30th Annual Meeting of the Division of Particles and Fields of the Mexican Physical Society (DPyC 2016) and XV Mexican Workshop on Particles and Fields (MWPF 2015)

Journal of Physics: Conference Series Volume 761

Puebla, Mexico
23 – 25 May 2016

Mazatlan, Mexico
2 – 6 November 2015

Editors:

Eduard de la Cruz Burelo **Arturo Fernandez Tellez**
Pablo Roig

ISBN: 978-1-5108-3180-3
ISSN: 1742-6588

Printed from e-media with permission by:

Curran Associates, Inc.
57 Morehouse Lane
Red Hook, NY 12571

Some format issues inherent in the e-media version may also appear in this print version.

Printed by Curran Associates, Inc. (2016)

For permission requests, please contact the Institute of Physics
at the address below.

Institute of Physics
Dirac House, Temple Back
Bristol BS1 6BE UK

Phone: 44 1 17 929 7481
Fax: 44 1 17 920 0979

techtracking@iop.org

Additional copies of this publication are available from:

Curran Associates, Inc.
57 Morehouse Lane
Red Hook, NY 12571 USA
Phone: 845-758-0400
Fax: 845-758-2633
Email: curran@proceedings.com
Web: www.proceedings.com

TABLE OF CONTENTS

HISTORICAL CONTRIBUTIONS

XXX DPYC MEETING (PUEBLA)

XV MWPF (MAZATLAN)

ADDITIONAL INVITED CONTRIBUTIONS

XIV MWPF (OAXACA)

Preface

Eduard de la Cruz Burelo[1], **Arturo Fernández Téllez**[2], **Pablo Roig**[1]
[1] Cinvestav-IPN, [2]FCFM-BUAP
E-mail: eduard@fis.cinvestav.mx, afernand@fcfm.buap.mx, proig@fis.cinvestav.mx

The 30th Annual Meeting of the Division of Particles and Fields (DPyC) of the Mexican Physical Society was held at Edificio Carolino, Benemérita Universidad Autónoma de Puebla, Mexico, on 23-25 May 2016. 125 participants and 83 presentations (20 posters and 63 talks including 47 contributed, 15 invited and a closing round table discussion) of high scientific quality certify the success of this meeting and witness the good health of the DPyC. Excited and encouraged by this fact, we decided to coordinate the edition of a proceedings volume commemorating and celebrating this anniversary, including the communications corresponding to this meeting as well as those of the recent XV Mexican Workshop on Particles and Fields, held at Playa Mazatlán Beach Hotel, Mazatlán, Mexico on 2-6 November 2015, where 46 works were discussed.

All presenters of these meetings were invited to contribute and the majority of them accepted -in spite of many other pressing simultaneous commitments. Additionally, we decided to invite researchers who were not able to attend these conferences, despite they used to be very involved in the activities of our community. With these invitations we also tried to reduce the inequality between male and female contributors and to enable researchers from the most isolated places in Mexico to participate, as well.

Given the commemorative nature of this volume, we wished to have some historical perspectives on specific aspects of our community which were not addressed in previous texts on the birth of the DPyC [1, 2, 3, 4]. We acknowledged our colleagues preparing essays on the creation of research networks and their interactions with DPyC, on the development of accelerator physics in Mexico and on the experimental and theoretical-research summers on high-energy physics, giving our best undergrads the support to spend a summer in a world-leading institution making their first steps in research.

The scientific program of the 2015 Workshop and 2016 Meeting covered a wide range of topics, including physics beyond the Standard Model (theory and experimental searches, astroparticles, cosmology, dark matter and dark energy, strings and QFT fundamentals), hadronic matter (perturbative QCD, jets and diffractive physics, hadronic structure, soft QCD, spectroscopy, heavy ion collisions and soft physics at hadron colliders and lattice simulations), flavor physics (quark masses and mixing, electroweak symmetry breaking, heavy quark physics, neutrino physics, CP violation, CKM and rare decays) and science and technology of accelerators and high performance computing. These are all reflected in this volume celebrating our 30th anniversary and wishing for the best to the DPyC and to its members in the future.

[1] A. Mondragón, "Early days of particle physics at the Institute of Physics UNAM," AIP Conf. Proc. **623** (2002) 117. Proceedings of the 8th Mexican Workshop on Particles and Fields: Zacatecas, Mexico, November 14-20, 2001. Editors: J.Lorenzo Díaz-Cruz (Puebla U., Mexico), Jürgen Engelfried (San Luis Potosí U.), Mariana Kirchbach (Zacatecas U.) and Myriam Mondragón (Mexico U., CEN). 2002, 416 pages.

[2] A. García and M. Á. Pérez, "High energy physics in Mexico: Historical sketch and implications," AIP Conf. Proc. **857B** (2006) 3. Proceedings of the 10th Mexican Workshop on Particles and Fields (MWPF 2005), Part A and B : Morelia, Mexico, November 6-12, 2005. Editors: Adnan Bashir, Víctor Villanueva, Luis Villaseñor (IFM-UMSNH, Michoacan), Miguel Á. Pérez (CINVESTAV, IPN) and Luis F. Urrutia (Mexico, ININ). 2006, 346 pp.

[3] J. L. Lucio and A. Zepeda, "Remembrances on the origin of the Mexican School of Particles and Fields," AIP Conf. Proc. **857B** (2006) 11. Proceedings of the 10th Mexican Workshop on Particles and Fields (MWPF 2005), see [2].

[4] L. M. Lederman, "Fermilab and Latin America," AIP Conf. Proc. **857B** (2006) 355. Proceedings of the 10th Mexican Workshop on Particles and Fields (MWPF 2005), see [2].

Journal of Physics: Conference Series **761** (2016) 011002 doi:10.1088/1742-6596/761/1/011002

Acknowledgments

The editors would like to thank Iraís Bautista Guzmán, Lorenzo Díaz Cruz and Mario Rodríguez Cahuantzi (FCFM-BUAP) for their excellent task organizing the 30th Annual Meeting of the Division of Particles and Fields (DPyC) of the Mexican Physical Society. We also acknowledge Pedro Podesta Lerma, Isabel Domínguez Jiménez, Edgar Alejandro León Espinoza and Juan Antonio Nieto García (UAS) for planning such an enjoyable XV Mexican Workshop on Particles and Fields. We are grateful as well to the members of the national organizing committee of both events.

We are indebted to CONACYT financial support through RED-FAE and SMF. We acknowledge the assistance and help of the Secretariats of both events.

Special words of gratitude are deserved to Mauro Napsuciale Mendívil (DCI-Univ. Gto.) who took the reins of the DPyC and RED-FAE in difficult times and managed to improve the situation of our Community, an endeavor in which the current (and hopefully future) representatives will pursue.

Committees

Local organizing committee of the XV Mexican Workshop on Particles and Fields:

- Pedro Podesta Lerma (FCFM-UAS, Chairman).
- Isabel Domínguez Jiménez (FCFM-UAS).
- Edgar Alejandro León Espinoza (FCFM-UAS).
- Juan Antonio Nieto García (FCFM-UAS).

Local organizing committee of the 30th Annual Meeting of the Division of Particles and Fields of the Mexican Physical Society:

- Iraís Bautista Guzmán (FCFM-BUAP).
- J. Lorenzo Díaz Cruz (BUAP).
- Arturo Fernández Téllez (FCFM-BUAP, Co-chair).
- Mauro Napsuciale Mendívil (DCI-Univ. Gto., Chair).
- Mario Rodríguez Cahuantzi (FCFM-BUAP).

National organizing committee of both events:

- Mauro Napsuciale Mendívil (DF-UG, Chairman)
- Arturo Fernández Téllez (FCFM-BUAP).
- Pedro Podesta Lerma (FCFM-UAS).
- Ildefonso León Monzón (FCFM-UAS).
- David Delepine (DF-UG).
- J. Lorenzo Díaz Cruz (BUAP).
- Eduard de la Cruz Burelo (CINVESTAV-DF).
- Mariana Kirchbach (IF-UASLP).
- Alfredo Aranda (UCOL).
- Saúl Ramos (IF-UNAM).
- Adnan Bashir (IFM-UMSNH).
- Alexis Aguilar Arévalo (ICN-UNAM).
- Salvador Carrillo Moreno (UIA).
- Karen Salomé Caballero Mora (UNACH).
- María Elena Tejeda Yeomans (UNISON).

List of participants

XV Mexican Workshop on Particles and Fields

(i) AGUILAR-ARÉVALO, Alexis, ICN-UNAM.
(ii) AGUILERA, José, UAS.
(iii) AHMAD, Aftab, UMSNH.
(iv) ARELLANO CELIZ, Concepción, BUAP.
(v) ARTEAGA VELÁZQUEZ, Juan Carlos, UMSNH.
(vi) BARRADAS PALMEROS, Rodolfo Enrique, BUAP.
(vii) BARRANCO, Juan, Univ. de Gto.
(viii) BASHIR, Adnan, UMSNH.
(ix) BAUTISTA, Iraís, BUAP.
(x) BEDOLLA, Marco Antonio, UMSNH.
(xi) BELLO, Héctor, BUAP.
(xii) BENCEDI, Gyula, Wigner RCP, Hungary.
(xiii) BIETENHOLZ, Wolfgang, ICN-UNAM.
(xiv) BURSCHE, Albert, University Zurich - Phyics Institute.
(xv) CAMPORESI, Tiziano, CERN.
(xvi) CARRAMIÑANA, Alberto, INAOE.
(xvii) CAÑAS ORDUZ, Blanca Cecilia, Cinvestav-IPN.
(xviii) COBOS-MARTÍNEZ, Javier, UMSNH.
(xix) CONCHA-SÁNCHEZ, Yajaira, UMSNH.
(xx) CORTÉS MALDONADO, Ismael, FCFM-UMSNH/CONACyT.
(xxi) COURTOY, Aurore, Conacyt-Cinvestav.
(xxii) DAS, Siba Prasad, BUAP.
(xxiii) DE LA CRUZ BURELO, Eduard, Cinvestav-IPN.
(xxiv) DOMÍNGUEZ JIMÉNEZ, Isabel, UAS.
(xxv) DURÁN, Cecilia, Cinvestav-IPN.
(xxvi) ERLER, Jens, IFUNAM.
(xxvii) ESCOBAR RUIZ, Carlos, ICN-UNAM.
(xxviii) ESPINOSA, Rafael, UDLAP.
(xxix) ESPINOZA, Catalina, IFUNAM.
(xxx) EZEQUIEL, Rodríguez, UNISON.
(xxxi) FERNÁNDEZ RANGEL, Luis Albino, UMSNH.

(xxxii) FERRO-HERNÁNDEZ, Rodolfo, Univ. Gto.
(xxxiii) FORTUNA MONTECILLO, Fabiola Elena, Univ. Gto.
(xxxiv) GALLARDO GALAVIZ, Daniela, UAS.
(xxxv) GÓMEZ BOCK, Melina, UDLAP.
(xxxvi) GONZÁLEZ HERNÁNDEZ, Emma, BUAP.
(xxxvii) GONZÁLEZ MACÍAS, Vannia, Univ. California Riverside.
(xxxviii) GONZÁLEZ QUAGLIA , Rodrigo, UDLAP.
(xxxix) GUEVARA, Adolfo, Cinvestav.
(xl) GUTIÉRREZ GUERRERO, Laura Xiomara, UMSNH.
(xli) HENTSCHINSKI, Martin, ICN-UNAM.
(xlii) HEREDIA DE LA CRUZ, Iván, Cinvestav-Conacyt.
(xliii) HERNÁNDEZ, Haydee, Univ. Gto.
(xliv) HERNÁNDEZ-TOMÉ, Gerardo, BUAP.
(xlv) HERNÁNDEZ, Jaime, BUAP.
(xlvi) IZAGUIRRE GÁMEZ, Samuel Doeg, UAS.
(xlvii) JUÁREZ, Pavel, Univ. Gto.
(xlviii) KIRCHBACH, Mariana, IF-UASLP.
(xlix) LÓPEZ, Lao Tsé, UAEH.
(l) MARTÍNEZ HERNÁNDEZ, Mario Iván, BUAP.
(li) MEDRANO, Santiago, UMSNH.
(lii) MILLÁN, Fernando, Univ. Gto.
(liii) MIRANDA HERNÁNDEZ, Jess Alejandro, Univ. Gto.
(liv) MIRANDA PALACIOS, Luis Salvador, ICN-UNAM.
(lv) MURILLO QUIJADA, Javier Alberto, Univ. Birmingham.
(lvi) NAPSUCIALE, Mauro, Univ. Gto.
(lvii) ORDUZ-DUCUARA, Javier-Andrés, UNAM.
(lviii) ORTIZ VELASQUEZ, Antonio, ICN-UNAM.
(lix) PEINADO, Eduardo, IF-UNAM.
(lx) PICENO, Elena, BUAP.
(lxi) PODESTA LERMA, Pedro Luis Manuel, UAS.
(lxii) RAMÍREZ ZAVALETA, Fernando, UMSNH.
(lxiii) RAYA, Khépani, UMSNH.
(lxiv) REYES, Humberto, IFUNAM.
(lxv) REYES, Marco A., Univ. Gto.
(lxvi) RICALDE HERRMANN, Daniel Enrique, UAS.
(lxvii) RIVERA, David, UMSNH.
(lxviii) RODRÍGUEZ CAHUANTZI, Mario, BUAP.
(lxix) ROIG, Pablo, Cinvestav-IPN.
(lxx) SAHU, Sarira, ICN-UNAM.
(lxxi) SALDAÑA, Ulises, Cinvestav-IPN.

(lxxii) SÁNCHEZ VÉLEZ, Ricardo, BUAP.
(lxxiii) TOLEDO, Genaro, IFUNAM.
(lxxiv) TOSTADO ROBLEDO, Sergio, Cinvestav-IPN.
(lxxv) URRUTIA, LUIS F., ICN-UNAM.
(lxxvi) VALERIO LIZARRAGA, Cristhian Alfonso, UAS.
(lxxvii) VARGAS MAGAÑA, Mariana, IFUNAM.
(lxxviii) VÁZQUEZ-JAUREGUI, Eric, IFUNAM.
(lxxix) VILLASEÑOR, Luis, UMSNH.
(lxxx) YEBRA, Azarael, Univ. Gto.

30th Annual Meeting of the Division of Particles and Fields of the Mexican Physical Society:

(i) AMAYA ALMAZÁN, Raúl Antonio, UAS.
(ii) AMEZCUA GERARDO, Lucero, BUAP.
(iii) ARANDA SÁNCHEZ, Jorge Isidro, UMSNH.
(iv) ARAUJO MARTÍNEZ, Aurora Cecilia, Univ. Gto.
(v) ARROYO, Jorge Armando, IFUNAM.
(vi) BARRADAS PAMEROS, Rodolfo Enrique, BUAP.
(vii) BARRANCO MONARCA, Juan, Univ. Gto.
(viii) BASHIR, Adnan, UMSNH.
(ix) BAUTISTA, Iraís, BUAP.
(x) BEDOLLA, Marco Antonio, UMSNH.
(xi) BERMÚDEZ, Rocío, UNISON.
(xii) BESPROSVANY, Jaime, IFUNAM.
(xiii) BOLAÑOS, Azucena, UV.
(xiv) CABALLERO MORA, Karen Salomé, UACH.
(xv) CABANILLAS NORIS, Juan Carlos, UAS.
(xvi) CARRAMIÑANA, Alberto, INAOE.
(xvii) CAÑAS ORDUZ, Blanca Cecilia, Cinvestav.
(xviii) CECILIA, Durán, Cinvestav.
(xix) CERVANTES MARTÍNEZ, Juan Mario, IF-UASLP.
(xx) CHAMORRO, Sindy, Cinvestav.
(xxi) COBOS-MARTÍNEZ, Javier, UMSNH.
(xxii) CORTÉS MALDONADO, Ismael, UMSNH.
(xxiii) COTZOMI PALETA, Jorge, BUAP.
(xxiv) CUAUTLE, Eleazar, ICN-UNAM.
(xxv) DE LA CRUZ BURELO, Eduard, Cinvestav.
(xxvi) DE LA FUENTE ACOSTA, Eduardo, Univ. Guadalajara.
(xxvii) DEL RAZO REYES, Josu I., UDLAP.
(xxviii) DÍAZ FURLONG, ALFONSO, BUAP.

(xxix) DÍAZ-CRUZ, Lorenzo, BUAP.
(xxx) DÍAZ, Enrique, BUAP.
(xxxi) DÍEZ-TEJEDOR, Alberto, Univ. Gto.
(xxxii) DOMÍNGUEZ ROSAS, Edgar, ICN-UNAM.
(xxxiii) DORANTES GARCÍA, Valentín, BUAP.
(xxxiv) DÍAZ NARANJO, Roberto R., ICN-UNAM.
(xxxv) ELIZALDE, Karen, BUAP.
(xxxvi) ENGELFRIED, Jürgen, IF-UASLP.
(xxxvii) ESPINOSA, David, UMSNH.
(xxxviii) ESPINOSA, Rafael, BUAP.
(xxxix) ESTRADA TRISTÁN, Nora Patricia, IF-UASLP.
(xl) EYSERMANS, Jan, BUAP.
(xli) FABELA ENRÍQUEZ, Brenda, UAZ.
(xlii) FERNÁNDEZ RANGEL, Luis Albino, UMSNH.
(xliii) FERNÁNDEZ TÉLLEZ, Arturo, BUAP.
(xliv) FERNÁNDEZ NÁJERA, Moisés, ICN-UNAM.
(xlv) FIERRO ROJAS, Pablo, BUAP.
(xlvi) FLORES SANDOVAL, Luis Jorge, Cinvestav.
(xlvii) FÉLIX BELTRAN, Olga, BUAP.
(xlviii) GAITÁN LOZANO, Ricardo, UNAM.
(xlix) GARCÉS, Estela A., UNAM.
(l) GÓMEZ BOCK, Melina, UDLAP.
(li) GONZÁLEZ CANALES, Félix Francisco, Cinvestav.
(lii) GONZÁLEZ HERNÁNDEZ, Emma, BUAP.
(liii) GUEVARA, Adolfo, Cinvestav.
(liv) GUTIÉRREZ GUERRERO, Laura Xiomara, UMSNH.
(lv) GUTIÉRREZ, Raúl, Univ. Gto.
(lvi) GÓMEZ, Selim, Univ. Gto.
(lvii) HECTOR, Bello Martínez, BUAP.
(lviii) HENTSCHINSKI, Martin, BUAP.
(lix) HERNÁNDEZ ÁLVAREZ, Octavio Emmanuel, UDLAP.
(lx) HERNÁNDEZ VILLANUEVA, Michel, Cinvestav.
(lxi) HERNÁNDEZ-PINTO, Roger J., UAS.
(lxii) HERNÁNDEZ-TOMÉ, Gerardo, BUAP.
(lxiii) HERRERA, Karla, Univ. Gto.
(lxiv) JARAMILLO ÁVILA, Benjamín Raziel, Univ. Gto.
(lxv) LARIOS, Bryan, BUAP.
(lxvi) LEÓN MONZÓN, Ildefonso, UAS.
(lxvii) LÓPEZ, Lao Tse, UAEH.

(lxviii) LÓPEZ OSORIO, María Alicia, Cinvestav.
(lxix) LÓPEZ, Alejandra, BUAP.
(lxx) MARCO, Arroyo, BUAP.
(lxxi) MARTÍNEZ-HUERTA, HUMBERTO, Cinvestav.
(lxxii) MARTÍNEZ, Tania, BUAP.
(lxxiii) MARTÍNEZ, Óscar, BUAP.
(lxxiv) MARTÍNEZ HERNÁNDEZ, Mario Iván, BUAP.
(lxxv) MARTÍNEZ VEGA, Alfredo, Univ. Gto.
(lxxvi) MARÍN OCHOA, Bibiana Marcela, Cinvestav.
(lxxvii) MAURY CUNA, Humberto, Univ. Gto.
(lxxviii) MEJÍA, Jhovanny, Cinvestav.
(lxxix) MEZA ALDAMA, Óscar, BUAP.
(lxxx) MIRANDA PALACIOS, Luis Salvador, UNAM.
(lxxxi) MOJICA MOLINA, Hebert Rodrigo, BUAP.
(lxxxii) MONDRAGÓN, Myriam, IFUNAM.
(lxxxiii) MONTEJO, Sebastián, BUAP.
(lxxxiv) MONTES DE OCA YEMHA, José Halim, UNAM.
(lxxxv) MORALES, Alma, BUAP.
(lxxxvi) MORENO PALACIOS, Óscar Eduardo, Univ. Gto.
(lxxxvii) MORENO, Eduardo, BUAP.
(lxxxviii) MOYOTL-ACUAHUITL, Agustín, Cinvestav.
(lxxxix) MÓNICA, Salinas Ibáñez, BUAP.
(xc) NAPSUCIALE, Mauro, Univ. Gto.
(xci) NIZ, Gustavo, Univ. Gto.
(xcii) NOVALES, Héctor, BUAP.
(xciii) ORDUZ-DUCUARA, Javier Andrés, UNAM.
(xciv) OROPEZA BARRERA, Cristina, Univ. Iberoamericana.
(xcv) ORTIZ VELÁSQUEZ, Antonio, ICN-UNAM.
(xcvi) PARRA, Alejandra, FCM-BUAP.
(xcvii) PEDRAZA MORALES, María Isabel, BUAP.
(xcviii) PÉREZ SUÁREZ, FERMÍN, Univ. Gto.
(xcix) PODESTA LERMA, Pedro Luis Manuel, UAS.
(c) PONCE, Epifanio, BUAP.
(ci) PÉREZ BOLAÑOS, Edgar Javier, UDLAP.
(cii) PÉREZ DE LEÓN, Mario Aldair, BUAP.
(ciii) PÉREZ MARTÍNEZ, Adriana, UNAM.
(civ) RAMOS-SÁNCHEZ, Saúl, UNAM.
(cv) RAMOS, Raymundo, William and Mary.
(cvi) RAMÍREZ GARCÍA, Mateo, Cinvestav.
(cvii) RAMÍREZ ZAVALETA, Fernando, UMSNH.

(cviii) RAYA, Khépani, UMSNH.
(cix) RENDÓN, Javier, Cinvestav.
(cx) REYES HERRERA, Juan, IFUNAM.
(cxi) RIVERA AGUDELO, Diana Carolina, Cinvestav.
(cxii) RODRÍGUEZ GALVÁN, Valentina, UASLP.
(cxiii) RODRÍGUEZ PÉREZ, David, UAS.
(cxiv) RODRÍGUEZ BECERRA, Gerardo de Jesús, Univ. Gto.
(cxv) RODRÍGUEZ CAHUANTZI, Mario, BUAP.
(cxvi) ROIG GARCÉS, Pablo, Cinvestav.
(cxvii) ROJAS CIOFALO, Diana del Carmen, IFUAP.
(cxviii) RUBALCAVA GARCÍA, Iraís, BUAP.
(cxix) SALAZAR, Humberto, BUAP.
(cxx) SALDAÑA-SALAZAR, Ulises J., BUAP.
(cxxi) SERGIO ARTURO, Iga Buitrón, ICN-UNAM.
(cxxii) SIERRA, Ángel, BUAP.
(cxxiii) SÁNCHEZ GARCÍA, Marxil, IFUNAM.
(cxxiv) TEJEDA-YEOMANS, María Elena, UNISON.
(cxxv) TIRADO BUENO, Eduardo, BUAP.
(cxxvi) TOLEDO, Genaro, IFUNAM.
(cxxvii) TOSTADO ROBLEDO, Sergio, Cinvestav.
(cxxviii) TUTUTI, Eduardo S., UMSNH.
(cxxix) URIBE ESTRADA, Cecilia, BUAP.
(cxxx) VALERIO LIZARRAGA, Cristhian Alfonso, UAS.
(cxxxi) VARELA, Enrique, BUAP.
(cxxxii) ZEPEDA, Héber, Cinvestav.
(cxxxiii) ÁNGELES MARTÍNEZ, René, Univ. Manchester.
(cxxxiv) Sosa Chávez Arturo, UMSNH.
(cxxxv) Gallegos Medina, Talhía, ICN-UNAM.

Peer review statement

All papers published in this volume of *Journal of Physics: Conference Series* have been peer reviewed through processes administered by the proceedings Editors. Reviews were conducted by expert referees to the professional and scientific standards expected of a proceedings journal published by IOP Publishing.

Journal of Physics: Conference Series **761** (2016) 012001 doi:10.1088/1742-6596/761/1/012001

Mexican High Energy Physics Network

J C D'Olivo[1], M Napsuciale[2], M A Pérez-Angón[3]

[1]Instituto de Ciencias Nucleares, Universidad Nacional Autónoma de México, Apdo. Postal 70-543, 04510, CDMX, Mexico

[2]Departamento de Física, Universidad de Guanajuato, León, Guanajuato, Mexico

[2]Departamento de Física, CINVESTAV-IPN, Apdo. Postal 14-740, 07000, CDMX, Mexico

Abstract. The Mexican High Energy Physics Network is one of CONACYT's thematic research networks, created with the aim of increasing the communication and cooperation of the scientific and technology communities of Mexico in strategic areas. In this report we review the evolution, challenges, achievements and opportunities faced by the network.

1. Historical overview

The Mexican community of high energy physics (HEP) originated from several efforts done by different institutions. Some of its first members came from the nuclear physics group at the Physics Institute of UNAM, who eventually moved to higher energies. Others came from the recruitment of foreign researchers in the field by the Department of Physics of CINVESTAV, like Jean Pestieau and Jerzi Plebañski. Although not all of them settled permanently in Mexico, they trained some students who after pursuing a Ph.D. abroad, returned to our country consolidating the initial group. Other pioneers in the field were trained abroad in processes not related to the above institutions and, when they returned to Mexico, joined one of the emerging groups in the early 70's. We can say with certainty that, by the end of that decade, there were small but stable groups in both of the aforementioned institutions. Later on, some of their members migrated to other institutions in Mexico City (UAM) or to state universities (BUAP, Univ. of Guanajuato, UMSNH, UASLP), leading the formation of new groups.

In the early 80's systematic process for training of human resources in HEP started with the first editions of the Mexican School of Particles and Fields (MSPF) and the Mexican Workshop of Particles and Fields (MWPF). The foundation of the Division of Particles and Fields of the Mexican Physical Society (DPyC-SMF) in 1984 made possible to work in a coherent and institutional environment towards to the consolidation of the Mexican HEP community. At a later stage, researchers working in more formal aspects of the field, along with those focused on understanding the nature of the gravitational interaction, founded the Division of Gravitation and Mathematical Physics (DGFM-SMF). In an effort to promote the scientific cooperation in the region, in November 1996, the DPyC organized the first edition of the Latin American Symposium on High Energy Physics (SILAFAE). Since then, this symposium has been held without interruption every two years, in different Latin American countries, and has become one of the most important and traditional events on HEP in the region.

In the 70's the interest of the HEP Mexican community focused on theoretical aspects, but in the late 80's the formation of experimental groups begun. Presently, particle physics is a mature discipline in our country, with researchers distributed among the most important Mexican universities and research centers. Currently, there are about 130 active researchers doing work in understanding the various phenomena related to elementary particles, approximately one third of which are experimental physicists. To them we must add roughly the same number of graduate students, astrophysicists and scientists from other disciplines with an interest in high-energy physics phenomena. The experimental component has been growing up steadily over the past 25 years. Its activity evolved from data analysis to the design and construction of parts of the detectors used in huge experiments (Pierre Auger Observatory, ALICE, CMS) and, more recently, to science and technology of particle accelerators and the development of new detection systems. Today, several frontier experiments in high energy physics and astrophysics around the world have a significant participation of Mexican research groups. Meanwhile, those engaged in theoretical research have been gaining greater international recognition for their work on phenomenological data analysis, model building for physics beyond the Standard Model or non-perturbative aspects of QCD.

Although the HEP community participates actively in the regular meetings organized by the DPyC-SMF, in general terms, its progress has been more a consequence of the efforts of individuals or small groups with specific interests than the result of a systematic planning of the community at the national level. The most important attempt undertaken in this regard has been done within the HEP Thematic Network, known as Red FAE for its Spanish acronym.

2. Mexican High Energy Physics Network

On August 31, 2006, the National Council of Science and Technology (CONACYT) of Mexico, issued a Call for Proposals for the realization of mega projects, with the aim of identifying projects with the potential to strength the search of new knowledge focused towards the solution of national problems and to promote the development of basic and applied research. After the evaluation process, 79 proposal passed to a second round and received some support. These proposals, and some others brought subsequently to the consideration of CONACYT, were integrated, in September 2007, within a set of 14 strategic thematic lines, named CONACYT's Thematic Research Networks (RTIC), with the purpose of providing high impact solutions for the country's development and the welfare of its population. This idea was conceived by Dr. José Antonio de la Peña, who at that time was Director for Scientific Development of CONACYT. One of the thematic lines corresponds to the HEP Network and brought six of the approved mega projects together:

(i) HAWC, Mexico observatory of gamma rays.
(ii) National network of air showers detectors.
(iii) Mexican multidisciplinary underground laboratory.
(iv) Development of particle accelerators. Synchrotron light source.
(v) Implementation of an advanced light source (UV-VIS-IR-UVV) based on short-pulse lasers.
(vi) Sensors for detecting IR radiation.

After a year of discussions and work with participation of representatives of various mega projects, in September 2008, the Technical and Administrative Committee of the Institutional Fund approved the Guidelines for the Formation and Consolidation of the RTIC. This marks the beginning of the HEP Network, whose Technical and Academic Committee (CTA) dedicated itself to the task of elaborating the first general working-plan. By then, the Thematic Networks Directorate (DRT) was created by CONACYT, being its first director Dr. Tomás Viveros García.

The activities of the thematic networks started in February 2009 with the First Congress of CONACYT's Thematic Networks, organized by the DRT, in the city of Querétaro. The First National Congress of the HEP Network took place shortly after, from 4 to 7 March 2009, in the city of Taxco, Guerrero. In this meeting, the basis for operation were established and a number of specific objectives were identified. Despite the long time required to launch the networks, during the congress it became evident the great interest in the new program. In the Call for Integration of RTCI 2009-01, issued by the DRT, on August 10, 2009, 156 researchers joined the HEP Network. The formal integration agreement was signed in February 2010 and the first resources were provided in May 2010. During 2011 the DRT launched a Call for Adherence to Thematic Networks and 68 new members joined the HEP Network, which by December 2010 had a total of 224 members, including scientists, postdocs and graduate students of 25 higher education institutions, distributed over 16 states of the Mexican Republic. Such numbers have suffered some variation and presently the network consists of 171 researchers from 22 institutions and a similar number of graduate students. Network coordination has always been in charge of the representative of the CTA, a responsibility that during the first stage was taken over by Juan Carlos D'Olivo (ICN-UNAM). Subsequently, the coordinators were Heriberto Castilla (CINVESTAV-IPN), from 2012 to 2014, and Mauro Napsuciale (Univ. of Guanajuato), from 2014 to 2016. At present, the representative of the CTA is Eduard de la Cruz Burelo (CINVESTAV-IPN).

The beginnings of the HEP Network leaded by Juan Carlos D'Olivo were promising and opened new opportunities, despite the vicissitudes of the networks program within CONACYT. To a greater or lesser extent, during this period there were achievements or progress in the various initial goals, both in supporting the activities of experimental and theoretical groups as well as in the planning of the development of the country in HEP. The support provided for mobility and organization of events opened new possibilities for group collaborations both at the national and international level. A large number of meetings, conferences, and other academic events in Mexico and abroad were sponsored by the HEP Network, several of which could not have been done without its support. Also, while most of the projects were already going on before the foundation of the network, they were benefited not only by the financial aid provided, including equipment, but also by the support received when negotiating funds for the international collaborations. Network membership facilitated the communication between the national leaders of such projects and the funding agencies, mainly CONACyT.

By its very genesis, the thematic lines of the network had considerable dispersion and, seeking a greater cohesion, three general lines were defined: i) Physics with accelerators and accelerator physics, ii) Detectors Physics, and iii) Astroparticle Physics. A team was integrated for each of these lines with the task of preparing a diagnostic of the situation in Mexico and making proposals for its development in the short, medium and long term. At the same time, taking into account the importance assigned to the popularization of science, a fourth team addressed this topic. By the end of 2012 the four teams produced documents containing information relevant to an assessment of the "state of the art" and the challenges and opportunities for Mexico in HEP. In addition, a catalogue of infrastructure was completed and, thanks to a collaboration with the Thematic Network on Complexity Science and Society, a Catalog of Human Resources was elaborated.

The successive editions of the National Congresses of the network made evident the interest generated within the HEP community. The second one was held in the city of Tlaxcala, from 20 to 22 January 2011, with a larger audience than the first edition. During this meeting, there were intense and open discussions about the direction of the network, its priorities and badge projects, the future of the discipline in Mexico and other issues of great importance. An important conclusion of this and other similar exercises is that the community of high energy and its network are able and willing to face great challenges and pursue ambitious undertakings, provided they are the result of agreements and consensus within the community.

The second period of the HEP Network leaded by Heriberto Castilla coincided with changes in the management of CONACyT. Initially, the new administrators do not attach great importance to the program of thematic networks and, accordingly, considerably reduced the budget allocated

Journal of Physics: Conference Series **761** (2016) 012001 doi:10.1088/1742-6596/761/1/012001

to this. Despite the adverse situation, the experimental groups working in international collaborations were supported by the HEP Network, as well as some initiatives of the theoretical groups and popularization of science. The scarce money available caused a lack of interest of most of the members not directly involved in HEP. The situation fortunately improved by the end of 2013, and from 23 to 28 January 2014, the HEP Network organized in the city of Guanajuato two academic events of great importance: The Third National Congress and the Thematic Workshop "Particle Physics and Astrophysics: Challenges and Opportunities in Mexico and Latin America", the later with the sponsorship of CONACYT and the Ministry of Public Education (SEP). These meetings had particular relevance, since it represented the re-launching of the activities of the network. During the course of them, the achieved goals were examined and future lines of action were discussed. It is worth noting that the main purposes that motivated the realization of the thematic workshop were the integration to the network of Mexican colleagues living abroad, and the seeking for the establishment of links with networks and similar organizations in Mexico an other countries in the region, a still pending issue.

Something worthy to remark is that, up to this point, the HEP network was an entity legally belonging to CONACyT and the administration of the networks was done by the DRT, which eventually caused some difficulties. Under this situation, CONACyT decided to change the administration scheme and switched to conventional group projects for thematic networks, managed by the higher education institutions appointed for this purpose. It was at this point that the original and promising idea that the networks could serve as thematic advisory entities for the administration of national science by CONACyT was lost.

During the third period leaded by Mauro Napsuciale an inclusive and horizontal exercise of planning was done. Although formally the CTA included only a few members, in practice a wider advisory committee (working group) was formed including members of all experimental groups and representatives of the most active theoretical groups in different topics. In an initial meeting at León Guanajuato on August 2, 2014, the working group conceived a route for an intensive planning exercise changing the focus to the by then conventional frontiers (energy, intensity and cosmos) but including new categories necessary for the Mexican community (conceptual, outreach and education and innovation and technological development). The documents generated during the 2010-2012 period served as background for this intensive planning exercise realized from August to December 2014. An important outcome of this exercise was the formulation of a National Plan for Research in High Energy Physics (PNIFAE), whose content was validated in the Third General Meeting of the HEP Network, held in the city of Manzanillo, Colima, on December 6-7 of 2014. In addition to this important exercise, the working group for every frontier realized topical meetings across the country and all the participating experimental groups received support for their international collaborations. Along the conclusions of the PNIFAE, the 2015 project was organized about working groups that maintained their activity according to their own dynamics and were responsible for applying the funds granted by the project. The Fourth General Meeting of the HEP Network was held in the city of Guanajuato at the beginning of December 2015.

The Mexican community of high energy physics has increased substantially and diversified its activity over the last twenty years, taking increasing responsibility in the international collaborations and generating the critical mass to undertake and lead national projects in the area, in particular the design and operation of a High Energy National Laboratory. The economic importance of promoting high energy projects in our country has been highlighted and the HEP Network is the natural space for their discussion and conceptualization. In this concern, it is important to remark that the academic interest of the HEP Network is focused on research in theoretical or experimental physics of elementary particles and gravitation and their ultimate goal is to understand the structure of nature at the microscopic level, the forces driving this structure and the implications for the macrocosm. Therefore, it is oriented to basic science. However, it is well known that research at the frontier of knowledge generates cutting edge technology developed for scientific purposes only, but which eventually produces social benefits. In order to explore the phenomena of micro and macrocosm a variety of prototype designs are developed in universities research laboratories, whose massive production is granted to industry usually requiring it to improve existing production processes and generating new opportunities related to the new technologies developed for the industrial production of such prototypes. Another important feature of the activity of the HEP community is an intensive international cooperation. Groups of researchers, technicians, engineers and students in the network take part in multinational collaborations doing work for large laboratories and observatories, such as CERN, Fermilab, Riken, Observatory Pierre Auger and SNOLAB, which carry out high-energy experiments using particles either produced in astrophysical phenomena or produced in colliders.

In the PNIFAE an inventory of the various experimental and theoretical contributions of the HEP Mexican community is done, a review of the state of art of the different frontiers is carried on and the possibilities for technological development projects related to the activities of this community are highlighted. Also the challenges in outreach, education, spin off and cutting edge technologies are unfolded. As a result of this diagnosis we have identified numerous areas of opportunity for the work of the network, in the short, medium and long term. In addition to the recommendations for the frontiers related to their specific problems, the document contains recommendations transverse to the community which are the points of basic agreement for the future work of the HEP community in Mexico. These recommendations were divided into three categories to consider the following aspects of the activities of the HEP Network: i) Strengthening, consolidation and integration of the HEP community; ii) Promoting international leadership of the HEP Mexican groups; iii) Improving the social impact of the Mexican HEP Network. In addition to the HEP community, these recommendations have the potential to target all sectors of the public administration related to the development of science, technology and innovation in Mexico. These include the head of the sector, CONACyT, their state counterparts, the State Councils of Science and Technology, the legislative entities of the federal order, the Congress and the House of Representatives, and state and municipal, local Congress and Councils.

The implementation of the program of Thematic Networks has been a process not without difficulties, especially when it is an innovative idea on how to do science and to boost technological development in Mexico. For a couple of years, after the Dr. de la Peña era in CONACYT, the continuity of the networks was seriously questioned, which was accompanied by a cut in both the amount and the type of the support provided. However, the program's success, as evidenced by the progress achieved and the high adhesion of the scientific community, prevented its cancellation. In this process, the HEP Network has gained presence as a meeting place that fosters the planning and execution of joint activities. Today, the HEP Network is a must in high energy physics and related disciplines in Mexico. However, there are pending issues that need to be addressed by the present leadership, such as:

- To consolidate and institutionalize the network, giving clear operational rules that are known to all its members and which are contained in an Operations Manual.
- To improve and facilitate the administrative management of the network so that resources flow faster.
- Transcend the role of small funding agency and become a true cooperative venture.
- To establish priorities of the network and its badge projects and to work out plans to translate them into short, medium and long term goals.
- To become an essential partner in the definition of national scientific policies related to high energy physics.

This should be accompanied by a change of attitude concerning the role of huge scientific projects in Mexico, not only by the members of the network, but also by CONACYT itself. The challenge is great, but so are the benefits.

References

[1] García A and Pérez M A, High energy physics in Mexico: historical sketch and implications, AIP Conf. Proc. 857B, 3-10 (2006).
[2] Lucio J L, The Mexican School of Particles and Fields, AIP Conf. Proc. 857B, 11-14 (2006).
[3] Comité Técnico Académico (D'Olivo J C et al.), Programa General de Trabajo 2012-2013, Red FAE, CONACYT (2012).
[4] Grupo de Trabajo, (Napsuciale, M et al.), Plan Nacional de Investigación en Física de Altas Energías, Red FAE, CONACYT (2015).
[5] Aguilar Arévalo A A, Castilla Valdés H, D'Olivo Saez J C and Napsuciale M, eds., Workshop on Particle Physics and Astrophysics: Challenges and Opportunities in Mexico and Latin America (UNAM, Mexico, 2015).
[6] Contreras J G and M. Napsuciale M, Creation of a group on particle accelerator science and technology in Mexico, arXiv:1608.07456.
[7] De La Cruz Burelo E, Hernández Villanueva M and Molina Gonzalez N W, Estructura de la Red de Fisica de Altas Energias en Mexico. In preparation.

Creation of a group on particle accelerator science and technology in Mexico

J. G. Contreras[1], **M. Napsuciale**[2]
[1]Faculty of Nuclear Sciences and Physical Engineering, Czech Technical University in Prague, Czech Republic
[2]Departamento de Física, Universidad de Guanajuato, León, Guanajuato, México

E-mail: jgcn@mail.cern.ch,mauro@fisica.ugto.mx

Abstract. The story-line of the activities that have produced the first group of scientists in Mexico working on particle accelerator science and technology is briefly presented.

1. Introduction

In Mexico, there is a strong community of physicists working on High-Energy Physics (HEP). Initially, most of the practitioners were working on theoretical subjects, but by the early 2000s there was already a strong group of experimentalists working on international collaborations at CERN, DESY and FERMILAB [1]. To many of us working in HEP, it seemed that the natural next step had to be to contribute also to the accelerator side of our field.

There were some previous attempts to develop accelerator science and technology in Mexico. The Institute of Physics of Guanajuato University (IFUG) founded in 1986 was conceived as a centre for the development of accelerator science and technology under the auspices of FERMILAB and other national laboratories in the USA. Unfortunately, this project did not succeed, in 1992 the accelerator project was closed and new groups working on other topics started a new era of this institute.

An opportunity to explore again the formation of groups in this area came from the National Council for Science and Technology of Mexico (CONACyT), which in 2006 opened a call for proposals to develop large scientific and technological projects. One of the proposed projects was the development of accelerator science and technology with the ultimate end of building a light source in Mexico [2].

After evaluating the different proposals, CONACYT decided to create nation-wide networks in specific topics such that the scientist could organise themselves, work-out a set of priorities and develop the ideas and projects as one community. Among the original networks was the Topical Network on High-Energy Physics (Red-FAE), which started working in 2008. One of the conclusions in the original diagnostic of the field presented during the First General Meeting of the Red FAE held at the city of Taxco in 2009, was the urgency to develop accelerator science and technology for different purposes in Mexico. Within this topic, one important item was the development of human resources in the different disciplines involved in the field.

In this context, several groups got interested in the process, to be described below. On the other hand, in 2010 the then Deputy General Direction for Scientific Development of CONACYT,

Dr. J. A. de la Peña, created an ad-hoc panel with the mandate to explore the possibility of large scale projects in the area of accelerators. The panel recommended to support projects in two areas: hadron therapy and light sources [3]. One of the issues emphasised in the report, was the need to form human resources in all related fields.

The development of human resources for accelerator science and technology in Mexico clustered initially around two independent initiatives to be described in the next sections.

2. The collaboration with CERN

One of the outcomes of the discussions leading to the proposal [2] to build a light source, was that, given that most of us had little or no experience in frontier research in accelerators, it was clear that we had to start looking for students interested in this field to train them at the highest level, in order that after some time they would be the driving force of this type of projects. Then, after the submission of the proposal we started to look for interested students.

In 2007, the first student willing to participate in this adventure was found. Following the advice of Dr. E. Elsen, one of us (JGC) contacted Dr. Koutchouk towards the end of 2007. Dr. Koutchouk was finalising the preparation of an European project (EuCARD) under the FP7 program. After some discussion of the situation, both in Mexico and in Europe, we decided that the best contact to help us train this first student would be Dr. F. Zimmermann from CERN.

In 2008 the student spent the Summer at CERN working with Dr. Zimmermann and later on this work was the basis of his 2009 M.Sc. thesis. The student continued the collaboration with Dr. Zimmermann working on electron cloud effects in the LHC [4, 5] and obtained his Ph. D. degree in September of 2013. The success of this first student motivated us to send more students to CERN and Dr. Zimmermann helped in finding adequate supervisors for them.

From the very beginning it was clear that funding the stays of students at CERN would be a problem. Having no local community with experts in this field meant that there was no clear way to successfully submit projects to obtain grants for the students. Furthermore, the disparities in the cost of living in Mexico and in the Geneva area meant that the minimal funding required at CERN was substantially larger than the norm in Mexico.

The funding for the initial stays was obtained from many sources. In Mexico the home institutes of the students, the RedFAE and CONACYT contributed funds; the rest of the money was obtained from the EuCARD , HELEN and E-Planet projects.

In parallel we explored how to solve permanently the problem. Since 1998 there is a cooperation agreement between CERN and CONACYT. There was a possibility to write an addendum to this agreement to take care specifically of the collaboration in the area of accelerator science. This was thought to be the right way to proceed, but it took a long time to crystallise. Finally in 2015 such an addendum was signed by the representatives of both institutions and a formal way to request funding was finally available. The project was called BEAM and was de facto in operation already before the formal signature of the addendum.

Under the auspices of BEAM three students have obtained a Ph.D. in accelerator sciences. In addition to the work on electron cloud effects, they have also contributed to our current knowledge on Crab cavities [6] and the design and construction of the LINAC4 [7]. Also three students got their M. Sc. degree from Mexican institutions, with the research work performed at CERN.

Currently there are five Mexicans working in the field of accelerator sciences at CERN. One is a CERN fellow, three are doing the research towards their Ph.D. and one is working towards his M. Sc. degree. They are working in a variety of fields covering synchrotron radiation at the LHC and future colliders, design testing and operation of superconducting RF cavities and crab cavities and optics and design of the interaction regions for the future colliders (TLEP, FCC-ee).

3. The collaboration with JLAB

During the trip back to León from the Taxco meeting in 2009, a discussion with Gerardo Moreno (first high energy physics experimentalist in Mexico, participant of the original IFUG project and presently retired) triggered the interest in the field by one of us (M.N.). Following the conclusion on the need to develop accelerator science and technology, Gerardo Moreno argued that a starting point to this endeavour could be to involve students in projects for the development of compact light sources which received a boost in terms of financial support with the incorporation of a possibility to generate X-rays by Inverse Compton Scattering (ICS). The calculation of Compton scattering is one of the goals of every course in QED, hence highly standard, the novelty here is the kinematical regime when the electron is moving in whose case it transfers energy to the photon increasing its energy. The technical problem that has prevented the development of efficient machines by ICS is the luminosity of the beams and those related to the small size of a compact light source.

By January of 2009, a master student at IFUG (Alejandro Castilla Loeza) got interested in the topic and after six months he learned QED and calculated ICS in the kinematical region of interest having a complete picture for the design. There were no facilities for the experimental part so a collaboration was established with the time-group at the National Centre of Metrology (CENAM) to make a feasibility demonstration experiment with old TV electron guns to produce X-rays from visible light.

By the end of 2009, a Mexican researcher working at the Thomas Jefferson National Laboratory (JLAB), Carlos Hernandez García, was invited by the DPyC to deliver one of the lectures at the XII Mexican Workshop on Particles and Fields held in Mazatlán. This was the opportunity to establish a fruitful collaboration between Mexican institutions and JLAB. At this very same meeting a program was designed for the Ph.D. program of Alejandro Castilla Loeza, focused on the development of crab cavities for the projected upgrade to 12 GeV of JLAB. After his return to JLAB, Carlos Hernández got a formal invitation from JLAB to the DPyC to discuss the possibilities for collaboration with Mexican institutions. Since Guanajuato had already an initial collaboration with JLAB, one of us (M.N.) was designed by the DPyC to attend JLAB's invitation.

The visit to JLAB was a highly motivating experience and we will always be in debt with Carlos for the preparation in every detail of this visit which included discussions with the engineers in charge of every part of the accelerators at CEBAF, the living experience of the free electron laser group in their everyday duties, highly supportive meetings with the Director of the Accelerator Division (Andrew Hutton) and the general director of JLAB (Hugh E. Montgomery), as well as a general meeting to explore the modalities to support the development of accelerator science and technology in Mexico. Many ideas emerged from this meeting, among them the need to organise a periodic school on accelerators in Mexico, a permanent space for Mexican students in the JLAB summer program and in the United States Particle Accelerator School (USPAS), the possibility of joint Ph.D. programs between Mexican universities and Old Dominion University (close to JLAB) or the direct participation of Mexican students in the collaborations working in the forthcoming upgrade to 12 GeV of JLAB. On our side we insisted in the need of non-conventional training programs for the first Mexican PhD students, who would require not only skills on simulations and design of components but also the experience of building, testing and operating them in order to have the knowledge to eventually lead groups on the corresponding topics.

As a product of this collaboration a student got his Ph D (Alejandro Castilla Loeza) working on the design, building and testing of a crab cavity for the 12 GeV upgrade of JLAB; three students got his/her Master degree (Karim Hernández Chain, Celene Cuevas and Christian López) and at least seven students participated in the JLAB summer program (Christian Valerio, Salvador Sosa, Luis Medina, Ernesto Barrientos, Saúl Cuen, Cesar Serna y Gabriel Palacios up to 2013). Presently, Salvador Sosa is about to finish his Ph D program working at JLAB and another student (Daniel Chávez Valenzuela) is working on the design of the magnets for the upgrade JLAB in a collaboration with Texas A& M University and Guanajuato University.

Most of the work done by Mexican students at JLAB had finantial support from both, Mexican institutions and JLAB. However, unlike the CERN collaboration, CONACyT only contributed with the national grants and the "Becas Mixtas" program to this process and eventually the crisis in the USA reached scientific projects including those of JLAB making impossible to support further the work of Mexican students. Under this situation some students got their master at JLAB and continued working in the field at CERN under the auspices of the BEAM project.

4. The Mexican Particle Accelerator School.

As part of the initiatives to trigger the development of accelerator science and technology in Mexico by the ends of 2010 the then Deputy General Direction for Scientific Development of CONACYT, Dr. J. A. de la Peña, created another panel with people interested in having a Mexican Synchrotron to work on several aspects considered important before starting formally the process, mainly a meeting of Mexican users of synchrotron facilities to have an idea of the size of this community and the formulation of a methodological route for the process. To this end the group received a grant from CONACyT leaded by one of us (M.N.). As part of the activities of this group we organised the First Mexican Particle Accelerator School (MEPAS I) on November 2011 at the city of Guanajuato.

The program of MEPAS-I was designed by our colleagues at JLAB under the leadership of Carlos Hernández and included talks by leaders of the accelerator divisions of several synchrotron facilities in the world and also some high energy physics facilities. In particular, it was very important the participation of Frank Zimmerman from CERN, which helped to close the loop of connections of the Mexican community with the accelerator international community.

It took a long time to organise the second Mexican Particle accelerator School (MEPAS-2) which was held on November 2015 at Guanajuato city. However, during these four years between the first and the second edition, the number of students involved in the field grow up considerably as we will review in the next section, and in addition to leaders of several organisations (remarkably William Barletta director of the United States Particle Accelerator School, Liu Lin from the Brazilian Synchrotron and Robert Hettel from SLAC) we had an active participation of the first Mexicans graduated in the field and PhD students working at JLAB, KEK, CERN and other facilities, as instructors. The second MEPAS had financial support mainly from the Red-FAE.

The third edition of MEPAS will be held this year under the auspices of the Mesoamerican Center for Theoretical Physics (MCTP) and will be completely organised by the Mexican Particle Accelerator Community of (CMAP).

5. Mexican Particle Accelerator Community

In addition to the JLAB and CERN initiatives during the past five years other institutions have been involved in the development of Mexican human resources in the field. One of them is the Texas A& M Magnet Laboratory where a student (Daniel Chávez Valenzuela) got his Master and is presently pursuing hid Ph. D. working on the design of the magnets for the upgrade to 12 GeV of JLAB. Other students got involved in accelerator projects at ALBA, DESY, FERMILAB, ALS, TRIUMF and Liverpool University.

During the past two years Mexican students working at different laboratories in the world and the first Mexican PhD in the field have been discussing the need to focus efforts and as a first step they have organised in the Mexican Community of Particle Accelerators. This community has about 25 members and although most of them are graduate students working for their Ph D degree at several institutions, they are already committed to the conceptual design and operation of the first Linear Accelerator (LINAC) built in Mexico [8, 9].

6. Accelerator Projects in Mexico

There is no doubt that the project that triggered the whole process was the Mexican Synchrotron Light Source, which has been pushed by different groups in the past. After the original proposal for this light source for the CONACyT call for proposals of large scientific and technological projects in 2006, the idea evolved with the grant to a different group in 2010 described above which boosted the training of students and lead to the formation of the Mexican Synchrotron Radiation Users Meeting which has had an active role in pushing the project. A new grant was provided by CONACyT and the state of Morelos in 2014 within the "Fondos Mixtos" program to a group which now included representatives of the user's community. The goal is to have a conceptual design report for the Mexican Synchrotron Light Source this year aiming to construct it in the state of Morelos.

In parallel, some smaller projects involving accelerator science and technology have been proposed by different groups. Among them, an Electron Beam Irradiation Facility for "cold pasteurisation" of fruit has been pushed by the state of Michoacán and Michoacán State University (UMSNH); an Excellence Centre for Nuclear Medicine aiming to produce radio-pharmaceutical products for imagenology and cancer therapy has been proposed in Puebla and some applications on national security are being implemented in the state of Puebla.

References

[1] Garcia A and Perez M A 2006 *AIP Conf. Proc.* **857B** 3–10
[2] Contreras J G, Herrera G, Lopez R, Moreno M *et al.* 2007 Desarrollo de la tecnología de aceleradores: Fuente de Luz de Sincrotón, Ideas for megaprojects proposal 55329, Conacyt
[3] Celis M A, Hernández-Calderón I, Contreras J G, Lárraga J M and Moreno M 2010 Aceleradores para el desarrollo científico, tecnológico y social de México, Report for Conacyt
[4] Maury Cuna H, Contreras J G and Zimmermann F 2012 *Phys. Rev. ST Accel. Beams* **15** 051001
[5] Dominguez O, Li K, Arduini G, Metral E, Rumolo G, Zimmermann F and Maury Cuna H 2013 *Phys. Rev. ST Accel. Beams* **16** 011003
[6] Yee-Rendon B, Lopez-Fernandez R, Barranco J, Calaga R, Marsili A, Tomás R, Zimmermann F and Bouly F 2014 *Phys. Rev. ST Accel. Beams* **17** 051001
[7] Valerio-Lizarraga C A, Leon-Monzon I and Scrivens R 2015 *Phys. Rev. ST Accel. Beams* **18** 080101
[8] Valerio-Lizarraga C 2016 Actividades de la comunidad mexicana de aceleradores de partículas XXX Annual meeting of the Mexican Division of Particles and Fields, Mexican Physical Society
[9] Maury-Cuna H 2016 Mexican linear particle accelerator XXX Annual meeting of the Mexican Division of Particles and Fields, Mexican Physical Society

Journal of Physics: Conference Series **761** (2016) 012003 doi:10.1088/1742-6596/761/1/012003

"Emergence and Development of experimental particle physics in Mexico"

Salvador Carrillo Moreno, Universidad Iberoamericana

Birth of an idea:

Thanks to two interviews to Drs. Miguel Ángel Pérez Angón and Matías Moreno (Sep/2016), I can present to you how the idea of forming researchers in experimental particle physics in Mexico started.

There is a connection with the spread of elementary particle physics in Europe: in 1983, CERN (European Centre for Nuclear Research) supports the inclusion of Spain as a member state, as it is the case three years later with Portugal. This boosts the development, in both countries, of high energy physics. In parallel, Nobel Prize winner (1988) and second Fermilab (Fermi National Accelerator Laboratory) director (1978 to 1989), Dr. Leon Max Lederman, decides to promote the growth of experimental high energy physics in Latin America.

Fermilab is created in 1967 close to the city of Batavia (USA), and it is the equivalent, in our American continent, of CERN. Leon Lederman visits Mexico and other Latin American countries at the beginning of the eighties. By this time, only two institutions are dedicated to particle physics: Physics Institute of UNAM (IF-UNAM) and Physics Department of Cinvestav. With the active participation of Clicerio Avilés (Clic) a first "Panamerican Simposium on Physics and Technology" is held in Oaxtepec, Morelos in 1982. Both Matías and Miguel Ángel agree on the fact that this meeting became a milestone in the emergence of experimental particle physics in Mexico. It is important to highlight that the idea of encouraging stays abroad of Mexican researchers (and even the first of such stays) was conceived before the Mexican Division of Particles and Fields of the Mexican Physical Society (DPyC-SMF) was created.

Dr. Matías Moreno IF-UNAM

Nobel Prize winner
Leon Lederman

Dr. Miguel Ángel Pérez Cinvestav

"As a result of these gatherings, some colleagues from UNAM and CINVESTAV got interested in driving the training of young Mexican researchers in this area. From IF-UNAM, Jorge Flores, Clicerio Avilés and Matías Moreno decided to start this process with

Journal of Physics: Conference Series **761** (2016) 012003 doi:10.1088/1742-6596/761/1/012003

the 'conversion' of active physicists working on high-energy physics (FAE) theoretical and phenomenological research in experimental physicists. Clicerio Avilés himself managed to succeed on this goal through long stays at Fermilab and, on his return to Mexico, created the Institute of Physics of Guanajuato University. From CINVESTAV, Augusto García, Arnulfo Zepeda and Miguel Ángel Pérez decided to took a different strategy and got several CINVESTAV students excited with the idea of perfecting his doctoral training in experimental high energy physics via thesis advised by Fermilab, CERN and DESY (Deutsches Elektronen Synchrotron, near Hamburg, in Germany) researchers [1]. This strategy was also chosen by some Brazilian colleagues, as Joao dos Anjos and Alberto Santoro."(Miguel Ángel Perez, 2016).

First stays abroad and a sad tragedy at CERN

Finally, the first visits are defined by the groups at IF-UNAM and CINVESTAV. Also within these institutions, the idea of creating both a Workshop and a School of the Mexican DPyC-SMF arises; but this is another story, that we leave to his protagonists. Then, in the middle of the eighties, before the foundation of the DPyC-SMF, the proposal of research stays abroad and the corresponding interaction with Fermilab (and a little later with CERN) start.

A first mention of the contest for stays abroad on high-energy physics appears in the bulletin of the SMF (Volume 4, Number 1, January-April 1990). Later on, it is included as a formal call (Volume 7, Number 3, July-September 1993); specifically, for Scientific Summers abroad, under the name Fellowship "Bruno González" in the case of the student appointed to visit CERN.

Physical engineer Bruno González, graduated at Iberoamerican University and Ph. D. student at Cinvestav, passes away during his visit at CERN as a consequence of a crash with a car (he was hit by a car while he was running his bike) very close to the lab in 1988. Since this incident, CERN commits to support every year a Mexican student in his memory. Bruno González, only child, outstanding student, who had been an excellent (very demanding) lecturer on Theoretical Mechanics, is still missed in our Community. Personally, I keep a clear remembrance of his burial, this sad loss shook profoundly all Iberoamerican University and Cinvestav students.

From my point of view, this extremely unlucky event, boosted the organizers to enlarge the call for research summers in order to allow more Mexican students to visit Fermilab and CERN and, by means of their work, honour the memory of Bruno González.

Formalization and development of the call in its current format

At the 2011 edition of the annual meeting of the DPyC-SMF and during the gathering of the General Assembly, a group of (mostly very young) experimental physicists proposes to take the reins of the contest and promote its realization outside Mexico City. Drs. Alexis Aguilar Arévalo (ICN-UNAM), Eduard de la Cruz Burelo (CINVESTAV Phys. Dept.), Pedro Podesta (Universidad Autónoma de Sinaloa) and Salvador Carrillo (Universidad Iberoamericana) make up a first National Committee, which is approved in the General Assembly. I remember a very good acceptance of this idea by the attendants and both Drs. Matías and Miguel Ángel, express very punctually the route and way the Call should follow from then on, and ask us for a yearly report of this activity at every annual meeting of the DPyC-SMF.

Since then, the contest has been repeated annually, intending to join the best applicants, for their preparation close to the centre of Mexico and in the outer states of the Republic, alternatively (the purpose of the latter is to foster the scientific development of these areas). Along its history, the call has been held in Puebla, Sinaloa, Faculty of Sciences at UNAM, Guadalajara, Univ. Iberoamericana México, Chiapas, and this year in Universidad Autónoma del Estado de Hidalgo. The presidency of the national committee has been taken and rotated among its members voluntarily.

The format of the call has recently changed with the intention of covering the most important abilities required for the type of work that will be done during the stay. It is established in a mini-workshop format, with the participation of at least 6 people responsible, each one in charge of a course: 3 experimental lectures, two on computing and one in statistics with an introduction to experimental high-energy physics. Experimental courses intend to use particles physics detectors and are distributed in parallel sessions, in such a way that these allow evaluating the students, who are collaborating in small teams, on their aptitudes and experimental abilities. Computing sessions use software such as ROOT or PYTHIA under Unix environment. Participation of the local researchers is encouraged, including students from the state of the hosting institutions or neighbouring states.

The current format of the contest is the listed below:

1) Elaboration of a pdf poster, which is advertised from the web site of the event.
2) Promotion in social networks. The Facebook page of the event was visited by more than 1500 people during the first hours and it has more than 2000 followers by now. As the call is made, a pre-registration is required.
3) Pre-registration, where students general data are collected. This allows the organizers to quantify the required expenses and organize students transportation, mainly. Also, a pre-selection of the students is made at this stage.
4) A deadline is established for delivering the documents. After it, in less than a week, the appointed candidates start their training.
5) The students competition starts. It consists of a mini-workshop held during the two-days courses. It is important to emphasize that all lecturers deliver, at the end of the second day, a list with the marks achieved by every student.
6) Marks are averaged together on the third day and a ranking is established. Depending on the number of available stays, a list with the selected candidates for accessing the final stage of the competition is established. This final phase consists in an interview, which is carried on in English. At the end of this 10-minutes talk, time is allotted for the judges to tell freely their opinion to the contestant. The students are also asked about their preferred lab to visit. In case they are selected, this opinion is certainly taken into account. I would like to emphasize the importance of a good English level, although no proficiency is expected at this point. The visiting students will have close contact with foreign colleagues during their stays and this will strength naturally their English skills, helping their needed interaction with other collaborators along the stay.

Journal of Physics: Conference Series **761** (2016) 012003 doi:10.1088/1742-6596/761/1/012003

7) A final ranking is made including all marks obtained in the lectures and the interview, and the joint average determines the final selection of visiting students. This process is open to the scrutiny of all judges and invited participants.

During this procedure, we were fortunate to count with –at least once- the participation of a Mexican scientist coming from one of our hosting institutions: Dr. Carlos García Hernández, from Jefferson Lab (USA). This time, Dr. Carlos García appointed directly the student who would work in this stay and also helped us to consolidate Jefferson Lab's support. We would like that this type of collaboration be promoted by other labs in the future.

The bitter with (all) the sweet

I consider that the new format has been very much successful and it has been an important achievement of the National Committee. However, the topic that remains difficult to tackle every year is the financial support for making the contest. The National Network of High-Energy Physics (RED-FAE, which is a national thematic network of CONACyT) and the same DPyC have offered support, but it has not been always enough and, in several instances, the National Committee members had to restore to funding from their own institutions, which is not always easy to obtain.

Perspectives

Unfortunately, there has not been yet any woman among the National Committee members. I cannot avoid mentioning that Drs. Elsa Fabiola Vázquez and Cristina Oropeza have participated in the three-day workshops a couple of times each. Moreover, on the 30th annual meeting of the DPyC-SMF, Dra. Karen Caballero (UNACH) and Isabel Pedraza (BUAP) were invited directly to join the National Committee. The only requested condition is that they participate continuously for at least two or three contests so that they have the chance of presiding the Committee that is appointed for simple voluntary service and in agreement with the members of the National Committee in every general Assembly of the DPyC-SMF.

A final topic that I consider worth mentioning deals with the path which needs to be promoted in the future of science (physics) in Mexico. Unfortunately, this issue has created some controversy inside our Community. I consider that independently of that, we must appreciate the labour started, more than 30 years ago, by a group of theoreticians and phenomenologists, who fostered scientific stays abroad on experimental high-energy physics. Dr. Matías is concerned with the possible brain drain affecting our most talented students, however, the result of this activity has been –beyond any doubt- beneficial for the development of experimental high-energy physics in Mexico. All members of the DPyC-SMF, we shall learn the lessons from this example and apply them to develop other related areas.

Luckily, the detailed tracking made by Dr. Jurgen Engelfried (UASL) in the period 1982-2000 includes the record of the first students benefitting of the experimental stays on high-energy physics. Later on, Dr. Marco Reyes (IF- Univ. Gto.) continued briefly this activity, but we lack some information from the early 2000's until 2011, when the members of the new group resumed this activity.

1982				
Fermilab				
	Juan Carlos D'Olivo	CINVESTAV	Ph. D.	Theory, ICN-UNAM
	José Mustre de León	CINVESTAV	Master Th.	Material Sc., Cinvestav
1984				
Fermilab				
E715	Antonio Morelos Pineda	CINVESTAV	Master Th.	Professor, IF-UASLP
E715	Gerardo Moreno López	CINVESTAV	Master Th.	Professor, IFUG
E715	Albino Hernández Galeana	CINVESTAV	Master Th	IPN
1986/7				
Fermilab				
	Bruno González	CINVESTAV	Master Th.	
	Juan Manuel Álvarez Haro	CINVESTAV	Master Th.	Economy, SHCP
	Juan José Alvarado Gil	CINVESTAV	Master Th.	Photoacustics, CINVESTAV UM
1988				
CERN				
	Bruno González	CINVESTAV	Ph. D.	***Passed away, CERN accident**
	Heriberto Castilla (replacing)	CINVESTAV	Ph. D.	CINVESTAV
1989				
CERN				
	Agustín Enciso Muñoz	CINVESTAV	Master Th.	Professor, U Zacatecas
1990				
CERN				
UA1	Marco Octavio Lanzagorta Saldaña	UNAM	Intern	Computer Graphics, US Navy
1991				
CERN				
	Alejandro Corichi Rodríguez	UNAM	Intern	Gravitation, ICN-UNAM
1992				
CERN				
LHC	J. Guillermo Contreras	CINVESTAV	Master Th.	CINVESTAV UM
LHC	Leonel Magaña	CINVESTAV	Master Th.	PostDoc, ALICE, Torino
Fermilab				
SELEX	Emilio Esparza Coss	CINVESTAV	Master Th.	Med. Phys., IFUG y CINVESTAV
E761	J. Flores	CINVESTAV	Ph. D.	
E791	Marco Reyes Santos	CINVESTAV	Ph. D.	Professor, UMSH

1993

CERN

LHC	Salvador Carrillo Moreno	CINVESTAV	Master Th.	Ph. D., FOCUS, CINVESTAV
LHC	- - - - -	UNAM	Intern	*Abandons CERN stay
DESY				
H1	Fabiola Vázquez Valencia	CINVESTAV	Master Th.	Ph. D., FOCUS, CINVESTAV
Fermilab				
E791	Rubén Flores Mendieta	CINVESTAV	Master Th.	Theory, IF-UASLP
E791	Alberto Sánchez Hernández	CINVESTAV	Master Th.	CINVESTAV

1994				
CERN				
	Alberto Rubio Ponce	CINVESTAV UM	Master Th.	Solid State, CINVESTAV
	Jesús Martínez	ESFM-IPN	Intern	Ph. D., ESCOM, IPN
DESY H1	Javier Espinoza	CINVESTAV UM	Master Th.	PostDoc, Sync. Rad., Seattle
Fermilab				
E791	Luis Manuel Montaño Zetina	CINVESTAV	Master Th.	PostDoc, Med. Phys., Torino
E791	Paulo César Lozano	CINVESTAV	Master Th.	Ph.D., Aeronautics, MIT

1995				
DESY				
H1	Oscar Ramírez	ESFM-IPN	Intern	Computing, Germany
Fermilab				
SELEX	Ricardo López Fernández	IF-UASLP	Master Th.	Ph.D., ALEPH, Grenoble
E791	Armando Encinas Oropesa	IF-UASLP	Master Th.	Solid State Physics.

1996				
CERN				
FELIX	Benjamín Gutiérrez	UNAM	Intern	Master Th., UNAM
NA49	Manuel Calderón de la Barca	CINVESTAV	Propedeutics	Ph. D., STAR, Yale
DESY				
H1	Miguel Mondragón	CINVESTAV	Master Th.	Ph. D., H1, Dortmund
	Horacio Cantú Quirino	IF-UASLP	Master Th.	
Fermilab				
DZERO	Ivette Fuentes	FC-UNAM	Intern	Imperial College, London
Lab D	Liliana Velasco	FC-UNAM	Intern	Ph. D., Oxford (UK)
	Mónica Borunda Pacheco	FC-UNAM	Intern	Ph. D., UC Davis

1997				
CERN				
	Marciano Vargas Treviño	FCFM-BUAP		
NA49	Víctor Tapio Rangel Kuoppa	ESFM-IPN	Master Th.	Ph. D. Tampere, Finland
	Julio Barreiro			
DESY				
H1	Gabriela Murguía Romero	UNAM	Intern	Ph. D., UNAM
	Miguel Francisco Villoro-Valdés	UNAM	Intern	Astrophysics, IA-UNAM
Fermilab				
CDF	Enrique Moreno Méndez	UNAM	Intern	Master Th., UNAM
DZERO	Ma.Magdalena González Sánchez	UNAM	Intern	Astronomy, U. Wisconsin
FOCUS	Javier González Beltrán	UNAM		Ph. D., UNAM
DZERO	Ricardo Hernández Pérez	ESFM-IPN		
CDF	Alfredo Raya Montaño*	EFM-Morelia		
SUNY				
	Bettina Delgadillo Lozano	ITESM		
	Erick Rosas	CIO		

Journal of Physics: Conference Series **761** (2016) 012003 doi:10.1088/1742-6596/761/1/012003

1998				
DESY				
H1	Andrea Vargas	BUAP	Intern	Master Th., Auger, Cinvestav
HERA-B	Erika Yvonne Carrasco Zavala	FC-UNAM		Ph .D., Theory, Cambridge (UK)
Fermilab				
CMS	Erika Álvarez	UNAM		
CDF	Rogelio Cuevas Saavedra	UNAM		
DZERO	Homero Díaz Marín	U. Michoacana		

1999				
CERN				
	Juan Barranco M.	UDLA	Intern	
	Alejandro Reyes C.	FC-UNAM	Intern	
DESY				
H1	Cristo Yee (Manuel) Rendón	UA Sinaloa	Intern	
	Jimmy Hernández Bello	UDLA	Intern	
H1	J. L. Gamboa	CINVESTAV UM	Master Th.	Master Th., CINVESTAV
Fermilab				
HyperCP	Zaida del Rosario Urrutia del Cid	IFUG	Master Th.	Master Th., E690, IFUG
Axion	Diana Eva Bedolla Orozco	FC-UAEM	Intern	Degree, FC-UAEM
SDSS	Ángel Mancar*	UNAM		

2000				
CERN				
HARP	Alejandro Castillo Ramírez	CINVESTAV		
ALICE	Arturo Avelino Huerta	FCFM-BUAP		
DESY				
HERA-B	Beatriz Arellano Bautista	UDLA		
HERA-B	Eric Martínez Pascual	FCFM-BUAP		
HERMES	Eric Vázquez Jáuregui	UV	Master Th.	Postgrade, IF-UASLP
HERMES	Jorge Antonio Félix Hernández	CINVESTAV	Master Th.	Master Th., IF-UNAM
Fermilab				
CKM	Belinka Fernández González	FCFM-BUAP	Master Th.	
SUNY	Adrián Pérez Galván	Iberoamericana		
	Ignacio Francisco Gallardo Guerra	UNAM		

This record was kept by Dr. Jurgen Engelfried UASL

2001		
CERN		
	Argelia Bernal Bautista	CINVESTAV
	Carlos Alberto Chávez Barajas	CINVESTAV
	Jorge Mercado Pérez	CINVESTAV
DESY		
	Edén Figueroa Barragán	ITESM
	Jacinta Grajales Cravioto	BUAP
	Joaquín Miranda Mena	CINVESTAV
	Lao Tse López Lozano	CINVESTAV
Fermilab		
	Marié López del Puerto de la Garza	UDLA
SUNY		

Julio Vargas Mena
Salvador Hernández Oliver

This record was kept by Dr. Marco Reyes IF-UGTO

The following record is being kept by the new Committee of the scientific summers on experimental high-energy physics

2012		
CERN	Ramón Heberto Martínez	U Sonora
JLab	Cesar Serna	U Gto.
DESY	Omar Fernando Sosa	BUAP
Fermialb	Grecia Guijarro	UA Sinaloa
2013		
CERN	Karla J. Peña R.	U Sonora
DESY	Juan Manuel Grados L.	BUAP
Fermilab	J. Javier Rendón C.	UA Sinaloa
JLab	Gabriel G. Palacios S.	UAM
2014		
CERN	Diego Berdeja S.	Iberoamericana
DESY	Juan L. Cuspinera C.	BUAP
DESY	Arturo A. Santaella O.	UNAM
Fermilab	Oscar Meza A.	BUAP
JLab	Andrés Sánchez P.	CUCEI
Berkeley	Luis A. Zazuleta R.	U Sonora
2015		
CERN	Felipe Gilberto Ortega G.	ITESM
DESY	Pamela Patricia Ornlas S.	U G
DESY	Saulo Hernández L.	U Sonora
Fermilab	Gonzalo Sánchez G.	ESFM IPN
JLab	Edgar Abarca M.	UNAM
2016		
CERN	Edna Loredana Ruiz V.	UNAM
Fermilab	María Martínez C.	ITESM
DESY	Tania Martínez C.	BUAP
DESY	Laila Vleeschower	UNACH
JLab	Adriana Ivonne Canales R.	UNAM
SNOLab	José Guillermo Lara	UNAM

References:

[1] A more complete versión of this process can be found in A. García y M.Á. Pérez, "High energy physics in Mexico: historical sketch and implications", AIP Conf. Proc. 857, 3-10 (2006).

Theoretical–research summer:

For a new generation of experts on high energy physics

Saúl Ramos-Sánchez

Instituto de Física, Universidad Nacional Autónoma de México, POB 20-364, Cd.Mx. 01000, México

E-mail: `ramos@fisica.unam.mx`

Abstract. Motivated by the need to strengthen the comprehensive training of young Mexican physicists interested in theoretical high energy physics, the *Theoretical–research summer on high energy physics* program was conceived. This program, that celebrates its sixth anniversary, consists in a yearly, nationwide challenging contest in which a board of experts identify the best undergraduate contestants to support them during short research stays in high–energy–theory groups of prestigious international institutions. Out of 80 contestants, the eight awarded students have demonstrated their skills, producing highly advanced (and publicly available) reviews on particle physics, field theory, cosmology and string theory, and a published paper.

1. The origins

During more than 30 years, the (experiment–oriented) *Verano Científico en Laboratorios Extranjeros* program has had great success and acceptance among young students interested in becoming experts on experimental high energy physics. Unfortunately, targeting a different population, many talented undergraduate students devoted to theory could not be supported by that program.

Looking for a mechanism to identify the future generation of theoretical high energy physicists, in 2010 the *Theoretical–research summer on high energy physics* contest was conceived by a group of Mexican high energy researchers, with support of the Division of Particles and Fields of the Mexican Physical Society (DPyC) and the CONACyT National Network on High Energy Physics (RedFAE). The main purpose of the program is to promote the integral professional training of future researchers on theoretical high energy physics, by encouraging international contact in early stages of their education. Evidently, reaching this goal shall lead to an even more successful new generation of Mexican experts on theoretical high energy physics.

In this program, talented Mexican students contest to earn the opportunity to contribute to the development of theoretical research in highly recognized institutions in a foreign environment. The best contestants are awarded with a full two–month scholarship for a research stay at the theoretical high energy physics departments of the International Centre for Theoretical Physics (ICTP) or the Thomas Jefferson National Accelerator Facility (JLab).

The first call for applications took place on 2011 and, since then, its success has demonstrated the need of such a program. During the short life span of this program, not only top–quality reviews and reports have been produced, but also a published article and a rich exchange of ideas and collaborations as well as the seeds for future research have been the result of the effort of the successful contestants. Furthermore, since the awarded students are motivated to share their experience and knowledge after their stays with the younger generations, the new applicants can profit from the discussions that take place during the contest.

2. The selection process

The call for applications of the contest is published nationwide every year in summer, through posters and emails that are shared among the Mexican experts on high energy physics and directors of scientific institutions. Applicants are expected to be young students, attending lectures of the last two semesters of their B.Sc. or the first two semesters of their M.Sc.. That is, contestants should be between 20 and 25 years old. They should also have demonstrated in their career a clear tendency towards theoretical physics. Beside knowledge on particle physics and its mathematical and conceptual basics (quantum mechanics, relativity, differential geometry, group theory, etc.), contestants should have some skills on analytical and numerical calculation methods.

The applications should include *i)* the full CV of the applicant, *ii)* her/his academic transcript, *iii)* a concise statement of interests and scientific prospects, and *iv)* two letters of recommendation (to be sent independently by the referees). The complete applications are submitted by the end of November through the electronic portal

http://stringpheno.fisica.unam.mx/veranosteoricos/

which contains many details of the program as well as a historic overview of its progress.

The applications are evaluated by a selection committee composed by five to eight members, who vary from year to year in order to incorporate different perspectives into the examination process. The selection committee comprises recognized scientists, including some experimentalists, of various disciplines within high energy physics. Some of the members of previous selection committee are A. Aguilar (UNAM), J. Barranco (Guanajuato U.), A. Bashir (UMSNH), K. Caballero-Mora (Chiapas U./MCTP), E. Castellanos (Chiapas U./MCTP), J. Erler (UNAM), A. García-Zenteno (UNAM), A. Güijosa (UNAM), I. Huet (Chiapas U./MCTP), R. Linares (UAM-I), M. Mondragón (UNAM), G. Murguía (UNAM), L. Patiño (UNAM), M. Tejeda-Yeomans (Sonora U.), G. Toledo (UNAM), and L. Valencia-Palomo (Chiapas U./MCTP).

The selection process has three stages. In the first stage, after the candidates have submitted their documents and recommendation letters, the selection committee evaluates them and, based on this information, must decide who may be the most qualified students. The successful candidates in this first stage are the finalists, who are then invited to participate in the next stages, which are hosted at a Mexican institution, where theoretical high energy physics is developed.

In the second stage, the candidates attend a couple of crash courses on essential topics (such as quantum field theory, QCD and QED particle physics, relativity, cosmology and string theory) and talks held by previously awarded students and experts of the area. The language of lectures and seminars is English; questions and discussions are encouraged. After these sessions, which may take up to three days, students are examined in a three-hour-long written test, partly based on the topics discussed on the crash courses and the talks, and also on elementary concepts of high energy physics.

The third and last stage gives the selection committee the opportunity to gain an overall perception of the students' skills to have an academic discussion. In this stage, the experts interview the best candidates about their previous knowledge and experiences in theoretical high energy physics, their research interests, and their expectations at the research center where they would like to make a stay. Some questions should invite the candidate to exhibit the technical abilities (s)he has in her/his favorite research topic. The interviews are personal and conducted in English.

All examination elements acquired during the contest let easily identify the best candidates. The names of the awarded contestants are announced in the closure ceremony, where all finalists also receive their certificates.

The *Theoretical–research summer on high energy physics* program has received about 100 applications from which over 60 students have excelled in the first stage of the contest. The selection committees have made difficult decisions to identify among them the best qualified contestants. Until the summer of 2016, eight exceptional students have been awarded with a full scholarship to develop a self–contained project on theoretical physics. The topics they have investigated range from low–energy meson and baryon physics to higher–order corrections of string theories. They have led to works of remarkable quality [1] and even to published papers.

The ICTP and the JLab have provided financial support and academic guidance to the awarded contestants. Scientists of all levels, including e.g. the director of the ICTP, have participated advising the projects of the students. Besides, students have been assigned an office and have been incorporated in the group activities. Particularly, they were included in seminars, workshops and schools organized at the institutions, such as the Hampton University Graduate School (HUGS) at the JLab and the various summer schools at the ICTP.

After their research stays, the successful candidates are invited to report the main results of their work and to share their achievements in public seminars, attended by the new contestants and active researchers of the area. The goal of these activities is twofold. First, the seminars and reports motivate young students who often see the research topics of the awarded students for the first time in there, and find out that they could be presenting similar findings, should they pull ahead in the contest. Second, holding professional talks before a critical academic audience is a limestone of a comprehensive training as a scientist, complementing thereby the purpose of the program.

3. Awarded students

In the following, we summarize some of the achievements of the eight awarded young scientists.

Lucía Gómez Córdova

In December of 2011, Lucía (21y) was a 7th semester B.Sc. student at the University of Guanajuato, campus León, working under the supervision of Oscar Loaiza-Brito. During her stay at the ICTP, in the summer of 2012, she developed the project *Moduli stabilization and inflation in type IIB supergravity*, supervised by Michele Cicoli. Now, she is a Ph.D. student, supervised by Pedro Vieira and Robert Myers, at the Perimeter Institute for Theoretical Physics. She works on quantum field theories and has published three papers.

David Theurel Lambert

In December of 2012, David (24y) was an undergraduate student at the UNAM, working on his B.Sc. thesis on an attempt towards quantum gravity. During his stay at the ICTP, in the summer of 2013, he was supervised by Gabrijela Zaharijas and Emiliano Sefusatti on the project *A bound on the short-wavelength extrapolation of the dark matter power spectrum*, that led to the publication of the paper entitled *Extragalactic gamma-ray signal from dark matter annihilation: an appraisal* [2]. Since 2013, he is a Ph.D. student at the MIT, where till 2015 he studied non–geometric theories of gravity; he is now working on computational neuroscience.

Juan Carlos De Haro Santos

In December of 2012, Juan Carlos (23y) was an undergraduate student at the University of Guanajuato, working on his B.Sc. thesis under the supervision of David Delepine. During his stay at the JLab, in the summer of 2013, he developed the project *Noether's theorem for dynamical symmetries*, supervised by José Goity. After defending successfully his M.Sc. thesis *Commissioning of CUORE Fast Cooling System* in 2016, he does now a Ph.D. at the U. Guanajuato, supervised by D. Delepine. He works on CP violation in mesons.

Mariana Carrillo González

In January of 2014, Mariana (22y) was an undergraduate student at the University of Morelos, working on her B.Sc. thesis, advised by Gabriel Germán. During her stay at the JLab, in the summer of 2014, she was supervised by José Goity on the project *Spin-flavor symmetry for large N baryons in a boosted frame.* After completing her M.Sc. at the Perimeter Institute in 2015, she became a Ph.D. student at the University of Pennsylvania under the supervision of Mark Trodden. She works on particle cosmology and has published five papers.

Diego Vidal Cruz Prieto

In January of 2014, Diego (23y) was an undergraduate student at the UNAM, working on his B.Sc. thesis under the supervision of David Vergara. During his stay at the ICTP, in the summer of 2014, he worked on *D-brane action corrections through S matrix computations at disk level*, under the direction of Hesan Hatefi. In 2015, he started his M.Sc. studies at UNAM, supervised by D. Vergara.

Ricardo Espíndola Romero

In December of 2014, Ricardo (22y) was an undergraduate student at the UNAM, working on his B.Sc. thesis under the direction of Antonio García Zenteno. During his stay at the ICTP, in the summer of 2015, he was supervised by Leopoldo Pando Zayas and worked on *Localization in Field Theory and Matrix Models.* He is now an M.Sc. student at UNAM, working on field–theoretical aspects of the AdS/CFT correspondence with his B.Sc. advisor. He is co–author of a paper.

Marco Antonio Carrillo Bernal

In December of 2015, Marco (22y) was an undergraduate student at the University of Veracruz, working on his B.Sc. thesis with the advise of Carlos Vargas and Genaro Toledo. He had already co–authored three papers on different topics. During his stay at the JLab, in the summer of 2016, he worked on the project *Extracting the chiral anomaly from $\gamma\pi \to \pi\pi$ through lattice QCD*, under the direction of Raúl A. Briceño. In autumn 2016, he shall start his M.Sc. studies at UNAM, supervised by Wolfgang Bietenholz, specializing on the topics he developed at JLab.

Uriel Adrián Luviano Valenzuela
In December of 2015, Uriel (21y) was a 7th semester B.Sc. student at the UNAM. During his stay at the ICTP, in the summer of 2016, he was supervised by Luis Aparicio and Fernando Quevedo, who guided him in his research on *α' corrections to supersymmetry breaking in type IIB string compactifications*. He shall conclude his B.Sc. studies by the winter of 2016.

4. Outlook

The *Theoretical–research summer on high energy physics* program has identified gifted undergraduate students and ignited their development as future experts on theoretical high energy physics. However, there are various challenges that must be tackled in order to reach the final goal of our program. One of them is the possibility to offer scholarships at other prestigious institutions, such as FermiLab, DESY and IPMU. Discussions between the organizing committee of this program and the administration of those institutions are conducted and could well lead to enlarge the current program in few years.

Further, even though this program has achieved Mexican institutions with few or no experts on high energy physics, it is clear that the contest by itself cannot influence much the training process of future experts on theoretical high energy physics at those places. Perhaps, a coordination with other programs, such as the yearly *Escuela en física fundamental*, could facilitate the success of the candidates from those institutions.

Finally, the *Theoretical–research summer on high energy physics* program has welcome several talented students from large and small universities across the country, who unfortunately could not be awarded by the program. Besides, there are great institutions in Mexico with specialists on high energy physics, who could well host the brilliant students in similar nationwide research stays, to perform research of international quality. With this purpose and with the aid of this program, suitable candidates and institutions shall be identified and supported to achieve their common scientific goals.

A new generation of talented Mexican experts on theoretical high energy physics is in development. At this stage, it is not possible to know the achievements that they could bring, but we are confident that their effort and a proper guidance can let them reach the greatest of all purposes in physics: a better understanding of laws of Nature. Hopefully, this program contributes to fulfill this aim.

Acknowledgments

I am especially grateful to F. Quevedo, M. Pennington, J.L. Goity, and the ICTP and JLab administrations, whose support and advice make possible this project. I am also indebted to A. Aguilar, J. Barranco, A. Bashir, K. Caballero-Mora, E. Castellanos, J. Erler, A. García-Zenteno, A. Güijosa, I. Huet, R. Linares, M. Mondragón, G. Murguía, M. Napsuciale, L. Patiño, M. Tejeda-Yeomans, G. Toledo, L. Valencia-Palomo, A. Zepeda, among others, for their enthusiastic support as members of the organizing and selection committees at various stages of this project. Al last, but not at least, I would like to thank to L. Aparicio, R.A. Briceño, M. Cicoli, J.L. Goity, H. Hatefi, R. Myers, L. Pando-Zayas, F. Quevedo, E. Sefusatti, P. Vieira, and G. Zaharijas for their time and remarkable skills to advise the young students awarded by this program. This project has been partly supported by CONACyT grants 151234, RedFAE2015, PNPC2013, PAPIIT grants IB101012 and IA101811, and by the DGDC-UNAM. I would like to thank the ICTP for the kind hospitality and support received through its Junior Associateship Scheme during the time I wrote this report.

References

[1] Description and project reports on `http://stringpheno.fisica.unam.mx/veranosteoricos/winners.html`
[2] Sefusatti E, Zaharijas G, Serpico P D, Theurel D and Gustafsson M 2014 *Mon. Not. Roy. Astron. Soc.* **441** 1861–1878 (*Preprint* 1401.2117)

Science and Technology of Accelerators

Cristhian Valerio Lizarraga[1], Alejandro Castilla Loaeza[2], Gerardo Guillermo Cantón[3], Carlos Duarte[1], Daniel Chavez Valenzuela[2], Karim Hernández Chahín[2], Humberto Maury Cuna[2], Luis Medina Medrano[2], Juan Reyes Herrera[4], Salvador Sosa Güitrón[5], Alan Valdivia García[2], Bruce Yee Rendón[6]

[1]Universidad Autónoma de Sinaloa,
[2]Universidad de Guanajuato,
[3]Centro de Investigación y de Estudios Avanzados del I.P.N.
[4]Universidad Nacional Autónoma de México,
[5]Old Dominion University.
[6]The High Energy Accelerator Research Organization.
cvalerio@uas.edu.mx

Abstract. The Mexican Particle Accelerator Community (CMAP) was created in 2015 and currently its members participate in different experiments around the world. Using their expertise, they are working in develop the particle accelerators area in Mexico. This paper provides a resume of the research done by its members and presents the preliminary design of an electron linear particle accelerator (eLINAC). This proposal will be the first accelerator designed and created in Mexico.

1. Introduction

Since 2007, the Division of Particles and Fields of the Mexican Physical Society (DPyC-SMF) sent a group of students to different laboratories to learn about particle accelerators [1], from this endeavor the CMAP was created. The good performance of their students allowed to consolidated the collaborations of this field between Mexican Universities and international Labs such as European Organization for Nuclear Research (CERN), Jefferson Lab, Berkeley Lab and ALBA.
Additionally, the two Mexican Particle Accelerators Schools (MEPAS) were a key factor to support the growth of the accelerator community. The MEPAS represents a relevant resource for the training of the students. Indeed, thanks to that experience learned they continue its research in the area, moreover, CMAP members start as students in these schools.
Nowadays, several laboratories uses particles accelerators for their research around the world, in addition, the industry and government are also users of these machines. In Mexico, the user community of accelerator is large, however, the numbers of facilities and experts are insufficient to cover the demand. The CMAP is developing the project of eLINAC , which will be the first accelerator designed and created by Mexicans. The main goals of this proposal is to cover part of the necessities of the users and the consolidation of the particle accelerator area in Mexico, [3].

2. CMAP members Research

This section we explain some of the research made by the CMAP members. The beam propagation through the accelerator requires a deep understanding of the interaction of electromagnetic fields with charged particles in order to estimate the behavior of the charged particles. Moreover, the collective effects created by the interaction of the particle beam itself, such as the space charge, when the density is high needs to be taken in to account as well as the relativistic effects when the particles travel close to the speed of light or they lose energy due the synchrotron radiation.
These studies in the machine required to run some complex simulations using several variables, codes as Methodical accelerator Design (MADX) [4] can estimate the propagation of the beam for long circumference like the LHC ring.
On the other hand, some quantum effects are needed to understand the beam formation in the plasma at the beam sources and the generation of secondary particles due electromagnetic fields or particle collisions.
Finally, another relevant subject is the beam losses. The particles lost in the ring are unavoidable, therefore, it is important to understand the mechanism that produced the losses and mitigate its effect in the acceleration chain in order to protect the workers safety and the integrity of the machine components.
In the next subsection we resume the research expertise of some members of CMAP group:

2.1 Gerardo Guillermo Canton

Synchrotron Radiation (SR) Studies using numeric method tools (montecarlo) are important for the good performance of the LHC. SR photons are simulated inside an accelerator's beam pipe, they are tracked from generation until final absorption, though all reflections in the vacuum chamber. The tool used is Synrad3D developed at Cornell University by D. Sagan which uses a mathematical model by G. Dugan.

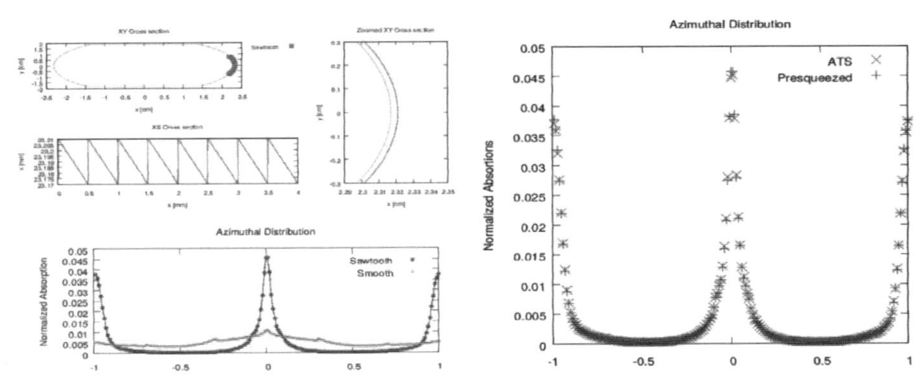

Figure 1: Effect of the sawtooth in the chamber of LHC and the effect of ATS optics in HLLHC

We have studied the difference between sudden (step function) changes of shape in the vacuum chamber and smooth changes regarding synchrotron absorption distribution.

Finally, the behavior of SR in a novel beam pipe (Figure 2) designed by R. Kersevan for FCC-hh which presents two slits in its equator.

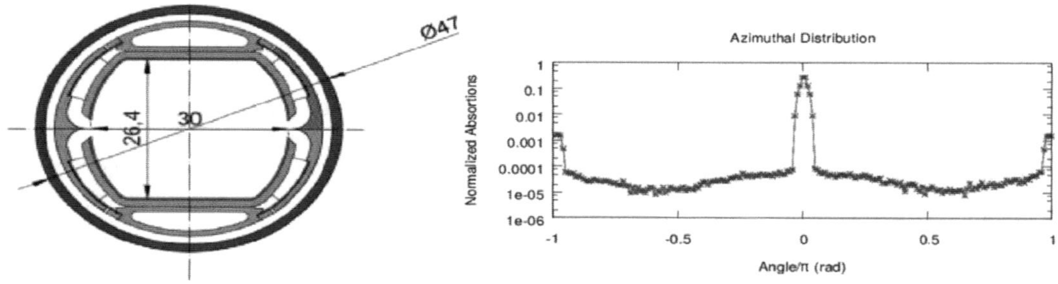

Figure 2: Enhanced geometry for synchrotron radiation

2.2 Luis Medina Medrano

The dynamic aperture (DA) is a key parameter of an accelerator for the evaluation of its performance, since it is related to the stability of the trajectories of the particles in the beam. Estimations of the DA of the two current lattices of the Future Circular Collider - ee (FCC-ee) at the top pair energy have been obtained. Both lattices, for which significant progress has been made, have asymmetric interaction regions (IRs) with β_x*/ β_y =50 cm/mm.mrad and $\varepsilon_x/\varepsilon_y$ =2nm/2pm, and fulfil the geometric constrains arising from the compatibility with the FCC-hh machine.[5] The DA at the interaction point (IP) has been computed in the four-dimensional phase space for different energy deviations with MAD-X and PTC, and the results exhibits --in the on-momentum case a peak of 20σ-28σ, for the different proposals. The review[6] of the DA with synchrotron motion, calculated using SAD and including different effects such as RF, tapering and radiation damping, shows that the requirement of 15σ for particles on-momentum, and 5σ for particles with up to 2% energy deviation, is almost satisfied. A statistical approach implementing the random effect of radiation, and the inclusion of errors and misalignments in the magnets are being explored for more realistic estimations of the DA.

2.2.1 *Performance of HL-LHC scenarios.* The baseline of the HL-LHC, an approved upgrade of the LHC aiming at the increase of the integrated luminosity, [7] has been studied --with the latest set of parameters--, along with its main alternative scenarios: the 8b+4e filling scheme, the 200 MHz RF system, and two scenarios fore-seeing the inability to use crab cavities, [8] and their performance compared in terms of the respective integrated luminosities. The model for the simulation of a typical fill assumes a β* levelling technique (demonstrated at low intensity) for a constant pile-up of 140 in the two main IPs, the reduction of the beam intensity due to burn-off, and the evolution of emittances including radiation damping and intrabeam scattering. New criteria, such as the inclusion of penalty steps to consider, for example, the lost time during the change between optics, and the adoption of a step-based luminosity levelling, have been included as well.

Results for the baseline in its first stage, that is, with only partial compensation of the crossing angle by crab cavities, demonstrates that flat optics restores the nominal performance (260fb^{-1}) that is lost with the use of round optics. The 200MHz option provides a similar performance to the nominal, while adding the advantage of electron cloud suppression. For the scenarios without crab cavities, regarded as a backup for operation, a drop of 8% in performance is found provided that, an increase of 20% or 60%, for the cases with and without wire compensation, respectively, of pile-up density is accepted. [9] It has been realized that adopting a levelling technique will require a large number of optics for the HL-LHC for the case of small luminosity steps; therefore a 5% step seems to be reasonable, and with a negligible reduction of integrated luminosity. The different scenarios at a pile-up of 200 events/mm deliver a similar but more pronounced behaviour, as expected.

2.2.2 Beta-beating due to beam-beam effects. Studies on the β-beating due to beam-beam (BB) interactions were conduced for the LHC, and will be extended for the HL-LHC to aid in the optimization of its lattice design and, ultimately, its performance. Peak amplitudes of 8% and 6% in the horizontal and vertical planes, respectively, taking into account both head-on (HO) and long-range (LR) effects, have been found. [10]

To correct the β-beating due to BB-LR, a rematch of the IR with the closest quadrupole magnets to the IP (excluding the inner triplet) was implemented: scanning the strengths of Q4 and Q5 at the left and right of the IP (so that the identity map between them is preserved) around their nominal values, a reduction of the 2% peak β-beating along the machine by a factor of ten was found. Rematching the optics at the start and end of the IR, as well at the IP, involving different configurations of the quadrupole strengths of Q4 to Q7, was the strategy adopted to correct the effect induced by HO-BB. The correction achieved by the adjustment of a single quadrupole, such as Q4 or Q5, proved to be the most efficient option: the peak and rms β-beating are reduced by a factor of 4 in both planes, and the beating at exactly the IP is decreased at least by a factor of 2.5. In both cases, beams 1 and 2 were studied separately, with similar results. Future work for the HL-LHC contemplates the study of the BB effects in the different optics (and its corresponding parameters) from the levelling process, and the corresponding means of correction; experimental verification to validate the correction proposal for the LHC is foreseen as well.

2.3 Bruce Yee Rendon

2.3.1 Electron cloud at j-parc main ring. The Japan Proton Accelerator Research Complex (J-PARC) is a set of high power proton accelerators: Linac, Rapid Cycling Synchrotron (RCS) and Main Ring. The presence of Electron Cloud (EC) instability at Main ring during Slow extraction (SX) mode, represents a threat for the good performance of the high intensity beam.
Several surveys are under developed to detect EC signal, study its properties, dependencies and find a way to mitigate it (See Figure 3)[11].

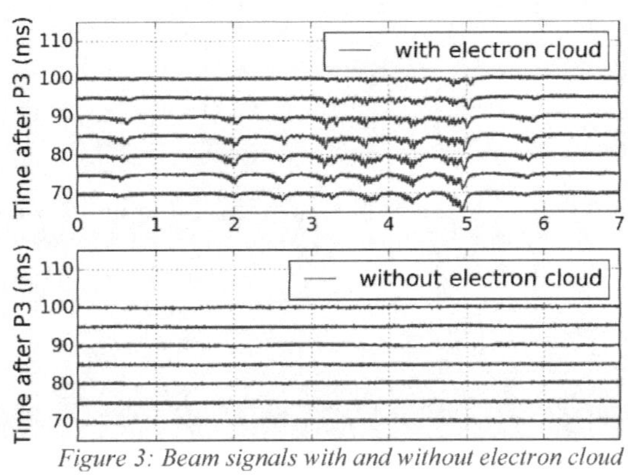

Figure 3: Beam signals with and without electron cloud

2.3.2 SIMULATIONS OF THE BEAM LOSS DISTRIBUTION AT J-PARC MAIN RING. At Main Ring of J-PARC, for some detectors the recorded signal is less than the one measured by hand dosimeters. Thus, using the software SAD and Geant 4, a model is created to explain the differences. The Figure 4 shows the results comparison between the simulations and values registered for the Beam Loss Monitors (BLM) [12].

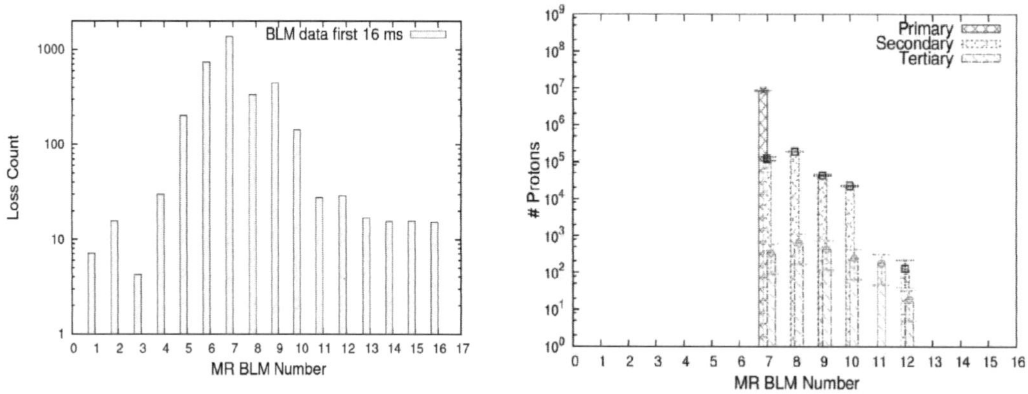

Figure 4: The comparison between the loss count for the P-BLMs (top) and the lost particles obtained by using the programs SAD and Geant 4 (bottom) at the INS A of the MR. The statistical errors are included.

2.4 *Juan Reyes Herrera*

Development of Particle Induced X-ray Emission (PIXE)(Figure 5) and X-ray optics, beamlines designing and plasma treatment of optical surfaces in the ALBA light source.

Planning of the first Mexican synchrotron and research in synchrotron light industrial applications.

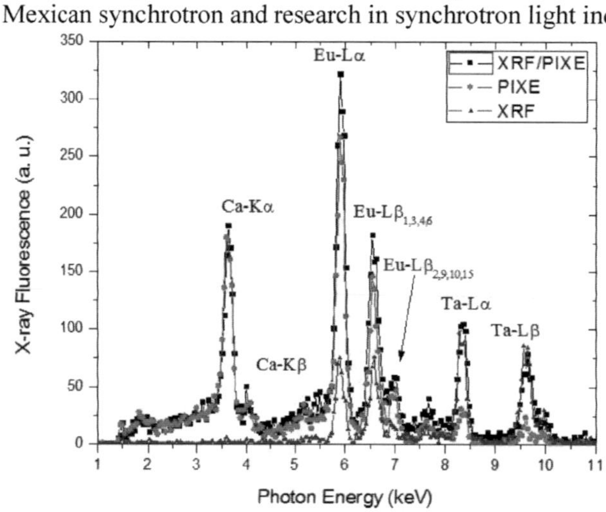

Figure 5: Characteristic X-ray emission spectra from a same sample, using XRF, PIXE and both simultaneously

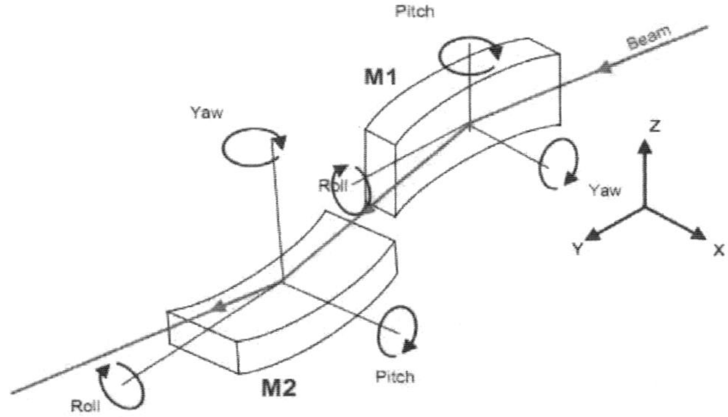

Figure 6: X-ray beam under a Kirkpatrick-Baez focusing system.

2.5 Cristhian Alfonso Valerio Lizarraga.

2.5.1 The space charge compensation (SCC) is one of the beam instabilities sources in the Low energy beam region. Studies at CERN explain how these instabilities can be predicted and propose a solution to set limits depending of the beam pressure [13].
2.5.2 Extraction systems design consists in the design of ion sources with its extraction system to enhance the emittance of the beam at the time of its creation, this work was applied in the linear accelerator LINAC4 at CERN where one of the designs is used to deliver beam in the actual accelerator.

2.6 Humberto Maury Cuna

2.6.1 Electron Cloud (EC) Effects in the Large Hadron Collider.

It is a group of electrons produced inside the vacuum chamber of an accelerator and is capable of interfering with the proper operation of the machine and/or degrade the quality of the particle beam. There are two main triggering process of an electron cloud: residual gas ionization and photoemission due to the beam-induced synchrotron radiation. This phenomenon has been a big concern for the LHC performance because the EC has prevented to reach the LHC nominal bunch spacing from its operation in 2010. An EC is able to produce beam instabilities, pressure rise and become an additional heat-load source to the LHC cryogenic systems. In order to improve the understanding of these effects simulations were performed, a review can be found in references [14, 15]. As part of my PhD thesis work, it was important to know the heat load due to the EC, in Figure 1 we can observe the average heat load for a 25 ns bunch spacing as a function of secondary emission yield.

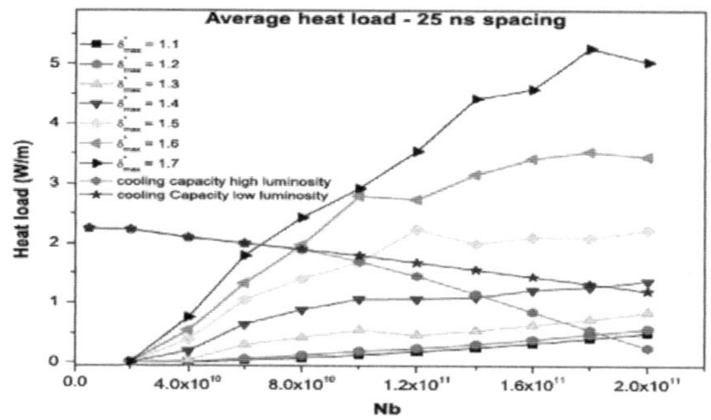

Figure 7: Average heat load as function of bunch population and different secondary emission yield (δ^{*}_{max}).

2.6.2 Mapping of the beam-induced synchrotron radiation at the LHC. The synchrotron radiation (SR) due to the beam in the LHC is a key ingredient to electron cloud build-up codes. Employing codes as SYNRAD3D can help to map the distribution of the SR photons along the LHC. In Figure 8, we are reporting the distribution of absorption sites of the photons coming from beam-induced SR around the LHC arc beam pipe [16].

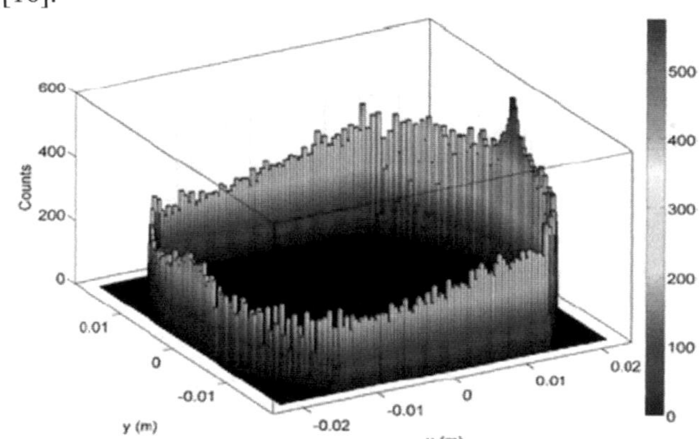

Figure 8 Photon distribution absorption sited of the beam-induced synchrotron radiation.

He is currently coordinating the particle accelerator group at Universidad de Guanajuato. Physics and engineer students integrate the group. They are working on the design of mechanical pieces for accelerator.

2.7 Karim Gibran Hernandez Chahin

Based on a combination of simulation and measurement results from different cavities with transverse deflecting for the high Luminosity LHC (3 different designs): the 400 MHz (CRAB) niobium cavity, 704MHz 5-cell elliptical niobium accelerating cavity for the High Gradient project and two 1.3 GHz Monocell elliptical niobium cavities.

The goal is to define and validate the preparation and operational techniques that permit high gradient or high performance operation of bulk Niobium SRF cavities of both elliptical and non-elliptical geometry.

As part of the goal, quantitatively assesses the performance factors in bulk niobium superconducting RF cavities operating in superfluid helium baths. Focus is given to a detailed consideration of cavity surface preparation and operational techniques that can affect the overall cavity performance.

To understand the cavity preparation, optimization in the electro-polish was done using different cathodes. The high-pressure rinse (HPR) is under study evaluating the performance as function of nozzles designs and operational pressures. Also studies to understand the different sources of contamination during the procedure of cavity manipulation and assembly in the cleanroom (based on the analysis of particulates by energy dispersive X-ray spectroscopy).

Diagnostic measurement and analysis techniques are under development to characterize the thermal loading and quench behavior of bare cavities, with an aim to separate and optimize the RF performance, material defect and geometrical factor effects in the cavity performance. This characterization leads to detailed list of induced cavity failure modes and quench types.

3. The eLinac project

A preliminary proposal for the Mexican eLINAC consists in a sequence of three 3-m in length FODO cells. Tree RF cavities are interspersed among the three cells. In order to minimize the beam size along the line, a 60° phase advance per cell is selected, thereby minimizing the maximum value of the beta functions; the emittances are assumed to be constants. Magnetic gradients for the three pairs of quadrupoles (from the source to end of the beamline) are 0.09 T/m, 0.5 T/m and 0.9 T/m, respectively. The maximum size of the beam is around 2 mm, therefore, that a transport tube with an inner radius an order of magnitude bigger, requires a 10-σ admittance. The last improvements on the design involves adding a solenoid magnet at the beginning of the Linac and incorporating a dipole magnet at the end of the beam line in order to bend the beam. The dipole will be employed to produce the synchrotron radiation and steering the beam to the target. In Table 1, the main machine and beam parameters are listed.

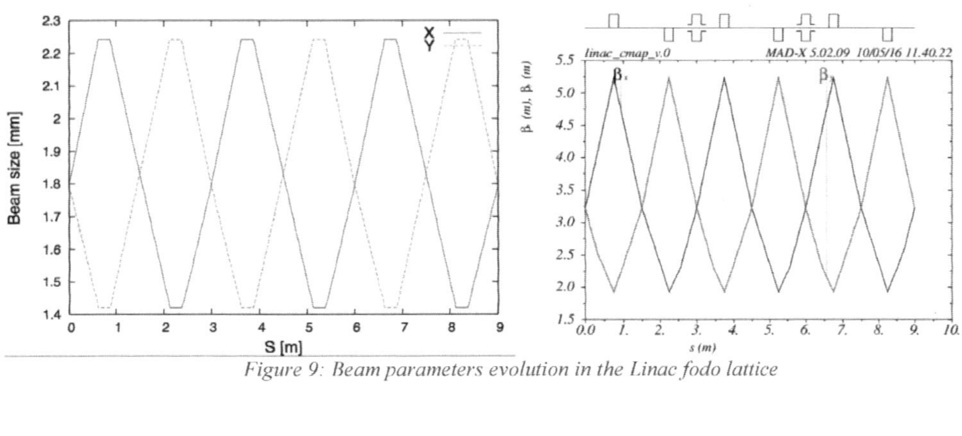

Figure 9: Beam parameters evolution in the Linac fodo lattice

Table 1. Basic parameters of the electron Linac

Beam Parameter	Value
Beam Energy	100 MeV
Transversal Emittance	1mm.mrad
Current	15 mA
Fodo cells	3
Number of Cavities	3

4. Final remarks

Mexican society has begun to have a critical mass of accelerator physicists and engineers capable of designing, building and operating an eLINAC able to provide two operation modes of beam: electron and Gamma radiation beams. With these two modes it can cover a wide range of applications from industrial, medical and basic research applications. This would improve technology development of the country and also boost the area along with the training of highly qualified scientists and engineers. In addition, the machine can serve as the first accelerating chain of a more complex machine.

References

[1] http://dpyc.smf.mx/dpyc-es/
[2] https://mepas2015.wordpress.com/
[3] Ramos lara, Maria de la Paz,2004 Experiencia mexicana en aceleradores de partículas(Mexico, UNAM siglo XXI).
[4] Methodical Acelerator Design, MAD, https://mad.web.cern.ch/mad/
[5] K. Oide *et al.* "Design of beam optics for the FCC-ee collider",in *Proc. 7th Int. Particle Accelerator Conf. (IPAC'16)*, Busan, Korea, May 2016, pp. 3821–3823.
[6] L. Medina "FCC-ee dynamic aperture at 175 GeV", FCC Week, Rome, Apr. 2016. http://indico.cern.ch/event/438866/contributions/1084987/
[7] G. Apollinari, I. Béjar Alonso, O. Brüning, M. Lamot, and L. Rossi, *"High-luminosity Large Hadron Collider (HL-LHC) Technical Design Report"*, CERN-2015-005, Dec. 2015, p. 3, 22.
[8] R. Tomás *et al.*, "HL-LHC alternative scenarios", in *Proc. Of Chamonix 2014 Workshop on LHC Performance*, CERN-2015-002, pp. 217–224.
[9] L. Medina, and R. Tomás, "Performance and operational aspects of HL-LHC scenarios", in *Proc. 7th Int. Particle Accelerator Conf. (IPAC'16)*, Busan, Korea, May 2016, pp. 1516–1519.
[10] X. Buffat *et al.*, "Dynamic beta and beta-beating effects in the presence of the beam-beam interactions", in *Procc. of HB2016*, Malmö, Sweden, Jul. 2016, to be published.
[11] B. Yee-Rendon et al., "Electron Cloud Measurements at J-PARC Main Ring", Proceedings of the International Particle Accelerator Conference (2016).
[12] B. Yee-Rendon et al., "SIMULATIONS OF THE BEAM LOSS DISTRIBUTION AT J-PARC MAIN RING", Proceedings of the International Particle Accelerator Conference (2016).
[13] Cristhian A. Valerio-Lizarrxcaga, Ildefonso Leon-Monzon, and Richard Scrivens, Phys. Rev. ST Accel. Beams **18**, 080101 – Published 19 August 2015
[14] Domínguez, O., Maury Cuna, H., Li, K., Arduini, G., Metral, E., Rumolo, G., and Zimmermann, "First electron-cloud studies at the Lage Hadron Collider" F., Phys. Rev. ST Accel. Beams 16, 011003, Jan 2013.
[15] H. Maury Cuna, J.G. Contreras and F. Zimmermann, "Simulations of electron-cloud heat load

for the cold arcs of the CERN Large Hadron Collider and its high-luminosity upgrade scenarios" , Phys. Rev. ST Accel. Beams 15, 051001, May 2012.
[16] G. H. I. Maury Cuna G. Dugan D. Sagan F. Zimmermann, "Synchrotron-radiation photon distribution for highest energy circular colliders", Proceedings of IPAC 2013, Vol., Pag.1340-1342.

Acknowledgments
Authors express their gratitude to the Mexican National Council for Science and Technology (**CONACYT**) for financing its researches.

Hard X-ray Sources for the Mexican Synchrotron Project

Juan Reyes-Herrera
Instituto de Física, Universidad Nacional Autónoma de México, Circuito Investigación Cientíca S/N, Ciudad Universitaria, 04510 México, D.F., Mexico

E-mail: jureyherrera@gmail.com

Abstract. One of the principal tasks for the design of the Mexican synchrotron was to define the storage ring energy. The main criteria for choosing the energy come from studying the electromagnetic spectrum that can be obtained from the synchrotron, because the energy range of the spectrum that can be obtained will determine the applications available to the users of the future light source. Since there is a public demand of hard X-rays for the experiments in the synchrotron community users from Mexico, in this work we studied the emission spectra from some hard X-ray sources which could be the best options for the parameters of the present Mexican synchrotron design. The calculations of the flux and the brightness for one Bending Magnet and four Insertion Devices are presented; specifically, for a Superconducting Bending Magnet (SBM), a Superconducting Wiggler (SCW), an In Vacuum Short Period Undulator (IV-SPU), a Superconducting Undulator (SCU) and for a Cryogenic Permanent Magnet Undulator (CPMU). Two commonly available synchrotron radiation programs were used for the computation (XOP and SRW). From the results, it can be concluded that the particle beam energy from the current design is enough to have one or more sources of hard X-rays. Furthermore, a wide range of hard X-ray region can be covered by the analyzed sources, and the choice of each type should be based on the specific characteristics of the X-ray beam to perform the experiments at the involved beamline. This work was done within the project Fomix Conacyt-Morelos "Plan Estratégico para la construcción y operación de un Sincrotrón en Morelos" (224392).

1. Introduction

A typical storage ring light source consists of a source, an injector, transport lines between accelerators, a storage ring, and a collection of surrounding beamlines with their experimental stations. A basic storage ring scheme with some synchrotron radiation sources is shown in the Figure 1.

In the present, there are three main groups of synchrotron sources based on storage rings. These are classified by the electron beam energy, low energy storage rings (<2 GeV) for VUV and soft X-rays radiation, high energy (6-8 GeV) for hard X-rays radiation, and the intermediate-energy light sources (2.5 GeV to 3.5 GeV); the Mexican project has been planned to be an intermediate-energy light source (ILS). Besides the good profitability of the ILS [2], there are other reasons for the popularization of the ILS around the world. This popularity is largely based on the high performance of these machines, backed up by recent technology developments like better systems to control instabilities at high beam current, higher harmonic RF-cavities

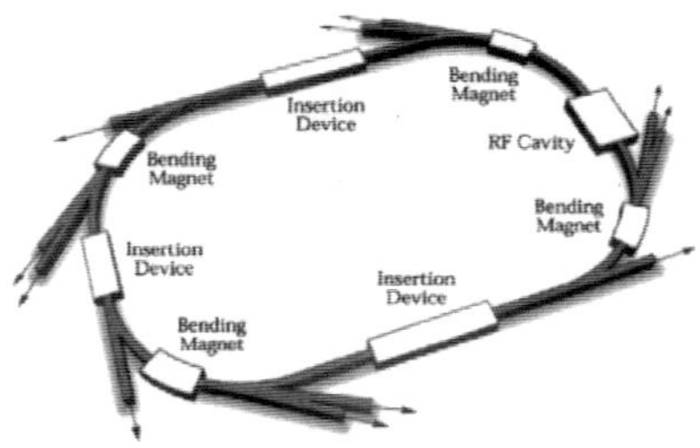

Figure 1. Basic components of a storage ring light source. The injection system is not shown [1].

to extend beam life time, between others [3]. One of the most important technological improvements was the development of insertion devices (IDs) to generate plentiful hard X-rays at relatively low electron beam energy [1]. This progress allows ILS to deliver hard X-rays (5-50 keV) [3].

There are many synchrotron radiation sources (bending magnets and IDs) that can provide hard X-rays in an ILS facility. The main target of this work is to study these sources under the electron beam parameters of the Mexican synchrotron design.

2. Hard X-ray sources for ILS

The radiation from bending magnet (BM) and wiggler sources is a continuum spectrum with a critical photon energy ε_c,

$$\varepsilon_c(\text{keV}) = 0.665B(\text{T})E^2(\text{GeV}), \tag{1}$$

where E is the electron beam energy and B is the magnetic field of the bending or wiggler magnet. The high magnetic field of the wiggler can extend the radiation to higher photon energy compared to a dipole. The radiation spectrum from an undulator magnet consists of a series of discrete harmonic peaks with the photon energy of ε_n [5],

$$\varepsilon_n[\text{keV}] = \frac{0.95nE^2[\text{GeV}]}{\lambda_u[\text{cm}](1+K^2/2)}, \tag{2}$$

where λ_u is the undulator period (Figure 2), n is the radiation harmonics, and $K = eB\lambda_u/2\pi\beta mc$ is the undulator magnetic strength parameter, where β is the relative speed of the electron with respect to the light.

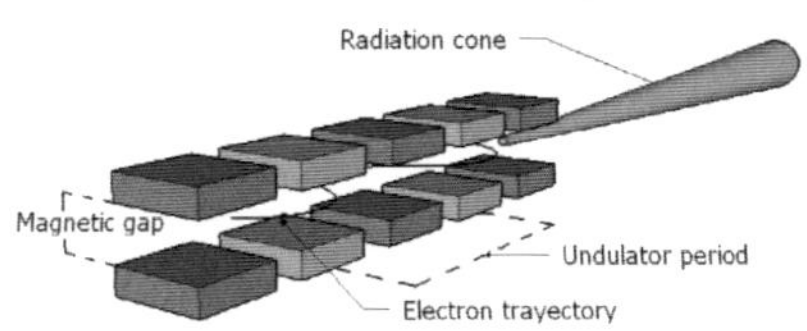

Figure 2. Magnetic structure of an Undulator.

Two important figures of merit for synchrotron radiation are the spectral photon flux F, defined as the photons emitted by electron beam per second and in 0.1% bandwidth, and the spectral brightness $\mathbf{B}$, defined as the photon flux per unit source area and per unit solid angle of the radiation cone [4],

$$F = \frac{N_{ph}}{0.1\%\Delta\omega/\omega}, \tag{3}$$

$$\mathbf{B} = \frac{N_{ph}}{4\pi^2 \sigma_{Tx}\sigma_{Tx'}\sigma_{Ty}\sigma_{Ty'} 0.1\%\Delta\omega/\omega}, \tag{4}$$

where N_{ph} is the number of photons per second and $\sigma_{Tx}\sigma_{Tx'}\sigma_{Ty}\sigma_{Ty'}$ are the effective photon beam sizes and divergences in transverse planes.

It can be noticed from the equations (1) and (2) that in order to obtain higher X-ray energies, it is necessary to increase either the electron beam energy (E), the magnetic field (B) or, to decrease the λ_u in the case of an Undulator. Since the cost of a synchrotron facility notably increases at higher electron beam energies, in the present there exists a good variety of technologies aimed to have IDs and BMs with stronger magnetics fields and shorter undulator periods as hard X-ray sources. The devices that could have good potential (as hard X-rays sources) for the Mexican light source project are

(i) High field bending magnets or Super-Bending Magnets (SBMs).
(ii) SuperConducting Wigglers (SCWs).
(iii) In Vacuum Short Period Undulators (IV-SPUs).
(iv) Superconducting Undulators (SCUs).
(v) Cryogenic Permanent Magnets Undulators (CPMUs).

2.1. High field bending magnets or Super-Bending Magnets (SBMs)

Some light sources around the world are using, or planning to use, SBMs, these SBMs can produce high magnetic fields based on super-conducting materials. For instance, there is a proposal in the SIRIUS (Brazil) designed to use SBMs of 2 T to generate hard X-rays [6]. Another case is in the SSRF (China) where the researchers are planning an upgrade of the sources by replacing normal BMs by SBMs [7]. Since there is a intrinsic spreading in a cone of the radiation from a magnetic dipole, this kind of sources can be used in experiments where hard X-rays are required but not needed with high brightness.

2.2. Super-Conducting Wigglers (SCWs)

There are many ILS using SCWs to generate hard X-rays, some of them are shown in the table 1. It is worth to mention that this SCWs were built in the Budker Nucler Physics Institute, Novosibirsk, Russia [8].

Table 1. Examples of SCWs installed at some ILS: the Canadian Light Source (CLS), the English Source (DLS), the Italian (Elettra) and the Spanish synchrotron (ALBA). Here E_e is the electron beam energy and E_p is the photon energy.

Synchrotron	E_e (GeV)	Beamline	E_p (keV)
CLS	2.9	BMIT	20-100
DLS	3	I12	50-150
Elettra	2.4	XRD2	8-30
ALBA	3	BL04 MSPD	8-50

Due to its low operating temperature (4 K), the manufacturing and operation of the SCWs are significantly higher than those based on permanent magnets. For these reasons, there are many efforts to build a Wiggler with high magnetic fields using permanent magnets only [9].

2.3. In-Vacuum Short Period Undulators (IV-SPUs)

The Undulators (hybrids or pure structure) are the workhorses of every third generation synchrotron facility. However most of them have long undulator period and big magnetic gaps, making impossible to generate high brightness hard X-rays for an ILS. To reduce the magnetic gap of these devices, the magnetic structure can be placed inside the vacuum chamber (In-vacuum Undulators). Therefore, in principle, the magnetic gap is limited only by the electron beam acceptance. Additionally, if the Undulator period is short, this ID can be an improvement of the hard X-rays brightness in a ILS [10]. There is an excellent explanation of the advantages and disadvantages of the use of IV-SPUs in the work by Hwang *et al* [10].

2.4. SuperConducting Undulators (SCUs)

The use of Undulators as hard X-rays sources carries some costs, first, reducing λ_u reduces K proportionally, resulting in fewer emitted photons with high energy. Both electromagnet and permanent-magnet based Undulators are material limited in their maximum magnetic field strength B, so there is no obvious way to compensate for this reduction in K. There are also some practical challenges. To maintain sufficient magnetic field strengths, the undulators air gap (between top and bottom magnet arrays) must also be scaled down proportionally. This narrowed gap demands strict beam alignment and low beam emittance in order to pass the electrons through the structure without beam scraping [11]. These difficulties can be avoided by the use of superconducting materials to get high magnetic fields. Since 1990 many proposals appeared, about the replacement of permanent magnets by superconducting wires or coils [12] y [13]. There is a complete study about NbTi and Nb_3Sn SCUs made by J. Bahrdt and Y. Ivanyushenkov [14], in this work they concluded that the SCU technology permits an excellent spectral performance, and the SCUs will become the preferred devices as soon as operational issues have been demonstrated in a multi-user facility.

2.5. Cryogenics Permanent Magnets Undulators (CPMUs)

Another option is to use short period Undulators based on permanent magnets, for short Undulator periods ($\lambda \leq 9$ mm) the CPMUs can reach the same magnetic field intensity than the SCUs [14]. Since the operation temperature of the CPMUs is around the liquid nitrogen temperature, the operation costs and the heating problems are fewer than with the SCUs. Another advantage of the CPMUs over the SCUs is that the field correction techniques developed for conventional permanent magnet Undulators can be directly applied to the CMPUs [15]. Currently, there are many CPMUs devices under development at some light sources in the world, for example the ones at SOLEIL [16] and the ESRF [17]. There are promising results and it seems like the CPMUs will dominate the short period undulator technology for the next years [14], for storage rings [18] and for free electron lasers (FELs) [19].

3. Spectra calculations from the sources

The parameters of the storage ring design are shown in the table 2.

Using the codes XOP (X-ray Oriented Programs, see 2.3) [20] and SRW (Synchrotron Radiation Workshop, see 3.92) [21] the flux and brightness from some sources are shown in the figures 3 and 4, respectively.

4. Conclusions

The study of different hard X-ray sources for the Mexican synchrotron project has been done. From the calculated spectra it can be concluded that the parameters of the present storage ring design are enough to have one or more hard X-rays sources for the users. The hard X-rays region can be covered from the studied sources, each of these has some advantages and disadvantages and the election will depend on the specifications needed from the corresponding beamline.

Table 2. Electron beam parameters of the Mexican source project, corresponding to the straight sections of the storage ring.

Beam Energy	E_e= 3 GeV
Circumference	300 m
Beam current	I= 0.250 A
Horizontal emittance	ε_x= 1.1 nm rad
Vertical emittance	ε_y= 0.161 pm rad
Horizontal β function	β_x= 2.06 m
Vertical β function	β_y= 1.32 m

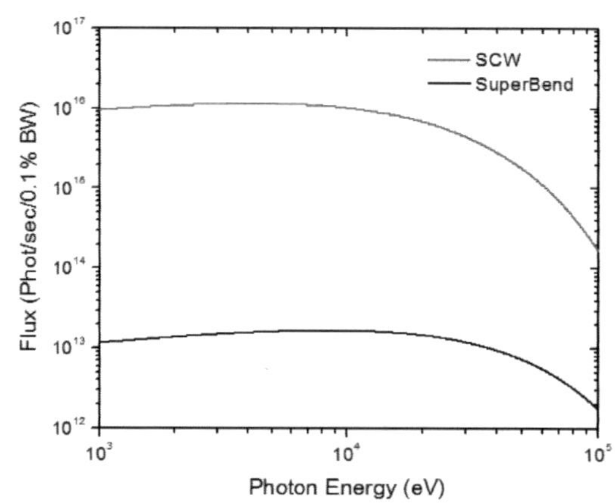

Figure 3. Flux from a 5 T Super bend (ALS [22]) and a SCW (Elettra [23]).

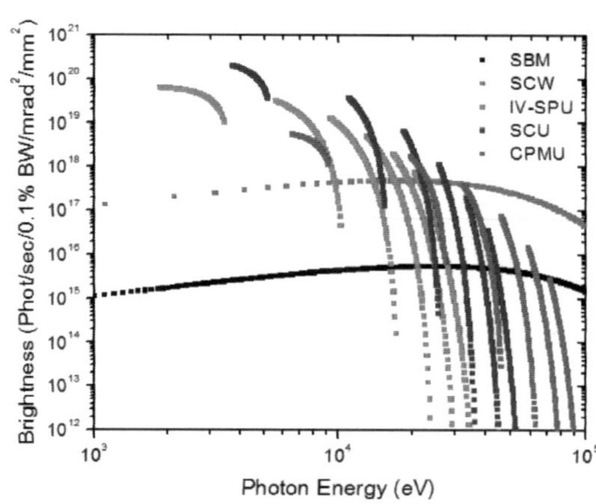

Figure 4. Brightness spectra from a 5 T Super-Bend (ALS [22]), a SCW (Elettra [23]), a IV-SPU (Diamond [24]), a SCU (APS [25]) and a CPMU (under development at the HZB [26]).

Acknowledgments
The author would like to thank Dr. Alain Flores Tlalpa (IFUNAM) and Dr. Armando Antillón Daz (ICF-UNAM) for providing the storage ring parameters of the Mexican synchrotron project.

The author acknowledges the support of the Fomix Conacyt-Morelos project (224392), as well as the encouragement given by Dr. F. Matías Moreno Yntriago (IFUNAM).

[1] J. Corbett, T. Rabedeau 2003 *Nucl. Instrum. Methods* A **500** 11
[2] Z.T. Zhao 2010 *Reviews of Accelerator Science and Technology Vol. 3* (World Scientific Publishing Company) 57
[3] W. A. Barletta, H. Winick 2003 *Nucl. Instrum. Methods* A **500** 1
[4] H. Wiedemann 2002 *Handbook of Accelerator Physics and Engineering* (World Scientific) 184
[5] H. Onuki, P. Elleaume 2003 *Wigglers, Undulators ant their Applications* (Taylors and Francis, USA, Canada)
[6] L. Liu, X. R. Resende, R. D. Rodrigues 2010 *Proccedings of IPAC'10* Kyoto, Japan, 2481
[7] Tian Shun-Qiang *et al* 2014 *Nucl. Science and Tech.* **25** 010102
[8] S. Khruchechev *et al* 2014 *Proccedings of IPAC2014* Dresden, Germany, 4103
[9] J. Chavanne, P. Van Vaerenbergh, P. Elleaume 1999 *Nucl. Instrum. Methods* A **421** 352
[10] C.S. Hwang, J.C. Jan, S.D. Chen, C.H. Chang, and T.M. Uen 2011 *Phys. Rev. ST Accel. Beams* **14** 044801
[11] B. A. Peterson *et al* 2014 *Physics Procedia* **52** 36
[12] I. Ben-Zvi, Z.Y. Jiang, G. Ingold, L.H. Wu 1990 *Nucl. Instrum. Methods Phys. Res.* A **297** 301.
[13] H.O. Moser, B. Krevet, H. Holzapfel, *Research Center Karlsruhe* German Patent No. DE 4101094
[14] J. Bahrdt, Y. Ivanyushenkov 2013 *J. Phys.: Conf. Ser.* **425** 032001
[15] T. Hara, T. Tanaka, H. Kitamura, T. Bizen, X. Marchal, T. Seike, T. Kohda, Y. Matsuura 2004 *Phys. Rev. ST Accel. Beams* **7** 050702
[16] C. Benabderrahmane *et al* 2013 *J. Phys.: Conf. Ser.* **425** 032019
[17] J. Chavanne, M. Hahn, R.Kersevan, C.Kitegi, C.Penel, F.Revol 2008 *Proceeding of EPAC Conf.* Genova, Italy 2243
[18] F. Holy *et al* 2014 *Phys. Rev. ST Accel. Beams* **17** 050704
[19] F.H. O'Shea, G. Marcus, J. B. Rosenzweig, M. Scheer, J. Bahrdt, R. Weingartner, A. Gaupp, F. Grner *Phys. Rev. ST Accel. Beams* **13** 070702
[20] M. Sanchez del Rio, R.J. Dejus 2004 *SPIE Proceedings Vol.* **5536** 171
[21] O. Chubar, P. Elleaume 1998 *Proc. of the EPAC98 Conference* 1177
[22] D. Robin, *et al* 2002 *Proceedings of EPAC 2002* Paris, France 215
[23] L. Tosi, C. Knapic, D. Zangrando 2004 *Proceedings of EPAC 2004* Lucerne, Switzerland 390
[24] J. A. Clarke, N. Bliss, D. G. Clarke, F. E. Hannon, C. Hill, A. A. Muir and D. J. Scott 2003 *Proceedings of the 2003 Particle Accelerator Conference* 1029
[25] Y. Ivanyushenkov *et al* 2015 *Phys. Rev. ST Accel. Beams* **18** 040703
[26] F. Holy *et al* 2014 *Phys. Rev. ST Accel. Beams* **17** 050704

Some Standard model problems and possible solutions

J. Barranco

Departamento de Física, División de Ciencias e Ingenierías, Campus León, Universidad de Guanajuato, León 37150, México

E-mail: jbarranc@fisica.ugto.mx

Abstract. Three problems of the standard model of elementary particles are studied from a phenomenological approach. *(i)* It is shown that the Dirac or the Majorana nature of the neutrino can be studied by looking for differences in the ν-electron scattering if the polarization of the neutrino is considered. *(ii)* The absolute scale of the neutrino mass can be set if a four zero mass matrix texture is considered for the leptons. It is found that $m_{\nu_3} \sim 0.05$ eV. *(iii)* It is shown that it is possible -within a certain class of two Higgs model extensions of the standard model- to have a cancelation of the quadratic divergences to the mass of physical Higgs boson.

1. A beautiful model

Louis de Broglie once said that if someone try to summarize the essential characteristics of physics in three words, those words would be: corpuscular, relativistic and quantum [1]. It was later that century that the concept of corpuscular was replaced with the idea of fields. These two concepts, fields and quantum were united in a single theory that now we call the quantum field theory (QFT). Thus we may say, paraphrasing de Broglie, that the essential of physics may be encompassed by QFT. In this spirit, it is natural that fundamental questions such as *What is the universe made of?* or *what are the fundamental interactions of nature?* should be tackled within the QFT formalism. Indeed, the Standard model of elementary particles (SM), the current theory that contains the fundamental building blocks of the ordinary matter, is a gauge quantum field theory. The internal symmetry that defines the SM is the local $SU(3) \times SU(2) \times U(1)$ gauge symmetry and the most general Lagrangian that describes the dynamics of the fields involves only 19 parameters. The last one of those parameters was finally measured by the LHC and it was announced on July 4th, of 2012: The mass of the Higgs boson which is 125.3 GeV [2, 3]. The other 18 parameters are: 6 quark masses, 3 masses of the charged leptons, 3 mixing angles in the quark sector and one CP violating phase (that give rise to the Cabibbo-Kobayashi-Maskawa, CKM, matrix), 3 coupling constants, the vacuum expectation value and the Θ_{QCD}.

It is almost incredible that with only those 19 parameters and within the formalism of QFT is possible to compute a huge number of processes involving elementary particles. Those processes cover energies from few MeV to the highest energies reached in both particle accelerators: man made or astrophysical. All those reactions can be computed and once the numerical values of those parameters are introduced, all the theoretical predictions match the experimental values within the one sigma error, except for few anomalies that are currently under study.

Table 1. Ratio of the solar neutrino events measured in different experiments over the expected number of events predicted by the Solar Standard Model.

Experiment	data/SSM	Experiment	data/SSM
^{37}Cl	0.310 ± 0.030	^{71} Ga	0.540 ± 0.039
Super-K	0.413 ± 0.015	SNO^{Scatt}	0.413 ± 0.05

Nevertheless such success, which in some cases corresponds to a precision in a billion of some computed quantities, such as the gyromagnetic moment of the muon, the task is not complete. Although there is a machinery that allow us to compute such variety of phenomena, there are some new insighta that tell us that the SM is not the final answer in the description of the fundamental blocks of the ordinary matter.

2. The ν-SM

The first indication that the SM need an extension came from the Sun. It was called the Solar Neutrino Problem and it consisted in a deficit between the expected number of neutrinos and the measured number of events at terrestrial detectors. Indeed, the ratio of the measured neutrinos over the expected number of neutrinos, predicted by the Solar Standard Model (SSM) was always different of one in all the experiments designed to measure the neutrinos coming from the Sun as can be seen in table 1.

Similar phenomena were also observed in atmospheric neutrinos and in neutrinos from reactor experiments. It took over 30 years to understand the origin of such deficit and the current solution to the problem is called the neutrino oscillation mechanism and it means that neutrinos change their flavor due to the fact that they are massive and their flavor states are a mixture of their mass states i.e. $|\nu_\alpha\rangle = \sum_a U^*_{\alpha a}|\nu_a\rangle\,, a = 1,2,3, \alpha = e,\mu,\tau$. Here the matrix U_{PMNS} is the mixing matrix between flavor states and PMNS is an abbreviation for Pontecorvo-Maki-Nagakawa-Sakata since this matrix was introduced by Ziro Maki, Masami Nakagawa and Shoichi Sakata [4] to explain the neutrino oscillations and because neutrino oscillation was first predicted by Bruno Pontecorvo in 1957 [5]. These two hypothesis on the neutrino sector imply that SM needs to incorporate:

(i) Three neutrinos with (at least two non-vanishing) masses m_1, m_2, m_3 and
(ii) three mixing angles $\theta_{12}, \theta_{13}, \theta_{23}$ and at least one CP violating phase δ.

That is, we have to change from 19 parameters to at least 26 parameters. Many of them still unknown. The probability that neutrinos change their identity is a function of those new parameters: $P(\theta_{12}, \theta_{13}, \theta_{23}, \delta, \Delta m^2_{12}, \Delta m^2_{13})$. As it can be seen, P is not a function of the absolute masses of the neutrinos m_i but a function of the differences $\Delta m^2_{ij} = m^2_j - m^2_i$. There exists an experimental program trying to determine with the highest precision the values of those parameters. It involves the detection of neutrinos produced in reactor, accelerators or astrophysical sources as the Sun and active galactic nuclei or supernova remnants that produce ultra-high energetic neutrinos. All this effort has cornered the allowed region of parameters as follows: $7.02 \times 10^{-5}(eV)^2 \leq \Delta m^2_{12} \leq 8.09 \times 10^{-5}(eV)^2$, $0.270 \leq \sin^2\theta_{12} \leq 0.344$, $2.317 \times 10^{-3}(eV)^2 \leq \Delta m^2_{13} \leq 2.607 \times 10^{-3}(eV)^2$, $0.382 \leq \sin^2\theta_{23} \leq 0.643$ at the 3σ C.L. [6].

Thus, the leptonic sector, after the discovery of neutrino oscillations, imply some questions that remain to be answered.

- What is the absolute scale of the neutrino masses?: Since neutrino oscillation data only allows the knowledge of the mass difference Δm^2_{ij}, the absolute values of $m_{\nu 1}, m_{\nu 2}, m_{\nu 3}$ remains a mystery. The strongest constraints on m_i come from Cosmology: $\sum_i m_i < 0.136$ eV [7].

- What is the hierarchy of masses in the neutrino sector?: From the neutrino data, it is not possible to determine the sign in Δm_{ij}^2, that is, either possibilities m_1 bigger or lower that m_2 are still allowed.
- Is there CP violation in the neutrino sector?: So far, the value of δ is feeble bounded, $\delta = 306^{+39}_{-70}$ degrees at the 1σ C.L., while at 3σ it can be from $0^o - 360^o$ [6].
- Why is the hierarchy of the CKM matrix so different to the one observed in the PMNS matrix?: Indeed, the phenomenon of mixing in the neutrino sector is very similar to the mixing in the quark sector but the structure of the mixing matrices are very different.
- *Is the neutrino a Dirac or a Majorana particle?*

The last question is not directly related with the oscillation of neutrinos, but it is a question about the fundamentals of the theory behind elementary particles. Grand Unification Theories (GUT) predict a Majorana nature for the neutrinos. Furthermore, if neutrinos are Majorana, the dynamical equation will not be the Dirac equation any more. Thus, changing our model where all fermions can be described as Dirac spinors. For a long time, it has been considered that the neutrinoless double β decay is the best way to test the nature of the neutrino [8].

Thus, it seems that most of the previous un-answered questions in the leptonic sector require an experimental effort. Nevertheless, one may wonder if from the theoretical or phenomenological approach it is possible to get some insight by using available data. The idea of this conference proceedings is to show that this is the case, and to motivate others to look for ideas that may help us to answer some of these questions.

3. The nature of the neutrino

It was shown some time ago that Majorana and Dirac neutrino scatter different on electrons or nucleons [9]. Nevertheless, since neutrinos have a very small mass and due to the left-handed nature of the charged weak interactions, neutrinos are born almost fully polarized. To account for such polarization, it is needed an extra state preparation factor $(1-\gamma_5)/2$ which is usually added and the resulting amplitudes for both Dirac or Majorana neutrino scattering become identical. In any case, the calculation of the $\nu - e$ scattering for both cases, Dirac and Majorana neutrinos by including as a free variable the neutrino polarization can be performed. The result for the Dirac neutrino in the center of mass system is [9, 10]:

$$\begin{aligned}
\frac{d\sigma^D}{d\Omega} &= \frac{G_F^2}{8\pi^2 s}((m_e^2(E_\nu - p^2\cos\theta)(g_A^{\ell\,2} - g_V^{\ell\,2}) + (E_\nu E_e + p^2)(g_V^\ell + g_A^\ell)^2 \\
&+ (E_\nu E_e + p^2\cos\theta)^2(g_V^\ell - g_A^\ell)^2 - p[s^{1/2}(E_\nu E_e + p^2)s_{||}(g_V^\ell + g_A^\ell)^2 + (E_\nu E_e + p^2\cos\theta) \\
&\times ((E_e + E_\nu\cos\theta)s_{||} + m_\nu s_\perp \sin\theta\cos\phi])\,(g_V^\ell - g_A^\ell)^2 \\
&+ m_e(E_\nu(1-\cos\theta)s_{||} - m_\nu|s_\perp|\sin\theta\cos\phi)(g_A^{\ell\,2} - g_V^{\ell\,2}))) \,, \qquad (1)
\end{aligned}$$

while for the Majorana case it is given by:

$$\begin{aligned}
\frac{d\sigma^M}{d\Omega} &= \frac{G_F^2}{4\pi^2 s}(((E_\nu E_e + p^2)^2 + (E_\nu E_e + p^2\cos\theta)^2 + m_\nu(E_\nu^2 - p^2\cos\theta))(g_V^{\ell\,2} + g_A^{\ell\,2}) \\
&+ m_e^2(E_\nu^2 - p^2\cos\theta + 2m_\nu^2)(g_A^{\ell\,2} - g_V^{\ell\,2}) - 2g_V^\ell g_A^\ell p(2E_\nu E_e + p^2(1+\cos\theta)) \\
&\times (E_\nu s_{||}(1-\cos\theta) - m_\nu|s_\perp|\sin\theta\cos\phi)) \,. \qquad (2)
\end{aligned}$$

Here E_e, E_ν are the energy of the electron and the neutrino, θ, ϕ the scattering angles and p the neutrino momentum and s the Mandelstam variable. Furthermore, the incident neutrino polarization vector is defined as $s_\nu = (0, s_\perp, 0, s_{||})$ in the neutrino rest frame, and

$g_V^\ell = -\frac{1}{2} + 2\sin^2\theta_W + \delta_{le}$, $g_A^\ell = -\frac{1}{2} + \delta_{le}$, $\ell = e, \mu, \tau$ are the coupling constants. Both cross sections are equal in the limit of $m_\nu \to 0$ and $s_{||} = -1$, that is, a fully polarized neutrino. If the neutrino polarization evolution is considered, $\nu - e$ scattering may shed some light into the Dirac or Majorana nature. Consider a change in the neutrino polarization that may be achieved by strong magnetic fields in astrophysical environments. Any particle, even neutral particles, possessing a magnetic moment, as the neutrino may have due to the fact that it is now a massive particle, may interact with external electromagnetic fields and consequently, its spin polarization may change.

Thus, as it can be seen from eqs. (1) and (2), even in the case of a very small neutrino mass, if the change in $S_{||}$ is big enough, there will be appreciable differences between the scattering of a Dirac or a Majorana neutrino regardless the energy of the incoming neutrino. Thus, it will be another possible way to observe differences between Dirac or Majorana neutrinos.

4. The absolute mass of the neutrino

As we have mentioned, the oscillation of neutrinos gives only information on δm_i^2. The absolute values of m_i are still unknown. So far, cosmology gives the strongest bound on the sum of all neutrinos. Currently, PLANCK data constraint on the mass of neutrinos is $\sum_i m_i < 0.136$ eV [7]. Other limit on the absolute neutrino mass come from the analysis of the Kurie plot in β-decays. In particular, the Tritium β-decay analysis done by the Troitsk Collaboration sets an upper limit on the electron neutrino mass of $m_\nu < 2.2$ eV. [11]. It looks like a matter of time when high precision experiments will eventually give the value of the absolute neutrino mass. Nevertheless, from a phenomenological approach it is possible to set the scale of the neutrino mass. For instance, unification models may shed light on the absolute value of the neutrino mass. Let us make an example of this possibility. The ideas of supersymmetry or grand unifications may set relations between the fermion masses and the mixing angles that arise in the context of left-right symmetric models which predict a Yukawa matrix texture.

Let's assume that the Yukawa matrices texture is a hermitian four-zero texture in the flavor basis, i.e.

$$M^f = \begin{pmatrix} 0 & C_f & 0 \\ C_f' & D_f & B_f \\ 0 & B_f' & A_f \end{pmatrix}. \quad (3)$$

To test this texture we have to do the following steps:

(i) Define a real and symmetric matrix $\tilde{H} = M^f M^{f\dagger}$, with M^f the mass matrix with a texture. In our case, a four-zero texture eq. (3). By writing $B(B') = be^{i\phi_B(\phi_{B'})}$, $C(C') = ce^{i\phi_C(\phi_{C'})}$, $D = de^{i\phi_D}$ and $A = ae^{i\phi_A}$, separate the phases of the non-diagonal elements with the help of a matrix $P = e^{-\frac{i}{2}\Xi}\mathrm{diag}\left(e^{\frac{i}{2}\Xi}, e^{i(\phi_C - \phi_D)}, e^{i(\phi_C + \phi_{B'} + \Xi)}\right)$. Thus, the new matrix H with no phases will be:

$$H = P^\dagger \tilde{H} P. \quad (4)$$

(ii) The elements of H are real and can be inverted with the help of the invariants:

$$\begin{aligned} \mathrm{Tr}(\tilde{H}) &= m_1^2 + m_2^2 + m_3^2, \\ \mathrm{Tr}^2(\tilde{H}) - \mathrm{Tr}(\tilde{H}^2) &= 2m_1^2 m_2^2 + 2m_1^2 m_3^2 + 2m_2^2 m_3^2, \\ \mathrm{Det}(\tilde{H}) &= m_1^2 m_2^2 m_3^2, \end{aligned} \quad (5)$$

into functions of the masses $\tilde{m}_j = m_j/m_3$. This leaves only one free parameter and one phase for each mass matrix. Let's denote as a_u, a_d, a_ℓ, a_ν and $\phi_u, \phi_d, \phi_\ell, \phi_\nu$ those parameters.

Quarks	$0.993 < a'_u < 0.997$	$0.9982 < a'_d < 1.0$	$\phi_{\rm bfp} = 1.5723$	$\chi^2_{\rm min} = 1.27$
Leptons	$a_\nu = 0.6750$	$a_l = 0.0375$	$\Phi_1 = 2.80233$	$m_{\nu_3} = 4.96 \times 10^{-2}$ eV^2

Table 2. Allowed values of the parameters a'_u and a'_d at 68% C.L. and the best fit for the free parameters of quark mass matrices that reproduce the CKM matrix. It is also shown the best fit points for the parameters a_l, a_ν, Φ_1 and the unknown mass m_{ν_3}. The best fit that reproduces the PMNS matrix gives a preferred value of the heaviest neutrino mass of the order of $m_{\nu_3} = 0.05$ eV.

It is convenient to normalize them: $a' = \frac{a}{m_3}$, $b' = \frac{b}{m_3}$, $c' = \frac{c}{m_3}$ and $d' = \frac{d}{m_3}$. The other parameters are given in terms of the masses of the fermions. Indeed, for instance:

$$\tilde{m}_1 \le a' \le \tilde{m}_3 \qquad \begin{cases} b' &= \sqrt{\frac{(a'-\tilde{m}_1)(a'+\tilde{m}_2)(\tilde{m}_3-a')}{a'}} \\ d' &= -a' + \tilde{m}_1 - \tilde{m}_2 + \tilde{m}_3 \\ c' &= \sqrt{\frac{\tilde{m}_1\tilde{m}_2\tilde{m}_3}{a'}} \end{cases} \tag{6}$$

(iii) Diagonalize H with the help of the matrices O_f constructed with the eigen-vectors of H. The theoretical mixing matrix CKM and PMNS are expressed in terms of such matrices. Indeed:

$$V^{\rm th}_{\rm CKM} = O_u^T P^{u-d} O_d\,, \qquad U^{\rm th}_{\rm PMNS} = O_l^T P^{l-\nu} O_\nu\,, \tag{7}$$

where $P^{u-d} = {\rm diag}[1, e^{i(\phi_1^u-\phi_1^d)}, e^{i(\phi_2^u-\phi_2^d)}]$ and in a similar way $P^{l-\nu} = {\rm diag}[1, e^{i(\phi_1^\ell-\phi_1^\nu)}, e^{i(\phi_2^\ell-\phi_2^\nu)}]$.

(iv) Compare the theoretical mixing matrices with the latest experimental values and find the allowed regions of the free parameters $a'_u, a'_d, a'_\ell, a'_\nu$ and the combinations $\phi \equiv \phi_1^u - \phi_1^d = \phi_2^u - \phi_2^d$ and $\Phi_1 \equiv \phi_1^\ell - \phi_1^\nu = \phi_2^\ell - \phi_2^\nu$.

In the leptonic case, since the absolute value of the neutrino masses are unknown, nevertheless, from neutrino oscillation data the difference Δm^2_{ij} are known and we can express the neutrino mass as:

$$\widetilde{m}_{\nu_1} = \sqrt{1 - \frac{(\Delta m^2_{32}+\Delta m^2_{21})}{m^2_{\nu_3}}}, \ \widetilde{m}_{\nu_2} = \sqrt{1 - \frac{\Delta m^2_{32}}{m^2_{\nu_3}}}. \tag{8}$$

Thus, in addition to the parameters a'_ℓ, a'_ν and Φ_1 we have to include in the fit the neutrino mass m_{ν_3} as a free parameter. By doing a χ^2 analysis, it is possible to find the allowed values of the parameters a'_u and a'_d and the best fit for the free parameters of quark mass matrices that reproduce the CKM matrix. Furthermore, in the leptonic case, similar regions are found a_l, a_ν, Φ_1 and, the unknown mass m_{ν_3}. Results are shown in table 2. As it can bee seen, the analysis shows a preferred value for m_{ν_3} to be around 0.05 eV. [12, 13]. It is important to mention the role of θ_{13} in the previous analysis. The confirmation that θ_{13} is different from zero reduce the possibilities and the allowed regions for m_{ν_3}. Once the true value of θ_{13} is introduced, the fit of the theoretical $U^{\rm Th}_{\rm PMNS}$ fixes $m_{\nu_3} \sim 0.05$ eV, as it can bee seen in Fig. 1 of [14].

5. The hierarchy problem

An aesthetic criterion for a satisfactory theory is that the dimensionless ratios between the free parameters of the Lagrangian should take values *of order of unity*. This criterion is called as the naturalness of the theory. This is not the case of the standard model of particles. The parameters that appear on the fundamental Lagrangian vary orders of magnitude between them. Some of the naturalness problems of the standard model are:

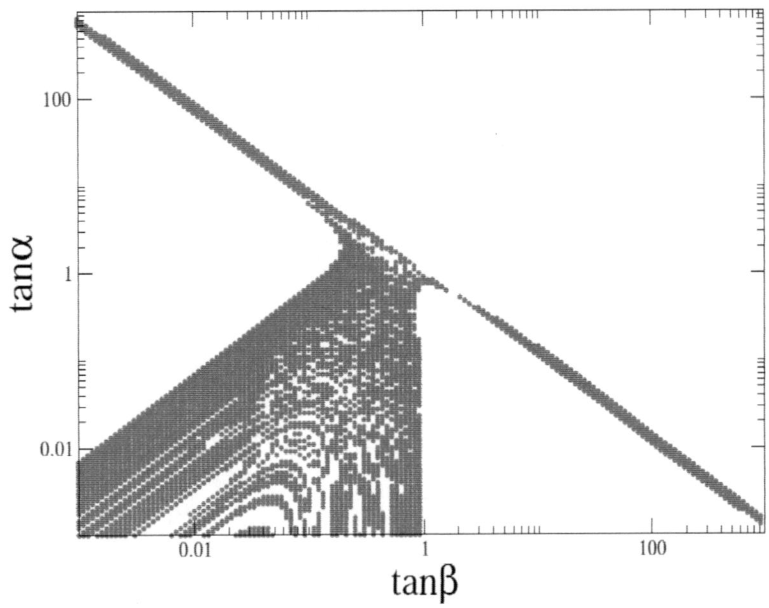

Figure 1. Some values of $\tan\alpha$ and $\tan\beta$ that fulfill the Newton-Wu conditions for reasonable masses of the scalar particles and that sets as one of those scalar the mass of the particle found by the LHC $m_1 = 125.3$ GeV.

- Neutrino masses are too small compared with the quark top mass.
- The Higgs mass is too small compared with the Planck mass.

Related with the problem in the hierarchy of the Higgs mass compared with the Planck mass is the possibility that large quantum contributions to the square of the Higgs boson mass would inevitably lead to quadratic divergences. In 1981, Veltman realized that such quadratic divergences can be cancelled if the following relation is fulfilled [15]:

$$\frac{3}{2}M_W^2 + \frac{3}{4}M_z^2 + \frac{3}{4}m_H^2 = \sum_f N_f m_f^2 . \tag{9}$$

Here f runs for all fermions in the Standard Model and $N_f = 3$ for quarks and $N_f = 1$ for leptons since N_f is the number of color degrees of freedom. LHC discovery [2, 3] sets a mass for the Higgs boson of ~ 125 GeV. Thus, the relation (9) is not valid.

Furthermore, as we have seen in the previous section, the neutrino mass will be of order of few electronvolts. While the mass of the quark top is ~ 174 GeV. In the SM there is only one scalar, the Higgs boson that gives mass to all fermions. Thus, the Yukawa coupling of the neutrino should be around 12 orders of magnitude smaller than the Yukawa of the top quark. In order to reduce this fine-tuning problem, in 1973 T. D. Lee introduced one of the simplest and economical extensions of the SM: to introduce another doublet Φ_2 [16]. Many realizations of the Two Higgs doublet model are possible. In one of them, Φ_2 gives mass to the 'heavier' quarks u, c, t and Φ_1 to d, s, b and to the leptons. Thus reducing the naturalness problem in the hierarchy of the Yukawa couplings. Is it possible to find a solution for the cancellation of the quadratic divergences of the Higgs mass within the THDM? It is possible. In 1994 Newton and Wu computed the quadratic divergences in a two Higgs extension of the SM [17]. They found the quadratic divergences as functions of the gauge, the Yukawa and the quartic scalar couplings and conditions between the masses of the fermions, the bosons and the Higgses to cancel the quadratic divergences. We will call them the Newton-Wu conditions. A realization of the Newton-Wu conditions in a THDM was done in [18]. In the following, we will re-evaluate the work of [18] with the new information that we have about the mass of one of the Higgses.

Consider the following Higgs potential for two doublets $\Phi_{1,2} = (\phi^+_{1,2}, \phi^0_{1,2})$:

$$\begin{aligned} V &= \mu_1^2\Phi_1^\dagger\Phi_1 + \mu_2^2\Phi_2^\dagger\Phi_2 + \mu_{12}^2(\Phi_1^\dagger\Phi_2 + \Phi_2^\dagger\Phi_1) + \frac{1}{2}\lambda_1(\Phi_1^\dagger\Phi_1)^2 + \frac{1}{2}\lambda_2(\Phi_2^\dagger\Phi_2)^2 \\ &+\lambda_3(\Phi_1^\dagger\Phi_1)(\Phi_2^\dagger\Phi_2) + \lambda_4(\Phi_1^\dagger\Phi_2)(\Phi_2^\dagger\Phi_1) + \frac{1}{2}\lambda_5(\Phi_1^\dagger\Phi_2)^2 + \frac{1}{2}\lambda_5^*(\Phi_2^\dagger\Phi_1)^2. \end{aligned} \tag{10}$$

Here the vacuum expectation values $\langle\phi^0_{1,2}\rangle = v_{1,2}$, fulfill $v^2 = v_1^2 + v_2^2$, and it is defined $\tan\beta = v_2/v_1$ as usual.

Thus, the Newton-Wu conditions reduce to [18]:

$$\begin{aligned} \frac{3}{2}M_W^2 + \frac{3}{4}M_Z^2 + \frac{v^2}{2}(3\lambda_1 + 2\lambda_3 + \lambda_4) &= \frac{1}{\cos^2\beta}\sum_{f_1} N_{f_1} m_{f_1}^2, \\ \frac{3}{2}M_W^2 + \frac{3}{4}M_Z^2 + \frac{v^2}{2}(3\lambda_2 + 2\lambda_3 + \lambda_4) &= \frac{1}{\sin^2\beta}\sum_{f_2} N_{f_2} m_{f_2}^2, \end{aligned} \tag{11}$$

For simplicity, let us assume that $\mu_{12} = 0$ and λ_5 real. We have five physical scalar particles: two charged Higgses $m_{H^\pm}$, one axial Higgs M_A, and two Boson Higgs particles m_1, m_2. Furthermore, we have five quartic scalar couplings in the potential $\lambda_{1...5}$. By rewriting the quartic scalar couplings $\lambda_{1...5}$ in terms of the masses of the physical scalars it is possible to rewrite the Newton-Wu conditions eqs. (11) in terms of the masses of the Higgses. Direct searches of other scalar particles have been done, and lower limits on the masses of those particles have been set. In particular, the Charged Higgs boson should be bigger than 80 GeV, and there are weaker limits for the axial scalar. By fixing $m_1 = 125.3$ GeV as the Higgs recently discovered by the LHC it is found a region of parameters that fulfills the Newton-Wu conditions. Examples of that region are shown in Fig. 1.

6. Conclusions

Although the SM is a beautiful model with a high predictive power, few unknowns prevail as well as some problems. Most of the unknowns are in the leptonic sector and in particular the neutrino sector. One of them is to determine the absolute scale of the neutrino mass or the nature of the neutrino, either it is a Dirac or a Majorana particle. In this note we have naively tried to find alternative ways of solving those problems from a phenomenological approach.

Another problems of the SM can be considered aesthetic problems: Why is Θ_{QCD} so small?, why is the Yukawa coupling of the neutrino so small compared with the Yukawa coupling of the quark top? Nevertheless, those questions can be related with other problems such as the cancellation of the quadratic divergences in the Higgs mass vacuum correction. Indeed, the two Higgs model extension of the SM may reduce the hierarchy problem and as shown in [18] and to cancel quadratic divergences in a class of two Higgs doublet models. In this note we have updated [18] with the constraint that one of the scalar Higgses being a scalar with the mass of $m_1 = 125.3$ GeV as found by LHC [2, 3].

Acknowledgments
This work has been supported by CONACyT SNI-Mexico.

[1] de Broglie, *El papel de las matemáticas en el desarrollo de la física teórica contemporanea.* Antologia de Matemáticas 2. UNAM 1971.
[2] G. Aad *et al.* [ATLAS Collaboration], Phys. Lett. B **716**, 1 (2012)
[3] S. Chatrchyan *et al.* [CMS Collaboration], Phys. Lett. B **716**, 30 (2012)
[4] Z. Maki, M. Nakagawa and S. Sakata, Prog. Theor. Phys. **28**, 870 (1962). doi:10.1143/PTP.28.870
[5] B. Pontecorvo, Sov. Phys. JETP **7**, 172 (1958) [Zh. Eksp. Teor. Fiz. **34**, 247 (1957)].
[6] M. C. González-García, M. Maltoni and T. Schwetz, Nucl. Phys. B **908**, 199 (2016) doi:10.1016/j.nuclphysb.2016.02.033 [arXiv:1512.06856 [hep-ph]].
[7] E. Di Valentino, E. Giusarma, M. Lattanzi, O. Mena, A. Melchiorri and J. Silk, Phys. Lett. B **752**, 182 (2016) doi:10.1016/j.physletb.2015.11.025 [arXiv:1507.08665 [astro-ph.CO]].
[8] J. Schechter and J. W. F. Valle, Phys. Rev. D **25** (1982) 2951.
[9] B. Kayser and R. E. Shrock, Phys. Lett. B **112** (1982) 137.
[10] J. Barranco, D. Delepine, V. G. Macias, C. Lujan-Peschard and M. Napsuciale, Phys. Lett. B **739**, 343 (2014) [arXiv:1408.3219 [hep-ph]].
[11] V. N. Aseev *et al.* [Troitsk Collaboration], Phys. Rev. D **84**, 112003 (2011) doi:10.1103/PhysRevD.84.112003 [arXiv:1108.5034 [hep-ex]].
[12] J. Barranco, F. González Canales and A. Mondragón, Phys. Rev. D **82**, 073010 (2010) doi:10.1103/PhysRevD.82.073010 [arXiv:1004.3781 [hep-ph]].
[13] J. Barranco, D. Delepine and L. López-Lozano, Phys. Rev. D **86**, 053012 (2012) doi:10.1103/PhysRevD.86.053012 [arXiv:1205.0859 [hep-ph]].
[14] J. Barranco, D. Delepine and L. López-Lozano, J. Phys. Conf. Ser. **485**, 012062 (2014). doi:10.1088/1742-6596/485/1/012062
[15] M. J. G. Veltman, Acta Phys. Polon. B **12**, 437 (1981).
[16] T. D. Lee, Phys. Rev. D **8**, 1226 (1973).
[17] C. Newton and T. T. Wu, Z. Phys. C **62**, 253 (1994). doi:10.1007/BF01560241
[18] E. Ma, Int. J. Mod. Phys. A **16**, 3099 (2001) doi:10.1142/S0217751X01004414 [hep-ph/0101355].

One-loop radiative corrections to the trilinear Higgs self-coupling

A. Moyotl, S. Chamorro and M. A. Pérez.
Departamento de Física, CINVESTAV, Apdo. Postal 14-740, 07000 México, D. F., México.
E-mail: amoyotl@fis.cinvestav.mx
E-mail: mchamorro@fis.cinvestav.mx
E-mail: mperez@fis.cinvestav.mx

Abstract. The one-loop corrections to the triple Higgs self-coupling are reported in the Littlest Higgs model with T-parity. Our results indicate that these corrections are no sensitive to high values of the symmetry breaking scale and the magnitude of one of momentum of one off-shell scalar boson legs.

1. Introduction

A measurement of the Higgs boson self-coupling will determine the structure of the Higgs potential, and will confirm that the observed scalar boson with mass of 125 GeV at the LHC [1, 2] really corresponds to the Higgs boson predicted by the Standard Model(SM). However, due to the high uncertainties on the hh cross sections measurements in both CMS and ATLAS [3], it is not clear if a meaningful measurement of the Higgs self-coupling could be possible in the latter case. In this sense, the Higgs self-coupling hhh was analyzed in processes such as gg double-Higgs fusion ($gg \to hh$), VV double-Higgs fusion ($qq' \to hhqq'$), double Higgs-strahlung ($q\bar{q}' \to Vhh$), associated production with top-quarks $q\bar{q}/gg \to t\bar{t}hh$[4] and high energy photon-photon collisions via the $\gamma\gamma \to t\bar{t}hh$ process [5]. Nevertheless, in these analyses the coupling hhh is independent of the 4-momentum of any Higgs boson leg. On other hand, it has been found that the Higgs boson self-coupling hhh at one loop level is very sensitive to new physics scenarios. In the SM, the top quark contribution is the dominant correction and this has a value of approximately of 9.8% [6, 7]. Another corrections to the self-coupling hhh were reported in models such as Two Higgs Doublet Models [6], and Minimal Supersymmetric Model (MSSM) with a light stop [7]. In this report we present the results to the Higgs self-coupling hhh at one loop level in the Littlest Higgs model (LHM) with T-parity. The LHM with T-parity is characterized by a global $SU(5)$ and a gauge $[SU(2)_1 \times U(1)_1] \times [SU(2)_2 \times U(1)_2]$ symmetries [8], and they are spontaneously broken at scale $f \sim \mathcal{O}(TeV)$. An important aspect of this model is that it offers an alternative to the hierarchy problem .

The present report is organized as follows. In the section 2 we review the theoretical set-up of LHM with T-parity necessary for our analysis. Analytic expressions for self-coupling hhh at one-loop level are presented in section 3. The numerical results for SM and LHM with T-parity are presented in section 4. Finally concluding remarks and an outlook are presented in section 5.

2. Littlest Higgs Model with T parity

The LHM is an effective theory based in the collective symmetry breaking approach, where in the first stage the global group $SU(5)$ is broken to $SO(5)$ at a scale f in the TeV range, via the symmetric tensor of vacuum expectation:

$$\Sigma_0 = \begin{pmatrix} \mathbf{0}_{2\times2} & \mathbf{0}_{2\times1} & \mathbf{1}_{2\times2} \\ \mathbf{0}_{1\times2} & 1 & \mathbf{0}_{1\times2} \\ \mathbf{1}_{2\times2} & \mathbf{0}_{2\times1} & \mathbf{0}_{2\times2} \end{pmatrix}.$$

Simultaneously the gauged subgroup $[SU(2)_1 \times U(1)_1] \times [SU(2)_2 \times U(1)_2]$ of $SU(5)$ is broken to the electroweak SM group $SU(2) \times U(1)$. Finally the gauged group $SU(2) \times U(1)$ is broken to $U(1)_{\rm em}$ via the usual Higgs mechanism; however, this Higgs potential corresponds to Coleman-Weinberg potential which is generated by one-loop radiative corrections. On other hand, from the global symmetry breaking of $SU(5)/SO(5)$ we generated 14 Nambu-Golstone bosons and four of these are absorbed by heavy gauge bosons ($W_H^{\pm}$, Z_H, A_H), and the remaining ten Nambu-Golstone bosons are parametrized by the nonlinear sigma model

$$\mathcal{L}_\Sigma = \frac{f^2}{8}\mathrm{Tr}|\mathcal{D}_\mu\Sigma|^2,$$

where $\Sigma = e^{i\Pi/f}\Sigma_0 e^{i\Pi^T/f}$, while the field Π and the covariant derivative $\mathcal{D}_\mu$ are given in [9, 10]. The implementation of T-parity on the gauge fields consists in exchanging the two $SU(2) \times U(1)$ factors; consequently the gauge coupling of the two $SU(2) \times U(1)$ factors are equal and therefore the number of free parameters is reduced. The T-parity in the fermion sector is introduced by implementing a $SU(2)_1$ doublet and another $SU(2)_2$ doublet, where under the T-parity the even linear combination is associated to the SM $SU(2)$ doublet, while the T-odd combination is associated with the called mirror fermion. The mirror fermions acquire mass through $SU(5)$ and T invariant Yukawa interaction:

$$\mathcal{L}_{\rm mirror}^{\rm LHM+T} = -\kappa_{ij} f\big(\bar{\Psi}_2^i\xi + \bar{\Psi}_1^i\Sigma_0\Omega\xi^\dagger\Omega\big)\Psi_R^j, \tag{1}$$

here κ_{ij} is a mixing matrix and it is different for each mirror fermion, $\xi = e^{i\Pi/f}$, $\Omega =$ diag(1,1,1-1,1,1), $\Psi_{1,2,R}$ is a multiplet with fermions doublets into its components. Then after expanding the Lagrangian (1) at $\mathcal{O}(v^2/f^2)$, the masses of the mirror up-quarks, mirror neutrinos and charged mirror leptons are given by:

$$m_{d_H} = m_{\ell_H} = \sqrt{2}\kappa_{ii} f, \tag{2}$$

$$m_{u_H} = m_{\nu_H} = \sqrt{2}\kappa_{ii} f\Big(1 - \frac{v^2}{8f^2}\Big). \tag{3}$$

Where the mirror down-quark (mirror charged lepton) receive only corrections at $\mathcal{O}(v^3/f^3)$. Moreover, the Higgs couplings to the mirror up-quark and heavy mirror neutrino are given by

$$h\bar{u}_H u_H = h\bar{\nu}_H\nu_H \sim \frac{i\kappa_{ii}}{2\sqrt{2}}\frac{v}{f}. \tag{4}$$

On other hand, the Higgs boson does not have direct coupling to mirror down-quarks and charged mirror leptons.

3. One loop correction to self-coupling hhh

In analogy with the SM, in LHM with parity T there are contributions arising from heavy mirror fermions, heavy bosons and scalars bosons. However, these contributions are suppressed by the energy scale f[1], but since the contributions of the SM fermions are proportional to the fourth power of its mass, it is logical to expect that the main contributions arise from mirror fermions in the LHM with T-parity. Therefore for this work we consider only the contributions of mirror fermions. Then, from the couplings (4) and masses (3) of the mirror fermions, the renormalizable expression for this result is given by:

$$\lambda_{f_H}^{\mathrm{LHM+T}}(q) = \sum_{i=1}^{3} \frac{\kappa_{ii}^2 m_{f_H} m_W N_c}{3\pi^2 2^{\frac{11}{2}} 3! g m_h^2} \frac{v^3}{f^3}\left[18 - \int_{x=0}^{1}\int_{y=0}^{1-x} \Xi_f^{SM}(x, y, s_{f_H}, s_q)\mathrm{d}x\mathrm{d}y\right], \qquad (5)$$

where N_c is the number of colors, $s_{f_H}^2 = m_{f_H}^2/m_h^2$ and $s_q^2 = q^2/m_h^2$, with f_H a heavy mirror up-quark or heavy mirror neutrino, and the sum comprises three families of mirror fermions and Ξ_f^{SM} is a dimensionless function .

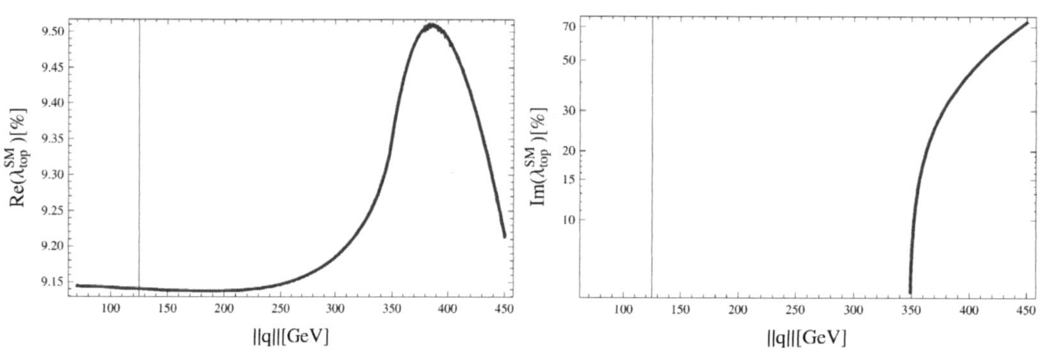

Figure 1. Real (left) and imaginary (right) contributions of top-quark to $\lambda_{\rm hhh}$ factor, as function of the 4-momentum magnitude of the off-shell scalar boson. There is no induced imaginary part for $||q|| < 2m_t$ and $\lambda_{\rm top}^{\rm SM} \simeq 9.14049$ % in the Higgs boson resonance region. The vertical line indicate the Higgs boson resonances ($||q|| = 125$ GeV).

4. Numerical results and discussion

4.1. Standard framework

To start, we analyzed the Higgs boson resonance region and we find $\lambda_{\rm top}^{\rm SM} = 9.14049\%$ for the top-quark, whereas for the bottom-quark ($\lambda_b^{\rm SM} \simeq 3.442 \times 10^{-6}\% + i6.4810 \times 10^{-9}\%$) and tau lepton ($\lambda_\tau^{\rm SM} \simeq 2.535 \times 10^{-8}\% + i7.962 \times 10^{-12}\%$) the respective contributions are very suppressed. Then, in the Higgs boson resonance region the main contribution to the $\lambda_{\rm hhh}$ factor comes from the top-quark. Thus, in the next analysis we consider only the top-quark. In order to compare our results with previous work, the Eq. (1) of [6] gives $\lambda_{hhh}^{\rm eff}(\mathrm{SM}) \simeq 9.8221\%$ and this result was obtained by the diagrammatic approach. On other hand, $\lambda_{\rm top}^{\rm SM} \simeq 9.14693\%$ can be obtained from $\Delta\hat{\Gamma}_H^3$ of the Eq. (32) in the reference [11]. These contributions are very similar to our results, but it is important to mention that these contributions were obtained with some approximations for the 4-momentum magnitude of the off-shell scalar boson. On other hand, for the 4-momentum

[1] The masses of heavy bosons and heavy scalar are directly proportional to the symmetry breaking scale f, as well as the mirror fermion masses [8].

magnitude of the off-shell scalar boson outside of Higgs boson resonance region; figure 1 shows the real and imaginary part of $\lambda_{\rm hhh}$ factor, as function of the 4-momentum magnitude of the off-shell scalar boson.

We obtained that there are no imaginary parts for $||q|| > 2m_t$. Nevertheless, we will show that the imaginary part increases considerably for higher values of $||q||$ and the real part remains basically stable, although it has a small maximum after $||q|| = 2m_t$. A high contribution to $\lambda_{\rm hhh}$ factor (for higher values of $||q||$) can violate unitarity, but the gauge bosons and self-coupling can reduce this large imaginary contribution.

4.2. Littlest Higgs Model with T parity

In the LHM with parity T there are two different contributions of the mirror fermion, one for mirror up-quarks and another for heavy mirror neutrinos. Consequently there are two different types of mixing matrices κ_{ii}, but for our analysis we will consider that such matrices have the same order of magnitude e.g. $\kappa_{i\ell_H} \sim \kappa_{iq_H} \equiv \kappa_{ii}$. Then, only two free parameters of the LHM with parity T are involved in the contribution (5), the symmetry breaking scale f and the mixing matrix κ_{ii}. Constraints on these parameters are discussed in Ref. [10], where it is considered that the mirror up-quarks are heavier than all the heavy gauge bosons. This corresponds to values of $\kappa_{ii} \gtrsim 0.45$, which makes the decay $q_H \to V_H q$ ($V_H = W_H^{\pm}, Z_H$) kinematically allowed, while for $\kappa_{ii} \lesssim 0.45$ the only kinematically allowed decay of mirror quark is $q_H \to A_H q$, and finally for $\kappa_{ii} \lesssim 0.1$ the mirror quarks are stable. Thus, for $0.1 \lesssim \kappa_{ii} \lesssim 0.45$ and the results from the 8 TeV run at the LHC, the combined analyses of electroweak precision physics and Higgs precision physics give a lower bound $f \gtrsim 694$ GeV at 95% C.L., while $f \gtrsim 638$ GeV was obtained from direct searches in $pp \to q_H q_H$ and $pp \to q_H A_H$ processes at 95% C.L. Therefore in the following analysis we will consider $\kappa_{ii} = 0.45$, and the region between 500 GeV and 2000 GeV for the symmetry breaking scale f. Moreover, for simplicity we consider that the masses of the three mirror families are the same, then the sum over family is replaced by a factor of three.

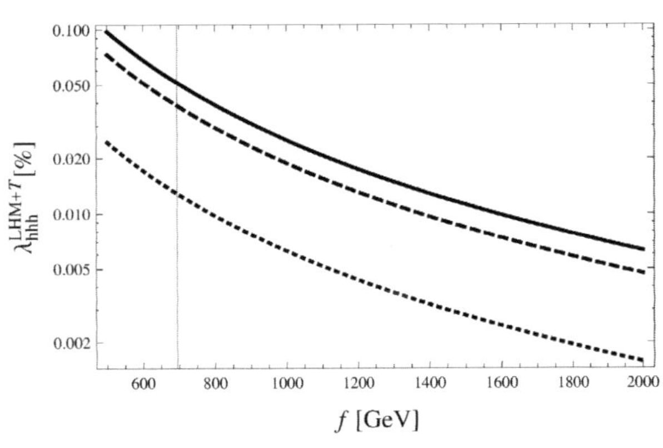

Figure 2. Prediction to $\lambda_{\rm hhh}$ factor in the LHM with T-parity, as a function of the symmetry breaking scale f. We have used the Higgs resonance region and $\kappa_{ii} = 0.45$, to obtain the heavy mirror up-quark (dashed line), heavy mirror neutrino (dotted line) and the sum of both contributions (black line). The vertical line corresponds to the lower bound $f = 694$ GeV.

In the Fig. 2 we plot the mirror up quark, mirror neutrino and the sum of both contributions to the $\lambda_{\rm hhh}$ factor, as a function of the symmetry breaking scale f for $\kappa_{ii} = 0.45$ and for the Higgs resonance region. We also include the lower bound $f \gtrsim 638$ GeV, which is given by the vertical line. Mathematically speaking, the difference between the contributions of heavy mirror up-quarks and heavy mirror neutrinos is the number of color N_c, consequently the mirror up-quark correction is the dominant contribution. We note that the corrections decrease softly with increasing symmetry breaking scale f, where the sum of both contributions for $f = 500$ GeV is the $\lambda_{\rm hhh}^{\rm LHM+T} \simeq 0.0977\%$ and for $f = 2000$ GeV is the $\lambda_{\rm hhh}^{\rm LHM+T} \simeq 0.0062\%$, while that for the lower bound $f = 694$ GeV is the $\lambda_{\rm hhh}^{\rm LHM+T} \simeq 0.0512\%$. It is important to note that there is no imaginary part in the regime of the Higgs resonance region, but for high 4-momentum of the off-shell scalar boson an imaginary part is induced, Figures 3 and 4 show this situation.

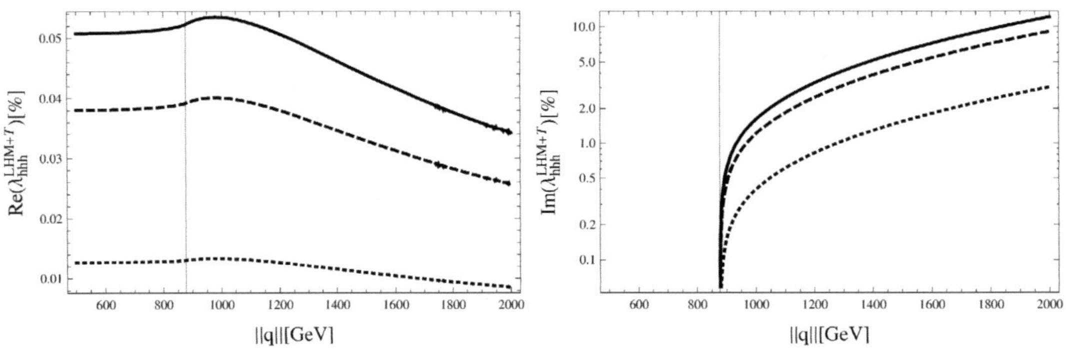

Figure 3. Real (left) and imaginary (right) contribuctions of the $\lambda_{\rm hhh}$ factor, as function of the 4-momentum magnitude of the off-shell scalar boson. We have used $f = 700$ GeV and $\kappa_{ii} = 0.45$, to obtain the heavy mirror up-quarks (dashed line), heavy mirror neutrinos (dotted line) and the sum of both contributions (black line). The vertical line corresponds to $||q|| = 2m_{f_H} \simeq 877$ GeV.

For $f = 700$ GeV and $\kappa_{ii} = 0.45$ we show the real and imaginary part of $\lambda_{\rm hhh}^{\rm LHM+T}$ factor, as function of the 4-momentum magnitude of the off-shell scalar boson in the Fig 3. The numerical behavior is similar to the fermionic SM contribution (see Fig. 1), but in a smaller scale. There is a small maximum in the real part after of $2m_{f_H} \simeq 877$ GeV and where also the imaginary part is induced. Moreover, the imaginary part increases with the 4-momentum magnitude of the off-shell scalar boson, such as the top quark in the SM, which is considerably reduced by contributions of heavy gauge bosons and self-coupling contributions. In particular, we have $\mathrm{Re}(\lambda_{\rm hhh}^{\rm LHM+T}) \simeq 0.0507\%$ and $\mathrm{Re}(\lambda_{\rm hhh}^{\rm LHM+T}) \simeq 0.0341\%$ for the real part of the sum of both contributions, in $||q|| = 500$ GeV and $||q|| = 2000$ GeV respectively.

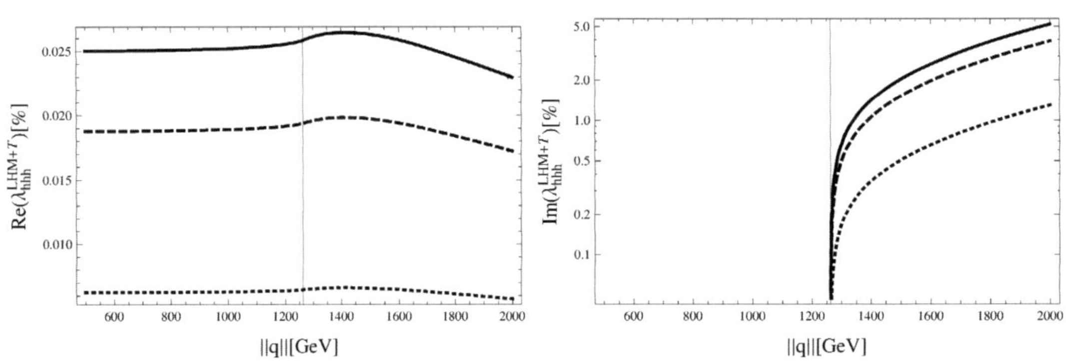

Figure 4. The same situation as in figure 3 but for $f = 1000$ GeV, where now $2m_{f_H} \simeq 1263$ GeV.

If the symmetry breaking scale is increased to $f = 1000$ GeV, the corrections decrease softly and we have the same behavior than the previous case, but now with $2m_{f_H} \simeq 1263$ GeV. In this case we have used the same values of $||q||$ than the previous case, to obtain $\mathrm{Re}(\lambda_{\mathrm{hhh}}^{\mathrm{LHM+T}}) \simeq 0.0250\%$ and $\mathrm{Re}(\lambda_{\mathrm{hhh}}^{\mathrm{LHM+T}}) \simeq 0.0229\%$ for the sum of both contributions respectively. Thus, we can see that the correction to the λ_{hhh} factor in LHM with parity T is not very sensitive to higher values of the symmetry breaking scale f, and of the 4-momentum magnitude of the off-shell scalar boson $||q||$.

5. Conclusions

We have studied the one loop correction to the self-coupling hhh in the Littlest Higgs Model with T-parity, which is induced by heavy mirror up-quarks and heavy mirror neutrinos. For this purpose, we consider the symmetry breaking scale f between 500 and 2000 GeV. Also, the constraint value of $\kappa_{ii} = 0.45$ was obtained from direct searches in $pp \to q_H q_H$ and $pp \to q_H A_H$ processes. In the Higgs resonance region, our result shows that the self-coupling correction is $\lambda_{\mathrm{hhh}}^{\mathrm{LHM+T}} \simeq 0.0977\%$ for $f = 500$ GeV, while that for $\lambda_{\mathrm{hhh}}^{\mathrm{LHM+T}} \simeq 0.0062\%$ for $f = 500$ GeV. Moreover, because of the larger mass of mirror fermions, there are not imaginary parts in the Higgs resonance region. However, for high values of 4-momentum magnitude of the off-shell scalar boson $||q||$ imaginary contributions may arise. It is important to note that mirror fermion correction is two orders of magnitude lower than the top-quark SM contribution, although is comparable with the gauge bosons contribution of SM. Finally, we show that the self-coupling correction $\lambda_{\mathrm{hhh}}^{\mathrm{LHM+T}}$ increased when the κ_{ii} value is increased.

Acknowledgement

We acknowledge support from CONACyT and SNI (Mexico).

References

[1] ATLAS Collaboration, *Observation of a new particle in the search for the Standard Model Higgs boson with the ATLAS detector at the LHC.* Phys. Lett. B **716**, 1-29 (2012).

[2] CMS Collaboration, *Observation of a new boson at a mass of 125 GeV with the CMS experiment at the LHC.* Phys. Lett. B **716**, 30-61 (2012).

[3] ATLAS Collaboration, *Searches for Higgs boson pair production in the* $hh \to bb\tau\tau, \gamma\gamma WW^*, \gamma\gamma bb, bbbb$ *channels with the ATLAS detector.* Phys. Rev. D **92**, 092004 (2015).

[4] Baglio, J. and Djouadi, A. and Grber, R. and Mhlleitner, M. M. and Quevillon, J. and Spira, M. *The measurement of the Higgs self-coupling at the LHC: theoretical status*, JHEP **04**, 151 (2013).

[5] Gutierrez-Rodriguez, A. and Peressutti, Javier and Sampayo, O. A. *Higgs Boson Self-Coupling at High Energy* $\gamma\gamma$ *Collider*, J. Phys. G **38**, 095002 (2011).

[6] Kanemura, Shinya and Kiyoura, Shingo and Okada, Yasuhiro and Senaha, Eibun and Yuan, C. P. *New physics effect on the Higgs selfcoupling*, Phys. Lett. B **558**, 157-164 (2003)

[7] Kanemura, Shinya and Okada, Yasuhiro and Senaha, Eibun. *Electroweak baryogenesis and quantum corrections to the triple Higgs boson coupling*, Phys. Lett. B **606**, 361-366 (2005).

[8] Han, Tao and Logan, Heather E. and McElrath, Bob and Wang, Lian-Ta *Phenomenology of the little Higgs model*, Phys. Rev. D **67**, 095004 (2003).

[9] Blanke, Monika and Buras, Andrzej J. and Duling, Bjoern and Poschenrieder, Anton and Tarantino, Cecilia. *Charged Lepton Flavour Violation and (g − 2)(mu) in the Littlest Higgs Model with T-Parity: A Clear Distinction from Supersymmetry*, JHEP **05**, 013 (2005).

[10] Reuter, Jrgen and Tonini, Marco and de Vries, Maikel. *Littlest Higgs with T-parity: Status and Prospects*, JHEP **02**, 053 (2014).

[11] Hollik, Wolfgang and Penaranda, Siannah. *Yukawa coupling quantum corrections to the selfcouplings of the lightest MSSM Higgs boson*, Eur. Phys. J. C **23**, 163-172 (2002).

Hadronic light-by-light contribution to the muon $g - 2$

Adolfo Guevara

Departamento de Física, Cinvestav del IPN, Apdo. Postal 14-740, 07000 Mexico DF. Mexico

E-mail: aguevara@fis.cinvestav.mx

Abstract. We have computed the hadronic light-by-light (LbL) contribution to the muon anomalous magnetic moment a_μ in the frame of Chiral Perturbation Theory with the inclusion of the lightest resonance multiplets as dynamical fields (RχT). It is essential to give a more accurate prediction of this hadronic contribution due to the future projects of J-Parc and FNAL on reducing the uncertainty in this observable. We, therefore, computed the pseudoscalar transition form factor and proposed the measurement of the $e^+e^- \to \mu^+\mu^-\pi^0$ cross section and dimuon invariant mass spectrum to determine more accurately its parameters. Then, we evaluated the pion exchange contribution to a_μ, obtaining $(6.66 \pm 0.21) \cdot 10^{-10}$. By comparing the pion exchange contribution and the pion-pole approximation to the corresponding transition form factor (πTFF) we recalled that the latter underestimates the complete πTFF by (15-20)%. Then, we obtained the $\eta^{(\prime)}$ TFF, obtaining a total contribution of the lightest pseudoscalar exchanges of $(10.47 \pm 0.54) \cdot 10^{-10}$, in agreement with previous results and with smaller error.

1. Introduction

Ever since the measurement of the electron magnetic moment in the splitting of the ground states of deuterium and molecular hydrogen [1], the anomalous magnetic moment has been an ever more stringent test of the underlying theory governing the interactions among elemental particles; giving us the lead from a way of renormalizing QED [2] to an outstanding confirmation of QFT with QED contributions [3] up to order $\left(\frac{\alpha}{\pi}\right)^6$. In this spirit, the a_μ has been seen as a very stringent test of beyond standard model physics (BSM). With the most recent measurements [4], a deviation from the Standard Model (SM) results would imply a contribution from BSM with a scale $\sim$ 100 TeV (assuming an interaction ~ 1). The current discrepancy [5] of 3.6σ and future plans on measuring more accurately this observable force theorists to make more precise predictions of SM contributions to the μ anomalous magnetic moment.

Within the SM, the contributions to the a_μ that have a greater uncertainty are the hadronic ones [5]. This is due to the fact that the underlying theory cannot be taken perturbatively in the whole energy range of the quark loop integrals, forcing theorists to compute these contributions using effective field theories (EFT) based on symmetries of Quantum Chromodynamics (QCD). This hadronic contribution can be splitted into two sub-contributions, the Hadron Vacuum Polarization (HVP) and the Hadronic Light-by-Light (HLbL), shown in Fig. 1. We analize the latter one by studying the $P\gamma^*\gamma^*$ interaction through its form factor (also called P transition form factor, PTFF), which gives the leading contribution to the HLbL through a pseudoscalar exchange diagram shown in Fig. 2. At low energies (*i.e.* in the chiral limit), the prediction for the πTFF has been confirmed by the measured rate of $\pi^0 \to \gamma\gamma$ decays [5]. On the other hand,

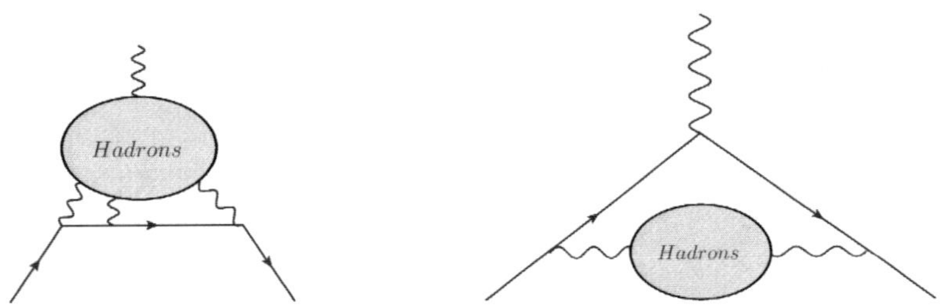

Figure 1. The two hadronic contribution to a_μ: Hadronic Light-by-Light (left) and Hadron Vacuum Polarization (right).

the prediction for a nearly on-shell photon and one with very large virtuality seems to be at odds with measurements at B-factories [6, 7]. These two limits have ruled the way of constructing the form factor to describe interactions in the intermediate energy region, where hadronic degrees of freedom play a crucial role. The EFT we use to compute the TFF is Resonance Chiral Theory (RχT) [8, 9], which makes use of short-distance QCD predictions to obtain the parameters of the theory in terms of known constants. In this work, we fit one of the parameters in the πTFF with the B-factories data and, using this information, we predict the $\eta^{(\prime)}$TFF and then obtain the contribution to the a_μ using these form factors.

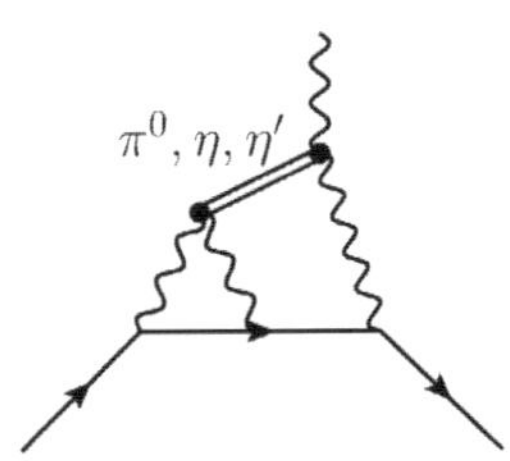

Figure 2. Pseudoscalar exchange contribution to the HLbL.

2. Theoretical Framework

Chiral Perturbation Theory (χPT) [10] is the EFT dual to QCD at low energies [11]. It is based on an expansion in powers of momenta and masses of the lightest pseudoscalar mesons over the chiral symmetry breaking scale ($\Lambda \sim 1$ GeV). Thus, the theory fails to be reliable at energies around 1 GeV; furthermore, when other mesons become relevant degrees of freedom ($\Lambda \gtrsim M_\rho$) the theory is no longer applicable. A generalization of χPT is obtained using $1/N_C$ as an expansion parameter [12] to include resonances as dynamical degrees of freedom. The theory that incorporates these elements is Resonance Chiral Theory (RχT) [8, 9], which requires unitary symmetry for the resonance multiplets. No *a priori* assumptions are made with respect to the role of resonances in this theory, therefore one obtains naturally Vector Meson Dominance [13] as a dynamical result of the theory [8]. The final ingredient of the theory comes from QCD behavior at short distances, which constraints a great amount of free parameters in the theory.

3. The $\pi\gamma^*\gamma^*$ form factor in RχT

In this framework, the form factor we obtain [14] is

$$\mathcal{F}_{\pi^0\gamma^*\gamma^*}(p^2,q^2,r^2) = \frac{2r^2}{3F}\left[-\frac{N_C}{8\pi^2 r^2} + 4F_V^2\frac{d_3(p^2+q^2)}{(M_V^2-p^2)(M_V^2-q^2)r^2} + \frac{4F_V^2 d_{123}}{(M_V^2-p^2)(M_V^2-q^2)}\right.$$
$$+\frac{16F_V^2 P_3}{(M_V^2-p^2)(M_V^2-q^2)(M_P^2-r^2)} - \frac{2\sqrt{2}}{M_V^2-p^2}\left(\frac{F_V}{M_V}\frac{r^2 c_{1235} - p^2 c_{1256} + q^2 c_{125}}{r^2} + \frac{8P_2F_V}{(M_P^2-r^2)}\right)$$
$$\left. + (q^2 \leftrightarrow p^2)\right]. \quad (1)$$

It contains contributions from the pseudoscalar resonances (as can be seen through the couplings P_2 and P_3) which need to be taken into account to obtain consistent short distance constraints [9, 15]. All the parameters, but one, can be obtained through these constraints. P_3 cannot be obtained requiring high energy constraints, therefore it is fitted using the combined analyses of $\pi(1300) \to \gamma\gamma$ and $\pi(1300) \to \rho\gamma$ decays as given in references [9, 14]. The consistent short distance constraints on the resonance couplings in the odd-intrinsic parity sector can be seen in refs [14, 15]. Thus, we obtain

$$P_3 = (-1.2 \pm 0.3)\cdot 10^{-2}\ \mathrm{GeV}^2. \quad (2)$$

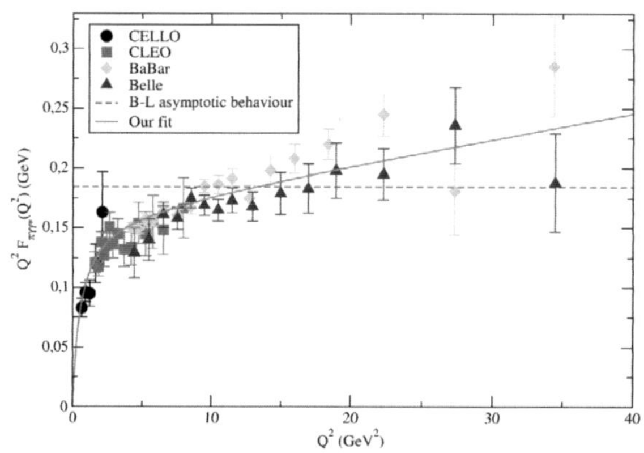

Figure 3. Our best fit compared to CELLO, CLEO, BaBar and Belle data for the πTFF.

On the other hand, the πTFF does not fit very well experimental data [6, 7] when P_2 is constrained by the short distance prediction. Therefore we allowed for it a small variation in a fit to Babar and Belle data of this form factor, where they measure the πTFF spectrum in a kinematical configuration that ensures that one of the photons is on-shell and the other is virtual. The form factor for such a configuration is given by taking [1] $p^2 \to 0$ and $Q^2 = -q^2$ in eq. (1)

$$\mathcal{F}_{\pi^0\gamma^*\gamma}(Q^2) = -\frac{F}{3}\frac{Q^2(1+32\sqrt{2}\frac{P_2F_V}{F^2}) + \frac{N_C}{4\pi^2}\frac{M_V^4}{F^2}}{M_V^2(M_V^2+Q^2)}. \quad (3)$$

We keep a very conservative 10% uncertainty from the asymptotic value of F_V around its predicted value [15] of $\sqrt{3}F$. Fig. 3 shows our best fit, with which we obtain

$$P_2 = (-1.13 \pm 0.12)\cdot 10^{-3}\ \mathrm{GeV}, \qquad \chi^2/dof = 1.01\,. \quad (4)$$

[1] By the kinematical configuration in which the process is chosen to be measured, the momenta of both photons are space-like.

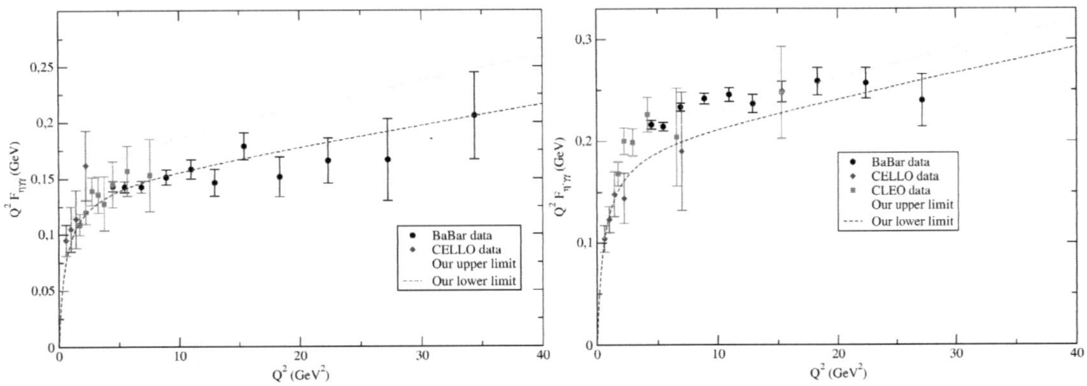

Figure 4. Our prediction of the ηTFF (left) and η'TFF (right) compared to CELLO, CLEO and BaBar data and using the parameters found with the πTFF.

4. The pseudoscalar exchange contribution to the a_μ^{HLbL}

Once all the parameters in the πTFF are determined, we insert the full off-shell TFF in the relations given in [16] obtaining thus

$$a_\mu^{\pi^0 LbL} = (5.75\pm0.06)\cdot10^{-10} \text{ on-shell } \pi^0 \qquad a_\mu^{\pi^0 LbL} = (6.66\pm0.21)\cdot10^{-10} \text{ whole } \pi^0\text{TFF}. \quad (5)$$

Table 1. Our result compared with other results obtained through different methods.

$a_\mu^{\pi^0 LbL} \cdot 10$	Model and Reference
5.58 ± 0.05	Extended Nambu-Jona-Lasinio [17]
5.56 ± 0.01	Naive VMD [18]
5.8 ± 1.0	Large N_C with two vector multiplets π-pole [16]
7.2 ± 1.2	π-exchange contribution [19]
6.54 ± 0.25	Holographic models of QCD [20]
6.58 ± 0.12	Lightest pseudoscalar and vector resonance saturation [9]
6.49 ± 0.56	Rational approximants [21]
5.0 ± 0.4	Non-local chiral quark model [22]
5.75 ± 0.06	Our result with on-shell π^0 [14]
6.66 ± 0.21	Our result whole π^0TFF [14]

This clearly shows that assuming an on-shell pion in the a_μ^{HLbL} underestimates the contribution in $\sim 15\%$, and the error by a factor of 4. The uncertainty comes mainly from the error in F_V, P_3 and in a chiral correction from very-low energy physics. We compare our result with previous results in table 1. The form factor for the η and η' can be obtained with the πTFF through eq. (6) with the minus sign for the case of the η.

$$\mathcal{F}_{\eta^{(\prime)}\gamma^*\gamma^*} = \left(\frac{5}{3}C_q^{(\prime)} \mp \frac{\sqrt{2}}{3}C_s^{(\prime)}\right)\mathcal{F}_{\pi^0\gamma^*\gamma^*} \qquad (6)$$

With this, we can compute the whole pseudoscalar exchange contribution to the a_μ^{HLbL}, shown in table 2 which includes other sub-leading HLbL contributions.

Table 2. Comparison of our contributions of the full a_μ^{HLbL} to previous determinations.

$a_\mu^{HLbL} \cdot 10^{10}$	Contributions
11.6 ± 4.0	F. Jegerlehner and A. Nyffeler [19]
10.5 ± 2.6	Prades, De Rafael and Vainshtein[2] [23]
11.8 ± 2.0	Our contribution [14]

5. Genuine probe of πTFF

All the experimental observables available to fit the parameters in the πTFF so far need an on shell photon and/or have photons with space-like momenta, while the HLbL contribution to the a_μ has both photons with time-like momenta. The photons in the process we study in this section, namely $\sigma(e^+e^- \to \gamma^* \to \pi^0\gamma^* \to \mu^+\mu^-\pi^0)$, both have time-like momenta and can be measured at very high photon virtualities by KLOE collab. for $q^2 \sim 1$ GeV^2 and Belle-II collab. for $q^2 \sim 10.5$ GeV^2. With the πTFF parameters fully determined, the prediction we obtain is shown in Fig. 5

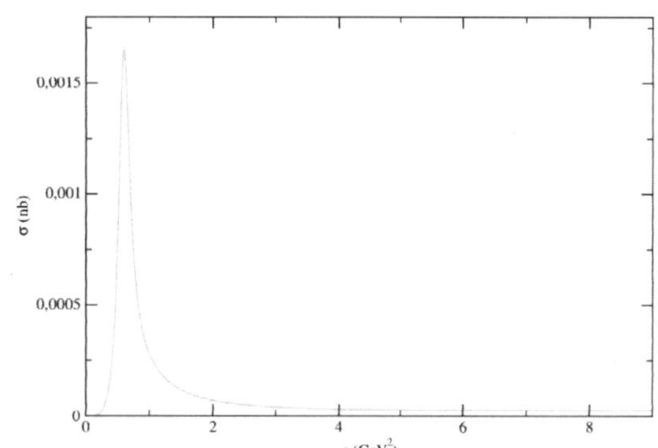

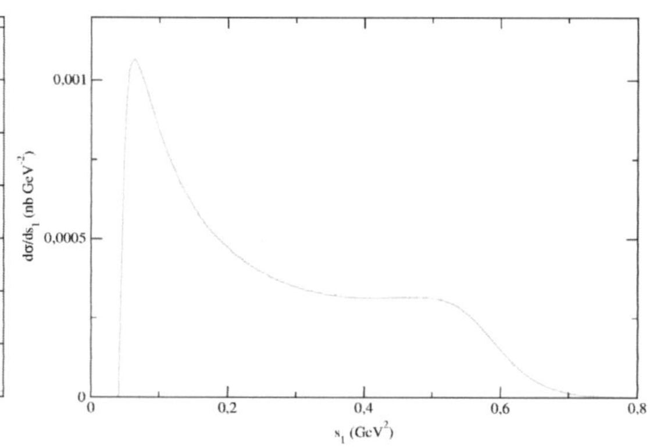

Figure 5. Our prediction for $\sigma(s)$ (left) and for $\frac{s\sigma}{ds_1}$ with $s = 1.02$ GeV^2, the error bands cannot be appreciated in these plots.

6. Conclusion

We found the pseudoscalar exchange contribution to the a_μ^{HLbL} with a very competitive uncertainty and consistent with other theoretical models; improving the analysis by including high energy constraints not realized in the reference [9] and also using Belle data released after the reference was published. Our error estimate is also more robust, since in addition to the errors of the resonance couplings, we have also included the uncertainty due to the value of the πTFF at very low energies.

We also obtained the first prediction for the cross section $\sigma(e^+e^- \to \mu^+\mu^-\pi^0)$, which might be measured in KLOE-2 and Belle-II. The measurement of this observable would be an interesting way of trying to reduce the error in the parameters of the πTFF. This, may also help to reduce the uncertainty on the mixing parameters between the η and η' mesons.

Acknowledgments
The author wishes to thank the DPyC and Cinvestav for financial support.

References

[1] Nafe J E, Nelson E B and Rabi I I 1947 Phys. Rev. **71** 914.
[2] Schwinger J S 1948 Phys. Rev. **73** 416.
[3] Kinoshita T and Nio M 2006 Phys. Rev. D**73** 053007.
[4] Bennett G W *et al.* [Muon g-2 Collab.] 2006 Phys. Rev. D**73** 072003.
[5] Olive K A *et al.* [PDG] 2014 Chin. Phys. C**38**.
[6] Aubert B *et al.* [BABAR Collab.] 2009 Phys. Rev. D**80** 052002.
[7] Shen C P *et al.* [Belle Collab.] 2013 Phys. Rev. D**88** 052019.
[8] Ecker G, Gasser J, Pich A and de Rafael E 1989 Nucl. Phys. **B321** 311; Ecker G, Gasser J, Leutwyler H, Pich A and de Rafael E 1989 Phys. Lett. B **223** 425; Cirigliano V, Ecker G, Eidenmuller M, Kaiser R, Pich A and Portolés J 2006 Nucl. Phys. **B753** 139.
[9] Kampf K and Novotný J 2011 Phys. Rev. D**84** 014036.
[10] Weinberg S 1979 Physica (Amsterdam) **96A** 327, Gasser J and Leutwyler H 1984 Ann. Phys. **158** 142; 1985 Nucl Phys **B250** 465.
[11] Coleman S R, Wess J and Zumino B 1969 Phys. Rev. **177** 2239; Callan C, Coleman S R, Wess J and Zumino B 1969 Phys. Rev. **177** 2247.
[12] 't Hooft G 1974 Nucl. Phys. **B72** 461; Nucl. Phys. **B75** 461.
[13] Sakurai J J 1969 *Currents and Mesons, Chicago Lectures in Physics* University of Chicago Press Chicago.
[14] Roig P, Guevara A and López Castro G 2014 Phys Rev D**89** 073016.
[15] Roig P and Sanz-Cillero J J Phys. Lett. B **733** (2014) 158.
[16] Knecht M and Nyffeler A 2002 Phys. Rev. D**65** 073034.
[17] Bijnens J, Fayyazuddin A and Prades 1995 J Phys. Lett. B**379** 209
[18] Hayakawa M, Kinoshita T and Sanda A I 1995 Phys. Rev. Lett. **75** 790
[19] Jegerlehner F and Nyffeler A 2009 Phys. Rep. **477** 1
[20] Cappiello L, Cata O and D'Ambrosio G 2011 Phys. Rev. D**83** 093006
[21] Masjuan P and Vanderhaeghen M 2015 J. Phys. G42 **12** 125004
[22] Dorokhov A E, Radzhabov A E and Zhevlakov A S (2012) Eur. J. Phys. C**72** 2227
[23] Prades J, de Rafael E and Vainshtein A Advanced series on directions in high energy physics (2009) Vol. 20 303

$\Delta\kappa'_V$ and ΔQ_V ($V=\gamma$, Z) form factors in the Georgi-Machacek model

M.A. Arroyo-Ureña, G. Hernández-Tomé, and G. Tavares-Velasco
Facultad de Ciencias Físico Matemáticas, Benemérita Universidad Autónoma de Puebla, Apartado postal 1152, 72001 Puebla, Pue., México

E-mail: gerardoht@alumnos.fcfm.buap.mx

Abstract. The CP-even static form factors $\Delta\kappa'_V$ and ΔQ_V ($V=\gamma$, Z) associated with the WWV vertex are studied in the context of the Georgi-Machacek model (GMM), which predicts nine new scalar bosons accommodated in a singlet, a triplet and a fiveplet. General expressions for the one-loop contributions to $\Delta\kappa'_V$ and ΔQ_V arising from neutral, singly and doubly charged scalar bosons are obtained in terms of both parametric integrals and Passarino-Veltman scalar functions, which can be numerically evaluated. It is found that the GMM yields 15 (28) distinct contributions to $\Delta\kappa'_\gamma$ and ΔQ_γ ($\Delta\kappa'_Z$ and ΔQ_Z), though several of them are naturally suppressed.

1. Introduction

Models with scalar triplet representations have attracted considerable attention due to their appealing features, such as the possibility of implementing the see-saw mechanism to endow the neutrinos with naturally light Majorana masses (the so called type-II see-saw), the appearance of the $H^\pm W^\mp Z$ coupling at the tree level, and the presence of doubly charged scalar particles. In this respect, the Georgi-Machacek model (GMM) [1, 2] is one of the most attractive Higgs triplet models as it preserves the relationship $\rho=1$ at the tree level via an $SU(2)$ custodial symmetry. The physical scalar spectrum of the GMM is given by the SM-like Higgs boson h and one extra CP-even singlet H, one scalar triplet H_3 (H_3^0, $H_3^\pm$), and one scalar fiveplet H_5 (H_5^0, $H_5^{\pm\pm}$, $H_5^\pm$). All of these multiplets are mass degenerate as a result of the custodial symmetry. Even if there is not enough energy available to produce the new scalar particles predicted by the GMM, one can search for their virtual effects through some observables. Particular interest has been put on the radiative corrections to the WWV ($V=\gamma, Z$) vertex, which represents a very sensitive scenario to search for any NP effects and test the gauge sector of the SM.

The on-shell WWV vertex can be written in terms of four form factors that define the CP-even and CP-odd static properties of the W boson. The two CP-odd form factors $\Delta\widetilde{\kappa}'_V$ and $\Delta\widetilde{Q}_V$ are absent up to the one-loop level in the SM and are thus expected to be negligibly small. As far as the CP-even form factors $\Delta\kappa'_V$ and ΔQ_V are concerned, they arise at the one-loop level in the SM and any other renormalizable theory, thereby being highly sensitive to NP effects.

The vertex function that determines the WWV CP-conserving coupling can be written as

$$\Gamma_V^{\mu\alpha\beta} = ig_V\Big\{A\left[2p^\mu g^{\alpha\beta}+4\left(Q^\beta g^{\mu\alpha}-Q^\alpha g^{\mu\beta}\right)\right]$$

MWPF IOP Publishing
Journal of Physics: Conference Series **761** (2016) 012010 doi:10.1088/1742-6596/761/1/012010

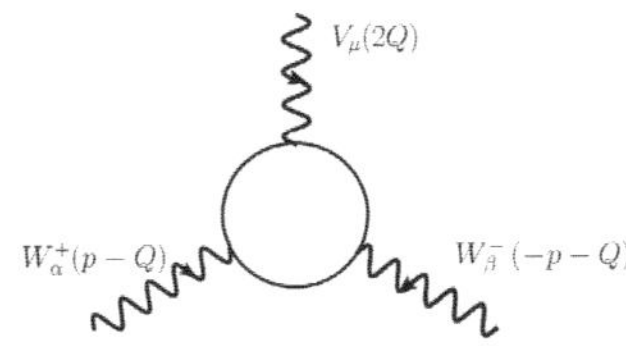

Figure 1. The WWV vertex function. The circle denotes radiative contributions.

$$+2\Delta\kappa_V'\left(Q^\beta g^{\mu\alpha}-Q^\alpha g^{\mu\beta}\right)+\frac{4\Delta Q_V}{m_W^2}\left(p^\mu Q^\alpha Q^\beta-\frac{1}{2}m_V^2 p^\mu g^{\alpha\beta}\right)\Bigg\}, \quad (1)$$

where g_V stand for the WWV tree-level coupling constant (in the SM $g_\gamma = g s_W$ and $g_Z = g c_W$) and we have used the convention employed in [3] for the external momenta, as shown in Fig. 1.

2. $\Delta\kappa_V'$ and ΔQ_V form factors in the GMM

The new scalar one-loop contributions from the GMM to the $\Delta\kappa_V'$ and ΔQ_V form factors arise from generic triangle diagrams (the bubble diagrams do not contribute) that can be classified according to the number of distinct particles circulating into the loop. Each type of diagram involves trilinear vertices between gauge bosons only or between scalar bosons and gauge bosons. In Fig. 2 we show a set of Feynman diagrams that contribute to both the $WW\gamma$ and WWZ vertices.

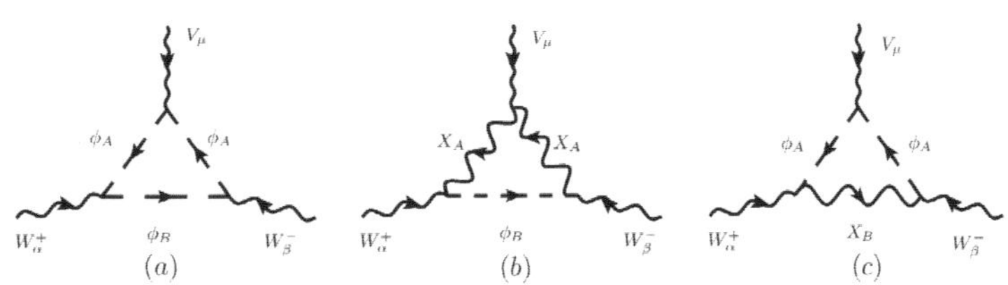

Figure 2. Generic Feynman diagrams for the new scalars contributions to both the $WW\gamma$ and WWZ vertices involving only two distinct virtual particles. The arrows stand for the directions of the 4-momenta. The possible combinations of internal particles are given by the vertices allowed in each particular model. For instance, when $V=\gamma$, the following electric charges of the internal particles are possible in the GMM, in units of the positron charge: if $Q_A=-1$ then $Q_B=0$, if $Q_A=1$ then $Q_B=-2$, if $Q_A=-2$ then $Q_B=1$.

Contrary to the couplings of the photon to a pair of charged scalar bosons, which can only be of diagonal type due to electromagnetic gauge invariance, the Z boson can have nondiagonal couplings to a pair of neutral or charged scalar bosons. Therefore, in addition to the diagrams of Fig. 2, the $\Delta\kappa_Z'$ and ΔQ_Z form factors can receive extra contributions from the Feynman diagrams shown in Fig. 3, which have three distinct particles circulating into the loop. Below we will present the contributions to the $\Delta\kappa_V'$ and ΔQ_V form factors for all these type of diagrams.

Before presenting our results, some remarks about our calculation are in order:

- The Feynman diagrams were evaluated via the unitary gauge. In order to make a cross check of our results we used both, the Feynman parametrization technique and the Passarino-Veltman method to solve the loop integrals.

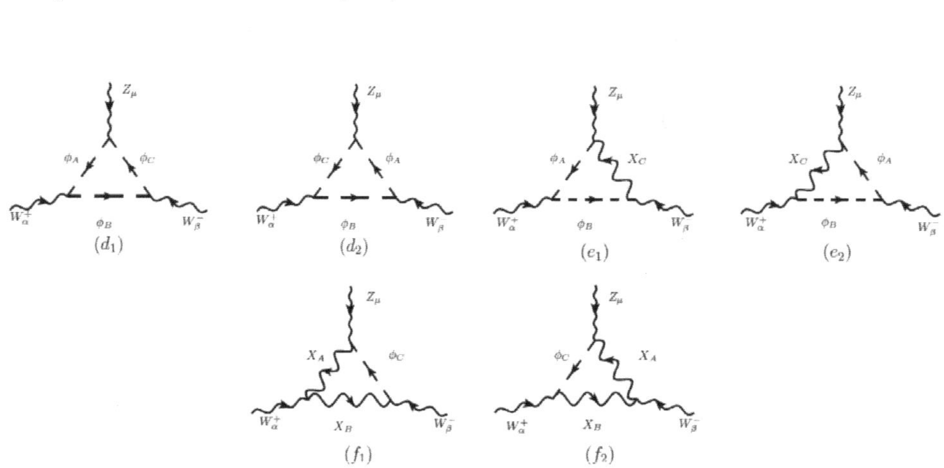

Figure 3. Extra contributions to the $\Delta\kappa'_Z$ and ΔQ_Z form factors from nondiagonal couplings.

- We verified that all the contributions of bubble diagrams to the $\Delta\kappa'_V$ and ΔQ_V form factors involving quartic vertices with two scalar bosons and two gauge bosons vanish, and thus the only contributions arise from triangle diagrams.
- The mass shell and transversality conditions for the gauge bosons enabled us to make the following replacements

$$Q^2 = \frac{m_V^2}{4}, \qquad p \cdot Q = 0, \qquad p^2 = m_W^2 - \frac{m_V^2}{4}, \tag{2}$$

and

$$p_\alpha \to Q_\alpha, \quad p_\beta \to -Q_\beta, \quad p_\mu \to 0, \tag{3}$$

which results in a considerable simplification of the calculation.
- Instead of dealing with the calculation of the $WW\gamma$ and WWZ vertices separately, we performed instead the calculation of the general WWV vertex, with V a massive neutral gauge boson. We have exploited the fact that there are only three generic trilinear vertices involved in the one-loop contributions to the WWV vertex and thus a model independent calculation was done using the generic Feynman rules of Fig. 4. The result for the contribution of each type of Feynman diagram will be presented in terms of loop functions, given as parametric integrals and also in terms of Passarino-Veltman scalar integrals, times a factor involving all the generic coupling constants associated with each vertex participating in the particular diagram. The contribution to the form factors of the $WW\gamma$ and WWZ vertices follow easily from our general expressions after taking the appropriate mass limits and substituting the corresponding coupling constants of the GMM or any other extension model.
- We corroborated all the contributions to the $\Delta\kappa'_V$ and ΔQ_V form factors are free of ultraviolet divergences.

Once the amplitude for each Feynman diagram is written down, the Feynman parametrization technique and the Passarino-Veltman method can be applied straightforwardly, along with some cumbersome algebra. Thereafter one can express the contributions to the $\Delta\kappa'_V$ and ΔQ_V form factors for each type of Feynman diagram of Fig. 2 as follows

$$\Delta\kappa_V'^i = -\frac{C_V^i}{16\pi^2} I_\kappa^{V-i}(x_A, x_B, x_V), \tag{4}$$

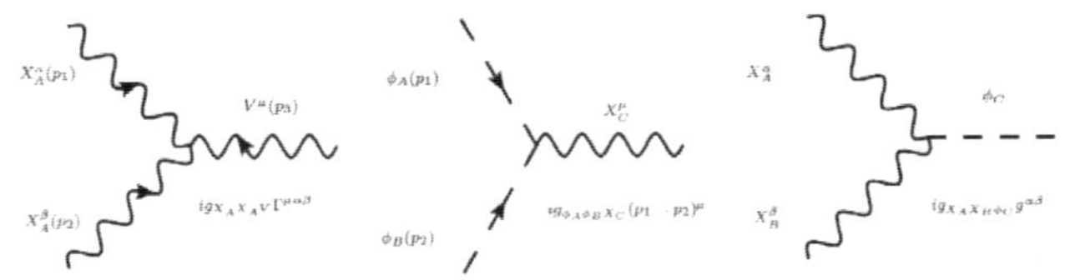

Figure 4. Generic Feynman rules for the relevant vertices involved in our calculation. The arrows stand for the direction of the 4-momenta and $\Gamma_{\mu\alpha\beta} = g_{\mu\alpha}(p_1 - p_3)_\beta + g_{\alpha\beta}(p_2 - p_1)_\mu + g_{\beta\mu}(p_3 - p_2)_\alpha$. $V = \gamma, Z$, ϕ_I $(I = A, B, C)$ denote a neutral singly, or doubly charged scalar boson, and X_J $(J = A, B)$ stands for a neutral or charged gauge boson.

$$\Delta Q_V^i = -\frac{C_V^i}{16\pi^2} I_Q^{V-i}(x_A, x_B, x_V), \tag{5}$$

for $V = Z, \gamma$ and $i = a, b, c$. We have introduced the scaled variable $x_I = m_I^2/m_W^2$ $(I = A, B)$, with m_A and m_B denoting the masses of the particles circulating into each type of diagram. A word of caution is in order here as m_A and m_B, and thereby x_A and x_B, are distinct for each type of contribution, which in turn is denoted by the superscript i $(i = a, b, c)$. As for the loop functions I_κ^{V-i} and I_Q^{V-i}, they are in terms of parametric integrals and Passarino-Veltman scalar integrals and can be found in [5], while the C_V^i factors are given in term of the coupling constants of the vertices involved in each Feynman diagram.

As explained above, the $\Delta\kappa_\gamma'^i$ and ΔQ_γ^i form factors can be obtained from the general expressions (4) -(5) by taking the $m_V \to 0$ limit. We have verified that these expressions are in agreement with the results presented in Ref. [4], where the $WW\gamma$ vertex was studied in the context of little Higgs models.

As far as the Feynman diagrams of Fig. 3 are concerned, they only contribute to the WWZ vertex and the respective form factors depend now on three distinct internal masses. They can be written as follows

$$\Delta\kappa_Z'^i = -\frac{C_Z^i}{16\pi^2} I_\kappa^{Z-i}(x_A, x_B, x_C, x_Z), \tag{6}$$

$$\Delta Q_Z^i = -\frac{C_Z^i}{16\pi^2} I_Q^{Z-i}(x_A, x_B, x_C, x_Z). \tag{7}$$

This time the superscript $i = d, e, f$ stands for the whole contributions of diagrams i_1 and i_2, and again the loop functions I_κ^{Z-i} and I_Q^{Z-i} can be found in [5].

Once the general expressions for the different kinds of contributions are obtained, we can compute the total contribution of the scalar sector of a given model by simple adding up all the partial contributions. We will present below a numerical analysis of the contributions of the GMM. For the numerical evaluation we computed the parametric integrals via the Mathematica numerical routines. A cross check was done using the results obtained by evaluating the results given in terms of Passarino-Veltman scalar functions with the help of the LoopTools routines.

2.1. $\Delta\kappa_\gamma'$ and ΔQ_γ form factors

Excluding the pure SM contributions, in the GMM, the $\Delta\kappa_\gamma'$ and ΔQ_γ form factors receive 10 contributions of the type-(a) diagrams, 3 of the type-(b) diagrams, and 2 of the type-(c) diagrams. Notice that all the new scalar bosons participate in the type-(a) diagrams, whereas the type-(b) diagrams only receive contributions from the singlet and the fiveplet scalar bosons, and the type-(c) diagrams from the fiveplet scalar bosons only. We examine the general behavior of $\Delta\kappa'_\gamma$ and ΔQ_γ as functions of the masses of the scalar bosons. For the type-(b) and type-(c) contributions we show in Fig. 5 the form factors as a function of the mass of the scalar boson circulating into the loop, whereas for type-(a) diagram we consider two scenarios: when both scalar bosons are degenerate and when one scalar boson mass is fixed and the other one is variable.

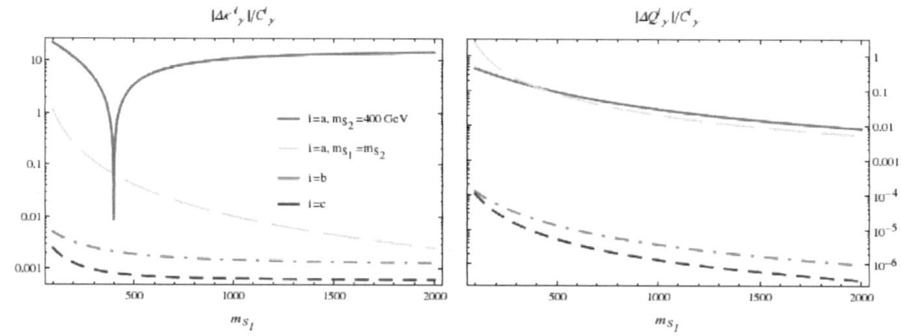

Figure 5. Behavior of the $\Delta\kappa'_\gamma$ and ΔQ_γ form factors as functions of the masses of the scalar bosons circulating into the loops of each type of contribution divided by the C^i_γ coefficient and in units of $a = g^2/(96\pi^2)$. While type-(a) contribution depends on two scalar boson masses m_{S_1} and m_{S_2}, type-(b) and type-(c) diagrams depends on only one scalar boson mass m_{S_1}.

2.2. $\Delta\kappa'_Z$ and ΔQ_Z form factors.

We will now analyze the $\Delta\kappa'_Z$ and ΔQ_Z form factors, for which we will follow a similar approach to that used above. We thus start by studying the general behavior of the distinct types of contributions. Apart from the diagrams of Fig. 2, there is additional contributions due to the diagrams of Fig. 3. As for the contributions of type (a), (b) and (c), their behavior is quite similar to that observed in Fig. 5, so here we will thus focus on the analysis of the extra contributions, whose behavior will turn out to be rather similar to that of contributions type (a), (b) and (c). There are 7 contributions of type (d), 4 of type (e), and 3 of type (f) to the $\Delta\kappa'_Z$ and ΔQ_Z form factors. Although our general results allow us to calculate type-(d) contributions with three distinct scalar boson masses m_{S_1}, m_{S_2} and m_{S_3}, in the GMM all the masses of the same multiplet are degenerate. It means that type-(d) contributions arise only from diagrams with al least two degenerate scalar bosons. Also, although type-(e) contribution arise from diagrams that can have two distinct scalar bosons, their masses are degenerate and there is dependence on one mass only, and this is also true for type-(f) contributions. Therefore, we expect that type-(d) contributions will be dominant as long as there is a large mass splitting between the scalar boson masses, whereas type-(e) and type-(f) contributions will only be important for a relatively light scalar boson mass. This is depicted in Fig. 6, where we show the behavior of the $\Delta\kappa'_Z$ and ΔQ_Z form factors for all the scenarios allowed in the GMM. For type-(d) contributions we consider three scenarios: m_{S_3} fixed and $m_{S_2} = m_{S_1}$ variables, $m_{S_3} = m_{S_2}$ fixed and m_{S_1} variable, and the three scalar boson masses degenerate $m_{S_3} = m_{S_2} = m_{S_1}$. On the other hand, for type-(e) contributions we only consider the case when the two scalar bosons are degenerate. In Fig. 6 we observe that $\Delta\kappa_Z$ and ΔQ_Z have a similar behavior to that of the $\Delta\kappa'_\gamma$ and ΔQ_γ form factors. In particular, the largest contributions to $\Delta\kappa'_\gamma$ are reached when there is a large mass splitting or when all the scalar bosons masses circulating into each loop are relatively light.

However, the decrease of $\Delta\kappa'_Z$ for large m_{S_1} is now less quick than in the case of $\Delta\kappa'_\gamma$. Again, the C^i_Z factor is proportional to v, so the values shown in the plots for type-(e) and type-(f) contributions will increase by two orders of magnitude. As for ΔQ_Z, it will reach its large value for the smallest allowed scalar boson masses as in the case of ΔQ_γ. When the scalar bosons are very heavy, they will be approximately degenerate, in which case ΔQ_Z will decrease significantly. Extra suppression for both form factors can arise from the C^i_Z coefficients and from potential cancellations between the distinct contributions.

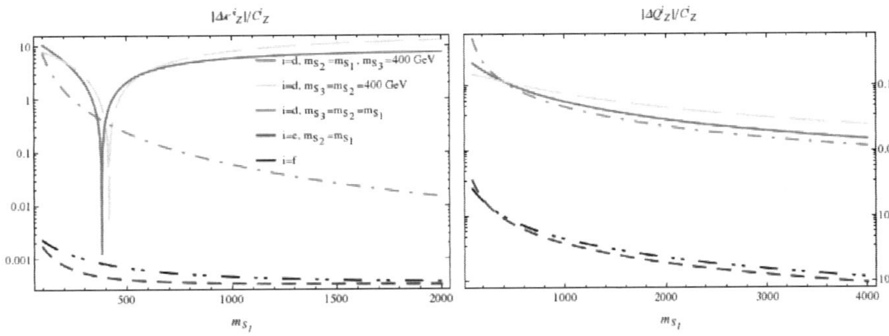

Figure 6. Behavior of the $\Delta\kappa'_Z$ and ΔQ_Z form factors as a function of the masses of the scalar bosons circulating into the loops of each type of contribution divided by the C^i_γ coefficient. We only consider the possible scenarios arising in the GMM.

3. Conclusions

A model independent calculation was done via both the Feynman parameter technique and the Passarino-Veltman reduction scheme. Our general results are expressed in terms of a six generic contributions to $\Delta\kappa'_V$ and ΔQ_V from scalar bosons that can be used to calculate the corrections arising from models with an extended scalar sector predicting new neutral, singly, and doubly charged scalar bosons. For the numerical analysis we have focused on the GMM, this model predicts 9 new scalar bosons accommodated in a singlet, a triplet and a fiveplet, which yield 15 new contributions to $\Delta\kappa'_\gamma$ and ΔQ_γ, whereas $\Delta\kappa'_Z$ and ΔQ_Z receive 28 distinct contributions. The general behavior of the $\Delta\kappa'_V$ and ΔQ_V form factors was analyzed and it was found that $\Delta\kappa'_V$ reaches values of the order of $a = g^2/(96\pi^2)$, with the largest values arising from the diagrams with two nondegenerate scalar bosons provided that there is a large splitting between their masses. On the other hand ΔQ_V reaches values of the order of $10^{-2}a$, with the largest contributions arising from diagrams with relatively light degenerate scalar bosons. Both form factors decrease rapidly when all the scalar boson masses are heavy. The values for $\Delta\kappa'_V$ and ΔQ_V predicted by the GMM are competitive with the ones predicted by other weakly coupled SM extensions, but a very high precision still would be necessary to detect such effects.

References

[1] H. Georgi and M. Machacek, Nucl. Phys. B **262**, 463 (1985). doi:10.1016/0550-3213(85)90325-6
[2] J. F. Gunion, R. Vega and J. Wudka, Phys. Rev. D **42**, 1673 (1990). doi:10.1103/PhysRevD.42.1673
[3] W. A. Bardeen, R. Gastmans and B. E. Lautrup, Nucl. Phys. B **46**, 319 (1972). doi:10.1016/0550-3213(72)90218-0
[4] A. Moyotl and G. Tavares-Velasco, J. Phys. G **37**, 105012 (2010) doi:10.1088/0954-3899/37/10/105012 [arXiv:1003.3230 [hep-ph]].
[5] M.A. Arroyo-Ureña, G. Hernández-Tomé, and G. Tavares-Velasco, *WWV* ($V = \gamma, Z$) vertex in the Georgi-Machacek model, Work in progress.

Journal of Physics: Conference Series **761** (2016) 012011 doi:10.1088/1742-6596/761/1/012011

Brief description of the flavor-changing neutral scalar interactions at two-loop level

R. Gaitán[1] **and J. A. Orduz-Ducuara**[2]
Departamento de Física,
FES-Cuautitlán Izcalli, UNAM
C.P. 54740,
Estado de México, México.
E-mail: [1]rgaitan@unam.mx, [2]jaorduz@ciencias.unam.mx

Abstract. In this letter we show a general description about flavor-changing neutral currents (FCNC) mediated by scalars. The analysis is extended at two-loop level for the Two-Higgs Doublet Model type-III because others models have strong constraints on its parameters, even at high orders of the perturbation. For this letter we focus on the standard model, calculating the amplitude for the $h \to \gamma\gamma$ process and discussing the results briefly.

To be published in Journal of Physics Conference Series (IOP). Joint Proceedings of the XV Mexican Workshop on Particles and Fields & the XXX Annual Meeting of the Division of Particles and Fields of the Mexican Physical Society

1. Introduction

The standard model (SM) of particles has awesome results: the experimental measurements and the theoretical results match very well. The observables are calculated using the Feynman diagrams. For instance, to obtain the decay width or the cross section, we need the Lagrangian or the Feynman rules (vertices and propagators of the model). In general, the SM extensions increase the vertex and propagators number. SM extensions are built to explore New Physics (NP) and describe unexplored phenomena. NP includes proposals as beyond standard model (BSM), it involves new particles, new interactions and new dimensions; for explaining or responding to the unsolved questions.

We can extend the gauge, fermion or scalar sector. Focusing on the last one, the simplest possibility to extend the SM is the Two-Higgs Doublet Model (THDM), which introduces a doublet scalar field plus the SM doublet. Theoretical motivations to enlarge the scalar sector could explain, e.g., the CP violation and the flavor-changing neutral currents (FCNC) in the gauge sector [1].

The rich phenomenology for the THDM has been widely studied by different theoretical and phenomenological groups. Some reports show interesting results for the Flavor-changing (FC) [2–4], an analysis at one-loop level for the pseudoscalar appears in [5], and at two loops in the lepton sector is found in [6]. Several experimental reports are dedicated to the exotic physics searching for new particles coming from a variety of models [7–13].

The purpose of this paper is to further explore the FC mediated by scalar bosons at loop-level. We were inspired by the recent experimental results (e.g., refs. [14, 15]) and for theoretical motivations on the scalar sector. We are aware of the strong constraints on loop-level and its low contributions in the processes and, even, in the loop-level with FC mediated by scalar bosons; but we shall expect a more general description for the neutral scalar interactions and their constraints [16, 17], which are so-called flavor-changing neutral scalar interactions (FCNSI), though they can be found as flavor-changing scalar currents (FCSC). We shall analyze the amplitudes at loop-level to have a new perspective for these processes. As known, GIM mechanism controls the neutral currents at tree-level. It means that processes with flavor-changing mediated by vector bosons are constrained because of the orthogonality of the CKM matrix [18, 19]. We would like to know whether it is similar for flavor-changing mediated by scalar bosons.

Our goal is to provide an analysis on the equation of the amplitud at two-loop level in general and for the SM and THDM to explain the FCNSI and its contribution, if any, for these kind of processes. We are motivated by the next generation of colliders where it would be possible to explore the couplings in the $\gamma\gamma$-processes [20]. We will use the THDM-III because it has flavor-changing neutral current (FCNC) transitions mediated by scalar bosons even at tree-level. We expect that this document be the first of a series of papers about the flavor-changing and scalars at high-loop level.

We organize this paper as follow: section 2 describes the lagrangian for the model and its systematic implementation by the computer. In section 3, we show the results at two-loop level for the amplitude and in section 4 we leave a brief discussion and conclusions.

2. Model

There are two versions of THDM, labeled as type I and type II, with invariance under Z_2 discrete symmetry which ensures CP conservation in the scalar sector. In the first case, all quarks acquire mass through one doublet whereas in type II one doublet gives mass to the up-type quarks while the other doublet gives mass to the down-type quarks. In the type III both doublets simultaneously give masses to all quarks. In the THDM-III, it is possible to have flavor-changing considering parameters, which may induce FCNSI; those parameters are free. There is FCNSI as long as the diagonalization of the fermion mass matrices does not ensure the diagonalization of each of the Yukawa matrices [1, 21].

2.1. Lagrangian for the THDM

For the THDM-III, the general potential is given by [3]:

$$
\begin{aligned}
V(\Phi_1\Phi_2) &= \lambda_1(\Phi_1^\dagger\Phi_1 - v_1^2)^2 + \lambda_2(\Phi_2^\dagger\Phi_2 - v_2^2)^2 + \lambda_3\left[(\Phi_1^\dagger\Phi_1 - v_1^2) + (\Phi_2^\dagger\Phi_2 - v_2^2)\right]^2 \\
&+\lambda_4\left[(\Phi_1^\dagger\Phi_1)(\Phi_2^\dagger\Phi_2) - (\Phi_1^\dagger\Phi_2)(\Phi_2^\dagger\Phi_1)\right] + \lambda_5\left[Re(\Phi_1^\dagger\Phi_2) - v_1v_2\right]^2 \\
&+\lambda_6\left[Im(\Phi_1^\dagger\Phi_2)\right]^2
\end{aligned}
$$

where λ_i are real, v_i's are the vaccuum expectation values and Φ_i's are the Higgs doublets.

The Yukawa sector for the THDM-III is

$$
\mathcal{L}_{YS}^{THDM-III} = Y_1^u\overline{Q}_L^0\tilde{\Phi}_1u_R^0 + Y_2^u\overline{Q}_L^0\tilde{\Phi}_2u_R^0 + Y_1^d\overline{Q}_L^0\Phi_1d_R^0 + Y_2^d\overline{Q}_L^0\Phi_2d_R^0 + h.c.
$$

where Y_i are the Yukawa couplings and $Q_L^0 = \begin{pmatrix} u_L \\ d_L \end{pmatrix}, \overline{Q}_L^0 = (\overline{u}_L, \overline{d}_L), \Phi_1 = \begin{pmatrix} \phi_1^\pm \\ \phi_1 \end{pmatrix}, \Phi_2 =$

$\begin{pmatrix}\phi_2^{\pm}\\ \phi_2\end{pmatrix}$, $\tilde{\Phi}_j = i\sigma_2\Phi_j^* = \begin{pmatrix}\phi_j{}^*\\ -\phi_i^{\mp}\end{pmatrix}$ and $\phi_i = \frac{1}{\sqrt{2}}(v_i + \phi_i^0 + i\chi_i)$. Then we obtained,

$$\mathcal{L}_{YS}^{THDM-III} = \mathcal{L}_{YS-NS}^{THDM-III} + \mathcal{L}_{YS-CS}^{THDM-III} + h.c.$$

where the neutral-sector is

$$\mathcal{L}_{YS-NS}^{THDM-III} = Y_1^u \overline{u}_L \phi_1^{0*} u_R + Y_2^u \overline{u}_L \phi_2^{0*} u_R + Y_1^d \overline{d}_L \phi_1^0 d_R + Y_2^d \overline{d}_L \phi_2^0 d_R \tag{1}$$

and the charged-sector is

$$\mathcal{L}_{YS-CS}^{THDM-III} = +Y_1^u \overline{d}_L(-\phi_2^-)u_R + Y_2^u \overline{d}_L(-\phi_1^-)u_R + Y_1^d \overline{u}_L \phi_1^+ d_R + Y_2^d \overline{u}_L \phi_2^+ d_R. \tag{2}$$

We will use the neutral sector for this paper. In this sector the interaction between fermions and scalar is given by the Yukawa couplings (Y_f); in general, this is

$$g_{hff} \sim Y^f \sim m_f \tag{3}$$

where g_{hff} is the coupling, which represents the vertex between the Higgs and pair of fermions. In the SM the coupling is given by

$$g_{hff}^{SM} = Y_{SM}^f = \sqrt{2}\frac{m_f}{v} \tag{4}$$

where m_f is fermion mass and v is the vaccuum expectation value (VEV). For the THDM, after the spontaneous symmetry breaking the mass matrix is given by

$$m_f = \frac{1}{\sqrt{2}}\left(v_1 Y_1^f + v_2 Y_2^f\right) \tag{5}$$

where we have two VEV's v_1 and v_2, which are relationated by: $\frac{v_2}{v_1} = \tan\beta$. In a general form, the eq. (5) is non-diagonal, and it can be made by:

$$(U_L^f)^\dagger m_f (U_R^f) = \widetilde{m}_f \tag{6}$$

It is important to note that the widetilde over the quantities means the flavor basis. Besides, in general, U is ortogonal and Y^f could be complex. So it is possible to have mixing between the mass (fermions) eigenstates at tree level. Considering the Yukawa matrices Hermitian, the mass eigenstates are

$$f = U_f^\dagger f' \tag{7}$$

where f can be $u, d-$type quarks and l leptons. Then the mass matrix is given by

$$\widetilde{m}_f = \frac{1}{\sqrt{2}}\left(v_1 \widetilde{Y}_1^f + v_2 \widetilde{Y}_2^f\right) \tag{8}$$

where, $\widetilde{Y}_i^f = U_L^\dagger Y_i^f U_R$.

In order to reduce the free parameters, one can rewrite the eq.(5) as:

$$\widetilde{Y}_1^f = \sqrt{2}\frac{m_f}{v\cos\beta} - \widetilde{Y}_2^f \tan\beta. \tag{9}$$

Sometimes Yukawa couplings are defined in terms on $\widetilde{\chi}_{ij}$ parameters [5]; namely,

$$\widetilde{Y}_{ij}^f = \sqrt{2}\frac{\sqrt{m_i m_j}}{v}\widetilde{\chi}_{ij}^f, \tag{10}$$

where m_i and m_j are the fermion masses, and $\widetilde{\chi}_{ij}^f$ are free parameters. This specific pattern is known as Cheng & Sher ansatz [22]. Through this mechanism it is possible to have FCNSI at tree-level. This relation has been used in different papers; e.g. [2, 5, 24].

In the next subsection we show the scheme to obtain the amplitude.

Journal of Physics: Conference Series **761** (2016) 012011 doi:10.1088/1742-6596/761/1/012011

2.2. Methods

We calculate the amplitude at two-loop level. We represent the scheme in fig. 1.

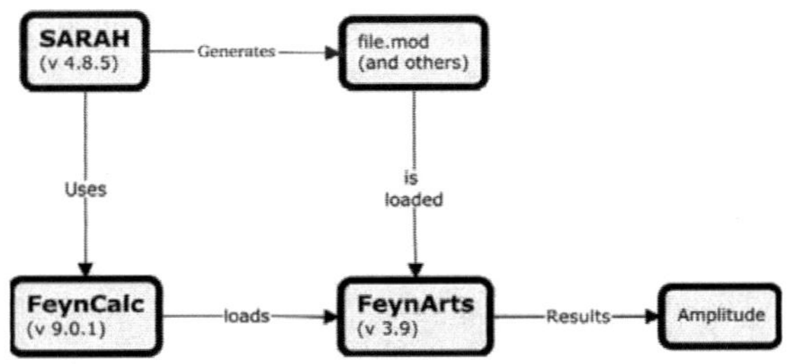

Figure 1. This scheme shows the path to follow for obtaining the amplitude at two-loop level, using SARAH [25–27], FeynCalc and FeynArts [28]. We implemented the THDM-III and FV and use different commands to exclude field points, isolating a process with Higgs boson, which has $m_h \sim 126$ GeV, and mediated by virtual Higgs boson.

Next section shows the results for the $h \to \gamma\gamma$ process at two-loop level, considering the flavor-changing mediated by a SM-like scalar boson.

3. Results

We implemented the THDM-III with the flavor-changing in SARAH for the process represented by fig. 2.

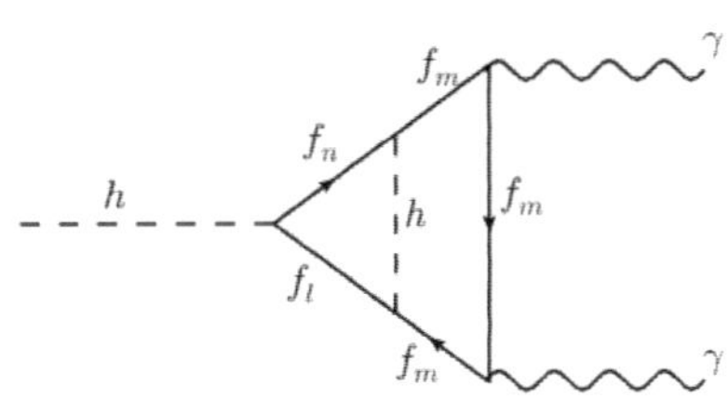

Figure 2. The Feynman diagram for the process $h \to \gamma\gamma$ at two-loop level; note FC in the loop mediated by the scalar. We do not draw the adjoint process.

The amplitude is given by

$$\mathcal{M}_2(\phi \to V'V) = -\frac{i}{256\pi^8 m_h^{12} \prod_{\lambda=1}^{6} P_\lambda} \mathrm{Tr}\left(M_2^{\mu_1\mu_2}\right)\epsilon^*_{\mu_1}(p_2)\epsilon^*_{\mu_2}(p_3) \tag{11}$$

where the product of the propagators is

$$\prod_{\lambda=1}^{6} P_\lambda = (r_{q_1} - r_l)(r_{q_2} - r_m)(r_{p_3\bar{q}_2} - r_m)(r_{q_1q_2} - 1)(r_{p_2p_3q_1} - r_n)(r_{p_2p_3\bar{q}_2} - r_m) \tag{12}$$

with $r_i = \frac{m_i^2}{m_h^2}$, $r_{p_i\bar{p}_j} = \frac{(p_i-p_j)^2}{m_h^2}$, $r_{q_iq_j} = \frac{(q_i+q_j)^2}{m_h^2}$, $r_{p_ip_jq_k} = \frac{(p_i+p_j+q_k)^2}{m_h^2}$ and $r_{p_ip_j\bar{q}_k} = \frac{(p_i+p_j-q_k)^2}{m_h^2}$. p_1 is the momentum for the scalar boson, $p_{2,3}$ are the momentum of the particles in the final state and $q_{1,2}$ are the momentum for the loops. The tensorial amplitud is

$$\begin{aligned} M_2^{\mu_1\mu_2} &= \kappa^{\rm SM}\big(m_m + \gamma_{\nu_1} P^{\nu_1}_{q_2\bar{p}_2\bar{p}_3}\big)\gamma^{\mu_1}(P_L + P_R) \\ &\quad \big(m_m + \gamma_{\nu_2} P^{\nu_2}_{q_2\bar{p}_3}\big)\gamma^{\mu_2}(P_L + P_R) \\ &\quad \big(m_m + \gamma_{\nu_3} q_2^{\nu_3}\big)(P_L + P_R) \\ &\quad \big(m_l - \gamma_{\nu_4} q_1^{\nu_4}\big)(P_L + P_R) \\ &\quad \big(m_n + \gamma_{\nu_5} P^{\nu_5}_{\bar{p}_2\bar{p}_3\bar{q}_1}\big)(P_L + P_R) \end{aligned} \tag{13}$$

where $\kappa^{\rm SM} = (-ie)^2\Big(-\frac{iem_f}{2M_W S_W}\Big)^3$ and $P^{\nu_1}_{q_2\bar{p}_2\bar{p}_3} = (q_2 - p_2 - p_3)^{\nu_1}$, $P^{\nu_2}_{q_2\bar{p}_3} = (q_2 - p_3)^{\nu_2}$, and $P^{\nu_5}_{\bar{p}_2\bar{p}_3\bar{q}_1} = (-p_2 - p_3 - q_1)^{\nu_5}$. In eq. (13) is possible to reduce the $(P_R + P_L)$ terms, however these could contain the model-dependent parameters. It is shown simply the general expression for the SM, and the results for the THDM will be reported soon elsewhere.

4. Discussion and conclusions

The discovery of the Higgs boson and exploration of its properties generate great interest in the scientific community. Theoretically, there are challenges to calculate at two- or multi-loop level associated to ultraviolet, infrared and mass singularities. Besides the integration for multi-loop, considering massive or, maybe, non-scalar integrals could also be awkward calculation. Several technical calculations at one-loop have been developed [29–31], even there are some numerical methods to calculate at this level [32–34]; however there are unexplored models with FCNSI at high-loop level. We expect that high-loop level studies can be interesting tests for the NP.

In this paper, we discuss $h \to \gamma\gamma$ decay. This kind of process may be explored by the future generation of colliders. The enhanced measurements could show more information about the FC mediated by scalars and the Yukawa couplings.

Acknowledgments

JO thanks to postdoctoral scholarships at DGAPA-UNAM and RED-FAE 2015, and we also acknowledge support from CONACYT-SNI (Mexico) and the PAPIIT-IN113916 project.

References

[1] Branco G C, Ferreira P M, Lavoura L, Rebelo M N, Sher M and Silva J P 2012 *Phys. Rept.* **516** 1–102 (*Preprint* 1106.0034)
[2] Arroyo, Marco and Diaz-Cruz, J Lorenzo and Diaz, Enrique and Orduz-Ducuara, Javier A 2013 (*Preprint* 1306.2343)
[3] Hernandez-Sanchez J, Moretti S, Noriega-Papaqui R and Rosado A 2013 *JHEP* **07** 044 (*Preprint* 1212.6818)
[4] Diaz-Cruz J L, Noriega-Papaqui R and Rosado A 2005 *Phys. Rev.* **D71** 015014 (*Preprint* hep-ph/0410391)
[5] Diaz-Cruz J L, Honorato C G, Orduz-Ducuara J A and Perez M A 2014 *Phys. Rev.* **D90** 095019 (*Preprint* 1403.7541)
[6] Harnik R, Kopp J and Zupan J 2013 *JHEP* **03** 026 (*Preprint* 1209.1397)
[7] CMS Collaboration (CMS) 2016 **CMS-PAS-EXO-16-017**
[8] Aad, Georges and others (ATLAS) 2015 *Eur. Phys. J.* **C75** 79 (*Preprint* 1410.5404)
[9] CMS Collaboration (CMS) 2016 **CMS-PAS-HIG-16-006**
[10] ATLAS collaboration 2015 **ATLAS-CONF-2015-061**
[11] CMS Collaboration (CMS) 2016 **CMS-PAS-EXO-16-008**
[12] Aad, Georges and others (ATLAS) 2013 *Phys. Lett.* **B719** 242–260 (*Preprint* 1210.6604)
[13] Aad, Georges and others (ATLAS) 2015 *JHEP* **07** 157 (*Preprint* 1502.07177)
[14] Khachatryan, Vardan and others (CMS) 2015 *Phys. Lett.* **B749** 337–362 (*Preprint* 1502.07400)
[15] CMS Collaboration (CMS) 2016 **CMS-PAS-HIG-16-005**
[16] Kopp J 2016 *51st Rencontres de Moriond on EW Interactions and Unified Theories La Thuile, Italy, March 12-19, 2016* (*Preprint* 1605.02865)
[17] Orduz-Ducuara, J A and Gaitán, R Manuscript in process

[18] Olive K A *et al.* (Particle Data Group) 2014 *Chin. Phys.* **C38** 090001
[19] Quang H K and Pham X Y 1998 *Elementary particles and their interactions: concepts and phenomena*
[20] Ginzburg I and Krawczyk M 2013 (*Preprint* 1310.5881)
[21] Sher M 2000 *Phys. Lett.* **B487** 151–154 (*Preprint* hep-ph/0006159)
[22] Cheng T P and Sher M 1987 *Phys. Rev.* **D35** 3484
[23] Gaitán R, Martinez R and Montes de Oca J H 2015 (*Preprint* 1503.04391)
[24] Gomez-Bock M and Noriega-Papaqui R 2006 *J. Phys.* **G32** 761–776 (*Preprint* hep-ph/0509353)
[25] Staub F 2015 *Adv. High Energy Phys.* **2015** 840780 (*Preprint* 1503.04200)
[26] Staub F 2014 *Comput. Phys. Commun.* **185** 1773–1790 (*Preprint* 1309.7223)
[27] Vicente A 2015 (*Preprint* 1507.06349)
[28] Hahn, Thomas 2001 *Comput. Phys. Commun.* **140** 418–431 (*Preprint* hep-ph/0012260)
[29] Catani, Stefano and Gleisberg, Tanju and Krauss, Frank and Rodrigo, German and Winter, Jan-Christopher 2008 *JHEP* **09** 065 (*Preprint* 0804.3170)
[30] Gonsalves, Richard J 1983 *Phys. Rev.* **D28** 1542
[31] Feynman R P 2000 *World Sci. Ser.20th Cent. Phys.* **27** pp.1–600
[32] Borowka, S and Heinrich, G and Jones, S P and Kerner, M and Schlenk, J and Zirke, T 2015 *Comput. Phys. Commun.* **196** 470–491 (*Preprint* 1502.06595)
[33] Buchta, Sebastian and Chachamis, Grigorios and Draggiotis, Petros and Rodrigo, German 2015 (*Preprint* 1510.00187)
[34] Cutkosky R E 1960 *J. Math. Phys.* **1** 429–433

An Introduction to the Massive Helicity Formalism with applications to the Electroweak SM

J. Lorenzo Díaz-Cruz[1], Bryan O. Larios-López[2] and O. Meza-Aldama[3]

[1] Facultad de Ciencias Físico - Matemáticas, BUAP
Apdo. Postal 1364, C.P. 72000, Puebla, Pue. México

E-mail: [1]ldiaz@ifuap.buap.mx, [2]bryanlarios@gmail.com, [3]oscar.meza@email.ucr.edu

Abstract. The power of the helicity formalism has been appreciated recently from its application to the massless case, where plenty of formal aspects of multi-legs amplitudes have been derived. However, in order to extend the formalism to the realistic cases, such as the electroweak Standard Model and QCD with massive quarks, some extra inputs are needed. We discuss first the formalism needed to evaluate amplitudes expressed with the modern notation for massive fermions and vectors based on a thorough treatment of Proca vector fields. Then, we present some examples of elementary processes where it is shown how the formalism leads to tremendous simplifications, these include 2-body decays $Z \to ff$, $h \to ff$ and $h \to W^-W^+$ as well as the 3-body decay $h \to Vf'\bar{f}$.

1. Introduction

The spinor helicity formalism (SHF) is a tool for calculating scattering amplitudes much more efficiently than the traditional approach. Its versatility is based on the fact that all objects appearing in the Feynman rules for gauge theories (QED, QCD, EW and SUSY) can be written in terms of two-component Weyl spinors (and Pauli sigma matrices). Dotted and undotted Weyl spinors are used, and in fact they are more fundamental than Minkowski four-vectors since they form irreducible representations of the Lorentz group, and unlike Dirac spinors they are not a "mixture" (i.e. a direct sum) of two different representations, furthermore Weyl spinors are easier to handle.

When dealing with a massless fermion, the solutions of the momentum-space Dirac equation have only two non-vanishing components, in that case the distinction between a 4-component Dirac spinor and a 2-component Weyl spinor essentially disappears. These Weyl spinors, which are helicity eigenstates in the massless case, can then be used to rewrite the Feynman rules for external legs (both fermionic and bosonic), vertices and propagators. The expressions obtained for the helicity amplitudes are in general very simple, and they can be squared directly, without any need of spinor completeness relations and Casimir tricks. This is one of the most marvelous advantages of both massless and massive SHF.

The SHF is usually introduced only at the massless level (see references [1]-[5]), while the massive formalism is less studied in the current literature. The only notable exception is Ref. [6], although it is written in a somewhat old-fashioned notation. Other papers, such as [7]-[12], study scattering amplitudes of several theories using mostly BCFW recursion (Britto-Cachazo-Feng-Witten) [13], creating a huge gap between the well-understood massless SHF and the work done in the massive SHF even for "simpler" theories such as QED.

The SHF brings huge simplifications (as compared to the usual Feynman's completeness-relations-trace-evaluation method) to the calculation of scattering amplitudes. In the massive case, although the simplifications are not that large, the method is in general a little bit more tricky than the massless case [16]. Finally, we must say that one of the main applications of the modern amplitude calculation techniques is to QCD gluon processes. Using recursion relations such as BCFW, one can calculate multi-gluon scattering amplitudes which would be impossible (in practice) to obtain using "traditional" methods. Of course, gluons are massless, so one does not need to extend the SHF to include massive particles in this case.

Computational tools have been developed to deal with both massless and massive spinor helicity amplitudes (see Refs. [14] and [15]), which of course comes very handy when dealing with complicated processes. However, one can notice that a self-contained theory of massive helicity amplitudes and detailed applications to simple processes of realistic theories is lacking, which is one of the motivations of this paper.

In this work we show some practical examples where we have applied the massive SHF to different processes (at tree level) in the Electroweak SM *without* using BCFW or other on-shell (or off-shell) recursion scheme, nor are we using symbolical or numerical evaluation methods.

This paper is structured as follows: Section 2 contains the solutions of the Dirac equation (massive case), as well the solutions of the Maxwell and Proca equations in order to derive the Feynman rules for external fermionic and bosonic lines, then we use the so-called light cone decomposition (LCD) [15, 16, 17] to express massive fermions in terms of massless ones. In Section 3 we present some calculations in Electroweak SM. Finally in Section 4 we present our conclusions and final comments of this work.

2. Helicity Method for QED

Consider a theory with a massive Dirac fermion, which is described by Dirac equation. Writing the momentum space Dirac spinor u in terms of two Weyl spinors the equation of motion takes the form

$$(\not{p}+m)u(p) = \begin{pmatrix} m & p_\mu \sigma^\mu_{a\dot{a}} \\ p_\mu \bar{\sigma}^{\mu\dot{a}a} & m \end{pmatrix} \begin{pmatrix} \chi_a(p) \\ \xi^{\dot{a}}(p) \end{pmatrix} = 0. \tag{1}$$

Through all the paper we shall use the following spacetime metric $g^{\mu\nu} = \text{diag}(-1,1,1,1)$, also we have used the base where the Dirac matrices take the following form

$$\gamma^\mu = \begin{pmatrix} 0 & \sigma^\mu \\ \bar{\sigma}^\mu & 0 \end{pmatrix}, \tag{2}$$

with $\sigma^\mu = (1, \vec{\sigma})$ and $\bar{\sigma}^\mu = (1, -\vec{\sigma})$.

Weyl spinor ($\xi^{\dot{a}}$, χ_a) could be expressed in terms of bracket labels, additionally the spinor indices could be omitted, this reads as $\chi = |r]$, $\xi = \frac{m}{\langle rq\rangle}|q\rangle$, where $|r]$ and $|q\rangle$ are 2-component Weyl spinors linked to the light-like momenta r^μ and q^μ respectively, these arise when one uses LCD to express a massive momentum (p^μ) in term of 2 massless momenta ($p^\mu = r^\mu + \frac{p^2 q^\mu}{2r\cdot q}$), this approach will be used throughout the paper to express massive spinors in terms of massless ones.

Equation (1) has 8 solutions, (see Ref. [16] for a complete direction), these are:

$$u_-(p) = |r] + \frac{m}{\langle rq\rangle}|q\rangle \quad , \quad u_+(p) = \frac{m}{[rq]}|q] + |r\rangle; \tag{3}$$

$$v_+(p) = |r] - \frac{m}{\langle rq\rangle}|q\rangle \quad , \quad v_-(p) = -\frac{m}{[rq]}|q] + |r\rangle; \tag{4}$$

$$\bar{u}_-(p) = \frac{m}{[qr]}[q| + \langle r| \quad , \quad \bar{u}_+(p) = [r| + \frac{m}{\langle qr\rangle}\langle q|; \tag{5}$$

$$\bar{v}_+(p) = -\frac{m}{[qr]}[q| + \langle r| \quad , \quad \bar{v}_-(p) = [r| - \frac{m}{\langle qr\rangle}\langle q|, \tag{6}$$

Spinor products satisfy $[rq] \equiv -[qr]$, similarly $\langle rq\rangle = -\langle qr\rangle$. For real momenta we have $\langle pk\rangle \equiv [kp]^*$. When $m = 0$, one can verify that $\bar{u}_\pm = \bar{v}_\mp$. Equations (3)-(6) will allow us to compute scattering amplitudes in a more practical way than the traditional approach. For massless spin-1 particles, one can write the transversal polarization vectors as follows $\epsilon_+^\mu(p) = -\frac{\langle q|\gamma^\mu|p]}{\sqrt{2}\langle qp\rangle}$ and $\epsilon_-^\mu(p) = -\frac{\langle p|\gamma^\mu|q]}{\sqrt{2}[qp]}$, see Ref.[1, 2], where p is the spin-1 momentum, and q is a reference momentum defined such that $q^2 = 0$. Note that $(\epsilon_\pm^\mu)^*(p) = -\epsilon_\mp^\mu(p)$. The slashed versions of these polarization vectors will also be useful to compute scattering amplitudes in the examples presented later, these are $\not{\epsilon}_+(p) = \frac{\sqrt{2}}{\langle qp\rangle}(|p]\langle q| + |q\rangle[p|)$ and $\not{\epsilon}_-(p) = \frac{\sqrt{2}}{[qp]}(|p\rangle[q| + |q]\langle p|)$.

Finally, for massive spin-1 bosons (see Ref. [15, 16]) making the identification $\epsilon_{1,2,3}^\mu(p) \equiv \epsilon_{-,+,0}^\mu(p)$, we can write $\epsilon_+^\mu(p) = \frac{\langle r|\gamma^\mu|q]}{\sqrt{2}[rq]}$, $\epsilon_-^\mu(p) = \frac{\langle q|\gamma^\mu|r]}{\sqrt{2}\langle rq\rangle}$ and finally the longitudinal mode $\epsilon_0^\mu(p) = \frac{1}{m}r^\mu + \frac{m}{2p\cdot q}q^\mu$, where now $p^2 = -m^2$, the momenta r and q are related with the momentum p by LCD.

3. Elementary Processes in the Standard Model

In this section we will use the SHF to compute several elementary processes in the Standard Model of particle physics. Considering massive particles but using LCD, we shall express the massive 4-component Dirac spinor in terms of the massless 2-component Weyl spinors. This will allow us to exploit all the available identities from the massless SHF.

3.1. 2-body Higgs decay $h(p_1) \to W^+(p_2)W^-(p_3)$

The helicity amplitude (HA) for the process $h \to W^+W^-$ is as follows

$$\mathcal{M}_{\lambda_2\lambda_3}(p_1, p_2, p_3) = \frac{2M_W^2 g_{\mu\nu}}{v}\epsilon_{\lambda_2}^\mu(p_2)\epsilon_{\lambda_3}^\nu(p_3), \tag{7}$$

where $p_1^2 = -M_h^2$, $p_2^2 = p_3^2 = -M_W^2$ and the λ's represent the helicity of the particles. We shall use simultaneous light cone decomposition (SLCD) to express the massive momenta p_2 and p_3 in terms of massless momenta (r_2, q_2, r_3, q_3). The massive momenta take the form; $p_2 = r_2 - \frac{M_W^2}{2r_2\cdot q_2}q_2$, and $p_3 = q_2 - \frac{M_W^2}{2r_2\cdot q_2}r_2$, here SLCD imposes that $r_3 = q_2$ and $q_3 = r_2$. For this process there are 3^2 HA's, $\mathcal{M}_{++}$, $\mathcal{M}_{+-}$, $\mathcal{M}_{-+}$, $\mathcal{M}_{--}$, $\mathcal{M}_{0+}$, $\mathcal{M}_{0-}$, $\mathcal{M}_{-0}$, $\mathcal{M}_{+0}$ and $\mathcal{M}_{00}$. Using SLCD to fix the massless momenta will be crucial to reduce the number of HA's, in this case HA's: $\mathcal{M}_{+-}$, $\mathcal{M}_{-+}$, $\mathcal{M}_{0-}$, $\mathcal{M}_{-0}$, $\mathcal{M}_{+0}$, $\mathcal{M}_{0+}$ vanish. The nonzero HA's are shown in Table 1.

$\lambda_2\lambda_3$	$\mathcal{M}_{\lambda_2\lambda_3}$
$--$	$\frac{2M_W^2}{v}\frac{\langle q_3q_2\rangle}{[q_3q_2]}$
$++$	$\frac{2M_W^2}{v}\frac{[q_3q_2]}{\langle q_3q_2\rangle}$
00	$-\frac{(2M_W^2)(s_{q_2q_3}^2+M_W^4)}{vM_W^2 s_{q_2q_3}}$

Table 1. Helicity Amplitudes for the 2-body Higgs decay $h \to W^+W^-$.

In the expressions of Table 1 we have used the following spinor relations

$$[q_2|\gamma^\mu|q_3\rangle = \langle q_3|\gamma^\mu|q_2], \tag{8}$$

$$[q_2|\gamma^\mu|q_3\rangle^* = [q_3|\gamma^\mu|q_2\rangle, \tag{9}$$

as well as Fierz identity:

$$\langle q_2|\gamma^\mu|r_2]\langle r_3|\gamma_\mu|q_3] = 2\langle q_2 r_3\rangle[r_2 q_3]. \tag{10}$$

The averaged and squared amplitude for the 2-body Higgs decay ($h \to W^+W^-$) is:

$$\langle|\mathcal{M}|^2\rangle = 2|\mathcal{M}_{++}|^2 + |\mathcal{M}_{00}|^2, \tag{11}$$

$$= \left(\frac{2M_W^2}{v}\right)^2 \left(2 + \left(\frac{1}{2M_W^2 s_{q_2q_3}}\right)^2 (s_{q_2q_3}^2 + M_W^4)^2\right), \tag{12}$$

$$= \frac{M_h^4}{v^2}\left(1 - 4x^2 + 12x^4\right), \tag{13}$$

with $s_{q_2q_3} = -(q_2+q_3)^2 = -2q_2\cdot q_3$ and $x = \frac{M_W}{M_h}$, the momenta q_2 and q_3 are defined as follows [15, 16].

$$q_2 = \frac{(\mathrm{sgn}(p_2\cdot p_3)\sqrt{\Delta} + p_2\cdot p_3)p_3 - p_3^2 p_2}{2\,\mathrm{sgn}(p_2\cdot p_3)\sqrt{\Delta}} \quad\text{and}\quad q_3 = \frac{(\mathrm{sgn}(p_2\cdot p_3)\sqrt{\Delta} + p_2\cdot p_3)p_2 - p_2^2 p_3}{2\,\mathrm{sgn}(p_2\cdot p_3)\sqrt{\Delta}}, \tag{14}$$

with $\Delta = (p_2\cdot p_3)^2 - p_2^2 p_3^2$. Furthermore we have used in Eq. (11) that $|\mathcal{M}_{--}|^2 = |\mathcal{M}_{++}|^2$. From Eq. (13) we find the decay width Γ for the process $h \to W^+W^-$:

$$\Gamma(h \to W^+W^-) = \frac{\lambda(1,x,x)^{1/2}}{16\pi M_h}\langle|\mathcal{M}|^2\rangle = \frac{\alpha_W M_h}{16x^2}\left(1 - 4x^2 + 12x^4\right)\sqrt{1-4x^2}, \tag{15}$$

where $\alpha_W = \frac{M_W^2}{v^2\pi}$ and the term $\sqrt{1-4x^2}$ is the W velocity in the Higgs rest reference frame.

3.2. 2-body Higgs decay $h(p_1) \to f(p_2)\bar{f}(p_3)$

The HA for the process $h \to f\bar{f}$ is the following:

$$\mathcal{M}_{\lambda_2\lambda_3}(p_1,p_2,p_3) = \frac{1}{v}\bar{u}_{\lambda_2}(p_2)v_{\lambda_3}(p_3), \tag{16}$$

where $p_1^2 = -M_h^2$, $p_2^2 = p_3^2 = -m_f^2$. There are 2^2 HA's, $\mathcal{M}_{+-}$, $\mathcal{M}_{-+}$, $\mathcal{M}_{--}$ and $\mathcal{M}_{++}$, but using SLCD for momenta p_2 and p_3, $\mathcal{M}_{-+}$ and $\mathcal{M}_{+-}$ vanish. The nonzero HA's are shown in Table 2.

$\lambda_2\lambda_3$	$\mathcal{M}_{\lambda_2\lambda_3}$
$--$	$\frac{m_f}{v[q_2q_3]}(s_{q_2q_3} - m_f^2)$
$++$	$\frac{m_f}{v\langle q_2q_3\rangle}(s_{q_2q_3} - m_f^2)$

Table 2. Helicity Amplitudes for the Higgs decay $h \to f\bar{f}$.

The averaged squared amplitude is then:

$$\langle|\mathcal{M}|^2\rangle = 2|\mathcal{M}_{--}|^2 = 2|\mathcal{M}_{++}|^2 = \frac{2m_f^2}{v^2 s_{q_2q_3}}(s_{q_2q_3} - m_f^2)^2 = \frac{y^2}{v^2}(1-4y^2), \tag{17}$$

with $y = \frac{m_f}{M_h}$. Then the decay width Γ goes as follows

$$\Gamma(h \to f\bar{f}) = \frac{\alpha_W M_h y^2}{8}(1-4y^2)^{3/2}. \tag{18}$$

3.3. 2-body Z boson decay $Z(p_1) \to f(p_2)\bar{f}(p_3)$

The HA for the process $Z \to f\bar{f}$ is given as follows

$$\mathcal{M}_{\lambda_1\lambda_2\lambda_3} = \frac{1}{2} g_Z \epsilon_{\mu\lambda_1}(p_1)\bar{u}_{\lambda_2}(p_2)\gamma^\mu(v_f - a_f\gamma^5)v_{\lambda_3}(p_3). \tag{19}$$

We shall assume that $m_f \ll M_Z \equiv M$, that could not true when the fermion is the quark top, while for the other cases in practice it will be equivalent to consider massless fermions ($p_2^2 = p_3^2 = 0$). Then the amplitude will vanish unless f and $\bar{f}$ have opposite helicities. We choose the arbitrary reference momentum $q_1 = p_2$, then by momentum conservation ($\sum_{i=1}^{n}[qi]\langle ik\rangle = 0$)

$$\langle p_3 r_1\rangle[r_1 p_2] = 0, \tag{20}$$

Using Eq.(20), the independent HA's are shown in Table 3.

$\lambda_1\lambda_2\lambda_3$	$\mathcal{M}_{\lambda_1\lambda_2\lambda_3}$
$++-$	$\frac{1}{2} g_Z \frac{\langle p_2\|\gamma_\mu\|r_1]}{\sqrt{2}\langle r_1 p_2\rangle}(v_f - a_f)[p_2\|\gamma^\mu\|p_3\rangle = \frac{g_Z(v_f - a_f)}{\sqrt{2}}\frac{\langle p_2 p_3\rangle[r_1 p_2]}{\langle r_1 p_2\rangle}$
$0+-$	$\frac{1}{2} g_Z \left(\frac{1}{M} r_{1\mu} + \frac{M}{2p_{12}} p_{2\mu}\right)(v_f - a_f)[p_2\|\gamma^\mu\|p_3\rangle = \frac{g_Z(v_f - a_f)}{\sqrt{2}M}\langle r_1 p_3\rangle[r_1 p_2] = 0$
$-+-$	$\frac{1}{2} g_Z \frac{\langle r_1\|\gamma_\mu\|p_2]}{\sqrt{2}[p_2 r_1]}(v_f - a_f)[p_2\|\gamma^\mu\|p_3\rangle = 0$

Table 3. Helicity Amplitudes for the Higgs decay $h \to f\bar{f}$.

The rest of the HA's are obtained by complex conjugation. Besides, because of the γ^5 matrix, the sign of the coefficient a_f will change. This is because $\gamma^5 v_-(p) = v_-(p)$ but $\gamma^5 v_+(p) = -v_+(p)$. Then we have

$$\mathcal{M}_{--+} = \frac{g_Z(v_f + a_f)}{\sqrt{2}}\frac{[p_2 p_3]\langle r_1 p_2\rangle}{[r_1 p_2]}. \tag{21}$$

Taking the square moduli of the nonzero amplitudes we obtain

$$|\mathcal{M}_{++-}|^2 = -g_Z^2\left(|v_f|^2 + |a_f|^2 - 2\mathrm{Re}(v_f a_f^*)\right)p_{23}, \tag{22}$$

$$|\mathcal{M}_{--+}|^2 = -g_Z^2\left(|v_f|^2 + |a_f|^2 + 2\mathrm{Re}(v_f a_f^*)\right)p_{23}. \tag{23}$$

From momentum conservation $p_1 = p_2 + p_3$, then we obtain: $p_2 \cdot p_3 = p_{23} = -\frac{1}{2}M^2$. Finally

$$\left\langle |\mathcal{M}|^2 \right\rangle = \frac{1}{3}\left(|\mathcal{M}_{++-}|^2 + |\mathcal{M}_{--+}|^2\right) = \frac{g_Z^2 M^2}{3}\left(|v_f|^2 + |a_f|^2\right). \tag{24}$$

The decay width for this channel is then:

$$\Gamma\left(Z \to f\bar{f}\right) = \frac{g_Z^2 M}{48\pi}\left(|v_f|^2 + |a_f|^2\right). \tag{25}$$

3.4. 3-body Muon Decay $\mu(p_1) \to \bar{\nu}_{e^-}(p_2)\,\nu_\mu(p_3)\,e^-(p_4)$

The HA for the process $\mu \to \bar{\nu}_{e^-}\,\nu_\mu e^-$ is as follows

$$\mathcal{M}_{\lambda_1\lambda_2\lambda_3\lambda_4} = \left(\frac{g_W}{\sqrt{8}M_W}\right)^2 \left[\bar{u}_{\lambda_3}(p_3)\gamma^\mu(1-\gamma_5)u_{\lambda_1}(p_1)\right]\left[\bar{u}_{\lambda_4}(p_4)\gamma_\mu(1-\gamma_5)v_{\lambda_2}(p_2)\right], \tag{26}$$

$$= \left(\frac{g_W}{\sqrt{8}M_W}\right)^2 \mathcal{A}^\mu_{\lambda_3\lambda_1}\mathcal{B}_{\mu\lambda_4\lambda_2}, \tag{27}$$

Journal of Physics: Conference Series **761** (2016) 012012 doi:10.1088/1742-6596/761/1/012012

where $p_1^2 = -m_\mu^2$, $p_4^2 = -m_e^2$, $p_2^2 = 0$ and $p_3^2 = 0$. We have defined $\mathcal{A}^\mu_{\lambda_3\lambda_1}$ and $\mathcal{B}_{\mu\lambda_4\lambda_2}$ as follows

$$\mathcal{A}^\mu_{\lambda_3\lambda_1} = 2\bar{u}_{\lambda_3}(p_3)\gamma^\mu \hat{P}_L u_{\lambda_1}(p_1), \tag{28}$$

$$\mathcal{B}_{\mu\lambda_4\lambda_2} = 2\bar{u}_{\lambda_4}(p_4)\gamma_\mu \hat{P}_L v_{\lambda_2}(p_2), \tag{29}$$

From equations (28) and (29) one obtains the following HA's; $\mathcal{A}^{\mu--} = 0$, $\mathcal{A}^{\mu-+} = 0$, $\mathcal{A}^{\mu+-} = \frac{2m_\mu}{\langle r_1 q_1\rangle}[p_3|\gamma^\mu|q_1\rangle$, $\mathcal{A}^{\mu++} = 2[p_3|\gamma^\mu|r_1\rangle$, and $\mathcal{B}_\mu^{++} = 0$, $\mathcal{B}_\mu^{--} = 0$, $\mathcal{B}_\mu^{-+} = \frac{2m_e}{[q_4 r_4]}[q_4|\gamma_\mu|p_2\rangle$, $\mathcal{B}_\mu^{+-} = 2[r_4|\gamma_\mu|p_2\rangle$, in Table 4 we show the nonzero products for the partial helicity amplitudes $\mathcal{A}^{\mu\lambda_3\lambda_1}$ and $\mathcal{B}_\mu^{\lambda_4\lambda_2}$.

$\lambda_1\lambda_2\lambda_3\lambda_4$	$\mathcal{A}^{\mu\lambda_3\lambda_1}$	$\mathcal{B}_\mu^{\lambda_4\lambda_2}$	$\mathcal{A}^{\mu\lambda_3\lambda_1}\mathcal{B}_\mu^{\lambda_4\lambda_2}$	$\mathcal{M}$ $(p_2 = q_1,\ p_3 = q_4)$
$+-+-$	$2[p_3\|\gamma^\mu\|r_1\rangle$	$2[r_4\|\gamma_\mu\|p_2\rangle$	$4\langle p_2 r_1\rangle\langle p_3 r_4\rangle$	$\left(\frac{g_W}{\sqrt{2}m_\mu}\right)^2 \langle p_2 r_1\rangle[p_3 r_4]$
$+++-$	$2[p_3\|\gamma^\mu\|r_1\rangle$	$\frac{2m_e}{[q_4 r_4]}[q_4\|\gamma_\mu\|p_2\rangle$	$4\frac{m_e}{[q_4 r_4]}\langle p_2 r_1\rangle\langle p_3 q_4\rangle$	0
$--++$	$\frac{2m_\mu}{\langle r_1 q_1\rangle}[p_3\|\gamma^\mu\|q_1\rangle$	$2[r_4\|\gamma_\mu\|p_2\rangle$	$4\frac{m_\mu}{\langle r_1 q_1\rangle}\langle p_2 q_1\rangle[p_3 r_4]$	0
$-++-$	$\frac{2m_\mu}{\langle r_1 q_1\rangle}[p_3\|\gamma^\mu\|q_1\rangle$	$\frac{2m_e}{[q_4 r_4]}[q_4\|\gamma_\mu\|p_2\rangle$	$4\frac{m_\mu m_e}{\langle r_1 q_1\rangle[q_4 r_4]}\langle p_2 q_1\rangle[p_3 q_4]$	0

Table 4. Helicity Amplitudes for Muon decay ($\mu \to e^-\ \bar{\nu}_{e^-}\ \nu_\mu$) with the momentum assignment $r_2 = q_1$ and $r_3 = q_4$.

We have used in the fourth column of the Table 4 the reflection property Eq. (8) and Fierz identity Eq. (10), while in the last column it was chosen $r_2 = q_1$ and $r_2 = q_4$, this reduces all the helicity amplitudes to one ($\mathcal{M}^{+-+-}$). The squared and averaged amplitude for the muon decay is:

$$\langle|\mathcal{M}^{+-+-}|^2\rangle = \frac{1}{2}|\mathcal{M}^{+-+-}|^2 = 2\left(\frac{g_W}{M_W}\right)^4 (p_1\cdot p_2)(p_3\cdot p_4) \tag{30}$$

From this result we can arrive to the decay width, which agrees with result of textbooks.

3.5. 3-body Higgs decay $h(p_1) \to Z(p_2)f(p_3)\bar{f}(p_4)$

The HA for this process is the following

$$\mathcal{M}_{\lambda_2\lambda_3\lambda_4} = \frac{M_Z^2}{v}\frac{e}{c_W s_W}\epsilon^\mu_{\lambda_2}(p_2)\left(g_{\mu\nu} + \frac{k_\mu k_\nu}{M_Z^2}\right)\frac{1}{k^2 + M_Z^2}\bar{u}_{\lambda_3}(p_3)\gamma^\nu(v_f - a_f\gamma^5)v_{\lambda_4}(p_4). \tag{31}$$

We will consider just the case when the fermions are leptons $(e, \mu, \tau, \nu_e, \nu_\mu, \nu_\tau)$, with this approximation the leptons are consider as massless, so they must have opposite helicities. Then, remembering that $k = p_3 + p_4$, we have

$$k_\nu \bar{u}_+(p_3)\gamma^\nu(v_f - a_f\gamma^5)v_-(p_4) = (v_f - a_f)[p_3|(\not{p}_3 + \not{p}_4)|p_4\rangle = 0, \tag{32}$$

$$k_\nu \bar{u}_-(p_3)\gamma^\nu(v_f - a_f\gamma^5)v_+(p_4) = (v_f + a_f)\langle p_3|(\not{p}_3 + \not{p}_4)|p_4] = 0; \tag{33}$$

it is very hard to find this kind of simplification (Equations (32) and (33)) with the traditional approach of computing scattering amplitudes, but now the amplitude (31) becomes simply

$$\mathcal{M}_{\lambda_2\lambda_3\lambda_4} = \mathcal{C}\bar{u}(p_3)\not{\epsilon}(p_2)(v_f - a_f\gamma^5)v(p_4), \tag{34}$$

where we have defined $\mathcal{C} \equiv \frac{M_Z^2}{v}\frac{e}{c_W s_W}\frac{1}{k^2+M_Z^2}$. Choosing $q_2 = p_3$, we get

$$\not\epsilon_+(p_2) = \frac{\sqrt{2}}{\langle p_3 r_2\rangle}(|r_2]\langle p_3| + |p_3\rangle[r_2|), \tag{35}$$

$$\not\epsilon_-(p_2) = \frac{\sqrt{2}}{[r_2 p_3]}(|p_3]\langle r_2| + |r_2\rangle[p_3|), \tag{36}$$

$$\not\epsilon_0(p_2) = \frac{1}{M_Z}\not r_2 + \frac{M_Z}{2p_{23}}\not p_3. \tag{37}$$

The nonzero HA's are show below in the Table 5.

$\lambda_2\lambda_3\lambda_4$	$\mathcal{M}_{\lambda_2\lambda_3\lambda_4}$
$++-$	$\mathcal{C}[p_3\|\frac{\sqrt{2}}{\langle p_3 r_2\rangle}\|r_2]\langle p_3\|(v_f - a_f\gamma^5)\|p_4\rangle = \sqrt{2}\mathcal{C}(v_f - a_f)\frac{[p_3 r_2]\langle p_3 p_4\rangle}{\langle p_3 r_2\rangle}$
$--+$	$\mathcal{C}\langle p_3\|\frac{\sqrt{2}}{[r_2 p_3]}\|r_2\rangle[p_3\|(v_f - a_f\gamma^5)\|p_4] = \sqrt{2}\mathcal{C}(v_f + a_f)\frac{\langle p_3 r_2\rangle[p_4 p_3]}{[p_3 r_2]}$
$0+-$	$\mathcal{C}[p_3\|\left(\frac{1}{M_Z^2}\not r_2 + \frac{M_Z}{2p_{23}}\not p_3\right)(v_f - a_f\gamma^5)\|p_4\rangle = \mathcal{C}\frac{v_f - a_f}{M_Z}[r_2 p_3]\langle r_2 p_4\rangle$
$0-+$	$\mathcal{C}\langle p_3\|\left(\frac{1}{M_Z^2}\not r_2 + \frac{M_Z}{2p_{23}}\not p_3\right)(v_f - a_f\gamma^5)\|p_4] = \mathcal{C}\frac{v_f + a_f}{M_Z}\langle r_2 p_3\rangle[r_2 p_4]$

Table 5. Helicity Amplitudes for the 3-body Higgs decay $h \to Zf\bar{f}$.

Therefore

$$\langle|\mathcal{M}|^2\rangle = 8\mathcal{C}^2\left(|v_f|^2 + |a_f|^2\right)\left[-p_{34} + \frac{(r_2\cdot p_3)(r_2\cdot p_4)}{M_Z^2}\right], \tag{38}$$

From momentum conservation we can obtain; $2p_{23} = M_Z^2 - M_H^2 - 2p_{24} - 2p_{34}$. Substituting everything in Eq.(38) we obtain

$$\left\langle|\mathcal{M}|^2\right\rangle = \sum|\mathcal{M}|^2 = 8\mathcal{C}^2\left(|v_f|^2 + |a_f|^2\right)\left[p_{24} - p_{34} - \frac{2p_{24}}{M_Z^2}\left(p_{24} + p_{34} + \frac{M_H^2}{2}p_{24}\right)\right]. \tag{39}$$

4. Conclusions

In this paper we have presented a summary of the basic formulae of the SHF. In order to appreciate the value of the methods, we studied the phenomenology of the Electroweak sector of the Standard Model of Particle Physics, including the evaluation of the decays: $Z \to ff$, $h \to ff$, $h \to W^-W^+$ and $h \to Vf'\bar{f}$. Although these results are well known, it can be appreciated that the simplification obtained by using SHF, make it worth to use then in teaching of Particle Physics with a modern approach.

5. Acknowledgements

The authors wish to thank economic support from CONACYT, L.D.C. was supported under the grant 220498, B.L. thanks to Fernando Febres Cordero for his valuable comments at the initial stage of this work.

References

[1] M. Srednicki, *Quantum Field Theory* (Cambridge University Press, 2007).
[2] H. Elvang and Y. Huang, *Scattering Amplitudes in Gauge Theory and Gravity* (Cambridge University Press, 2015), [arXiv:1308.1697v2 [hep-th]].
[3] M. D. Schwarz, *Quantum Field Theory and the Standard Model* (Cambridge University Press, 2014).

[4] E. Witten, Commun. Math. Phys. **252** 189-258 (2004), arXiv:hep-th/0312171.
[5] M. Peskin, http://arxiv.org/pdf/1101.2414.pdf.
[6] S. Dittmaier, Phys. Rev. D **59** 016007 (1998), hep-ph/9805445.
[7] S. Badger, E. N. Glover, V. Khoze, and P. Svrcek, JHEP **0507** 025 (2005), arXiv:hep-th/0504159.
[8] S. Badger, E. N. Glover, and V. V. Khoze, JHEP **0601** 066 (2006), arXiv:hep-th/0507161.
[9] K. Ozeren and W. Stirling, Eur. Phys. J. C **48** 159 (2006), arXiv:hep-ph/0603071.
[10] A. Hall, Phys.Rev. D **77** 025011 (2008), arXiv:0710.1300 [hep-ph].
[11] R. Boels, JHEP **1001** 010 (2010), arXiv:0908.0738 [hep-th].
[12] L. J. Dixon, J. M. Drummond, C. Duhr and J. Pennington, JHEP **1406** 116 (2014), arXiv:1402.3300 [hep-th].
[13] R. Britto, F. Cachazo, B. Feng, and E. Witten, Phys. Rev. Lett. **94** (2005) 181602, [hep-th/0501052].
[14] D. Maître and P. Mastrolia, Comput. Phys. Commun. **179** 501-574 (2008), arXiv:0710.5559 [hep-ph].
[15] J. Kuczmarski, arXiv:1406.5612 [hep-ph].
[16] Bryan O. Larios, O. Meza, Work in progress.
[17] D. A. Kosower, Phys. Rev. D **71**, 045007 (2005), [arXiv:hep-th/0406175].
[18] L. H. Ryder, *Quantum Field Theory* (Cambridge University Press, 1996).
[19] M. E. Peskin and D. V. Schroeder, *An Introduction to Quantum Field Theory* (Addison Wesley, 1995).
[20] J. L. Díaz Cruz, B. Larios, O. Meza Aldama and J. Reyes Pérez, Rev. Mex. Fis. E **61(2)** (2015) 104. English version: arXiv:1511.07477 [physics.gen-ph].
[21] C. Schwinn and S. Weinzierl, [On-shell recursion relations for all Born QCD amplitudes,] JHEP **0704** (2007) 072, arXiv:hep-ph/0703021 [HEP-PH].

MWPF IOP Publishing
Journal of Physics: Conference Series **761** (2016) 012013 doi:10.1088/1742-6596/761/1/012013

Flavored axion-monodromy inflation

Raymundo Ramos

High Energy Theory Group, Department of Physics, College of William and Mary, Williamsburg, VA 23187-8795

E-mail: `raramos@email.wm.edu`

Abstract. In this talk we consider the breaking of a flavor-symmetric potential as the origin of the pseudo-Goldstone bosons responsible for inflation. The breaking of flavor symmetry generates the fermion mass hierarchy while the breaking of accidental symmetries leads to pseudo-Goldstone bosons with an axion-monodromy potential appropriate for inflation. We deal with models where the inflaton is a linear combination of two fields. We will show that the mechanism by which inflation ends depends on the choice of parameters.

1. Introduction

The discussion presented here is based on the work published in [1]. As successful as big bang cosmology is, it fails to explain some important details. Two of them, of particular interest in this work, are the horizon and the flatness problems. In the first one, the regions that compose the cosmic microwave background (CMB) keep a high degree of homogeneity, even among regions that were far apart enough to be causally disconnected. The second one is the need for a very fine-tuned initial condition to explain the flatness of the current universe. The most widely accepted solution to these problems is called inflation, a period of exponential growth that took place in the early universe. The scalar responsible for driving inflation is called the inflaton. It provides the universe with a nearly constant energy density [2]. In this talk we discuss the possibility that accidental continuous global symmetries, present in models of flavor with discrete symmetries, may lead to pseudo-Goldstone bosons with the appropriate potential for an inflaton.

In models of flavor, Yukawa couplings arise from discrete flavor symmetries by adding flavons to the theory. Consider the flavor symmetry $\mathbf{Z}_N$ and a flavon Φ transforming as $\Phi \to \omega\Phi$, with $\omega = \exp(2\pi i/N)$. For standard model (SM) fermions charged under $\mathbf{Z}_N$ we can write higher-dimension operators of the form

$$\frac{1}{M_F^p}\bar{Q}_L H \phi^p D_R + \text{H.c.}\,, \tag{1}$$

where H is the standard model Higgs doublet and the power p depends on the flavor charge of the standard model fermions. The scale M_F will be fixed to the reduced Planck mass $M_* \equiv M_P/\sqrt{8\pi}$. The flavor symmetry is broken when the flavon field acquires vev and we are left with Yukawa couplings proportional to $(\langle\phi\rangle/M_*)^p$.

We will consider a class of inflation models that bears some resemblance to natural inflation scenarios [3] but provides cosmological predictions consistent with Planck data [4] while keeping sub-Planckian vevs: two-field axion-monodromy inflation [5–12]. The two-field structure requires that we consider two copies of the flavor symmetry mentioned before. Let the symmetry group be $\mathbf{Z}_p^\Phi \times \mathbf{Z}_r^\chi$ and the new flavons, Φ and χ, transform only under the corresponding symmetry as

$$\Phi \to \omega_\Phi \Phi \quad \text{and} \quad \chi \to \omega_\chi \chi, \tag{2}$$

where $\omega_\Phi = \exp(2\pi i/p)$ and $\omega_\chi = \exp(2\pi i/r)$, and p and r are integers. The requirement of a nearly flat potential for the inflaton demands that we forbid renormalizable terms made out with powers of only Φ and/or χ fields. This is achieved by assuming that $p, r \geq 5$. The resulting renormalizable flavon potential is

$$V(\Phi, \chi) = -m_\Phi^2 \Phi^\dagger \Phi + \frac{\lambda_\Phi}{2}(\Phi^\dagger \Phi)^2 - m_\chi^2 \chi^\dagger \chi + \frac{\lambda_\chi}{2}(\chi^\dagger \chi)^2 + \lambda_p \Phi^\dagger \Phi \chi^\dagger \chi. \tag{3}$$

It is easy to see that this potential has an accidental U(1)×U(1) global symmetry. Consider the following reparameterization

$$\Phi = \frac{\phi_0 + f_\theta}{\sqrt{2}} \exp(i\theta/f_\theta) \quad \text{and} \quad \chi = \frac{\chi_0 + f_\rho}{\sqrt{2}} \exp(i\rho/f_\rho) \tag{4}$$

where $f_\theta/\sqrt{2} \equiv \langle \Phi \rangle$ and $f_\rho/\sqrt{2} \equiv \langle \chi \rangle$. By substituting Eq. (4) in Eq. (3) we can check that the resulting $V(\rho, \theta)$ is exactly flat. The fields ϕ_0 and χ_0 will acquire heavy masses and can therefore be ignored in what follows.

2. Axion-monodromy inflation potential

Triangle diagrams from anomalous symmetries can generate modifications to the flat potential of the fields θ and ρ. Consider the additional gauge symmetries SU(N_1)×SU(N_2) and new fermions charged under these symmetries:

$$A_L, A_R \sim (\mathbf{N_1}, \mathbf{1}) \quad \text{and} \quad B_L^{(i)}, B_R^{(i)}, C_L, C_R \sim (\mathbf{1}, \mathbf{N_2}). \tag{5}$$

The new fermions, together with the flavons Φ and χ contribute the following Yukawa terms to the Lagrangian

$$\mathcal{L} \supset h_1 \bar{A}_R A_L \chi + \sum_{i=1}^{n} h_2^{(i)} \bar{B}_R^{(i)} B_L^{(i)} \chi + h_3 \bar{C}_R C_L \Phi^* + \text{ H.c. } . \tag{6}$$

These terms present anomalous accidental U(1) symmetries. The corresponding triangle diagrams result in the following interactions between the ρ and θ fields and the gauge bosons [8]

$$\frac{g_1^2}{32\pi^2}\left(\frac{\rho}{f_\rho}\right) F_1 \widetilde{F}_1 + \frac{g_2^2}{32\pi^2}\left(\frac{n\rho}{f_\rho} - \frac{\theta}{f_\theta}\right) F_2 \widetilde{F}_2 . \tag{7}$$

By integrating over instanton configurations, together with a simple redefinition of the (ρ, θ) space origin, we find the following terms for the potential $V(\rho, \theta)$ [8]

$$V(\rho, \theta) = \Lambda_1^4 \left[1 + \cos\left(\frac{\rho}{f_\rho}\right)\right] + \Lambda_2^4 \left[1 - \cos\left(\frac{n\rho}{f_\rho} - \frac{\theta}{f_\theta}\right)\right], \tag{8}$$

where the scales Λ_j are determined by the dynamics of each SU(N_j).

As complex as this potential may seem, we can describe some features relevant to inflation. A depiction of the potential, with ρ and θ as cylindrical coordinates, shows that the first cosine term creates a maximum at the origin. As ρ grows, the potential descends to a minimum similar to that of a Mexican-hat potential. The second cosine term draws a spiraling trench on the surface described by the first one. As the inflaton follows the trench around the θ direction, it moves forward in the ρ direction. It is precisely this monodromy what allows the number of e-folds to grow without the need for super-Planckian initial field values.

To simplify the potential in Eq. (8) we choose $f_\rho = f_\theta \equiv f_1$ and define $f_1/n \equiv f_2$. We also assume $f_1 \gg f_2$. Moreover, we simplify the second cosine term, the one responsible for the trench, by rotating the fields with $\rho = c\,\tilde{\rho} + s\,\tilde{\theta}$ and $\theta = c\,\tilde{\theta} - s\,\tilde{\rho}$, where

$$c = \frac{f_1}{\sqrt{f_1^2 + f_2^2}} \qquad \text{and} \qquad s = \frac{f_2}{\sqrt{f_1^2 + f_2^2}} \, . \tag{9}$$

The final form of the potential is

$$V(\tilde{\rho}, \tilde{\theta}) = \Lambda_1^4 \left[1 + \cos\left(\frac{c\tilde{\rho} + s\tilde{\theta}}{f_1} \right) \right] + \Lambda_2^4 \left[1 - \cos\left(\frac{\tilde{\rho}}{f} \right) \right], \tag{10}$$

where $f = f_1 f_2 / \sqrt{f_1^2 + f_2^2}$. To find the coordinates of the trench we just have to solve $\partial V/\partial\tilde{\rho} = 0$, or the equivalent equation

$$\sin\left(\frac{\tilde{\rho}}{f} \right) - s\,c\,\frac{\Lambda_1^4}{\Lambda_2^4} \sin\left(\frac{c\tilde{\rho} + s\tilde{\theta}}{f_1} \right) = 0. \tag{11}$$

The inflaton will be the linear combination of the fields $(\tilde{\rho}, \tilde{\theta})$ that moves along the trench. The end of inflation will be determined by the violation of the slow-roll conditions or when the condition $\partial V/\partial\tilde{\rho} = 1$ cannot be met. In the later case, inflation ends by a mechanism analog to the waterfalling in models of hybrid inflation [13].

3. Inflationary trajectories

The final value of the inflaton field is determined by the slow-roll conditions or by the violation of Eq. (11). The slow-roll parameters as functions of the potential and its derivatives are

$$\epsilon = \frac{1}{2}\left(\frac{V'}{V} \right)^2, \qquad \eta = \frac{V''}{V} \qquad \text{and} \qquad \gamma = \frac{V'V'''}{V^2}, \tag{12}$$

where the derivatives are taken with respect to the inflaton field. Units where $M_* = 1$ are used. In our case, the inflaton is a combination of the ρ and θ fields. Let a represent the inflaton, we can write the nth derivative of the potential as

$$\frac{d^n V}{da^n} = \left[\left(1 + \frac{d\tilde{\rho}}{d\tilde{\theta}} \right)_{\text{tr}}^{-1/2} \frac{d}{d\tilde{\theta}} \right]^n V\left(\tilde{\theta}, \tilde{\rho}(\tilde{\theta})_{\text{tr}} \right), \tag{13}$$

where $da = (d\tilde{\rho}^2 + d\tilde{\theta}^2)^{1/2}$ is the differential distance traveled along the trench and the function $\tilde{\rho}(\tilde{\theta})_{\text{tr}}$ is the solution to Eq. (11). The subscript "tr" inside the square brackets indicates that $\tilde{\rho}$ and $\tilde{\theta}$ values are taken along the trench.

The value of the inflaton at the end of inflation is determined by the slow-roll condition $\epsilon(a_f) = 1$ or the moment the condition $\partial V/\partial\tilde{\rho} = 0$ cannot be met, whatever happens first along the path of the inflaton. To find the initial value we require enough e-folds of inflation to solve the horizon problem. The e-folds are given by

$$N = \int_{a_i}^{a_f} \frac{1}{\sqrt{2\epsilon}}\, da. \tag{14}$$

In our numerical calculations we require $N = 60$. We show results for two sets of parameters to illustrate the difference between the two mechanism that end inflation.

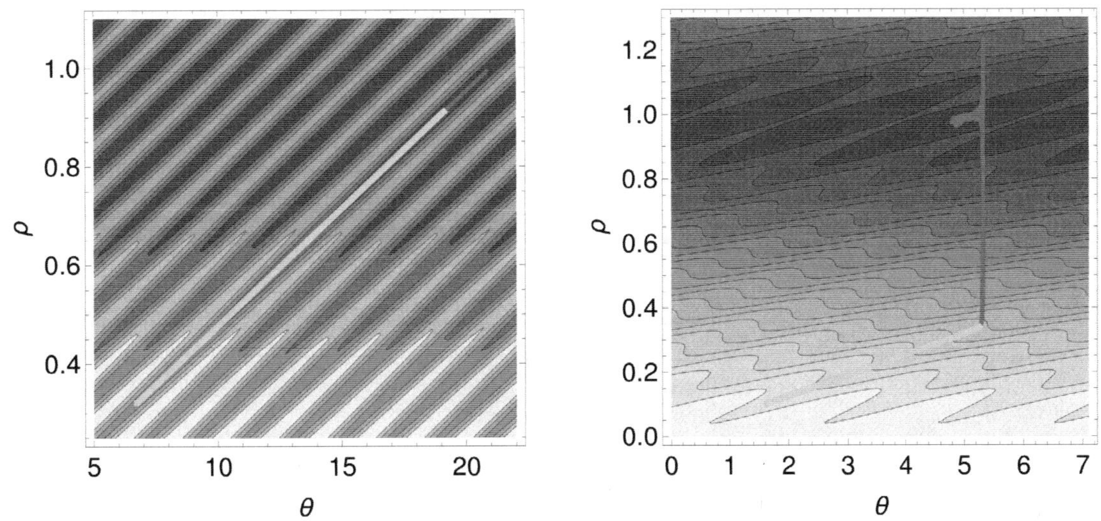

Figure 1. The path followed by the inflaton (green) for the cases described in Secs. 3.1 (left) and 3.2 (right) during the 60 e-folds of inflation. The red line shows the path followed until it settles in the corresponding minimum of the potential. In the right-hand side plot we see the point where inflation is terminated by an instability (see discussion in Sec 3.2.) Units where $M_* = 1$ were used.

3.1. Termination without a waterfall.

A choice of parameters where inflation ends when $\epsilon = 1$ is

$$f_1 = 0.22\sqrt{2}, \qquad f_2 = f_1/21, \qquad \Lambda_1 = \Lambda_2 = 0.006, \tag{15}$$

where units with $M_* = 1$ are used. This corresponds to the choice $n = 21$. After solving the field values from $\epsilon = 1$ and $N = 60$, we find the following initial and final values for the trajectory of the inflaton

$$(\tilde{\rho}, \tilde{\theta})_i = (6.04 \times 10^{-4}, 6.74) \qquad \text{and} \qquad (\tilde{\rho}, \tilde{\theta})_f = (1.50 \times 10^{-4}, 19.14), \tag{16}$$

We obtain predictions from the model by evaluating the slow roll parameters and the potential at the initial values of the fields. The cosmological parameters we determine are: the spectral index $n_s = 1 - 6\epsilon + 2\eta$, the ratio of tensor-to-scalar amplitudes $r = 16\epsilon$, the running of the spectral index $n_r = 16\epsilon\eta - 24\epsilon^2 - 2\gamma$ and the scalar amplitude $\Delta_R^2 = V/(24\pi^2\epsilon)$. Our predictions for these parameters are:

$$n_s = 0.96, \qquad r = 0.060, \qquad n_r = -0.00046, \qquad \Delta_R^2 = 2.2 \times 10^{-9}. \tag{17}$$

We compare these predictions with the latest results from the Planck experiment [4] to determine the consistency of this choice of parameters: $n_s = 0.968 \pm 0.006$, $r < 0.12$ (95% C.L.), $n_r = -0.003 \pm 0.007$ and $\Delta_R^2 = 2.19 \pm 0.08 \times 10^{-9}$. (Note that Δ_R^2, is called A_s in Ref. [4].)

3.2. Termination with a waterfall.

As mentioned before, inflation may end if the inflaton reaches a point where there is no solution to Eq. (11). This means we have reached an unstable point, suddenly accelerating the evolution

Q_{1L}	Q_{2L}	Q_{3L}	u^c_R	c^c_R	t^c_R	d^c_R	s^c_R	b^c_R
6	5	3	2	-1	-3	-1	-2	-2

L_{1L}	L_{2L}	L_{3L}	e^c_R	μ^c_R	τ^c_R	ν^c_{1R}	ν^c_{2R}	ν^c_{3R}
0	0	0	5	3	1	-3	-3	-3

Table 1. $\mathbf{Z}_{21}^{\Phi}$ charge assignments q, where the group transformation is defined by $\exp(2i\pi q/21)$. The Higgs doublet is a singlet under the flavor symmetry.

of the fields in an orthogonal direction [7]. This can be interpreted as the fields leaving the spiraling motion to rapidly fall to the minimum of the potential.

In this case, to determine the value of the fields at the end of inflation we have to solve

$$\left.\frac{\partial^2 V}{\partial \tilde{\rho}^2}\right|_{tr} = 0 \ , \tag{18}$$

then, similarly to the previous case, we will find the initial values using Eq. (14) to integrate back the 60 e-folds of inflation. Our choice of parameters is

$$f_1 = 0.22\sqrt{2}, \qquad f_2 = f_1/17, \qquad \Lambda_1 = 3.38\times 10^{-3}, \qquad \Lambda_2 = 1.61\times 10^{-3}. \tag{19}$$

which means that this time $n = 17$. Our analysis yields the following initial and final points for the inflationary trajectory:

$$(\tilde{\rho},\tilde{\theta})_i = (6.83\times 10^{-3}, 1.63) \qquad \text{and} \qquad (\tilde{\rho},\tilde{\theta})_f = (0.0281, 5.2970), \tag{20}$$

respectively. As before, we evaluate the slow roll parameters and the potential at the initial values. The predictions for the cosmological parameters in this case are:

$$n_s = 0.96, \qquad r = 0.0078, \qquad n_r = -7.2\times 10^{-5}, \qquad \Delta^2_{\mathcal{R}} = 2.2\times 10^{-9}. \tag{21}$$

Again, our results fall within those allowed by Planck's measurements. The path followed by the inflaton is displayed in Fig. 1 (left-hand side.) We can see the point where the fields leave the trench and fall to the minimum of the potential.

4. Standard model flavor

Here we aim to construct operators of the form of Eq. (1). Depending on the particular charge assignments of each SM fermion, the size of the Yukawa couplings will be given by powers of $\langle\Phi\rangle/M_*$ and $\langle\chi\rangle/M_*$. We choose $\mathbf{Z}_{21}^{\Phi}$ as the flavor model with Φ as the only flavon. We give in Table 1 the choices for SM fermion charges, where we also included a set of right-handed neutrinos. We construct $\mathbf{Z}_{21}$-invariant higher-order operators and set $\langle\Phi\rangle/M_* = 0.22 = \lambda$ the Cabibbo angle. For each Yukawa coupling we consider only contributions from lowest order operators. We find the following quark and charged fermion Yukawa matrices (with order-one coefficients ignored)

$$Y_u = \begin{pmatrix} \lambda^8 & \lambda^5 & \lambda^3 \\ \lambda^7 & \lambda^4 & \lambda^2 \\ \lambda^5 & \lambda^2 & 1 \end{pmatrix}, \qquad Y_d = \begin{pmatrix} \lambda^5 & \lambda^4 & \lambda^4 \\ \lambda^4 & \lambda^3 & \lambda^3 \\ \lambda^2 & \lambda & \lambda \end{pmatrix}, \qquad Y_e = \begin{pmatrix} \lambda^5 & \lambda^3 & \lambda \\ \lambda^5 & \lambda^3 & \lambda \\ \lambda^5 & \lambda^3 & \lambda \end{pmatrix}, \tag{22}$$

Is easy to show that these matrices achieve the mass ratios $m_u/m_t \sim \lambda^8$, $m_c/m_t \sim \lambda^4$, $m_d/m_b \sim \lambda^4$, and $m_s/m_b \sim \lambda^2$, as desired. For the neutrino case, the charges assigned to right-handed neutrinos lead to a mass matrix of the form $[\langle H\rangle^2/\Lambda_R]Y_\nu$, via the see-saw mechanism, where Λ_R represents the right-handed neutrino mass scale and $\langle H\rangle$ is the Higgs vev. The entries of the matrix Y_ν are of order λ^0 times a combination of order one coefficients that we can adjust according to the desired results.

5. Conclusions

In this talk we put together features from two different sectors: the Goldstone bosons from accidental continuous global symmetries in flavor models and the fields responsible for driving inflation in two-field axion monodromy models. The discrete symmetry choice allows us to successfully reproduce the mass ratios from fermion masses while keeping the appropriate conditions for an inflaton potential. The relation between inflation, the flavon vevs and the small flavor-symmetry-breaking parameter motivates the use of two-field axion monodromy inflation. This type of inflation models makes possible to predict the data on the CMB [4] while retaining a sub-Planckian vev.

Acknowledgments

This work was supported by the NSF under Grant No. PHY-1068008. The author thanks Lorenzo Díaz-Cruz for the invitation and support during the XXX Annual Meeting of the Division of Particles and Fields of the Mexican Physical Society.

References

[1] C. D. Carone, R. Ramos and Z. Wang, Phys. Rev. D **92**, no. 11, 116011 (2015) doi:10.1103/PhysRevD.92.116011 [arXiv:1509.08013 [hep-ph]].
[2] A. H. Guth, Phys. Rev. D **23**, 347 (1981); A. D. Linde, Phys. Lett. B **108**, 389 (1982); A. Albrecht and P. J. Steinhardt, Phys. Rev. Lett. **48**, 1220 (1982).
[3] F. C. Adams, J. R. Bond, K. Freese, J. A. Frieman and A. V. Olinto, Phys. Rev. D **47**, 426 (1993) [hep-ph/9207245]; K. Freese, J. A. Frieman and A. V. Olinto, Phys. Rev. Lett. **65**, 3233 (1990).
[4] P. A. R. Ade *et al.* [Planck Collaboration], arXiv:1502.02114 [astro-ph.CO].
[5] L. McAllister, E. Silverstein and A. Westphal, Phys. Rev. D **82**, 046003 (2010) [arXiv:0808.0706 [hep-th]].
[6] M. Berg, E. Pajer and S. Sjors, Phys. Rev. D **81**, 103535 (2010) [arXiv:0912.1341 [hep-th]].
[7] C. D. Carone, J. Erlich, A. Sensharma and Z. Wang, Phys. Rev. D **91**, no. 4, 043512 (2015) [arXiv:1410.2593 [hep-ph]].
[8] J. McDonald, arXiv:1407.7471 [hep-ph];
[9] J. McDonald, arXiv:1412.6943 [hep-ph].
[10] G. Barenboim and W. I. Park, arXiv:1412.2724 [hep-ph]; Phys. Rev. D **91**, no. 6, 063511 (2015) [arXiv:1501.00484 [hep-ph]].
[11] T. Li, Z. Li and D. V. Nanopoulos, arXiv:1507.04687 [hep-th]; Phys. Rev. D **91**, no. 6, 061303 (2015) [arXiv:1409.3267 [hep-th]]; arXiv:1412.5093 [hep-th].
[12] J. Erlich, J. Olsen and Z. Wang, arXiv:1509.06781 [hep-ph].
[13] A. D. Linde, Phys. Rev. D **49**, 748 (1994) [astro-ph/9307002].

One–loop corrections for Higgs–portal dark matter

J. Armando Arroyo & Saúl Ramos-Sánchez

Instituto de Física, Universidad Nacional Autónoma de México, POB 20-364, Cd.Mx. 01000, México

E-mail: `armandoarroyo@fisica.unam.mx, ramos@fisica.unam.mx`

Abstract. Models endowed with Higgs portals can probe into the hidden sectors of particle physics while providing stable dark matter candidates. Previous tree–level computations in such scenarios have shown that experimental bounds constrain dark matter to a very narrow region in parameter space. Aiming at improving the study of the implications of those constraints, we inspect one–loop corrections to the annihilation cross section for scalar dark matter into observable fermions. We find that these loop contributions might be enough to drastically change those results by deforming in about 10% the allowed parameter space for dark matter particles with masses even below 1 TeV. These findings encourage further investigation.

1. Introduction

Since the early days of cosmology [1], the quest for an understanding of dark matter (DM) and its origin has led to several proposals. These include massive astrophysical compact halo objects (MACHOs) [2], weakly interacting massive particles (WIMPs) [3] and axion-like particles (ALPs) [4]. WIMPs have been considered as optimal DM candidates mainly because of their natural appearance in many models that could explain other longstanding issues of particle physics. Their theoretical appeal has triggered an exhaustive experimental search by, among others, the XENON100 and LUX collaborations, that has resulted in severe constraints on the WIMP parameter space [5, 6]. Verifying that existing models comply with these strict bounds requires now that all their predicted observables, such as the production and decay rates, be computed with maximal precision.

On the other hand, the recent confirmation of the existence of the Higgs particle with a relatively large mass opens up a new set of possibilities. Especially, WIMPs might well couple directly to the Higgs field, establishing a connection known as *a Higgs portal* [7] between the *dark sector* and the Standard Model (SM). If there were such couplings, WIMPs would e.g. decay into visible matter via the exchange of a Higgs boson, affecting direct and indirect detection signals.

Beyond DM, Higgs portals may be relevant to address other intriguing questions, such as the nature of cosmological inflation [8], the instability in the electroweak vacuum [9], and more recently, the 750 GeV diphoton excess reported at the LHC [10]. Despite their great potential, in this work we shall only focus on the features of DM in these models.

The additional particles required in Higgs portals are considered to be *hidden* in the sense that they do not carry SM charges. They are dynamical (real) scalars S, spinors ψ or vectors V_μ, confined to interact with the SM via Lagrangian couplings with the Higgs field H, such as

$$S^2H^\dagger H\,, \qquad \bar{\psi}\psi H^\dagger H\,, \qquad V_\mu V^\mu H^\dagger H\,, \tag{1}$$

and affected by their own (in principle, arbitrary) potential energy density. The scalar case clearly requires the inclusion of a $\mathbb{Z}_2$ symmetry (or another similar mechanism) to avoid tadpole contributions and to ensure DM stability. These different particles are good DM candidates as long as they comply with experimental bounds. In this respect, fermions are disfavored because they do not only yield a non–renormalizable model, but they are already excluded by XENON data when considering only thermal DM [11].

The phenomenological implications of Higgs portals with scalars and vectors as DM have been studied in [11, 12] (and a recent global analysis for the scalar case has been done in [13]), where only tree–level contributions to the decay and production rates have been considered. It was shown in those works that experimental bounds of WMAP [14] and XENON100 [5] allow for a very reduced set of values in the parameter space of the models, so that radiative corrections may become important to falsify them. Although one–loop corrections to DM–nucleon scattering have been already analyzed [15], the corresponding study for DM annihilation processes is still missing. To motivate and to exhibit the importance of completing this task is one of the aims of the present work.

Based on the results obtained in [16], we discuss here the main aspects of Higgs portals and explore the importance of one–loop corrections to the cross sections of the processes associated with DM detection. The structure of this contribution is as follows. In section 2, we give a brief review of the scalar and vector Higgs portals; in section 3 we present the one–loop analysis for the cross section in the context of the scalar portal done with the aid of FeynArts [18] and FormCalc [19] and finally, in section 4 we discuss the obtained results and ongoing work.

2. Scalar and vector Higgs portals

The scalar and vector Higgs portals have been studied previously in [11, 12] and here we will only revise the main features.

The simplest example of these models is the scalar portal. It is achieved by introducing a (real) massive scalar field S subject to a $\mathbb{Z}_2$ symmetry, under which S is odd and all SM fields are even. The full renormalizable Lagrangian then reads

$$\mathcal{L} \;=\; \mathcal{L}_{SM} + \frac{1}{2}\partial_\mu S\,\partial^\mu S - \frac{1}{2}\mu_S^2 S^2 - \frac{\lambda_{hs}}{2}S^2 H^\dagger H - \frac{\lambda_S}{4}S^4\,, \tag{2}$$

where λ_{hs}, μ_S and λ_S are the portal and the two self–interaction real parameters. After electroweak symmetry breaking (EWSB), the Higgs–portal potential becomes

$$V(h,S) \;=\; \frac{1}{2}m_S^2 S^2 + \frac{\lambda_{hs}v}{2}S^2\,h + \frac{\lambda_{hs}}{4}S^2\,h^2 + \frac{\lambda_S}{4}S^4\,, \tag{3}$$

where $m_S^2 = \mu_S^2 + \lambda_{hs}v^2/2$ is the physical mass of S, $v \approx 246\,GeV$ denotes the electroweak vacuum expectation value (VEV) of the Higgs field H, such that $H(x) \to (v + h(x))/\sqrt{2}$ (and S has vanishing VEV) in the vacuum, with the small Higgs field perturbation h.

Similarly, for the vector Higgs portal we introduce a vector field V_μ, probably stemming from a U(1) gauge symmetry, whose Higgs–portal potential is given by

$$\mathcal{L} \supset \frac{1}{2}\mu_V^2 V_\mu V^\mu + \frac{\lambda_{hv}}{2}V_\mu V^\mu H^\dagger H + \frac{\lambda_V}{4}\left(V_\mu V^\mu\right)^2\,. \tag{4}$$

Since the mass term $V_\mu V^\mu$ is not gauge invariant if V_μ arises from a gauge symmetry,[1] some gauge–restoring mechanism, such as the Stueckelberg mechanism,[2] must be invoked. After EWSB, the physical mass of the vector is given by $m_V^2 = \mu_V^2 + \lambda_{hv}v^2/2$.

[1] V_μ may have a different origin. E.g. it could be a composite field, whose effective potential is eq. (4).

[2] For a review on the Stueckelberg mechanism, see [20]

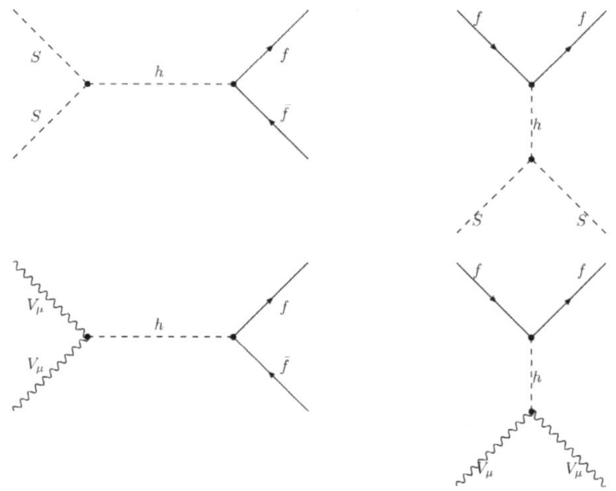

Figure 1: The left-hand-side diagrams depict DM annihilation into SM fermions, while those on the right, DM–fermion scattering for the scalar and vector cases.

Not only does EWSB generate the physical masses m_S and m_V of the DM particles, but, as eq. (3) evidences, it also "produces" the cubic interactions $h\,S^2$ and $h\,V_\mu V^\mu$. These interaction terms allow for the possibility of the *invisible* decay of the Higgs particle into DM. LHC bounds [21] on the associated decay width $\Gamma(h \to SS)$ do not constrain DM with masses over few hundred GeV [13]; hence, this decay shall not be investigated in this work.

Additionally, as depicted in Fig. 1, the hSS vertex allows for DM annihilation into SM fermions as well as DM–fermion scattering. The thermally–averaged cross section of the first process, $\langle \sigma_{DM} v_{rel} \rangle$ (with $_{DM}$ referring either to S or V_μ), relates to Planck's measurement of DM relic abundance [22], $\Omega_{CDM}h^2 = 0.1198 \pm 0.0015$, roughly by

$$\Omega_{CDM}h^2 \sim 3 \cdot 10^{-27} cm^3\, s^{-1} / \langle \sigma_{DM} v_{rel} \rangle\,, \tag{5}$$

where v_{rel} denotes the DM relative velocity or Møller velocity. On the other hand, DM–fermion scattering is important for direct DM detection, because the spin–independent DM–nucleon scattering cross section σ_{SI} can be computed from it under certain assumptions, allowing for a comparison with LUX and XENON100 data.

For scalar and vector Higgs portals, the newest version of micrOMEGAs [23] performs the numerical computation of σ_{SI} up to one–loop level and compares it with the LUX and XENON100 bounds. Therefore, only the computation of radiative corrections to $\langle \sigma_{DM} v_{rel} \rangle$ may be relevant if those contributions are sizable, as we shall show that they turn out to be.

At tree–level, the thermally–averaged annihilation cross sections are given by

$$\langle \sigma_S v_{rel} \rangle = \frac{\lambda_{hs}^2 m_f^2}{16\pi} \frac{(1 - m_f^2/m_S^2)^{3/2}}{(4m_S^2 - m_h^2)^2}\,, \tag{6}$$

$$\langle \sigma_V v_{rel} \rangle = \frac{\lambda_{hv}^2 m_f^2}{48\pi} \frac{(1 - m_f^2/m_V^2)^{3/2}}{(4m_V^2 - m_h^2)^2}\,, \tag{7}$$

for the scalar and vector case, respectively. In these eqs., m_f is the mass of the fermions in the final state. Note that the cross section eq. (6) (eq. (7)) depends solely on λ_{hs} (λ_{hv}) and m_S (m_V). It then follows that Planck's data bound on Ω_{CDM} leads to constraints on these parameters.

Journal of Physics: Conference Series **761** (2016) 012014 doi:10.1088/1742-6596/761/1/012014

3. One–loop corrections in the scalar portal

Sizable one–loop contributions to DM cross sections in Higgs portals may become important as they may close further the narrow region of allowed parameters, thereby finding the model already falsified by data.

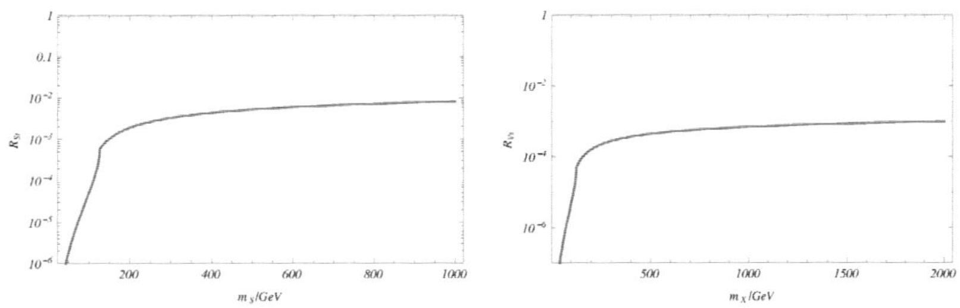

Figure 2: Approximate comparison between the amplitudes at one–loop and at tree–level for the process $SS \to t\bar{t}$ (left) and $V_\mu V^\mu \to t\bar{t}$ (right).

In a previous analysis [16], we studied the WIMP mass evolution of the ratio between the simplest one–loop amplitude correction and the tree–level amplitude for DM annihilation into t quarks,

$$R_{DMt} \equiv \frac{|\mathcal{M}_{\text{only 1-loop}}(DM\,DM \to t\bar{t})|^2}{|\mathcal{M}_{\text{tree}}(DM\,DM \to t\bar{t})|^2}, \tag{8}$$

where, as before, DM stands for either S or V_μ WIMP candidates. This ratio lets us estimate how subdominant one–loop corrections are. As it is evident from Fig. 2 (left), in the scalar case this ratio shows that, for DM masses around 1 TeV and larger, the simplest one–loop contribution analyzed can be as large as $\sim$ 1% of the tree–level amplitude. This preliminary result encourages further study, since other one–loop contributions may enlarge this already sizable correction. In the vectorial case, this approach reveals (see Fig. 2 (right)) that one–loop contributions may not be so relevant. Thus, we can safely focus on the corrections to the scalar Higgs portal.

We have extended that previous analysis in two ways. First, we completed the computation of the one–loop cross section σ_S for annihilation into t's, including all possible contributions. Secondly, we further studied the complete cross section at one–loop level for the annihilation of

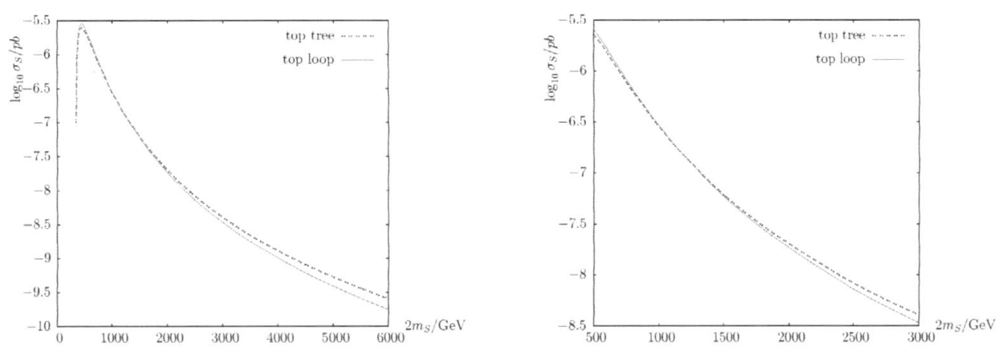

Figure 3: Annihilation cross section of scalar DM into t quarks for $\lambda_{hs} = 10^{-3}$ and m_S between 0.25 and 3 TeV and a zoomed–in interval. We compare the tree–level (blue/dashed line) contribution vs. the full up–to–one–loop cross sections (red/solid line). For $m_S \gtrsim 600$ GeV, the one–loop contribution becomes negative and can be (for $m_S \sim 3$ TeV) as large as 30% of the tree–level cross section.

scalar DM into t, b and c quarks and μ and τ leptons. The Feynman rules were generated with LanHEP [17]; the computation of amplitudes, one loop integrals and the cross sections have been performed by using the public numerical and analytical tools FeynArts [18], LoopTools and FormCalc [19].

Our results are displayed in Figs. 3 and 4. Fig. 3 shows the comparison of the tree–level $\sigma_{S,\text{tree}}$ and full up–to–one–loop $\sigma_{S,\text{1-loop}}$ cross sections for scalar DM annihilation into t quarks and $\lambda_{hs} = 10^{-3}$. We note that for $2m_S \gtrsim 1.2$ TeV, this contribution is negative and significantly larger than 1%, becoming as large as 20% for scalar Higgs–portal WIMPs with $m_S \sim 2$ TeV. A similar comparison has been done for DM annihilation into light quarks and heavy leptons (Fig. 4), to confirm our expectation that top contributions are dominant. For different values of λ_{hs}, we find similar results.

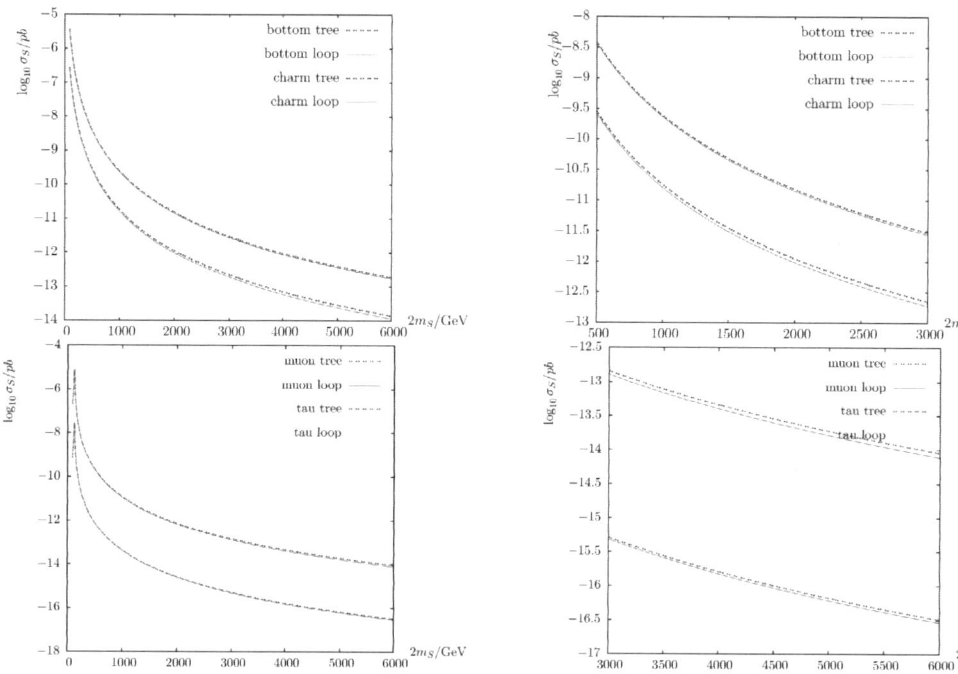

Figure 4: Annihilation cross section of scalar DM for $\lambda_{hs} = 10^{-3}$ and m_S between 45 GeV and 3 TeV (left panels) and a zoomed–in interval (right panels). The upper panels depict the tree–level (blue/dashed line) and the full up–to–one–loop (red/solid line) contributions with b and c quarks in the final states, whereas the lower panels display the same contributions with μ and τ leptons in the final states.

4. Punchline and outlook

We have computed the one–loop corrections to the annihilation cross section of Higgs–portal DM into SM fermions. Our preliminary results stress the need to carefully determine the influence of radiative corrections in the computation of the cross sections in models with Higgs portals. We notice that, particularly for the scalar Higgs portal and certain parameter values, the one–loop contributions can be negative and as large as 13% (21%) of the tree level cross section for admissible DM with masses of about 1 TeV (2 TeV).

The resulting thermally averaged cross sections $\langle\sigma_S v_{rel}\rangle$ are functions of the two parameters of the Higgs portal, m_S and λ_{hs}, such that for each value of m_S an interval of λ_{hs} is still allowed by Planck's data, according to eq. (5). Since the computed cross section σ_S is reduced by a sizable amount for heavy WIMPs, so is $\langle\sigma_S v_{rel}\rangle$. Thus, the predicted relic abundance increases, stressing the tension between Higgs portals and measurements by WMAP, LUX and XENON100, when the one–loop corrections to the DM–nucleon scattering amplitude σ_{SI} are additionally included. The detailed report of this analysis shall be presented elsewhere [24].

Acknowledgments

We are grateful for useful discussions with T. Hahn and A. Ibarra. This work was partly supported by CONACyT grant 151234. S. R-S. would like to thank the ICTP for the kind hospitality and support received through its Junior Associateship Scheme during the realization of this work.

References

[1] Zwicky F 1933 *Helvetica Physica Acta* **6** 110–127
[2] Bird S, Cholis I, Muñoz J B, Ali-Haïmoud Y, Kamionkowski M, Kovetz E D, Raccanelli A and Riess A G 2016 *Phys. Rev. Lett.* **116** 201301 (*Preprint* 1603.00464)
[3] Bertone G, Hooper D and Silk J 2005 *Phys. Rept.* **405** 279–390 (*Preprint* hep-ph/0404175)
[4] Sikivie P 2008 *Lect. Notes Phys.* **741** 19–50 [,19(2006)] (*Preprint* astro-ph/0610440)
[5] Aprile E *et al.* (XENON100) 2012 *Phys. Rev. Lett.* **109** 181301 (*Preprint* 1207.5988)
[6] Akerib D S *et al.* (LUX) 2016 *Phys. Rev. Lett.* **116** 161301 (*Preprint* 1512.03506)
[7] Patt B and Wilczek F 2006 (*Preprint* hep-ph/0605188)
[8] Gross C, Lebedev O and Zatta M 2016 *Phys. Lett.* **B753** 178–181 (*Preprint* 1506.05106)
[9] Falkowski A, Gross C and Lebedev O 2015 *JHEP* **05** 057 (*Preprint* 1502.01361)
[10] Choi S M, Kang Y J and Lee H M 2016 (*Preprint* 1605.04804)
[11] Djouadi A, Lebedev O, Mambrini Y and Quevillon J 2012 *Phys. Lett.* **B709** 65–69 (*Preprint* 1112.3299)
[12] Lebedev O, Lee H M and Mambrini Y 2012 *Phys. Lett.* **B707** 570–576 (*Preprint* 1111.4482)
[13] Han H and Zheng S 2015 *JHEP* **12** 044 (*Preprint* 1509.01765)
[14] Komatsu E e a 2011 *Astrophys.J.* **192** 18 (*Preprint* 1001.4538)
[15] Hisano J, Nagai R and Nagata N 2015 *JHEP* **05** 037 (*Preprint* 1502.02244)
[16] Arroyo J A 2015 *DM candidates from Higgs portals (in Spanish)* B.Sc. Thesis UNAM URL http://stringpheno.fisica.unam.mx/thesis/ArmandoArroyo.pdf
[17] Semenov A 2009 *Comput. Phys. Commun.* **180** 431–454 (*Preprint* hep-ph/0805.0555)
[18] Hahn T 2001 *Comput. Phys. Commun.* **140** 418–431 (*Preprint* hep-ph/0012260)
[19] Hahn T and Pérez-Victoria M 1999 *Comput. Phys. Commun.* **118** 153–165 (*Preprint* hep-ph/9807565)
[20] Zhang Y, Wang S Z and Wang Q 2008 *JHEP* **03** 047 (*Preprint* 0803.1275)
[21] Aad G *et al.* (ATLAS Collaboration and CMS Collaboration) 2015 *Phys. Rev. Lett.* **114**(19) 191803 URL http://link.aps.org/doi/10.1103/PhysRevLett.114.191803
[22] Ade P *et al.* (Planck) 2015 (*Preprint* 1502.01589)
[23] Bélanger G, Boudjema F, Pukhov A and Semenov A 2015 *Comput. Phys. Commun.* **192** 322–329 (*Preprint* 1407.6129)
[24] Arroyo J A and Ramos-Sánchez S 2016, To appear

Dark Matter candidate in Inert Doublet Model with additional local gauge symmetry $U(1)$

R. Gaitán
Departamento de Física, FES-Cuautitlán, UNAM, C.P. 54770, Estado de México, México
E-mail: rgaitan@unam.mx

E. A. Garcés
Instituto de Física, Universidad Nacional Autónoma de México, Apdo. Postal 20-364, 01000Ciudad de México, México
E-mail: egarces@fisica.unam.mx

J.H. Montes de Oca
Departamento de Física, FES-Cuautitlán, UNAM, C.P. 54770, Estado de México, México
E-mail: josehalim@gmail.com

L. G. Cabral-Rosetti
Centro Interdisiplinario de Investigación y Docencia en Educacin Técnica (CIIDET), Unidad de Ciencias Básicas, Av. Universidad No. 282 Pte., Col. Centro, C.P. 76000, Querétaro, Qro., México
E-mail: cabralrosetti@gmail.com

Abstract. We consider the Inert Doublet Model (IDM) with an additional local gauge symmetry $U(1)$ and a complex singlet scalar to break the symmetry $U(1)$. The continuous symmetry $U(1)$ is introduced to control the CP-conserving interaction instead of some discrete symmetries as usually. We present the mass spectrum for neutral scalar and gauge bosons and the values of the charges under $U(1)$ for which the model could have a candidate to dark matter.

1. Introduction

The existence of Dark matter (DM) is now essentially established [1]. The most convincing evidence for DM came from the observation that luminous objects such as stars, gas clouds, globular clusters, or entire galaxies move faster than one would expect if they only felt the gravitational attraction of other visible objects [2, 3]. DM is estimated to constitute about 23 % of the total matter in the universe. However, the origin of DM still remains a mystery.

The Standard Model (SM) of particle physics successfully explains experimental results observed in colliders, but none of the SM particles can be a good candidate for the dark matter. One of the main reasons is the lack of particles with scarce or null interactions with other SM particles. One of the first proposals for DM candidate within the content of SM particles were neutrinos but these have been discarded because they do not fit the description of hot DM. Therefore, it is necessary to look for new physics beyond the SM. In the literature there are various proposals as dark matter candidates, being a weakly interacting massive particle (WIMP) a promising candidate; in fact, the WIMP relic density is around the observed value (Planck Collaboration) [4]:

$$\Omega h^2_{CDM} = 0.1199 \pm 0.0027. \tag{1}$$

In the case of WIMPs as DM candidates in models beyond the standard model we comment some of them below. In supersymmetric models, one candidate is the lightest neutralino as a spin-1/2 WIMP dark matter [5]. Spin-1 WIMP dark matter has been studied in the context of models with extra dimensions [6]. Some extensions are focused to only include scalar fields in different representations under $SU(2)_L$ [7, 8, 9]; if these scalars are stable, they can in principle account for a density of DM. These last proposals have been taken into account more seriously because current experimental constraints and possible future observations for scalar particles.

Several new physics scenarios with extended scalar fields and extra Higgs doublets have been considered to accommodate new particles as DM candidates. For example, the Inert Two Higgs Doublet Model (IDM) [10], where a DM candidate coming from the doublet with Vacuum Expectation Value (VEV) equal to zero and its mass has been constrained to the order of GeV [11, 12]. Recently a two Higgs doublet model (2HDM) with a $U(1)$ gauge symmetry, including a gauge boson Z' with mass of order of GeV scale or below, has been proposed, that is the case of the D2HDM [13, 14]. However, the word "Dark" in references [13, 14] is used due to the invisible light gauge boson.

In the D2HDM a continuous symmetry $U(1)$ is the mechanism to keep very suppressed the Yukawa couplings between fermions and scalars, while in the IDM this is achieved by imposing a Z_2 discrete symmetry. In this work we consider as starting point the D2HDM in order to propose a model with WIMP-like DM candidate particle.

2. The Inert Doublet Model-$U(1)_X$

The considered gauge group for the model is defined as $G_{IDMU_X} = G_{SM} \otimes U(1)_X$ where G_{SM} is the gauge group symmetry of the SM. We name this model as Inert Doublet Model with $U(1)_X$ (IDM-$U(1)_X$) and the coupling constant associated to $U(1)_X$ is denoted as g_X. From here on we will explain the scalar and gauge bosons sectors such that a dark matter candidate can be viable in the IDM-$U(1)_X$.

2.1. Scalar sector

The scalar sector in the IDM-$U(1)_X$ contains two complex scalar doublets with hypercharge $Y = 1$, which is the content of the IDM. One doublet plays the role of the doublet of the SM, meanwhile the second one is the inert doublet with null VEV. Usually, the particles with very little or no interaction are referred as inert particles. Additional to these two doublets, we consider one complex singlet scalar field Φ_x with VEV denoted by v_x. The singlet Φ_x is introduced to break down the symmetry G_{IDMU_X} to the G_{SM}. As it happens in SM, the doublet with non zero VEV is responsible for the spontaneous electroweak symmetry breaking.

The models with discrete symmetry, for instance the Z_2 symmetry, are able to have an inert particle by imposing that its doublet is transformed in no trivial form under Z_2. However, our proposal is based in the continuous symmetry $U(1)_X$ instead of a discrete symmetry, which also can control interaction between neutral scalars and other particles. Scalar fields are denoted as Φ_1 and Φ_2 for doublets and Φ_x for singlet, as previously introduced. These scalar fields have quantum number under G_{IDMU_X} defined as

$$\Phi_1 \sim (1, 2, 1/2, x_1), \tag{2}$$

$$\begin{aligned}\Phi_2 &\sim (1,2,1/2,x_2),\\ \Phi_x &\sim (1,1,0,x),\end{aligned}$$

where two first entries denote the representation under $SU(3)_C$ and $SU(2)_L$ meanwhile the hypercharge and charge $U(1)_X$ are written in the last two entries. Then, the most general, renormalizable and gauge invariant potential is

$$\begin{aligned}V &= \mu_1^2\Phi_1^\dagger\Phi_1 + \mu_2^2\Phi_2^\dagger\Phi_2 + \mu_x^2\Phi_x^*\Phi_x + \left[\mu_{12}^2\Phi_1^\dagger\Phi_2 + h.c.\right]\\ &+\lambda_1\left(\Phi_1^\dagger\Phi_1\right)^2 + \lambda_2\left(\Phi_2^\dagger\Phi_2\right)^2 + \lambda_3\left(\Phi_1^\dagger\Phi_1\right)\left(\Phi_2^\dagger\Phi_2\right)\\ &+\lambda_4\left|\Phi_1^\dagger\Phi_2\right|^2 + \left[\lambda_5\left(\Phi_1^\dagger\Phi_2\right)^2 + \lambda_6\left(\Phi_1^\dagger\Phi_1\right)\left(\Phi_1^\dagger\Phi_2\right)\right.\\ &\left.+\lambda_7\left(\Phi_2^\dagger\Phi_2\right)\left(\Phi_1^\dagger\Phi_2\right) + h.c.\right] + \lambda_x\left(\Phi_x^*\Phi_x\right)^2\\ &+\left(\Phi_x^*\Phi_x\right)\left[\lambda_{1x}\left(\Phi_1^\dagger\Phi_1\right) + \lambda_{2x}\left(\Phi_2^\dagger\Phi_2\right)\right]\\ &+\left[\lambda_{12x}\left(\Phi_1^\dagger\Phi_2\right)\left(\Phi_x^*\Phi_x\right) + h.c.\right] + \left[\lambda_{1x}'\left(\Phi_1^\dagger\Phi_1\right)\Phi_x\right.\\ &\left.+\lambda_{2x}'\left(\Phi_2^\dagger\Phi_2\right)\Phi_x + \lambda_x'\left(\Phi_1^\dagger\Phi_2\right)\Phi_x + h.c.\right].\end{aligned} \tag{3}$$

The μ_1^2, μ_2^2, μ_x^2, λ_1, λ_2, λ_3, λ_4, λ_x, λ_{1x}, λ_{2x} are real parameters and the rest of the parameters are complex. The potential (3) is not only CP non conserving but it has terms involving odd power which do not allow the stability of the DM candidate, for instance the terms proportional to λ_{2x}' or λ_x'. We set the values of the $U(1)_X$ charges x_1, x_2 and x in order to suppress the terms in the potential that do not allow the stability for DM candidate. In the potential the terms of the form $\Phi_1\Phi_2$ are allowed by gauge invariance when the $U(1)_X$ charges satisfy $x_1 = x_2$. If $x_1 \neq x_2$, the parameters μ_{12}, λ_5 and λ_{12x} must be equal to zero in order to guarantee the gauge invariance. We also assume $x \neq 0$ in order to involve the singlet scalar in the $U(1)_X$ symmetry breaking mechanism. Therefore, the parameters λ_{1x}' and λ_{2x}' must also be null to guarantee the gauge invariance. The term $\left(\Phi_1^\dagger\Phi_2\right)\Phi_x$, which induces decay of the DM candidate, can be eliminated of the scalar potential if we assume $x_2 - x_1 \neq x$.

Under the last assumptions the scalar potential is simplified to

$$\begin{aligned}V &= \mu_1^2\Phi_1^\dagger\Phi_1 + \mu_2^2\Phi_2^\dagger\Phi_2 + \mu_x^2\Phi_x^*\Phi_x + \lambda_1\left(\Phi_1^\dagger\Phi_1\right)^2\\ &+\lambda_2\left(\Phi_2^\dagger\Phi_2\right)^2 + \lambda_3\left(\Phi_1^\dagger\Phi_1\right)\left(\Phi_2^\dagger\Phi_2\right) + \lambda_4\left|\Phi_1^\dagger\Phi_2\right|^2\\ &+\lambda_x\left(\Phi_x^*\Phi_x\right)^2 + \left(\Phi_x^*\Phi_x\right)\left[\lambda_{1x}\left(\Phi_1^\dagger\Phi_1\right) + \lambda_{2x}\left(\Phi_2^\dagger\Phi_2\right)\right].\end{aligned} \tag{4}$$

After spontaneous symmetry breaking, the scalar fields acquire VEV such that

$$\langle\Phi_1\rangle = \begin{pmatrix} 0 \\ \frac{v}{\sqrt{2}}\end{pmatrix}, \tag{5}$$

$$\langle\Phi_2\rangle = \begin{pmatrix} 0 \\ 0\end{pmatrix} \tag{6}$$

and

$$\langle\Phi_x\rangle = \frac{v_x}{\sqrt{2}}. \tag{7}$$

The masses of the physical scalars are

$$M_{h_1}^2 = 2\lambda_1 v^2, \tag{8}$$

$$M^2_{h_2,h_3} = \mu_2 + \frac{1}{2}v^2(\lambda_3+\lambda_4) + \frac{1}{2}v_x^2\lambda_x, \tag{9}$$

$$M^2_{H^\pm} = \mu_2 + \frac{1}{2}v^2\lambda_3 + \frac{1}{2}v_x^2\lambda_x, \tag{10}$$

$$M_x^2 = 2\lambda_x v_x^2. \tag{11}$$

The extremal conditions for (4) are

$$2\mu_1^2 + 2\lambda_1 v^2 + \lambda_{1x}v_x^2 = 0 \tag{12}$$

and

$$2\mu_x^2 + 2\lambda_x v_x^2 + \lambda_{1x}v^2 = 0. \tag{13}$$

In this case the mass matrix for neutral scalars is diagonal matrix and there is degeneration in the masses of the neutral scalars h_2 and h_3 due to the non diagonal elements that could arise from the terms proportional to the parameters that were taken equal to zero. The $U(1)_X$ charges for scalars can be taken with values such as to generate this mixing between neutral scalars. However, the only combination that retains the stability of the neutral scalar from Φ_2 is considered above, $(\ x_1,\ \ x_2,\ \ x_3\ ) = (\ 0,\ \ -x,\ \ x\ )$ with $x \neq 0$.

2.2. Gauge-scalar interactions

The interactions between the scalar and gauge bosons are given by

$$\mathcal{L}_{\text{gauge-scalar}} \;=\; |D_\mu\Phi_1|^2 + |D_\mu\Phi_2|^2 + |D_\mu\Phi_x|^2, \tag{14}$$

where the covariant derivative D_μ is defined as

$$D_\mu = \left(\partial_\mu + ig'Y\hat{B}_\mu + igT_3\hat{W}_{3\mu} + ig_{Z'}Q'\hat{Z}'_{0\mu}\right). \tag{15}$$

The hat notation in the vector bosons refers to states before a transformation is applied to eliminate the kinetic mixing between the $U(1)_Y$ and $U(1)_X$ tensor fields which is given by

$$\mathcal{L}_{\text{Kinetic}} = -\frac{1}{4}\hat{B}_{\mu\nu}\hat{B}^{\mu\nu} + \frac{1}{2}\frac{\varepsilon}{\cos\theta_W}\hat{B}^{\mu\nu}\hat{Z}'_{0\mu\nu} - \frac{1}{4}\hat{Z}'_{0\mu\nu}\hat{Z}_0'^{\mu\nu}, \tag{16}$$

where, $\hat{B}^{\mu\nu}$ and $\hat{Z}_0'^{\mu\nu}$ are the field strength tensors of the $U(1)_Y$ and $U(1)'$ gauge bosons, respectively. If the transformation

$$\begin{pmatrix} Z'_{0\mu} \\ B_\mu \end{pmatrix} = \begin{pmatrix} \sqrt{1-\varepsilon^2/\cos^2\theta_W} & 0 \\ -\varepsilon/\cos^2\theta_W & 1 \end{pmatrix}\begin{pmatrix} \hat{Z}'_{0\mu} \\ \hat{B}_\mu \end{pmatrix} \tag{17}$$

is applied, the kinetic term with mixing in (16) is removed. The transformation is also applied in (14) and after spontaneous symmetry breaking this Lagrangian is written as

$$\mathcal{L} = \mathcal{L}_{masses} + \mathcal{L}_{interactions}, \tag{18}$$

where

$$\mathcal{L}_{masses} = \frac{1}{2}m^2_{Z^0}Z^0Z^0 + \frac{1}{2}m^2_{Z'^0}Z'^0Z'^0 - \Delta^2 Z'^0Z^0 + ... \tag{19}$$

with

$$m^2_{Z^0} = \frac{1}{4}g_Z^2 v^2, \tag{20}$$

$$m^2_{Z'^0} = g^2_x(v^2 + v^2_x) + \frac{\varepsilon}{\cos\theta_W} g_x g'^2 + \frac{1}{4}\left(\frac{\varepsilon}{\cos\theta_W}\right)^2 g'^2 v^2, \tag{21}$$

$$\Delta^2 = \frac{1}{2} g_x g_Z v^2 + \frac{1}{4}\frac{\varepsilon}{\cos\theta_W} g_Z g'^2. \tag{22}$$

Additionally to the mass terms for the gauge bosons, the mixing term $(Z_{0\mu} Z_0'^{\mu})$ appears. This mixing term can be eliminated through the following rotation

$$\begin{pmatrix} Z \\ Z' \end{pmatrix} = \begin{pmatrix} \cos\xi & -\sin\xi \\ \sin\xi & \cos\xi \end{pmatrix} \begin{pmatrix} Z^0 \\ Z'^0 \end{pmatrix}, \tag{23}$$

where the mixing angle ξ satisfies the expression $\tan 2\xi = \frac{2\Delta^2}{m^2_{Z^0} - m^2_{Z'^0}}$, and has been constrained to $|\xi| < 10^{-3}$ [15]. The detailed description for $\mathcal{L}_{interactions}$ can be reviewed in references [13, 14].

3. Discussion

We considered a model with new particles that arise though extending the scalar sector and a neutral gauge boson from the additional symmetry $U(1)_X$. The mass spectrum for these neutral scalar and gauge bosons was presented in Eq. (9) and Eq. (21). These equations show the importance of the free parameters λ_X, v_X. The model presents a degeneration in the masses of the neutral scalars $h_{2,3}$ due to the elimination of the parameter λ_5. The λ_5 coupling must vanish to guarantee the gauge invariance when the values of x were restricted in order to control the interactions. This means that the model can provide a scenario for two dark matter candidates with degenerate masses. Other scenario where the DM candidate is identified with the singlet in a model with additional symmetry $U(1)$ was studied previously [16]. The next step is to take into account the limits reported for observable quantities involving these new particles.

Acknowledgments

This work was supported by projects PAPIIT-IN113916 in DGAPA-UNAM, PIAPIVC07 in FES-Cuautitlán UNAM and *Sistema Nacional de Investigadores* (SNI) México. E. A. Garcés thanks CONACYT postdoctoral grant.

References

[1] G. Bertone, D. Hooper and J. Silk, *Phys. Rept.* **405**, 279 (2005), hep-ph/0404175.
[2] M. Roos, astro-ph.CO/1001.0316.
[3] N. Arkani-Hamed, D. P. Finkbeiner, T. R. Slatyer, N. Weiner, *Phys. Rev.* D**79**, 015014 (2009), hep-ph/0810.0713.
[4] P. A. R. Ade *et al.* [Planck Collaboration], *Astron. Astrophys.* (2014), astro-ph.CO/1303.5076.
[5] G. Jungman, M. Kamionkowski and K. Griest, *Phys. Rept.* **267**, 195 (1996), hep-ph/9506380.
[6] G. Servant and T. M. P. Tait, *Nucl. Phys. B* **650**, 391 (2003), hep-ph/0206071.
[7] J. McDonald, *Phys. Rev. D* **50**, 3637 (1994), hep-ph/0702143.
[8] V. Silveira and A. Zee, *Phys. Lett. B* **161**, 136 (1985).
[9] C. P. Burgess, M. Pospelov and T. ter Veldhuis, *Nucl. Phys. B* **619**, 709 (2001), hep-ph/0011335.
[10] E. M. Dolle and S. Su, *Phys. Rev. D* **80**, 055012 (2009), hep-ph/0906.1609.
[11] L. Lopez Honorez, E. Nezri, J. F. Oliver and M. H. G. Tytgat, *JCAP* **0702**, 028 (2007), hep-ph/0612275.
[12] L. López Honorez and C. E. Yaguna, *JCAP* **1101**, 002 (2011), hep-ph/1011.1411.
[13] H. Davoudiasl, H. -S. Lee and W. J. Marciano, *Phys. Rev. D* **85**, 115019 (2012), hep-ph/1203.2947.
[14] H. S. Lee and M. Sher, *Phys. Rev. D* **87**, no. 11, 115009 (2013), hep-ph/1303.6653.
[15] S.A. Abel, M. D. Goodsell, J. Jaeckel, V. V. Khoze and A. Ringwald, *J. High Energy Phys.* 07 (2008) 124, hep-ph/0803.1449.
[16] Gaitán, R. and Garcés, E. A. and de Oca, J. H. Montes, hep-ph/1410.5462.

Three- and four-body systems with the Functional Renormalization Group

Benjamín Raziel Jaramillo Ávila

División de Ciencias e Ingenierías, Universidad de Guanajuato, León, México

E-mail: `raziel@fisica.ugto.mx`

Abstract. The Efimov effect arises in three-body systems near the unitary limit. Some of its features are universal, while others are not. This article uses a Functional-Renormalization-Group approach to discuss the Efimov effect and four-body systems. In this context, the Efimov effect appears as a consequence of the Renormalization-Group flow of couplings. On the four-body system, we find three tetramers below each Efimov trimer, and no evidence of four-body universality breaking. Two of these tetramers are in agreement with quantum-mechanical calculations and experimental results.

1. Introduction

Different physical systems are formed by nonrelativistic particles with interactions that have a large scattering length. There, the range of interparticle interactions can be neglected when two conditions are met: the range is much smaller than the scattering length, and interactions occur at sufficiently low energy. When this happens, an interaction is solely described by one parameter, its scattering length.

Under these conditions, there is independence of the detailed form of the interparticle interaction. This can produce universality in the sense that different systems might display the same low-energy behavior. This happens even though the particles (and interactions) that form these systems are very different. For example, particles can be atoms in some cases and nucleons in others. A few parameters (including the scattering length) might be enough to describe these systems. References [1, 2] review systems with these interactions.

The particle-particle interaction is the starting point to describe these systems. This work is restricted to systems of identical scalar particles, hence there is only one particle-particle scattering length. When the absolute value of the scattering length is large, the interaction is close to a zero-energy two-particle bound state. There are two possible cases. **a)** If the scattering length is large and positive, the two-body system has a weakly bound state with energy $-\frac{1}{m\,a^2}$ (a is the scattering length and m is the mass of each particle). **b)** On the other hand, if a is very negative, the two-body system does not support a bound state, but the interaction is almost strong enough to do so. There, two-particle subsystems play an important role in the dynamics of larger systems even though there are no two-particle bound states. A limit sits between these cases: when the scattering length approaches either $+\infty$ or $-\infty$, the system supports a zero-energy bound state. This is know as the unitary (or resonant) limit, which can also be expressed as $1/a \to 0$.

Two physical systems have the features described above, ultracold gases of atoms and low-energy nucleon systems. Experimentalists have an impressive amount of control over ultracold gases of atoms, where temperatures reach $\lesssim 10^{-4}$ Kelvin. In particular, it is possible to control the atom-atom interaction (and its scattering length) in carefully prepared systems. This is achieved by setting an external magnetic field close to a Feshbach resonance [3]. There, the scattering length can be set very close to the resonance, both with positive and negative scattering lengths. Because of this, ultracold gases of atoms provide a benchmark to study low-energy systems and some of their universal behavior.

Low-energy nucleon systems also have some of the features described in this introduction. When the kinetic energy of each nucleon is around or below a few tens of MeV, it is not necessary to use pion-mediated nucleon interactions. Instead, the dynamics can be described by a pionless Effective Field Theory [4, 5]. There, a contact vertex describes the two-nucleon interaction. The S-wave component of this interaction (which dominates at low-energy) has two channels: each channel controlled by a scattering length. The spin-singlet channel $^1S_0$ has scattering length $a_s = -23.7\,\text{fm}$, and the triplet channel $^3S_1$ has $a_t = +5.4\,\text{fm}$. These scattering lengths are larger than the range of interaction, which is roughly 1 fm. Low-energy Effective Field Theories are useful to describe both few- [4, 5] and many-nucleon systems [6]. Furthermore, corrections due to the finite range of the interactions can be added as a series in (r_{eff}/a), where r_{eff} is the effective range of the interaction [7, 8, 4].

In 1970, V. Efimov showed that some three-body systems have a large sequence of weakly bound states, and that the energies of these states are arranged in a geometric sequence [9, 10]. Furthermore, the dimensionless constant in the geometric sequence depends on the ratios of the masses of the particles. For particles of equal mass, this constant is an universal number, $E_{i+1}^{(3)}/E_i^{(3)} = 515.0$.

The Efimov effect arises as a consequence of a large separation between two physical scales. Particularly, a separation between a (very small) two-body interaction range and a (very large) scattering length. Efimov states, or trimers, appear in scales between this range and scattering length. For simplicity, assume we are dealing with nonrelativistic systems of indistinguishable scalar particles with mass m. In natural units, the two-body interaction range, R, sets a *high*-energy scale,[1] $E_R = \frac{R^{-2}}{m}$, and the scattering length, a, sets a *low*-energy scale, $E_a = \frac{a^{-2}}{m}$. If $R \ll |a|$, then $E_R \gg E_a$. All Efimov states have energy between these two scales, $E_R \gg |E_i^{(3)}| \gg E_a$.

Since the energies of Efimov states appear in a geometric sequence, the energy of the lowest state determines the energies of the rest. This introduces a dimensionful, three-body physical parameter, the Efimov parameter. Its value is determined by the detailed form of the microscopic interactions, and in this low-energy framework it is an independent parameter.

The Efimov effect breaks a scale-invariance symmetry. This is because, if Efimov states did not exist, there would not be any physical parameters to set a scale between E_R and E_a. There, the system would be independent of the detailed form of the interactions, and the behavior of the three-body system would be universal. However, the Efimov effect breaks the continuous scaling symmetry into a discrete symmetry realized by the Efimov sequence. In that sense, the Efimov parameter breaks the universality of the three-body system.

Four-body systems are also influenced by the Efimov effect. Numerical quantum-mechanical calculations find two four-body states for each Efimov three-body state (or trimer) [11, 12, 13] and no four-body universality-breaking parameter. Experiments on ultracold gases of atoms also reveal the presence of two four-body states [14] close to an Efimov trimer.

The Functional-Renormalization-Group approach sketched in this article finds three

[1] We work in systems where even though E_R is the high-energy scale, it is sufficiently low to use a nonrelativistic framework.

tetramers for each Efimov trimer [15, 16], but one of them might be an artifact of the approach. The energies of the other two tetramers are in agreement with those in quantum-mechanical calculations [12, 13] and with experimental results [14]. Furthermore, we find no evidence of four-body universality-breaking parameters.

2. Low-energy systems and the Functional Renormalization Group

The Functional Renormalization Group (FRG) is a nonperturbative Quantum-Field-Theoretical tool to integrate low-momentum quantum fluctuations. It was developed by C. Wetterich [17, 18] based on ideas by Wilson. Reference [19] and Chapter 2 in [20] are reviews on the topic.

The FRG uses artificial mass terms, called regulators, to suppress low-momentum quantum fluctuations. The regulators depend on an unphysical momentum scale, k, and they suppress fluctuations with momentum below k.

The FRG starts with an (unphysical) *running action* at an ultra violet scale, K_0. Fluctuations below K_0 are suppressed from this action, but physical information from scales above K_0 is included. A flow equation describes the change in the running action as k changes. We use this equation to evolve the running action, Γ_k, from $k = K_0$ to the physical limit, $k \to 0$. In this limit there are no suppressed fluctuations and the running action becomes the effective action. The latter contains the full dynamics, including all quantum fluctuations.

The flow equation,

$$\partial_k \Gamma_k = -\frac{\mathrm{i}}{2}\,\mathrm{STr}\left(\partial_k \mathbf{R} \cdot \left(\Gamma_k^{(2)} - \mathbf{R}\right)^{-1}\right),$$

depends on the regulators, $\mathbf{R}$, and on the second derivatives of the running action with respect to the fields, $\Gamma_k^{(2)}$. This is a second-order functional differential equation, and there are no general techniques to solve it. However, in some cases, if we use an ansatz for the running action, Γ_k, the flow equation reduces to a system of ordinary differential equations. The functions that solve this system of equations are k-dependent parameters, known as running functions; their k-dependence is given by the FRG flow. In the physical limit, these functions contain dynamical information about the system.

The use of an ansatz instead of solving the full FRG flow equation means that the results are not exact. However, the FRG can provide sensible physical results if the running action ansatz retains the main features of the (unknown) full effective action. Furthermore, FRG results are nonperturbative in the sense that they are obtained without using a series expansion around a small parameter.

3. Running action and three-body physics

To study two-, three-, and four-particle systems we use a running action ansatz that contains nonrelativistic kinetic terms and contact-interaction terms. The ansatz contains a one-particle field, $\psi(x)$; a two-particle (or dimer) field, $\phi(x)$; and a three-particle (or trimer) field, $\chi(x)$. All of these fields have nonrelativistic kinetic terms. Dimer kinetic terms approximately represent the energy dependence of two-particle subsystems, and trimer kinetic terms do a similar job for three-particle subsystems.

Dimer fields are introduced at the UV via a Hubbard-Stratonovich transformation [21]. At the UV, dimers are auxiliary (non dynamical) fields, but the FRG flow gives them dynamics. Trimer fields can also be introduced at the UV as non dynamical fields.

The explicit form of the running action ansatz is

$$\Gamma_k = \int_x \Big\{ \psi^\dagger \Big(\mathrm{i}\,\partial_0 + \frac{\nabla^2}{2m} \Big) \psi + Z_d\, \phi^\dagger \Big(\mathrm{i}\,\partial_0 + \frac{\nabla^2}{4m} - \frac{u_d}{Z_d} \Big) \phi + Z_t\, \chi^\dagger \Big(\mathrm{i}\,\partial_0 + \frac{\nabla^2}{6m} - \frac{u_t}{Z_t} \Big) \chi$$
$$- \frac{g}{2} (\phi^\dagger\, \psi\, \psi + \psi^\dagger\, \psi^\dagger\, \phi) - \lambda\, \phi^\dagger\, \psi^\dagger\, \phi\, \psi - h(\chi^\dagger\, \phi\, \psi + \phi^\dagger\, \psi^\dagger\, \chi)$$
$$- \frac{w}{4}\, \phi^\dagger\, \psi^\dagger\, \psi^\dagger\, \phi\, \psi\, \psi - \frac{v_d}{4} (\phi^\dagger\, \phi^\dagger\, \phi\, \psi\, \psi + \phi^\dagger\, \psi^\dagger\, \psi^\dagger\, \phi\, \phi) - \frac{u_{dd}}{2} (\phi^\dagger\, \phi)^2$$
$$- \frac{v_t}{2} (\phi^\dagger\, \psi^\dagger\, \psi^\dagger\, \chi\, \psi + \chi^\dagger\, \psi^\dagger\, \phi\, \psi\, \psi) - \frac{u_{dt}}{2} (\phi^\dagger\, \phi^\dagger\, \chi\, \psi + \chi^\dagger\, \psi^\dagger\, \phi\, \phi)$$
$$- u_{tt}\, \chi^\dagger\, \psi^\dagger\, \chi\, \psi \Big\}.$$

This contains several k-dependent functions which are determined by solving the FRG flow equation. The objects Z_d, u_d, g are parameters associated with two-body systems. Z_d is the k-dependent dimer field renormalization, u_d is a k-dependent dimer field bilinear, and g is a constant coupling between particles and dimers. The quantities Z_t, u_t, λ, h are three-body parameters, and each of them is k-dependent. Z_t and u_t are a renormalization and a bilinear for the trimer (respectively), whereas λ and h are couplings. Finally, w, v_d, u_{dd}, v_t, u_{dt}, u_{tt} are k-dependent four-body coupling constants.

The flow of two-body parameters is analytically solvable, and it must satisfy one condition in the physical limit: two-particle scattering must produce the correct scattering length [1, 2, 15, 21]. Explicitly, two-particle scattering amplitude at zero energy must yield

$$T_{\psi\psi} \to \frac{8\,\pi\, a}{m}.$$

Using the running action ansatz, the same amplitude yields

$$T_{\psi\psi} = \frac{-g^2}{u_d(k \to 0)}.$$

The two expressions for the amplitude match when

$$u_d(k \to 0) = -\frac{g^2\, m}{8\pi a},$$

which is a boundary condition on u_d.

The flow of three-body running functions reflects the Efimov effect. In the unitary limit, these functions oscillate periodically in $t = \log(k/K_0)$. The periodicity in t corresponds to geometric scaling in units of energy (or momentum). When the flow of couplings oscillates indefinitely in a cycle we say there is a limit cycle. This is in contrast with flows that approach a fixed point.

In the present approach, with the regulators in [15], the Efimov period in t is $\frac{50\pi}{\sqrt{535}} \simeq 6.79$. These cycles are larger than the physical ones, which have period 6.25. In FRG approaches, the exact value of this period depends on an unphysical choice: the explicit form of the regulators, $\mathbf{R}$. References [15, 21] and Chapter 3 of [22] study these three-body systems in an FRG approach.

Out of the unitary limit, the finite scattering length breaks the Efimov periodicity in the flow. The cycles stop when $k \sim 1/|a|$ (in natural units). For fluctuations with momenta smaller than $1/|a|$, the three-body functions settle.

4. Four-body physics

The flow of four-body systems inherits the Efimov periodicity of three-body systems. In the unitary limit, the four-body couplings display an infinite sequence of peaks inside each Efimov cycle [15, 22]. This is known as super Efimov effect, and it arises from the structure of the flow equations. However, not all these peaks correspond to physical four-body resonances.

Out of the unitary limit we find three four-body peaks inside each Efimov cycle. They are labeled as (0), (1), and (2), where (0) is the deepest one, and (2) is extremely shallow. At zero energy these states appear when the two-particle scattering length is

$$\frac{a_4^{(0)}}{a_3} = 0.438 \qquad \frac{a_4^{(1)}}{a_3} = 0.877 \qquad \frac{a_4^{(2)}}{a_3} = 0.997,$$

where a_3 is the scattering length that produces the closest Efimov three-body state above the four-body resonances. The scattering lengths for the deepest of these peaks, (0) and (1) are in 5% agreement with quantum-mechanical calculations [12, 13], and they are also consistent with experimental results [14]. The shallow resonance, (2), quickly merges with the particle-trimer threshold and might be an artifact of using an ansatz. Figure 1 shows a plot for the energies and scattering lengths that produce the four-body resonances.

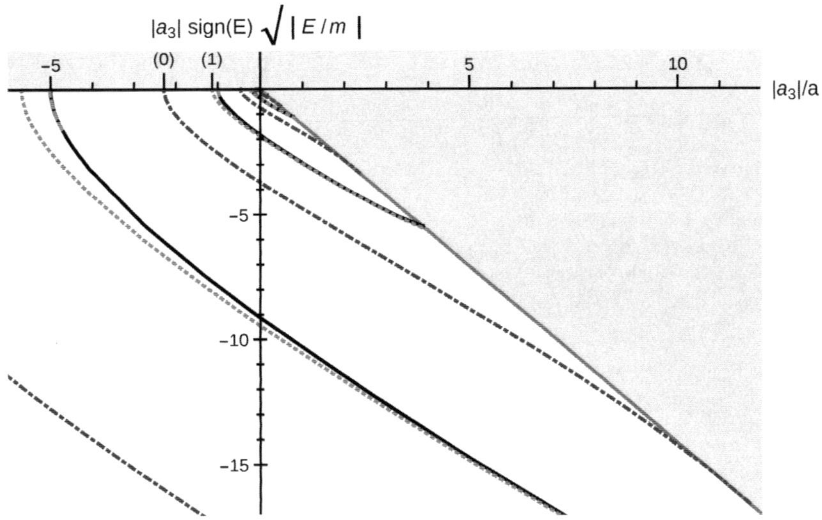

Figure 1. This plot shows the features of the four-body system that appear in the present approach. It displays four-body resonances and particle-trimer thresholds of several Efimov cycles. The blue straight lines below the shaded region represent the dimer-dimer threshold. The black solid curves are the particle-trimer thresholds. The red dot-dashed curve is the lowest four-body resonance below its corresponding Efimov trimer, (0). The green short-dashed curve is the four-body resonance (1). The magenta long-dashed curve is the shallow tetramer, (2), which quickly merges with the particle-trimer threshold. The horizontal axis is the inverse of the two-particle scattering length, $1/a$, but it is normalized by (the absolute value of) the inverse of a two-particle scattering length that produces a trimer, $1/|a_3|$. The vertical axis is proportional to $\text{sign}(E)\sqrt{|E|/m}$, which (in natural units) has units of inverse length. This axis is also normalized by $1/|a_3|$. On the horizontal axis, the ratio between four-body resonances should be 29.8 in our truncation (the physical value is $22.7 \simeq \sqrt{515.0}$). Instead, this number was artificially replaced by 5 to show several Efimov cycles in one plot.

5. Conclusions

The FRG can describe few-body systems with reasonable accuracy. However, to improve these results, it might be necessary to use a more general energy dependence in the running action. It is also necessary to study how much do FRG results depend on unphysical choices in the FRG, in particular, the choice of regulators.

The FRG can be a powerful and computationally-cheap tool to study many-body systems, and few-body FRG studies can provide foundations for many-body studies.

References

[1] Braaten E and Hammer H W 2006 *Phys. Rept.* **428** 259
[2] Platter L 2009 *Few Body Syst.* **46** 139
[3] Chin C, Grimm R, Julienne P and Tiesinga E 2010 *Rev. Mod. Phys.* **82** 1225
[4] Bedaque P F and van Kolck U 2002 *Annu. Rev. Nucl. Part. Sci.* **52** 339
[5] Epelbaum E, Hammer H W and Meißner U G 2009 *Rev. Mod. Phys.* **81** 1773
[6] Drews M and Weise W 2015 *Phys. Rev.* C **91** 035802
[7] Birse M C and McGovern J A and Richardson K G 1999 *Phys. Lett.* B 464 169
[8] Kaplan D B, Savage M J and Wise M B 1998 *Nucl. Phys.* B **534** 329
[9] Efimov V 1970 *Phys. Lett.* B **33** 563
[10] Efimov V 1971 *Sov. J. Nucl. Phys.* **12** 589
[11] Hammer H W and Platter L 2007 *Eur. Phys. J.* A **32** 113
[12] von Stecher J, D'Incao J P and Greene C H 2009 *Nature Phys.* **5** 417
[13] Deltuva A 2010 *Phys. Rev.* A **82** 040701
[14] Ferlaino F, Knoop S, Berninger M, Harm W, D'Incao J P, Nägerl H C and Grimm R 2009 *Phys. Rev. Lett.* **102** 140401
[15] Jaramillo Ávila B and Birse M C 2013 *Phys. Rev.* A **88** 043613
[16] Jaramillo Ávila B and Birse M C 2015 *Phys. Rev.* A **92** 023601
[17] Wetterich C 1991 *Nucl. Phys.* B **352** 529
[18] Wetterich C 1993 *Phys. Lett.* B **301** 90
[19] Berges J, Tetradis N and Wetterich C 2002 *Phys. Rept.* **363** 223
[20] Schwenk A and Polonyi J (editors) 2012 *Renormalization Group and Effective Field Theory Approaches to Many-Body Systems* (Berlin Springer-Verlag)
[21] Schmidt R and Moroz S 2010 *Phys. Rev.* A **81** 052709
[22] Jaramillo Ávila B 2015 *Functional Renormalisation Group and Nuclear Matter* (University of Manchester PhD thesis)

Journal of Physics: Conference Series **761** (2016) 012017 doi:10.1088/1742-6596/761/1/012017

The Belle II experiment: fundamental physics at the flavor frontier

Ivan Heredia de la Cruz
for the Belle II Collaboration
CONACYT – Physics Department, Centro de Investigación y de Estudios Avanzados del IPN (CINVESTAV-IPN) – Mexico City, Mexico
E-mail: iheredia@fis.cinvestav.mx

Abstract. After the major success of B-factories to establish the CKM mechanism and its proven potential to search for new physics, the Belle II experiment will continue exploring the physics at the flavor frontier over the next years. Belle II will collect 50 times more data than its predecessor, Belle, and allow for various precision measurements and searches of rare decays and particles. This paper introduces the B-factory concept and the flavor frontier approach to search for new physics. It then describes the SuperKEKB accelerator and the Belle II detector, as well as some of the physics that will be analyzed in Belle II, concluding with the experiment status and schedule.

1. Introduction

The first generation of B-factories, PEP-II at SLAC and KEKB at KEK, provided e^+e^- collisions at the BaBar and Belle experiments during the last decade. In particular, KEKB reached the world highest instantaneous luminosity of about 2×10^{34} cm^{-2}s^{-1}. Collisions took place at centre-of-mass energies equal to the mass of the $\Upsilon(nS)$, mainly at $m_{\Upsilon(4S)} = 10.58$ GeV, and also at “off-resonance” energies. BaBar and Belle collected an integrated luminosity of about 0.5 and 1 fb^{-1}, respectively. In Belle, 1 fb^{-1} corresponds to nearly 772 million $B\bar{B}$ and 900 million $\tau\bar{\tau}$ pairs.

The discovery potential of B-factories is beyond all possible doubt. For instance, Belle discovered charge-parity violation (CPV) in the B meson system [1, 2] and found direct CPV in $B \to K\pi$ decays [3, 4]. Belle also measured mixing in charm [5] and observed several B and τ rare decays (e.g., see section 4.8), as well as many predicted and unexpected new states, such as the $X(3872)$ [6] and the tetraquark candidate $Z(4430)^+$ [7].

One of the greatest achievements of B-factories was the experimental confirmation of the CKM mechanism, which led to the Nobel Prize in Physics in 2008 awarded to M. Kobayashi and T. Maskawa [8]. This result, normally summarized in the so called unitary triangles as the one in figure 1, shows the impressive agreement of the Standard Model (SM) global fit.

In the light of so many impressive achievements, a second generation of B-factories is eagerly awaited. SuperKEKB and Belle II will supersede KEKB and Belle to extend the exploration of fundamental physics at the flavor frontier.

Journal of Physics: Conference Series **761** (2016) 012017 doi:10.1088/1742-6596/761/1/012017

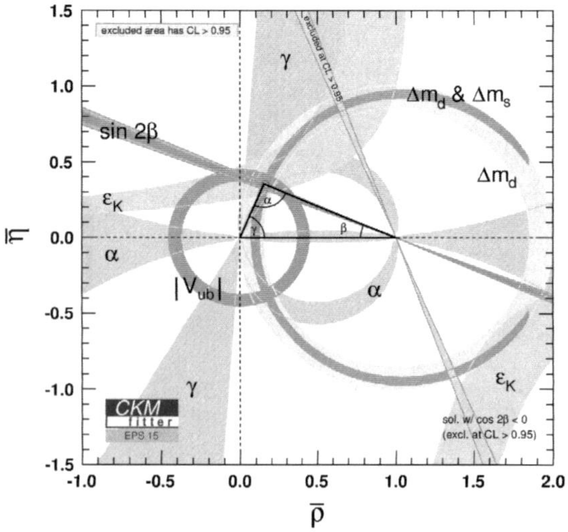

Figure 1. Constrains on the CKM $(\bar{\rho}, \bar{\eta})$ plane from the global SM CKM fit [9].

2. The flavor frontier

There exist two approaches at particle colliders to search for new physics (NP), namely, the energy and flavor (or intensity) frontiers. The first produces directly new particles and is only limited by the beam energy. The second can reveal beyond SM virtual particles appearing in loop diagrams through precise measurements of well known and rare processes. CMS and ATLAS at the LHC, for example, explore the energy frontier using 13 TeV centre-of-mass energy pp collisions, while B-factories (and also LHCb) can test indirectly a mass scale up to 100 TeV within the flavor frontier. If there is NP at a scale of a few TeV, effects will emerge in B, D and τ decays and, thus, will be measured by the LHC experiments. If not, flavor measurements will provide unique ways to find NP. This complementarity is crucial to understanding the need to upgrade Belle to Belle II, and KEK to SuperKEKB. The well defined initial state energy in Belle II allows also for efficient detection of inclusive decay modes (e.g. $b \to s\ell\ell$) and final states with neutral particles and missing energy (e.g. $B \to \tau\nu$), which are complementary features to the LHCb experiment's specialty of detecting decays to charged final states (e.g. $B \to \mu^+\mu^-$).

A new generation of (super) B-factories will make it possible to measure with higher precision almost all the CKM matrix elements through B and D decays. They will also be key to testing beyond SM models by measuring or looking for mixing, CPV, lepton flavor and number violation (LFV and LNV) and rare decays of B, D and τ particles. Furthermore, it is possible to search directly for light particles such as sterile neutrinos, dark photons, light Higgs bosons and axions, and to gain a better understanding of quantum chromodynamics (QCD) by a more complete characterization of open charm, light meson and baryon states, and quarkonium and quarkonium-like exotic states.

There are unique features in B-factories which make them attractive and a real alternative to the energy frontier approach. The most obvious is the much lower background produced at lepton colliders with respect to hadron colliders, implying lower track multiplicity and detector occupancy, and in turn resulting in higher B, D and τ reconstruction efficiencies and easier detection of neutral particles. This point is illustrated in figures 2 and 3. Moreover, there is no need of a strict event filtering (trigger selection) that can bias a measurement and lead to intricate corrections or further systematic uncertainties. The asymmetric beam energies permit a more precise study of time dependent CPV, while the tuning of the beams to specific Υ resonances results in a straightforward bottomonium exploration. Another nice feature is the ability to identify a signal decay (B_{sig}) as the complement of the decay products of the other B meson (B_{tag}) in the event.

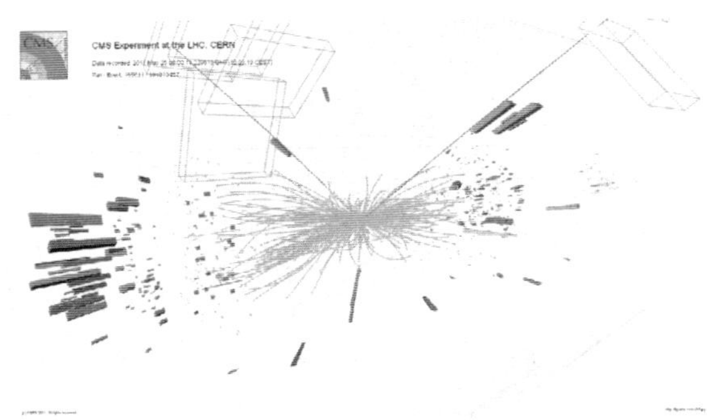

Figure 2. High-multiplicity environment observed in pp collisions. Event recorded by the CMS experiment [10].

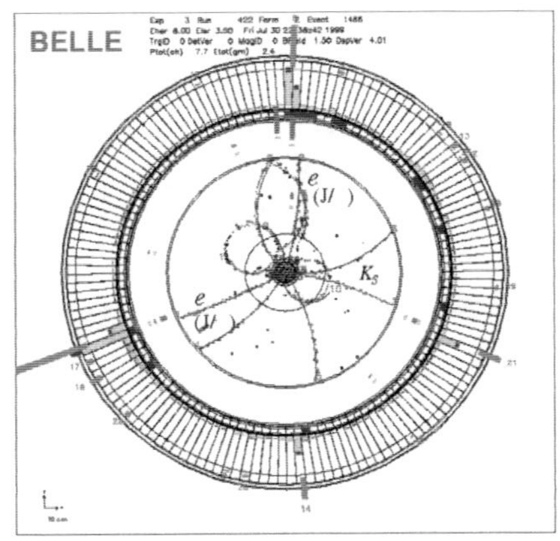

Figure 3. Low-multiplicity environment from e^+e^- collisions. Event recorded by the Belle experiment [11].

3. SuperKEKB and Belle II

The SuperKEKB collider and Belle II detector are shown in figures 4 and 5, and described below.

3.1. *SuperKEKB collider*

SuperKEKB is located at KEK (The High Energy Accelerator Research Organization) in Tsukuba, Japan, and represents a major upgrade to the KEKB accelerator. It consists of two 3 km rings equipped with radiofrequency (RF) systems which accelerate electron and positron beams to 4 and 8 GeV, respectively, and make them collide in the center of the Belle II detector. With respect to KEKB, SuperKEKB has longer dipoles and redesigned magnet lattice to squeeze the emittance, and includes additional (or modifies) RF systems for higher ($\times 2$) beam currents. SuperKEKB reduces ($\times 20$) the beam spot size using a nano-beams scheme achieved with new superconducting final focusing quadrupole magnets near the interaction region, and further increases the luminosity by introducing crab cavities. The last impart a kick to restore head-on collisions for an initial large crossing angle, avoiding long-range beam-beam collisions. The upgraded collider will deliver an integrated luminosity of 50 ab^{-1} and reach a peak luminosity of 8×10^{35} cm^{-2}s^{-1}, about 50 and 40 times KEKB, respectively, meaning about 10^{10} BB or $\tau\tau$ pairs per year.

3.2. *The Belle II detector*

The Belle II detector consists of several subdetectors that, with respect to Belle, were upgraded or replaced for improved detection performance at higher luminosity. A larger tracker improves the impact parameter and secondary vertex resolutions, increases the K_S^0 and pion efficiencies, and provides better flavor tagging. A smaller beryllium double-wall beam pipe ($r = 1.5 \to 1.0$ cm), combined with an innermost silicon pixel layer much closer to the interaction region ($r = 1.4$ cm), improve the impact parameter resolution along the beam line ($\sigma_{d_z} \sim 60 \to 20$ μm). An upgraded

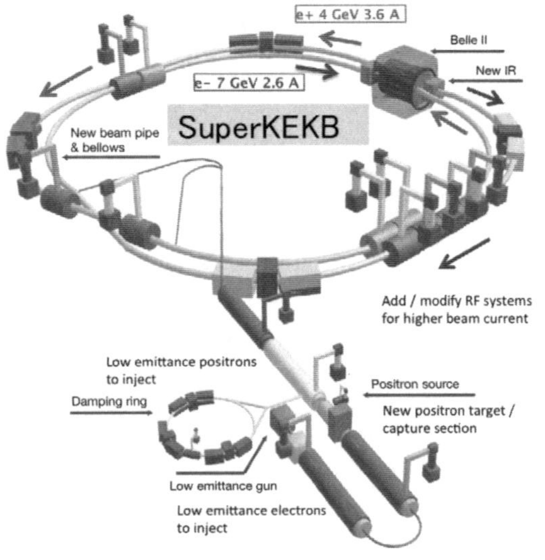

Figure 4. The SuperKEKB collider.

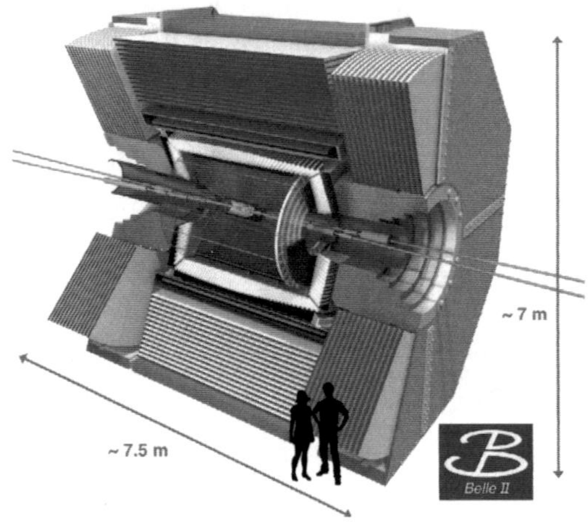

Figure 5. The Belle II detector.

time-of-propagation (TOP) counter and ring-imaging Cherenkov counters with aerogel radiator (A-RICH), together with a faster and more hermertic K_L^0 and μ (KLM) detector, allow a better particle identification and further enhance the flavor tagging and background rejection. Faster and more reliable trigger and data acquisition (DAQ) systems and algorithms permit the operation of Belle II at a much higher event rate.

3.3. Tracking system

The tracking system is formed by the vertex detector (VXD) and the central drift chamber (CDC). The VXD provides precise measurements of the impact parameter of tracks and secondary vertex of short-lived particles, while the CDC complements the VXD for finer track reconstruction and measurement of long-lived particles. The tracker is designed to cope with the particle multiplicity emerging from SuperKEKB collisions (consisting in average of 11 charged tracks, 5 neutral pions and 1 neutral kaon) and to measure the soft momentum of tracks.

The VXD comprises two layers of DEPFET-based pixel detectors (PXD) and four layers of double sided silicon strip detectors (SVD). The energy loss of tracks traversing the PXD and SVD is very low; in terms of radiation lengths, the material budget is about 0.2% and 0.8% X_0, respectively. With about 8 million pixels, the PXD has excellent spacial granularity ($\sigma < 15$ μm); also, the excellent timing ($2-3$ ns) of the 187 SVD sensors provides fast readout. The VXD production is expected to be completed in 2017 and fully installed in spring 2018.

The CDC is made up of 14,336 drift cells with sense wires seated in 56 concentric layers within $r = 16.8 - 111.1$ cm from the interaction point, forming 9 axial/stereo alternating superlayers. Axial layers are parallel to the beam line, while stereo layers are slightly twisted to enable 3-dimensional reconstruction. The CDC was completed in 2015 and commissioning with cosmics rays continues to date. Full CDC operation is expected for the 2017 data taking period.

3.4. Particle identification system

Two Cherenkov counter technologies exist in Belle II to enable particle identification (PID) in the barrel (central) and endcaps (forward) regions. The TOP counter in the barrel region consists of 16 modules located in the gap between the CDC and the calorimeter. A TOP module is composed of two quartz bars instrumented with an array of photodetectors, a mirror and an extension prism that collect the Cherenkov radiation of the charged particle at the edge of the bars. Photodetectors are micro channel plate photomultipliers (PMT) which provide ultra-fast (~ 50 ps) and hight sensitivity photon detection.

Both, the time of propagation of the Cherenkov light throughout the TOP bar and the detection angle (PMT channel), are needed to create the space-time image that classifies between pions and kaons. In the forward region, in contrast, PID is obtaind directly by the Cherenkov rings created by particles passing through the A-RICH radiator. The last consists of two adjacent layers of aerogel with different refractive index to increase the photon yield (by overlapping the Cherenkov rings) detected by hybrid avalanche photo detectors (HAPD).

The TOP and A-RICH counters are already built, and will be installed and operational in Belle II by 2017.

3.5. Calorimeter, K_L^0 and muon detectors
Already installed in Belle II, an electromagnetic calorimeter (ECL) is located inside a superconducting solenoid coil that provides a 1.5 T magnetic field. The ECL uses CsI(Tl) crystals and CsI in the barrel and endcaps regions, respectively, and is equipped with faster readout electronics with respect to Belle in order to compensate for larger backgrounds. Outside of the solenoid coil, an iron flux-return is instrumented to detect K_L^0 mesons and to identify muons (KLM). The KLM consists mainly of resistive plate chambers (RPC) in the barrel region and of scintillators in the endcaps region. K_L^0 mesons can shower hadronically in the absorptive iron plates and be detected in the KLM active detectors.

3.6. Trigger and data acquisition systems
Bhabha scattering and $\gamma\gamma$ processes, as well as intrabeam (Touschek) scattering, are expected to be major sources of background in Belle II. To deal in general with the much higher physics rate, a two-stage trigger system based on hardware (L1) and software (HLT) will be implemented to reduce the rate to less than 10 kHz, recording only collision events of interest. The DAQ system gathers the information from the subdetectors and produces events of about 200 kB, implying a storage rate around 1.5 GB/s. Belle II makes use of a Grid computing model based on various levels, called tiers, to distribute the raw data storage, processing, Monte-Carlo simulation production and further reprocessing to several data centers and computer cluster sites in more than 17 countries.

4. Physics cases
Belle II has a rich physics program that is well documented elsewhere [12] and is currently divided in nine topics developed by corresponding working groups: (1) leptonic and semileptonic B decays; (2) radiative and electroweak penguins; (3) $\phi_{1,2}$; (4) ϕ_3; (5) charmless hadronic B decays; (6) open charm decays; (7) quarkonium and quarkonium-like states; (8) τ, low multiplicity and electroweak; and (9) NP. Some analyses are presented in the following sections. No attempt is made to describe them in detail, but only to show the discovery potential of Belle II.

4.1. $B \to \ell\nu_\ell$
The decay of a B^+ ($u\bar{b}$) meson to a lepton and neutrino takes place by the exchange of a W^+ boson; however, it could also be mediated by a charged Higgs boson inspired from two-Higgs-doublet (2HDM) or Supersymmetry models [13, 14]. Using inputs from other decay modes and lattice QCD, it is found that $\mathcal{B}(B^+ \to \tau^+\nu_\tau) = 0.76^{+0.08}_{-0.06} \times 10^{-4}$ [9]. This prediction can be modified by a factor that depends on the ratio of the Higgs doublet vacuum expectation values,

normally called $\tan\beta$. Measurements of $\mathcal{B}(B^+ \to \tau^+\nu_\tau)$ by BaBar and Belle are consistently higher than expected, with a current average of $0.848^{+0.036}_{-0.055} \times 10^{-4}$ [9]. The measurement by Belle II is expected to be more precise than this average and, together with the decay $B^+ \to \mu^+\nu_\mu$, will impose tight constrains on charged Higgs models with large $\tan\beta$ and relatively small mass m_{H^+} of the charged Higgs.

4.2. $B \to D^{(*)}\ell\nu_\ell$

The ratios $R(D^{(*)}) = \mathcal{B}(B \to D^{(*)}\tau\nu)/\mathcal{B}(B \to D^{(*)}\ell\nu)$, where $\ell = \mu, e$, can also constrain charged Higgs models through lepton universality violation (LUV) tests. While BaBar found incompatible values of $x = \tan\beta/m_{H^+}$ for $R(D)$ and $R(D^*)$ [15], largely disfavoring the 2HDM hypothesis, Belle found agreement at $x \approx 0.5$ GeV^{-1} [16]. A recent measurement by LHCb [17] is in good agreement with both BaBar and Belle results, but their combination is in disagreement at a level 4.0σ with the SM [18], as shown in figure 6. Belle II will reduce the uncertainty of these measurements by an order of magnitude, potentially resolving between the SM and 2HDM models.

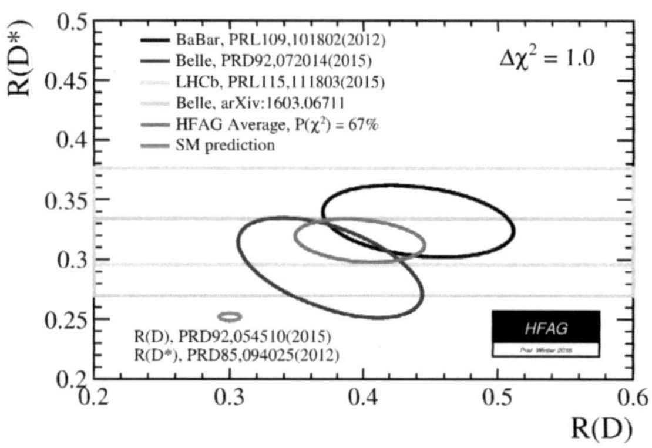

Figure 6. SM prediction and measured values of $R(D)$ and $R(D^*)$ by BaBar, Belle and LHCb [18].

4.3. Second-class current decays

It is worth noting that the measurement of yet undiscovered second-class currents [19] was proposed during this Workshop to improve the bounds on a charged Higgs, certainly within reach of Belle II through $\tau^- \to \pi^-\eta^{(\prime)}\nu_\tau$ decays [20, 21].

4.4. Direct CPV in $B \to K\pi$ decays

Tension in differences between CP asymmetries of B decay modes, e.g. $\Delta A = A_{CP}^{B^0\to K^+\pi^-} - A_{CP}^{B^+\to K^+\pi^0} = -0.122 \pm 0.022$ [22], could be a hint of NP in the electroweak penguin operator, or just a result of hadronic effects. To investigate this issue, a model independent sum rule can be used to test the SM expectations [23]. This requires the measurement of A_{CP} in neutral final state modes (full isospin analysis), where Belle II will have better sensitivity than LHCb by about a factor of two.

4.5. Unitary triangle

Belle II will further over-constrain the unitary triangle (UT) in figure 1 and test the CKM mechanism for CPV at the percent level, reducing the world average values of the UT angles by about a factor of four.

4.6. Analysis of the inclusive process $b \to s\ell^+\ell^-$
Recent measurements by LHCb of the observables in the angular distribution of flavor changing neutral current (FCNC) exclusive decays, $B \to X_s\ell^+\ell^-$, present several discrepancies which can be explained by a wide range of NP scenarios [24]. Either the confirmation of these anomalies or the discrimination among NP models require more precise measurements. Belle II can take advantage of its ability to reconstruct inclusive decays to perform the measurement of the inclusive branching fraction of the process $b \to s\ell^+\ell^-$ and the forward-backward asymmetry of the final state leptons.

4.7. Quarkonium
Systems composed of a quark and antiquark of the same flavor ($q\bar{q}$), called quarkonia, have been extensively studied by Belle and many other experiments using several production mechanisms. A quarkonium is an excellent laboratory where perturbative and non-perturbative QCD models can be tested reliably. The large data sample to be collected by Belle II represents a new opportunity to continue this research field, including the search for and characterization of states clearly containing a $q\bar{q}$ quark pair, but with properties that deviate from theoretical expectations for quarkonium spectroscopy. Such non-conventional or "exotic" states have been subject of extensive research in recent years (since the discovery by Belle II of the X(3872) in 2003 [6]) and point to the existence of tetraquarks, molecules, hybrids, etc.

4.8. Lepton flavor violation
Lepton flavor violating (LFV) decays are highly suppressed in the SM and, therefore, represent a clean null test of the SM. Several NP scenarios increase the branching fraction of LFV decays to order of magnitudes that are accesible to current and coming experiments. Belle II sensitivity for LFV decays is over 100 times greater than its predecessor for the cleanest channels (e.g. $\tau \to 3\ell$) and about 10 times for channels where irreducible backgrounds play a major role (e.g. $\tau \to \ell\gamma$). Figure 7 summarizes the status of searches for LFV in τ decays.

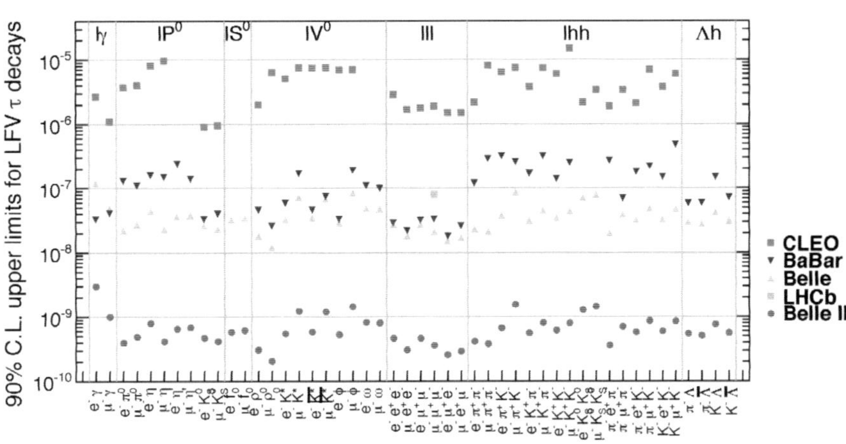

Figure 7. Upper limits on τ LFV branching fractions [12].

5. The Belle II collaboration and experiment schedule
The Belle II collaboration consist of about 650 collaborators from 100 institutions and 23 countries. A joint theory-experiment effort, called the Belle II Theory Interface Platform

(B2TIP), was established to study and advance the prospects of the Belle II physics program.

The Belle II schedule can be divided in three phases. Beams started circulating in SuperKEKB in 2016 as part of Phase-1, with a commissioning detector, named BEAST II, to measure the beam backgrounds. During Phase-2 (2017) SuperKEKB will initiate and tune collisions, while Belle II will operate without the VXD. Finally, in Phase-3 (2018) the complete Belle II detector will start taking physics data. The instantaneous luminosity is expected to evolve over time and to reach its peak value in 2022. The goal of Belle II is to collect an integrated luminosity of 50 ab^{-1} by 2024.

6. Summary

The first generation of B-factories were key to establishing the CKM mechanism. In the process, several hints of NP and unexpected states were found. In the following years, the Belle II and the LHC experiments will complement and verify each other to find the much-expected NP. Belle II is not only Belle with higher luminosity; there are also many improvements in the collision scheme, detector, trigger system, reconstruction algorithms, etc. Belle II will start operations in 2017 without the VXD, and a fully functional detector is expected in 2018 that will take data until collecting 50 ab^{-1}.

Acknowledgments

IHC would like to thank to Red-FAE-CONACyT and the CINVESTAV Physics Department for the funding provided to attend to the XV Mexican Workshop on Particles and Fields, and to CONACyT and SEP for the funding provided by means of projects FOINS-296 and CB-254409.

References

[1] Abe K *et al.* (Belle) 2002 *Phys. Rev.* **D66** 071102 (*Preprint* hep-ex/0208025)
[2] Abe K *et al.* (Belle) 2001 *Phys. Rev. Lett.* **87** 091802 (*Preprint* hep-ex/0107061)
[3] Lin S W *et al.* (Belle) 2008 *Nature* **452** 332–335
[4] Chao Y *et al.* (Belle) 2004 *Phys. Rev. Lett.* **93** 191802 (*Preprint* hep-ex/0408100)
[5] Staric M *et al.* (Belle) 2007 *Phys. Rev. Lett.* **98** 211803 (*Preprint* hep-ex/0703036)
[6] Choi S K *et al.* (Belle) 2003 *Phys. Rev. Lett.* **91** 262001 (*Preprint* hep-ex/0309032)
[7] Choi S K *et al.* (Belle) 2008 *Phys. Rev. Lett.* **100** 142001 (*Preprint* hep-ex/0708.1790)
[8] Nobel Media AB 2014 *Nobelprize.org*
URL http://www.nobelprize.org/nobel_prizes/physics/laureates/2008/press.html
[9] Charles J *et al.* 2015 *Phys. Rev.* **D91** 073007 (*Preprint* hep-ph/1501.05013)
[10] Taylor L and McCauley T 2011 CMS collision events (7 TeV): candidate ZZ to 2e+2mu
URL https://cds.cern.ch/record/1374433
[11] Belle Collaboration URL http://www2.kek.jp/ja/news/archive/kekb-00jan-j.html
[12] Golob B, Trabelsi K and Urquijo P 2015 BELLE2-NOTE-0021
URL https://belle2.cc.kek.jp/ twiki/pub/B2TiP/WebHome/belle2-note-0021.pdf
[13] Hou W S 1993 *Phys. Rev.* **D48** 2342–2344
[14] Akeroyd A G and Recksiegel S 2003 *J. Phys.* **G29** 2311–2317 (*Preprint* hep-ph/0306037)
[15] Lees J P *et al.* (BaBar) 2013 *Phys. Rev.* **D88** 072012 (*Preprint* hep-ex/1303.0571)
[16] Huschle M *et al.* (Belle) 2015 *Phys. Rev.* **D92** 072014 (*Preprint* hep-ex/1507.03233)
[17] Aaij R *et al.* (LHCb) 2015 *Phys. Rev. Lett.* **115** 111803 (*Preprint* hep-ex/1506.08614)
[18] Heavy Flavor Averaging Group (HFAG) Winter 2016
URL http://www.slac.stanford.edu/xorg/hfag/semi/winter16/winter16_dtaunu.html
[19] Weinberg S 1958 *Phys. Rev.* **112**(4) 1375–1379
[20] Roig P 2016 *15th Mexican Workshop on Particles and Fields (MWPF 2015) Mazatlan, Mexico, November 2-6, 2015* (*Preprint* hep-ph/1608.02538)
[21] Escribano R, Gonzalez-Solis S and Roig P 2016 *Phys. Rev.* **D94** 034008 (*Preprint* hep-ph/1601.03989)
[22] Heavy Flavor Averaging Group (HFAG) July 2016
URL http://www.slac.stanford.edu/xorg/hfag/rare/July2016/acp/OUTPUT/TABLES/acp.pdf
[23] Gronau M 2005 *Phys. Lett.* **B627** 82–88 (*Preprint* hep-ph/0508047)
[24] Descotes-Genon S, Hofer L, Matias J and Virto J 2016 *JHEP* **06** 092 (*Preprint* hep-ph/1510.04239)

Lepton number violation in $\tau^- \to \pi^+\mu^-\mu^-\nu_\tau$ at Belle II experiment

D. Rodríguez Pérez, P. L. M. Podesta Lerma, Isabel Domínguez Jiménez

Universidad Autónoma de Sinaloa, Facultad de Ciencias Físico-Matemáticas, Av. de las Américas y Blvd. Universitarios, Cd. Universitaria, CP 80000, Cln, Sinaloa, México

E-mail: davidrope@uas.edu.mx

Abstract. The nature of neutrino masses is one of many questions opened about the Standard Model (SM) of elementary particles. The existence of lepton number violating processes, where it changes by two units implies the presence of a (heavy) Majorana neutrino. High precision measurements are essential to this kind of searches. In the B factory experiment Belle II is possible to find these kind of processes, we implemented the simulation and reconstruction of

$$\tau^- \to \pi^+\mu^-\mu^-\nu_\tau$$

using belle2 software bas2f to investigate the feasibility of this channel.

1. Introduction

The SM is the current theory of fundamental particles and their interactions, which was created based on observations and theoretical predictions through several years. In this theory all the elementary matter particles are fermions, interacting among themselves by the exchange of gauge bosons and the mass is generated by the Higgs mechanism.

But, despite its tremendous success in describing the basic forces of nature many fundamental questions remain unanswered within the SM. Thus, considering the open questions that in the SM remain unanswered, it is fair to conclude that the present theory is an extremely successful but phenomenological description of subatomic processes at the energy scales up to O(1 TeV), several issues remain to be explained such as the mass hierarchy, the nature of the neutrino, the nature of dark matter and dark energy, the absence of hierarchy and the validity of the SM up to higher energy scales, several approaches had been proposed such as supersymmetry, WIMPs, and Majorana neutrinos, the last one is the one we will use to try to extend our understanding of neutrinos.

At the energy frontier the main representatives are the ATLAS and the CMS experiments at the Large Hadron Collider (LHC) at CERN [1]. The second is the rate/precision frontier, exemplified by the LHCb experiment at the LHC and the Belle II experiment at SuperKEKB [2].

At the energy frontier, the LHC experiments will be able to discover new particles produced in proton-proton collisions at a center-of-mass energy of up to 14 TeV if there is any. Since the constituent gluons or quarks interact in these collisions, only a fraction of the center-of-mass energy available to produce such new particles, and the mass reach is limited to O(1 TeV/c^2).

Sensitivity to the direct production of a specific new particle depends on the cross section and on the size of the data sample.

At the rare/precision frontier, the observable signatures of new particles or processes can be obtained through measurements of flavor physics reactions at lower energies and evidence of a deviation from the SM prediction. An observed discrepancy can be interpreted in terms of NP models. This is the approach of Belle II.

2. Lepton number violation

Experiments with neutrinos [3] [4] have questioned the SM, since this one considers that neutrinos are massless. Solar neutrino experiments [5] show that they oscillate from a definite flavor to another through long distances, thus the most reasonable explanation is that neutrinos should have mass. However, these neutrino oscillations don't give information about absolute neutrino masses. Thus, there exist different mechanisms to obtain neutrino masses. The requirement of lepton number violation is necessary to generate neutrino masses in the case of Majorana nature.

The double-β decay process without neutrinos [6] is the most studied process that violates the lepton number by two units. Until now, it has not been observed. But besides of the neutrinoless double beta decay, other processes had been searched at LHCb at CERN [1], BaBar at the SLAC [7] and Belle at KEKB [8]. They had imposed limits in the branching fraction to different decays, see Figure 1.

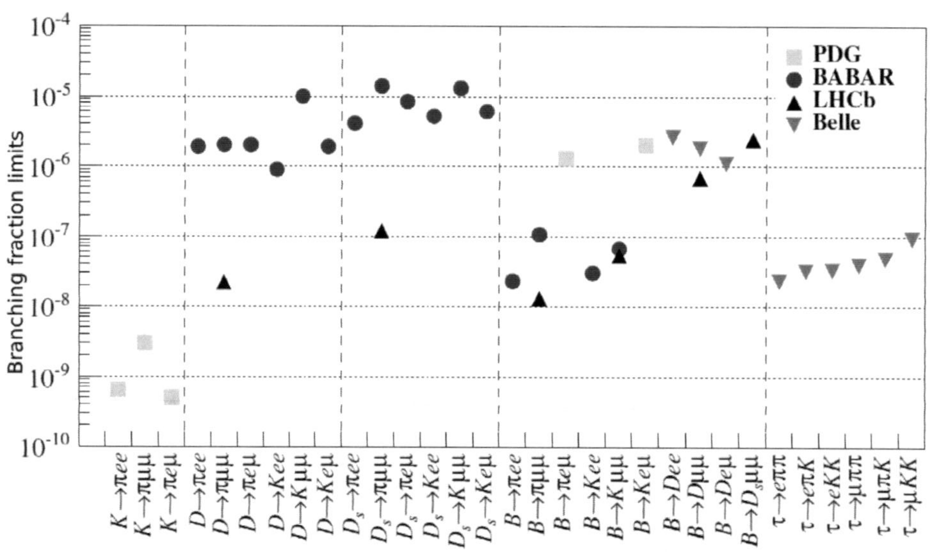

Figure 1. Branching fraction limits in channels where $\Delta L = 2$.

The Majorana neutrino mass (m_N) in the range of $\sim$ (100 GeV - 1 TeV) can explain the suppressed channels. However, the m_N is not determined and other cases can exist. In particular, when m_N in the range of $\sim$ (MeV to few GeV) the branching fractions can be enhanced for decays to four bodies (for more details see [9]). We focus in semi-leptonic decays, specifically on $\tau^- \to \pi^+\mu^-\mu^-\nu_\tau$ channel. We will study the feasibility of searching this channel in the Belle II detector.

Decay processes that violate the total lepton number $\Delta L = 2$ can be induced by the exchange of Majorana neutrinos. We consider a scenario where these decays are dominated by the exchange of only one heavy neutrino, which produces an enhancement of the decay amplitude via the resonant mechanism and the high luminosity in the Belle II experiment allows to study the $\tau^- \to \pi^+\mu^-\mu^-\nu_\tau$ channel. The resonant contribution of a heavy Majorana neutrino to $\Delta L = 2$ four-body decays of τ lepton provides additional information and complementary constrains on the parameter space of Majorana neutrino.

3. Belle II

The B factory experiment Belle II at the SuperKEKB [2] accelerator in Tsukuba, Japan, has confirmed the Kobayashi-Maskawa mechanism of CP violation in the SM. Furthermore, it has a rich physics program of τ leptons. The aim of e^+e^- colliders is to accumulate 50 ab^{-1}, corresponding to about 47 billion $\tau^+\tau^-$ events, by the year 2022. This corresponds to 50 times the luminosity of its predecessor KEKB.

3.1. Accelerator

SuperKEKB uses the same two rings that were used by KEKB: a low-energy ring (LER) for positrons and a high-energy ring (HER) for electrons. These are located side by side 11 meters below the ground level in the TRISTAN tunnel, which has a circumference of around 3 km. However, the beam energies have been changed from the present values of 3.5 and 8.0 GeV to 4.0 and 7.0 GeV. In SuperKEKB the Nano-Beam scheme is used [2], in which the emittance growth due to intra-beam scattering and the short beam lifetime due to the Touschek effect and both are serious problems, particularly in the LER. The increase in the beam energy of the LER from 3.5 to 4.0 GeV helps mitigate these problems. The decrease in the beam energy of the HER from 8.0 to 7.0 GeV allows a lower emittance.

In the interaction region, the 7 GeV electrons in the HER and the 4 GeV positrons in the LER collide at one interaction point (IP) with a non-zero crossing angle of 83 mrad.

4. Luminosity

The center-of-mass (CM) energy corresponds to the mass of the $\Upsilon(4S)$ resonance,

$$E_{CM} = 2\sqrt{E_{HER}E_{LER}} = 10.58GeV \sim M_{\Upsilon(4S)}\,. \tag{1}$$

The production cross section for $\Upsilon(4S)$ (bound $b\bar{b}$ state) at the CM is 1.1 nb, 1.3 nb for $c\bar{c}$ pair production, and 2.1 nb for $q\bar{q}$ pair production. Where q is a u, d, or s quark. Cross sections of various processes are summarized in Table 1.

Table 1. Cross-section for various processes in e^+e^- collisions at $\sqrt{s} = 10.58$ GeV. QCD refers to Bhabha and radiative Bhabha processes.

Process	σ[nb]
$b\bar{b}$	1.1
$c\bar{c}$	1.3
$q\bar{q}$ $(q = u, d, s)$	2.1
$\tau^+\tau^-$	0.93
QED $(25.551° < \theta < 159.94°)$	37.8
$\gamma\gamma$	11.1

The rate of interaction in e^+e^- collisions is proportional to the interaction cross section, $\sigma_{interaction}$:

$$\frac{dN_{\text{interaction}}}{dt} = \mathcal{L}_{e^+e^-}\sigma_{\text{interaction}}\,, \tag{2}$$

where the coefficient $\mathcal{L}_{e^+e^-}$ is the luminosity of the accelerator and is a measure of the number of particles per unit area per unit time at the IP,

$$\mathcal{L}_{e^+e^-} = \nu n \frac{N_{e^+}N_{e^-}}{A} \tag{3}$$

here, n is the number of bunches in the storage ring composed of N_{e^+} positrons and N_{e^-} electrons that collide ν times per second and A is the overlapping area of the two colliding beams. The luminosity goal of SuperKEKB is $8 \times 10^{35}\text{cm}^{-2}\text{s}^{-1}$, which is around 50 times as large as the peak luminosity achieved by the KEKB collider.

5. Reconstruction

The main problem in reconstructing the channel is the background: we have two kinds of background combinatorial and contamination from other physics decays that have similar topology as a first approximation we generated three samples one for signal and two for background. We used the **KKMC** as the generator and the Belle II software called **basf2** to simulate the detector response. We generate 50K events for the each one of the channels listed in table. The cuts we used are invariant mass window of the tau candidate 1.0 GeV $< M_\tau <$ 1.9 GeV probability of identification in this case is the global probability, for the muon $PID_\mu >$ 20% and for the pion $PID_\pi >$ 80%, since the muons have to travel all the detector and need to be identified in the KLM. And finally a beam constraint $M_{bc} \leq$ 4 GeV. The reconstructed candidates for each channel are also showed in the table 2. In the figure 2 we showed the result for the signal for some variables. Similar results are obtained for the other two channels. With this we see that combinatorial background is quite low as expected, since this is the ideal case.

Channel	Reconstructed events	expected BR
$\tau^- \to \pi^+\mu^-\mu^-\nu_\tau$	11680	$< 10^{-5}$%
$\tau^- \to \pi^+\mu^-\mu^+\nu_\tau$	10834	$\sim 10^{-6}$%
$\tau^- \to \pi^+\pi^-\pi^-\nu_\tau$	11116	9.31 ± 0.06%

Table 2. Reconstructed events; mass, PID and M_{bc} cuts are used.

We used the background events and look for contamination, we called them **false signals**.The used cuts in these searches were the same that we used to find the simulated signal: 1.0 GeV $< M <$ 1.9 GeV invariant mass window to reconstruct τ's; PID at $\geq$ 80% and $\geq$ %20 for π's and μ's, respectively; and $M_{bc} \leq$ 4.0 GeV. The results are in Figure 3

6. Conclusions

In Belle II will be possible to determine the neutrino nature. If neutrinos are Majorana particles we can use it to search violation of leptonic number process.

We simulated the $\tau^- \to \pi^+\mu^-\mu^-\nu_\tau$ which is a decay to four bodies and its branching fraction is enhanced with respect to decays to three bodies. The feasibility study of this decay in Belle II has been started and looks like very promising, more studies need to be done to find all sources of background and the understanding of them will determine the reach of this search for new physics.

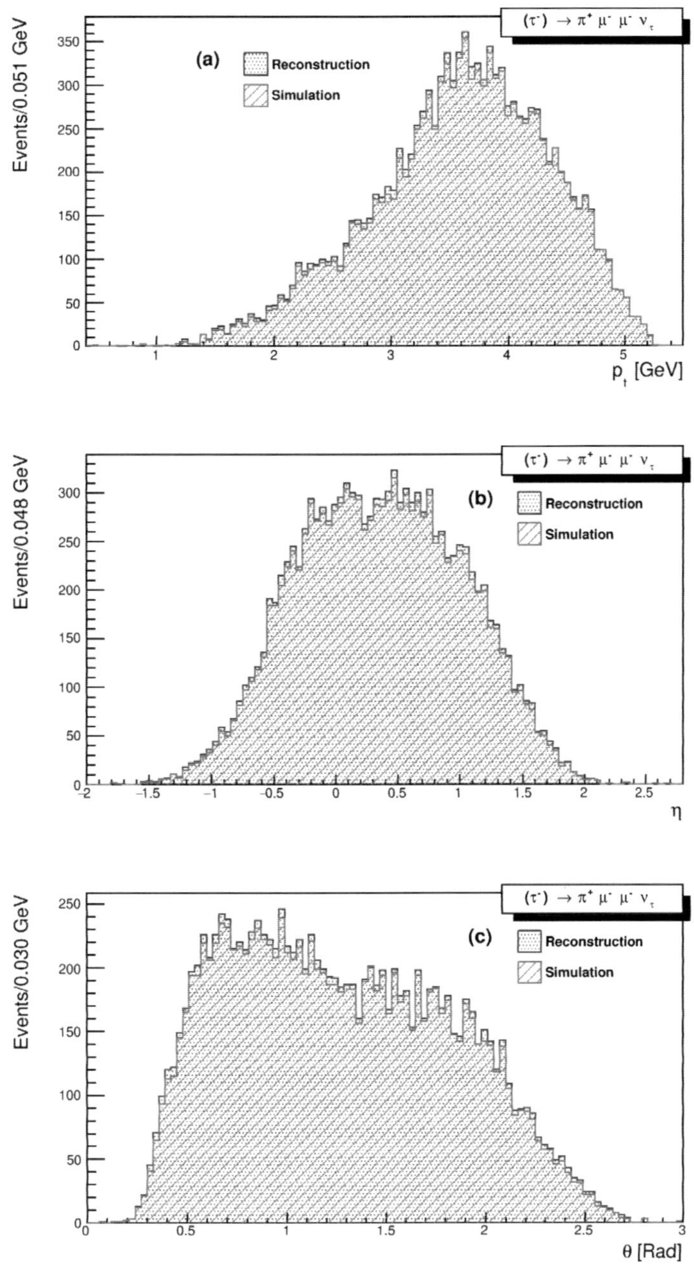

Figure 2. Transverse momentum (top), pseudorapidity (middle) and polar angle (bottom) for the decay channel $\tau^- \to \pi^+\mu^-\mu^-\nu_\tau$ with an invariant mass window of 1.0 GeV $< M_\tau <$ 1.9 GeV, PID ($\mu >$ 20% $\pi >$ 80%), $M_{bc} \leq$ 4 GeV, using 11,991 reconstructed events (blue) and 11,680 simulated events (red).

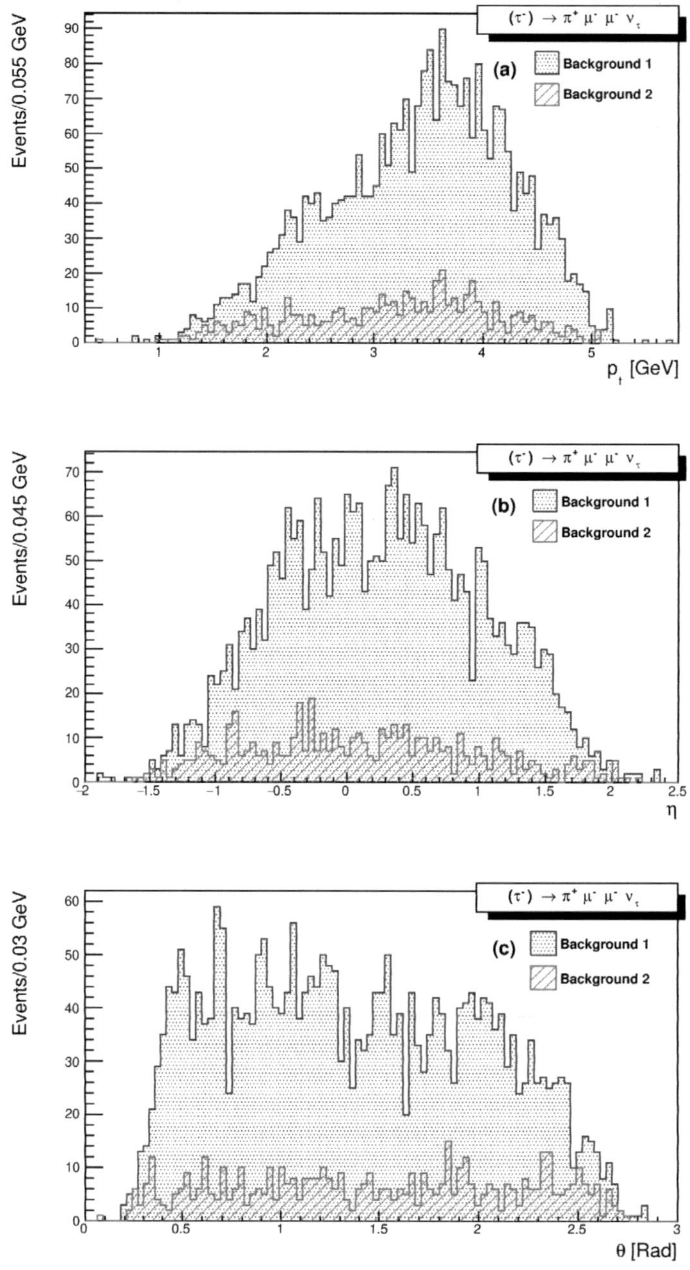

Figure 3. Transverse momentum (top), pseudorapidity (middle) and polar angle (bottom) for the decay channel $(\tau^-) \rightarrow \pi^+\mu^-\mu^-\nu_\tau$ with an invariant mass window of 1.0 GeV $< M_\tau <$ 1.9 GeV, PID ($\mu > 20\%$ $\pi > 80\%$), $M_{bc} \leq 4$ GeV, using 2,825 for background 1 (blue) and 537 for background 2 (red).

Journal of Physics: Conference Series **761** (2016) 012018 doi:10.1088/1742-6596/761/1/012018

References

[1] Brning O S, Collier P, Lebrun P, Myers S, Ostojic R, Poole J and Proudlock P 2004 *LHC Design Report* (Geneva: CERN) URL https://cds.cern.ch/record/782076
[2] Belle group P D I 2010 *Belle II, Technical Design Report*
[3] Mann A K and Primakoff H 1977 *Phys. Rev. D* **15**(3) 655–665 URL http://link.aps.org/doi/10.1103/PhysRevD.15.655
[4] Frati W, Gaisser T K, Mann A K and Stanev T 1993 *Phys. Rev. D* **48**(3) 1140–1149 URL http://link.aps.org/doi/10.1103/PhysRevD.48.1140
[5] Kuo T K and Pantaleone J 1990 *Phys. Rev. D* **41**(12) 3842–3845 URL http://link.aps.org/doi/10.1103/PhysRevD.41.3842
[6] Bilenky S M 2010 690–715 ISSN 1063-7796 (*Preprint* 1001.1946) URL http://arxiv.org/abs/1001.1946
[7] Aubert B *et al.* (BaBar) 2002 *Nucl. Instrum. Meth.* **A479** 1–116 (*Preprint* hep-ex/0105044)
[8] Sagawa H 1996 *Il Nuovo Cimento A* **109** 1055–1060 ISSN 0369-3546 URL http://dx.doi.org/10.1007/BF02823646
[9] Quintero N 2014 *Estudios de violación del número leptónico en procesos resonantes inducidos por un neutrino de mayorana* Ph.D. thesis

B_s lifetime measurement in the CP-odd channel $B_s^0 \to J/\psi f_0(980)$ by the D0 experiment

Michel Hernández Villanueva, on behalf of the D0 collaboration
CINVESTAV, Mexico City 07360, Mexico

E-mail: emhernand@fis.cinvestav.mx

Abstract. The lifetime measurement of the B_s^0 meson in the decay channel $B_s^0 \to J/\psi\ f_0(980)$ is reported, which is a CP-odd state. Using 10.4 fb^{-1} of data collected with the D0 detector in Run II of the Tevatron, the lifetime of the CP-odd component of the B_s^0 meson is measured to be $\tau(B_s^0) = 1.70 \pm 0.14$ (stat) $\pm\, 0.05$ (syst) ps.

1. Introduction

The $B_s^0 \to J/\psi f_0(980)$ channel is a pure CP-odd eigenstate decay. Indeed, the B_s^0 is a spin 0 state, the resonance $f_0(980)$ has $J^{PC} = 0^{++}$ and the J/ψ has $J^{PC} = 1^{--}$. For this reason, this channel is a very good alternative to the well known golden channel $B_s^0 \to J/\psi\phi$ to study the B_s^0 mixing, because the last one requires an angular analysis to obtain the CP-even and CP-odd contributions of the eigenstates [1]. The $B_s^0 \to J/\psi f_0(980)$ decay has been observed by the LHCb, Belle, CDF and D0 collaborations [2, 3, 4, 5]. Due to its CP-odd nature, in the absence of CP violation in mixing, the final state can be produced only by the decay of the B_s^0 heavy mass eigenstate, and the measurement of the $B_s^0 \to J/\psi f_0(980)$ lifetime can be translated into a measurement of the heavy mass eigenstate lifetime. In 2011, the CDF collaboration reported a first measurement of the lifetime [4], obtaining $c\tau = (510 \pm 36 \pm 9)$ μm. After that, the LHCb collaboration reported a lifetime of $c\tau = (510 \pm 12 \pm 8)$ μm [6].

In this talk, a measurement by the D0 collaboration of the B_s^0 lifetime in the CP-odd channel $B_s^0 \to J/\psi f_0(980)$ is reported.

2. B_s^0 mass and CP eigenstates

The B_s^0 (with quark content $\bar{b}s$) and $\bar{B}_s^0$ (quark content $b\bar{s}$) mesons are produced as flavor eigenstate at colliders. However, neither conservation law prevents B_s^0 and $\bar{B}_s^0$ from having transitions. Therefore, they oscillate between themselves. After the production, they propagate as mass eigenstates

$$|B_L\rangle = p|B_s^0\rangle + q|\bar{B}_s^0\rangle; \quad |B_H\rangle = p|B_s^0\rangle - q|\bar{B}_s^0\rangle, \tag{1}$$

being p and q complex parameters with the normalization condition $|p|^2 + |q|^2 = 1$. The state $|B_H\rangle$ corresponds to the heavy mass eigenstate and the state $|B_L\rangle$ to the light one.

On the other hand, CP transformation exchanges the B_s^0 and $\bar{B}_s^0$ states

$$CP|B_s^0\rangle = e^{i\xi}|\bar{B}_s^0\rangle; \quad CP|\bar{B}_s^0\rangle = e^{-i\xi}|B_s^0\rangle. \tag{2}$$

Journal of Physics: Conference Series **761** (2016) 012019 doi:10.1088/1742-6596/761/1/012019

Based on this, eingenstates of CP can be defined

$$|B_{\pm}\rangle = \frac{1}{\sqrt{2}}(|B_s^0\rangle \pm e^{i\xi}|\bar{B}_s^0\rangle), \tag{3}$$

calling $|B_+\rangle$ and $|B_-\rangle$ the even and odd eigenstates, respectively. If CP violation is neglected in mixing, it can be shown that $q/p = \pm e^{i\xi}$, giving

$$CP|B_H\rangle = -|B_H\rangle; \quad CP|B_L\rangle = |B_L\rangle. \tag{4}$$

That is, the mass eigenstates are also eigenstates of CP [7]. The heavy mass eigenstate coincides with the odd eigenstate of CP, and the light mass eigenstate corresponds to the even eigenstate of CP.

So, a measurement of the B_s^0 lifetime in a pure CP eigenstate gives information about the lifetime of the heavy or light mass eigenstates, which are important parameters to improve the understanding of the mixing process in the B_s^0 system.

3. The D0 detector

The D0 detector is one of the two large particle physics experiments at Fermilab's Tevatron collider. The D0 collaboration began in 1983 and involved hundreds of scientists from 77 institutions. The experiment was upgraded from 1996 to 2001 and ran until the Tevatron ceased operations in 2011.

The D0 detector is described in detail elsewhere [8]. The detector components most relevant to this analysis are the central tracking and the muon systems. Preshower detectors and electromagnetic and hadronic calorimeters surround the tracker. The muon system is located outside the calorimeter, and consists of multilayer drift chambers and scintillation counters inside 1.8 T iron toroidal magnets, and two similar layers outside the toroids.

4. Data selection

To reconstruct the B_s^0 candidates, first a reconstruction of $J/\psi \to \mu^+\mu^-$ candidates is performed looking for a pair of muon tracks with opposite charge and common vertex. Each J/ψ candidate must have a transverse momentum greater than 1.5 GeV/c and a mass inside the 2.80 – 3.35 GeV/c^2 window. After that, a search is made for $\pi^+\pi^-$ tracks forming a common vertex with the J/ψ candidate. The invariant mass of the two pions is required to be in the mass interval 880 – 1080 MeV/c^2, consistent with the $f_0(980)$ resonance. The B_s^0 candidates are reconstructed by a constrained fit to a common vertex for the four charged tracks, requiring an invariant mass inside the range 5.1 – 5.8 GeV/c^2 and a P_T greater than 6.0 GeV/c.

Figure 1 shows the invariant mass distribution of the B_s^0 candidates after the selection and cuts, in which several background sources are identified.

- Cross-feed contamination: mainly produced by the combination of J/ψ mesons from b hadron decays with another particles.
- Combinatorial background: coming from the random combinations of J/ψ mesons with additional tracks.
- B^+ contamination: $B^+ \to J/\psi K^+$ decays in which the kaon is misidentified as a pion, and an extra track accidentally forms a vertex with the $J/\psi K^+$.

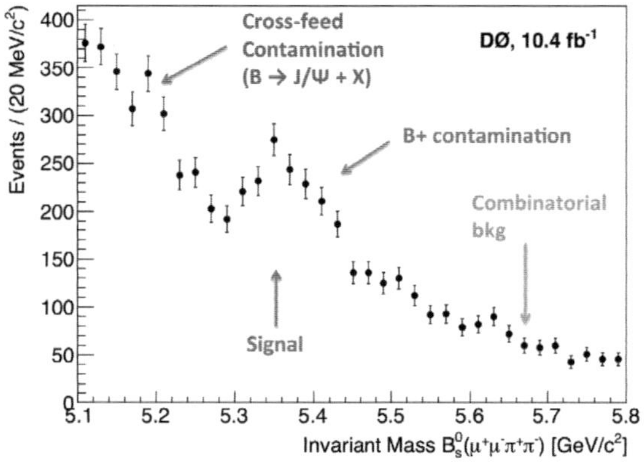

Figure 1. Invariant mass distribution of the B_s^0 candidates after the selection and cuts. Several background sources around the B_s^0 peak are identified.

5. B_s^0 lifetime measurement

The lifetime measurement is based on the transverse decay length method. The transverse decay length is defined as

$$L_{xy} = \vec{L}_{xy} \cdot \vec{P}_T / P_T, \tag{5}$$

where $\vec{P}_T$ is the transverse momentum vector of the B_s^0, and $\vec{L}_{xy}$ is defined as figure 2 shows. The proper transverse decay length λ is obtained applying a boost to L_{xy}. If β and γ are the Lorentz factors

$$\lambda = \frac{L_{xy}}{(\beta\gamma)_T} = L_{xy}\frac{M_B}{P_T}, \tag{6}$$

where M_B is the mass of the B_s^0 taken from the world average reported in the Particle Data Group [9].

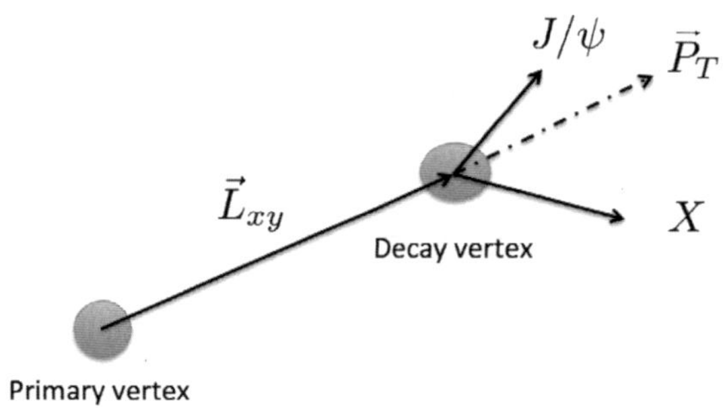

Figure 2. The decay vertex is located as the point where the tracks of the final state particles coincide. The vector, transverse to the beam, between the primary vertex (where the candidate is produced) and the decay vertex is labeled as $\vec{L}_{xy}$.

To measure the lifetime, a simultaneous likelihood fit to the mass and proper decay length distributions is performed. The mass distribution of the signal is modeled by a Gaussian. In the case of the mass distribution of the background components, the cross-feed contamination is modeled by a Gaussian function, the combinatorial background by an exponential function, and the B^+ contamination is modeled as a template determined by the following way: the events in the data sample are reconstructed as $B^+ \to J/\psi K^+$ to identify a peak associated with a B^+, the events inside the mass window are selected and reconstructed, adding an extra pion, as $B_s^0 \to J/\psi\pi^+\pi^-$, and the shape of the distribution is determined by a template. Details about the likelihood function and the models are described in Ref. [10].

The fit yields 494 ± 85 events and a lifetime of $c\tau = 504 \pm 42$ μm for the signal component. Figures 3 and 4 show the distributions of the data with the fit result superimposed. To confirm the presence of the $f_0(980)$ resonance, the events within one σ of the B_s^0 mass are selected, and the mass distribution of the $\pi^+\pi^-$ pairs is fit with a Flatté model [11].

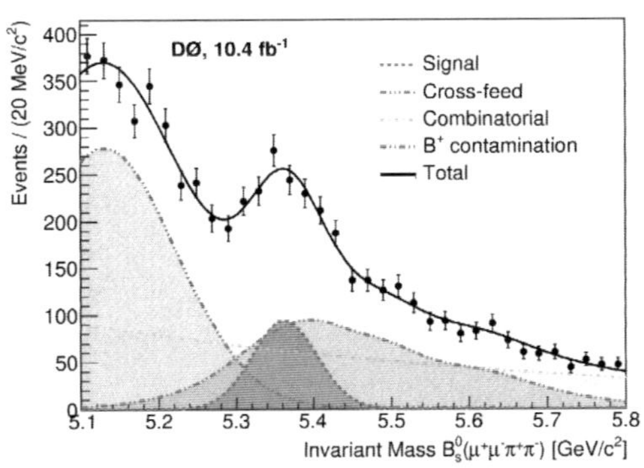

Figure 3. Invariant mass distribution for B_s^0 candidates. The model obtained from the fit and each one of its components are superimposed [10].

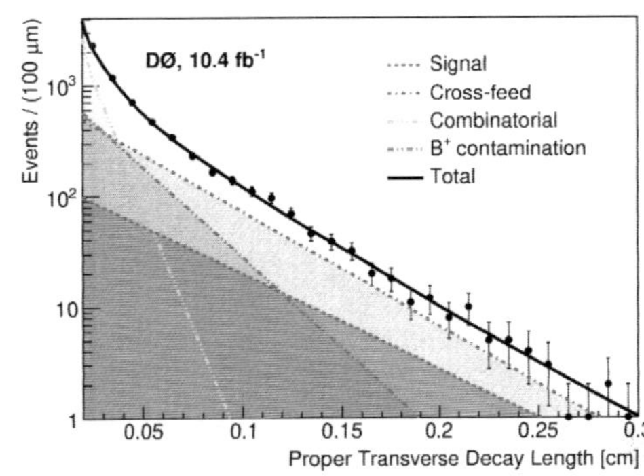

Figure 4. Proper transverse decay length distribution for B_s^0 candidates. The model obtained from the fit and each one of its components are superimposed [10].

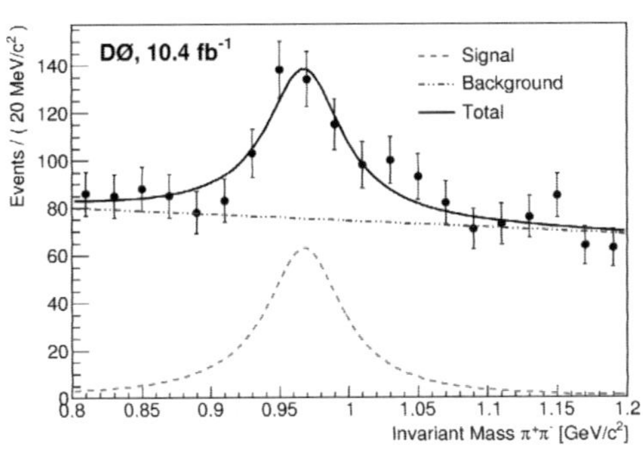

Figure 5. Invariant mass distribution of the $\pi^+\pi^-$ pairs coming from events within $\pm 1\sigma$ of the B_s^0 mass [10].

6. Systematic uncertainties

Several sources of systematic uncertainty are considered. Table 1 summarizes the uncertainties with its source and the total uncertainty, determined by combining individual uncertainties in quadrature. The uncertainty related to alignment has been previously determined. The $\pi^+\pi^-$ invariant mass window has is varied from its nominal width of 200 MeV$/c^2$ to between 160 and 240 MeV$/c^2$. The modeling and fitting method is tested with data generated in pseudoexperiments, in which a bias of -4.4 μm arises for an input lifetime of 500 μm and 500 signal events. This bias is corrected in the nominal result. Finally, the parametrization of the models used to describe the distributions are changed, and the variation with respect to the nominal result is taken as a systematic uncertainty.

Table 1. Summary of systematic uncertainties.

Source	Variation (μm)
Alignment	5.4
$\pi^+\pi^-$ invariant mass window	8.0
Fit bias	4.4
Distribution models	12.5
Total (combining in quadrature)	**16.4**

7. Conclusions

In summary, the lifetime of the B_s^0 is measured [10], obtaining:

$$c\tau(B_s^0) = 508 \pm 42 \text{ (stat)} \pm 16 \text{ (syst) } \mu\text{m}, \tag{7}$$

from which:

$$\tau(B_s^0) = 1.70 \pm 0.14 \text{ (stat)} \pm 0.05 \text{ (syst) ps}, \tag{8}$$

in the decay channel $B_s^0 \to J/\psi\pi^+\pi^-$ with $880 \leq M_{\pi^+\pi^-} \leq 1080$ MeV/c^2. Neglecting CP violation in this decay, the measurement can be translated into the width of the heavy mass eigenstate of the B_s^0

$$\Gamma_H = 0.59 \pm 0.05 \text{ (stat)} \pm 0.02 \text{ (syst) ps}^{-1}. \tag{9}$$

Our result is in good agreement with previous measurements from LHCb and CDF collaborations, and provides an independent confirmation of the longer lifetime for the CP-odd eigenstate of the $B_s^0/\bar{B}_s^0$ system.

8. Acknowledgements

The D0 Collaboration thanks the staffs at Fermilab and collaborating institutions, and acknowledges support of the many international funding agencies which made these experimental programs possible.

References

[1] V. M. Abazov, *et al.* (D0 Collaboration) (2005). *Measurement of the B_s^0 lifetime in the exclusive decay channel $B_s^0 \to J/\psi\phi$.* Phys. Rev. Lett. **94**(4), 042001.

[2] R. Aaij *et al.* (LHCb Collaboration) (2011). *First observation of $B_s^0 \to J/\psi f_0(980)$ decays*, Phys. Lett. B **698**, 115.

[3] J. Li *et al.* (Belle Collaboration) (2011). *Observation of $B_s^0 \to J/\psi f_0(980)$ and Evidence for $B_s^0 \to J/\psi f_0(1370)$*, Phys. Rev. Lett. **106**, 121802.

[4] T. Aaltonen *et al.* (CDF Collaboration) (2011). *Measurement of branching ratio and B_s^0 lifetime in the decay $B_s^0 \to J/\psi f_0(980)$ at CDF*, Phys. Rev. D **84**, 052012.

[5] V.M. Abazov *et al.* (D0 Collaboration) (2012). *Measurement of the relative branching ratio of $B_s^0 \to J/\psi f_0(980)$ to $B_s^0 \to J/\psi\phi$*, Phys. Rev. D **85**, 011103(R).

[6] R. Aaij *et al.* (LHCb Collaboration) (2012). *Measurement of the $\bar{B}_s^0$ Effective Lifetime in the $J/\psi f_0(980)$ Final State*, Phys. Rev. Lett. **109**, 152002.

[7] G. C. Branco, L. Lavoura, & J. P. Silva (1999). *CP violation* (No. 103). Oxford University Press.

[8] V.M. Abazov *et al.*, *The upgraded D0 detector*, Nucl. Instrum. Methods A **565**, 463 (2006).

[9] K. A. Olive *et al.* (Particle Data Group) (2014). *Review of Particle Physics*, Chin. Phys. C **38**, 090001.

[10] V.M. Abazov *et al.* (D0 Collaboration) (2016). *B_s^0 lifetime measurement in the CP-odd decay channel $B_s^0 \to J/\psi\ f_0(980)$*, Phys. Rev. D **94**, 012001.

[11] S. M. Flatté (1976). *Coupled-channel analysis of the $\pi\eta$ and $K\bar{K}$ systems near $K\bar{K}$ threshold*, Phys. Lett. B **63**, 224.

Electromagnetic corrections to $B^- \to V^0$ semileptonic transitions

S.L. Tostado[1]

[1] Departamento de Física, Centro de Investigación y de Estudios Avanzados del Instituto Politécnico Nacional, Apdo. Postal 14-740, 07000 México D.F., México

E-mail: stostado@fis.cinvestav.mx

Abstract. We evaluate the $O(\alpha)$ electromagnetic radiative corrections to the $B^- \to V^0 \ell^- \bar{\nu}_\ell$ (V is a vector meson and $\ell = e, \mu$) decay rates. Intermediate resonance states in the real-photon amplitude are considered in order to extend the region of validity of the soft-photon approximation. The virtual corrections are regularized by the cutoff scale Λ, which is used to match the long- and short-distance regimes. The radiative corrections to the integrated decay rate are presented and we discuss its impact in the precise determination of the b quark mixing parameters.

1. Introduction

In the Standard Model (SM) of weak interactions, the elements of the Cabbibo-Kobayashi-Maskawa matrix are fundamental parameters to describe the quark-mixing [1]. In particular, improved measurements of the magnitude of V_{ub} and V_{cb} are needed to test the SM mechanism of CP violation [2, 3]. The two methods to extract these elements in B decays are based on the study of inclusive and exclusive semileptonic decays. However, these values differ from each other by $\Delta|V_{xb}| = |V_{xb}|_{\rm incl} - |V_{xb}|_{\rm excl} = (1.13 \pm 0.36) \times 10^{-3}$, and $(2.7 \pm 1.1) \times 10^{-3}$ [4], for $x = u$ and c, respectively. The values extracted from exclusive decays are commonly obtained from $B \to P \ell \nu$ decays (P is a pseudoscalar meson) since they are widely known theoretically [5, 6]. Isospin breaking corrections are useful to test the consistency of experimental data and theoretical calculations, as shown in superallowed Fermi transitions (SFT) and Kaon semileptonic ($K_{\ell 3}$) decays [7]. In this work we are concerned with the calculation of electromagnetic radiative corrections to $B^\pm \to V^0$ semileptonic transitions, a useful input in testing isospin symmetry in B decays.

2. Tree level amplitude for $B^- \to V^0$ semileptonic transitions

At the tree-level, the decay amplitude is:

$$\mathcal{M}^0 = \frac{G_F}{\sqrt{2}} V_{qb} c_V W_\nu(P_V, P) L^\nu, \tag{1}$$

where G_F is the Fermi constant, V_{qb} ($q = u, c$) the relevant CKM matrix element, $L^\nu = \bar{u}_\ell \gamma^\nu (1 - \gamma_5) v_\nu$ represents the leptonic current, and $c_V = 1/\sqrt{2}$ is the Clebsh-Gordan coefficient for $V = (\rho^0, \omega)$ mesons. P (P_V) is the four momentum of the decaying (daughter) meson. The

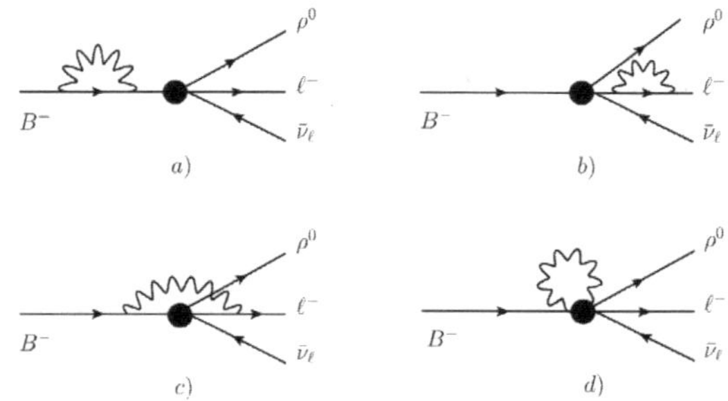

Figure 1. Virtual photon corrections to $B^- \to \rho^0 \ell^- \bar{\nu}_\ell$. Figs. *a)* and *b)* correspond to the self-energies of charged particles and *c)*, *d)* to vertex contributions.

hadronic matrix element can be parametrized in terms of four $q^2 = (P - P_V)^2$-dependent form factors, which we choose as (V, A_1, A_2, A)

$$\begin{aligned} W_\nu(P_V, P) &= \frac{2V^{B\to V}}{M + m_V}\epsilon_{\nu\alpha\beta\gamma}\varphi^\alpha P^\beta P_V^\gamma - i\frac{2m_V A^{B\to V}}{q^2} q\cdot\varphi q_\nu \\ &\quad -i(M+m_V)A_1^{B\to V}\varphi_\nu + i\frac{A_2^{B\to V}}{M+m_V} q\cdot\varphi(P+P_V)_\nu \, . \end{aligned} \tag{2}$$

The four-vector φ denotes the vector meson polarization, with the orthogonality condition $P_V \cdot \varphi = 0$. The form factors are specific to the $B \to V$ transition [8].

3. Virtual radiative corrections

The charged meson and lepton self-energies (Fig. 1 *a* and *b*) contribute to the decay amplitude as

$$\mathcal{M}^1_{a+b} = \mathcal{M}^0 \frac{1}{2}(\delta Z_\ell + \delta Z_B)\, , \tag{3}$$

where

$$\begin{aligned} \delta Z_\ell &= \frac{\alpha}{4\pi}\Big(2 - \mathrm{B}_0[m^2, 0, m^2] + 4m^2 \mathrm{B}_0'[m^2, \lambda^2, m^2]\Big) \\ \delta Z_B &= \frac{\alpha}{4\pi}\Big(2\mathrm{B}_0^M[M^2, 0, M^2] + 4M^2 \mathrm{B}_0'^M[M^2, \lambda^2, M^2]\Big)\, , \end{aligned}$$

which contribute to the wave function renormalization of the charged particles. $\mathrm{B}_0^{(M)}[\cdots]$ and $\mathrm{B}_0'^{(M)}[\cdots]$ are the Passarino-Veltman functions corresponding to the scalar two-point integral and its derivative [9]. We will isolate the infrared (IR) singularity in $\mathrm{B}_0'^{(M)}$ by providing a fictitious mass λ to the photon, and regulate the ultraviolet (UV) singularity of B_0 by using the cutoff Λ as the maximum scale at which the long-distance approximation is expected to be valid. By long-distance (LD) corrections we mean, emission/absorption of real and virtual photons with small momenta ($k < \Lambda$) such that they cannot resolve the structure of hadrons; at these low photon momenta, the point-like approximation for hadrons and the use of scalar QED for describing photon-meson interactions should be a good approximation.

We will follow Sirlin's prescription [10] to separate the model-independent and model-dependent contributions from Figure 1c,d. For the model-independent part of the virtual corrections we obtain (here $u = (P-p)^2$)

$$\mathcal{M}^1_{c+d,\mathrm{MI}} = \frac{\alpha}{4\pi}\mathcal{M}^0\Big\{\mathrm{B}_0^M[M^2, 0, M^2] + 4p\cdot P C_0[m^2, M^2, u, m^2, \lambda^2, M^2]$$

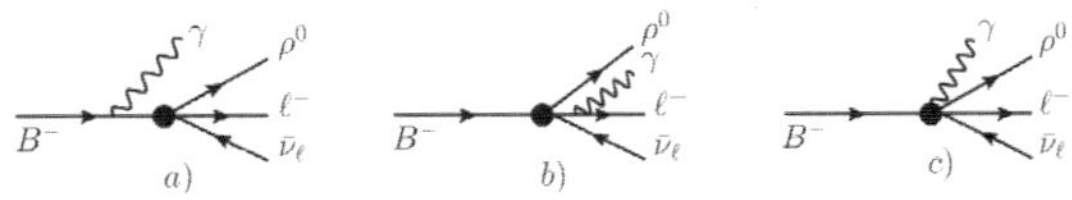

Figure 2. Feynman diagrams for real photon emission.

$$+2M^2\mathrm{C}_1[m^2,M^2,u,m^2,0,M^2]+4P\cdot p\mathrm{C}_2[m^2,M^2,u,m^2,0,M^2]\Big\}$$
$$-\frac{\alpha}{4\pi}\mathcal{M}_{\mathrm{NF}}\mathrm{C}_2[m^2,M^2,u,m^2,0,M^2]\;, \tag{4}$$

where $\mathrm{C}_i[m^2,M^2,u,m^2,\lambda^2,M^2]$ $(i=0,1,2)$ denote the Passarino-Veltman functions. We have used again λ to isolate the IR-divergence contained in C_0, while $\mathrm{C}_{1,2}$ are evaluated in the limit $\lambda\to 0$. The last term in Eq.(4) involves $\mathcal{M}_{\mathrm{NF}}=\frac{G_F}{\sqrt{2}}mV_{ub}W^\mu(P_V,P)\bar{u}_\ell\, \not{P}\gamma_\mu(1-\gamma_5)v_{\bar{\nu}_\ell}$, which corresponds to a non-factorizable (NF) amplitude, it is IR safe and is suppressed for light charged leptons.

The short distance (SD) corrections to semileptonic decays are finite in the electroweak theory [11, 12]. In the dominant logarithmic approximation the $O(\alpha)$ corrections depend upon the separation scale Λ as follows (m_Z is the Z boson mass)

$$\delta^1_{\mathrm{SD}}=\frac{2\alpha}{\pi}\ln\left(\frac{m_Z}{\Lambda}\right)\;. \tag{5}$$

It is customary to set the lower cutoff of SD corrections as the mass of the decaying particle; however, we will keep it explicitly to study the cutoff scale dependence of the full radiative corrections.

4. Real photon emission

By taking into account the contributions in Fig. 2 one gets the decay amplitude in the Low's expansion [13]

$$\mathcal{M}^{\mathrm{Low}} = e\frac{G_F}{\sqrt{2}}V_{qb}c_V\Big\{W_\mu(P_V,P)\bar{u}_\ell\left[-\frac{P\cdot\varepsilon}{P\cdot k}+\frac{2p\cdot\varepsilon+\not{\varepsilon}\,\not{k}}{2p\cdot k}\right]\gamma^\mu(1-\gamma_5)v_{\nu_\ell}$$
$$+L^\mu D^\lambda W_{\mu\lambda}-2P_V\cdot D\;L^\mu\frac{\partial W_\mu(P_V,P)}{\partial q^2}\Big\}, \tag{6}$$

where ε denotes the polarization four-vector of the real photon, with $k\cdot\varepsilon=0$. We have defined $D^\lambda=(\varepsilon\cdot P/k\cdot P)k^\lambda-\varepsilon^\lambda$, L and W are the same as in Eq. (1), and

$$W_{\mu\lambda} = -\frac{2}{M+m_V}V\epsilon_{\mu\lambda\nu\alpha}\varphi_V^\nu P_V^\alpha+i\frac{q\cdot\varphi}{M+m_V}A_2\delta_{\mu\lambda}$$
$$+i\frac{A_2}{M+m_V}(P+P_V)_\mu\varphi_\lambda-2i\frac{m_VA}{q^2}q_\mu\varphi_\lambda-2i\frac{m_VA}{q^2}q\cdot\varphi\delta_{\mu\lambda}\;. \tag{7}$$

As required by soft photon theorems [13], Eq. (6) contains terms up to $\mathcal{O}(k^0)$ which are fixed by gauge-invariance and depend only upon the form factors of the non-radiative decay amplitude. The partial derivative in the last term of Eq. (6) must be carried over explicit q^2-dependent form factors in W. We will consider the evaluation of the last two terms in Eq. (6) as a part of the model-dependent contribution.

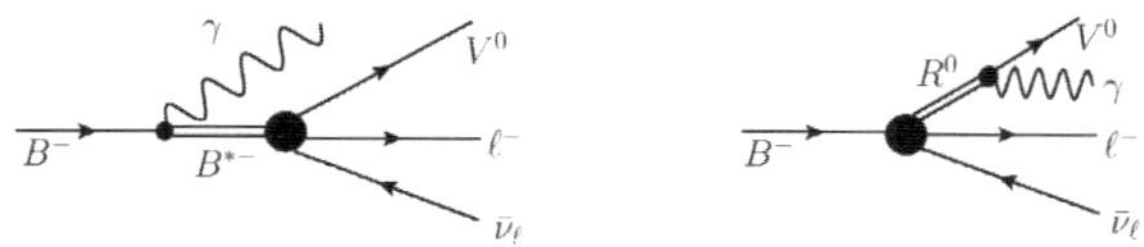

Figure 3. Model-dependent real photon corrections; R^0 denotes a pseudoscalar resonance.

4.1. Model-dependent contribution

For relatively low photon energies, [6] points out that the diagrams shown in Figure 3 must be taken into account in $B \to D\ell\nu_\ell\gamma$ decays since on-shell $D^*(\to D\gamma)$ intermediate states can have an important contribution as non-radiative events. For the decays considered in this paper $B^\pm \to (\rho^0, \omega, D^{*0})\ell^\pm\nu_\ell$, there are not nearby narrow intermediate states decaying into $V^0\gamma$ states that can produce a similar important effect. In our study, we evaluate the model-dependent (MD) contributions in Fig. 3 for moderately energetic photons following closely Ref. [6].

The general form of the MD decay amplitude becomes

$$\mathcal{M}^{MD} = e\frac{G_F}{\sqrt{2}}V_{qb}c_V\epsilon^\mu\,(V^{MD}_{\mu\nu} - A^{MD}_{\mu\nu})L^\nu\ . \tag{8}$$

From Figure 3 we get [6]

$$V^{MD}_{\mu\nu} - A^{MD}_{\mu\nu} = \tag{9}$$
$$\frac{\langle V^0|(V_\nu - A_\nu)|B^{*-}(P_1)\rangle\langle B^{*-}(P_1)|J_\mu|B^-\rangle}{P_1^2 - m^2_{B^*} + i\epsilon} + \frac{\langle V^0|J_\mu|R^0(P_2)\rangle\langle R^0(P_2)|(V_\nu - A_\nu)|B^-\rangle}{P_2^2 - m^2_R + i\epsilon}\ .$$

where $P_1 = P - k$ and $P_2 = P_V + k$ denote the four momenta of the corresponding resonances, and $m_{B^*}(m_R)$ the mass of $B^{*-}(R^0)$ (if the R resonance is produced on-shell, we have to use $\epsilon \to m_R\Gamma_R$). In the above expression J_μ denotes the electromagnetic current and $(V_\nu - A_\nu)$ denotes the usual V-A weak currents [6]. Details about the expressions for the matrix elements in Eq. (9) can be found in [14].

5. Results

In this section we present the results of radiative corrections to $B^- \to V^0\ell^-\bar{\nu}_\ell$, where $V^0 = (\rho^0, \omega, D^{*0})$. The integrals involved are evaluated numerically, and we include the LD radiative corrections obtained in the previous sections and the SD correction in Eq. (5).

In the case of $V = \rho^0$ meson, the total radiative corrections for the two leptonic channels are

$$\delta^1_T(\ell) = \begin{cases} (1.62 \pm 0.10 \pm 0.04)\%\ , & \text{for } \ell = \mu \\ (1.63 \pm 0.11 \pm 0.04)\%\ , & \text{for } \ell = e \end{cases}\ . \tag{10}$$

The first uncertainty is associated to the variation in the scale Λ between $m_\rho/2$ and 2 GeV, while the central value is taken for $\Lambda = m_\rho$. The remaining uncertainty corresponds to the error in the numerical evaluation and the use of different models for the form factors [8]. We do not found any relevant difference between this case and $V^0 = \omega$, given the near equality of the masses and the matrix elements.

A similar analysis can be applied to the case with $V^0 = D^{*0}$, where we obtain

$$\delta^1_T(\ell) = \begin{cases} (1.53 \pm 0.06 \pm 0.04)\%, & \text{for } \ell = \mu \\ (1.53 \pm 0.08 \pm 0.04)\%, & \text{for } \ell = e \end{cases}\ . \tag{11}$$

The values are taken for $\Lambda = m_{D^{*0}}$, the first uncertainty corresponds to variations between $m_{D^{*0}}/2$ and $2m_{D^{*0}}$, and the second one to the numerical estimation as in the previous case.

We have found that the total radiative corrections of Eqs. (10) and (11) are dominated by the SD radiative corrections since the LD are 4 (7) times smaller, for $V^0 = \rho^0$ (D^{*0}). This is due to partial cancellations of the contributions coming from the three and four body region of the phase space. The MD corrections play a subleading role because their contributions were estimated to be at most ten times smaller that the MI ones, which is in agreement with the results obtained for radiative $K_{\ell 3}$ decays [15]. The effect of LD radiative corrections to specific points of the Dalitz plot has been studied in [14].

6. Conclusions

We have evaluated numerically the long-distance (LD) electromagnetic corrections to the charmless and charmfull semileptonic $B^- \to V^0\ell^-\bar{\nu}_\ell$ decays involving the ground state vector mesons V^0 and $\ell = \mu$ and e flavored leptons. We have associated an uncertainty to the radiative corrections in the decay rate due to the cutoff dependence by allowing it to vary between $m_V/2$ and approximately $2m_V$. The effects of using different models for the weak form factors are estimated to play a subleading role compared to the dependence upon the separation scale Λ. We have found that there exist large cancellations between the contributions of LD corrections to the integrated decay rates coming from the three- and four-body regions of the Dalitz plot. We found that the MD contributions are negligible since they are smaller than the quoted error. Our results can be useful for future/improved measurements of the different charged decay channels of $B \to V\ell^-\bar{\nu}_\ell$ branching ratios in order to test consistency of data with isospin symmetry. Actually, the determination of the V_{ub} and V_{cb} CKM elements at the few percent level requires the consideration of the full radiative corrections of $O(\alpha)$.

Acknowledgements

The author is grateful to G. López-Castro for suggesting this problem and several conclusions. The author would like to thank Conacyt (México) for partial financial support under project 236394, and the organizing committee for financial support.

References

[1] N. Cabibbo 1963 Phys. Rev. Lett. **10** 531; M. Kobayashi and T. Maskawa 1973 Prog. Theor. Phys. **49** 652.
[2] P. del Amo Sánchez et al [BaBar collaboration] 2011 Phys. Rev. D**83** 032007.
[3] A. Sibidanov et al [Belle Collaboration] 2013 Phys. Rev. D**88** 0.32005.
[4] K.A. Olive et al. [Particle Data Group] 2014 Chin. Phys. C, **38** 090001.
[5] V. Cirigliano and D. Pirjol 2005 Phys. Rev. D **72** 094021.
[6] D. Becirevic and N. Kosnik 2010 Acta Phys. Polon. Supp. **3** 207.
[7] I. S. Towner and J. C. Hardy 2015 Phys. Rev. C **91** 025501; V. Cirigliano, G. Ecker, H. Neufeld, A. Pich and J. Portolés 2012 Rev. Mod. Phys. **84** 399.
[8] P. Ball and R. Zwicky 2005 Phys. Rev. D**71** 014015; L. Del Debbio *et al.* (UKQCD collaboration) 1998 Phys. Lett B **416** 392; I. Caprini, L. Lellouch and M. Neubert 1998 Nucl. Phys. B **530** 153.
[9] G. Passarino and M. Veltman 1979 Nucl. Phys. B **160** 151.
[10] A. Sirlin 1967 Phys. Rev. **164** 1767.
[11] A. Sirlin 1978 Rev. Mod. Phys. **50** 573; A. Sirlin 1974 Nucl. Phys. **B71** 29.
[12] W. J. Marciano and A. Sirlin 1988 Phys. Rev. Lett. **61** 1815, 1993 *ibid.* **71** 3629.
[13] F.E. Low 1958 Phys. Rev. **110**, 974; S.L. Adler and Y. Dothan 1966 Phys. Rev. **151** 1267; H. W. Fearing and E. Fischbach 1970 Phys. Rev. D **2** 542.
[14] S.L. Tostado and G. López-Castro *arXiv* 1510.08020 [hep-ph].
[15] T. C. André 2007 Annals Phys. **322** 2518.

NLO cross sections in 4 dimensions without DREG

R. J. Hernández-Pinto[1]*, F. Driencourt-Mangin[2], G. Rodrigo[2] and G. F. R. Sborlini[2]

[1] Facultad de Ciencias Físico-Matemáticas, Universidad Autónoma de Sinaloa, Ciudad Universitaria, CP 80000, Culiacán, Sinaloa, México.
[2] Instituto de Física Corpuscular, Universitat de València - Consejo Superior de Investigaciones Científicas, Parc Científic, E-46980 Paterna, Valencia, Spain.

E-mail: `roger@uas.edu.mx`

Abstract. In this review, we present a new method for computing physical cross sections at NLO accuracy in QCD without using the standard Dimensional Regularisation. The algorithm is based on the Loop-Tree Duality theorem, which allow us to obtain loop integrals as a sum of phase-space integrals; in this way, transforming loop integrals into phase-space integrals, we propose a method to merge virtual and real contributions in order to find observables at NLO in $d = 4$ space-time dimensions. In addition, the strategy described is used for computing the $\gamma^* \to q\bar{q}(g)$ process. A more detailed discussion related on this topic can be found in Ref [1].

1. Introduction

Theoretical tools for describing physical observables at the LHC require to be very precise for the determination of the properties of new particles and the description of new physics processes. In general, the formalism used to obtain accurate theoretical predictions is based in the so-called Dimesional Regularisation [2] (DREG). In DREG, real and virtual integrals are implemented in $d = 4-2\epsilon$ instead of $d = 4$ space-time dimensions and singularities are manifested as poles in the ϵ parameter. However, cancellation of singularities on *infrared-safe observables* is guaranteed by the KLN [3] theorem. These ideas have been studied in terms of the substraction techniques [4] and the developments along them have had an enormous impact on the theoretical predictions for the LHC. In this document, we will describe a new method based on the Loop-Tree Duality (LTD) theorem [1, 5, 6, 7, 8, 9, 10, 11, 12, 13, 14, 15, 16]; we will present the general algorithm for the mapping between real and virtual variables so the cancelation of singularities is achieved at integrand level, meaning that integrals can be done in $d = 4$ dimensions; finally, we will present a realisation of the method in the $\gamma^* \to q\bar{q}(g)$ process at NLO in QCD.

2. The Loop-Tree duality

LTD presents a new paradigm for writing loop integrals, at one-loop and beyond, as a sum of phase-space integrals, the so-called dual integrals. In order to illustrate the theorem at one-loop, let's consider a generic N-particle scalar one-loop integral with massive particles, i.e.

$$L^{(1)}(p_1, \ldots, p_N) = \int_\ell \prod_{i=1}^{N} G_F(q_i) \ , \tag{1}$$

* Speaker

where $G_F(q_i) = (q_i^2 - m_i^2 + \imath 0)^{-1}$. Then, the dual representation consists of the sum of N dual integrals:

$$L^{(1)}(p_1, \ldots, p_N) = -\sum_{i=1}^{N} \int_\ell \tilde{\delta}(q_i) \prod_{j=1, j\neq i}^{N} G_D(q_i; q_j) \,, \tag{2}$$

where $G_D(q_i; q_j) = (q_j^2 - m_j^2 - \imath 0\, \eta \cdot k_{ji})^{-1}$ are dual propagators, i, j label the available internal lines and $\tilde{\delta}(q_i) \equiv 2\pi \imath\, \theta(q_{i,0})\, \delta(q_i^2 - m_i^2)$. Masses and momenta of the internal lines are denoted m_i and $q_{i,\mu} = (q_{i,0}, \mathbf{q}_i)$, respectively, where $q_{i,0}$ is the energy and $\mathbf{q}_i$ are the spatial components. Internal lines can be written in terms of the loop momentum ℓ and the outgoing four-momenta of the external particles p_i, as $q_i = \ell + k_i, k_i = p_1 + \ldots + p_i$, where momentum conservation holds, i.e. $k_N = 0$. Finally, Eq. (2) shows explicitly a relation between a one-loop integral with N external legs to N phase-space integrals, where the change in sign in the $\imath 0$ prescription has to be considered carefully. This procedure can also be used for topologies with more than one-loop and it has been studied previously [7].

3. Unsubtraction in $d = 4$

The general procedure for describing *next-to-leading order* (NLO) cross sections involves the calculation of real and virtual corrections to the Born cross-section. Let's consider a process with m particles in the final state; the NLO cross-section is written as

$$\sigma^{\rm NLO} = \int_m d\sigma_{\rm V}^{(1,{\rm R})} + \int_{m+1} d\sigma_{\rm R}^{(1)} \,, \tag{3}$$

where $d\sigma_{\rm R}^{(1)}$ denotes the real correction and $d\sigma_{\rm V}^{(1,{\rm R})}$ contains all the information related to the renormalised virtual correction; indeed, using the LTD, virtual correction can be written as

$$d\sigma_{\rm V}^{(1,{\rm R})} = \sum_{i=1}^{N} \int_\ell 2\,{\rm Re}\langle \mathcal{M}_N^{(0)} | \mathcal{M}_N^{(1,{\rm R})}(\tilde{\delta}(q_i))\rangle\, \mathcal{O}_N(\{p_i\}) \,, \tag{4}$$

where $\mathcal{M}_N^{(0)}$ represents the *leading order* (LO) amplitude with N-legs, $\mathcal{M}_N^{(1,{\rm R})}$ is the renormalised one-loop scattering amplitude, and $\mathcal{O}_N(\{p_i\})$ is a measure function which is defined for a given observable. It is worth mentioning that $\mathcal{M}_N^{(1,{\rm R})}$ also contains the self-energy contributions of external legs, even if they become zero in the massless case. Besides, the renormalised amplitude includes counter-terms which are needed in order to subtract UV singularities locally, this aspect is widely discussed in Ref. [1]. On the other hand, the real correction is given by

$$\int_{m+1} d\sigma_{\rm R}^{(1)} = \sum_{i=1}^{N} \int_{m+1} |\mathcal{M}_{N+1}^{(0)}(q_i, p_i)|^2 \mathcal{R}_i(q_i, p_i) \mathcal{O}_{N+1}(\{p'_j\}) \,, \tag{5}$$

where the external momenta p'_j, the phase-space and the tree-level scattering amplitude, $\mathcal{M}_{N+1}^{(0)}$, are rewritten in terms of the loop three-momentum and the external momenta p_i of the Born process.

One of the basic ideas in order to build a four dimensional representation of the scattering amplitudes resides on a proper mapping between real and virtual variables in order to merge both contributions at integrand level. Following these ideas, let's consider the first parton as the *emitter* and the second as the *spectator*; then, in order to reconstruct the kinematics associated to the real emission, we propose to map the loop three-momentum and the four momenta of the emitter and the spectator into a region of the real phase-space where the twin of the emitter decays to two partons in a soft and collinear configuration; it means,

$$p_r'^{\mu} = q_i^{\mu}, \qquad p_i'^{\mu} = p_i^{\mu} - q_i^{\mu} + \alpha_i\, p_j^{\mu}, \qquad p_j'^{\mu} = (1-\alpha_i)\, p_j^{\mu}, \qquad p_k'^{\mu} = p_k^{\mu}, \qquad k \neq i, j \,, \tag{6}$$

where $\alpha_i = (q_i - p_i)^2/(2p_j\cdot(q_i-p_i))$, p_i is the momenta of the final-state emitter, q_i is the internal on-shell momentum prior to the emitter, p_j is the momentum of the final-state spectator, and p'_r is the momentum of the extra radiated particle. In addition, it is straightforward to show that momentum conservation is satisfied in the mapping. A final piece in the method consists on the segmentation of the real phase-space. In Eq. (5), we have introduced a complete partition of the real phase-space

$$\sum \mathcal{R}_i(p_i, q_i) = \sum \prod_{jk\neq ir} \theta(y'_{jk} - y'_{ir}) = 1 \ , \tag{7}$$

which is similar to divide the phase-space as a function of the minimal dimensionless two-body invariants $y'_{ir} = s'_{ir}/s$. This separation of the real phase-space takes into account the relation between the real and virtual kinematics, such that, the real correction is actually contributing over the region where the soft and collinear divergences are generated. Hence, IR singularities are cancelled at the integrand level when both real and virtual corrections are simultaneously added. On the other hand, the UV divergences are subtracted locally by suitable counter-terms [1]. Finally, the implementation of the NLO cross-section can be implemented in four-dimensions due to the absence of divergences at integrand level, i.e. $\epsilon = 0$ in DREG.

4. Unsubtraction applied to $\gamma^* \to q\bar{q}(g)$ at NLO in QCD

In this section we present an example where the method can be tested. NLO correction to the process $\gamma^* \to q\bar{q}(g)$ includes, for the virtual correction, the self-energy and the triangle topologies, the most simple topologies where the LTD can be applied. Considering that internal loop momenta are $q_1 = \ell + p_1$, $q_2 = \ell + p_{12}$ and $q_3 = \ell$, we parametrise them as $2q_i^\mu = \sqrt{s_{12}}\xi_{i,0}(1, 2\sqrt{v_i(1-v_i)}\mathbf{e}_{i,\perp}, 1-2v_i)$ when $q_i^2 = 0$. Then, proceeding with the method described in the previous section, we obtain by combining real and virtual correction the dual cross-sections

$$\begin{aligned}
\widetilde{\sigma}_1^{(1)} &= \sigma^{(0)}\frac{\alpha_S}{\pi}C_F \int_0^1 d\xi_{1,0}\int_0^{1/2} dv_1\, \mathcal{R}_1(\xi_{1,0}, v_1)\left[2\left(\xi_{1,0} - (1-v_1)^{-1}\right) - \frac{\xi_{1,0}(1-\xi_{1,0})}{(1-(1-v_1)\xi_{1,0})^2}\right], \\
\widetilde{\sigma}_2^{(1)} &= \sigma^{(0)}\frac{\alpha_S}{2\pi}C_F \int_0^1 d\xi_{2,0}\int_0^1 dv_2\, \mathcal{R}_2(\xi_{2,0}, v_2)\,(1-v_2)^{-1}\left[\frac{2\,v_2\,\xi_{2,0}\,(\xi_{2,0}(1-v_2)-1)}{1-\xi_{2,0}}\right. \\
&\left. -1 + v_2\,\xi_{2,0} + \frac{1}{1-v_2\,\xi_{2,0}}\left(\frac{(1-\xi_{2,0})^2}{(1-v_2\,\xi_{2,0})^2} + \xi_{2,0}^2\right)\right],
\end{aligned} \tag{8}$$

where the segmentation of the real phase-space is defined as

$$\mathcal{R}_1(\xi_{1,0}, v_1) = \theta(1-2v_1)\,\theta\left(\frac{1-2v_1}{1-v_1} - \xi_{1,0}\right), \quad \mathcal{R}_2(\xi_{2,0}, v_2) = \theta\left(\frac{1}{1+\sqrt{1-v_2}} - \xi_{2,0}\right) \ ; \tag{9}$$

and the remnant dual pieces are collected in a single element leading to

$$\begin{aligned}
\overline{\sigma}_V^{(1)} &= \sigma^{(0)}\frac{\alpha_S}{4\pi}C_F\int_0^\infty d\xi \int_0^1 dv \left\{ -2\,(1-\mathcal{R}_1(\xi,v))\,v^{-1}(1-v)^{-1}\frac{\xi^2(1-2v)^2+1}{\sqrt{(1+\xi)^2-4v\,\xi}} \right. \\
&+ 2\,(1-\mathcal{R}_2(\xi,v))\,(1-v)^{-1}\left[2\,v\,\xi\,(\xi(1-v)-1)\left(\frac{1}{1-\xi+\imath 0} + \imath\pi\delta(1-\xi)\right) - 1 + v\,\xi\right] \\
&+ 2\,v^{-1}\left(\frac{\xi(1-v)(\xi(1-2v)-1)}{1+\xi} + 1\right) - \frac{(1-2v)\,\xi^3\,(12 - 7m_{\rm UV}^2 - 4\xi^2)}{(\xi^2+m_{\rm UV}^2)^{5/2}} \\
&\left. - \frac{2\,\xi^2(m_{\rm UV}^2 + 4\xi^2(1-6v(1-v)))}{(\xi^2+m_{\rm UV}^2)^{5/2}}\right\}.
\end{aligned} \tag{10}$$

In Eq. (10) we have unified the integration variables, $\xi = \xi_{2,0} = \xi_{3,0} = \xi_{\rm UV}$ and $v = v_2 = v_3 = v_{\rm UV}$, while $(\xi_{1,0}, v_1)$ are expressed in terms of $(\xi_{2,0}, v_2)$ with the appropriate change of variables. The last two termns in Eq. (10) are the UV counter-terms of the vertex and self-energies. Finally, integrating these expressions, we obtain the well known result

$$\sigma = \sigma^{(0)} \left(1 + 3\, C_F \, \frac{\alpha_{\rm S}}{4\pi} + \mathcal{O}(\alpha_{\rm S}^2) \right) , \tag{11}$$

where $\sigma^{(0)} = \alpha\, e_q^2\, C_{\rm A}/2$. It is worth mentioning that using the method described in here, it was unnecessary to introduce any tensor reduction; Gram determinants are naturally avoided in LTD, and therefore the spurious singularities that the tensor reduction introduces leading to numerical instabilities in the integration over the phase-space.

5. Conclusions

In this review, we illustrated a new method for computing NLO cross sections with the application of the LTD theorem and without using DREG. The general algorithm is described and a practical realisation is implemented in the calculation of the $\gamma^* \to q\bar{q}(g)$ at NLO in QCD. The advantage on the use of the LTD is that the calculation of a full NLO correction is achieved with purely four-dimensional expressions, i.e. with the DREG parameter $\epsilon = 0$.

6. Acknowledgements

The work of RJHP is partially supported by CONACyT, México. This work has been supported by CONICET Argentina, by the Spanish Government and ERDF funds from the European Commission (Grants No. FPA2014-53631-C2-1-P and SEV-2014-0398) and by Generalitat Valenciana under Grant No. PROMETEOII/2013/007. FDM acknowledges support from Generalitat Valenciana (GRISOLIA/2015/035).

References

[1] G. F. R. Sborlini, F. Driencourt-Mangin, R. J. Hernández-Pinto and G. Rodrigo, arXiv:1604.06699.
[2] C. G. Bollini and J. J. Giambiagi, Nuovo Cim. B 12 (1972) 20. G. 't Hooft and M. J. G. Veltman, Nucl. Phys. B 44 (1972) 189. G. M. Cicuta and E. Montaldi, Lett. Nuovo Cim. 4 (1972) 329. J. F. Ashmore, Lett. Nuovo Cim. 4 (1972) 289.
[3] T. Kinoshita, J. Math. Phys. 3 (1962) 650.
[4] S. Catani and M. H. Seymour, Nucl. Phys. B 485 (1997) 291 [Erratum-ibid. B 510 (1998) 503]. S. Catani and M. H. Seymour, Phys. Lett. B 378 (1996) 287. S. Frixione, Z. Kunszt and A. Signer, Nucl. Phys. B 467 (1996) 399. A. Gehrmann-De Ridder, T. Gehrmann and E. W. N. Glover, JHEP 0509 (2005) 056. S. Catani and M. Grazzini, Phys. Rev. Lett. 98 (2007) 222002. V. Del Duca, C. Duhr, G. Somogyi, F. Tramontano and Z. Trócsányi, JHEP 1504 (2015) 036. M. Czakon, Phys. Lett. B 693 (2010) 259. R. Boughezal, C. Focke, X. Liu and F. Petriello, arXiv:1504.02131 [hep-ph]. J. Gaunt, M. Stahlhofen, F. J. Tackmann and J. R. Walsh, JHEP 1509 (2015) 058.
[5] S. Catani, T. Gleisberg, F. Krauss, G. Rodrigo and J. C. Winter, JHEP 0809 (2008) 065.
[6] I. Bierenbaum, S. Catani, P. Draggiotis and G. Rodrigo, JHEP 1010 (2010) 073.
[7] I. Bierenbaum, S. Buchta, P. Draggiotis, I. Malamos and G. Rodrigo, JHEP 1303 (2013) 025.
[8] I. Bierenbaum, P. Draggiotis, S. Buchta, G. Chachamis, I. Malamos and G. Rodrigo, Acta Phys. Polon. B 44 (2013) 2207.
[9] S. Buchta, G. Chachamis, P. Draggiotis, I. Malamos and G. Rodrigo, JHEP 1411 (2014) 014.
[10] S. Buchta, PhD thesis, Universitat de València, 2015, arXiv:1509.07167 [hep-ph].
[11] S. Buchta, G. Chachamis, P. Draggiotis and G. Rodrigo, arXiv:1510.00187 [hep-ph].
[12] R. J. Hernández-Pinto, G. F. R. Sborlini and G. Rodrigo, JHEP 1602 (2016) 044.
[13] G. F. R. Sborlini, R. Hernández-Pinto and G. Rodrigo, PoS EPS-HEP 2015 (2015) 479.
[14] G. F. R. Sborlini, PoS RADCOR 2015 (2015) 082.
[15] G. F. R. Sborlini, F. Driencourt-Mangin, G. Rodrigo, e-Print: arXiv:1608.01584 [hep-ph].
[16] G. Rodrigo, F. Driencourt-Mangin, G. F. R. Sborlini and R. J. Hernández-Pinto, PoS (LL2016) 037, arXiv:1608.01800 [hep-ph].

Top quark decays with flavor violation in extended models

J I Aranda[1], D E Gómez[1], I Cortés-Maldonado[2], F Ramírez-Zavaleta[1], E S Tututi[1]

[1] Facultad de Ciencias Físico Matemáticas, Universidad Michoacana de San Nicolás de Hidalgo, Avenida Francisco J. Mújica S/N, 58060, Morelia, Michoacán, México
[2] Cátedras Consejo Nacional de Ciencia y Tecnología, México

E-mail: `feramirez@umich.mx`

Abstract. We analyze the top quark decays $t \to cg$ and $t \to c\gamma$ mediated by a new neutral gauge boson, identified as Z', in the context of the sequential Z model. We focus our attention on the corresponding branching ratios, which are a function of the Z' boson mass. The study range is taken from 2 TeV to 6 TeV, which is compatible with the resonant region of the dileptonic channel reported by ATLAS and CMS Collaborations. Finally, our preliminary results tell us that the branching ratios of $t \to cg$ and $t \to c\gamma$ processes can be of the order of 10^{-11} and 10^{-13}, respectively.

1. Introduction

The Standard Model of elementary particles (SM) is the most successful physical theory developed so far because it has offered very precise predictions of numerous experimental results. The SM is a theory which includes the study of the three fundamental interactions, the electromagnetic, weak and strong; however, to date, it is known that the SM can not explain several questions such as the flavor problem; the phenomena of flavor violation can arise in many of its well-motived extensions. One interesting feature of most models beyond the SM is the presence of generalized neutral current sectors, which favor nondiagonal transitions mediated by neutral massive spin-1 particles. The presence of a new neutral massive gauge boson Z' is predicted in the context of numerous extensions of the SM. The simplest extended model that predicts the existence of the neutral weak gauge boson is based on the extended electroweak gauge group $SU_C(3) \times SU_L(2) \times U_Y(1) \times U'(1)$ [1].

Today it is well known the existence of neutrino oscillations, which implies that lepton flavor conservation is violated in nature. Moreover, only signals of lepton flavor violation coming from transitions between neutral leptons have been detected. Therefore, flavor-violating transitions between charged fermions constitute an interesting topic, since if they occur in nature are additional evidence of flavor violation. The purpose of this work is to study the effect of the flavor violation of the up-quark sector mediated by Z' through $t \to cg$ and $t \to c\gamma$ processes, particularly, our interest is to calculate the branching ratios of both processes, and then, to contrast these with the SM predictions.

Currently, exhaustive searches for the processes mentioned above are in progress at the LHC, but so far they have not been detected and only upper limits have been established [2, 3]. Up to now, the ATLAS Collaboration has imposed the most stringent bound over the $t \to cg$ decay, *i.e.*, $Br(t \to cg) < 2.7 \times 10^{-4}$ via single top production coming from the $pp \to qg \to t \to W(\to l\nu_l)b$ reaction [2]. In contrast, the CMS Collaboration establishes that $Br(t \to c\gamma) < 1.7 \times 10^{-3}$ through $pp \to qg \to t\gamma \to W(\to l\nu)b\gamma$ reaction, being the most stringent limit [3].

2. The extension model

We start from the most general renormalizable Lagrangian, which includes flavor violation mediated by a new neutral massive gauge boson Z'. This comes from any extended Lagrangian model or grand unification theories [1, 4], and can be written as follows:

$$\mathcal{L}_{NC} = \sum_{i,j} \left[\overline{f_i}\gamma^\alpha(\Omega_{Lf_if_j}P_L + \Omega_{Rf_if_j}P_R)f_j + \overline{f_j}\gamma^\alpha(\Omega^*_{Lf_if_j}P_L + \Omega^*_{Rf_if_j}P_R)f_i \right] Z'_\alpha \quad (1)$$

where f_i is any fermion of the SM, Z'_α is the neutral gauge boson predicted by several extensions of the SM and $P_{L,R} = \frac{1}{2}(1 \mp \gamma_5)$ are the chiral projectors. The $\Omega_{Lq_iq_j}, \Omega_{Rq_iq_j}$ parameters represent the strength of the $Z'_{q_iq_j}$ coupling, where q_i is any SM quark. The Lagrangian mentioned above includes couplings that conserve and violate flavor mediated by the Z' boson. The flavor-conserving couplings, $Q^{f_i}_{L,R}$ [4, 5], are related with the Ω parameters as follows: $\Omega_{Lf_if_i} = -g_2 Q^{f_i}_L$ and $\Omega_{Rf_if_i} = -g_2 Q^{f_i}_R$, where g_2 is the gauge coupling of the Z' boson. For the various extended models we are interested in, the gauge couplings of $Z's$ are of the form $g_2 = \sqrt{\frac{5}{3}} \sin\theta_W g_1 \lambda_g$, where $g_1 = g/\cos\theta_W$ and λ_g depends on the symmetry breaking pattern, being of $\mathcal{O}(1)$ [5]. In the sequential Z model, the gauge couplings fulfil $g_2 = g_1$. It should also be mentioned that g is the weak coupling constant and θ_W is the Weinberg angle.

3. $t \to cg$ and $t \to c\gamma$ processes

For the analysis of $t \to cg$ and $t \to c\gamma$ it is necessary to known the Feynman rules involved. The contributions of the vertices that violate flavor and those that respect it can be extracted from the renormalizable Lagrangian given in Eq. (1). Specifically, the Feynman rules associated with the couplings $Z'tc$ and $Z'tt$ are given as $\gamma^\alpha(\Omega_{Ltc}P_L + \Omega_{Rtc}P_R)$ and $-g_2(Q^t_L P_L + Q^t_R P_R)$, respectively. In the rest of the paper we will assume pure vector couplings (i.e. $\Omega_{Ltc} = \Omega_{tc}$ and $\Omega_{Rtc} = \Omega_{tc}$). The contribution of the $Z'tc$ vertex to the $t \to cg$ and $t \to c\gamma$ decays is computed by using the Feynman diagrams shown in Fig. 1. After applying the Passarino-Veltman reduction

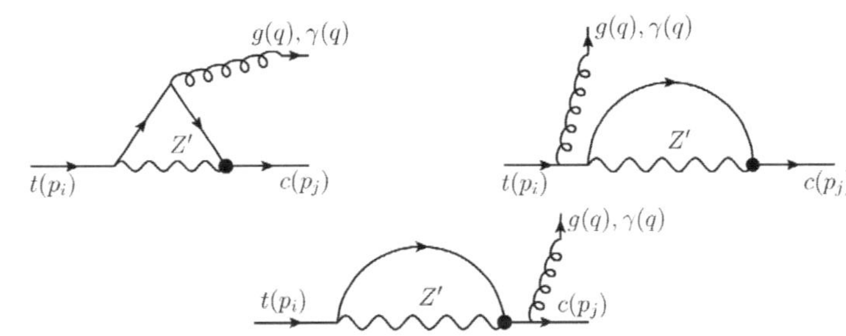

Figure 1. Feynman diagrams contributing to the $t \to cg, c\gamma$ decays.

scheme, the total amplitude of these processes takes the following form: the total amplitude of

$t \to cg$ decay is given by

$$\mathcal{M}(t \to cg) = \frac{igg_2}{64\pi^2 m_i}\bar{u}(p_j)\sigma^{\mu\alpha}q_\alpha\epsilon^a_\mu(q)\Bigg[F_1(Q^t_L - Q^t_R)(\Omega_{Ltc} - \Omega_{Rtc}) + F_2(Q^t_L\Omega_{Ltc} + Q^t_R\Omega_{Rtc}) + \Big(F_1(Q^t_L - Q^t_R)(\Omega_{Ltc} + \Omega_{Rtc}) + F_2(Q^t_L\Omega_{Ltc} - Q^t_R\Omega_{Rtc})\Big)\gamma^5\Bigg](t^a)_{kl}\, u(p_i), \tag{2}$$

where $(t^a)_{kl} = (\lambda^a/2)_{kl}$, λ^a are Gell-Mann matrices, g_s is the strong coupling constant and $\epsilon^a_\mu(q)$ is the gluon polarization vector. In contrast, the total amplitude for the $t \to c\gamma$ decay can be written as

$$\mathcal{M}(t \to c\gamma) = \frac{ieg_2Q_u}{64\pi^2 m_i}\bar{u}(p_j)\sigma^{\mu\alpha}q_\alpha\epsilon_\mu(q)\Bigg[F_1(Q^t_L - Q^t_R)(\Omega_{Ltc} - \Omega_{Rtc}) + F_2(Q^t_L\Omega_{Ltc} + Q^t_R\Omega_{Rtc}) + \Big(F_1(Q^t_L - Q^t_R)(\Omega_{Ltc} + \Omega_{Rtc}) + F_2(Q^t_L\Omega_{Ltc} - Q^t_R\Omega_{Rtc})\Big)\gamma^5\Bigg]u(p_i), \tag{3}$$

where e is the electric charge of the electron, $Q_u = 2/3$ and $\epsilon_\mu(q)$ is the photon polarization vector. The form factors F_1 and F_2 are explicitly given as

$$F_1 = \frac{m_i^2}{m_{Z'^2}} - 6\Bigg(B_0(0, m_i^2, m_{Z'}^2) - B_0(m_i^2, m_i^2, m_{Z'}^2)\Bigg), \tag{4}$$

$$F_2 = 2\Bigg[1 + 2\Bigg(C_0(m_i^2, 0, 0, m_i^2, m_{Z'}^2, m_i^2)m_i^2 + \frac{m_{Z'}^2}{m_i^2}\Bigg(B_0(0, m_i^2, m_{Z'}^2) - B_0(m_i^2, m_i^2, m_{Z'}^2)\Bigg)\Bigg)\Bigg], \tag{5}$$

where $C_0(m_i^2, 0, 0, m_i^2, m_{Z'}^2, m_i^2)$, $B_0(0, m_i^2, m_{Z'}^2)$ and $B_0(m_i^2, m_i^2, m_{Z'}^2)$ are Passarino-Veltman scalar functions. The amplitudes given in Eqs. (2) and (3) are free of ultraviolet divergences because the B_0 functions appear subtracting each other and respect gauge invariance.

4. Results and conclusions

Since our main goal is the numerical estimation of the branching ratios for the $t \to cg$ and $t \to c\gamma$ processes, we start from the definition of the decay width [6] and we found that it is given as

$$\Gamma(t \to cV) = \frac{1}{16\pi\, m_t}\,|\mathcal{M}(t \to cV)|^2 \tag{6}$$

where m_t is the top quark mass and $V = g, \gamma$. Based on this definition and squaring the amplitude, we can obtain the branching ratio associated with each process of our study

$$Br(t \to c\gamma) = \frac{\alpha g_2^2 Q_u^2}{4096\pi^4}\left[(|F_1(Q^t_L - Q^t_R) + F_2Q^t_L|^2 + |F_1(Q^t_R - Q^t_L) + F_2Q^t_R|^2)|\Omega_{tc}|^2\right]\frac{m_i}{\Gamma_t}, \tag{7}$$

$$Br(t \to cg) = \frac{\alpha_s g_2^2 C_F}{4096\pi^4}\left[(|F_1(Q^t_L - Q^t_R) + F_2Q^t_L|^2 + |F_1(Q^t_R - Q^t_L) + F_2Q^t_R|^2)|\Omega_{tc}|^2\right]\frac{m_i}{\Gamma_t}, \tag{8}$$

where $C_F = \frac{4}{3}$. Here $\alpha_s = \frac{g_s^2}{4\pi}$ and $\alpha = \frac{e^2}{4\pi}$ are the coupling constants of the strong and electromagnetic interactions, respectively. Since we only try to estimate the strength of flavor changing neutral currents (FCNC), it is assumed that $\alpha_s = 0.12$.

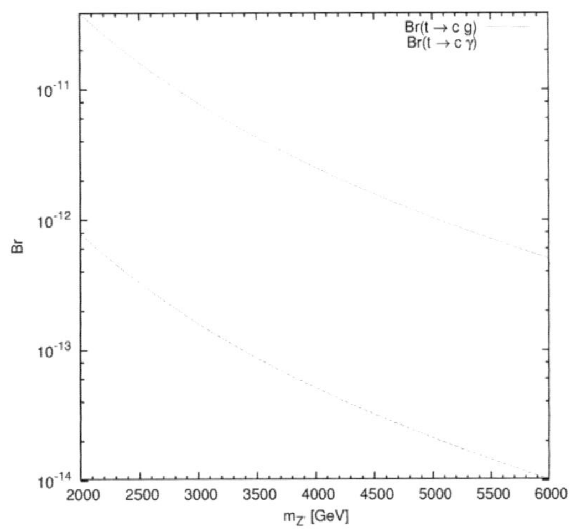

Figure 2. $\mathrm{Br}(t \to cg)$ ($\mathrm{Br}(t \to c\gamma)$) as a function of $m_{Z'}$.

In this work we use the LoopTools package for the numerical evaluation of the branching ratios [7]. Our numerical study will only refer to the sequential Z model. The behaviors of $Br(t \to cg)$ and $Br(t \to c\gamma)$ as a function of the Z' boson mass are shown in Fig. (2). The value of Ω_{tc} was set at 9.25×10^{-2}, which is consistent with the estimate obtained for the $Z'tc$ coupling in a previous work [8]. In both graphs we can see that as the Z' boson mass grows, the branching ratio of both processes decreases. For the mass interval 2.5 TeV$< m_{Z'} < 3.5$ TeV, it can be clearly appreciated that in the $t \to cg$ and $t \to c\gamma$ processes induced by FCNC, $Br(t \to cg) \sim 10^{-11}$ and $Br(t \to c\gamma) \sim 10^{-13}$, respectively, which is one order of magnitude above the SM predictions [9]. The explored region of the Z' boson mass is between 2.5 TeV$< m_{Z'} < 6$ TeV, which is consistent with the experimental limits on the Z' boson mass reported by the experimental collaborations ATLAS and CMS [10]. We stress that if two-loop SM contributions were considered, they are expected to be suppressed regarding the one-loop SM contributions, since these dominant contributions are very suppressed because the GIM mechanism, so the new physics contribution should be dominant.

Acknowledgments
This work has been partially supported by SNI and CIC-UMSNH (México).

References
[1] Langacker P, Rev. Mod. Phys. **81**, 1199 (2008); Durkin L S and Langacker P, Phys. Lett. B**166**, 436 (1986); Cvetic M and Langacker P, Proceedings of Ottawa 1992: Beyond the standard model III, 454-458, (1992).
[2] ATLAS Collaboration, Phys. Lett. B**712**, 351 (2012).
[3] Khachatryan V *et al.* [CMS Collaboration], JHEP**04**, 035 (2016).
[4] Aranda J I, Montaño J, Ramírez-Zavaleta F, Toscano J J and Tututi E S, Phys. Rev. D**86**, 035008 (2012).
[5] Robinett R W and Rosner J L, Phys. Rev. D**25**, 3036 (1982); Robinett R W, Phys. Rev. D**26**, 2388 (1982).
[6] Olive K A *et al.*, Chin. Phys. C**38**, 090001 (2014).
[7] Hahn T and Perez-Victoria M, Comput. Phys. Commun. **118**, 153 (1999).
[8] Aranda J I, Ramírez-Zavaleta F, Toscano J J and Tututi E S, J. Phys. G: Nucl. Part. Phys. **38**, 045006 (2011).
[9] Aguilar-Saavedra J A and Nobre B M Phys. Lett. B**553**, 251 (2003).
[10] Aad G *et al.* (ATLAS Collaboration), ATLAS-CONF-2015-070, (2015); The CMS Collaboration, CMS-PAS-EXO-12-015, (2015).

MWPF IOP Publishing
Journal of Physics: Conference Series **761** (2016) 012023 doi:10.1088/1742-6596/761/1/012023

Design, construction, characterization and use of a detector to measure time of flight of cosmic rays

A. C. Araujo[1], J. Félix[2]
[1,2]División de Ciencias e Ingenierías, Campus León, Universidad de Guanajuato
E-mail: [1]`araujoma2012@licifug.ugto.mx`, [2]`felix@fisica.ugto.mx`

Abstract. In the study of cosmic rays, measurements of time of flight and momentum have been used to identify incident particles from its physical properties, like mass. In this poster we present the design, construction, characterization, and operation of a detector to measure time of flight of cosmic rays. The device is comprised of three plates of plastic scintillator arranged in vertical straight line, they are coupled to one photomultiplier tube. The analogical output has been connected to a data acquisition system to obtain the number of digital pulses per millisecond. We present preliminary results.

1. Introduction

Cosmic rays at the top of the terrestrial atmosphere include all stable charged particles and nuclei with lifetimes of order 10^6 years or longer, when they collide with atmospheric atoms produce secondary particles [1]. Scintillation is a method for detecting this cosmic radiation. It uses photomultiplier tubes (PMT) coupled to scintillators that produce flash light when struck by radiation. These flashes are detected by the PMT, hence the output pulses of it contain information on the energy of the incident particle. Scintillation is applied in Time of Flight detectors (TOF) that are made to measure the velocity of the particles from the time difference between the pulses obtained from the scintillator counters [2]. Velocity can be combined with a measure of the momentum of particles to determine its mass, and therefore its identity [3].

2. Hypothesis

Particles can be identified by measuring its time of flight and momentum.

3. Objectives

3.1. General Objectives

Design, construct, operate, and characterize a detector to measure time of flight of cosmic rays.

3.2. Specific Objectives

- To detect particles from the flash light produced in plates of plastic scintillator in a particular range of time.
- Determine the velocity of particles that go through the detector.
- Calculate the mass of particles to identify them.

4. Design of the experimental prototype

The main parts of the detector are three plates of plastic scintillator produced by Sumitomo Chemical, arranged in a vertical straight line, separated by one meter with extensor aluminium plates, as is shown in Figure 1.

They are connected with optical fibers of 3.15 m long, and a guide of light to a photomultiplier tube from Hamamatsu (H10493-012), whose effective area is a circle of 25 mm of diameter [4]. The dimensions of the plastic scintillators are 5 cm length, 0.6 cm width, and 5 cm height. They are covered with aluminium plates to block out external light. The inside walls of the aluminium plates are polished to mirror level to reflect photons inside the box, and ensure that most of them are guided toward the PMT. Three pieces with a set square form were designed to set the plastic scintillators. The base for the photomultiplier tube is in the middle part of the detector, which is inside of an aluminium tube.

5. Construction

We cut, polished, and fixed with screws all the pieces. Finally, the detector was fixed to a cabinet using aluminium tape. This process is presented in Figure 2.

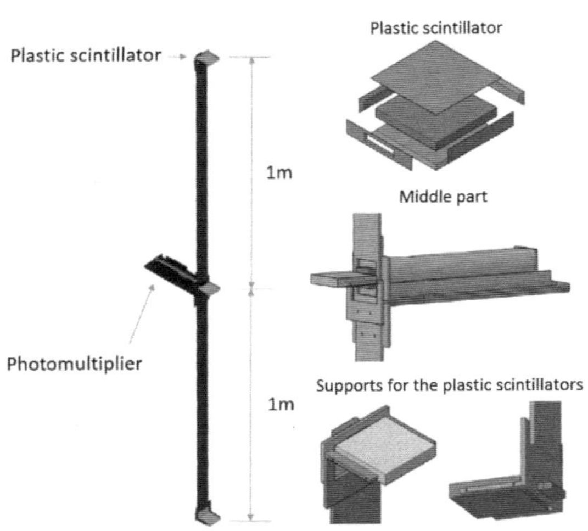

Figure 1. Designed parts of the detector.

Figure 2. Construction and assembling.

6. Tests

Tests were done at Fermi National Accelerator Laboratory (Fermilab) in Batavia, Illinois. The block diagram for the connection of the detector is shown in Figure 3. The operation voltages for the PMT applied were: Low Voltage Input (Red) +14.01 V, Low Voltage Input (Green) -14.04 V, and Vcont Input (White) +1.135 V.

The discriminator board was powered applying +5 V and -5 V to its terminals, the input signal was produced for the detector, and digital signal was connected with a BNC coaxial cable to a data acquisition system of 32 channels, that is the CompactRIO from National Instruments that was programmed to obtain the number of counts per millisecond.

7. Results

The analogical and digital outputs are shown in Figure 4. The configuration parameters for the oscilloscope were: Channel 1 (Digital) 2.00 V, Channel 2 (Analogical) 200 mV, and Time 1.00 μs. The chart Counts vs Time, and the histogram for the frequency distribution are shown in Figures 5, and 6.

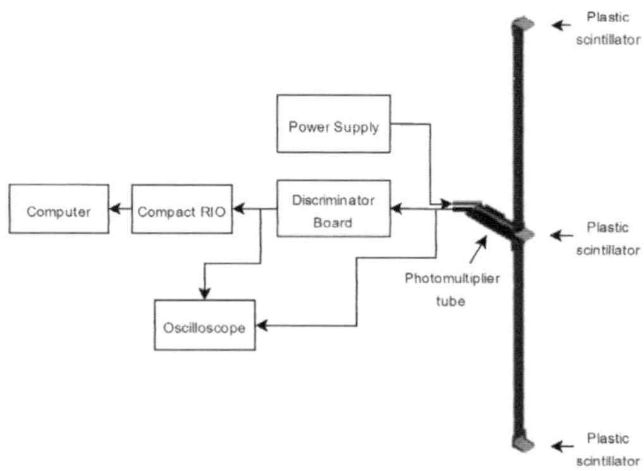

Figure 3. Block Diagram for the connection of the system.

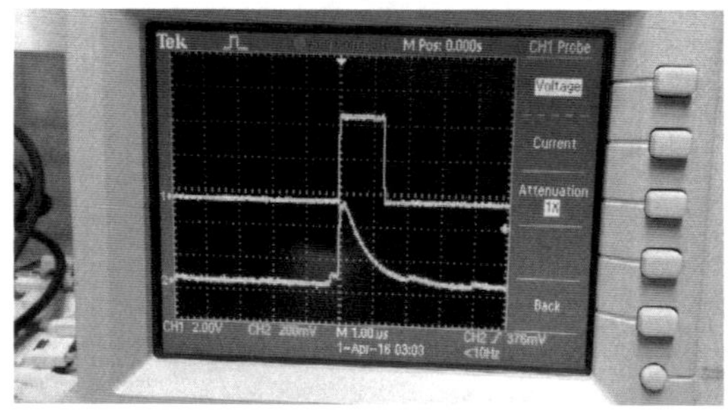

Figure 4. Oscilloscope display for the detector outputs. Digital output is channel 1 at the top. Analogical output is channel 2 at the bottom.

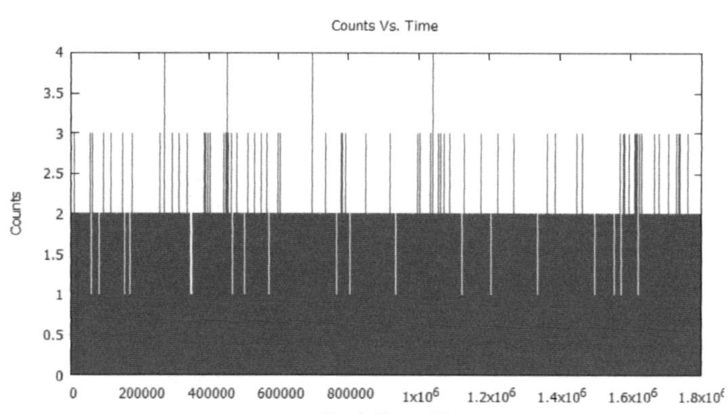

Figure 5. Counts per millisecond vs time in a recording of 30 minutes

8. Conclusion

Analogical signals of 400mV of amplitude in a time scale of 1 μs were obtained with the photomultiplier tube. A faster data acquisition system is required to see the difference of time in the three pulses produced by the plastic scintillators, since they should be in the range of nanoseconds. That would be necessary for measuring the time of flight of particle that come from cosmic rays.

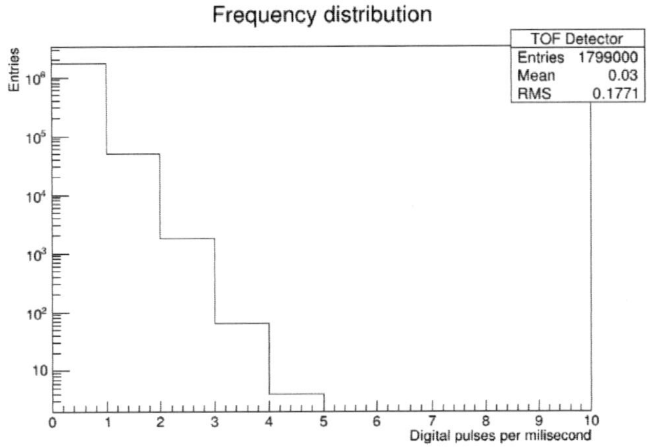

Figure 6. Frequency distribution of digital pulses per millisecond in a recording of 30 minutes.

9. Acknowledgements

To the project 223179, CB-2013-01, Fondo SEP CONACYT. And, to the working group of the Laboratorio de Partículas Elementales de la Universidad de Guanajuato.

References

[1] Particle Data Group, *Cosmic rays*. Available online from: http://pdg.lbl.gov/2014/reviews/rpp2014-rev-cosmic-rays.pdf
[2] Hamamatsu Photonics, *Photomultiplier Tubes. Basics and Applications*, Third Edition, pp 136-139.
[3] CERN, *How detectors works*. Available online at: http://home.cern/about/how-detector-works
[4] Hamamatsu Photonics, Photosensor Module H10493-012 Data Sheet.

Design, construction and characterization of a mini cosmic ray spectrometer

Julian Félix
Department of Physics, Division of Science and Engineering Campus Leon, Guanajuato University
E-mail: `felix@fisica.ugto.mx`

Everardo Granados
Department of Physics, Guanajuato University, León
E-mail: `granadosve2012@licifug.ugto.mx`

Oscar Moreno
Department of Physics, Guanajuato University, León
E-mail: `morenopo2012@licifug.ugto.mx`

Abstract. This article describes the design and construction of a prototype called *mini spectrometer* to see if materials can be characterized using cosmic rays flux around the Earth. It started with common materials like water, air and oil. The mini spectrometer is implemented by two optical techniques: Cherenkov effect and scintillation. The interaction is observed with materials favoring the possibility of characterizing through a digital process.

1. Introduction

Cosmic rays are particles generated by Astrophysical sources, called "primary", those produced by interaction of primaries with interstellar gas are called "secondaries". A source of secondaries is the layer of Earth's atmosphere [1]. The objective is the use of secondary cosmic rays to develop a new technique for characterizing materials, therefore, improve and find applications. Using a natural source of radiation could avoid the spending of an artificial radiation source. Other studies that could be developed include the research for new materials to detect particles.

2. Design

The prototype have three modules located on vertical column, the top and bottom part plastic scintillators were put and the central module was used for introducing different materials. The plastic scintillators are of organic type "Saint-Gobain BC-404" [2] of volume equal to one cubic decimeter. In the back of each module an analogical photomultiplier Sens-Tech Photo-detector PC30CW5 [3] was put to measure the light produced by interaction of cosmic rays with plastic scintillator and the material. In the Fig 1 the plastic scintillator and the photomultiplier are shown.

The detector was made of aluminum owing to his electromagnetic light reflectance property. In Fig. 2 the general design is shown.

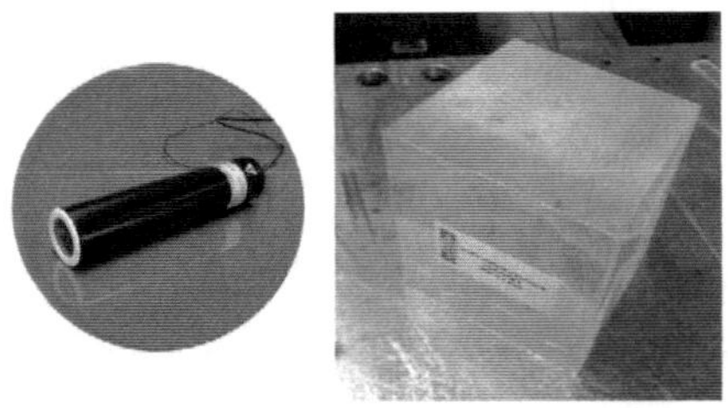

Figure 1. Analogical photomultiplier Sens-Tech Photodetector PC30CW5 (left picture). Plastic scintillator Saint-Gobain BC-404 (right picture).

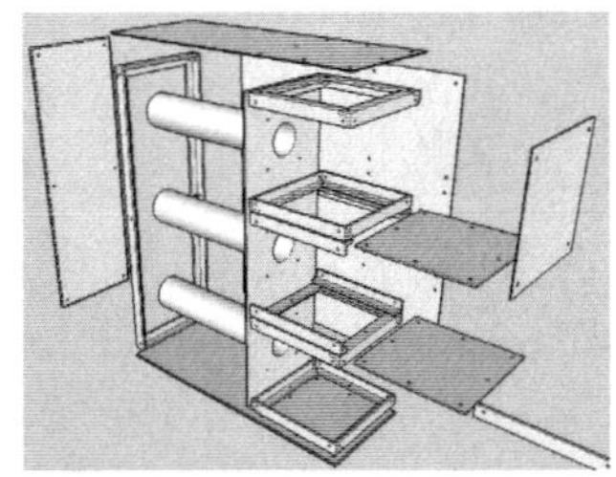

Figure 2. Internal parts of the prototype.

The photomultiplier Inputs, have a distinction color to his proper operation: Red, white and black. Operating ranges are: Red (5-8) V , white (0.3-1.8) V, black is the reference to 0 V.

The use of three photomultiplier requires at least three dual voltage sources, to avoid using many sources an electronic board of six channels output was designed to supply three photomultipliers using one dual voltage source. The three-dimensional circuit design is shown in Fig. 3

Figure 3. Six channels electronic board.

3. Construction

Aluminum parts were cut by "water jet" to have parts with precise measurements. Plastic scintillators were covered with mirrors to reflect the photons generated in the modules and reach the optical part of the photomultiplier. Center module is internally polished to mirror level for the same purpose. The process of building the prototype is shown in Fig. 4 - 7.

The LM3174 voltage regulator [4] was used to make the six-channel electronic board. The designed circuit was printed on a phenolic plate with serilustre ink to protect the copper ferric chloride. The building process is shown in Fig. 8 - 11

To observe the voltage in each channel a LCD Display Digital Panel Voltmeter [5] was used.

4. Test

The connection diagram is shown in the Fig. 12, photomultiplier outputs were connected to a data acquisition system to digitize the signal (CompactRIO). Counts were captured in a LabView program developed in laboratory, it takes the analogical counts generated every millisecond per module. Analogical and digital signal were checked by an oscilloscope as shown in Fig. 13. The first material used was potable water.

Figure 4. Prototype back.

Figure 5. Aluminium tube for each photomultiplier.

Figure 6. Side-View.

Figure 7. Plastic scintillator covered.

Figure 8. Phenolic plate in ferric chloride

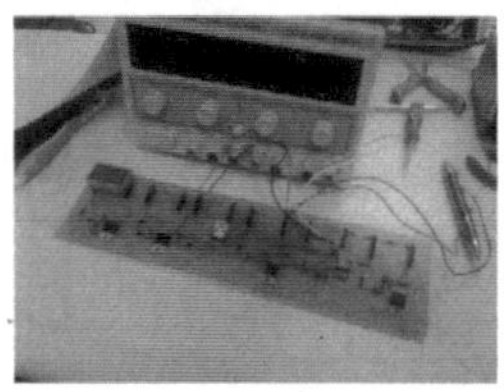

Figure 9. Phenolic plate test

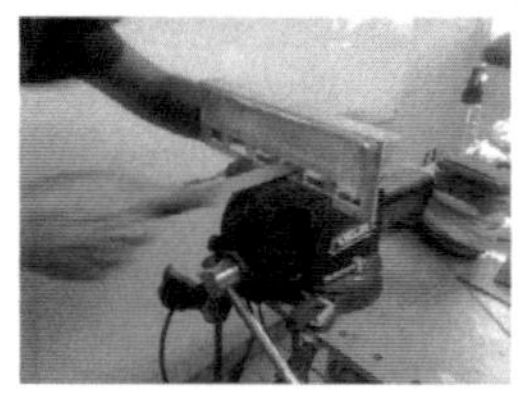

Figure 10. Chassis contruction

Figure 11. Six-Channels voltage source finished

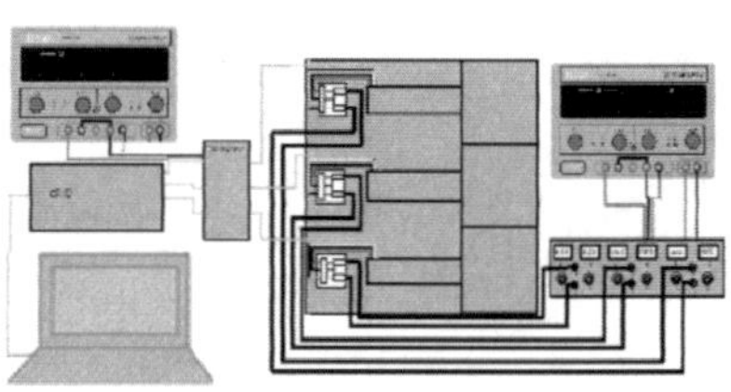

Figure 12. Connection Diagram

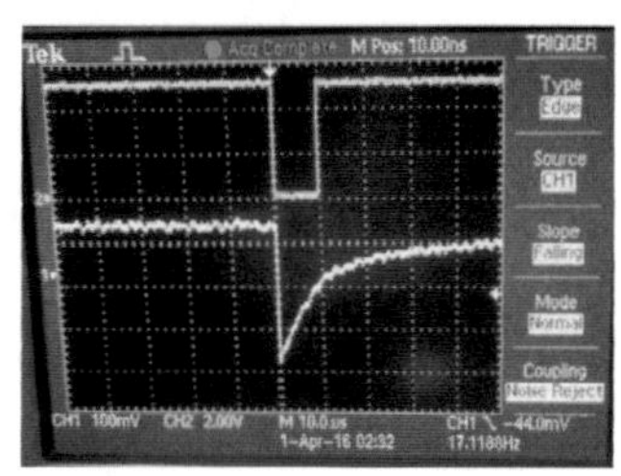

Figure 13. Analog and digital signal.

Several tests were taken to found the most stable region for each photomultiplier with a combination of allowable voltage ranges.

The stable regions obtained are shown in table 1.

Table 1. Stable regions of voltages for each photomultiplier.

Channel	Red (v)	White (V)
1	5.0	0.8
2	7.0	1.1
3	5.5	0.8

In table 1 "channel one" is the bottom module and "channel three" is the upper module. The first materials for test were: distilled water, mineral water and oil, each one was taken during 30 minutes.

5. Results

Data were plotted in counts versus time for each voltage, and the frequency distribution for channel two are shown in the next Figures for distilled water.

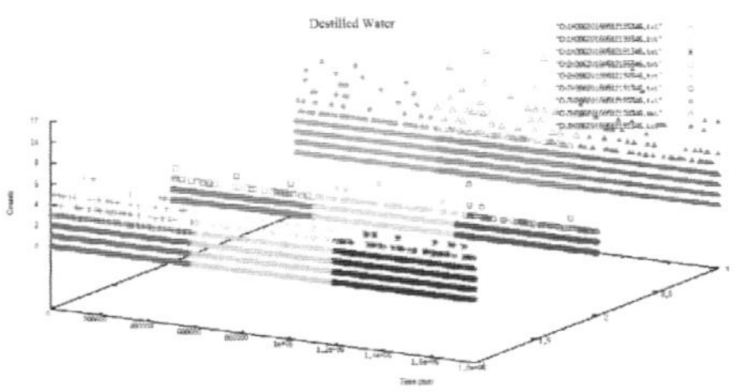

Figure 14. Example of graph counts vs time for each voltage, in the case of distilled water.

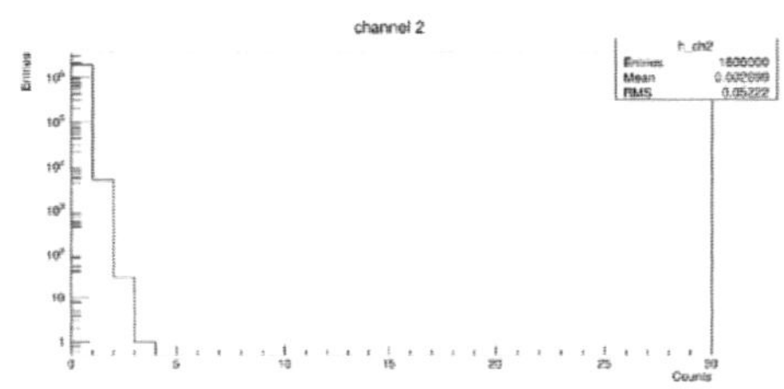

Figure 15. Frequency distribution.

To observe differences between materials, simultaneous pulses in all modules were captured by a program made in C, this ensures that a particle passed through the three modules avoiding dark counts. In table 2 is shown the total simultaneous pulses obtained for each material in a data taking of thirty minutes.

Table 2. Simultaneos counts.

Material	Counts
Distilled water	477
Oil	756
Mineral water	569

6. Conclusions

The amount of cosmic rays that hit in a determined area is not a constant,on the other hand, the frequency distribution depends on the material, therefore, it is possible to establish a technique for identifying materials. To improve the data collection accuracy it is indispensable to have a reference material. Finally the study keep going until new results are obtained for comparison.

References

[1] http://pdg.lbl.gov/2015/reviews/rpp2015-rev-cosmic-rays.pdf.
[2] Plastic scintillator data sheet
http://www.crystals.saint-gobain.com/uploadedFiles/SG-Crystals/Documents/SG.pdf.
[3] Photodetector data sheet
http://scipp.ucsc.edu/outreach/TW/kliewer/CRMD%20Manuals/P30CW5.pdf
[4] LM317 data sheet
http://www.ti.com/lit/ds/symlink/lm317.pdf.
[5] LCD Display Digital Panel Voltmeter data sheet.
http://www.farnell.com/datasheets/1484210.pdf.

ALICE Diffractive Detector Control System for RUN-II in the ALICE Experiment

J C Cabanillas[1], **M I Martínez**[2], **I León**[1]

[1] Universidad Autónoma de Sinaloa (UAS), Josefa Ortíz de Dominguez S/N Culiacan Rosales, Sinaloa, México, CP: 80013
[2] Benemérita Universidad Autónoma de Puebla (BUAP), Cd. Universitaria, 72570 Heróica Puebla de Zaragoza, Puebla, México

Email: juan.carlos.cabanillas.noris@cern.ch, mario.martinez.hernandez@cern.ch, ildefonso.leon.monzon@cern.ch

Abstract. The ALICE Diffractive (AD0) detector has been installed and commissioned for the second phase of operation (RUN-II). With this new detector it is possible to achieve better measurements by expanding the range of pseudo-rapidity in which the production of particles can be detected. Specifically the selection of diffractive events in the ALICE experiment which was limited by the range over which rapidity gaps occur. Any new detector should be able to take data synchronously with all other detectors and to be operated through the ALICE central systems. One of the key elements developed for the AD0 detector is the Detector Control System (DCS). The DCS is designed to operate safely and correctly this detector. Furthermore, the DCS must also provide optimum operating conditions for the acquisition and storage of physics data and ensure these are of the highest quality.

The operation of AD0 implies the configuration of about 200 parameters, as electronics settings and power supply levels and the generation of safety alerts. It also includes the automation of procedures to get the AD0 detector ready for taking data in the appropriate conditions for the different run types in ALICE. The performance of AD0 detector depends on a certain number of parameters such as the nominal voltages for each photomultiplier tube (PMT), the threshold levels to accept or reject the incoming pulses, the definition of triggers, etc. All these parameters affect the efficiency of AD0 and they have to be monitored and controlled by the AD0 DCS.

1. Introduction

ALICE (A Large Ion Collider Experiment) is an experiment dedicated to the study of strongly interacting matter, particularly to the study of heavy ion collisions at very high energies at the LHC in the European Organization for Nuclear Research (CERN) [1]. The new AD0 detector improves the detection capacity for diffractive events in proton-proton (p-p) and lead ion (Pb-Pb) collisions by adding four planes of particle counters at very small angles with respect to the beam direction. The detection of diffractive events in ALICE during the first phase of LHC operation (RUN-I) was limited by the range over which rapidity gaps were identified. AD0 will expand the range in which the production of particles (or lack thereof) can be detected [2]. Another benefit of the AD0 diffractive physics trigger is the larger amount of data recorded due to the fact that about 25% of the results of collision events are of the diffractive type [3]. To integrate AD0 to the ALICE data taking, it was necessary to design a new control architecture specific to this detector.

2. The AD0 Detector

The AD0 detector consists of two sub-detectors called ADA and ADC. Each of them comprises two detector layers, each one formed by four scintillator modules arranged around the LHC beam pipe. The ADA and ADC designations refer to the positions where they are installed at both ends of the ALICE experimental site with respect to the interaction point (IP). The positions of ADA and ADC in the ALICE reference frame and the nomenclature for layers and modules are shown in Fig. 1.

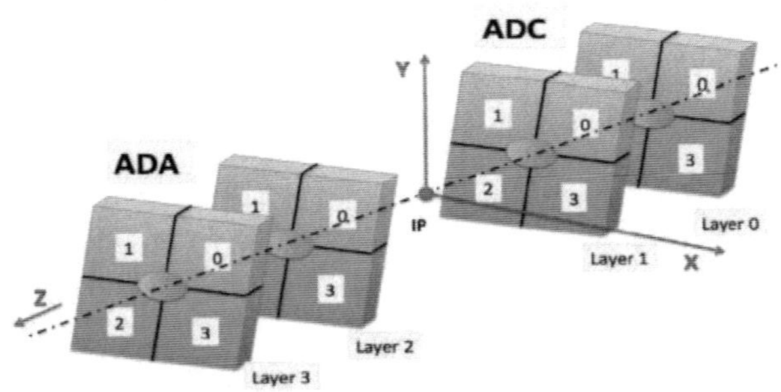

Figure 1. ADA and ADC sub-detectors nomenclature in the ALICE cavern

3. Detector Control System (DCS)

The DCS is responsible for controlling, monitoring and configuring the AD0 detector equipment among which there are commercial hardware devices like power supplies, voltage cards, crates, etc. as well as specific custom equipment like the Front-End Electronics (FEE). Also all relevant parameters for the offline analysis of physical data are periodically archived by the AD0 DCS in the ALICE Online Conditions Database (OCDB) [1]. These tasks are accomplished by sending commands and reading the status from the equipment. The control systems in ALICE are developed using a Supervisory Control and Data Acquisition (SCADA) platform called WinCC Open Architecture (OA) ® (formerly PVSS) from ETM Company. The ADO DCS is integrated to the global control system of the ALICE experiment [4].

3.1 AD0 DCS Software Architecture

The software architecture is a tree-like hierarchy that models the structure of the hardware sub-systems and devices. This tree structure is composed of nodes, each one having a single parent, except the top node. The performance and functionality of each node in the tree hierarchy are implemented as a Finite State Machine (FSM) [5-6]. Fig. 2 shows the simplified software architecture in the AD0 DCS where the main subsystems are shown.

3.2 AD0 DCS Hardware Architecture

AD0 has an architecture compatible with other hardware architectures in the ALICE experiment which are subdivided into three layers; a) supervision, b) process control and c) field layer [1]. The AD0 detector hardware architecture is described in Fig. 3.

3.3 Implementation of AD0 DCS

The Control Unit (CU) and Device Unit (DU) nodes in the control hierarchy of AD0 are implemented as finite state machines. The FSM tool built into the DCS software is based on the SMI++ (State Machine Interface) language. The FSM Framework component allows the implementation of the DCS as a collection of logical objects (CU's) whose state depends on the state of the objects below in the control hierarchy. The state of the DU's depends on the real status of the hardware devices. On the opposite way commands are sent down the hierarchy up to the DU's who drive the devices to actually execute the commands [1].

Journal of Physics: Conference Series **761** (2016) 012025 doi:10.1088/1742-6596/761/1/012025

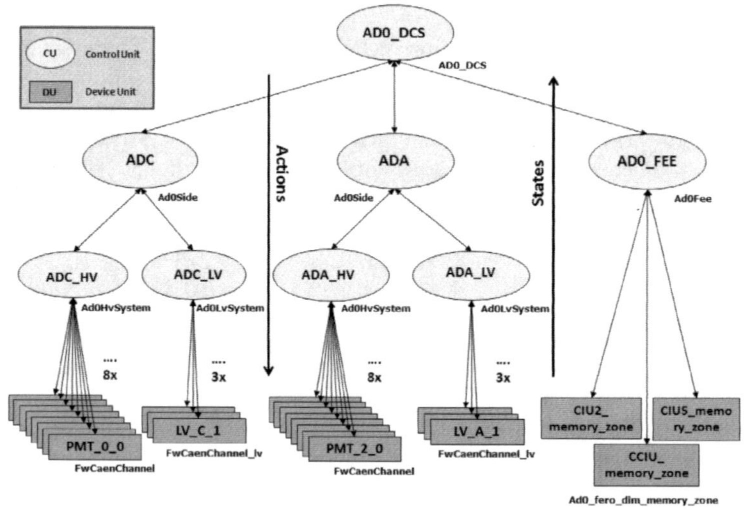

Figure 2. AD0 DCS Software Architecture.

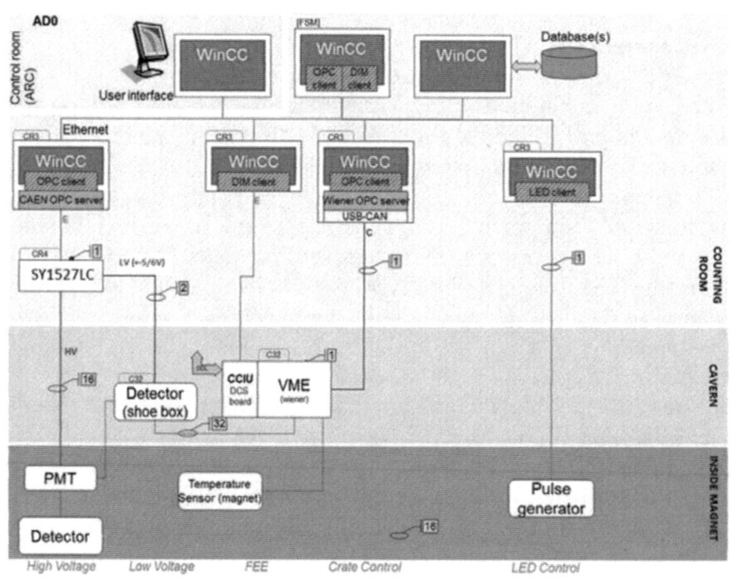

Figure 3. AD0 DCS Hardware Architecture.

3.4 FSM Node on the Top Level of the Hierarchy

At the highest level of the hierarchy the top node (AD0_DCS in Figure 2) is the main control unit. The FSM state diagram for this is shown in Fig. 4. This implementation of AD0 top node is based on the guidelines provided by the ALICE Control Coordination (ACC).

3.5 DCS User Interface (UI)

Custom user interfaces (panels) are associated with any CU or DU in the hierarchy. Commands can be sent, and states can be shown graphically from these interfaces. The user can also navigate through the hierarchy and display the operation panel for each node. The user interface of the AD0 DCS top node is shown in Fig. 5.

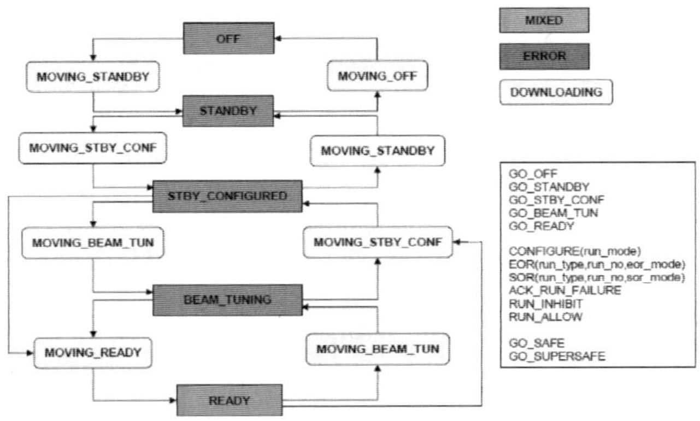

Figure 4. FSM state diagram of the AD0 top node.

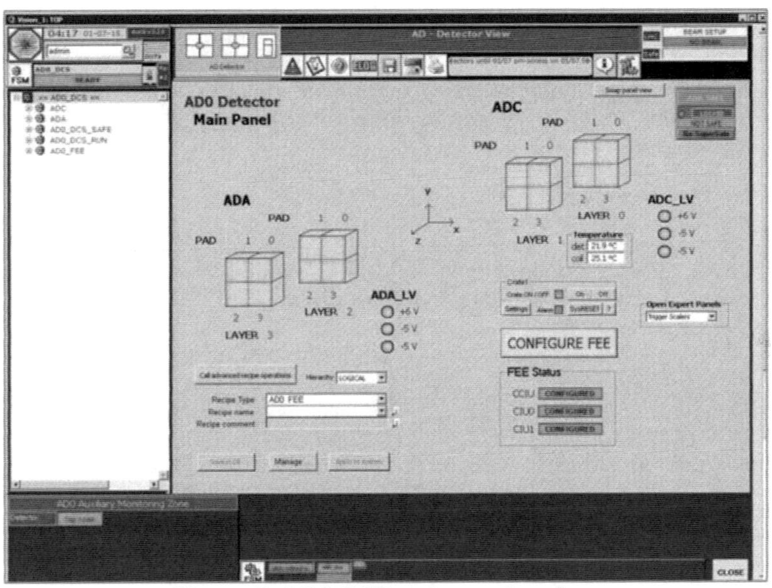

Figure 5. AD0 DCS main user interface.

4. Conclusions

The control system for AD0 detector was successfully developed and integrated in ALICE. AD0 detector is operational for data taking since March 2015 for the second running period (RUN II) of the LHC. AD0 DCS has standard and easy to operate interfaces for the detector expert. Also this control system is continuously maintained and updated following ALICE Control Coordination (ACC) requirements. Finally, AD0 control system has a stable operation within the ALICE central DCS.

References

[1] Aamodt K 2008 *JINST* **3** S08002
[2] Villalobos O and the ALICE Collaboration 2012 J. Phys.: Conf. Ser. **381** 012039
[3] Herrera G 2011 *AIP Conf.Proc.* **1350** 176-179
[4] Carena F et al 2005 *Proc. 10th ICALEPCS Int. Conf. on Accelerator & Large Expt. Physics Control Systems* (Geneva).
[5] Chochula P et al 2012 *Proc. of ICALEPCS2013* (San Francisco) p 1485
[6] Chochula P et al 2012 *Real Time Conference (RT) 18th IEEE-NPSS. IEEE* (Berkeley) p. 1-6

Auger Prime the new stage of the Pierre Auger Observatory, using Universality

Alejandra Parra, Oscar Martínez, Humberto Salazar
Facultad de Ciencias Físico - Matemáticas, (BUAP), Av. San Claudio y 18 Sur, Col. San Manuel, Edif. FM1-101B, C.P. 72570, Puebla, México
E-mail: alejandra.parra@alumno.buap.mx

Abstract. The Pierre Auger Observatory is currently in an update stage denominated AugerPrime. The Observatory will have scintillator detectors on top of each of the surface stations (WCD). The main goal of AugerPrime is to improve the studies on mass composition for ultra high energy cosmic rays, for this purpose AugerPrime will use Universality. The model will parameterize the signal in four principal components, the objective is an adequate discrimination of the muonic and electromagnetic components. We are interested in the discrimination of these two components using simulations. To do that, we are working with OfflineTrunk (the official software of the Collaboration). Our work is focused on the development of some modules for analysis and study of the signal from AugerPrime.

1. Introduction

The cosmic rays were measured for the first time by Victor Hess in 1912, he made several experiments in a balloon with an electroscope. He supposed that as the altitude increases the radiation measure will decrease, however the results showed the opposite [1]. From that moment on, searches to understand and explain this new radiation started.

In 1938 Pierre Auger discovered the extensive air showers (EAS), showers of secondary particles caused by the collision of primary high energy particles with nitrogen molecules from air. He had positioned particle detectors high in the Alps and noticed that two detectors located several meters apart had signals of the arrival of particles at the same time. The conclusion was that he was measuring a shower of 10^{15} eV approximately, never seen before.

In 1962, John Linsley detected the first 10^{20} eV cosmic ray, in New Mexico, US. He used a ground based array of detectors to measure this cosmic ray. His observations suggested that not all cosmic rays are confined to the galaxy. This was the first evidence of ultra-high energy cosmic rays. In 2003 the Pierre Auger Observatory is proposed, to be the largest cosmic ray air shower array in the world.

2. The Pierre Auger Observatory

The main objective of the Observatory is the study of the ultra high energy cosmic rays (UHECR), with energies between $10^{18} - 10^{20}$ eV. The Observatory is located at Malarge, Argentina, at 1400 m over the sea level. The Pierre Auger Observatory combines two different techniques, one is the detection of the fluorescence due to a cosmic ray crossing the atmosphere and the other one is using water Cerenkov stations (WCD). The fluorescence array (FD) consists of 4 buildings located surrounding the surface array. Each building has six telescopes, the telescopes only can work during moonless nights and good weather [2]. The combination of these two techniques allows the observatory for more precise measurement of the ultra-high energy cosmic rays.

Both arrays FD and SD have extensions to measure low energy cosmic rays, around 10^{17} eV. To the FD array the low energy extension is the High Elevation Auger Telescopes (HEAT) which is situated in front of Coihueco, one of the four sites of the FD array. In the case of the SD array, the extension is called Infill. This region consists in 63 surface stations, separated by 750 m, half the usual separation [3]. Both arrays are located in the same region, with the purpose of having simultaneous measurements.

Currently the Observatory is beginning a new stage, called AugerPrime. This update consists on placing a scintillator detector on top of each WCD station, this makes possible the separation of the electronic and muonic components of the showers. It is planned that AugerPrime works for the next ten years, during this time the Observatory will have twice the original data and will be able to collect additional information for important questions as:

- Higher certainty of the origin of the flux suppression at the highest energies, if it is due to propagation effects, and the maximum energy reached by astroparticle sources.
- A proton flux contribution at highest energies.
- A better understanding of the air shower cosmic rays and the interaction of the particles in the atmosphere. This can lead the Observatory to study the hadronic models and a deeper study of the particle physics involved, to very high energies.

3. AugerPrime

One of the principal motivations for this additional array of detectors is to add information about the particles in the air showers independent from the measurements did by the water-Cerenkov stations. To do that, the additional measurements should be done in the same position as the WCD with a detector with different response to the components of the EAS. The chosen detector is a plane of plastic scintillator positioned on top of the surface detector [4].

The Surface Scintillator Detectors (SSD) consist in one module of $\approx 4m^2$ extruded plastic scintillators which are read out by wavelength-shifting fibers coupled to a single photo detector. The active part of the scintillator is a plane made by 12 1.5 m long bars of extruded polystyrene scintillator. Each bar is 1 cm thick and 10 cm wide. The fibers are positioned following the grooves of the routers at both ends, in a "U" configuration that maximizes light yield and allows the use of a single photomultiplier [5].

The external detector enclosure is made from aluminum to guarantee light tightness, robustness for 10 years of operation in the field, and enough rigidity for transportation. A double aluminum roof is installed, separated by 2 cm, to allow air flow and reduce the temperature changes. The temperature control is very important not just for correcting the behavior of the electronics but also for the aging of the detector.

3.1. Status

In march of this year the site to establish the Engineering Array (EA) of AugerPrime was decided, the requirements for the site were:

- An accessible area, relatively near to the Central Station.
- Solid ground, accessible during all year.
- Landowners should agree with the installation of new detectors.
- Due to communication problems, the infill area was avoided.

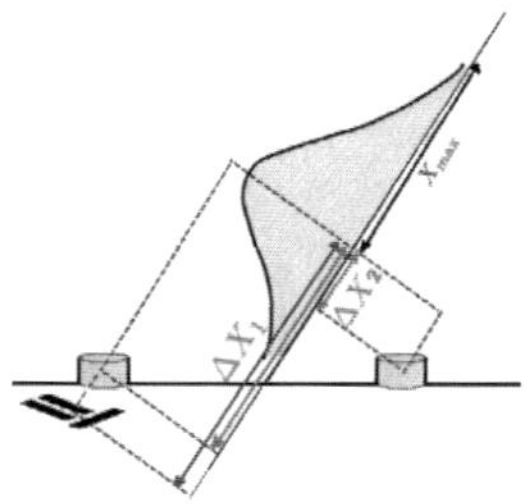

Figure 1. Integral along the shower axis of the density between the height at the projected detector position at the height of the shower maximum.

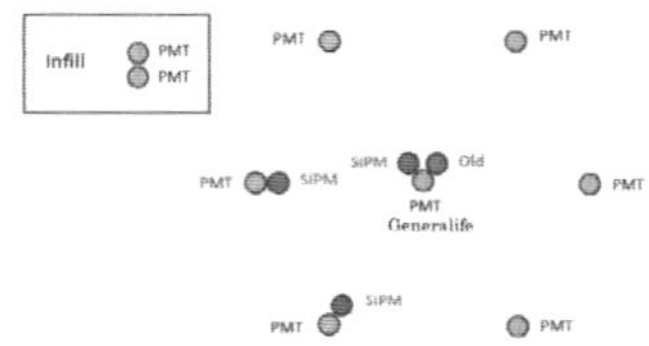

Figure 2. Layout of the Engineering Array. The different configurations of the detectors are shown.

The chosen place is around the tank called "Generalife". The EA will consist of 12 stations, ten of them will be allocated in a hexagon of the standard array. Two more will be placed in the infill area, with the objective of having more statistics. As show in Fig.2, the hexagonal array cell of 7 positions will include 4 single stations, 2 double and one triple station. The EA allows the comparison between the different proposed electronics for the SSD, and also between the old and new stations. According to planned, the EA start will be deployed in September of this year.

3.2. Universality

Additional to the SSD detectors, a new method to describe the air showers was proposed. This new method uses the shower universality, which states that the air shower can be entirely described by its energy, number of muons and maximum slant deep. According to this, an analytical parameterization of the signals was proposed. The signal will be separated in four components:

- The muonic component.
- The pure electromagnetic component.
- The electromagnetic component due to muon interactions and muon decay.
- The electromagnetic component due to low-energy hadrons (jet component).

Another important quantity used by this parameterization will be DX: the integral along the shower axis of the density between the height at the projected detector position and the height of the shower maximum, Fig.1. For each of the signal components the arrival time of the particles will be different and can also be parameterized.

So, the total expected signal in a surface station at certain distance r and with some relative depth DX, is:

$$S_{tot} = S_{em}(r, DX, E) + N_{\mu}^{rel}[S_{\mu}^{ref}(r, DX, E) + S_{em}^{\mu}(r, DX, E)] + (N_{\mu}^{rel})^{\alpha} S_{em}^{low-energy}(r, DX, E)\,. \tag{1}$$

The alpha parameter is obtained from simulations with different hadronic models, if $\alpha \approx 1$ it will be a very good approximation. The integrated signal as well as the time trace of each surface station is fitted by adjusting the parameters E, DX and N_{μ}^{rel} [6].

The primary energy and the number of muons are strongly related, and can not be determined independently. The approach used to solve this consists in parameterizing the number of muons as a function of the energy and X_{max} using hybrid events. With the number of muons, it is possible to reconstruct the energy E and the depth of the shower maximum X_{max}.

4. Software

Our principal interest is to work with this new detector, so the work should be done with simulations. The program used to simulate the air showers is CORSIKA (COsmic Ray SImulation for KAscade). CORSIKA allows us to fix the principal characteristics of the showers as energy, primary particle, magnetic field, first interaction. At the moment we count with a library of simulated events with the following characteristics:

- Primary energy: from 10^{18} to 10^{20} eV.
- Primary particle: proton and iron.
- Zenith angle: 0°, 25°, 45° and 60°.
- Azimuthal angle: random (0 - 360 degrees).

These simulations are being done in the LNS (Laboratorio Nacional de Supercómputo del Sureste de México), thanks to this it is possible to generate the data sets in reasonable times.

The OfflineTrunk (development version of the official software of the Collaboration) program will be used to know the response of the detector. The program already implemented the Universality method and allows to reconstruct the events using it. There are also some basic modules to simulate the SSD detector. The work in progress is mainly focused in these modules, in order to do an analysis of the signal.

We are working in a proper discrimination of the signal. As was mentioned, the signal will be separated in four components, which allows us a better understanding of the muonic and electromagnetic component and therefore of the mass composition of the showers. The main objective of the work is to have an adequate separation of the signal of each detector and also of the four components of the signal proposed by Universality. Once we reach this objective, an analysis of the different PMT's proposed for the SSD will also be done.

5. Conclusions

The main objective of AugerPrime is improving the studies on mass composition for the ultra high energy cosmic rays. It is planned to work for another ten years, in this way the Observatory Pierre Auger will have double the original data and will be able to do an event by event analysis.

Thanks to the new scintillator detector (SSD), it will be possible to discriminate the muonic and electromagnetic components of the air showers. The chosen method to do this is Universality. The new model proposes a parameterization of the signal in four components. An adequate separation of the signal will allow a better understanding of the ultra high energy spectrum. As a consequence of this, it will be possible to elucidate the possible astrophysics scenarios for the origin of the ultra high energy cosmic rays.

References

[1] A.M. Hillas, "Cosmic Rays", Serie "Selected readings in physics", Pergamon Press, (1972).

[2] Pierre Auger Collaboration, A. Aab et al., "The Pierre Auger Cosmic Ray Observatory", *Nuclear Instruments and Methods in Physics Research Section A: Accelerators, Spectrometers, Detectors and Associated Equipment* **798** (2015) 172-213.

[3] P. Abreu et al. (Pierre Auger Collaboration), Phys. Rev. Lett. 109 (2012) 062002 [arXiv:1208.1520] [hep-ex]

[4] "The Pierre Auger Observatory Upgrade: Preliminary Design Report",
`https://www.auger.unam.mx/Augeriki/UTF_Home?action=AttachFile\&do=get\&target=AugerUpgradePDR_17April2015.pdf`.

[5] "SSD Engineering Array review report, November 2015",
`https://www.auger.una.mx/AugerWiki/Useful_Documents?action=AttachFile\&co=get\&target=SSD_EA_Report_Nov15.pdf`.
[6] M.Ave, R.Engel,J.Gonzalez,D.Heck, T.Pierog,andM.Roth,"Extensive Air Shower Universality of Ground Particle Distributions", *Proc.of31st Int.Cosmic Ray Conf., Beijing* (2011) #1025.

Building a non-perturbative quark-gluon vertex from a perturbative one

Rocio Bermudez
Departamento de Investigación en Física, Universidad de Sonora, Blvd. Transversal S/N, 83000, Hermosillo, Mexico

E-mail: rabermudezr@gmail.com

Abstract. The quark-gluon vertex describes the electromagnetic and the strong interaction among these particles. The description of this interaction at high precision in both regimes, perturbative and non-perturbative, continues being a matter of interest in the context of QCD and Hadron Physics. There exist very helpful models in the literature that explain perturbative aspects of the theory but they fail describing non-perturbative phenomena, as confinement and dynamic chiral symmetry breaking.
In this work we study the structure of the quark-gluon vertex in a non-perturbative regime examining QCD, checking results with QED, and working in the Schwinger-Dyson formalism.

1. Introduction

Studies of the non perturbative quark propagator show that the mass function in the infrared limit $M(p^2)|_{p^2 \to 0}$ agrees with the mass of the constituent quarks, and its perturbative limit is well reproduced in the ultraviolet limit[1].
The quark propagator is intimately connected with the gluon propagator and with the quark-gluon vertex through the Schwinger-Dyson equations (SDE), and the Slavnov-Taylor identities. This is why it becomes necessary to study the behaviour of the gluon propagator and the vertex of this interaction.
The quark-gluon vertex is already described perturbatively at one-loop[2]. Nevertheless, this structure fails as we study mass generation via dynamical chiral symmetry breaking, which is a non-perturbative phenomenon.
In this work we are going to study the behaviour of the quark-gluon vertex, and we present a guide of the basic ideas needed to build this vertex.

2. Key elements

Even though the quark-gluon vertex is not an observable, it is necessary to preserve its gauge invariance. This is accomplished by using Ward-Slavnov-Taylor identity[3, 4]. This identity relates the vertex with the inverse of the quark propagator and the ghost propagator as follows

$$q^\mu \Gamma_\mu(p,k,q) = G(q^2)[S^{-1}(p)H(p,k,q) - \overline{H}(k,p,q)S^{-1}(k)] \tag{1}$$

In this equation, $\Gamma_\mu(p,k,q)$ is the quark-gluon vertex, $G(p^2)$ is a scalar function associated with the ghost propagator. Functions H and $\overline{H}$ involve the full quark-quark-ghost-ghost vertex, and

$S^{-1}(p^2)$ is the inverse quark propagator defined as

$$S^{-1}(p^2) = \frac{i\gamma \cdot p + M(p^2)}{F(p^2)} \tag{2}$$

Because it is the tree level in the perturbative expansion for small coupling, a guark-gluon vertex must have the same transformation properties of the bare vertex γ^μ under charge conjugation and parity operations. Moreover, we need the quark-gluon vertex to be free of kinematic singularities. This implies that the vertex has to have a unique and finite limit when $k^2 \to p^2$.

Finally, the SDEs are an infinite set of coupled equations. Then we need to assure the absence of overlapping divergences of these equations. We achieve this by imposing multiplicative renormalizability, which means that functions $F(p^2)$ and $M(p^2)$ have to be written as a power law in the quark propagator, i.e. $F(p^2) = (p^2)^\nu$, where ν is the anomalous dimension.

3. Procedure

We can separate the full quark-gluon vertex into two kinematic contributions using the Ward-Slavnov-Taylor identity, similar to the case of QED using Ward-Takahashi identity. Actually, WST identity can be split into two parts, abelian and non-abelian, where the abelian part corresponds to the Ward-Takahashi identity.

Thus, we can write the quark-gluon vertex as

$$\Gamma^\mu = \Gamma^\mu_L + \Gamma^\mu_T \tag{3}$$

where Γ^μ_L and Γ^μ_T are the longitudinal and the transverse part of the vertex, respectively. Moreover, $q_\mu \Gamma^\mu_T = 0$, meaning that the transverse part does not contribute to the WST identity.

Now, we extend the decomposition of the QED vertex to the case of QCD. We can write the longitudinal part of the vertex as follows

$$\Gamma^\mu_L(p,k) = \sum_{i=1}^{4} \lambda_i(p^2,k^2) L^\mu_i(p,k) \tag{4}$$

where $\lambda_i = \lambda_i^{(a)} + \lambda_i^{(b)}$, with (a) the abelian contribution and (b) the non-abelian one. For the transverse part we have

$$\Gamma^\mu_T(p,k) = \sum_{i=1}^{8} \tau_i(p^2,k^2) T^\mu_i(p,k) \tag{5}$$

Similar to the longitudinal form factor λ_i, we take $\tau_i = \tau_i^{(a)} + \tau_i^{(b)}$. We use the Ball and Chiu basis for both partitions, except for T^μ_6 which has the opposite sign[5].

For QED, the functions λ_i are defined in [5], and we can find perturbative definitions at one-loop for QCD in [2]. Nevertheless, in order to have a non-perturbative description, we need to extend these definitions[6, 7].

From Eq. (1), the function H can be expressed as

$$H(p,k,q) = \chi_0(p^2,k^2,q^2)\mathbb{I} + \chi_1(p^2,k^2,q^2)\gamma \cdot p + \chi_2(p^2,k^2,q^2)\gamma \cdot k + \chi_3(p^2,k^2,q^2)\sigma^{\mu\nu} p_\mu k_\nu \tag{6}$$

If we perform a perturbative expansion for the χ_i, it follows that χ_0 is the dominant term[2]. So, we can choose the longitudinal part of the quark-gluon vertex as

$$\Gamma^\mu_L(p,k) = G(q^2)\chi_0(p^2,k^2,q^2)\Gamma^\mu_{BC}(p,k) \tag{7}$$

where $G(q^2)$ and $\chi_0(p^2,k^2,q^2)$ are defined perturbatively in [2].

On the other hand, we need to analyse the behaviour of the transverse part in order to make a practical ansatz. There are basics requirements for τ_i coefficients that must be fulfilled. For instance, these coefficients have to be free of kinematics singularities. In general, τ_i need to satisfy the following:

- All τ_i must be symmetric under the exchange of $k^2 \leftrightarrow p^2$, except for τ_6 which must be antisymmetric under this exchange.
- Since the quark propagator is expected to be multiplicative renormalizable, then the quark-gluon vertex has to be a function of $F(p^2)$ and $M(p^2)$, because the longitudinal part contains them both. So, $\tau_i = \tau_i(F(p^2), M(p^2))$.
- Multiplicative renormalizability also implies that in the asymptotic limit $k^2 \gg p^2$, the transverse vertex must be unique.

About this last point, we perform the asymptotic limit over the perturbative definitions of τ_i, given in [2]. For the abelian (a) and the non-abelian (b) coefficients we find [8]

$$k^4\frac{\tau_1^a}{m} = \left(C_F - \frac{1}{2}C_A\right)\frac{\alpha_s}{4\pi}(3+\xi)\ln\left(\frac{p^2}{k^2}\right) \qquad k^4\frac{\tau_1^b}{m} = \frac{\alpha_s C_A}{48\pi}\left[12+7\xi-\xi^2\right]\ln\left(\frac{p^2}{k^2}\right)$$

$$k^4\tau_2^a = \left(C_F - \frac{1}{2}C_A\right)\frac{\alpha_s}{12\pi}(1-2\xi)\ln\left(\frac{p^2}{k^2}\right) \qquad k^4\tau_2^b = \frac{\alpha_s C_A}{16\pi}\left[1-\frac{3\xi}{2}+\frac{\xi^2}{6}\right]\ln\left(\frac{p^2}{k^2}\right)$$

$$k^2\tau_3^a = \left(C_F - \frac{1}{2}C_A\right)\frac{\alpha_s}{12\pi}(1+\xi)\ln\left(\frac{p^2}{k^2}\right) \qquad k^2\tau_3^b = \frac{\alpha_s C_A \xi}{32\pi}\left[1+\frac{\xi}{3}\right]\ln\left(\frac{p^2}{k^2}\right)$$

$$k^6\frac{\tau_4^a}{m} = \left(C_F - \frac{1}{2}C_A\right)\frac{\alpha_s}{12\pi}(\xi-1)\ln\left(\frac{p^2}{k^2}\right) \qquad k^6\frac{\tau_4^b}{m} = \frac{\alpha_s C_A}{48\pi}\left[2-\xi-\xi^2\right]\ln\left(\frac{p^2}{k^2}\right)$$

$$k^2\frac{\tau_5^a}{m} = \left(C_F - \frac{1}{2}C_A\right)\frac{\alpha_s}{24\pi}(\xi-1)\ln\left(\frac{p^2}{k^2}\right) \qquad k^2\frac{\tau_5^b}{m} = \frac{\alpha_s C_A}{96\pi}\left[22+11\xi+3\xi^2\right]\ln\left(\frac{p^2}{k^2}\right)$$

$$k^2\tau_6^a = \left(C_F - \frac{1}{2}C_A\right)\frac{\alpha_s}{12\pi}\left(1-\frac{\xi}{2}\right)\ln\left(\frac{p^2}{k^2}\right) \qquad k^2\tau_6^b = \frac{\alpha_s C_A}{64\pi}\left[1-\xi+\frac{2\xi^2}{3}\right]\ln\left(\frac{p^2}{k^2}\right)$$

$$k^4\frac{\tau_7^a}{m} = \left(C_F - \frac{1}{2}C_A\right)\frac{\alpha_s}{12\pi}(1-\xi)\ln\left(\frac{p^2}{k^2}\right) \qquad k^4\frac{\tau_7^b}{m} = \frac{\alpha_s C_A}{24\pi}\left[-1+\xi\right]\ln\left(\frac{p^2}{k^2}\right)$$

$$k^2\tau_8^a = \left(C_F - \frac{1}{2}C_A\right)\frac{\alpha_s}{4\pi}\ln\left(\frac{p^2}{k^2}\right) \qquad k^2\tau_8^b = -\frac{\alpha_s C_A}{16\pi}\left[1+\frac{3\xi}{2}+\frac{\xi^2}{2}\right]\ln\left(\frac{p^2}{k^2}\right)$$

where ξ is the gauge parameter, C_F and C_A are the color factors of the eigenvalues of the square Casimir operators in the fundamental and the adjoint representation, respectively.

In the asymptotic limit, the transverse quark-gluon vertex only has contributions of $T_3^\mu = -T_6^\mu$, i.e., $\Gamma^\mu \sim (\tau_3 - \tau_6)T_6^\mu$ when $k^2 \gg p^2 \gg m^2$. Thus, in this limit

$$\tau_3 - \tau_6 = -\alpha_s \frac{C_A(1+\xi) - 8\xi C_F}{64p^2\pi}\log\frac{p^2}{k^2} \tag{8}$$

In Fig. (1) we compare $\tau_3 - \tau_6$ obtained using Eq. (8) with the definitions from [2]. If we take the QED case, where $C_F = 1$ and $C_A = 0$, we obtain the corresponding limit[9].

A non-perturbative proposal for the quark-gluon vertex has to be gauge invariant[10]. Work in this direction will be reported elsewhere[11].

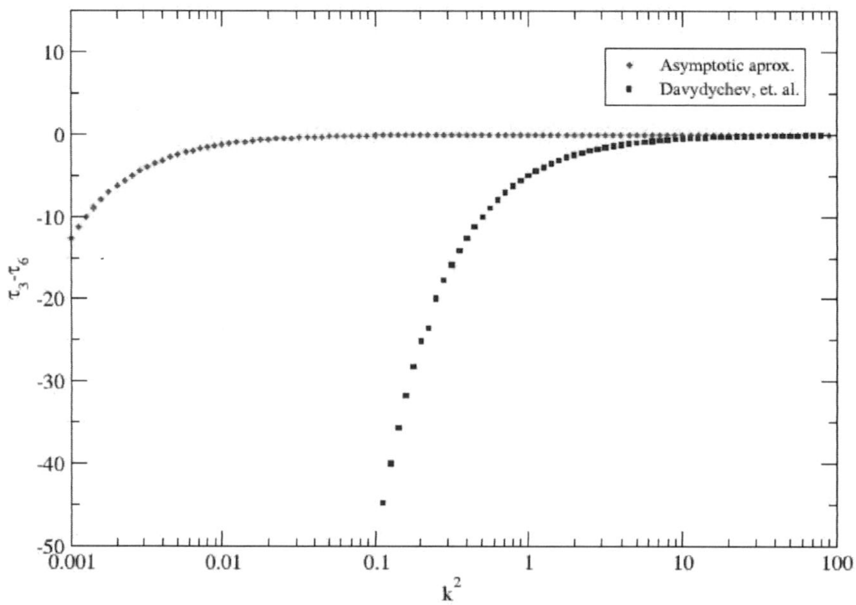

Figure 1. In this plot, the *asymptotic approx.* is obtained using eq. (8), while *Davydychev, et. al.* result is produced using the full definition in [2]. The plotting parameters are $\alpha_s = 0.118$ for the coupling, $m = 5$MeV for the quark mass and we computed in Landau gauge $\xi = 0$.

4. Conclusions

In this work we presented the basic requirements that a non-perturbative ansatz for the quark-gluon vertex must satisfy, and they agree with the literature of the area.

References

[1] A. Bashir, L. Chang, I.C. Cloet, *et al.*, Commun. Theor. Phys. **58**, 79 (2012)
[2] A.I. Davydychev, P. Osland and L. Saks, Phys. Rev. D **63**, 014022 (2001)
[3] A.A. Slavnov, Teor. Mat. Fiz. **10**, 153 (1972) Theor. Math. Phys. **10**, 99 (1972)
[4] J.C. Taylor, Nucl. Phys. B **33**, 436 (1971)
[5] J.S. Ball and T.W. Chiu, Phys. Rev. D **22**, 2542 (1980)
[6] A.C. Aguilar and J. Papavassiliou, Phys. Rev. D **83**, 014013 (2011)
[7] E. Rojas, J.P.B.C. de Melo, B. El-Bennich, *et al.* JHEP **1310** (2013) 193
[8] R. Bermudez, L.A. Fernandez-Rangel, L.X. Gutierrez-Guerrero, M.E. Tejeda Yeomans and A. Bashir, Manuscript in progress *Quark-gluon vertex: A perturbation theory primer and beyond* (2016)
[9] D. Curtis and M.R. Pennington, Phys. Lett. D **42**, 4165 (1990).
[10] A. Bashir, R. Bermudez, *et al.*, Phys. Lett. C **85**, 045205 (2012).
[11] R. Bermudez and M.E. Tejeda-Yeomans, Work in progress (2016).

Ultraviolet radiation detector to obtain the rate of particles at different heights

E Ponce, E Flores and R Conde.
Autonomous University of Puebla, Mexico.
E-mail: eponce@yandex.ru

Abstract.
The nature and origin of cosmic rays remains one of the greatest puzzles of modern astrophysics after more than 50 years since their first registration. Several ground experiments have reported the rate registered at its height of operation. To continue with the study of cosmic rays, we propose obtain the rate at different heights in the Earth's atmosphere, developing a small and portable ultraviolet radiation detector, consisting of a scintillation plastic, a PMT, and a fast DAQ system. In this work we present the design and construction of the UV detector and the rate recorded in the Sierra Negra Volcano near Puebla, Mexico (4200 m.a.s.l).

1. Introduction
The Earth's atmosphere is subject to constant arrival of atomic nuclei, mostly protons, known as primary cosmic radiation; during its trajectory to the surface, by interacting with atmospheric nuclei, produce a variety of elementary particles, identified as secondary cosmic rays. In order to obtain the rate of cosmic particles, we design a portable ultraviolet radiation detector, which consists of a scintillator plastic, thick 1 cm, radius 6 cm, threshold energy approximately 1 MeV; the interaction of high energy cosmic rays with scintillation plastic molecules, produces photons, the light is detected using a photomultiplier tube (PMT), the PMT converts photons into an electrical signal, finaly, it is processed by a DAQ system of high speed. In this paper, we present the portable ultraviolet radiation detector, and the laboratory calibration, the next step will be to test the detector performance, operating during the journey from university city of BUAP campus (2200 m.a.s.l) to Pico de Orizaba national park (4200 m.a.s.l) in Mexico. The data obtained will be compared with the rate registered by experiments operating at heights known as AUGER, HAWC, LAGO.

2. Ultraviolet radiation detector setup
To obtain the continuous rate of the cosmic rays at different heights, it is necessary a detector operating for each specific height; for this reason we developed a small and portable ultraviolet radiation detector, the main purpose of the experiment is obtain the cosmic rays rate of experimental form in the range of heights 2200-4200 m.a.s.l, approximately.

The UV detector [1] consists of a scintillator plastic as first transducer, see figure 1. Scintillation detectors make use of special materials, when exposed to ionizing radiation, they emit a small flash of light. This phenomenon occurs when radiation interacts with matter, exciting and ionizing one large number of atoms and molecules, when they return to their steady state, emit photons in the visible spectrum or around it.

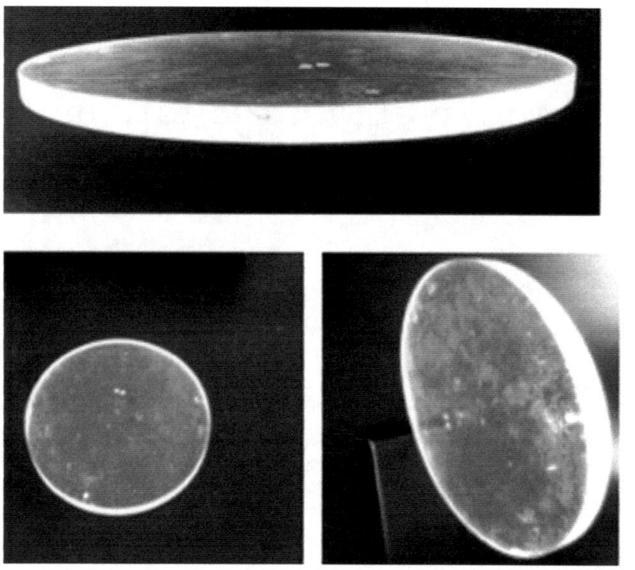

Figure 1. Scintillation plastic as a medium radiator, thick 1 cm, radius 6 cm, threshold energy approximately 1 MeV. Used in the UV radiation detector to obtain the rate of particles at different heights.

To obtain useful information from scintillation plastic, a photodetector is added [2], which converts the photons into an electrical signal. The photodetector most frequently used for a scintillator detector is a photomultiplier tube (PMT), see figure 2, this device, has a photocathode, an electron multiplier chamber and a polarization base as main parts; the first part of the PMT is a thin layer of a special material that emits electrons (more know as photoelectrons) when photons pass through the photocathode. The second part of the PMT, is an array of electrodes polarized with an active (semiconductor devices) or passive (resistive) base, to keep a gradual distribution of voltage in each electrode and achieve gains of the order up to 10^6, resulting a current pulse in the PMT output pin; connecting a load resistor in the output, a voltage is generated in the order of a few millivolts, which is processed by a DAQ system for further analysis, in order to obtain information about the incident radiation.

Figure 2. The UV detector contains one R1463 photomultiplier tube from the Hamamatsu company with an ultraviolet entrance window and a multialkali photocathode.

The DAQ system has as main computer an FPGA Xilinx, see figure 3, this controls the information of: a digitalization board every 10 ns; a GPS and a temperature-pressure sensor every second; a high voltage source to control the constant gain of the photomultiplier [3], every minute. The time intervals assigned, are with the purposes of autocalibrate the UV detector during the trajectory of operation in the journey.

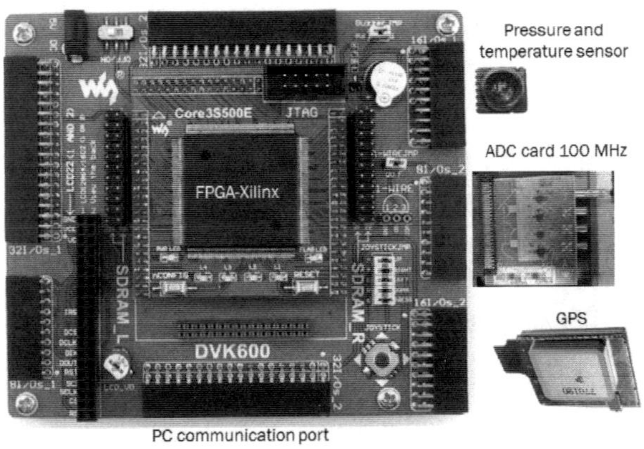

Figure 3. DAQ system of high speed processing, has as main computer an FPGA Spartan 3S500 Xilinx family, a Temperature-Pressure sensor HP03S, a digitalization board, a GPS EM408E and a communication port with the PC.

3. Ultraviolet radiation detector testing and preliminary results

As a first step in the calibration process of the UV detector, we measured the response of the PMTs to a single photoelectron (SPE) signals [4], figure 4. From these measurements we can obtain the signal shape for any event figures 5, 6.

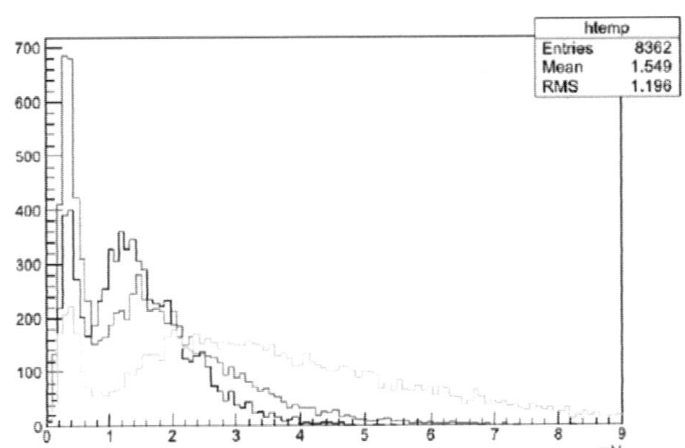

Figure 4. Single photoelectron response of PMT R1463, for different voltages: 1.25 kV (green signal), 1.30 kV (red signal) and 1.35 kV (black signal).

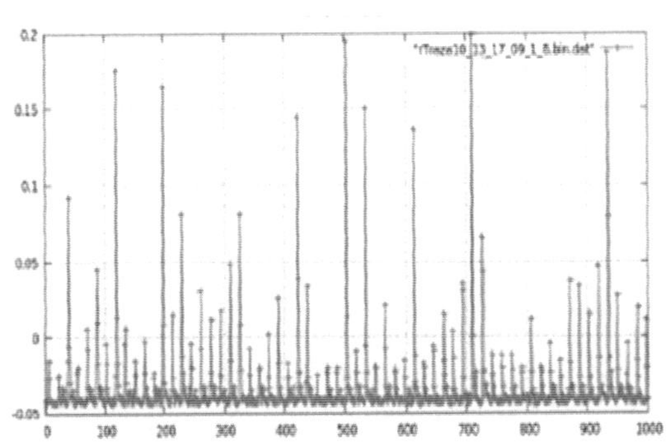

Figure 5. Traces recorded at a height around 4200 m.a.s.l.

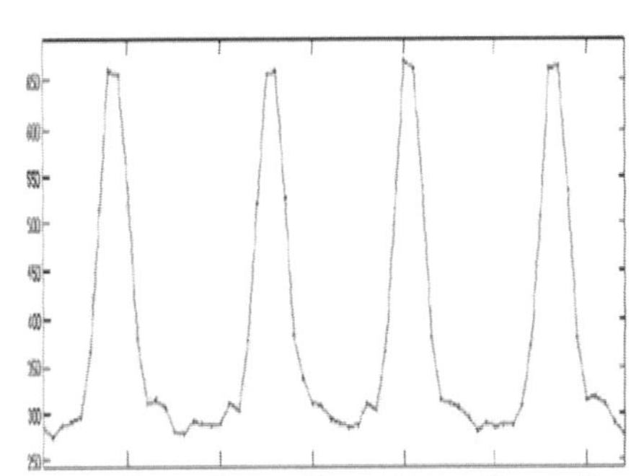

Figure 6. Zoom of some traces recorded.

Journal of Physics: Conference Series **761** (2016) 012028 doi:10.1088/1742-6596/761/1/012028

The UV detector was operating in Sierra Negra Volcano near Puebla, Mexico (around 4200 m.a.s.l), The rate recorded is presented in figure 7.

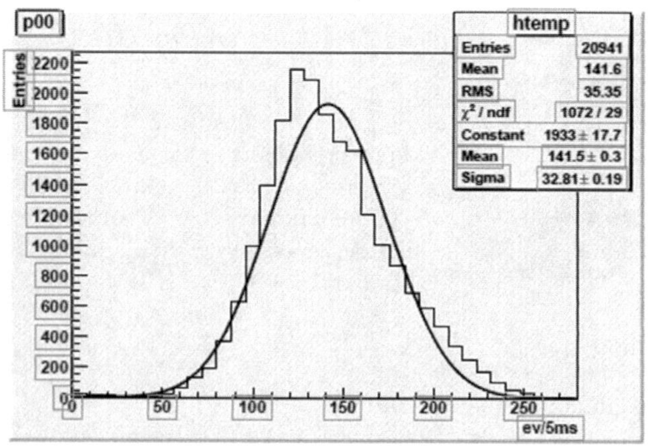

Figure 7. Histogram of particles recorded every 5 milliseconds with DAQ system. The maximun peak is around 140, for one second are needed 200 histograms, with this considerations the rate is around 28x10^3 particles/second.

HAWC located at 4100 meters above sea level, reports a flow of 26x10^3 particles per second. It is indicating that our detector operates properly.

4. Conclusions

In this paper, was presented the portable ultraviolet radiation detector, together with results obtained at a height of 4200 meters above sea level. The next step will be to test the detector performance, operating during the journey from university city of BUAP campus to Pico de Orizaba national park. The data obtained will compared with the number of registered particles per second, provided by experiments operating at heights known as AUGER, LAGO, etc. Will create a database with free access, recording the profile signal processed by DAQ system, generated by each particle, for further and off line analysis, by students or researchers interested.

5. Bibliography

[1] Dmitriev 2007 *UV radiation from the night-time atmosphere seen from the universitetsky TATIANA satellite* (AIP proceeding)

[2] Garipov 2011 *Program of transient UV event research at TATIANA-2 satellite* (Journal of Geophysical research A 115)

[3] E. Ponce 2013 *Pinhole camera for study of atmospheric UV flashes as a source of background in the TUS experiment* (NIMA 639)

[4] E. Ponce 2015 *Transient Luminous events registered with a ground pinhole* (PoS ICRC2015 437)

Forward-backward correlations in proton-proton collisions

E. Dominguez-Rosas, E. Cuautle and G. Paić
Instituto de Ciencias Nucleares, Universidad Nacional Autónoma de México, Apartado Postal 70-543, Ciudad de México 04510, México
E-mail: sidej.rex@gmail.com

Abstract. The correlation multiplicity in forward-backward pseudorapidity is really important for the overall event characterization and to extract information on non-pertubative hadronic processes. In this work we present a study of the correlation conditions at LHC. Subsequently, the study is extended using p+p simulations of events to different energies, the correlations in multiplicity are studied with different initial conditions of the events in order to show how it is related to the mechanisms of hadronization, as well as concerning multiple partonic interactions, and consequently the degree of hardness of the collision. The sensitivity to the QCD color reconnection strength is reported.

1. Introduction

Forward-backward (F-B) correlations are quantified by the strength of the factor, b_{corr}, which is the slope obtained from the relationship between the average multiplicity measured in the forward rapidity $\langle n_B \rangle_{n_F}$ and the multiplicity in the forward rapidity hemisphere, n_F. this relationship can be expressed as[1]:

$$\langle n_B \rangle_{n_F} = a + b_{corr} \cdot n_F. \tag{1}$$

This linear relation has been observed[2] in ALICE experiment. An alternative definition of the correlation coefficient is possible through the direct calculations of the corresponding correlation factor [3]:

$$b_{corr} = \frac{\langle n_B n_F \rangle - \langle n_B \rangle \langle n_F \rangle}{\langle n_F^2 \rangle - \langle n_F \rangle^2}. \tag{2}$$

This correlation factor has been studied in a variety of collisions systems, from the simplest to the most complicated, and what is certain is that at higher collision energy of the center of mass, larger strength is obtained for these correlations, then it is interesting to investigate the meaning of this multiplicity F-B correlations[4] using other like long-range multiplicity correlations[5]. One of the recent results to incorporate flow-like effects in p+p [6] using the so called Color reconnection (CR) among partons in the final state can produce important effects to this correlation factor.

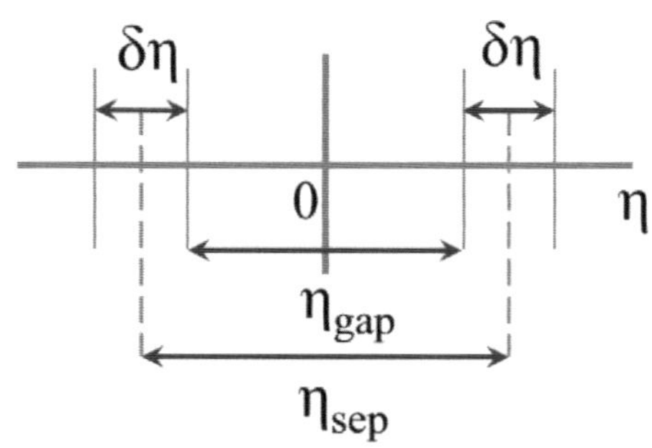

Figure 1. Sketch of the pseudorapidity variables used for the present simulations.

2. Results

The pseudorapidity values to extract n_F y n_b are shown on figure 1, The simulation data were Minimum Bias (MB) generated with PYTHIA 8.186 [7] tune Monash 2013 [4] at different energies: 0.9, 2.76 and 7 TeV, which allows us to compare with ALICE data[2]. The correlation factors have been calculated in the $|\eta| < 0.9$ and in the momentum range 0.3 GeV/$c \leq p_T \leq$ 1.5 GeV/c.

Variables used to estimate the correlation factor are: the window size is $\delta\eta$, in which the particles were counted, both in front and behind; η_{gap} is the separation between $\delta\eta$ in front and behind, *i. e.*, the distance between the windows.

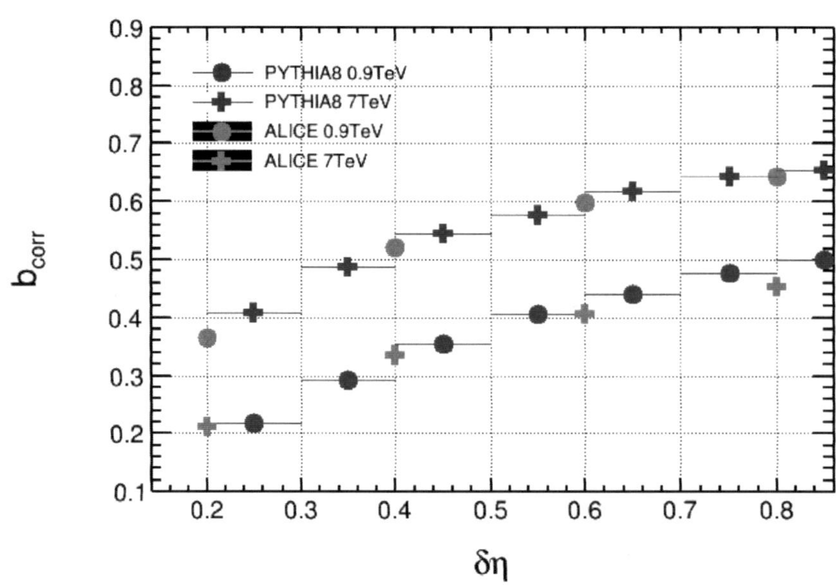

Figure 2. (Color online) Correlation factor for two different energies comparing ALICE data with the PYTHIA Monash 2013 generator in function of the width of the window $\delta\eta$, with $\eta_{gap} = 0$

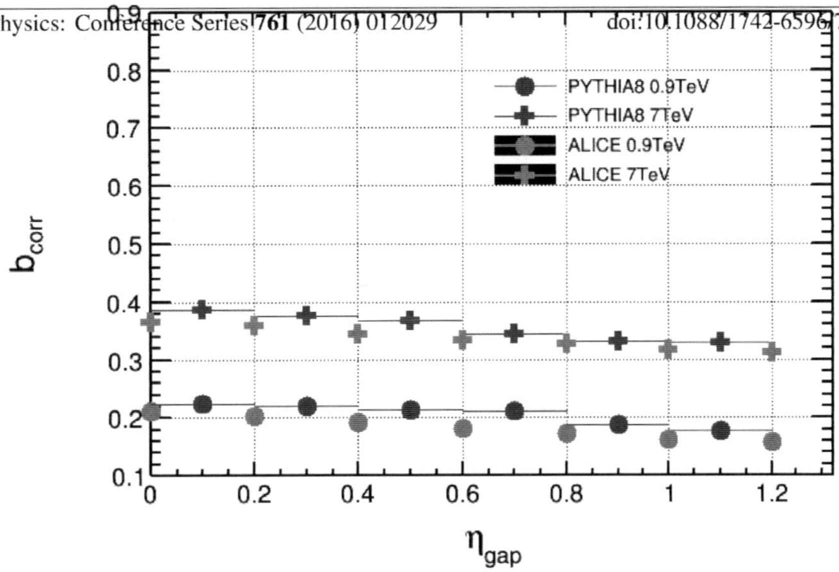

Figure 3. (Color Online) Correlation factor as function of $\eta_{\rm gap}$ for $\delta\eta = 0.2$. The results show the behavior for two different energies, comparing PYTHIA to ALICE data[2].

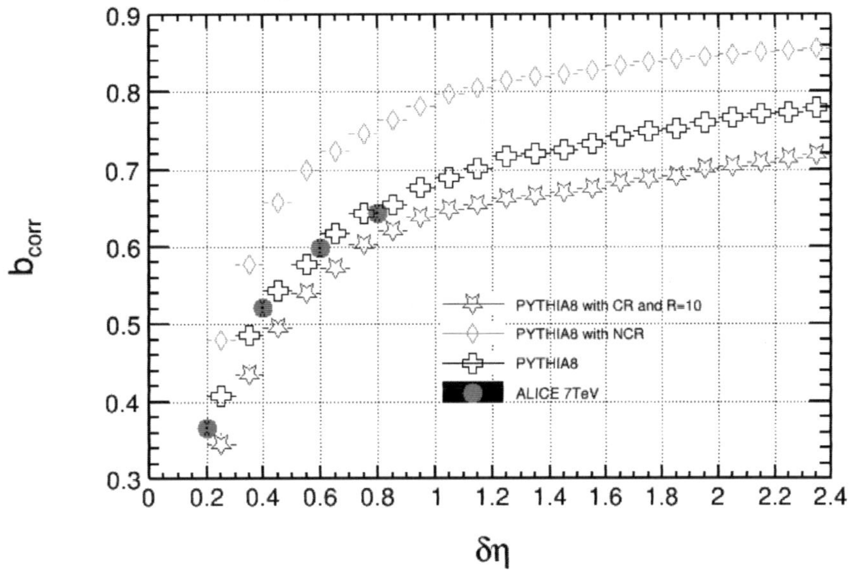

Figure 4. (Color online) Correlation factor as function of $\delta\eta$ for p+p at 7 TeV with PYTHIA Monash 2013, comparing model with three different ranges of color reconnection and with ALICE data.

Figure 2 shows how the F-B correlation factor increases as function of $\delta\eta$, and the collision energy. One can observe a complete agreement between ALICE data[2] and PYTHIA event generator.

The figure 3 shows the behavior of $b_{\rm corr}$ with respect to $\eta_{\rm gap}$ keeping $\delta\eta$ fixed to 0.2, it shows a downward trend, which is expected because we are changing the analyzed windows, from higher to lower number of particles to compute the correlations. Again a very good description with the existing data is observed.

Finally we wanted to investigate whether the color reconnection [6] model explains the observed correlations. In the figure 4 we present the results of the simulation with three different values of the color reconnection strength. Indeed one observes a very strong dependence of the correlation factor with the reconnection strength. With the extreme value of 10 for the CR one observes a much lower correlation factor at 7 TeV and with a value zero one undershoots the data. The default value of 1.8 for CR seems to reproduce well the data.

3. Conclusion

We have compared simulations with an up to date generator to the published FB correlations. The excellent agreement has led us to study the details of the influence of an important ingredient of the generator use *i.e* the color reconnection.

The present work links, to our knowledge, for the first time explicitly the QCD process of color reconnection in a very obvious manner indicating an interesting alley of investigation. It also suggests that FB correlations can be used to determine the intensity of the CR strength.

4. Acknowledgments

The authors acknowledge useful discussions with A. Ortiz. Support for this work has been received from CONACYT under the grant No. 270067 and from DGAPA-UNAM under grants IN108414.

5. References

[1] A. Capella and J. Tran Thanh van,Phys. Rev. **D29** (1984) 2512; G.J. Alner et al., Phys. Rep. **154** (1987) 247.
[2] ALICE Collaboration, JHEP **05** (2015) 097.
[3] A. Capella et al., Phys.Rep. **236**,225 (1994).
[4] I.G.Altsybeev, G.A. Feofilov, E.L. Gillies, J.Phys.Conf.Ser. **668 (2016)** 012034
[5] Asryan A., Feofilov G., Ivanov A., Grebenyuk A., Naumenko P. and Vechernin V., Proc. ISHEPP-XIX, vol.2, pp. 208-214, JINR, Dubna.
[6] Antonio Ortiz Velasquez, Peter Christiansen, Eleazar Cuautle Flores, Ivonne Maldonado Cervantes, Guy Paić, Phys. Rev. Lett. **111** (2013) 042001.
[7] Torbjörn Sjöstrand, Stephen Mrenna, Peter Skands, JHEP **05** (2006) 026.

MWPF IOP Publishing
Journal of Physics: Conference Series **761** (2016) 012030 doi:10.1088/1742-6596/761/1/012030

Available on CMS information server **CMS CR -2016/177**

The Compact Muon Solenoid Experiment

Conference Report

Mailing address: CMS CERN, CH-1211 GENEVA 23, Switzerland

03 August 2016 (v3, 25 August 2016)

Charged Higgs Analysis in CMS

Jan Eysermans and Isabel Pedraza for the CMS Collaboration

Abstract

In this talk an overview is given of the possible searches of the Charged Higgs Boson during run 2 of the LHC data taking period. The Charged Higgs boson emerges in several (minimal) Standard Model (SM) extensions such as the 2 Doublet Higgs Model, which predicts 5 physical Higgs bosons, consistent with the SM Higgs boson. Based on the main production and decay modes, the possible intermediate and final state particles are predicted for a Charged Higgs mass higher than the top quark mass ($m_{H^\pm} > m_t$). In particular, the dominant H to tau nu and H to tb channels are discussed in more detail together with their associated background.

Presented at *DPC-SMF2016 XXX Reunion Anual de la Division de Particulas y Campos de la SMF*

Journal of Physics: Conference Series **761** (2016) 012030 doi:10.1088/1742-6596/761/1/012030

Charged Higgs Analysis in CMS

Jan Eysermans[1], María Isabel Pedraza Morales[1] on behalf of the CMS Collaboration

[1]Facultad de Ciencias Físico-Matemticas, Benemérita Universidad Autónoma de Puebla

E-mail: jan.eysermans@cern.ch

Abstract. An overview is given of the possible searches of the Charged Higgs Boson during run 2 of the LHC data taking period. The Charged Higgs boson emerges in several (minimal) Standard Model (SM) extensions such as the 2 Doublet Higgs Model, which predicts 5 physical Higgs bosons, consistent with the SM Higgs boson. Based on the main production and decay modes, the possible intermediate and final state particles are predicted for a Charged Higgs mass higher than the top quark mass ($m_{H^\pm} > m_t$). In particular, the dominant $H^\pm \to \tau\nu$ and $H^\pm \to tb$ channels are discussed in more detail together with their associated background.

1. Introduction to the theory of Charged Higgs bosons

In the Standard Model (SM), gauge invariance requires the (heavy) W and Z gauge bosons to be massless, which is in contradiction with experiment. They acquire mass through the Higgs mechanism, based on spontaneous symmetry breaking of a Higgs doublet

$$\phi = \frac{1}{\sqrt{2}}\begin{pmatrix}\phi^+ \\ \phi^0\end{pmatrix},$$

within the Mexican hat shaped potential ($\mu^2 < 0, \lambda > 0$):

$$V(\phi) = \mu^2\phi^\dagger\phi - \lambda(\phi^\dagger\phi)^2.$$

Expanding $V(\phi)$ around its (non-zero) vacuum expectation value $\nu = \sqrt{-\mu^2/\lambda}$, a scalar CP even particle arises in the spectrum of the Lagrangian: the SM Higgs boson. The corresponding Higgs field interacts with the gauge fields giving the corresponding gauge boson mass (except the photon) whereas fermion masses are obtained from Yukawa couplings to this Higgs field.

A simple extension of the SM introduces a second Higgs doublet (ϕ_1, ϕ_2): the 2 doublet Higgs Model (2DHM). Expanding the potential $V(\phi_1, \phi_2)$ around the vacuum expectation values (ν_1 and ν_2 respectively) results in five physical Higgs bosons:

- two neutral CP-even scalars: h (SM Higgs) and H (heavy Higgs);
- two charged Higgs bosons $H^\pm$;
- one neutral CP-odd pseudoscalar A.

The free parameters of the model are the remaining masses of the Higgs bosons and the value of $\tan\beta = \nu_2/\nu_1$. In type-II 2DHM the fermions couple to ϕ_2 for the up quark and to ϕ_1 for the down quark and leptons. The minimal supersymmetric SM is a type II 2DHM.

2. Charged Higgs production and decay

The production of charged Higgs bosons in a pp collider can be distinguished in two mass regions. In the low mass region ($m_{H^\pm} < m_t$), the $H^\pm$ production is mainly through the decay of a top quark to $H^\pm b$ in $t\bar{t}$ production. In the high mass region ($m_{H^\pm} > m_t$), the production of charged Higgs boson is through fusion of top-bottom quarks. Two semi-final states, $H^\pm tb$ or $H^\pm t$ are possible depending whether the 4 or 5 flavor scheme is used (4FS, 5FS).

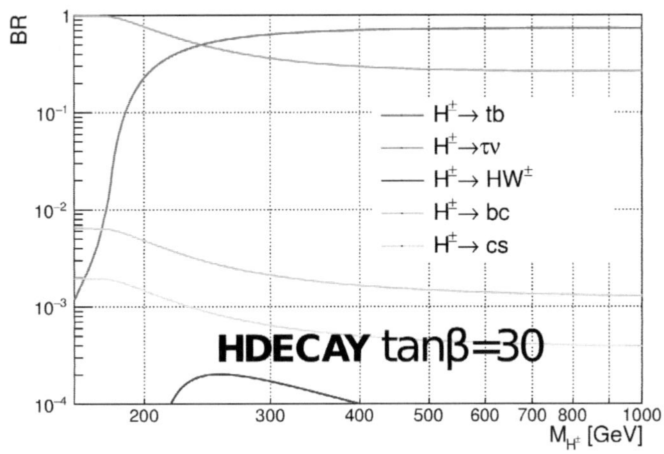

Figure 1. Branching ratios for the main $H^\pm$ decay modes for $\tan\beta = 30$ [3].

The decay of the charged Higgs boson depends mainly on its mass and to a lesser extent on the value of $\tan\beta$. In Figure 1, the most important decay channels with their branching ratios are shown. The most contributing channels in the high mass region are $H^\pm \to \tau\nu$ and $H^\pm \to tb$. Currently the analysis in CMS is focused on those channels, yet the $H^\pm \to hW$ and more exotic channels (e.g. $H^\pm \to$ SUSY, $H^\pm \to AW$ with A the neutral CP-odd pseudoscalar particle) are also under investigation. The channels $H^\pm \to cb, cs$ are more important in the low mass region and are also under investigation.

In the Compact Muon Solenoid (CMS) collaboration [1] the current searches for charged Higgs signals are both focussed on light and heavy Charged Higgs particles, involving all the channels as mentioned above as well as the $H^\pm \to WZ$ channel. Here only the $H^\pm \to \tau\nu$ and $H^\pm \to tb$ channels will be briefly discussed.

3. The $H^\pm \to \tau\nu$ channel

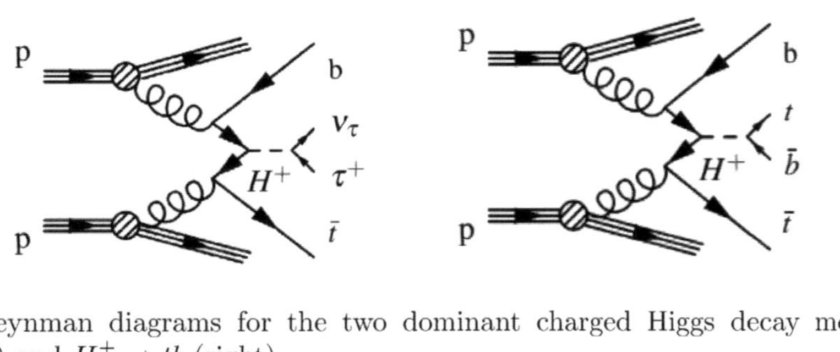

Figure 2. Feynman diagrams for the two dominant charged Higgs decay mode channels $H^\pm \to \tau\nu$ (left) and $H^\pm \to tb$ (right).

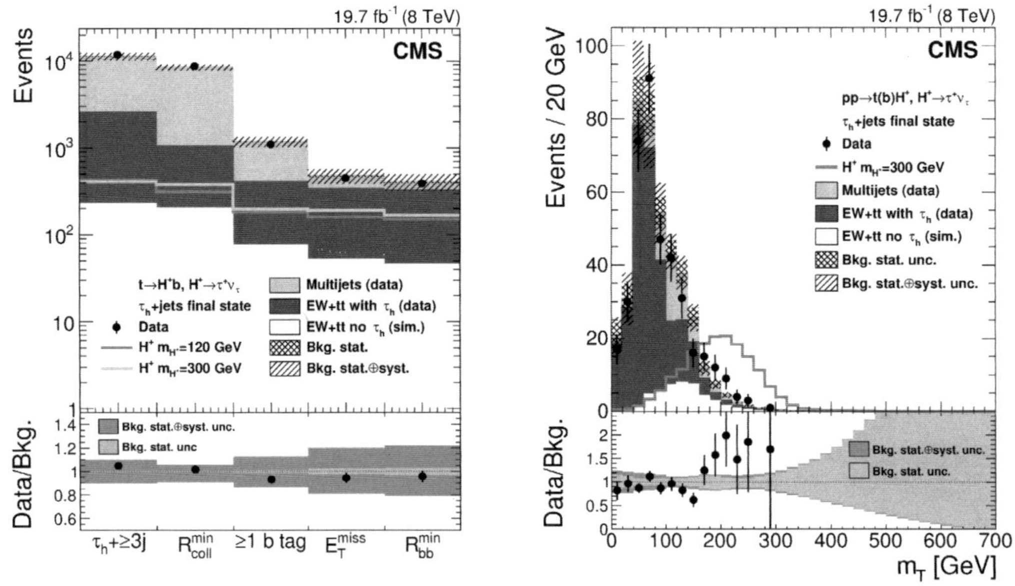

Figure 3. Left: Cutflow diagram for event selection of the $H^{\pm} \rightarrow \tau\nu$ channel. Right: invariant transverse mass distribution after event selection based on the leading τ and MET [2].

This channel is characterized with the $t(b)\tau\nu$ as semi-final state particles. The neutrino will be assigned as missing transverse energy (MET) whereas the τ can decay leptonically (35 %) or hadronically (65%) involving (charged) pions [4]. The decay of the top quark results in W+b-jet, where the W can again decay leptonically or hadronically producing jets. The main background events are multijets (QCD), dibosons (WW, WZ, ZZ), W+jets and $t\bar{t}$.

When restricting to hadronic decay modes (both for the W and the τ), the 2 neutrinos originate only from the charged Higgs and the transverse mass m_T can be reconstructed based on the leading τ (highest p_T) and MET. A shape analysis performed on the m_T distribution will be used to separate the signal from background events.

The analysis strategy is based on the 8 TeV experience [2]. Event cuts are applied on the final-state particles based on the required event topology as described above and additional cuts for background suppression:

- at least 1 τ, $p_T > 51$ GeV, $|\eta| < 2.1$;
- no (isolated) leptons;
- large MET > 130 GeV;
- at least 3 jets, one b-jet;
- angular cuts $R^{min}_{coll,bb}$ for multijet suppression.

The QCD multijet contribution is estimated via a τ_h-misidentification rate technique based on events passing up to the one b-jet requirement whereas the EWK+$t\bar{t}$ background is obtained by using an embedded technique using μ+jets events. The event cut-flow diagram for the 2014 data taking with $\sqrt{s} = 8$ TeV is shown in Figure 3 (left). After passing the full selection criteria, the invariant mass based on the leading τ and MET is calculated and is shown in Figure 3 (right). A signal sample with $m_{H^{\pm}} = 300$ GeV is superimposed for illustrative purposes.

For the $\sqrt{s} = 13$ TeV analysis the selection criteria will be optimized and refined based on the ratio $S/\sqrt{B}$ by varying the cut boundaries or by categorization of the events (e.g. on b-jets). The dominant backgrounds will be estimated based on data-driven techniques using events in control regions.

4. The $H^{\pm} \to tb$ channel

For this channel the semi-final state particles are $tt(b)b$. The top quarks will decay into W+b-jets, yielding at least 3 b-jets in the final state. Both W bosons can decay leptonically or hadronically. Current analyses in CMS are focused on one- or two-leptonic final states by selecting on a specific lepton trigger. To distinguish signal from background the scalar sum of jet transverse energies distribution H_T is used. The full hadronic final state (no leptons) is characterized by a high jet multiplicity: two jets for each W, and at least 3 b-jets. Therefore in the hadronic final state the b-jet multiplicity distribution is also used to distinguish the signal from background. The dominant backgrounds are W+jets, $t\bar{t}$, single t and QCD multijets. Dibosons, $Z/\gamma*$+jets and $t\bar{t} + W/Z$ have less contribution to the background.

The search strategy for the $H^{\pm} \to tb$ channel is also based on the 8 TeV experience [2]. Cuts are applied based on the amount of leptons, lepton p_T, MET and jet multiplicity. For the leptonic final states, the event spectrum is divided into different regions based on the amount of jets, both for muons and electrons:

- Control region ($2 \leq N_{jet} \leq 3$): low jet multiplicity used to derive background normalizations;
- Signal region ($N_{jet} \leq 4$): high jet multiplicity consistent with the signal signature.

Each region is further classified in the b-tag multiplicity, ranging from 0, 1 and ≥ 2 b-jets for the control region and 1, ≥ 2 b-jets for the signal region. For each signal region, the H_T distribution is obtained and the signal is compared to the background.

5. Results and conclusions

In the high mass region of the charged Higgs, the dominant channels are $H^{\pm} \to \tau\nu$ and $H^{\pm} \to tb$. Within the CMS collaboration, these channels are currently being investigated with the 13 TeV data as well as other channels in the low mass region and some more exotic channels. Several cuts are applied for background suppression based on the 8 TeV experience and the signal is distinguished from background based on the m_T, H_T and b-jet multiplicity distributions. For the 8 TeV analysis, a combined statistical analysis has been performed in the high mass region ($180 < m_{H^{\pm}} < 600$ GeV). The data agree with the SM prediction and upper limits of the Charged Higgs have been set, including systematic uncertainties.

References

[1] S. Chatrchyan *et al.* [CMS Collaboration], JINST **3** (2008) S08004. doi:10.1088/1748-0221/3/08/S08004
[2] V. Khachatryan *et al.* [CMS Collaboration], JHEP **1511** (2015) 018 doi:10.1007/JHEP11(2015)018 [arXiv:1508.07774 [hep-ex]].
[3] A. Djouadi, J. Kalinowski and M. Spira, Comput. Phys. Commun. **108**, 56 (1998) doi:10.1016/S0010-4655(97)00123-9 [hep-ph/9704448].
[4] K. A. Olive *et al.* [Particle Data Group Collaboration], Chin. Phys. C **38**, 090001 (2014). doi:10.1088/1674-1137/38/9/090001

Journal of Physics: Conference Series **761** (2016) 012031 doi:10.1088/1742-6596/761/1/012031

Event shape variables, a tool for QCD study

M. Fernández, E. Cuautle and G. Paić

Instituto de Ciencias Nucleares, Universidad Nacional Autónoma de México,
Apartado Postal 70-543, Ciudad de México 04510, México

Abstract. The proton-proton collisions at LHC energies reveals important new phenomena observed in heavy ions collisions and not completely understood in small systems. Different approaches and techniques have been implemented to study those processes, but still the details of the underlying event, which correspond at certain level to soft processes remains unknown. Here we present a study of proton-proton collisions using the event shape variable sphericity for identified particles combined with multiplicity bins. The results allow us to get important features of collisions, which could be used to explore in more details and extract phenomena of QCD processes.

1. Introduction

The event shape [1] variables provide a measure of the geometry of hadronic energy momentum flow, that offer the possibility to classify the events on pencil-like and isotropic[2]. The event shape variables had a very important role in the LEP accelerator to determine the strong coupling constant, for test models of corrections by hadronic effects and for the validation of the Monte Carlo event generators. The correction by hadronization is possible because the event shape variables are calculated with the information of the final hadrons of the event.

2. PYTHIA and PHOJET event generators

QCD perturbation theory, formulated in terms of quarks and gluons, is valid at short distances. At long distances, QCD becomes strongly interacting and perturbation theory breaks down. In this confinement regime, the colored partons are transformed into colorless hadrons, through hadronization or fragmentation processes. The PYTHIA event generator allows us to compute QCD processes to simulate high-energy physics events. These are simulation of collisions at high energies between elementary particles or hadrons such as e^+, e^-, p, $\bar{p}$ among others, and their different combinations[3]. The generator also contains phenomenological models to describe the soft physics aspects. It incorporates parton distributions functions, initial and final state interactions, multiparton interactions, decays and string fragmentation as the main hadronization mechanisms.

An alternative generator is PHOJET[4] whose main difference with others is that the hadronization processes described by Dual Parton Model (DPM) in terms of Reggeon and Pomeron exchanges combined with the QCD improved parton model[5]. PHOJET has in particular a detailed modeling for diffractive and soft interactions.

3. Event Shape Variable

The transverse sphericity S_{T} [6] is defined in terms of the eigenvalues $\lambda_1 > \lambda_2$ of the linearised version of the transverse momentum tensor:

$$\mathbf{S_{XY}}^{\mathbf{L}} = \frac{1}{\sum_i p_T^{(i)}} \sum_i \frac{1}{p_T^{(i)}} \begin{pmatrix} p_x^{(i)^2} & p_x^{(i)} p_y^{(i)} \\ p_y^{(i)} p_x^{(i)} & p_y^{(i)^2} \end{pmatrix} \tag{1}$$

where $\vec{p_T}^i = (p_x^{(i)}, p_y^{(i)})$ is the vector of transverse momentum of the *i-th* particle in a given acceptance. The transverse sphericity is given for:

$$S_T \equiv \frac{2\lambda_2}{\lambda_2 + \lambda_1}. \tag{2}$$

It has the property that it tends to 1 for isotropic and is 0 for pencil-like events. The definition is inherent in the multiplicity, for example S_{T} goes to 0 for events with a low multiplicity.

4. Results

The sphericity distribution given by Eq.(2) was calculated with PYTHIA 6.421 Perugia0 and PHOJET 1.12, generating 17.5×10^6 and 13.5×10^6 of inelastic p+p events at $\sqrt{s} = 7$ TeV respectively, in the central pseudorapidity region, $|\eta| < 0.8$, and transverse momentum $0.15 \leq p_{\mathrm{T}} \leq 10$ GeV/c, selecting only primary particles, which include strong decay products, plus weak decay products from heavy flavour.

Left panel of figure 1 shows the invariant transverse momentum distributions for charged hadrons comparing the two event generators normalized to the respective event numbers. The ratio between them is shown in the bottom; right panel of the same figure shows the comparison of the transverse sphericity and the ratio on the bottom.

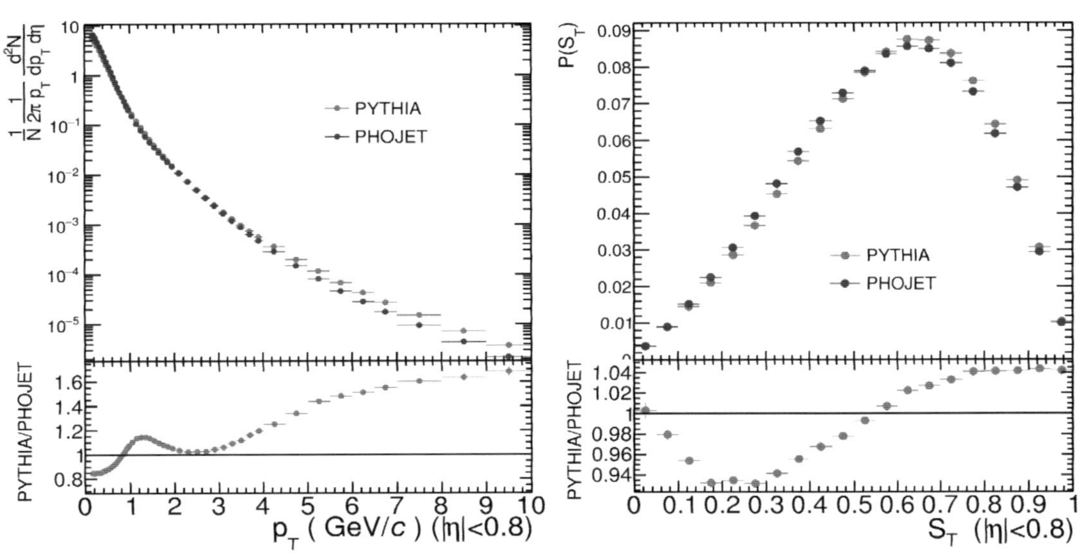

Figure 1. (Color online) The transverse momentum distribution for all charged hadrons (left) and transverse sphericity (right), comparing PYTHIA and PHOJET at $\sqrt{s} = 7$ TeV.

Figure 2 shows the sphericity distribution for soft and hard events, defined as those which have $p_{\mathrm{T}}^{leading}$ less than 2 GeV/c and largest than 5 GeV/c respectively. The main differences between

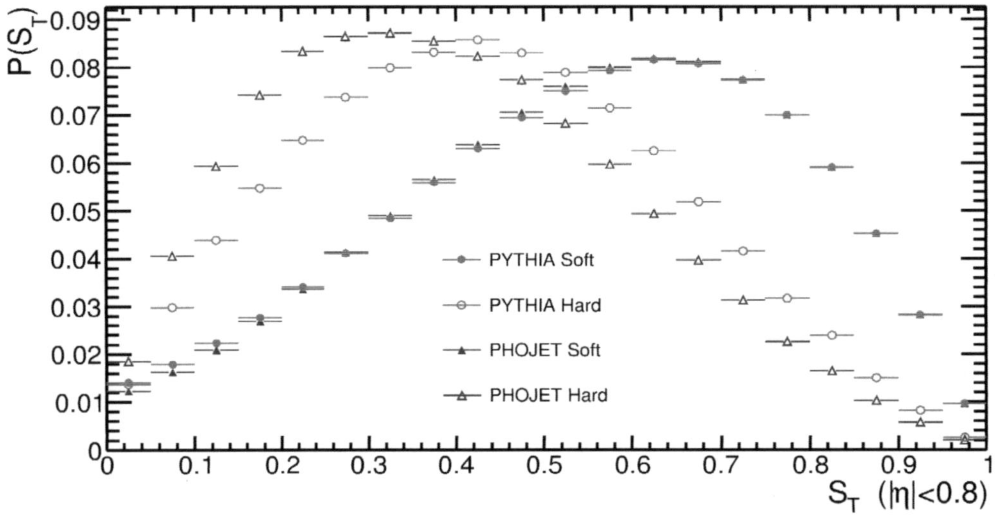

Figure 2. (Color online) Probability distribution of $S_{\rm T}$ for soft ($p_{\rm T}^{leading} < 2$ GeV/c) and hard events ($p_{\rm T}^{leading} > 5$ GeV/c) comparing two event generators.

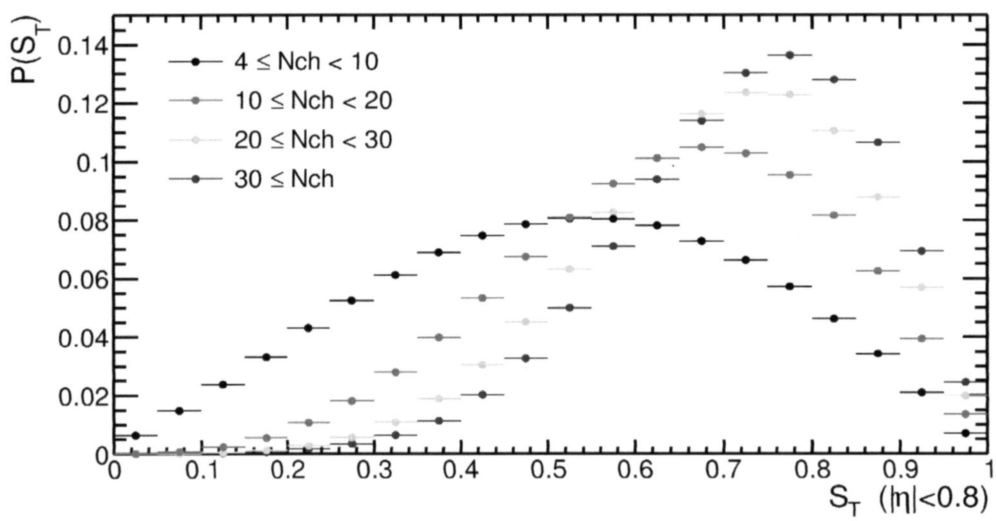

Figure 3. (Color online) Distribution of $S_{\rm T}$ in multiplicity bins, generated with PYTHIA.

two event generators are observed for hard processes, while the soft once seem to be compatible between them. It is important to mention that the distributions are normalized to 1, to compare the shape of the distributions.

Figure 3 illustrates how the hardness (softness) of the event could be measured using multiplicity bins of the sphericity distribution: higher (lower) multiplicity means more isotropic (jetty-like) events. This kind of studies could be extended to other variables, like the number of multiple parton interactions.

A detailed analysis of the sphericity in terms of the multiplicity reveals important features of the structure of the events. Figure 3 shows the transverse sphericity for four multiplicity bins, from low to high multiplicity we see that the events evolve from dijets to isotropic events.

5. Conclusions

The sphericity variable has been studied in proton-proton collisions, comparing two different event generator models, PYTHIA and PHOJET. Transverse momentum distributions used to compute sphericity present larger differences at higher values. The sphericity obtained shows the main differences for hard QCD processes, identified by the cuts on the $p_{\mathrm{T}}^{leading}$ variable. On the other side, lower and higher multiplicity bins allow to split between hard and soft sphericity distributions. A combination of the different variables (mutiplicity and $p_{\mathrm{T}}^{leading}$) could bring an insight of structure of the events.

6. Acknowledgments

The authors acknowledges useful discussions with Antonio Ortiz. Support for this work has been received from CONACYT under the grant No. 270067 and from DGAPA-UNAM under PAPIIT grants IN108414.

References

[1] G. Hanson *et al.*, Phys. Rev. Lett. **35**, 1609 (1975).
[2] A. Ortiz, G. Paić and E. Cuautle, Nucl. Phys. **A 941**, 78 (2015).
[3] T. Sjöstrand, Computer Phys. Commun **82**, 74 (1994).
[4] R. Engel and J. Ranft, Phys. Rev. D **54**, 4244 (1996).
[5] F. W. Bopp, R. Engel and J. Ranft, High energy physics. Proceedings, LAFEX International School, Session C, Workshop on Diffractive Physics, LISHEP'98, Rio de Janeiro, Brazil, February 16-20, 729-741 (1998).
[6] A. Banfi, G. P. Salam and G. Zanderighi, JHEP **1006**, 038 (2010).

MWPF IOP Publishing
Journal of Physics: Conference Series **761** (2016) 012032 doi:10.1088/1742-6596/761/1/012032

The $\Phi^0 \to \gamma Z$ decay in the Littlest Higgs Model

J I Aranda[1], I Cortés-Maldonado[2], S Montejo-Montejo[1], F Ramírez-Zavaleta[1], E S Tututi[1]

[1] Facultad de Ciencias Físico Matemáticas, Universidad Michoacana de San Nicolás de Hidalgo, Avenida Francisco J. Mújica S/N, 58060, Morelia, Michoacán, México
[2] Cátedras Consejo Nacional de Ciencia y Tecnología, México

E-mail: icortesm@cern.ch

Abstract. The existence of new particles with masses of the order of TeV's is a characteristic of several extended models. We study a heavy scalar particle decaying into γZ final states at the one-loop level in the Littlest Higgs Model. This process is analyzed in the context of the recent results reported by ATLAS and CMS Collaborations, concerning the observation of a resonance around 750 GeV. We analyze our results in a scenario where the energy scale f ranges from 2 TeV to 5 TeV. We found that the branching ratio for the $\Phi^0 \to \gamma Z$ decay is of the order of 10^{-7}, which mainly depends on a single parameter.

1. Introduction

The standard model of elementary particles (SM) is a theory whose predictions have been validated by current high energy physics experiments (HEP). However, it is not a complete theory because it does not offer solutions to certain fundamental problems such as: a consistent gravitational quantum theory, neutrino oscillations, matter-antimatter asymmetry, hierarchy problem, among others. In order to explain these shortcomings new theories beyond the SM are developed. The recent discovery of the Higgs boson by ATLAS and CMS experiments in the Large Hadron Collider (LHC) at CERN [1] opens the door to new questions and new projects in HEP. An example of these future projects is search for evidence of new physics beyond the SM where it involves the existence of particles whose masses would be within reach of the LHC. Recent preliminary observations suggest the presence of new scalar resonance decaying into two photons [2]. In fact, from the experimental analyses performed by ATLAS collaboration there are arguments to think that a new scalar resonance could be manifestating itself among 200 GeV and 1.7 TeV [2]. Thus, it is of particular interest the search for new neutral scalar bosons predicted by various Standard Model Extensions (SME). In this paper, we will concentrate on one of the so-called Little Higgs Models (LHM), namely, the Littlest Higgs model (LTHM). The LHM propose a solution to the hierarchy problem [3, 4, 5, 6, 7] and provide an alternative for the study of the electroweak symmetry breaking [3, 4] on the basis of dimensional deconstruction [8, 9], with feature of canceling quadratic divergences. The LTHM is particularly interesting since there are no new degrees of freedom beyond the SM under TeV scale along with its reduced spectrum of new scalar particles [4].

2. Littlest Higgs Model

The most economical version of the LHM is known as the LTHM [4, 5], which is based on a nonlinear sigma model with global symmetry $SU(5)$ alongside gauged group $[SU(2) \times U(1)]^2$ [4, 5]. The $SU(5)$ group breaks to $SO(5)$ at the energy scale f, which is constrained to be of the order of $2-4$ TeV [6]. At the same time, the $[SU(2) \times U(1)]^2$ group is broken to its subgroup $SU_L(2) \times U_Y(1)$, which corresponds to the SM electroweak gauge group. Owing to the breaking breaking pattern, 14 Goldstone bosons are generated, which transform under the $SU_L(2) \times U_Y(1)$ group as a real singlet $\mathbf{1}_0$, a real triplet $\mathbf{3}_0$, a complex doublet $\mathbf{2}_{\pm\frac{1}{2}}$ and a complex triplet $\mathbf{3}_{\pm 1}$ [4, 5]. The real singlet and the real triplet are absorbed by longitudinal components of the heavy gauge bosons, whereas the complex doublet and the complex triplet remain massless [5]. The complex triplet acquires a mass of the order of f by means of the Coleman-Weinberg potential when the global symmetry of the $SO(5)$ breaks down. The complex doublet is identifed as the SM Higgs field. The explicit form of the LTHM Lagrangian as well as the complete set of new particles, masses and new Feynman rules can be found in Ref. [5]. At the leading order in the LTHM, the masses of the new heavy scalar particles are degenerated, *i.e.*, $m_\Phi = \frac{\sqrt{2} m_H f}{\sqrt{1-y_v^2 v}}$, where Φ can be any of the six new scalar bosons, in particular, we are interested in the physics of the new neutral scalar, denoted as Φ^0. Here, m_H is the Higgs boson mass and $y_v = 4v'f/v^2$, where v' is the vacuum expectation value (VEV) of the complex triplet and v is the electroweak energy scale. Moreover, there are another relevant parameters, namely, the coupling constants of the $SU(2)$ and $U(1)$ groups, which are g_j and g'_j for $j = 1, 2$, respectively, where $c \equiv g_1/\sqrt{g_1^2 + g_2^2}$ and $c' \equiv g'_1/\sqrt{{g'_1}^2 + {g'_2}^2}$. Typical values for the c parameters can be found in Refs. [5, 7]. Here, we will use $0.1 < c < 0.9$ and $c' = 1/\sqrt{2}$.

3. $\Phi^0 \to \gamma Z$ decay

In the LTHM, the $\Phi^0 \to \gamma Z$ decay is absent at the tree level, so the process is induced at one-loop level. To perform this calculation we take into account only the triangular diagrams mediated by the SM fermions. The contributions due to the charged gauge bosons are not considered because their couplings to Φ^0 are null. In the LTHM this decay is only mediated by SM quarks and a new heavy top quark [10]. In Fig 1, the Feynman diagrams that contribute to the $\Phi^0 \to \gamma Z$ decay are shown.

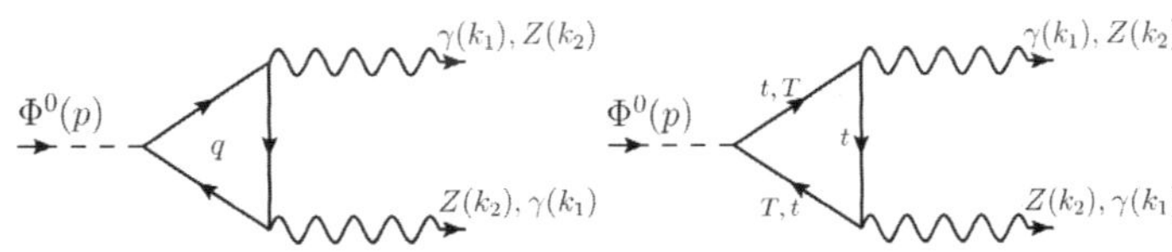

Figure 1. Feynman diagrams contributing to the $\phi^0 \to \gamma Z$ decay at the one-loop level. Here, q symbolizes any SM quark, t is the top quark and T represents the heavy top quark.

The amplitude for the $\Phi^0 \to \gamma Z$ process can be expressed as

$$\mathcal{M}_{\Phi^0 \to \gamma Z} = A^{\gamma Z} \left(k_1 \cdot k_2\, g^{\mu\nu} - k_1^\nu k_2^\mu\right) \epsilon_\mu(k_1) \epsilon_\nu(k_2), \tag{1}$$

where $\epsilon_\mu(k_1)$ and $\epsilon_\nu(k_2)$ are the polarization vectors for the photon and the Z boson, respectively. The $A^{\gamma Z}$ coefficient is given as

$$A^{\gamma Z} = \frac{e g g_V m_{\Phi^0}^2 v y_t \{C_a(y_Z - 1)(4t_t + y_Z - 1) + 2[(B_b - B_a + 1)y_Z - 1]\}}{4\sqrt{2}\, c_W f^2 (y_Z - 1)}, \tag{2}$$

where $B_a = B_0(m_{\Phi^0}^2, m_t^2, m_t^2)$, $B_b = B_0(m_Z^2, m_t^2, m_t^2)$, $B_c = B_0(m_{\Phi^0}^2, m_t^2, m_T^2)$, $B_a = B_0(m_Z^2, m_t^2, m_T^2)$ and $C_a = m_{\Phi^0}^2 C_0(m_{\Phi^0}^2, m_Z^2, 0, m_t^2, m_t^2, m_t^2)$ denote Passarino-Veltman scalar functions. Here, $y_t = m_t^2/m_{\Phi^0}^2$ and $y_T = m_T^2/m_{\Phi^0}^2$. It is worth mentioning that the $A^{\gamma Z}$ coefficient is free of ultraviolet divergences, whereas the amplitude satisfies gauge invariance. The numerical evaluation was performed by using the LoopTools package [11].

Finally, we get the following decay width for $\Phi^0 \to \gamma Z$ process

$$\Gamma(\Phi^0 \to \gamma Z) = \frac{e^2 g^2 v^2 (y_Z - 1)|A^{\gamma Z}|^2}{16\pi c_W^2 f^2 m_{\Phi^0}}. \tag{3}$$

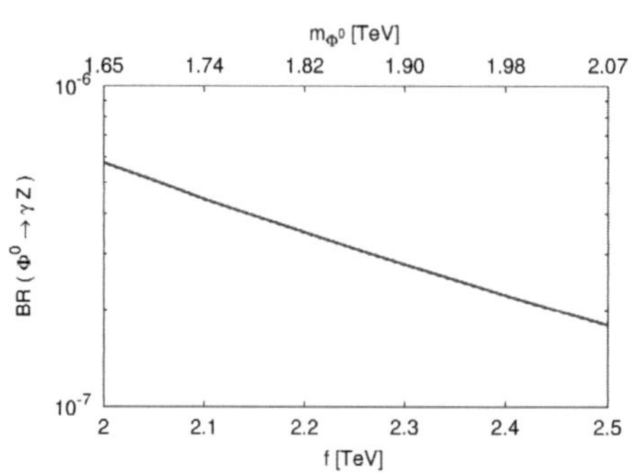

Figure 2. Branching ratio of the $\Phi^0 \to \gamma Z$ decay as a function of f parameter. On the upper x-axis, the value of the mass of the scalar particle Φ^0 is shown.

4. Results

Since the parameter space of the LTHM has been severely constrained by the Higgs discovery channels and electroweak precision observables [6], the $\Phi^0 \to \gamma Z$ decay could help to test this model. In this sense, a scenario where the mass of the scalar particle is assumed to be around 1 TeV is proposed. We begin our analysis by describing the behavior of the branching ratio for the $\Phi^0 \to \gamma Z$ process in the range 2 TeV $< f <$ 2.5 TeV, as it is shown in Fig. 2. For this interval of the energy scale f, the value of the mass of Φ^0 is $m_{\Phi^0} = 1.65$ TeV at $f = 2$ TeV and $m_{\Phi^0} = 2.07$ TeV with $f = 2.5$ TeV. The curve represents the branching ratio for the $\Phi^0 \to \gamma Z$ decay, which is of the order of 10^{-7}. For $f = 2$ TeV and $m_{\Phi^0} = 1.65$ TeV $Br(\Phi^0 \to \gamma Z) \approx 6 \times 10^{-7}$. It is important to mention that our results represent a preliminary study which must be enriched with a hypothetical analysis of the Φ^0 production at the LHC [10].

5. Conclusions

We have presented a brief description of the LTHM model. Our particular interest is the phenomenology of the scalar particle Φ^0 whose mass is assumed to be of the order of 1 TeV. However, although the space of parameters of this model has been severely restricted, there is a window to explore this scalar particle decaying into the final states γZ by using values for the parameter f between 2 TeV and 2.5 TeV. For these values of the energy scale f, the masses of Φ^0 are between 1.65 TeV and 2.07 TeV. For the process in question the corresponding branching ratio is of the order of 10^{-7}, which implies that this observable is very suppressed and it would be difficult to observe at the LHC in the near future.

Acknowledgments
This work has been partially supported by SNI and CIC-UMSNH (México).

References

[1] G. Aad *et al.* [ATLAS Collaboration], Phys. Lett. B **716** (2012) 1. S. Chatrchyan *et al.* [CMS Collaboration], Phys. Lett. B **716** (2012) 30.
[2] The ATLAS Collaboration, ATLAS-CONF-2015-081. The CMS Collaboration, CMS-PAS-EXO-15-004.
[3] Arkani-Hamed N, Cohen A G and Georgi H, Phys. Lett. **B513**, 232 (2001); Arkani-Hamed N, Cohen A G, Gregoire T and Wacker J G, JHEP**08**, 020 (2002); Arkani-Hamed N, Cohen A G, Katz E, Nelson A E, Gregoire T and Wacker J G, JHEP**08**, 021 (2002); Low I, Skiba W and Smith D, Phys. Rev. **D66**, 072001 (2002).
[4] Arkani-Hamed N, Cohen A G, Katz E and Nelson A E, JHEP**07**, 034 (2002).
[5] Han T, Logan H E, McÈlrath B and Wang L-T, Phys. Rev.**D67**, 095004, 2003.
[6] Reuter J and Tonini M, JHEP**02**, 077 (2013).
[7] Aranda J I, Cortés-Maldonado I, Ramírez-Zavaleta F and Tututi E S, J. Phys. G: Nucl. Part. Phys. **43**, 025002 (2016).
[8] Arkani-Hamed N, Cohen A G and Georgi H, Phys. Rev. Lett. **86**, 4757 (2001).
[9] Hill C T, Pokorski S and Wang J, Phys. Rev. **D64**, 105005 (2001).
[10] Aranda J I, Cortés-Maldonado I, Montejo-Montejo S, Ramírez-Zavaleta F and Tututi E S, work in preparation.
[11] Hahn T and Perez-Victoria M, Comput. Phys. Commun. **118**, 153 (1999).

Review of recent results on heavy-ion physics and astroparticle physics in ALICE at the LHC

Héctor Bello and Arturo Fernández
Facultad de Ciencias Físico Matemáticas BUAP, 1152, Puebla, México.

Antonio Ortiz
Instituto de Ciencias Nucleares, UNAM, Ciudad de México, 04510, México
E-mail: antonio.ortiz@nucleares.unam.mx

Abstract.
In this work we present a summary of the most relevant results on heavy-ion and astroparticle physics in ALICE. The summary includes a brief overview of the current status on the characterization of the hot and dense QCD medium created in the heavy-ion collisions produced at the LHC, as well as the intriguing finding of collective-like phenomena in small collision systems.

1. Introduction
ALICE (A Large Ion Collider Experiment [1]) is a dedicated heavy-ion detector to exploit the unique physics potential of nucleus-nucleus interactions at the Large Hadron Collider (LHC) energies. The main goal of the detector is to study the physics of strongly interacting matter at extreme energy densities, where we have evidence that a new phase of matter, the quark-gluon plasma (QGP), is formed [2, 3, 4, 5]. Although several important measurements have been carried out by the LHC high energy physics experiments (CMS, ATLAS and LHCb), the present work focuses on results from ALICE and it is based on a recent review on heavy-ion physics at the LHC [6]. In this paper, we discuss the discovery of QGP-like phenomena in small collision systems, proton-proton (pp) and proton-led (p–Pb) collisions. And also, we briefly discuss other contributions of ALICE in astroparticle physics.

2. The ALICE apparatus
Particle identification (PID) is an important tool to study the hot and dense matter created in heavy-ion collisions. Therefore, ALICE [7] is an experiment specialized in PID from low (≈ 0.15 GeV/c) up to high (20 GeV/c) transverse momenta. The central barrel of ALICE is placed inside a large solenoidal magnet which provides a magnetic field of 0.5 T. It is dedicated to detect hadrons, electrons, and photons produced at mid-pseudorapidity, $|\eta|<0.8$. It comprises an Inner Tracking System (ITS) of high-resolution silicon detectors, a cylindrical Time-Projection Chamber (TPC), and particle identification arrays of Transition-Radiation Detectors (TRD) and of Time-Of-Flight (TOF) counters. Additional central subsystems, not-covering full azimuth, are a ring-imaging Cherenkov detector for High-Momentum Particle IDentification (HMPID), and

two electromagnetic calorimeters: a high-resolution PHOton Spectrometer (PHOS) and a larger-acceptance ElectroMagnetic Calorimeter (EMCal). The muon arm detects muons emitted within $2.5 < \eta < 4$ and consists of a complex arrangement of absorbers, a dipole magnet, five pairs of tracking chambers, and two trigger stations. Several smaller detectors (VZERO, TZERO, FMD, ZDC, and PMD) for triggering, multiplicity measurements and centrality determination are installed in the forward region. On the three upper faces of the octagonal yoke of the solenoid, a dedicated cosmic ray detector (ACORDE [8]) is placed, which is made of 60 scintillator modules, covering 10% of solenoid upper faces. This subsystem produces a cosmic ray trigger signal when a single module fires (single muon cosmic ray event) or when more than two modules are activated by a cosmic ray shower.

3. Characterization of QGP

3.1. Soft probes

To learn about the early stage of the system, low transverse momentum ($p_{\rm T} < 2.2\,{\rm GeV}/c$) direct photons are studied. A temperature $T = 297{\pm}12^{\rm stat}{\pm}41^{\rm syst}$ MeV has been measured for the 0-20% Pb–Pb collisions at $\sqrt{s_{\rm NN}}$ =2.76 TeV [9]. Hence, the system at the LHC is hotter than that produced at RHIC, where an early temperature of $239{\pm}25^{\rm stat}{\pm}7^{\rm syst}$ MeV was measured for the same centrality class in Au–Au collisions [10]. It is worth noticing that such temperatures are already above the one predicted to achieve the QCD phase transition [11]. In addition, the average multiplicity per number of participant is twice that measured at RHIC [12].

The system expands and cools down, then, the inelastic interactions cease and the yields of particles reach fixed values. This is the stage of the so-called chemical freeze-out which is studied using the yields of identified hadrons. Within 20% particle ratios, e.g., the proton yield normalized to that of pions, are described by thermal models with a common chemical freeze-out temperature of $T_{\rm ch} \approx 156$ MeV [13]. However, large deviations are observed for protons and $\rm K^{*0}$; for the latter, this is not a surprise since its mean lifetime is smaller than that of the fireball ($\approx$10 fm/c) [14], and therefore the resonance yield may deviate from the expected values due to hadronic processes like re-scattering and regeneration [15].

On the other hand, the measurement of the spatial extent at decoupling time is obtained using intensity interferometry, a technique which exploits the Bose-Einstein enhancement of identical bosons emitted close by in phase space. This approach is known as Hanbury Brown-Twiss analysis (HBT) [16]. Using identical charged pions ALICE has measured a homogeneity volume of $\approx 300\,{\rm fm}^3$ (two times that reported at RHIC) and a decoupling time of $\approx 10\,{\rm fm}/c$ [14].

The transverse momentum distributions of identified hadrons contain valuable information about the collective expansion of the system ($p_{\rm T} \lesssim 2\,{\rm GeV}/c$), the presence of new hadronization mechanisms like quark recombination ($2 \lesssim p_{\rm T} \lesssim 8\,{\rm GeV}/c$) [17] and, at larger transverse momenta, the possible modification of the fragmentation due to the medium [18, 19]. ALICE has reported the transverse momentum spectra of charged pions, kaons and (anti)protons as a function of the collision centrality in broad transverse momenta intervals [20, 21, 22].

Figure 1 shows that for central Pb–Pb collisions the low $p_{\rm T}$ parts ($< 2\,{\rm GeV}/c$) of the spectra are well described by hydrodynamic models (within 20%), except the low $p_{\rm T}$ ($< 1\,{\rm GeV}/c$) proton yield [23, 24, 25, 26]. Models which best describe the data include hadronic rescattering with non-negligible antibaryon annihilation [25, 26]. The description of the results by hydrodynamic models is only observed in 0-40% Pb–Pb collisions, results for more peripheral collisions disagree with such prediction. This behavior has been recently studied for the average $p_{\rm T}$ in different colliding systems [27].

The analysis of the spectral shapes of the $p_{\rm T}$ distributions can be done using a blast-wave fit [28], which allows the extraction of the parameters related with the temperature at the kinetic freeze-out ($T_{\rm kin}$) and the average transverse expansion velocity ($\langle\beta_{\rm T}\rangle$). At the LHC, the radial flow in the most central collisions is found to be $\approx$ 10% higher than at RHIC, while the kinetic

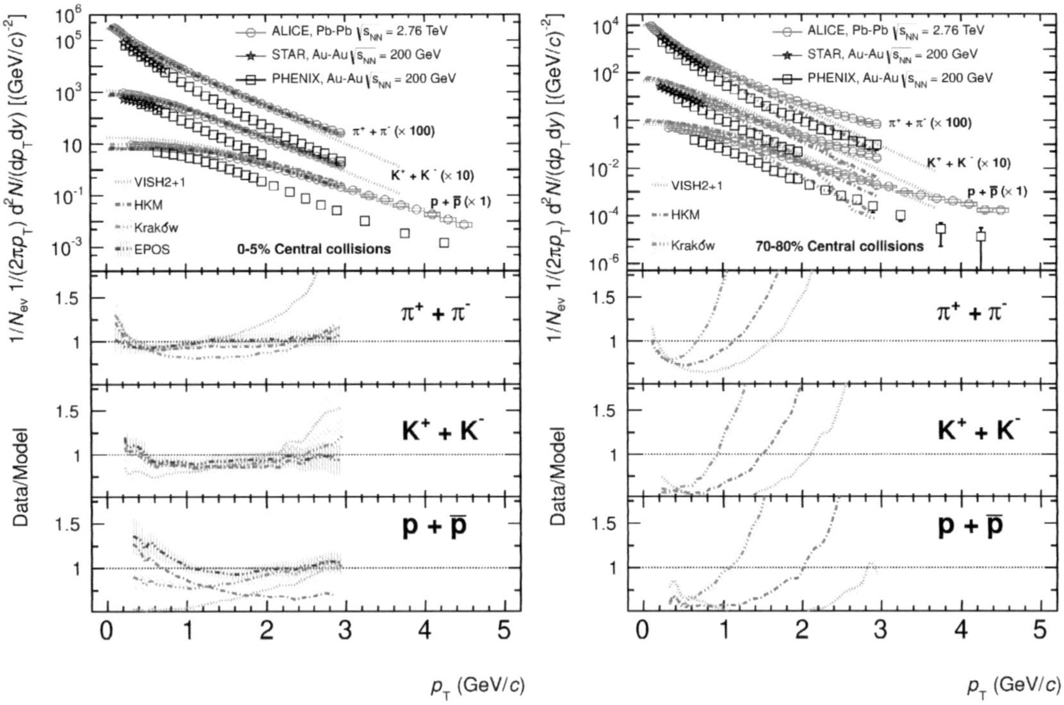

Figure 1. (Color online) Transverse momentum spectra of charged pions, kaons, and (anti)protons measured in central (left) and peripheral (right) Pb–Pb collisions at $\sqrt{s_{\rm NN}} = 2.76\,\text{TeV}$. The systematic and statistical uncertainties are plotted as color boxes and vertical error bars, respectively.

freeze-out temperature was found to be close to that measured at the RHIC, $T_{\rm kin} = 95\,\text{MeV}$ [20]. From the study of the low $p_{\rm T}$ particle production we conclude that at the LHC the created system is larger, hotter and longer-lived than that produced at RHIC.

The intermediate $p_{\rm T}$ is studied with the particle ratios as a function of $p_{\rm T}$ [22]. For central heavy-ion collisions, Fig. 2 shows a comparison of particle ratios with results from STAR and PHENIX at the RHIC measured in Au–Au collision at $\sqrt{s_{\rm NN}} = 200\,\text{GeV}$. The proton-to-pion peak at the LHC is approximately 20% larger than at the RHIC, which is consistent with an increased average radial flow velocity. Interestingly, in ALICE we also observe a peak in the kaon-to-pion ratio supporting the idea of a strong radial flow. To check the effect of radial flow the shapes of the $p_{\rm T}$ distributions of ϕ-meson and protons are compared. The results indicate that for central Pb–Pb collisions the shapes of the proton-to-pion and ϕ-to-pion ratios are the same. On the other hand, for $p_{\rm T} < 4\,\text{GeV}/c$, the ϕ-meson yield normalized to that of protons becomes flat going from the most peripheral to the most central Pb–Pb collisions. This suggests that the mass, and not the number of quark constituents, determines the spectral shape in central Pb–Pb collisions. This is in a good agreement with the hydrodynamical interpretation. Recently, it has been shown that the spectral shapes, studied with the average $p_{\rm T}$, exhibit a scaling with the hadron mass (number of constituent quarks) only in the 0-40% (40-90%) Pb–Pb collisions [27].

Using the scalar product method, the elliptic flow for identified hadrons has been

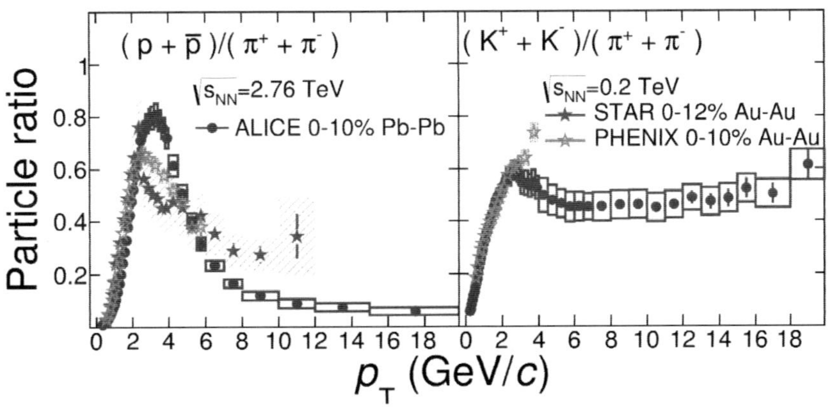

Figure 2. (Color online) ALICE (circles) results from $\sqrt{s_{\rm NN}} = 2.76\,$TeV Pb–Pb collisions compared with STAR and PHENIX results for $\sqrt{s_{\rm NN}} = 200\,$GeV Au–Au collisions. Left panel: the proton-to-pion ratio. Right panel: the kaon-to-pion ratio.

measured [29] over a broad $p_{\rm T}$ range. Going from central to semi-peripheral Pb–Pb collisions, the second order Fourier coefficient (v_2) increases as expected due to the eccentricity increase. For $p_{\rm T}$ below $2\,{\rm GeV}/c$ a mass ordering is observed indicating the interplay between elliptic and radial flow. For higher $p_{\rm T}$, the hadron-v_2 seems to be grouped into baryons and mesons, the exception is the v_2 of ϕ-mesons, which for central Pb–Pb collisions follows that for baryons. This observation indicates that the behavior of v_2 is driven by the hadron mass and not by the number of constituent quarks. ALICE has also reported the violation of the scaling of v_2 with the number of constituent quarks, such a observation is also against the scenario with quark recombination/coalescence.

3.2. Jet studies

The amount of suppression of inclusive particle production relative to pp collisions can be quantified with the nuclear modification factor, $R_{\rm AA}$, defined as

$$R_{\rm AA}(p_{\rm T}) = \frac{1}{\langle T_{\rm AA}\rangle}\frac{{\rm d}^2N^{AA}/{\rm d}y{\rm d}p_{\rm T}}{{\rm d}^2\sigma^{pp}/{\rm d}y{\rm d}p_{\rm T}} \qquad (1)$$

where $N^{\rm AA}$ ($\sigma^{\rm pp}$) is the yield (cross section) measured in AA (pp) collisions; and $\langle T_{\rm AA}\rangle$ is the nuclear overlap function [30].

Figure 3 shows $R_{\rm AA}$ of full jets with $R = 0.2$–0.3 in central Pb–Pb collisions at $\sqrt{s_{\rm NN}} = 2.76\,$TeV. The data exhibit a strong suppression of the jet production relative to pp collisions [31]. Comparing the nuclear modification factor of charged jets to that of charged particles at large $p_{\rm T}$ one can see that the amount of the suppression is similar although the underlying parton $p_{\rm T}$ scale is different for inclusive particles and jets.

In order to study details of path-length dependence of energy loss, ALICE performed studies of elliptic anisotropy of inclusive charged jets [32] and semi-inclusive distributions of recoil jets [33] which complement and further extend earlier studies of elliptic anisotropies of inclusive high-$p_{\rm T}$ particles and modification of away-side di-hadron correlations [34]. For collisional energy loss, the path length dependence is expected to be linearly proportional to the length traversed by the parton in medium, while for radiative energy loss, where in addition interference effects play a role, the dependence can be quadratic. In AdS/CFT class of models an even stronger dependence on path length traversed is predicted.

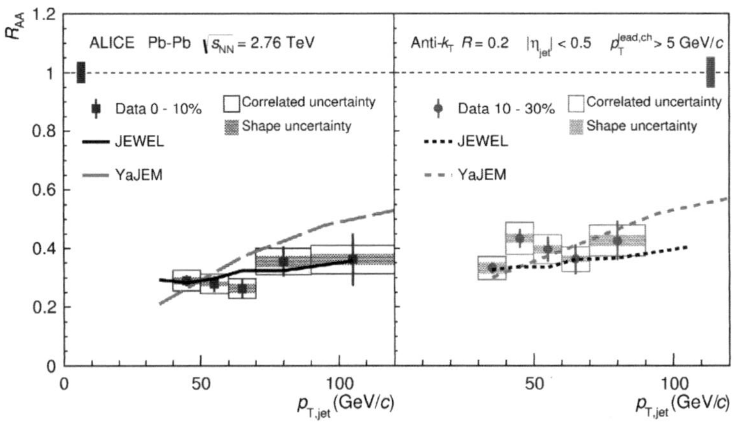

Figure 3. (Color online) Nuclear modification factor of full jets with $R = 0.2$ and p_T leading bias of 5 GeV/c in central Pb–Pb collisions [35].

In Figure 4 the measurement of elliptic anisotropy v_2 for charged jets with the resolution parameter $R = 0.2$ is shown in central and semi-central Pb–Pb collisions. The data show significant positive v_2 value in semi-central Pb–Pb collisions pointing to the path length dependence of jet suppression. In central collisions the current uncertainties on the measurement do not allow to draw a definite conclusion, although the v_2 magnitude is also positive. These data are also compared to v_2 for full jets measured by ATLAS [36] and inclusive charged particles [37, 38]. Although these measurements cannot be directly compared quantitatively due to different p_T scales and centrality selections, qualitatively they agree and provide a clear evidence of path-length dependent parton energy loss.

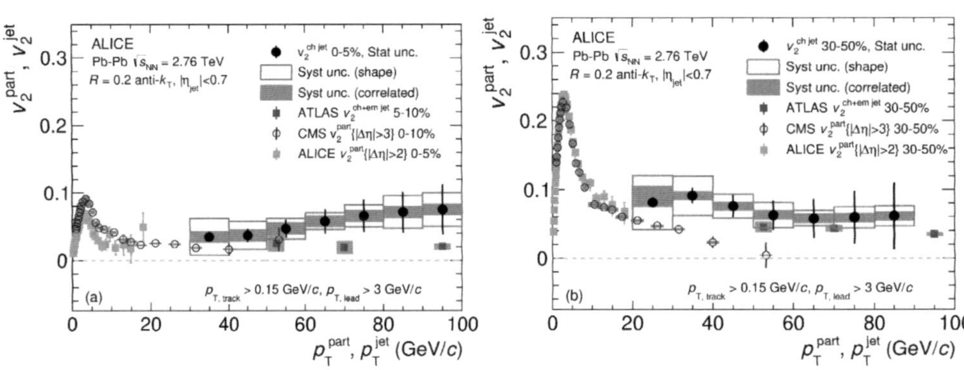

Figure 4. (Color online) Elliptic anisotropy (v_2) of inclusive charged particles and charged jets with the resolution parameter $R = 0.2$ in central (left) and semicentral (right) Pb–Pb collisions at $\sqrt{s_{NN}} = 2.76$ TeV measured by ALICE [32]. The data are compared with charged particle v_2 anisotropies measured by ALICE [37] and CMS [38] and calorimetric jets with $R = 0.2$ measured by ATLAS [36].

3.3. Open heavy flavour and quarkonia production

The nuclear modification factors of D mesons [39, 40] and electrons from heavy flavour decays [41] at mid-rapidity and of muons from heavy flavour decays at forward rapidity have been measured with ALICE. The results show a strong reduction of the yields (relative to pp) at large transverse momenta ($p_{\rm T} \geq 5\,{\rm GeV}/c$) in the most central collisions. Figure 5 (left) shows that, within uncertainties and in the measured $p_{\rm T}$ interval, the D meson nuclear modification factor is similar to that of charged pions and inclusive charged particles. It should be noted that the $R_{\rm AA}$ of D mesons and pions is also sensitive to the shape of the parton momentum distribution and their fragmentation functions. Model calculations including those effects and a colour-charge hierarchy in parton energy loss are able to describe the measurements [42].

In Fig. 5 (right), $R^{\rm D}_{\rm AA}$ as a function of the average number of participant nucleons [43] is shown. This measurement is compared with results from the CMS collaboration of non-prompt J/ψ [44] and theoretical predictions [45, 46]. For D mesons, a smaller suppression in peripheral than in central collisions is observed. A larger suppression in central collisions is seen for D mesons than for non-prompt J/ψ, indicating a different energy loss for charm and beauty quarks. This observation is supported by predictions from energy loss models, where the difference between the $R_{\rm AA}$ of D and B mesons arises from the different masses of c and b quarks.

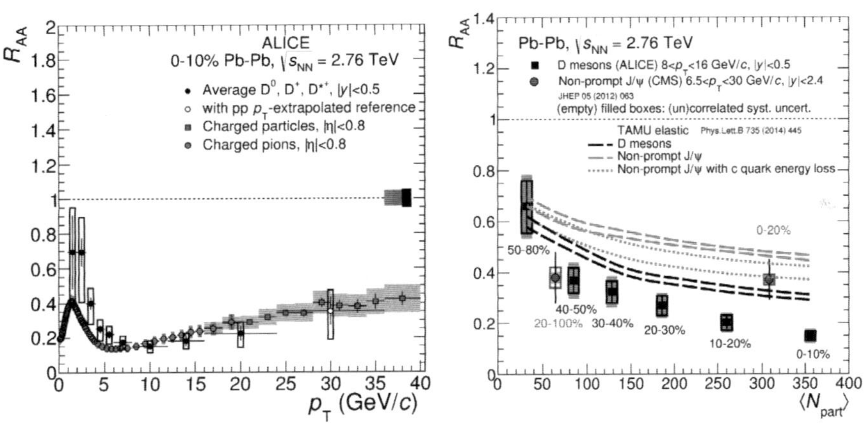

Figure 5. (Color online). D meson nuclear modification factor, $R_{\rm AA}$, in Pb–Pb collisions at $\sqrt{s_{\rm NN}} = 2.76\,{\rm TeV}$. Left: $R_{\rm AA}$ as a function of $p_{\rm T}$ compared to charged hadrons and pions. Right: $R_{\rm AA}$ as a function of $N_{\rm part}$ [43] compared to non-prompt J/ψ measured by the CMS collaboration [44].

The v_2 of prompt D^0, D^+ and D^{*+} mesons at mid-rapidity was measured in three centrality classes: 0-10%, 10-30% and 30-50% [47, 48]. The comparisons of the D meson v_2 as a function of $p_{\rm T}$ with analogous results for inclusive charged particles are shown in Fig. 6. The v_2 decreases from peripheral to central collisions as expected due to the decreasing initial geometrical anisotropy. The average of the v_2 of D^0, D^+ and D^{*+} in the centrality class 30-50% is larger than zero with 5σ significance in the range $2 < p_{\rm T} < 6\,{\rm GeV}/c$. A positive v_2 is also observed for $p_{\rm T} > 6\,{\rm GeV}/c$, which most likely originates from the path length dependence of the in-medium partonic energy loss, although the present statistics does not allow to reach a firm conclusion on this. The measured D meson v_2 is comparable in magnitude with that of the charged particles, which are mostly light-flavour hadrons. This result indicates that low-$p_{\rm T}$ charm quarks take part in the collective motion of the system.

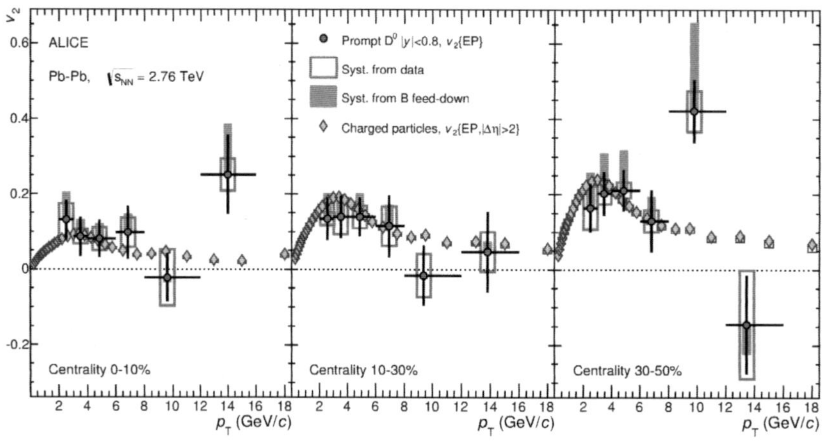

Figure 6. (Color online). D^0 meson v_2 as a function of p_T in three centrality ranges and compared with the v_2 of charged particles [47, 48].

R_{AA} and v_2 are two complementary measurements that can be used to better understand the connection between transport coefficients and the description of the in-medium energy loss [49].

3.4. Fluid-like behavior in small systems: pp *and p–Pb collisions*

The study of particle production in high-multiplicity pp and p–Pb collisions at the LHC has revealed unexpected new collective-like phenomena. The understanding of the phenomena is ongoing and different explanations have been proposed [50, 51, 52, 53, 54].

The CMS Collaboration studied the two-particle angular correlations of charged particles in pp collisions at $\sqrt{s} = 7\,\text{TeV}$ and discovered the ridge structure in small systems [55]. While the p_T-integrated correlation does not show any special feature, in the p_T range 1–3 GeV/c the near side long range angular correlation is clearly observed. Similar structures were also observed in p–Pb collisions at $\sqrt{s_{NN}} = 5.02\,\text{TeV}$ [56, 57, 58]. Furthermore, in high multiplicity events, non-zero second-order Fourier coefficients were extracted from the long-range correlations. Using the ALICE capabilities for particle identification, the proton v_2 was observed to be smaller than that for pions, up to about $p_T = 2\,\text{GeV}/c$ [59]. This effect is similar to the mass ordering of v_2 observed in heavy-ion collisions.

The transverse momentum spectra of charged pions, kaons and (anti)protons as a function of the event multiplicity have been measured up to 20 GeV/c [60]. At low p_T ($<$ 2–3 GeV/c) the spectra exhibit a hardening with increasing multiplicity, this effect is more important for heavy particles than for light particles. We are therefore observing features which resemble the radial flow effects well known from heavy-ion collisions [22] and which are well described when a hydrodynamical evolution of the system is considered. At the LHC [61] it was shown that for high multiplicity p–Pb events, the p_T spectra were described by the blast-wave function. Using the parameters obtained from the simultaneous fit to pion, kaon, proton and lambda p_T spectra the model is able to describe the multi-strange baryon p_T distributions ($p_T < 4\,\text{GeV}/c$) [62]. The feature is also observed in pp collisions simulated with PYTHIA 8 [63, 64], where no hydrodynamical evolution is included, instead multiple partonic interactions (MPI) and color reconnection are producing the effects. Some ideas have been proposed to understand the role of MPI in data [65, 66, 67, 68, 69]. The multiplicity dependence of the intermediate to high-p_T particle production is studied with the particle ratios [60]. The proton-to-pion ratio as a function of the event multiplicity exhibits a maximum (bump) at $p_T \approx 3\,\mathrm{GeV}/c$ and the size of the bump increases with increasing multiplicity. On the other hand, at higher transverse momenta ($p_T > 10\,\mathrm{GeV}/c$) the ratios return to the values measured for pp and Pb–Pb collisions. Any particle species dependence of the nuclear modification factor is therefore excluded.

In order to look for the presence of re-scattering effects in high multiplicity p–Pb collisions; the K^{*0} and ϕ relative to charged kaons production is studied as a function of the cube root of the average mid-rapidity charged particle density. In heavy-ion collisions the decreasing trend of K^{*0}/K with increasing fireball size has been explained as a consequence of a re-scattering of K^{*0} decay daughters in the hadronic phase. It is worth noticing that a similar trend is also observed in p–Pb collisions [70].

Finally, we discuss the latest results on multiplicity-dependent enhancement of strange and multi-strange hadron production in pp and p–Pb collisions [71]. Figure 7 shows a significant enhancement of strange to non-strange hadron production with increasing particle multiplicity in pp collisions. The behaviour observed in pp collisions resembles that of p–Pb collisions at a slightly lower centre-of-mass energy [62], both in the values of the ratios and in their evolution with the event activity. This suggests that the origin of strangeness production in hadronic collisions is driven by the characteristics of the event activity rather than by the initial-state collision system or energy. In the context of heavy-ion collisions, this effect (strangeness enhancement) has been considered a signature of the QGP formation. Recently, it has been pointed out that a perfect scaling of the particle ratios with the energy density holds for the different colliding systems [72] opening new possibilities for a better understanding of the QGP-like features in small systems.

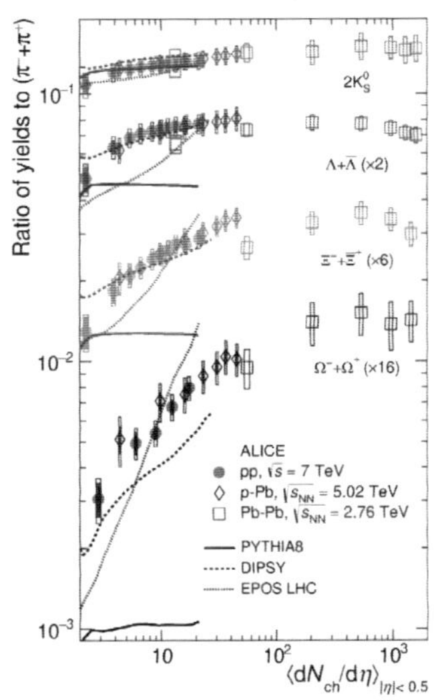

Figure 7. (Color online). Multiplicity dependence of strange and multi-strange hadrons yields normalized to that for charged pions. The empty and dark-shaded boxes show the total systematic uncertainty and the contribution uncorrelated across multiplicity bins, respectively. The values are compared to calculations from MC models and to results obtained in Pb–Pb and p–Pb collisions at the LHC.

4. Cosmic ray physics

Cosmic ray muons are created in Extensive Air Showers (EAS) following the interaction of cosmic ray primaries (protons and heavier nuclei) with nuclei in the upper atmosphere. Primary cosmic rays span a broad energy range, starting at approximately 10^9 eV and extending to more than 10^{20} eV. The use of high-energy physics detectors for cosmic ray physics was pioneered by ALEPH [73], DELPHI [74] and L3 [75] during the Large Electron-Positron (LEP) collider era at CERN. An extension of these earlier studies is now possible at the LHC, where experiments can operate under stable conditions for many years.

ALICE has also been used to perform studies that are of relevance to astroparticle physics [76]. Recent studies focus on events containing more than four reconstructed muons in the ALICE TPC, which we refer to as multi-muon events, stem from primaries with energy $E > 10^{14}$ eV. ALICE dedicated 30.8 days of cosmic ray data taking recording approximately 22.6 million events containing at least one reconstructed muon. ALICE has reported (see Fig. 8) 5 events with, $N_\mu > 100$ and zenith angles less than 50° (corresponding to an area density $\rho_\mu > 5.9\ m^2$ in a rate of 1.9×10^{-6} Hz).

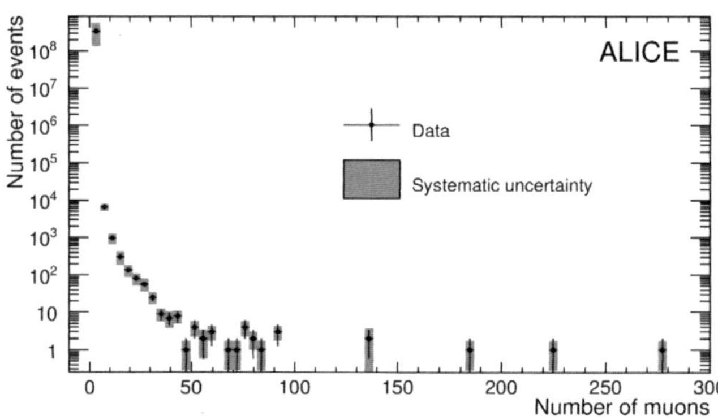

Figure 8. (Color online). Muon multiplicity distribution of the whole sample of data (2010-2013) corresponding to 30.8 days of data taking.

Figure 9 shows a comparison between data and MC simulations, the result suggests a mixed-ion primary cosmic ray composition with an average mass that increases with energy.

The ALICE measurement agrees with results from other experiments wich work at the same energy domain (knee). Following the successful description of the magnitude of the muon multiplicity distribution (MMD) in the low-to-intermediate range of muon multiplicities the same simulation framework was used to study the frequency of high muon multiplicity (HMM) events.

It has been found that the observed rate of HMM events is consistent with the rate predicted by CORSIKA 7350 using QGSJET II-04 to model the development of the resulting airshower, assuming a pure iron composition for the primary cosmic rays.

Only primary cosmic rays with an energy $E > 10^{16}$ eV were found to give rise to HMM events. The expected rate of HMM events is sensitive to assumptions made about the dominant hadronic production mechanisms in air shower development. The latest version of QGSJET differs from earlier versions in its treatment of forward neutral meson production resulting in a higher muon yield and has been retuned taking into account early LHC results on hadron production in 7 TeV proton-proton collisions. This is the first time that the rate of HMM events, observed at the relatively shallow depth of ALICE, has been satisfactorily reproduced using a conventional hadronic model for the description of extensive air showers; an observation that places significant constraints on alternative, more exotic production mechanisms [76].

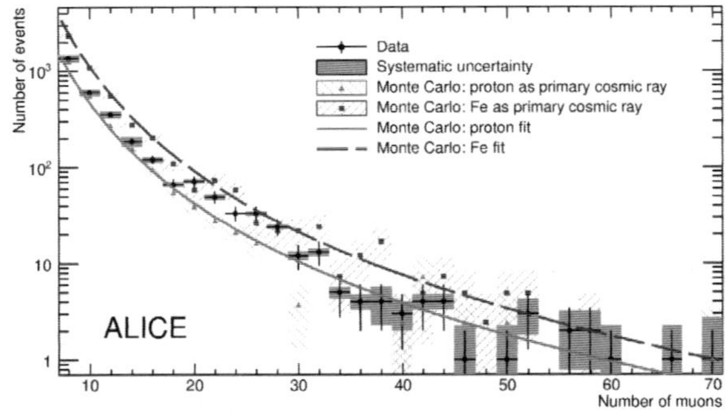

Figure 9. (Color online). The measured MMD compared with the values and fits obtained from CORSIKA simulations with proton and iron primary cosmic rays for 30.8 days of data taking. The errors are shown separately (statistical and systematic) for data, while for Monte Carlo they are the quadrature sum of the statistical and systematic uncertainties.

5. Outlook

In this paper we have presented the status and some highlights of the ALICE-LHC experiment, including the latest news on cosmic ray physics and the progress on the characterization of the hot and dense QCD medium created in the heavy-ion collisions. We have discussed several hot topics as the collective-like phenomena in small collision systems. More differential studies are needed in order to reveal the origin of the effects and their impact in our current understanding on the physics of heavy-ion collisions.

Acknowledgements

Support for this work has been received from CONACYT under the grant No. 260440 and grant No. 241408; from DGAPA-UNAM under PAPIIT grant IA102515.

References

[1] ALICE Collaboration (B. Alessandro *et al.*), *Jour. of Phys. G: Nucl. and Part. Phys.* **32** (2006) 1295.
[2] STAR Collaboration (J. Adams *et al.*), *Nucl. Phys.* **A757** (2005) 102.
[3] PHOBOS Collaboration (B. B. Back *et al.*), *Nucl. Phys.* **A757** (2005) 28.
[4] BRAHMS Collaboration (I. Arsene *et al.*), *Nucl. Phys.* **A757** (2005) 1.
[5] PHENIX Collaboration Collaboration (K. Adcox *et al.*), *Nucl. Phys.* **A757** (2005) 184.
[6] R. Bala, I. Bautista, J. Bielcikova and A. Ortiz, *Int. J. Mod. Phys.* **E25** (2016) 1642006, `arXiv:1605.03939 [hep-ex]`.
[7] ALICE Collaboration (B. B. Abelev *et al.*), *Int. J. Mod. Phys.* **A29** (2014) 1430044, `arXiv:1402.4476 [nucl-ex]`.
[8] A. Fernandez Tellez, ACORDE, The ALICE Cosmic Ray Detector, in *Proceedings, 30th International Cosmic Ray Conference (ICRC 2007): Merida, Yucatan, Mexico, July 3-11, 2007*, (2007), pp. 1201–1204.
[9] ALICE Collaboration (J. Adam *et al.*), *Phys. Lett.* **B754** (2016) 235.
[10] PHENIX Collaboration (A. Adare *et al.*), *Phys. Rev.* **C91** (2015) 064904.
[11] A. Ayala *et al.*, *Nucl. Phys.* **B897** (2015) 77.
[12] ALICE Collaboration (K. Aamodt *et al.*), *Phys. Rev. Lett.* **106** (2011) 032301.
[13] M. Floris, *Nucl. Phys.* **A931** (2014) 103.
[14] ALICE Collaboration (K. Aamodt *et al.*), *Phys. Lett.* **B696** (2011) 328.
[15] ALICE Collaboration (B. Abelev *et al.*), *Phys. Rev.* **C91** (2015) 024609.
[16] R. Hanbury Brown *et al.*, *Nature* **178** (1956) 1046.
[17] R. J. Fries *et al.*, *Ann. Rev. Nucl. Part. Sci.* **58** (2008) 177.

[18] S. Sapeta *et al.*, *Eur. Phys. J.* **C55** (2008) 293.
[19] R. Bellwied *et al.*, *Phys. Lett.* **B691** (2010) 208.
[20] ALICE Collaboration (B. Abelev *et al.*), *Phys. Rev.* **C88** (2013) 044910.
[21] ALICE Collaboration (B. Abelev *et al.*), *Phys. Lett.* **B736** (2014) 196.
[22] ALICE Collaboration (J. Adam *et al.*), *Phys. Rev.* **C93** (2016) 034913.
[23] P. Bozek *et al.*, *Phys. Rev.* **C85** (2012) 064915.
[24] I. A. Karpenko *et al.*, *Phys. Rev.* **C87** (2013) 024914.
[25] K. Werner *et al.*, *Phys. Rev.* **C85** (2012) 064907.
[26] C. Shen *et al.*, *Phys. Rev.* **C84** (2011) 044903.
[27] A. Ortiz, *Nucl. Phys.* **A943** (2015) 9.
[28] E. Schnedermann *et al.*, *Phys. Rev.* **C48** (Nov 1993) 2462.
[29] ALICE Collaboration (B. Abelev *et al.*), *JHEP* **06** (2015) 190.
[30] ALICE Collaboration (B. Abelev *et al.*), *Phys. Rev.* **C88** (2013) 044909, arXiv:1301.4361 [nucl-ex].
[31] ALICE Collaboration (B. Abelev *et al.*), *Phys. Lett.* **B722** (2013) 262.
[32] ALICE Collaboration (J. Adam *et al.*), *Phys. Lett.* **B753** (2016) 511.
[33] ALICE Collaboration (J. Adam *et al.*), *JHEP* **09** (2015) 170.
[34] ALICE Collaboration (K. Aamodt *et al.*), *Phys. Rev. Lett.* **108** (2012) 092301.
[35] ALICE Collaboration (J. Adam *et al.*), *Phys. Lett.* **B746** (2015) 1.
[36] ATLAS Collaboration (G. Aad *et al.*), *Phys. Rev. Lett.* **111** (2013) 152301.
[37] ALICE Collaboration (B. Abelev *et al.*), *Phys. Lett.* **B719** (2013) 18.
[38] CMS Collaboration (S. Chatrchyan *et al.*), *Phys. Rev. Lett.* **109** (2012) 022301.
[39] ALICE Collaboration (J. Adam *et al.*), *JHEP* **03** (2016) 081.
[40] ALICE Collaboration (J. Adam *et al.*), *JHEP* **03** (2016) 082.
[41] ALICE Collaboration (B. Abelev *et al.*), *Phys. Rev. Lett.* **109** (2012) 112301.
[42] M. Djordjevic, *Phys. Rev. Lett.* **112** (2014) 042302.
[43] ALICE Collaboration (J. Adam *et al.*), *JHEP* **11** (2015) 205.
[44] CMS Collaboration (S. Chatrchyan *et al.*), *JHEP* **05** (2012) 063.
[45] A. Andronic *et al.*, *Eur. Phys. J.* **C76** (2016) 107.
[46] M. Nahrgang *et al.*, *Phys. Rev.* **C89** (2014) 014905.
[47] ALICE Collaboration (B. Abelev *et al.*), *Phys. Rev. Lett.* **111** (2013) 102301.
[48] ALICE Collaboration (B. Abelev *et al.*), *Phys. Rev.* **C90** (2014) 034904.
[49] A. Ayala, I. Dominguez, J. Jalilian-Marian and M. E. Tejeda-Yeomans (2016) arXiv:1603.09296 [hep-ph].
[50] L. J. Gutay *et al.*, *Int. J. Mod. Phys.* **E24** (2015) 1550101.
[51] I. Bautista *et al.*, *Phys. Rev.* **D92** (2015) 071504.
[52] K. Dusling *et al.*, *Phys. Rev. Lett.* **108** (2012) 262001.
[53] K. Dusling *et al.*, *Phys. Rev.* **D87** (2013) 051502.
[54] K. Dusling *et al.*, *Phys. Rev.* **D93** (2016) 014034.
[55] CMS Collaboration (V. Khachatryan *et al.*), *JHEP* **09** (2010) 091.
[56] CMS Collaboration (S. Chatrchyan *et al.*), *Phys. Lett.* **B718** (2013) 795.
[57] ALICE Collaboration (B. Abelev *et al.*), *Phys. Lett.* **B719** (2013) 29.
[58] ATLAS Collaboration (G. Aad *et al.*), *Phys. Rev. Lett.* **110** (2013) 182302.
[59] ALICE Collaboration Collaboration (B. Abelev *et al.*), *Phys. Lett.* **B726** (2013) 164.
[60] ALICE Collaboration (J. Adam *et al.*), *Phys. Lett.* **B760** (2016) 720, arXiv:1601.03658 [nucl-ex].
[61] ALICE Collaboration Collaboration (B. Abelev *et al.*), *Phys. Lett.* **B728** (2014) 25.
[62] ALICE Collaboration (J. Adam *et al.*) (2015) arXiv:1512.07227 [nucl-ex].
[63] R. Corke *et al.*, *JHEP* **03** (2011) 032.
[64] A. Ortiz *et al.*, *Phys. Rev. Lett.* **111** (2013) 042001.
[65] ALICE Collaboration (B. Abelev *et al.*), *JHEP* **09** (2013) 049.
[66] ALICE Collaboration (B. Abelev *et al.*), *Phys. Lett.* **B741** (2015) 38.
[67] ALICE Collaboration (B. Abelev *et al.*), *Eur. Phys. J.* **C72** (2012) 2124.
[68] A. Ortiz *et al.*, *Nucl. Phys.* **A941** (2015) 78.
[69] A. Ortiz, G. Bencedi and H. Bello (2016) arXiv:1608.04784 [hep-ph].
[70] ALICE Collaboration (J. Adam *et al.*), *Eur. Phys. J.* **C76** (2016) 245, arXiv:1601.07868 [nucl-ex].
[71] ALICE Collaboration (J. Adam *et al.*) (2016) arXiv:1606.07424 [nucl-ex].
[72] Cuautle, E. and Paić, G. (2016) arXiv:1608.02101 [hep-ph].
[73] C. Grupen *et al.*, *Nucl. Instrum. Meth.* **A510** (2003) 190.
[74] DELPHI Collaboration (J. Ridky and P. Travnicek), *Nucl. Phys. Proc. Suppl.* **138** (2005) 295, [,295(2005)].
[75] L3 Collaboration (P. Ladron de Guevara), *Nucl. Phys. Proc. Suppl.* **143** (2005) 500.
[76] ALICE Collaboration (J. Adam *et al.*), *JCAP* **1601** (2016) 032, arXiv:1507.07577 [astro-ph.HE].

First year results of the High Altitude Water Cherenkov observatory

Alberto Carramiñana - for the HAWC Collaboration
Instituto Nacional de Astrofísica, Óptica y Electrónica,
Luis Enrique Erro 1, Tonantzintla, Puebla 72840, México

E-mail: alberto@inaoep.mx

Abstract. The High Altitude Water Cherenkov (HAWC) γ-ray observatory is a wide field of view (1.8 Sr) and high duty cycle (> 95% up-time) detector of unique capabilities for the study of TeV gamma-ray sources. Installed at an altitude of 4100m in the Northern slope of Volcán Sierra Negra, Puebla, by a collaboration of about thirty institutions of Mexico and the United States, HAWC has been in full operations since March 2015, surveying 2/3 of the sky every sidereal day, monitoring active galaxies and mapping sources in the Galactic Plane to a detection level of 1 Crab per day. This contribution summarizes the main results of the first year of observations of the HAWC γ-ray observatory.

1. High energy γ ray astronomy

The advent of new spectral windows for the study of the Universe during the second half of the XXth Century brought a remarkable expansion of our knowledge of the cosmos. Today, astronomy is performed across more than twenty decades of the electromagnetic spectrum, from tens of neV, the photon energy of 100 MHz radio-waves, to at least 100 TeV. Ranging from the MeV to the PeV, gamma rays represent the most extreme form of electromagnetic radiation. Its upper boundary may be even further in energy, having been sought up to the 10^{19} eV range with the Pierre Auger observatory [1]. From the observational point of view, γ rays are highly non-thermal radiation associated to astrophysical particle accelerators that produce cosmic rays. And, in the same way as high-energy γ-ray astronomy is complemented at other wavelengths, it complements searches in the dawn of the multi-messenger era.

Astronomical observations of γ rays are performed with space and ground-based instruments. There are three types of detectors with their own particular characteristics, as illustrated in the diagram 1. Their differences in energy coverage, field of view and sensitivity makes them highly complementary: in the same manner as space-borne pair production telescopes, like *Fermi*-LAT or *AGILE*, perform continuous observations of large portions of the sky in the GeV range, HAWC permanently surveys the sky transiting above it in the TeV range; both types of instruments provide useful information to atmospheric Cherenkov telescopes, which are highly sensitive but require to be pointed at specific parts of the sky, which are inaccessible part of the year.

Journal of Physics: Conference Series **761** (2016) 012034 doi:10.1088/1742-6596/761/1/012034

	GeV	TeV
Sr	**Pair production telescopes** 0.1 - 100 GeV Small effective area Background free Large FoV and high duty cycle All sky surveys & monitoring Transients (AGN, GRB) Extended diffuse emission	**Air shower arrays** 0.1 - 100 TeV Large effective area Good noise rejection Large FoV and high duty cycle Partial sky surveys & monitoring Transients (AGN, GRB) Extended diffuse emission
Deg		**Atmospheric Cherenkov telescopes** 30 GeV - 30 TeV Large effective area Excellent noise rejection Small FoV and low duty cycle Detailed studies of known sources Deep surveys of limited regions High resolution spectra

Diagram 1: three types of astronomical γ ray telescopes separated in terms of their energy coverage (*horizontally*) and field of view (FoV) (*vertically*).

The Large Aperture Telescope (LAT) on board of the *Fermi γ-Ray Space Telescope* has been one of the main driving forces in high energy astrophysics in the last years. Since its 2008 launch, *Fermi*-LAT has obtained a deeper and deeper exposure of the whole sky, leading to the finding of about 3000 sources of γ rays listed in various catalogs. In particular the 3FGL catalog is the deepest all-sky survey in the 0.1-300 GeV range performed to date, while the 1FHL catalog contains sources detected using photons of energies > 10 GeV [2, 3]. Of particular relevance for high energy studies is the 2FHL catalog of sources detected in the 50 GeV $-$ 2 TeV range, which offers the possibility of complementary studies with atmospheric Cherenkov telescopes and air shower arrays, in particular HAWC [4].

Atmospheric Cherenkov telescopes (ACTs) provide unique detailed information about very high-energy γ-ray sources [5]. They employ the atmosphere as the upper part of the detector, taking advantage of particle cascades triggered by the interaction of cosmic and γ rays with atmospheric nuclei. ACTs sample the Cherenkov light produced by the secondary particles of the cascade as they travel down the atmosphere faster than the speed of light in the air. As the emission cone of these particles is rather narrow, about $1.4°$, these telescopes have to be pointed to individual objects for their study. Furthermore, the emitted Cherenkov light is faint and can only be detected during night-time, on clear dark nights. To their advantage, ACTs are much more sensitive than air shower arrays: they can detect in a few minutes point sources that require more than one transit with HAWC. In the last couple of decades ACTs have found tens of TeV sources through pointed observations, mostly active galactic nuclei, in particular BL Lacs & flat spectrum radio quasars, and also starburst galaxies and Galactic sources, as listed in the TeVCat [6]. An important input to TeVCat is the Galactic Plane survey performed by the HESS collaboration. This is a very deep survey in the TeV band covering Galactic latitudes $|b| < 3.5°$, and longitudes $250° < \ell < 360°$ and $\ell < 65°$ [7]. As a consequence of the large

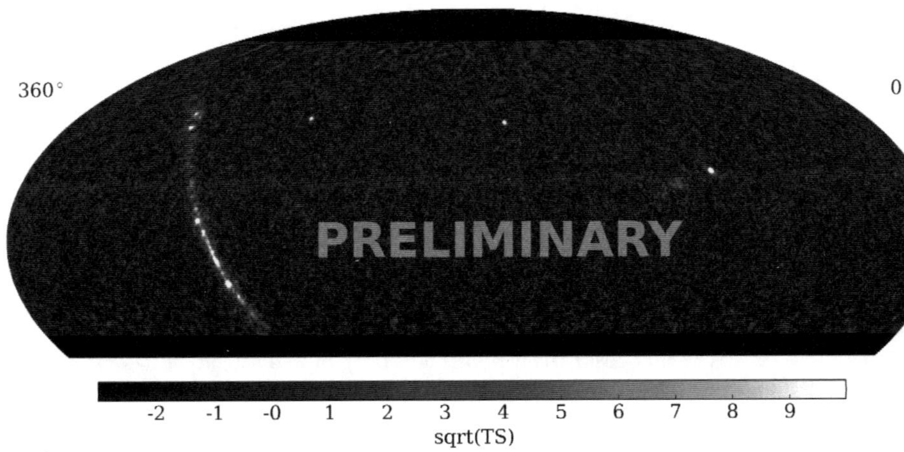

Figure 1. HAWC skymap of statistical significances above background, in celestial coordinates. The declination coverage is approximately $-30° < \delta < +70°$. The emission band at the first Galactic quadrant is clearly visible on the left. The three outstanding point sources from left to right, Mrk 501, Mrk 421 and the Crab Nebula. Next to the Crab, with lower significance, is Geminga.

number of Galactic sources found in the HESS Galactic Plane survey, the distribution of source types in TeVCat departs from the one derived from the 2FHL, highlighting the bias of (deep) pointed observations and highlighting the relevance of TeV wide field of view and high duty cycle observatories.

Extensive air shower (EAS) arrays can perform continuous survey and monitoring observations complementary to the deeper pointed ACT exposures. EAS arrays have been employed as cosmic-ray detectors for decades; the cosmic-ray experiment installed at Haverah Park in the 1960s is one of the earliest examples of water Cherenkov detectors [8]. EAS with the capability of separating photons and hadrons have been developed in United States (MILAGRO), Mexico (HAWC) and China (Tibet AS-γ and ARGO). In particular MILAGRO proved the feasibility of astronomical observations with the water Cherenkov technique by performing a successful multi-steradian survey of the γ-ray TeV sky. It detected photons of median energy 40 TeV from Galactic objects, in particular extended diffuse sources associated to GeV pulsars. At the Galactic anti-center MILAGRO detected the Crab Nebula and an extended source at the location of Geminga; the first Galactic quadrant showed prominent emission at the Cygnus region and the previously unreported source MGRO J1908+06. The BL Lac object Mrk 421 was also detected and sampled in timescales of months during the lifetime of the observatory [9].

2. The High Altitude Water Cherenkov γ-ray observatory

The success of MILAGRO prompted for the development of a water Cherenkov observatory of improved design located at higher altitude. The HAWC γ-ray observatory has been installed at the Northern base of Volcán Sierra Negra[1], in the Mexican state of Puebla, by a collaboration of about thirty institutions of Mexico and the United States formed in 2007. The site has an altitude of 4100m and a somewhat equatorial latitude of 19°N, which allows to cover the declination band $-26° \lesssim \delta \lesssim +64°$, or roughly two thirds of the sky for an instantaneous aperture of 45°. The site selected for HAWC is one kilometer North of the Large Millimeter

[1] named Tliltepetl, in Nahuatl.

Telescope Alfonso Serrano (LMT), a 50m diameter antenna for millimeter wave observations installed at 4563m [10]. The base infrastructure is shared between both facilities, as the road, electrical line and optical fiber of the LMT were extended to the HAWC site between 2008 and 2010. The project funding in place in early 2011 and the HAWC array platform was prepared between May and December 2011. After three years of installation, the complete HAWC array was inaugurated on March 2015 in a ceremony taking place at the 4100m site.

HAWC has four times the sampling area of MILAGRO and ten times higher muon detection area. The detector is segmented in 300 optically isolated detectors. Each one of these water Cherenkov detectors (WCD) consists of a metallic cylinder of 7.2m diameter and 5m height containing 180,000 liters of treated transparent water, with three 8 inch and one 10 inch up-facing photomultiplier tubes at its bottom. The array occupies an area of 22,500 m^2. The central Counting House hosts the data acquisition system, able to cope with a cosmic-ray event rate of 20 kHz. The relative time resolution is of the order of 1 nanosecond, to properly locate events in the sky; the recording of charge deposits allows for photon-hadron discrimination and energy estimation. A laser calibration system is routinely run from the Counting House.

HAWC started early science operations with a partial configuration of 111 detectors in August 2013. The array was upgraded to 250 WCDs in November 2014, four months prior to the start of full operations formalized with the inauguration event on equinox day 2015. The HAWC γ-ray observatory is currently exploring continuously the sky in the 0.1-100 TeV energy range with an instantaneous field of view of 1.8 Sr and the ability to reject 99.9% of the registered hadronic events, making it able to detect the Crab Nebula with a signal to noise of 5σ in a single 6h transit. The increasing understanding of the detector response results in an improved analysis software, currently at its Pass 4 version.

3. HAWC first year observations

A skymap of statistical significances, measured by $\sqrt{T_s}$, where T_s is the test statistic resulting of the comparison of a point source hypothesis versus the null hypothesis, was generated for declinations $-30°$ and $+70°$. The HAWC data, acquired between late November 2014 and December 2015, were selected with standard cuts that optimize the response for point sources with a Crab-like spectrum. The map shown in figure 1 shows distinct sources that will be described in detail in the 2HWC catalog, currently in preparation [11]. They are briefly summarized below.

3.1. The Galactic anti-center region

Three sources have been detected by HAWC in the direction of the Galactic anti-center. The most prominent is the Crab Nebula (M1), a well-known object studied all-through the electromagnetic spectrum. The Crab Nebula and its pulsar are the remnants of SN1054, a supernova recorded by several civilizations over 960 years ago. The proof of their association provided the first hard evidence of the link between supernovae and neutron stars [12]. The nebula, located at about 2 kpc as estimated through its apparent and physical expansion [13], is powered by the young and powerful 33ms pulsar PSR J0534+2200. The *Fermi*-LAT spectra of the Crab show the pulsed emission to be dominant between 100 MeV and a few GeV, to be rapidly out-shined by the nebular emission at higher energies [14]. The Crab pulsar has been detected with VERITAS and MAGIC up to ~ 2 TeV [15, 16]; however, the pulsed emission may be too faint for a HAWC detection, at least on the short term [17]. The Crab Nebula is a bright source of TeV γ rays: it is the brightest source detected by HAWC, with a significance $\gtrsim 80\sigma$ and small enough to be assumed a point source. It serves the TeV community as a calibrator: Crab observations by HAWC allow to compare simulations of its performance with actual data, and to optimize the analysis through improved γ/hadron cuts (fig.2, right panel).

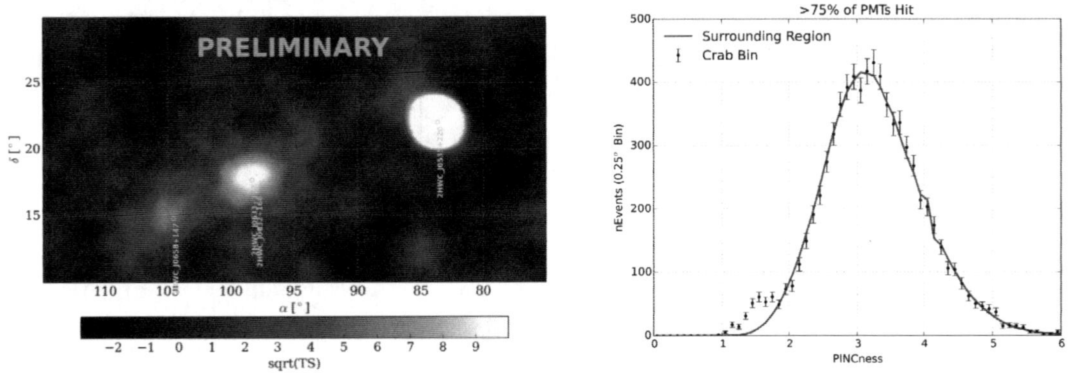

Figure 2. *Left*: Significance map of the Galactic anti-center region. The saturated source on the right corresponds with the Crab Nebula. The next bright emission region, at $\alpha \sim 98°$, coincides with the position of Geminga, previously reported which MILAGRO. On the left is a likely extended source matching the position of the GeV γ-ray pulsar PSR B0656+14. *Right:* distribution of the PINC parameter for γ/hadron discrimination on a region excluding the Crab (blue line) and centered on the Crab (black points).

One of the most topical current issues is the discovery of flaring activity of the emission of the Crab, reported by *AGILE* in 2011 [18]. HAWC has monitored the Crab between November 2014 and February 2016 on a daily basis, obtaining an unprecedented continuous 400-days light curve consistent with a constant flux. We note that, even thought ARGO-YBJ reported an enhancement of the TeV flux of the Crab in 2010, HESS did not find TeV emission associated to the > 100 MeV flaring event of March 2013, and that no MeV flares have been reported since November 2014 [19, 20]. HAWC monitoring of the Crab will be relevant to assess its TeV variability and also to explore its spectrum up to 100 TeV [21]. A dedicated Crab paper is currently in drafting by the HAWC collaboration.

Geminga is an older 237-ms pulsar, of dynamic age $\sim 300,000$ years, located at a distance of just 130 pc. It is a very bright GeV γ-ray point source, detected since the 1970-80s with the *SAS-II* and *COS-B* satellites [23, 24]. *ROSAT* and *CGRO*-EGRET data showed it contains a radio-silent and γ-ray loud pulsar [25], becoming the early prototype of this class of objects later unraveled by *Fermi*-LAT. The HAWC data are consistent with the MILAGRO observation of an extended emission matching the position of the Geminga pulsar [9].

Relatively close to Geminga is the pulsar PSR B0656+14 (PSRJ0659+1414), detected first in γ rays with *CGRO*-EGRET and COMPTEL. As in the case of Geminga, this is a relatively old ($\sim 110,000$ years) and nearby pulsar (300 pc), also detected in the radio and optical, but with intense X-ray emission. It is associated with the Monoceros SN remnant [26], itself associated to the EGRET source 2EG 0635+0521; this has been sought, but not detected, in TeV γ rays with HEGRA [27, 28]. HAWC sees an extended emission of a few degrees, difficult to map with ACTs, centered at about J0657+146. The HAWC data of the Geminga and Monoceros regions are currently undergoing detailed studies.

3.2. Galactic plane

HAWC scans every sidereal day close to 70% of Galactic plane, on longitudes $0° \lesssim \ell \lesssim 245°$. Aside from the Galactic anti-center region described above, HAWC has observed an extended ridge of γ-ray emission on the first quadrant of the Galactic Plane, specifically between

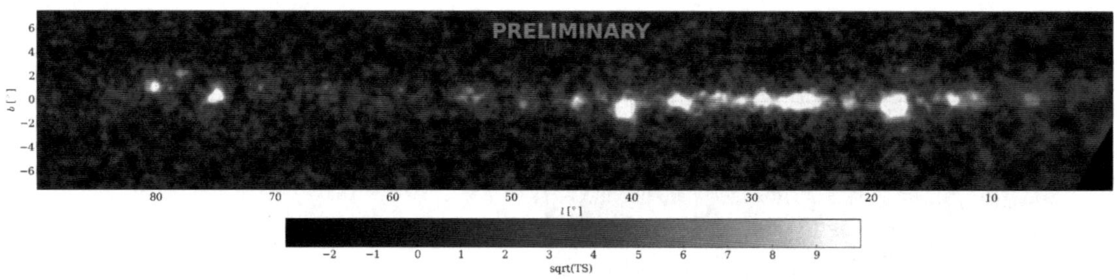

Figure 3. HAWC map of the first Galactic quadrant, displayed in Galactic coordinates. The Cygnus region is at $70° \lesssim \ell \lesssim 80°$; at $\ell \sim 54°$ is the triplet of sources related to SNR G054.1+00.3 and TeV J1930+188; and several prominent features at lower longitudes, in particular at 40° (MGRO J1908+06) and 18° (HESS J1841–055 and J1837–069).

$5° \lesssim \ell \lesssim 80°$. The Galactic Plane emission was present in the MILAGRO skymap [9], and is now been studied in more depth with HAWC. Most of the emission is confined to Galactic latitudes $|b| < 2°$, indicative of a young population of sources. Deconvolution with a point spread function provides tens of sources, a large fraction of which appear to be extended (fig. 3). The early HAWC-111 data showed already a good agreement with surveys performed with pointed observations by HESS and VERITAS [7, 29]; this is reinforced by the full-array data. Zones to highlight are: (i) the Cygnus region, which corresponds to a tangent spiral arm, where HAWC sees at least five distinct emissions which can be related to sources like TeV J2032+4130 (MGRO J2031+41), VER J2019+407, and MGRO J2019+37. These, in turn, correspond well with 3FGL sources; (ii) a complex zone of emission containing the SNR G054.1+00.3, coincident with TeV J1930+188, and two previously unreported sources; (iii) a large ridge of γ rays emission between $16° \lesssim \ell \lesssim 42°$, with several regions of intense emission coincident with MGRO J1908+06, HESS J1843–033, HESS J1837–069 and HESS J1825–137. This emission appears to continue down to the Galactic Center, where the response of HAWC decreases significantly. The Galactic Center itself has not been detected with HAWC, remaining a long term target requiring the full understanding of the high zenith angle response of the detector.

3.3. *Extragalactic sources*

The nearby BL Lac active galaxies Mrk 421 and Mrk 501, previously known to be sources of TeV photons, have been detected by HAWC at high very significant levels, $\sqrt{TS} \simeq 31$ and 22, respectively [11]. They are bright enough to be frequently detected on weekly timescales, and on single days when flaring. This has allowed to have one-year light curves for each of these

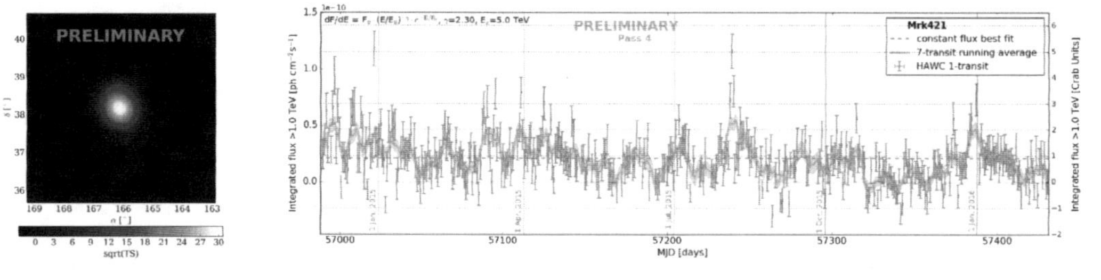

Figure 4. Detection and one-year light curve of Mrk421. Note the activity around MJD 57240.

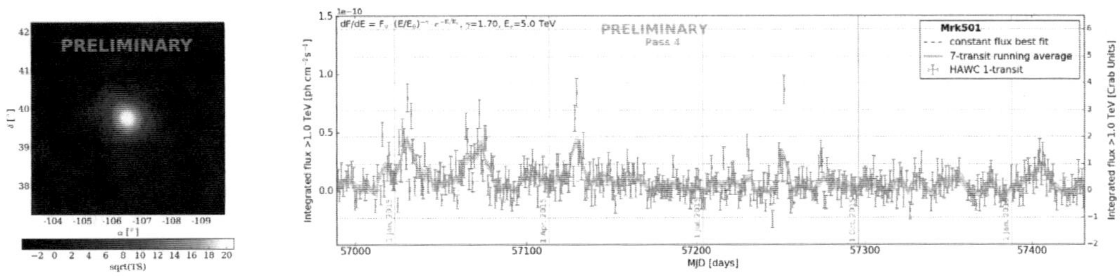

Figure 5. Skymap image and one-year light curve of Mrk 501. A large flare occurring two months after the last data point shown here was reported to the community.

Markarian objects with daily sampling (figs. 4 and 5). Both light curves are inconsistent with a constant flux with p-values below 10^{-10}.

Mrk 421 was discovered as a γ-ray source by *CGRO*-EGRET, instrument that established active galactic nuclei as the predominant sources of MeV-GeV gamma rays in the sky [30]. The detections with the Whipple ACT of TeV photons from Mrk 421 *only*, the nearest of the first sample of EGRET blazars, represented the first indication of the absorption of very high energy photons by the intervening extragalactic background light (EBL) [31, 32]. Figure 4 shows the light curve of Mrk 421 from November 2014 to February 2016. This is the first time that Mrk 421 has been sampled continuously for a whole year, as it cannot be observed by ACTs between July and October. When flaring, Mrk 421 can surpass the Crab and become the brightest object in the HAWC sky. The light curve on fig. 4 shows seven one-day transits with a flux $\geq 4\sigma$ above its average. Episodes of significant enhanced emission can be seen on timescales of $\sim$ 10 days. Note in particular the episode registered around August 2015, *i.e.* MJD 57240.

Mrk 501 was discovered as a source of TeV γ rays with the Whipple telescope, anticipating by over a decade its finding in GeV energies with *Fermi*-LAT [33, 34]. Although not as bright as Mrk 421, this source is detected with over 20σ and also monitored on daily sidereal basis. Its flux is inconsistency with a constant value, showing significant short timescale activity. Just a few months ago, on the 6th of April 2016, HAWC detected Mrk 501 with a flux 2.2 times larger than the Crab Nebula; this was promptly reported to the astronomical community [35].

One particularity of these two blazars is their proximity: with a redshift $z = 0.031$, Mrk 421 is at a luminosity distance $d_L = 134$ *Mpc*, while Mrk 501 has a slightly larger redshift $z = 0.033$ and corresponding to $d_L = 143$ Mpc. At larger distances the effect of the absorption of high-energy γ rays through the pair production process, $\gamma\gamma \rightarrow \mathrm{e}^+\mathrm{e}^-$, with the EBL is due to become important [32, 36]. The $\tau = 1$ opacity curve can be approximated as $(E/\mathrm{TeV})z \simeq 0.15$ in the low TeV band. HAWC is currently performing a search of active galactic nuclei with $z \lesssim 0.3$ to test absorption EBL models.

3.4. *Fundamental physics*

In addition to the search and study of TeV sources registered in its first year γ-ray skymap, the HAWC science case considers topics of fundamental relevance for physical sciences, some of which are very briefly highlighted below.

3.4.1. Cosmic rays – With an event rate of about 20,000 cosmic rays per second, HAWC is rapidly building a very large dataset for the study of cosmic particles. Early results confirming the anisotropy in the arrival directions of TeV cosmic rays have already been published [37]. As the cosmic-ray energy range in reach of HAWC is roughly from 1 TeV to 1 PeV, its data can link ground-based PeV observations with the multi-GeV energies reachable with space-borne experiment. The HAWC collaboration is currently working on spectral and composition studies of its extensive cosmic-ray data [38].

3.4.2. Gamma-ray bursts – The wide field of view and high duty cycle characteristics of HAWC are ideal for searching TeV emission from γ-ray bursts (GRBs). Even thought these explosive events have been studied in practically all wavebands, it remains to be proven that they do produce TeV photons. *Fermi*-LAT detected 35 GRBs in its first three years, some of them with photons in the 50-100 GeV range [39]. Even before the start of early science operations in August 2013, HAWC had been seeking to detect GRBs. The exceptional event GRB 130427A, in which a 94 GeV photon was detected by *Fermi*-LAT [40], happened under unfortunate circumstances for HAWC. Then in early phases of installation, the observatory was operating with 29 WCDs and the DAQ system in scaler mode only; the event had a zenith angle of 57° and setting. Given these non-optimal conditions, only a very rough upper limit could be set [41]. Just a few months later, better conditions allowed HAWC to start setting constraining upper-limits on later events [42]. HAWC is constantly searching for GRBs occurring in its field of view: the recent and notorious GRB 160509A, with a multi-GeV photon recorded by LAT, but also with an unfortunately high zenith angle for HAWC, was reported through the GCN system [43].

3.4.3. Dark matter searches – HAWC is well suited to search for γ rays produced by the decay of weakly interacting massive particle (WIMPs). The 5-year sensitivity of HAWC for the dark matter emission expected for the Virgo cluster, M31 and the Galactic Center were shown at the ICRC2015 [44]. The nearby Andromeda galaxy is probably the best candidate at low WIMP masses, while the Galactic Center may provide the best limits at higher energies, $M_\chi \gtrsim 30 - 100$ TeV, despite its unfavorable location relative to the Sierra Negra site [44].

3.4.4. Primordial Black Holes – Primordial black holes (PBHs) are remnants of the Big Bang that may have formed during phase transitions related to collapses of cosmic strings or domain walls, in the very early Universe [45]. According to quantum gravity theories, they evaporate through Hawking radiation following the formalism of classical thermodynamics extended to gravitation [46, 47]. PBHs radiate according to their mass on timescales $\tau(M) \propto M^3$, such that black holes with $M_{BH} \lesssim 5 \times 10^{14}$g should be evaporating now. Within the uncertainties on how the final stages of the evaporation process lead to an explosive event, PBHs may have a distinct signature observable in the GeV/TeV range [48]. The most constraining upper limits of the volume density of PBHs decaying has been set with MILAGRO and will be superseded by HAWC with its first year data, to be gradually improved as the HAWC database builds up [49].

3.4.5. Neutrino astrophysics – The IceCube neutrino observatory is providing a new fresh look at astrophysical phenomena in the high-energy regime. Astrophysical neutrinos are a powerful diagnostic of high-energy hadronic processes, which necessarily produce γ rays. HAWC is coordinating efforts with IceCube in the search for enhanced TeV photon emission related to multi-TeV and PeV neutrino events in real-time observations, and also in archival data [51, 52].

3.4.6. Gravitational wave events follow-up – The spectacular discovery of a gravitational wave emission event observed by LIGO on the 15th of September 2015, interesting in its own right as an astrophysical black hole merger, is a landmark in the study of the Universe [53]. Since 2014 the HAWC collaboration has worked in coordination with the LIGO/VIRGO collaboration to search for high energy electromagnetic counterparts of gravitational wave events. The GW150915 discovery event occurred towards the celestial South Pole, making unfeasible such a follow-up.

The localization by LIGO of the more recent GW151226 event, a coalescence of two black holes of masses 14 and 7 $M_\odot$ respectively [54], intersected the field of view of HAWC at the time of the event. The analysis of these HAWC data is on-going.

4. HAWC summary

The HAWC γ-ray observatory has become a prime survey instrument for high-energy astrophysics. Its first year data have provided a novel view of the TeV sky, with prominent sources in the Galactic anti-center, the first Galactic quadrant and outside our own galaxy. A deeper understanding of these, together with studies in fundamental topics like cosmic rays, GRBs, dark matter searches, PBHs and multi-messenger follow-ups, makes us look forward to further HAWC data. The HAWC collaboration is currently working on upgrading its facilities with a high bandwidth internet optical fiber connection from the Sierra Negra site to the INAOE, and a sparse 100,000 m^2 outrigger array to improve the response of the instrument above 10 TeV.

Acknowledgements

The High Altitude Water Cherenkov γ-ray observatory is a collaboration of over thirty institutions in Mexico and the United States, with current participation of Germany, Poland and Costa Rica. HAWC has been possible thanks to the generous support of the National Science Foundation, the US Department of Energy and the Consejo Nacional de Ciencia y Tecnología in Mexico.

This research has made use of the SIMBAD database, operated at CDS, Strasburg, France [55]; and of NASA's Astrophysics Data System [56].

References

[1] *The Pierre Auger and Telescope Array Collaborations:* Aab, A. et al., ApJ 794, 172 (2014).
[2] Acero, F., et al. ApJS 218, 23 (2015).
[3] Ackermann, M., et al. ApJS 209, 34 (2013).
[4] Ackermann, M., et al. ApJS 222, 5 (2016).
[5] Weekes, T.C., AIP Conf Ser 1085, 3 (2008).
[6] http://tevcat.uchicago.edu/
[7] The HESS Galactic Plane survey - arxiv/1307.4690 from ICRC 2013.
[8] Lawrence, M.A., Reid, R.J.O., Watson, A.A., J. Phys. G: Nucl. Part. Phys. 17, 733 (1991).
[9] Abdo, A.A., et al. ApJ 664, L91 (2007).
[10] Serrano Pérez-Grovas, A., RMAA SC 4, 67 (1996).
[11] *The HAWC Collaboration*: Abeysekara, A.U., et al. - in preparation
[12] Mayall, N.U. & Oort, J.M. PASP 54, 85 (1942).
[13] Trimble, V., AJ 73, 535 (1968).
[14] Abdo, A.A., et al. ApJ 708, 1254 (2010).
[15] *VERITAS Collaboration:* Aliu, E., et al. Science 334, 69 (2011).
[16] Ansoldi, S., et al. A&A 585, 133 (2016).
[17] Álvarez Ochoa, C., et al., *for the HAWC Collaboration*, PoS(ICRC2015)890 – arxiv/1508.03327.
[18] Tavani, M., et al. Science 331, 736 (2011).
[19] Aielli, G., et al. ATel 2921 (2010).
[20] *The HESS Collaboration:* Abramowski, A., et al., A&A 562, L4 (2014).
[21] Amenomori, M., et al. ApJ 813, 98 (2015).
[22] *The HAWC Collaboration*: Crab Nebula paper, in preparation.
[23] Thompson, D.J., et al., ApJ 213, 252 (1977).
[24] Swanenburg, B.N., et al. ApJ 243, L69 (1981).
[25] Halpern, J., Holt, S.S., Nature 357, 222(1992).
[26] Leahy, D.A., Naranan, S., Singh, K.P., MNRAS 220, 501 (1986).
[27] Kaaret, P., et al. ApJ 523, 197 (1999).
[28] Aharonian, F.A., et al. A&A 417, 973 (2004).
[29] *The HAWC Collaboration*: Abeysekara, A.U., et al., ApJ 817, 3 (2016).
[30] Fichtel, C. E., et al. ApJ SS 94, 551 (1994).

[31] Punch, M., et al., Nat 358, 477 (1992).
[32] de Jager, O.C., Stecker, F.W., Salamon, M.H., Nature 369, 294 (1994).
[33] Quinn, J., et al., ApJL 456, 83 (1996).
[34] Abdo, A.A., et al., ApJS 183, 46 (2009).
[35] Sandoval, A., Lauer, R., Wood, J. *on behalf of the HAWC Collaboration*, ATel 8922 (2016).
[36] Dominguez, A., et al. ApJ 770, 77 (2013).
[37] *The HAWC Collaboration*: Abeysekara, A.U., et al., ApJ 796, 108 (2014).
[38] Hampel-Arias, Z. for the *HAWC Collaboration*, APS April 2016 Meeting, S13.006 (2016).
[39] Ackermann, M., et al., ApJS 209, 11 (2013).
[40] Ackermann, M., et al, Science 343, 42 (2014).
[41] *The HAWC Collaboration*: Abeysekara, A.U., et al., ApJ 800, 78 (2015).
[42] Lennarz, D., Taboada, I., *for the HAWC Collaboration*, PoS(ICRC2015)715 – arxiv/1508.03327.
[43] Lennarz, D., at HAWC, GCN Circular 19423 (2016).
[44] Harding, J.P., Dingus, B., *for the HAWC Collaboration*, PoS(ICRC2015)1227 – arxiv/1508.04352.
[45] Carr, B.J., Proc. of "Inflating horizon of particle astrophysics and cosmology" – astro-ph/0511743 (2005).
[46] Hawking, S.W., Nature 248, 30 (1974).
[47] MacGibbon, J.H., et al. Proc. Fifth Fermi Symposium, eConf C121020.1 – arxiv/1503.01166 (2015).
[48] Ukwatta, T.N., et al. Astroparticle Physics 80, 90 (2016).
[49] Abdo, A.A., et al., Astroparticle Physics 64, 4 (2015).
[50] *The IceCube Collaboration:* Aartsen, M.G., et al. astro-ph/1510.05223.
[51] Ignacio Taboada, on behalf of the HAWC Collaboration, ATel 7868 (2015).
[52] Ignacio Taboada, GCN Circular 19361 (2016).
[53] *The LIGO Scientific Collaboration and VIRGO Collaboration:* Abbott, B.P., et al. Phys. Rev. Lett 116, 061102 (2016).
[54] *The LIGO Scientific Collaboration and VIRGO Collaboration:* Abbott, B.P., et al. Phys. Rev. Lett 116, 241103 (2016).
[55] Wenger, et al. A&AS 143, 9 (2000).
[56] http://www.adsabs.harvard.edu/

Vacuum Cherenkov radiation and photon decay rates from generic Lorentz Invariance Violation

H. Martínez-Huerta and A. Pérez-Lorenzana

E-mail: aplorenz@fis.cinvestav.mx
Physics Department, Centro de Investigacion y de Estudios Avanzados del IPN, Mexico City, DF, Mexico

Abstract.
Among the most studied approaches to introduce the breaking of Lorentz symmetry, the generic approach is one of the most frequently used for phenomenology, it converges on the modification of the free particle dispersion relation. Using this approach in the photon sector, we have calculated the squared probability amplitude for vacuum Cherenkov radiation and photon decay by correcting the QED coupling at tree level and first order in LIV parameters. For the lower order energy correction we calculate the emission and decay rate for each process.

1. Introduction

Cosmic rays are the most energetic phenomena known so far, reaching energies of several decades of EeV [1, 2]. Among the many particulars their study and understanding reveal, they also provide an energy window to test fundamental physics. Such is the case of the search for signatures of Lorentz Invariance Violation (LIV), mainly motivated by Quantum Gravity theories and string theories [3, 4, 5, 6]. Due the the nature of this symmetry, the derived physics from LIV tends to be unique and energy dependent [7, 8]. Therefore, its consequences at the highest energies and very long distances could be identified in the current observatories and experiments.

Following this spirit, in this paper we present a generic approach and a phenomenological first order correction to the production rates of two processes that could have a significant impact on cosmic particle propagation. The generic mechanism for introducing LIV, often found in the literature [7, 8, 9, 10, 11, 12, 13, 14], is summarized in an explicit not Lorentz invariant (LI) term added to the free particle Lagrangian density that will converge into the following generic correction to the dispersion relation:

$$S_a = E_a^2 - p_a^2 = m_a^2 \pm \alpha_{a,n} A^{n+2}, \tag{1}$$

where E_a and p_a stand for the four-momenta associated with an a particle species. Additionally, for particular models, A can take the form of E or p, however, for the ultra relativistic limit where $m_a \ll \{E, p\}$, any particular choice of A will be equivalent. The coefficient $\alpha_{a,n}$ in Eq. (1), parametrizes the particle species dependent LIV correction, where n expresses the correction order to the mass shell. It is common to associate a generic $\alpha_n \approx E_{QG}^{-n}$, where E_{QG} is the scale of Quantum Gravity or the scale of the expected new physics beneath. Several methods are used in the search for LIV signals, some of them can lead to lower limits to E_{QG} [15, 16, 17, 18], which is expected to be close to 10^{19} GeV .

In the next section, we have applied the generic correction shown in Eq. (1) for photons to derived the LIV corrected square amplitude at tree level from the diagrams in figure 1. It is worth stressing that such processes are forbidden in the standard theory by energy-momentum conservation, but under the LIV hypothesis they can be possible. Both of them are used in the search for LIV evidence. The first one is the emission of a single photon by a charged particle that propagates in vacuum and it is frequently named vacuum Cherenkov radiation [11, 12]. The second is LIV photon decay [11, 13], and it is motivated by the LIV extra term in Eq. (1) that can be read as an effective photon mass and it will depend on the LIV coefficients; the simplest process will produce an electron - positron pair. Once the corrected square amplitude has been derived, we used it to find the process rates.

2. LIV corrected rates

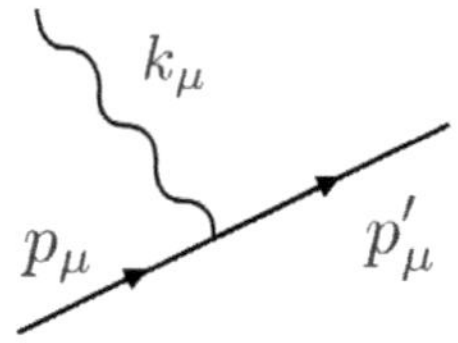

Figure 1. Diagram for LIV vacuum Cherenkov radiation.

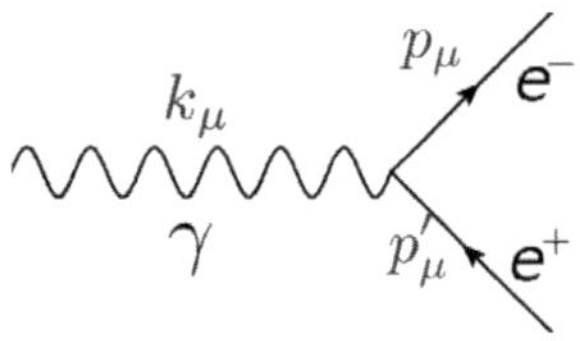

Figure 2. Diagram for LIV photon decay.

Following the diagrams in figures 1 and 2, let be (E_p, p) and $(E_{p'}, p')$ the four-momenta for the initial and final particle in the particular case in figure 1, but the four-momenta for the e^+-e^- pair in figure 2, as indicated. We will call (ω, k) the photon four-momenta in both the cases. With such a consideration, we can express the square probability amplitude for both processes with a single expression, given by

$$\begin{aligned}\frac{1}{2}\sum_{spin}|M|^2 &= (\not{p} - \not{k} + m)(e^2\gamma^\mu\gamma^\nu)(\not{p} + m)\epsilon_\mu\bar{\epsilon}_\nu \\ &= 2\sum_{spin}(p^\mu p^\nu - k^\mu p^\nu - p^\nu p^\mu + k^\nu p^\mu + p^\alpha p_\alpha g^{\mu\nu} - k^\alpha k_\alpha g^{\mu\nu} + m^2 g^{\mu\nu})\epsilon_\mu\bar{\epsilon}_\nu\ ,\end{aligned} \tag{2}$$

where Minkowski metric $g^{\mu\nu} = \eta^{\mu\nu}$ and Dirac gamma matrices standard properties under a Lorentz Invariant (LI) theory are assumed. Since the correction is taken at first LIV order correction, any term $\geq \alpha_n^2$ is negligible and the Ward identity is preserved.

From Eq. (1), $S_\gamma = k^\alpha k_\alpha = \alpha_n k^{n+2}$, and assuming the charged particle dispersion relation to be Lorentz conserving along the process, we note that

$$S_p = S_{p'} = (p-k)^\alpha(p-k)_\alpha = S_p + S_\gamma - 2k^\alpha p_\alpha, \tag{3}$$

then

$$2k^\alpha p_\alpha = S_\gamma = \alpha_n k^{n+2}, \tag{4}$$

where we have dropped out the particle index on the photon LIV parameter. Hence, Eq. (2) becomes:

$$\frac{1}{2}\sum_{spin}|M|^2 = e^2|4m_a^2 - \alpha_n k^{n+2}| \tag{5}$$

The absolute value is taking in order to ensure physical congruence due the limits and the generality in the sign of α_n.

Once the modified squared amplitude for both process is found, we have derived the modified decay and emission rates on an arbitrary preferential frame and for $n = 1$.

For vacuum Cherenkov radiation we have found that

$$\Gamma_{a\to a\gamma} = \frac{e^2}{4\pi}\frac{1}{4E_a}\int_0^{\pi}\frac{|4m_a^2 - \alpha_1 k_+^3|}{\omega(k_+,\alpha_1)}\frac{k_+^2 \sin\theta d\theta}{|p_a\cos\theta - k_+ - \frac{(1+\frac{3}{2}\alpha_1 k_+)}{\sqrt{1+\alpha_{(1)}k_+}}\sqrt{k_+^2 + E_a^2 - 2k_+ p_a \cos\theta}|}, \tag{6}$$

where the photon momenta modes from the corrected energy-momenta conservation are

$$k_\pm = \frac{1}{2p_a\cos\theta}\left(E_a^2 \pm \sqrt{E_a^4 + \frac{4p_a\cos\theta}{\alpha_1}(E_a^2 - p_a^2\cos^2\theta)}\right). \tag{7}$$

We have used only k_+, since we are looking for a phenomenological approach and k_- will not be physical.

On the other hand, for LIV photon decay we have:

$$\Gamma_{\gamma\to e^+e^-} = \frac{e^2}{4\pi}\frac{|4m_e^2 - \alpha_1 k^3|}{4\omega(k,\alpha_1)}\int_0^{\pi}\sum_{p=p_\pm}\frac{p^2\sin\theta d\theta}{|(k\cos\theta - p)E_e - p\sqrt{k^2 + E_e^2 - 2kp\cos\theta}|}, \tag{8}$$

where the momenta modes from the corrected energy-momenta conservation are

$$p_\pm = \frac{1}{2(\alpha_1 k + \sin^2\theta)}\left(\alpha_1 k^2\cos\theta \pm \sqrt{\alpha_1^2 k^4\cos^2\theta - 4(\sin^2\theta + \alpha_1 k)(1+\alpha_1 k)m_e^2}\right). \tag{9}$$

Above, e stands for the electron charge, θ is the angle between final particles, $p_a = \sqrt{E_a^2 - m_a^2}$ for the given particle a and $\omega(k,\alpha_1) = |k|\sqrt{1+\alpha_1 k}$. Notice that the formulae for photon decay are actually generic for any fermion pair in the final state, provided the corresponding mass is used. There exist vacuum Cherenkov radiation and photon decay rates obtained from different LIV approaches, for instance, from the minimal Standard-Model extension by the spontaneous breaking of Lorentz symmetry [19] and from the introduction of Lorentz violating operators of dimensions four and six can be found in [20, 21].

The last integration in eqs. (6) and (8) are numerically performed for both expressions. For $n = 1$, the results are depicted in figures 3 and 4. Since we are looking for an approach at the highest energies, we choose $m_a = m_{proton}$ in the first considered process due its relevance for cosmic rays [1, 22]. A small angle approach is taken for vacuum Cherenkov radiation in order to clarify the threshold and in LIV photon decay to allow the process in a large energetic window.

As it can be seen, the phenomena are sensitive to the LIV term. For vacuum Cherenkov emission, there is an energy threshold, proportional to (E_a, m_a, α_1), that preserves unaffected the physics below it. The threshold grows inversely with the value of the LIV photon parameter, α_1, and above it, the emission rate grows with charged particle energy. In the photon decay case, the rate decreases with the LIV photon parameter, $\alpha_{(1)}$, and grows with the photon momenta. There is also a threshold that turns off photon decay process at low energies that comes from energy-momentum conservation in the LI fermion sector and the corrected LIV photon.

3. Conclusions

We have found a generic first order LIV correction to the emission and decay rates for vacuum Cherenkov radiation and LIV photon decay. Both processes, kinetically forbidden in a LI theory, can be possible under LIV hypothesis. The possible consequences of both processes decrease while the LIV photon coefficient $\alpha_1 \to 0$. Both processes can lead to different expected physics at the most energetic scenarios, such as cosmic rays, but we will present such an analysis in a future work.

Journal of Physics: Conference Series **761** (2016) 012035 doi:10.1088/1742-6596/761/1/012035

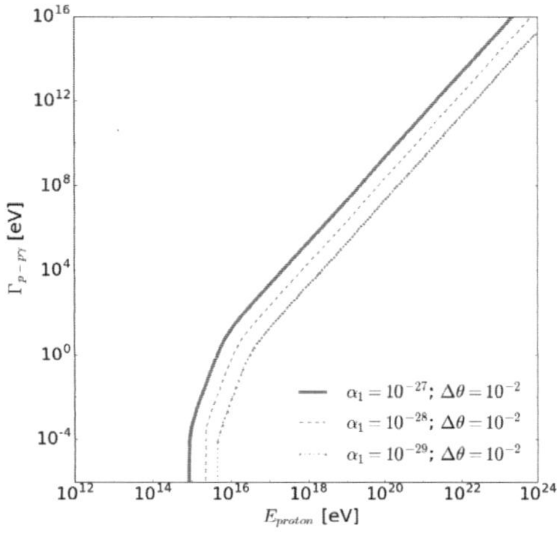

Figure 3. Emission rate for LIV vacuum Cherenkov radiation for n=1, and $m_a = m_{proton}$. Lower values of the LIV photon parameter lead to a higher energetic threshold.

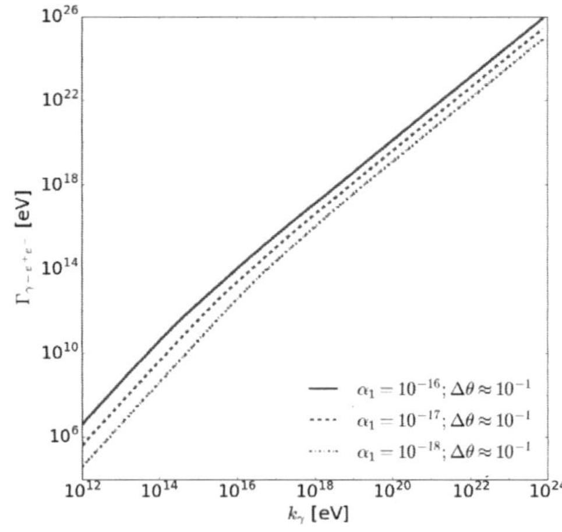

Figure 4. Decay rate for LIV photon decay into electron positron pairs, for n=1. Lower values of the LIV photon parameter lead to a lower decay rate.

Acknowledgments

This work was partially supported by Conacyt grant No. 237004.

References

[1] Aab A *et al.* (Pierre Auger Collaboration) 2014 *Phys. Rev.* D **90** 122005
[2] Abraham J *et al.* (Pierre Auger Collaboration) 2010 *Phys. Lett.* B **685** 239–246
[3] Ellis J, Mavromatos N E and Nanopoulos D V 2000 *Phys. Rev.* D **61** 027503
[4] Gambini R and Pullin J 1999 *Phys. Rev.* D **59** 124021
[5] Ellis J, Mavromatos N and Nanopoulos D 2000 *Gen. Rel. Grav.* **32** 127–144
[6] Ellis J, Mavromatos N, Nanopoulos D and Volkov G 2000 *Gen. Rel. Grav.* **32** 1777 – 1798
[7] Coleman S and Glashow S L 1999 *Phys. Rev.* D **59** 116008
[8] Amelino-Camelia G, Ellis J, Mavromatos N E, Nanopoulos D V and Sarkar S 1998 *Nature* **393** 763–765
[9] Amelino-Camelia G 2001 *Nature* **410** 10651069
[10] Ahluwalia D V 1999 *Nature* **398** 199
[11] Galaverni M and Sigl G 2008 *Phys. Rev. Lett.* **100** 021102
[12] Gagnon O and Moore G D 2004 *Phys. Rev.* D **70** 065002
[13] Galaverni M and Sigl G 2008 *Phys. Rev.* D **78** 063003
[14] Martínez H and Pérez-Lorenzana A 2013 *Journal of Physics: Conference Series* **468** 012005
[15] Abramowski A *et al.* (HESS Collaboration) 2011 *Astropart. Phys.* **34** 738747
[16] Vasileiou V, Jacholkowska A, Piron F, Bolmont J, Couturier C, Granot J, Stecker F W, Cohen-Tanugi J and Longo F 2013 *Phys. Rev.* D **87** 122001
[17] Xu H and Ma B Q 2016 *Phys. Lett.* B **760** 602–604
[18] Nellen L (HAWC) 2015 The potential of the HAWC Observatory to observe violations of Lorentz Invariance *Proceedings, 34th International Cosmic Ray Conference (ICRC 2015)* (*Preprint* 1508.03930)
[19] Colladay D and Kostelecky V A 1998 *Phys. Rev. D* **58** 116002
[20] Klinkhamer F and Schreck M 2008 *Phys. Rev. D* **78** 085026
[21] Grigory Rubtsov P S and Sibiryakov S 2012 *Phys. Rev. D* **86** 085012
[22] Aab A *et al.* (Pierre Auger Collaboration) 2014 *Phys. Rev.* D **90** 122006

Non-perturbative QCD and hadron physics

J J Cobos-Martínez

Instituto de Física y Matemáticas, Universidad Michoacana de San Nicolás de Hidalgo, Edificio C-3, Ciudad Universitaria, Morelia, Michoacán 58040, México.

E-mail: `j.j.cobos.martinez@gmail.com`

Abstract.
A brief exposition of contemporary non-perturbative methods based on the Schwinger-Dyson (SDE) and Bethe-Salpeter equations (BSE) of Quantum Chromodynamics (QCD) and their application to hadron physics is given. These equations provide a non-perturbative continuum formulation of QCD and are a powerful and promising tool for the study of hadron physics. Results on some properties of hadrons based on this approach, with particular attention to the pion distribution amplitude, elastic, and transition electromagnetic form factors, and their comparison to experimental data are presented.

1. Introduction

The quantitative understanding of the properties of strongly interacting matter in all its manifestations and forms in terms of the fundamental theory, Quantum Chromodynamics, is without a doubt one of the most challenging and exciting problems in modern science [1, 2, 3, 4, 5, 6, 7]. QCD is the fundamental theory of quarks, gluons, and their interactions. However, the QCD Lagrangian does not by itself explain the data on strongly interacting matter, and it is not clear how the observed bound states, the hadrons, and their properties arise from QCD. Neither confinement nor dynamical chiral symmetry breaking (DCSB) is apparent in QCD's lagrangian, yet they play a dominant role in determining the observable characteristics of QCD. The physics of strongly interacting matter is governed by emergent phenomena such as these, which can only be elucidated through the use of non-perturbative methods in QCD [4, 5, 6, 7]

The SDE-BSE are a well suited continuum approach to non-perturbative QCD and hadron physics [5, 6, 7] since they provide access to infrared as well as ultraviolet momenta, thus giving a clear connection with processes that are well understood because QCD is asymptotically free. The SDE-BSE form an infinite tower of coupled non-linear integral equations that must be truncated in order to define a tractable problem. Because QCD is asymptotically free the interaction kernel in each SDE is known within the perturbative domain and hence any model dependence is restricted to the long-range behavior of the kernels. The SDE-BSE connect observables with QCD's fields and parameters, and feedback between theory and experiment can then refine the statements and lead to an understanding of non-perturbative QCD and hadron physics. The mass spectrum of hadrons, hadron elastic and transition form factors, distribution functions, and the phase structure of hot and dense QCD, all contribute information that is critical to elucidate the non-perturbative interaction between quarks and gluons. Existing and future hadron physics laboratories around the world will be accumulating data so that for the first time we will be able to see the transition from non-perturbative physics (mesons and baryons) to the perturbative domain (quarks and gluons), governed by QCD. Providing predictions that can be tested experimentally is very important. An understanding of the emergent phenomena of confinement and DCSB from first principles will surely provide us the foundations for an understanding of hadron physics from QCD.

2. Electromagnetic structure of hadrons and pQCD predictions

The investigation of matter's substructure through electron scattering experiments is a well-proved technique since the electromagnetic (EM) probe is well known. Suppose that we want to study the EM charge distribution of a hadron, for example the pion of Fig 1. The experimental procedure is to measure the angular distribution of the scattered electrons and compare it to the known cross section for scattering electrons from a point charge

$$\frac{d\sigma}{d\Omega} = \left(\frac{d\sigma}{d\Omega}\right)_{\text{point}} \left|F(q^2)\right|^2, \tag{1}$$

where $q = k' - k = P' - P$ is the momentum transfer carried by the virtual photon and $F(q^2)$ is the hadron's EM form factor (FF). We then attempt to deduce the structure (and dynamics of its constituents) of the target from the measured FF. The amplitude for the Feynman diagram

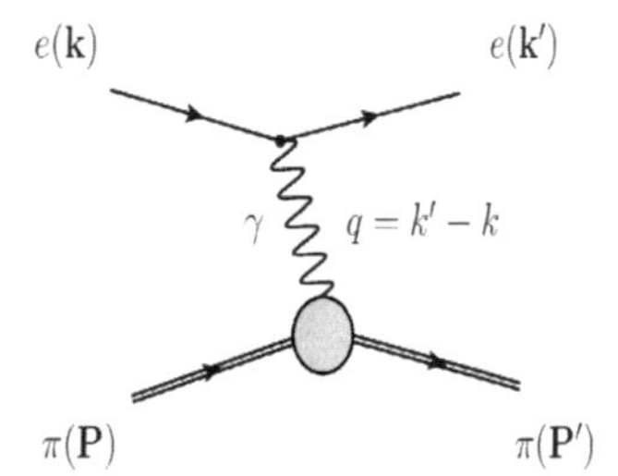

Figure 1. Electron-pion scattering.

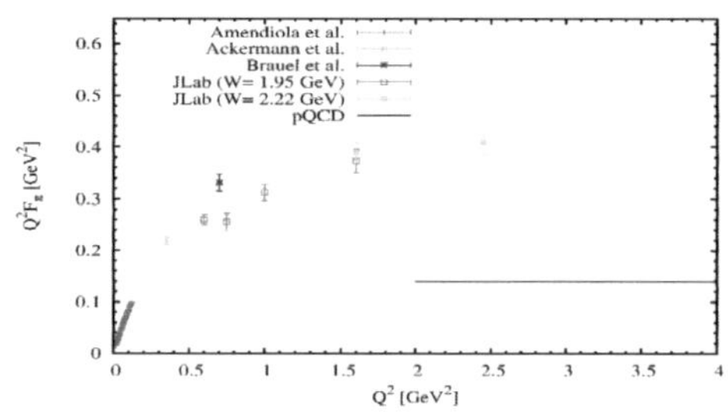

Figure 2. $F_\pi(Q^2)$ and pQCD prediction.

of Fig 1 is given by

$$\mathcal{M} = (-\mathrm{i}e)^2 \overline{u}(k')\gamma^\mu u(k) \frac{-\mathrm{i}}{q^2} \langle \pi(P')|J_\mu|\pi(P)\rangle, \tag{2}$$

where u, $\overline{u}$ are electron spinors, $|\pi(P)\rangle$ is the full pion bound state, and J_μ its EM current. The quantity $\langle \pi(P')|J_\mu|\pi(P)\rangle$ is the pion-photon vertex, whose structure we do not know and would like to understand in terms of the non-perturbative interactions between quarks, gluons, and photons, from their underlying theories. Although we do not know the details about $\langle \pi(P')|J_\mu|\pi(P)\rangle$ in terms of QCD and QED dynamics, we know that J_μ must be a Lorentz four-vector, and therefore the EM interaction of a spinless particle, like the pion, can be parametrized by a single FF

$$\langle \pi(P')|J_\mu|\pi(P)\rangle = (P' + P)_\mu F_\pi(Q^2), \tag{3}$$

where $F_\pi(Q^2)$ is the pion FF that appears in Eq. (1) and we have defined $Q^2 \equiv -q^2$ since q^2 is negative for electron scattering. The FF parametrizes our ignorance about the detailed structure of the pion and the dynamics of its substructure, represented by the blob in Fig 1. In principle, it can be calculated from perturbative QCD (pQCD) for large Q^2 because QCD is asymptotically free. However, it is not clear whether presently accessible momentum transfers are large enough to test predictions based on a perturbative analysis in QCD. For low-to-present-day momentum transfers, the contributions coming from the dynamics of strong QCD play an important, if non-dominant, role. In this regime, pQCD is inapplicable and other methods that incorporate these effects must be developed.

According to Brodsky and Lepage [8, 9], the large Q^2 limit of lepton-hadron scattering processes can be factorized into a soft part containing the non-perturbative longe-range dynamics, absorbed into a universal distribution amplitude (DA), and a hard scattering amplitude T_H, which encodes the quark-gluon subprocesses and can be calculated in pQCD, for the high-momentum exchange between the lepton and the valence quarks in the hadron. Thus $F_\pi(Q^2)$ can be written as [8, 9]

$$F_\pi(Q^2) = \iint \mathrm{d}x\mathrm{d}y \phi^*_\pi(x,Q^2) T_H(x,y;Q^2) \phi_\pi(y,Q^2), \tag{4}$$

where $\phi_\pi(x,Q^2)$ is the pion DA (PDA) and $x\,(y)$ is the fraction of the pion momentum $P\,(P')$ carried by the individual valence quarks, $0 < x, y < 1$. The hard scattering kernel is a sum over contributions from one-gluon exchange, two-gluon exchange, and so on. Due to the asymptotic freedom of QCD, in the $Q^2 \to \infty$ limit Eq. (4) is dominated by one-gluon exchange [8, 9]. Thus

$$F_\pi(Q^2) \overset{Q^2\to\infty}{=} 16\pi \frac{\alpha_s(Q^2)}{Q^2} \left| \frac{1}{3} \int \mathrm{d}x \frac{1}{x} \phi_\pi(x) \right|^2, \tag{5}$$

where α_s is QCD's coupling constant, and xyQ^2 is the "virtuality" of the exchanged gluon, which sets the scale for $\alpha_s(Q^2)$ and can be understood as a measure of the applicability of pQCD to the interaction. In the $Q^2 \to \infty$ limit the PDA evolves to [8, 9] $\phi^{\text{asy}}_\pi(x)$

$$\phi_\pi(x,Q^2) \overset{Q^2\to\infty}{\longrightarrow} \phi^{\text{asy}}_\pi(x) = 6 f_\pi x(1-x), \tag{6}$$

where $f_\pi = 92.2$ MeV is the pion decay constant. Using Equations (5) and (6) one gets the pQCD prediction [8, 9] for $F_\pi(Q^2)$

$$\lim_{Q^2\to\infty} Q^2 F_\pi(Q^2) = 16\pi f_\pi^2 \alpha_s(Q^2). \tag{7}$$

Brodsky and Lepage [8, 9] have also provided pQCD predictions for the transition FF $G_{\pi^0\gamma\gamma^*}(Q^2)$ that parametrizes the process $\gamma\gamma^* \to \pi^0$

$$\lim_{Q^2\to\infty} Q^2 G_{\pi^0\gamma\gamma^*}(Q^2) = 4\pi^2 \frac{1}{3} \int \mathrm{d}x \frac{1}{x} \phi_\pi(x,Q^2) = 4\pi^2 f_\pi, \tag{8}$$

where we have used the asymptotic form of $\phi_\pi(x,Q^2)$. Note that in the both pQCD predictions for $F_\pi(Q^2)$ and $G_{\pi^0\gamma\gamma^*}(Q^2)$, the quantity $\frac{1}{3}\int \mathrm{d}x \frac{1}{x}\phi_\pi(x,Q^2)$, or the universal PDA $\phi_\pi(x,Q^2)$, controls their normalization. The PDA plays an important role in the theoretical description of many QCD processes [10]. Equations (7) and (8) are just two examples. The PDA also plays a role in semileptonic weak decays $B \to \pi l\nu$ and weak hadronic decays $B \to \pi\pi$, both of which contribute to the determination of the parameters of the quark mixing matrix in the Standard Model; see, for example, Ref. [10]. The precise definition of $\phi_\pi(x,Q^2)$ is based on the representation as the matrix element of a non-local quark-antiquark operator on the light-cone formalism; its x-dependence cannot be computed from pQCD but its evolution with Q^2 can.

The experimental measurement of $F_\pi(Q^2)$ is a non-trivial task. To date, reliable experimental data for $F_\pi(Q^2)$ exist in the timelike (negative Q^2) region, for small spacelike values of Q^2, where $F_\pi(Q^2)$ is dominated by the ρ meson pole, and up to $Q^2 = 2.45$ GeV^2; see, for example, Ref. [11]. We are awaiting JLAB's 12 GeV upgrade to extend the above measurements up to $Q^2 = 6$ GeV^2. In Fig 2 the pQCD prediction for $F_\pi(Q^2)$ is compared to the available experimental data. As can be seen, for the highest Q^2 data available, there is a weak suggestion that $Q^2 F_\pi(Q^2) = \text{constant}$ is being approached. Since these data are several times larger than the pQCD prediction, non-perturbative effects are still dominant. A question that comes to mind is: Has $\phi_\pi(x, Q^2)$ reached its asymptotic value $\phi_\pi^{\text{asy}}(x)$ at present energies? For a comparison of the experimental data on $G_{\pi^0\gamma\gamma^*}(Q^2)$ to the pQCD prediction and recent SDE results, see Khépani Raya's contribution to these conference proceedings and also Ref. [12].

3. The Schwinger-Dyson equations

The Schwinger-Dyson equations are the equations of motion for the Greens functions of a quantum field theory (QFT). These are derived from the full generating functional of the theory, making no assumptions about the coupling constant. In the Euclidean space formulation of QCD, the renormalized SDE for the full quark propagator, see Fig 3, for a particular quark flavor is

$$S^{-1}(p) = Z_2 i\gamma \cdot p + Z_4 m(\mu) + Z_1 \int \frac{d^4q}{(2\pi)^4} g^2 D_{\mu\nu}(p-q) \frac{\lambda^a}{2} \gamma^\mu S(q) \Gamma^{a\nu}(p,q), \tag{9}$$

where g is QCD's coupling constant, $D_{\mu\nu}$ is the gluon propagator, $\Gamma^{a\nu}$ is the quark-gluon vertex, and $m(\mu)$ the renormalized current-quark mass; $Z_1(\mu, \Lambda)$ and $Z_2(\mu, \Lambda)$ are renormalization constants, which depend on the renormalization (μ) and regularization (Λ) mass scales. The quark propagator, the gluon propagator, and the quark-gluon vertex, also depend on μ; however observables do not. Both $D_{\mu\nu}$ and $\Gamma^{a\nu}$ satisfy their own SDE, which in turn are coupled to higher n-point functions and so on *ad infinitum*. Thus the quark SDE, explicitly shows that the SDEs form a coupled infinite set of non-linear integral equations. A tractable problem is defined once a truncation scheme has been specified. See, for example, Refs. [5, 6, 7] for more details on SDEs and their application to non-perturbative QCD and hadron physics. The general form

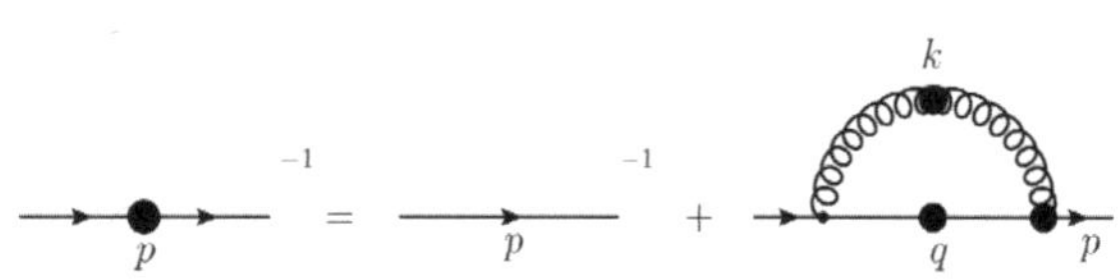

Figure 3. Quark Schwinger-Dyson equation

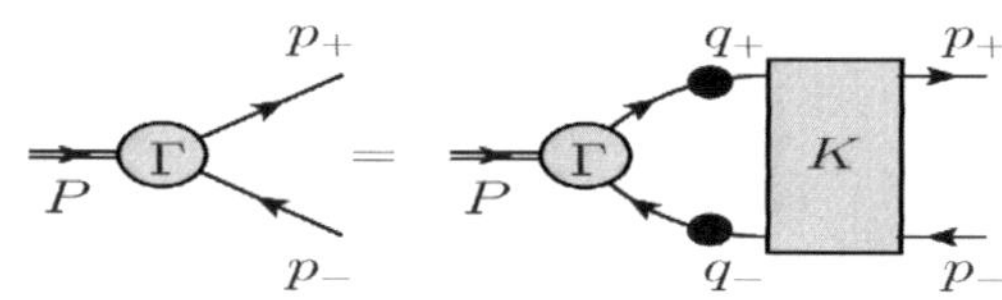

Figure 4. Meson Bethe-Salpeter equation

of the quark propagator is given in terms of two Lorentz-scalar dressing functions

$$S^{-1}(p) = i\gamma \cdot pA(p^2, \mu^2) + B(p^2, \mu^2) = Z^{-1}(p^2, \mu^2)\left(i\gamma \cdot p + M(p^2)\right). \tag{10}$$

In the last form, $Z(p^2, \mu^2)$ is known as the wave function renormalization and $M(p^2)$ is the quark mass function. The solution of Eq. (9) is further subject to a renormalization condition [7]. The best known truncation scheme of the SDEs is the weak coupling expansion. It is an essential tool for the investigation of large momentum transfer phenomena because QCD is asymptotically free. However, it excludes the possibility of obtaining information about the low-energy regime of the strong interactions [5, 6, 7]. As an example of this, consider the chiral limit in pQCD. In this case the theory is chirally symmetric and a perturbative evaluation [13, 14] gives

$$B(p^2) = m[1 - (\alpha/\pi)\ln(p^2/m^2) + \cdots], \tag{11}$$

where the ellipsis denote higher order terms in α. However, it is always true that at any order in pQCD $\lim_{m\to 0} B(p^2) = 0$ and hence DCSB is impossible in perturbation theory.

4. The Bethe-Salpeter equation

Meson bound states, whose flavor structure is given by a dressed quark-antiquark pair, are described by the Bethe-Salpeter equation (BSE), see Fig 4,

$$[\Gamma_H(p;P)]_{tu} = \int \frac{\mathrm{d}^4 q}{(2\pi)^4} [K(p,q;P)]_{tu;rs} \left[S^a(q_+) \Gamma_H(q;P) S^b(q_-) \right]_{sr}, \tag{12}$$

where $H = (a\bar{b})$ indicates the flavor structure; $\Gamma_H(p;P)$ is the meson Bethe-Salpeter amplitude (BSA) describing the bound state; $S^f(q_\pm)$ is the quark propagator, obtained from Eq. (9); $K(p,q;P)$ is the quark-antiquark scattering kernel; Latin indices indicate the color, flavor, and Dirac structure; Poincaré covariance and momentum conservation entail $q_+ = q + \eta P$, $q_+ = q - (1-\eta)P$, etc, with $P = p_+ - p_-$. The parameter $\eta \in [0,1]$ describes the meson's momentum (P) sharing between the quark-antiquark pair; observables, however, do not depend on this. The BSE is a homogeneous eigenvalue equation that admits solutions only for discrete values of the meson momenta $P^2 = -m_H^2$, where m_H is the meson mass [5, 6, 7]. In a particular channel, the lowest mass solution corresponds to the ground state. In the pseudoscalar channel the lowest mass solutions are the pion and kaon mesons. The general form of the BSA in this channel is given by

$$\Gamma_H(p;P) = \gamma_5 \left[\mathrm{i}E_H(p;P) + \gamma \cdot P F_H(p;P) + \gamma \cdot p (p \cdot P) G_H(p;P) + \sigma^{\mu\nu} p_\mu P_\nu H_H(p;P) \right]. \tag{13}$$

Since the BSE is homogeneous, the BSA has to be normalized by a separate condition [5, 6, 7]. For a more detailed presentation of the SDE-BSE and their application to hadron physics see, for example, Refs [5, 6, 7]. In Eq. (12), $K(p,q;P)$ is the fully-amputated, two-particle irreducible, quark-antiquark scattering kernel. It is a four-point Schwinger function, obtained formally as the sum of a countable infinity of skeleton diagrams. The complexity of $K(p,q;P)$ is one of the reasons why quantitative SDE-BSE studies employ modeling of $D_{\mu\nu}$ and $\Gamma^{a\nu}$, because $K(p,q;P)$ also appears in the SDE satisfied by $\Gamma^{a\nu}$. Despite their complexity, substantial progress has been made in unraveling the non-perturbative structure of $D_{\mu\nu}$ and $\Gamma^{a\nu}$ using their SDE [15, 16], as well as the lattice formulation of QCD [17, 18]. Ultimately, the detailed infrared behavior of these quantities should not materially affect the observable consequences of the quark SDE and meson BSE, as long as the infrared enhancement in the quark SDE implements the appropriate amount of dynamical chiral symmetry breaking and, explains the (pseudo)Goldstone character of the pion [5, 6, 7]. However, the lack of a full understanding of the interaction between quarks, through the complete knowledge of $K(p,q;P)$, does not prevent us from obtaining general results in hadron physics [19].

5. The rainbow-ladder truncation and the Maris-Tandy model

In QCD, chiral symmetry and its breaking pattern are the dominant effects in the low-energy regime, specially for light quarks. These are expressed through the axial-vector Ward-Takahashi identity (AxWTI) [19, 20, 21], which implies a relationship between the kernel in the BSE and that in the quark SDE (through the quark self-energy $\Sigma(k)$)

$$\int \frac{\mathrm{d}^4 q}{(2\pi)^4} K_{tu;rs}(k,q;P) \left[\gamma_5 S(q_-) + S(q_+)\gamma_5 \right]_{sr} = \left[\Sigma(k_+)\gamma_5 + \gamma_5 \Sigma(k_-) \right]_{tu}, \tag{14}$$

thus constraining the content of $K(p,q;P)$ if an essential symmetry and its breaking pattern are to be preserved. This relation must be satisfied by any viable truncation scheme of the SDE-BSE coupled system. In modeling the quark SDE kernel we follow [20, 21], and make use of the rainbow approximation for the self-energy

$$Z_1 \int \frac{\mathrm{d}^4 q}{(2\pi)^4} g^2 D^{\mu\nu}(k) \frac{\lambda^a}{2} \gamma_\mu S(q) \Gamma^{a\nu}(k,p) \to \int \frac{\mathrm{d}^4 q}{(2\pi)^4} \mathcal{G}(k^2) D^{\mathrm{free}\,\mu\nu}(k) \frac{\lambda^a}{2} \gamma_\mu S(q) \frac{\lambda^a}{2} \gamma_\nu, \tag{15}$$

where the phenomenological "effective" coupling $\mathcal{G}(k^2)$ contains information about the product $G(k^2, \mu^2)F^1(k, p, \mu)$, where $F^1(k, p, \mu)$ is the form factor associated with the γ_ν structure in the quark-gluon vertex and $G(k^2, \mu^2)$ that in the gluon propagator. From a practical point of view Eq. (14) provides a way of obtaining $K(p, q; P)$. Equations (14) and (15) give

$$K(p,q;P)_{tu;rs} = -\mathcal{G}(k^2) D^{\text{free}\mu\nu}(k) \left[\frac{\lambda^a}{2}\gamma_\mu\right]_{ts} \left[\frac{\lambda^a}{2}\gamma_\nu\right]_{ru}, \tag{16}$$

which define the rainbow-ladder (RL) truncation of the SDE-BSE complex. Constraints on $\mathcal{G}(k^2)$ come from the SDE satisfied by gluon propagator and the quark-gluon vertex. However, we know the behavior of $\alpha(k^2)$ in the ultraviolet, $k^2 >$ 2-3 GeV^2, is well described by pQCD, and therefore any model dependence on $\mathcal{G}(k^2)$ is restricted to the infrared. On the other hand, $\mathcal{G}(k^2)$ in the quark SDE should exhibit sufficient infrared enhancement capable of triggering DCSB and the generation of a momentum-dependent quark mass dressing function that connects the current-quark mass to a constituent-like quark mass [5, 6, 7]. Various models for the effective interaction $\mathcal{G}(k^2)$ combining the ultraviolet behavior known from pQCD and an ansatz for the infrared part have been designed in the past and have been applied to different detailed studies of hadron physics; see, for example, Refs. [5, 6, 7]. In choosing a form for $\mathcal{G}(k^2)$ we use the Maris-Tandy model [20, 21]. In Fig 5 we present numerical solutions for $M(p^2)$ for various

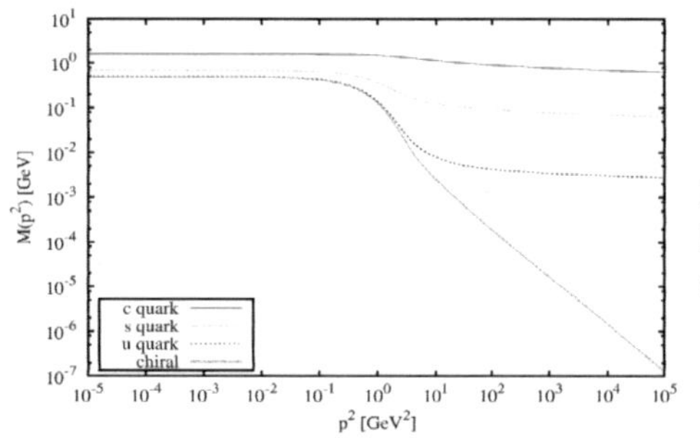

Figure 5. $M(p^2)$ in the RL truncation with the Maris-Tandy dressing function for various values of $m(\mu)$.

values of $m(\mu)$. It is evident that DCSB occurs in the rainbow truncation. At ultraviolet momenta, the magnitude of $M(p^2)$ is determined from $m(\mu)$. In the infrared, however, and specially for light-quarks, $M(p^2)$ is significantly enhanced. For light quarks, this enhancement is orders of magnitude larger than the mass present in the Lagrangian. The domain in which the effect of DCSB is relevant decreases as the $m(\mu)$ increases. We also note that the evolution from the current-quark mass to a constituent-like quark mass occurs at the scale of ≈ 1 GeV^2, as required from hadron phenomenology [22]. The behavior of $M(p^2)$ in Fig 5 has also been confirmed semi-quantitatively in lattice simulations of QCD [23, 24, 25].

6. SDE-BSE results on the PDA and form factors

The theoretical interest in the distribution amplitude (DA) of hadrons is due to their relevance in the description of exclusive reactions [8, 9, 26, 27] from QCD. The DAs describe probability amplitudes to find a hadron in a state with minimum number of Fock constituents and at small transverse separation. The scale dependence of the DA is given by the Efremov-Radyushkin-Brodsky-Lepage (ERBL) evolution equations [8, 9, 26] and can be calculated in pQCD, while the DA at a certain low scale provides the necessary non-perturbative input for the QCD treatment of exclusive reactions. In terms of quark propagators S and the pion BSA Γ_π, the pion DA is given by [28]

$$f_\pi \phi_\pi(x) = Z_2 N_c \text{Tr} \int \frac{d^4q}{(2\pi)^4} \delta(n \cdot q_+ - xn \cdot P)\gamma_5 \gamma \cdot n S(q_+) \Gamma_\pi(q;P) S(q_-). \tag{17}$$

Due to the appearance of δ in Eq. (17) it is easier to calculate the integer moments of $\phi_\pi(x)$, $\langle x^m \rangle = \int_0^1 dx x^m \phi_\pi(x)$ [28]. For the computation of moments, using Feynman integral techniques and a subsequent numerical integration over the Feynman parameters introduced, algebraic parametrizations of the RL numerical solutions for S and Γ_π are employed [28]. Once all needed moments are calculated, a novel reconstruction procedure is implemented to obtain the PDA [28]. Here, the PDA is expanded in terms of Gegenbauer polynomials of order α, instead with $\alpha = 3/2$ fixed. With $\alpha_- = \alpha - 1/2$ we write

$$\phi_\pi(x, \mu) = x^{\alpha_-}(1-x)^{\alpha_-}[1 + \sum_{j=2,4,\ldots} a_j^\alpha(\mu) C_j^\alpha(2x-1)]. \tag{18}$$

The parameter α and expansion coefficients $a_j^\alpha(\mu)$ are determined from the numerical moments obtained from Eq. (17). Once these are obtained one may project out the result onto an $\alpha = 3/2$ basis. However, this requires many nonzero coefficients $\{a_j^{3/2}\}$, and introduce spurious oscillations typical of Fourier-like approximations. The plot in Fig 6 shows the RL result with

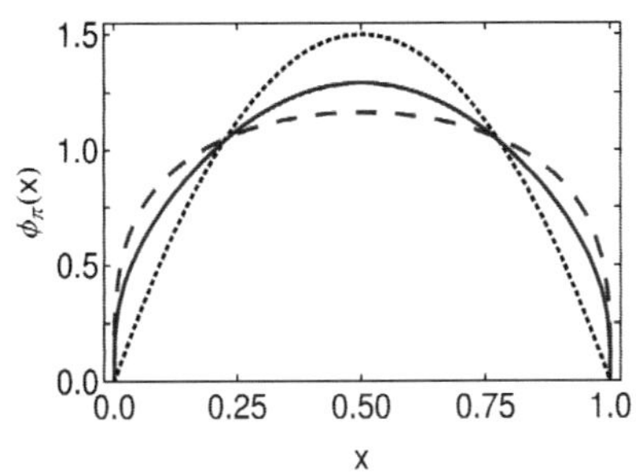

Figure 6. Computed distribution amplitude at $\mu = 2$ GeV. Curves: dashed, rainbow-ladder; and dotted asymptotic distribution.

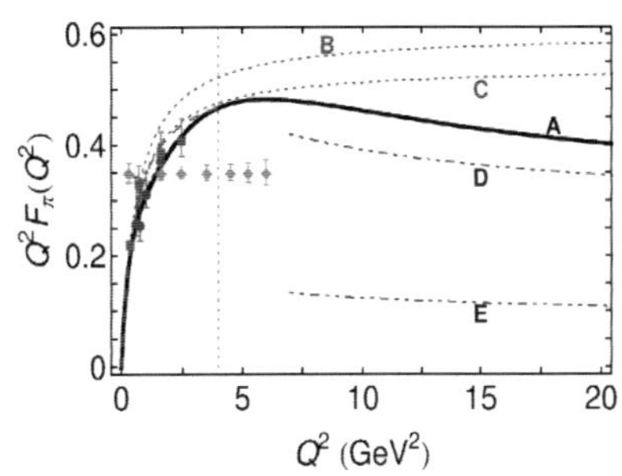

Figure 7. $Q^2F_\pi(Q^2)$. Curves: A, prediction from Ref. [29]; D, asymptotic result using the PDA from Ref. [28]; E is the pQCD result. See Ref. [29] for more details.

the Maris-Tandy interaction [28]. It is described with $\alpha^{\mathrm{RL}} = 0.79$, and $a_2^{\alpha^{\mathrm{RL}}} = 0.0029$. Projected onto a $\alpha = 3/2$ basis this corresponds to $a_2^{3/2} = 0.23, \ldots, a_{14}^{3/2} = 0.022$, etc; see Ref. [28] for more details on the computation and the conclusions drawn. Recall the quantity $\left|\frac{1}{3}\int dx \frac{1}{x}\phi_\pi(x)\right|^2$. The Gegenbauer expansion coefficients evolve with scale according to ERBL equations [8, 9, 26]. At leading-log accuracy it is necessary to evolve to $\mu = 100$ GeV before $a_2^{3/2}$ falls to 50 % of its value [29]. This means that the asymptotic value $\phi_\pi^{\mathrm{asy}}(x)$ lies at very large momenta. Furthermore using the asymptotic form for $\phi_\pi(x)$ we have $\left|\frac{1}{3}\int dx \frac{1}{x}\phi_\pi^{\mathrm{asy}}(x)\right|^2 = 1$. On the other hand, using the RL result for $\phi_\pi(x)$ we have $\left|\frac{1}{3}\int dx \frac{1}{x}\phi_\pi^{\mathrm{RL}}(x)\right|^2 = 3.2$. This means that the pQCD analysis result has to be multiplied by a factor of 3.2 and the asymptotic analysis of various models has to be compared to this new result [29], since the asymptotic domain given by pQCD lies indeed at very large momenta, far beyond our experimental capabilities; see curve E in Fig 7. Curve A in Fig 7 shows the $F_\pi(Q^2)$ computed in the RL truncation of the SDE-BSE system [29]. The magnitude of $Q^2F_\pi(Q^2)$ reflects the scale of dynamical chiral symmetry breaking. The analysis in Ref. [29] in shows that hard contributions to the pion form factor dominate for $Q^2 \geq 8$ GeV.

7. Conclusions

The SDE-BSE is a well founded continuum approach to non-perturbative QCD and hadron physics. They are a natural framework for the exploration of strong QCD since they provide access to infrared as well as ultraviolet momenta, thus giving a clear connection with processes that are well understood because QCD is asymptotically free.

$\phi_\pi^{\rm asy}$ is a poor approximation to $\phi_\pi(x)$ at all momentum-transfer scales that are either now accessible to experiments involving pion elastic or transition form processes, or will become so in the foreseeable future. Predictions for leading-order, leading-twist formulae involving $\phi_\pi^{\rm asy}$ are a misleading guide to interpreting and understanding contemporary experiments. At accessible energy scales a better guide is obtained by using the PDA described herein in such formulae [12, 28, 29, 30].

References

[1] Brodsky S J *et al.* 2015 QCD and Hadron Physics *Preprint* arXiv:1502.05728 [hep-ph]
[2] Heinz U *et al.* 2015 Exploring the properties of the phases of QCD matter-research opportunities and priorities for the next decade *Preprint* arXiv:1501.06477 [nucl-th].
[3] Alkofer R and von Smekal L (2001) *Phys. Rept.* **353**, 281
[4] Brambilla N *et al.* (2014) *Eur. Phys. J.* C **74**, no. 10, 2981
[5] A. Bashir, L. Chang, I. C. Cloet, B. El-Bennich, Y. X. Liu, C. D. Roberts and P. C. Tandy, (2012) *Commun. Theor. Phys.* **58**, 79 (2012)
[6] I. C. Cloet and C. D. Roberts (2014) *Prog. Part. Nucl. Phys.* **77**, 1
[7] C. D. Roberts (2016) *J. Phys. Conf. Ser.* **706**, 022003
[8] Lepage G P and Brodsky S J 1979 *Phys. Lett* B **87** 359
[9] Lepage G P and Brodsky S J 1980 *Phys. Rev.* D **22** 2157
[10] V. M. Braun, S. Collins, M. Gckeler, P. Prez-Rubio, A. Schfer, R. W. Schiel and A. Sternbeck *Phys. Rev.* D **92** 014504
[11] Cobos-Martinez J J (2010) *Static and dynamic properties of the pion from continuum modelling of strong QCD* Online: `http://etheses.dur.ac.uk/633/]`
[12] K. Raya, L. Chang, A. Bashir, J. J. Cobos-Martinez, L. X. Gutirrez-Guerrero, C. D. Roberts and P. C. Tandy, *Phys. Rev.* D **93** 074017
[13] Pagels H 1979 *Phys. Rev.* D **19** 3080
[14] Cornwall J M 1982 *Phys. Rev.* D **26** 1453
[15] Alkofer R, Fischer C S Llanes-Estrada F J and Schwenzer K 2009 *Annals Phys.* **324** 106
[16] Fischer C S 2003 *Non-perturbative propagators, running coupling and dynamical mass generation in ghost - antighost symmetric gauges in QCD Preprint* arxiv:hep-ph/0304233
[17] Skullerud J, Bowman P O, Kizilersu A, Leinweber D B and Williams A G 2005 *Nucl. Phys. Proc. Suppl.* **141** 244
[18] Cucchieri A and Mendes T 2007 *PoS LAT2007* 297
[19] Maris P, Roberts C D and Tandy P C 1998 *Phys.Lett.* B **420** 267
[20] Maris P and Roberts C D 1997 *Phys. Rev.* C **56** 3369
[21] Maris P and Tandy P C 1999 *Phys. Rev.* C **60** 055214
[22] Roberts C D 2016 Perspective on the origin of hadron masses *Preprint* arXiv:1606.03909 [nucl-th]
[23] Skullerud J, Leinweber D B and Williams A G 2001 *Phys. Rev.* D **64** 074508
[24] Bowman P O, Heller U M, Leinweber D B and Williams A G 2003 *Nucl. Phys. Proc. Suppl* **119** 323
[25] Bowman P O, Heller U M, Leinweber D B, Williams A G and Zhang J 2004 *Nucl. Phys. Proc. Suppl.* **128** 23
[26] Efremov A V and Radyushkin A V 1980 *Phys. Lett.* B **94** 245
[27] Farrar G R and Jackson D R 1979 *Phys. Rev. Lett.* **43** 246
[28] Chang L, Cloet I C, Cobos-Martinez J J, Roberts C D, Schmidt S M and Tandy P C 2013 *Phys.Rev.Lett.* **110** 132001
[29] Chang L, Cloet I C, Roberts C D, Schmidt S M and Tandy P C 2013 *Phys.Rev.Lett.* **111** 141802
[30] Cloet I C, Chang L, Roberts C D, Schmidt S M and Tandy P C 2013 *Phys.Rev.Lett.* **111** 092001

Global extraction of the parton-to-pion fragmentation functions at NLO accuracy in QCD

R. J. Hernández-Pinto[1]*, M. Epele[2], D. de Florian[3], R. Sassot[4] and M. Stratmann[5]

[1] Facultad de Ciencias Físico-Matemáticas, Universidad Autónoma de Sinaloa, Ciudad Universitaria, CP 80000, Culiacán, Sinaloa, México
[2] Instituto de Física La Plata, UNLP, CONICET Departamento de Física, Facultad de Ciencias Exactas, Universidad de La Plata, C.C. 69, La Plata, Argentina
[3] International Center for Advanced Studies (ICAS), UNSAM, Campus Miguelete, 25 de Mayo y Francia, (1650) Buenos Aires, Argentina
[4] Departamento de Física and IFIBA, Facultad de Ciencias Exactas y Naturales, Universidad de Buenos Aires, Ciudad Universitaria, Pabellón 1 (1428) Buenos Aires, Argentina
[5] Institute for Theoretical Physics, University of Tübingen, Auf der Morgenstelle 14, 72076 Tübingen, Germany

E-mail: roger@uas.edu.mx

Abstract.
In this review, we discuss the results on the parton-to-pion fragmentation functions obtained in a combined NLO fit to data of single-inclusive hadron production in electron-positron annihilation, proton-proton collisions, and lepton-nucleon deep-inelastic scattering. A more complete discussion can be found in Ref. [1].

1. Introduction

Fragmentation functions (FFs) are fundamental objects which describe the collinear transition of a quark i into a hadron H with a fraction z of its momentum, and it is usually named as $D_i^H(z)$. These FFs can only be obtained by performing global fits. However, since the FFs are non perturbative objects, they cannot be computed from first principles and they need to be extracted by fitting the experimental data of different kind of processes. However, the scale dependence of the FFs can be obtained in perturbative QCD (pQCD) and can be determined by renormalization group equations, similar to those for parton densities (PDF).

In here we present the results of Ref. [1] where we obtained an updated set of parton-to-pion FFs with the determination of their uncertainties by applying the Iterative Hessian (IH) method in light of all the newly available, precise experimental results in single-inclusive pion production in electron-positron annihilation (SIA), semi-inclusive deep-inelastic lepton-nucleon scattering (SIDIS), and pp collisions. This will allow us to scrutinize the consistency of the information on FFs extracted across the different hard scattering processes, i.e., to validate the fundamental notion of universality, which is at the heart of any pQCD calculation based on the factorization of short- and long-distance physics [2].

* Speaker

2. Functional Form and Fit Parameters

We parametrize the hadronization of a parton of flavor i into a positively charged pion at an initial scale of $Q_0 = 1\,\text{GeV}$ as

$$D_i^{\pi^+}(z, Q_0) = \frac{N_i z^{\alpha_i}(1-z)^{\beta_i}[1+\gamma_i(1-z)^{\delta_i}]}{B[2+\alpha_i, \beta_i+1] + \gamma_i B[2+\alpha_i, \beta_i+\delta_i+1]} \,. \tag{1}$$

Here, $B[a, b]$ denotes the Euler Beta-function, and the N_i in (1) are chosen in such a way that they represent the contribution of $zD_i^{\pi^+}$ to the momentum sum rule.

The improved experimental information now allows us to impose less constraints on the parameter space. More specifically, as before we still have to assume isospin symmetry, i.e.,

$$D_{\bar{u}}^{\pi^+} = D_d^{\pi^+} \,, \tag{2}$$

and we need to relate the total u-quark and d-quark FFs by a global, z-independent factor $N_{d+\bar{d}}$,

$$D_{d+\bar{d}}^{\pi^+} = N_{d+\bar{d}} D_{u+\bar{u}}^{\pi^+} \,, \tag{3}$$

which quantifies any charge symmetry violation found in the fit. The fragmentation of a strange quark into a pion is now related to the unfavored FFs in Eq. (2) by

$$D_s^{\pi^+} = D_{\bar{s}}^{\pi^+} = N_s z^{\alpha_s} D_{\bar{u}}^{\pi^+} \,, \tag{4}$$

rather than just using a constant as in the DSS analysis. Besides, the charm- and bottom-to-pion FFs no longer assume $\gamma_c = \gamma_b = 0$ in Eq. (1) but can now exploit the full flexibility of the ansatz. As in the DSS and all other analyses [3, 4, 5, 6, 7], we include heavy flavor FFs discontinuously as massless partons in the QCD scale evolution above their $\overline{\text{MS}}$ "thresholds", $Q = m_{c,b}$, with m_c and m_b denoting the mass of the charm and bottom quark, respectively.

In total we now have 28 free fit parameters describing our updated FFs for quarks, antiquarks, and gluons into positively charged pions. The corresponding FFs for negatively charged pions are obtained by charge conjugation and those for neutral pions by assuming $D_i^{\pi^0} = [D_i^{\pi^+} + D_i^{\pi^-}]/2$.

3. Data Selection

For the global fit, we use all the available experimental information on single-inclusive charged and neutral pion production in SIA, SIDIS, and hadron-hadron collisions.

Compared to the data sets already used in the DSS global analysis [3], we include the new results from BaBar [8] and Belle [9] in SIA at a c.m.s. energy of $\sqrt{S} \simeq 10.5\,\text{GeV}$. Both sets are very precise, with relative uncertainties of about $2-3\%$, and reach all the way up to pion momentum fractions z close to one, well beyond of what has been measured so far. As customary, we limit ourselves to data with $z \geq 0.1$ to avoid any potential impact from kinematical regions where finite, but neglected, hadron mass corrections, proportional to $M_\pi/(Sz^2)$, might become of any importance [3, 6, 7].

In case of SIDIS, we replace the preliminary multiplicity data from Hermes [10] by their released final results [11]. More specifically, we use the data for charged pion multiplicities as a function of momentum transfer Q^2 in four bins of z taken on both a proton and a deuteron target. In addition, we include the still preliminary multiplicity data for $\pi^\pm$ from the Compass Collaboration [12], which are given as a function of z in bins of Q^2 and the initial-state momentum fraction x. In addition, for the SIDIS data sets we do not have to impose any cuts on both data sets to accommodate them in the global analysis.

Finally, we add a couple of new sets of data for inclusive high-p_T pion production in pp collisions to the results from the Phenix experiment [13] already included in the DSS analysis.

Most noteworthy are the first results for neutral pions from the ALICE Collaboration at CERN-LHC [14], covering unprecedented c.m.s. energies of up to 7 TeV. In addition, we add STAR data taken at $\sqrt{S} = 200\,\text{GeV}$ in various rapidity intervals for both neutral and charged pion production and for the π^-/π^+ ratio [15, 16, 17, 18]. It is worth mentioning that it turns out that a good global fit of RHIC and LHC *pp* data, along with all the other world data, can only be achieved if one imposes a cut on the minimum p_T of the produced pion of about 5 GeV.

4. Fit Procedure and Uncertainty Estimates

The 28 free parameters describing the updated parton-to-pion FFs in Eq. (1) at the chosen input scale of 1 GeV are again determined from a standard χ^2 minimization where

$$\chi^2 = \sum_{i=1}^{N} \left[\left(\frac{1-\mathcal{N}_i}{\delta\mathcal{N}_i} \right)^2 + \sum_{j=1}^{N_i} \frac{(\mathcal{N}_i T_j - E_j)^2}{\delta E_j^2} \right], \tag{5}$$

for $i = 1, \ldots, N$ data sets, each contributing with N_i data points. E_j is the measured value of a given observable, δE_j the error associated with this measurement, and T_j is the corresponding theoretical estimate for a given set of parameters in Eq. (1). In this new fit, we derive the optimum normalization shifts analytically from the condition $\partial\chi^2/\partial\mathcal{N}_i = 0$, which yields

$$\mathcal{N}_i = \frac{\sum_{j=1}^{N_i} \frac{\delta\mathcal{N}_i^2}{\delta E_j^2} T_j E_j + 1}{1 + \sum_{j=1}^{N_i} \frac{\delta\mathcal{N}_i^2}{\delta E_j^2} T_j^2}\,. \tag{6}$$

Here, $\delta\mathcal{N}_i$ denotes the quoted experimental normalization uncertainty for data set i.

Now, in order to estimate the uncertainties we use the IH method [19]. The main idea of the method is to assume a quadratic behavior of the χ^2 hyper-surface of parameter displacements and to express the χ^2 increment from its minimum value in terms of combinations of fit parameters that maximize the variation. These sets correspond to fixed displacements along the eigenvector directions of the Hessian matrix. To define the eigenvector sets one has to choose a tolerance parameter $\Delta\chi^2$ for the increment in χ^2 which is still acceptable in the global fit. Here we proceed as follows: the tolerances for the eigenvector sets corresponding to 68% and 90% confidence level (C.L.) intervals are determined from the Gaussian probability density function for a χ^2 distribution with k degrees of freedom (d.o.f.):

$$P_k(x) = \frac{x^{k/2-1} e^{-x/2}}{\Gamma(k/2) 2^{k/2}}\,. \tag{7}$$

The $\Delta\chi^2$ related to the 68th and 90th percentiles are then obtained by solving $\int_0^{\chi^2+\Delta\chi^2} d\chi^2 P_k(\chi^2) = 0.68$ and 0.90, respectively.

Finally, we choose the next-to-leading order (NLO) set of PDFs from the MSTW group [20] and the corresponding uncertainty estimates in computations of the SIDIS and *pp* cross sections. For consistency, we also fix the strong coupling α_S to the values obtained in the MSTW fit.

5. Results

In this section we present and discuss the results of our global analysis of parton-to-pion fragmentation functions. First, we present the obtained fit parameters, normalization shifts, and individual χ^2 values. Next, the obtained $D_i^{\pi^+}(z, Q^2)$ and their uncertainties are shown and compared to the results of the DSS fit.

5.1. Parton-To-Pion Fragmentation Functions

Table 1 reveals already a notable difference to one of the findings of the DSS analysis which preferred an unexpectedly sizable breaking of the charge symmetry between $u+\bar{u}$ and $d+\bar{d}$ FFs of about 10% [3], within large uncertainties though. Now, with much improved experimental information on charged pion multiplicities both from HERMES [11] and COMPASS [12] and new data on the ratio π^-/π^+ in pp collisions from STAR [17], the parameter $N_{d+\bar{d}}$ in Eq. (3) prefers to stay very close to unity, i.e., very little or no breaking.

flavor i	N_i	α_i	β_i	γ_i	δ_i
$u+\overline{u}$	0.387	-0.388	0.910	7.15	3.96
$d+\overline{d}$	0.388	-0.388	0.910	7.15	3.96
$\overline{u}=d$	0.105	1.649	3.286	49.95	8.67
$s+\overline{s}$	0.273	1.449	3.286	49.95	8.67
$c+\overline{c}$	0.306	1.345	5.519	19.78	10.22
$b+\overline{b}$	0.372	-0.127	4.490	24.49	12.80
g	0.260	2.552	6.194	87.06	20.36

Table 1. Parameters describing the NLO FFs for positively charged pions, $D_i^{\pi^+}(z,Q_0)$, in Eq. (1) in the $\overline{\text{MS}}$ scheme at the input scale $Q_0 = 1\,\text{GeV}$. Results for the charm and bottom FFs refer to $Q_0 = m_c = 1.43\,\text{GeV}$ and $Q_0 = m_b = 4.3\,\text{GeV}$, respectively.

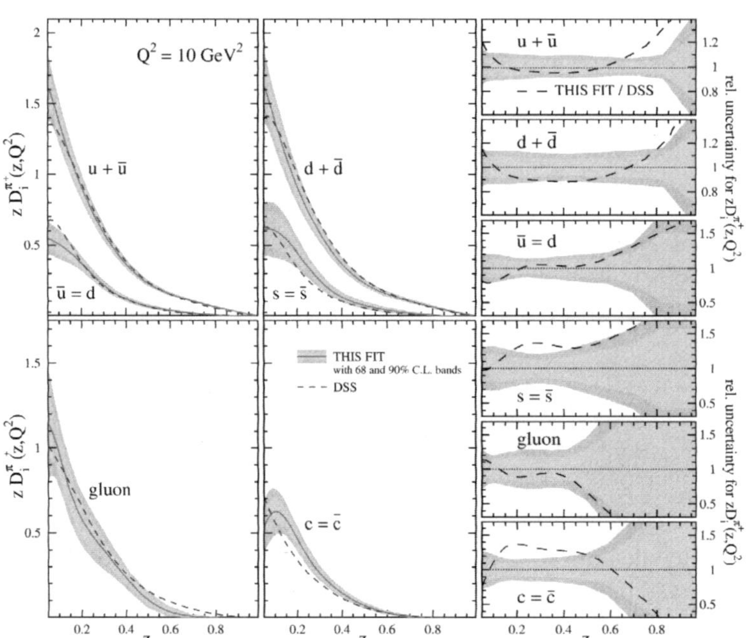

Figure 1. The individual FFs for positively charged pions $zD_i^{\pi^+}(z,Q^2)$ at $Q^2 = 10\,\text{GeV}^2$ along with uncertainty estimates at 68% and 90% C.L. indicated by the inner and outer shaded bands, respectively. The panels on the right-hand-side show the corresponding relative uncertainties. Also shown is a comparison to the previous global analysis by DSS [3] (dashed lines).

In Fig. 1 we present the new parton-to-pion FFs at $Q^2 = 10$ GeV^2. As can be inferred, for the light quark flavors the old DSS results are either close to the updated fit or within its 90% C.L. uncertainty band. The best determined pion FFs is $D^{\pi^+}_{u+\bar{u}}$, where the relative uncertainties are below 10% at 90% C.L. throughout most of the relevant z range. Only for $z \gtrsim 0.8$ the errors rapidly increase because of the lack of experimental constraints in this region. The corresponding uncertainties for $D^{\pi^+}_{d+\bar{d}}$ turn out to be slightly larger as they also include possible violations of SU(2) charge symmetry through Eq. (3). Bigger deviations from the DSS analysis are found for both the gluon and the charm FFs. In the latter case, this is driven by the greater flexibility of the functional form, five fit parameters rather than three. The significantly reduced $D^{\pi^+}_g$ as compared to the DSS fit is a result of the new ALICE pp data [14], which have a strong preference for less pions from gluon fragmentation for basically all values of z. Similar conclusions are obtained for $Q^2 = M_Z^2$.

experiment	# data in fit	χ^2
TPC [21]	44	37.1
TASSO [22]	18	54.7
SLD [23]	79	62.1
ALEPH [24]	22	23.0
DELPHI [25]	51	72.2
OPAL [26, 27]	46	192.3
BABAR [8]	45	44.0
BELLE [9]	78	46.8
HERMES [11]	128	181.1
COMPASS [12] prel.	398	369.9
PHENIX [13]	15	13.9
STAR [16, 18, 15, 17]	38	33.3
ALICE [14] 7 TeV	11	32.1
TOTAL:	973	1189.5

Table 2. Data sets used in our NLO global analysis.

The overall quality of the fit is summarized in Tab. 2 where we list all data sets included in our global analysis, along with their individual χ^2 values.

Firstly, it is worth mentioning that there is a more than twofold increase in the number of available data points as compared to the original DSS analysis [3]. Secondly, the quality of the global fit has improved dramatically from χ^2/d.o.f. $\simeq 2.2$ for DSS, see Tab. II in Ref. [3], to χ^2/d.o.f. $\simeq 1.2$ for the current fit. A more detailed comparison reveals that the individual χ^2 values for the SIA data [21, 22, 23, 24, 25, 26], which were already included in the DSS fit, have, by and large, not changed significantly. The biggest improvement concerns the SIDIS multiplicities from HERMES which, in their recently published version [11], are now described very well by the updated fit. Also, the preliminary charged pion multiplicities from COMPASS [12] and the new SIA data from BABAR [8] and BELLE [9] integrate nicely into the global analysis of parton-to-pion FFs. Finally, there is some tension among the pp data sets from RHIC and the LHC, which forced us to introduce a cut $p_T > 5$ GeV on the pion's transverse momentum in the current fit to accommodate both of them. The obtained individual χ^2 values are all reasonable, as can be inferred from Tab. 2, with the new ALICE data [14] being on the high side, which largely stems from the penalty for the still sizable normalization shift. This large shift reflects the preference of the new ALICE data for a smaller gluon-to-pion FF than extracted by the original DSS fit based on RHIC PHENIX data [13] alone. As a result of the p_T

cut, the number of pp data in the fit for RHIC has decreased as compared to the DSS analysis. Both the BRAHMS [28] and STAR [29] results at forward pseudo-rapidities do not pass the p_T cut anymore, and, hence, are excluded from the updated fit.

6. Summary

We have presented a new, comprehensive global QCD analysis of parton-to-pion fragmentation functions at next-to-leading order accuracy including the latest experimental information. The analyzed data for inclusive pion production in semi-inclusive electron-positron annihilation, deep-inelastic scattering, and proton-proton collisions span energy scales ranging from about 1 GeV up to the mass of the Z boson. The achieved, very satisfactory and simultaneous description of all data sets strongly supports the validity of the underlying theoretical framework based on pQCD and, in particular, the notion of factorization and universality for parton-to-pion fragmentation functions.

Acknowledgments

The work of RJHP is partially supported by CONACyT, México. This work was also supported in part by CONICET, ANPCyT, UBACyT, the Research Executive Agency (REA) of the European Union under the Grant Agreement number PITN-GA-2010-264564 (LHCPhenoNet) and the Institutional Strategy of the University of Tübingen (DFG, ZUK 63).

References

[1] D. de Florian, R. Sassot, M. Epele, R. J. Hernández-Pinto, M. Stratmann. Phys. Rev. D **91** 014035 (2015).
[2] See, e.g., J. C. Collins, D. E. Soper, and G. F. Sterman, Adv. Ser. Direct. High Energy Phys. **5**, 1 (1988).
[3] D. de Florian, R. Sassot, and M. Stratmann, Phys. Rev. D **75**, 114010 (2007).
[4] D. de Florian, R. Sassot, and M. Stratmann, Phys. Rev. D **76**, 074033 (2007).
[5] C. A. Aidala, F. Ellinghaus, R. Sassot, J. P. Seele, and M. Stratmann, Phys. Rev. D **83**, 034002 (2011).
[6] S. Kretzer, Phys. Rev. D **62**, 054001 (2000);
[7] B. A. Kniehl, G. Kramer, and B. Potter, Nucl. Phys. B **582**, 514 (2000); S. Albino, B. A. Kniehl, and G. Kramer, Nucl. Phys. B **725**, 181 (2005); *ibid.* **803**, 42 (2008); M. Hirai, S. Kumano, T. -H. Nagai, and K. Sudoh, Phys. Rev. D **75**, 094009 (2007).
[8] J. P. Lees *et al.* [BABAR Collaboration], Phys. Rev. D **88**, 032011 (2013).
[9] M. Leitgab *et al.* [BELLE Collaboration], Phys. Rev. Lett. **111**, 062002 (2013).
[10] A. Hillenbrand, "Measurement and Simulation of the Fragmentation Process at HERMES", Ph.D. thesis, Erlangen Univ., Germany, September 2005; private communications.
[11] A. Airapetian *et al.* [HERMES Collaboration], Phys. Rev. D **87**, 074029 (2013).
[12] N. Makke [COMPASS Collaboration], PoS DIS **2013**, 202 (2013).
[13] A. Adare *et al.* [PHENIX Collaboration], Phys. Rev. D **76**, 051106 (2007).
[14] B. Abelev *et al.* [ALICE Collaboration], Phys. Lett. B **717**, 162 (2012).
[15] J. Adams *et al.* [STAR Collaboration], Phys. Lett. B **637**, 161 (2006).
[16] B. I. Abelev *et al.* [STAR Collaboration], Phys. Rev. D **80**, 111108 (2009).
[17] G. Agakishiev *et al.* [STAR Collaboration], Phys. Rev. Lett. **108** (2012) 072302.
[18] L. Adamczyk *et al.* [STAR Collaboration], Phys. Rev. D **89**, 012001 (2014).
[19] J. Pumplin, D. R. Stump, and W. K. Tung, Phys. Rev. D **65**, 014011 (2001); J. Pumplin, D. Stump, R. Brock, D. Casey, J. Huston, J. Kalk, H. L. Lai, and W. K. Tung, Phys. Rev. D **65**, 014013 (2001).
[20] A. D. Martin, W. J. Stirling, R. S. Thorne, and G. Watt, Eur. Phys. J. C **63**, 189 (2009).
[21] H. Aihara *et al.* [TPC/TWO GAMMA Collaboration], Phys. Lett. B **184**, 299 (1987); Phys. Rev. Lett. **61**, 1263 (1988); X. -Q. Lu, Ph.D. thesis, Johns Hopkins University, UMI-87-07273, 1987.
[22] W. Braunschweig *et al.* [TASSO Collaboration], Z. Phys. C **42**, 189 (1989).
[23] K. Abe *et al.* [SLD Collaboration], Phys. Rev. D **59**, 052001 (1999).
[24] D. Buskulic *et al.* [ALEPH Collaboration], Z. Phys. C **66**, 355 (1995).
[25] P. Abreu *et al.* [DELPHI Collaboration], Eur. Phys. J. C **5**, 585 (1998).
[26] R. Akers *et al.* [OPAL Collaboration], Z. Phys. C **63**, 181 (1994).
[27] G. Abbiendi *et al.* [OPAL Collaboration], Eur. Phys. J. C **16**, 407 (2000).
[28] I. Arsene *et al.* [BRAHMS Collaboration], Phys. Rev. Lett. **98**, 252001 (2007).
[29] J. Adams *et al.* [STAR Collaboration], Phys. Rev. Lett. **97**, 152302 (2006).

Journal of Physics: Conference Series **761** (2016) 012038 doi:10.1088/1742-6596/761/1/012038

DIS at low x, high gluon densities, BFKL evolution and 3 particle correlations

Martin Hentschinski

Facultad de Ciencias Físico Matemáticas, Benemérita Universidad Autónoma de Puebla, Puebla 1152, Mexico

Instituto de Ciencias Nucleares, Universidad Nacional Autónoma de México

E-mail: hentschinski@correo.nucleares.unam.mx

Abstract. In this contribution we provide a short overview of some recent results in the study of Deep Inelastic Scattering (DIS) processes at low x. We focus in particular on recent applications of next-to-leading order Balitsky-Fadin-Kuraev-Lipatov (BFKL) evolution to inclusive DIS events and to the search for saturation effects in exclusive events. In particular we discuss in some detail the recently calculated 3 particle production process in DIS which – for DIS on a large nuclei – promises to be a suitable candidate for saturation searches.

1. Introduction

Understanding the high energy limit of strong interaction is a long standing challenge which predates the identification of Quantum Chromodynamics (QCD) as the correct theory of strong interactions. While QCD is generally strongly coupled and does not allow for a perturbative description, there exists a class of events where this is actually possible. Such perturbative or 'hard' events are characterized by a 'hard' scale M much bigger than the QCD confinement scale $\Lambda \sim 200$ MeV which – due to asymptotic freedom of QCD – implies a small strong coupling constant $\alpha_s(M) \ll 1$. With $\sqrt{s}$ the center of mass energy and M the hard scale of the process, the perturbative high energy or Regge limit corresponds then to the limit $x = M^2/s \to 0$ at fixed M. For such a scenario, even though $\alpha_s(M) \ll 1$, large logarithms $\ln 1/x$ can compensate for the smallness of the strong coupling, $\alpha_s \ln 1/x \sim 1$ and therefore invalidate the naïve perturbative expansion in α_s. To recover predictive power of the perturbative expansion in this kinematic limit, it is therefore necessary to re-arrange the perturbative series through a resummation of terms $(\alpha_s \ln 1/x)^n$, to all orders in α_s. For perturbative QCD amplitudes, this is achieved by the Balitsky-Fadin-Kuraev-Lipatov (BFKL) evolution equation [1], which is currently known up to next-to-leading order (NLO) w.r.t. an expansion in α_s [2].

2. BFKL evolution and DIS at small x

A first test of BFKL evolution has been possible at the HERA experiments which measured proton structure functions in Deep-Inelastic Scattering (DIS) reactions in ep (electron-proton) scattering, see Fig. 1. In such DIS events, the virtuality Q^2 of the photon, which is exchanged between electron and proton, provides a hard scale and allows therefore for the use of QCD perturbation theory. A first approximate solution of the BFKL evolution equation predicts a strong rise of perturbative cross-sections with energy, $\sigma \sim x^{-\lambda}$, with $\lambda = \alpha_s 12 \ln 2/\pi \simeq 0.52$

Journal of Physics: Conference Series **761** (2016) 012038 doi:10.1088/1742-6596/761/1/012038

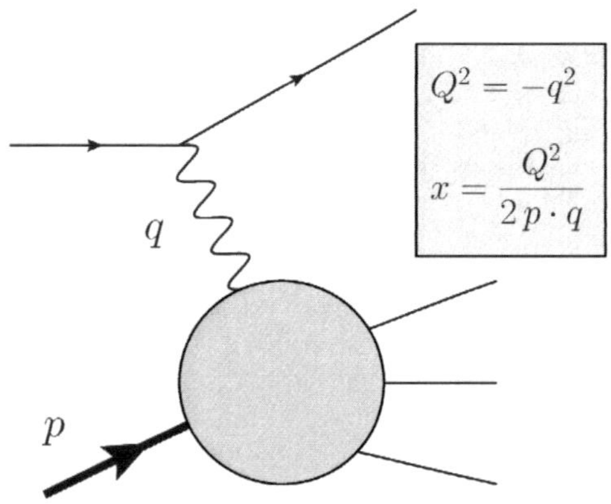

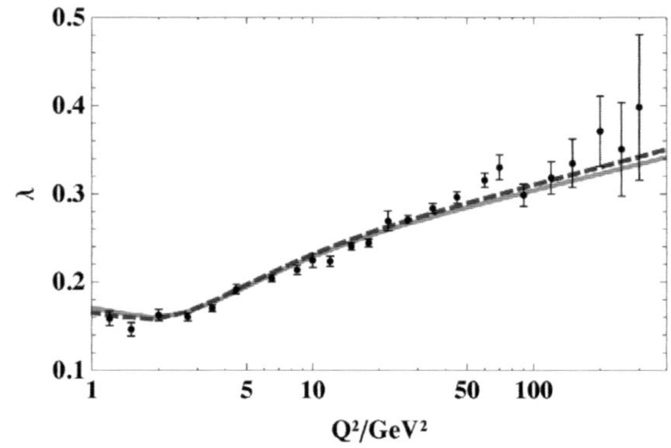

Figure 1. *Left:* Schematic DIS process: a virtual photon, emitted by the scattering electron, probes the proton at high virtuality Q^2. Bjorken x denotes the ratio of photon virtuality and center-of-mass energy squared of the virtual photon-proton scattering process. *Right:* The effective Pomeron intercept $\lambda(Q^2)$, obtained from combined HERA data, using the following parametrization of the proton structure function $F_2(x, Q^2) = c(Q^2)x^{-\lambda(Q^2)}$. The theory predictions correspond to two different NLO BFKL fits performd in [4] (plot originally published in [4]).

for $\alpha_s = 0.2$. This predicted rise is far stronger than the rise of non-perturbative cross-sections such as the total cross-sections for proton-proton scattering, where typical values extracted from experiments are of order $\lambda \sim 0.1$. The strong rise of proton structure functions at large photon virtuality Q^2, as measured by the HERA experiments, provided therefore first qualitatively evidence for BFKL evolution. While the above discussed approximate solution to BFKL evolution is not sufficient to describe the very precise combined HERA data[1], the agreement with data improves dramatically if a) BFKL evolution is solved exactly, not asymptotically and b) NLO corrections are included, see [3,4]. In the case of [4] this has been possible due to a resummation of collinear enhanced configurations within the NLO BFKL kernel, together with an optimal scale setting for the scale of the running coupling. To test these fits further, the natural choice is to study LHC cross-sections, which due to their large center-of-mass energies are well suited for such a purpose. Of particular interest are in this context hard production processes in the forward direction of one of the scattering protons, such as production of jets with [5] and without [6–8] rapidity gaps, Z-bosons [9,10] and bottom quarks [11], since they allow to probe the gluon distribution of the other proton at very small values of x, sometimes down to ultra-small values of $x \simeq 10^{-6}$. A particular interesting class of processes are in this context exclusive photo-production cross-sections of vector meson at the LHC, which are measured in ultra-peripheral pp (proton-proton) and pPb (proton-lead) collisions, where either the lead nucleus (pPb) or a proton (pp) serves as a photon source [12–15]. Combining LHC with HERA data [16–22], measured in ep collisions, such measurements allow to study directly the rise with energy of the photo-production cross-section. For photo-production of J/Ψ and Υ vector mesons, where the heavy charm and bottom masses provide a hard scale, it was recently shown [23] that this rise has an excellent description in terms of NLO BFKL fits of [4].

[1] In particular the asymptotic leading order BFKL 'intercept' $\lambda = \alpha_s(Q)12\ln 2/\pi$ is decreasing with decreasing $\alpha_s(Q)$ and therefore with increasing Q^2 in contrast to data which clearly find an intercept *increasing* with Q^2, see Fig. 1, right.

3. High gluon densities at small x

Even though NLO BFKL evolution provides a very good description of currently available low x data, there exist unitarity bounds for hadronic cross-sections which indicate that the power-like rise with increasing x cannot continue forever; instead the observed rise must eventually slow down and saturate. A dynamical mechanism which achieves perturbative unitarity of QCD cross-sections at low x is known under the term 'gluon saturation' as originally proposed by Gribov, Levin and Ryskin [24]. The framework of the Color Glass Condensate (CGC) provides then an effective action approach to gluon saturation [25]. It gives a description of the high energy hadron or nucleus as a weakly-coupled yet non-perturbative system of gluons with high occupancy number, characterized by a semi-hard scale Q_s, called the saturation scale. As a result one obtains a non-linear generalization of the BFKL evolution equation [26–28], which includes multiple-scattering effects; the latter provide the necessary slow down and saturation of the growth with energy, see [29] and references therein for a detailed discussion.

Since the CGC-framework assumes high occupation number of the gluonic target field, the latter can be no longer treated perturbatively. The resulting formalism represents therefore the DIS scattering process in the low x limit as the scattering of the virtual photon on a strong background color field, which represents the target. Defining light-cone vectors $n, \bar{n}$ through the four momenta of virtual photon and target, *i.e.* $n \sim p$ and $n \cdot \bar{n} = 1$ with $v^- \equiv n \cdot v, v^+ \equiv \bar{n} \cdot v$ for a generic four vector v, theory calculations are performed using quark and gluon propagators in the presence of the background field $A^{+,c}$. In particular one has for the propagators

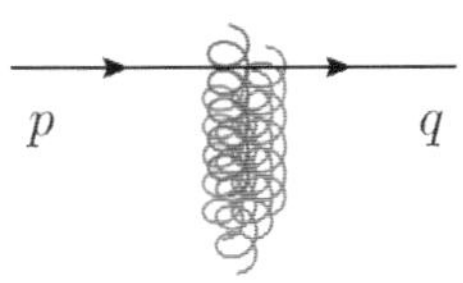

$$= (2\pi)^d \delta^{(d)}(p-q)\tilde{S}_F^{(0)}(p) + \tilde{S}_F^{(0)}(p) \quad p \quad q \quad \tilde{S}_F^{(0)}(q)\,,$$

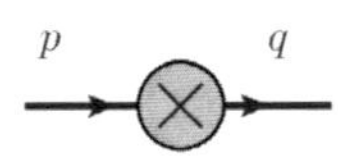

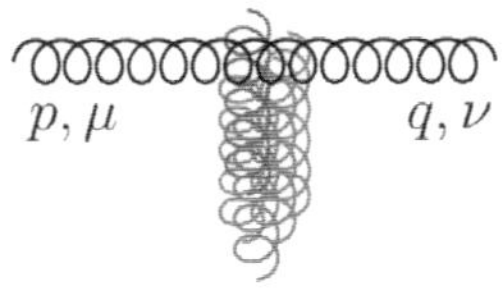

$$= (2\pi)^d \delta^{(d)}(p-q)\tilde{G}_{\mu\nu}^{(0)}(p) + \tilde{G}_{\mu\alpha}^{(0)}(p) \quad p \quad q \quad \tilde{G}_{\alpha\nu}^{(0)}(q)\,, \tag{1}$$

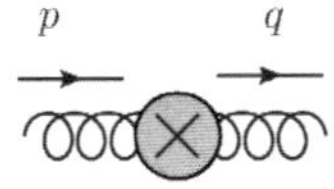

with the free propagators

$$\tilde{S}_F^{(0)}(p) = \frac{i\not{p} + m}{p^2 - m^2 + i0}\,, \qquad \tilde{G}_{\mu\nu}^{(0)}(p) = \frac{id_{\mu\nu}(p)}{p^2 + i0}\,, \qquad d_{\mu\nu}(p) = -g_{\mu\nu} + \frac{n_\mu^- p_\nu + p_\mu n_\nu^-}{n^- \cdot p}\,, \tag{2}$$

and the interaction terms

p q

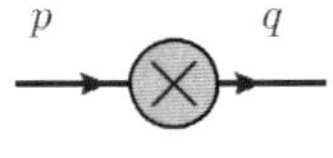

(3)

$$= 2\pi\delta(p^- - q^-)\not{n}^- \int d^{d-2}\boldsymbol{z} e^{-i\boldsymbol{z}\cdot(\boldsymbol{p}-\boldsymbol{q})} \left\{\theta(p^-)\big[V(\boldsymbol{z}) - 1\big] - \theta(-p^-)\big[V^\dagger(\boldsymbol{z}) - 1\big]\right\}\,,$$

p q

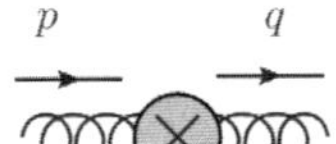

$$= -2\pi\delta(p^- - q^-)2p^- \int d^{d-2}\boldsymbol{z} e^{-i\boldsymbol{z}\cdot(\boldsymbol{p}-\boldsymbol{q})} \cdot \left\{\theta(p^-)\big[U(\boldsymbol{z}) - 1\big] - \theta(-p^-)\big[U^\dagger(\boldsymbol{z}) - 1\big]\right\}\,. \tag{4}$$

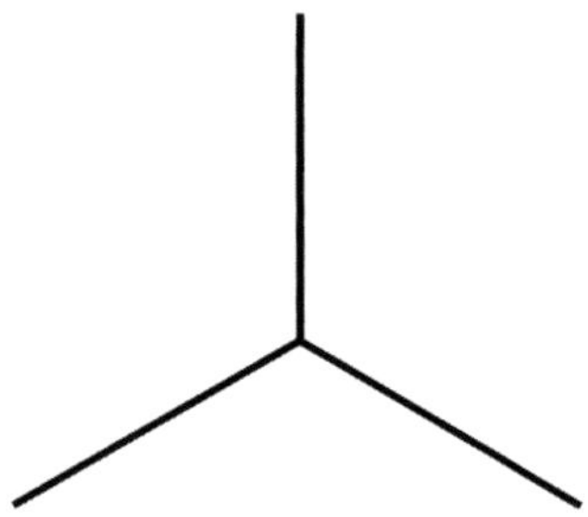

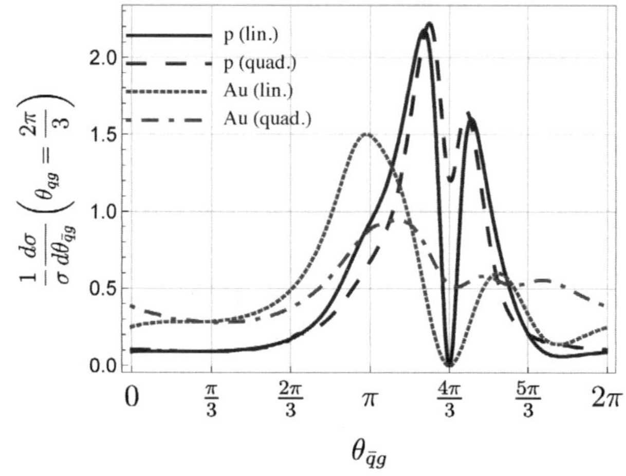

Figure 2. *Left:* Mercedes-Benz star configuration: all transverse momenta have equal absolute value, with relative angle $\Delta = 2\pi/3$. *Right:* Normalized cross-section for three parton production in the presence of large gluon densities against $\Delta\theta_{\bar{q}g}$ with $\Delta_{qg} = 2\pi/3$ for proton and gold up to linear and quadratic order in $N^{(2)}$. For the study the photon momentum fractions of the final state particles have been fixed to $z_1 = z_2 = 0.2$, $z_3 = 0.6$ while $|\boldsymbol{p}_1| = |\boldsymbol{p}_2| = |\boldsymbol{p}_3| = 2$ GeV and $Q^2 = 9$ GeV^2 (plot originally published in [31]).

The latter contain Wilson lines in the fundamental (V) and adjoint (U) representation which resum the interaction with the background field with

$$V(\boldsymbol{z}) \equiv V_{ij}(\boldsymbol{z}) \equiv \mathrm{P}\exp ig\int_{-\infty}^{\infty} dx^- A^{+,c}(x^-, \boldsymbol{z})t^c \tag{5}$$

and $U^{ba}(z_t) = 2\,\mathrm{tr}(t^a V(z_t) t^b V^\dagger(z_t))$; P denotes path ordering of the gluon field in the exponent while t^c is a SU(N_c) generator in the fundamental representation.

Current x values reached at today's collider experiments are most likely still too large to provide definite evidence for the presence of saturation effects. A first possibility to increase gluon densities and therefore the sensitivity to saturation is to consider scattering on large nuclei, where gluon densities are expected to be enhanced by a factor $A^{\frac{1}{3}}$, with A the mass number of the nucleon. A second possibility is to consider exclusive final states: while for inclusive observables the presence of high gluon densities leads mainly to a modification of the x dependence of gluon densities [30], exclusive production processes can provide matrix elements which differ for high gluon densities from their low density counter parts. They are therefore more sensitive to effects related to high gluon densities and provide a tool for their exploration. Such a study has been carried out in [31], where the differential cross-section for the process

$$\gamma^*(q) + \text{target}\,(p) \to q(p_1) + \bar{q}(p_2) + g(p_3) + X, \tag{6}$$

has been calculated for first time within the CGC formalism. For details on the calculation we refer to [31]. The results of a first numerical study are shown in Fig. 2. With the differential cross-section for the process Eq. (6) expressed in terms of dipole and quadrupole operators of Wilson lines Eq. (5),

$$S^{(2)}_{(\boldsymbol{x}_1\boldsymbol{x}_2)} \equiv \frac{1}{N_c}\,\mathrm{tr}\left[V(\boldsymbol{x}_1)V^\dagger(\boldsymbol{x}_2)\right], \quad S^{(4)}_{(\boldsymbol{x}_1\boldsymbol{x}_2\boldsymbol{x}_3\boldsymbol{x}_4)} \equiv \frac{1}{N_c}\,\mathrm{tr}\left[V(\boldsymbol{x}_1)V^\dagger(\boldsymbol{x}_2)V(\boldsymbol{x}_3)V^\dagger(\boldsymbol{x}_4)\right], \tag{7}$$

this study uses the large N_c and Gaussian approximation [33] to express the quadrupole in terms of the dipole operator. For the dipole profile itself a model motivated by recent rcBK[2] fits to HERA data [32] has been used,

$$S^{(2)}_{(\boldsymbol{x}_1\boldsymbol{x}_2)} = \int d^2\boldsymbol{l}\, e^{-i\boldsymbol{l}\cdot\boldsymbol{x}_{12}}\, \Phi(\boldsymbol{l}^2) = 2\left(\frac{Q_0|\boldsymbol{x}_{12}|}{2}\right)^{\rho-1} \frac{K_{\rho-1}(Q_0|\boldsymbol{x}_{12}|)}{\Gamma(\rho-1)},$$

$$\text{where} \qquad \Phi(\boldsymbol{l}^2) = \frac{\rho-1}{Q_0^2\pi}\left(\frac{Q_0^2}{Q_0^2+\boldsymbol{l}^2}\right)^{\rho}, \tag{8}$$

and Q_0 is a scale proportional to the saturation scale. To justify the dilute limit associated with the Gaussian approximation, we study the cross-section at large photon virtuality $Q^2 = 9$ GeV^2 and expand the differential cross-section up to quadratic order in $N^{(2)} = 1 - S^{(2)}$. We further use $\rho = 2.3$ and $Q_0^{\text{proton}} = 0.69$ GeV, as motivated by inclusive DIS fits at $x = 0.2 \times 10^{-3}$. Since we want to search for effects due to large gluon densities, we further simulate scattering on a gold nucleus through the re-scaling $Q_0^{\text{Au}} = A^{1/6} \cdot Q_0^{\text{proton}} = 1.67$ GeV. At linear order in $N^{(2)}$, the cross-section is proportional to the Fourier transform of the dipole, $\Phi\left((\boldsymbol{p}_1 + \boldsymbol{p}_2 + \boldsymbol{p}_3)^2\right)$ and therefore gives direct access to the gluon distribution in the target. At this order the cross-section is – up to the detailed shape and x-dependence of the dipole factor Eq. (8) – identical to a corresponding BFKL calculation. The configuration dominant in the 'collinear' limit corresponds on the other hand to the vanishing of the sum of transverse momenta $\boldsymbol{p}_1 + \boldsymbol{p}_2 + \boldsymbol{p}_3 = 0$, *i.e.* the transverse momentum transfer between projectile and target is zero. To study deviations from this configuration we first chose $|\boldsymbol{p}_1| = |\boldsymbol{p}_2| = |\boldsymbol{p}_3|$. For this choice the 'collinear' limit $\boldsymbol{p}_1 + \boldsymbol{p}_2 + \boldsymbol{p}_3 = 0$ is then given by the Mercedes-Benz star configuration, see Fig. 2. We observe vanishing of the partonic cross section at these 'collinear' configurations, Fig. 2, with a strong double peak. This vanishing is explained by the vanishing of the partonic matrix element at leading order in $N^{(2)}$ for zero momentum transfer between projectile and target, which itself is expected from gauge invariance. For a complete study at hadronic level, we expect the double peak to turn into a single peak. While in the case of the proton we find (besides the double-peak structure) mainly a smearing of the collinear peak at $\Delta_{\bar{q}g} = 4\pi/3$, the effect of a larger gluon saturation scale for a gold nucleus can be clearly seen: while the shape of the linear gold curve is similar to the proton curve, it is wider and depleted with respect to the former. Including further quadratic terms in $N^{(2)}$, we start to probe effects due to high gluon densities at the level of the matrix element beyond the mere modification of Eq. (8) and beyond the corresponding BFKL matrix element. In the case of the proton, these corrections are small, as expected for $x = 0.2 \cdot 10^{-3}$ and a relatively large photon virtuality $Q^2 = 9$ GeV^2. For a highly saturated gold nucleus, they are on the other hand sizable. While this apparently puts doubts on the reliability of the current expansion up to quadratic order in $N^{(2)}$, it clearly demonstrates the sensitivity of this processes to high and saturated gluon densities and identifies it as suitable tool in searching for such effects already at currently accessible collider energies.

Acknowledgments

Financial support by CONACyT-Mexico grant number CB-2014-241408 is gratefully acknowledged.

References

[1] L. N. Lipatov, Sov. J. Nucl. Phys. **23** (1976) 338, E. A. Kuraev, L. N. Lipatov, V. S. Fadin, Phys. Lett. B **60** (1975) 50, Sov. Phys. JETP **44** (1976) 443, Sov. Phys. JETP **45** (1977) 199. I. I. Balitsky, L. N. Lipatov, Sov. J. Nucl. Phys. **28** (1978) 822.

[2] The running coupling Balitsky-Kovchegov equation (rcBK) provides a non-linear generalization of the BFKL equation as discussed above, see [27,32] for details

[2] V. S. Fadin, L. N. Lipatov, Phys. Lett. B **429** (1998) 127, M. Ciafaloni, G. Camici, Phys. Lett. B **430** (1998) 349.
[3] H. Kowalski, L. N. Lipatov, D. A. Ross and G. Watt, Eur. Phys. J. C **70**, 983 (2010) [arXiv:1005.0355 [hep-ph]].
[4] M. Hentschinski, A. Sabio Vera and C. Salas, Phys. Rev. Lett. **110** (2013) no.4, 041601 [arXiv:1209.1353 [hep-ph]]; Phys. Rev. D **87** (2013) no.7, 076005 [arXiv:1301.5283 [hep-ph]].
[5] M. Hentschinski, J. D. Madrigal Martínez, B. Murdaca and A. Sabio Vera, Phys. Lett. B **735**, 168 (2014) [arXiv:1404.2937 [hep-ph]]; Nucl. Phys. B **887**, 309 (2014) [arXiv:1406.5625 [hep-ph]]; Nucl. Phys. B **889**, 549 (2014) [arXiv:1409.6704 [hep-ph]].
[6] M. Deak, F. Hautmann, H. Jung and K. Kutak, JHEP **0909**, 121 (2009) [arXiv:0908.0538 [hep-ph]].
[7] M. Hentschinski and A. Sabio Vera, Phys. Rev. D **85**, 056006 (2012) [arXiv:1110.6741 [hep-ph]]; G. Chachamis, M. Hentschinski, J. D. Madrigal Martínez and A. Sabio Vera, Phys. Rev. D **87**, no. 7, 076009 (2013) [arXiv:1212.4992].
[8] A. van Hameren, P. Kotko and K. Kutak, Phys. Rev. D **88**, 094001 (2013) Erratum: [Phys. Rev. D **90**, no. 3, 039901 (2014)] [arXiv:1308.0452 [hep-ph]].
[9] F. Hautmann, M. Hentschinski and H. Jung, Nucl. Phys. B **865**, 54 (2012) [arXiv:1205.1759 [hep-ph]].
[10] A. van Hameren, P. Kotko and K. Kutak, Phys. Rev. D **92**, no. 5, 054007 (2015) [arXiv:1505.02763 [hep-ph]].
[11] G. Chachamis, M. Deak, M. Hentschinski, G. Rodrigo and A. Sabio Vera, JHEP **1509**, 123 (2015) [arXiv:1507.05778 [hep-ph]].
[12] B. B. Abelev *et al.* [ALICE Collaboration], Phys. Rev. Lett. **113**, no. 23, 232504 (2014) [arXiv:1406.7819 [nucl-ex]].
[13] R. Aaij *et al.* [LHCb Collaboration], J. Phys. G **40**, 045001 (2013) [arXiv:1301.7084 [hep-ex]]; J. Phys. G **41**, 055002 (2014) [arXiv:1401.3288 [hep-ex]].
[14] R. Aaij *et al.* [LHCb Collaboration], JHEP **1509**, 084 (2015) [arXiv:1505.08139 [hep-ex]].
[15] CMS Collaboration [CMS Collaboration], "Measurement of exclusive Υ photoproduction in pPb collisions at $\sqrt{s_{\rm NN}} = 5.02$ TeV," CMS-PAS-FSQ-13-009.
[16] S. Chekanov *et al.* [ZEUS Collaboration], Eur. Phys. J. C **24**, 345 (2002) [hep-ex/0201043].
[17] S. Chekanov *et al.* [ZEUS Collaboration], Nucl. Phys. B **695**, 3 (2004) [hep-ex/0404008].
[18] C. Alexa *et al.* [H1 Collaboration], Eur. Phys. J. C **73**, no. 6, 2466 (2013) [arXiv:1304.5162 [hep-ex]].
[19] A. Aktas *et al.* [H1 Collaboration], Eur. Phys. J. C **46**, 585 (2006) [hep-ex/0510016].
[20] C. Adloff *et al.* [H1 Collaboration], Phys. Lett. B **483**, 23 (2000) [hep-ex/0003020].
[21] J. Breitweg *et al.* [ZEUS Collaboration], Phys. Lett. B **437** (1998) 432 [hep-ex/9807020].
[22] S. Chekanov *et al.* [ZEUS Collaboration], Phys. Lett. B **680**, 4 (2009) [arXiv:0903.4205 [hep-ex]].
[23] I. Bautista, A. Fernandez Tellez and M. Hentschinski, arXiv:1607.05203 [hep-ph].
[24] L. V. Gribov, E. M. Levin and M. G. Ryskin, Phys. Rept. **100**, 1 (1983).
[25] L. D. McLerran and R. Venugopalan, Phys. Rev. D **49**, 2233 (1994), Phys. Rev. D **49**, 3352 (1994).
[26] J. Jalilian-Marian, A. Kovner, L. D. McLerran and H. Weigert, Phys. Rev. D **55**, 5414 (1997); J. Jalilian-Marian, A. Kovner, A. Leonidov and H. Weigert, Nucl. Phys. B **504**, 415 (1997), Phys. Rev. D **59**, 014014 (1999), Phys. Rev. D **59**, 014015 (1999), Phys. Rev. D **59**, 034007 (1999), A. Kovner, J. G. Milhano and H. Weigert, Phys. Rev. D **62**, 114005 (2000); A. Kovner and J. G. Milhano, Phys. Rev. D **61**, 014012 (2000); E. Iancu, A. Leonidov and L. D. McLerran, Nucl. Phys. A **692**, 583 (2001), Phys. Lett. B **510**, 133 (2001); E. Ferreiro, E. Iancu, A. Leonidov and L. McLerran, Nucl. Phys. A **703**, 489 (2002); H. Weigert, Nucl. Phys. **A703**, 823-860 (2002); J. -P. Blaizot, E. Iancu, H. Weigert, Nucl. Phys. **A713**, 441-469 (2003).
[27] I. Balitsky, Nucl. Phys. B **463**, 99 (1996); Y. V. Kovchegov, Phys. Rev. D **60**, 034008 (1999), Phys. Rev. D **61**, 074018 (2000).
[28] Y. V. Kovchegov and E. Levin, Nucl. Phys. B **577**, 221 (2000) [hep-ph/9911523]; M. Hentschinski, H. Weigert and A. Schafer, Phys. Rev. D **73**, 051501 (2006) [hep-ph/0509272]; E. Levin and A. Prygarin, Phys. Rev. C **78**, 065202 (2008) [arXiv:0804.4747 [hep-ph]].
[29] E. Iancu and R. Venugopalan, In *Hwa, R.C. (ed.) et al.: Quark gluon plasma* 249-3363; H. Weigert, Prog. Part. Nucl. Phys. **55**, 461 (2005); J. Jalilian-Marian, Y. V. Kovchegov, Prog. Part. Nucl. Phys. **56**, 104-231 (2006); F. Gelis, E. Iancu, J. Jalilian-Marian, R. Venugopalan, Ann. Rev. Nucl. Part. Sci. **60**, 463-489 (2010); J. L. Albacete and C. Marquet, Prog. Part. Nucl. Phys. **76**, 1 (2014).
[30] J. Bartels and K. Kutak, Eur. Phys. J. C **53**, 533 (2008) [arXiv:0710.3060 [hep-ph]]; L. Motyka and M. Sadzikowski, Acta Phys. Polon. B **45**, no. 11, 2079 (2014) [arXiv:1411.7774 [hep-ph]].
[31] A. Ayala, M. Hentschinski, J. Jalilian-Marian and M. E. Tejeda-Yeomans, arXiv:1604.08526 [hep-ph].
[32] J. L. Albacete, N. Armesto, J. G. Milhano, P. Quiroga-Arias and C. A. Salgado, Eur. Phys. J. C **71**, 1705 (2011);
[33] F. Dominguez, C. Marquet, B. -W. Xiao and F. Yuan, Phys. Rev. D **83**, 105005 (2011).

Journal of Physics: Conference Series **761** (2016) 012039 doi:10.1088/1742-6596/761/1/012039

Pion transition form factor through Dyson-Schwinger equations

Khépani Raya

Instituto de Física y Matemáticas, Universidad Michoacana de San Nicolás de Hidalgo Edificio C-3, Ciudad Universitaria, C.P. 58040, Morelia, Michoacán, México.

E-mail: khepani@ifm.umich.mx

Abstract. In the framework of Dyson-Schwinger equations (DSE), we compute the $\gamma^*\gamma \to \pi^0$ transition form factor, $G(Q^2)$. For the first time, in a continuum approach to quantun chromodynamics (QCD), it was possible to compute $G(Q^2)$ on the whole domain of space-like momenta. Our result agrees with CELLO, CLEO and Belle collaborations and, with the well-known asymptotic QCD limit, $2f_\pi$. Our analysis unifies this prediction with that of the pion's valence-quark parton distribution amplitude (PDA) and elastic electromagnetic form factor.

1. Introduction

The neutral pion transition form factor is measured through electron-positron scattering. Although the available data of CELLO [1], CLEO [2], Babar [3] and Belle [4] collaborations agree in the domain of $Q^2 < 10$ GeV^2, the Babar and Belle data (the only data available above that point) are notoriously different. Moreover, how possibly the Babar data can reconcile with the asymptotic QCD limit, calculated by Brodsky and Lepage in [5], i.e., $2f_\pi$, remains unclear and unsatisfactory.

We have previous DSE input from [6], but because of the numerical methods developed by that time, it was not possible to arrive at momentum scales larger than $Q^2 > 4$ GeV^2. Complete understanding of $G(Q^2)$ demands simultaneously achieving correct asymptotic behavior, but also the essentially non perturbative Abelian anomaly, $2f_\pi G(Q^2 = 0) = 1$. Such features are achievable in the framework of DSEs. At the same time, we are able to connect our results with that of the pion's PDA, [7], and elastic electromagnetic form factor, [8]. The reference to our detailed published article is [9]. Most of the ingredients which have gone into this study are quite general and can be applied to other mesons and processes. In particular, the study of $\gamma^*\gamma \to \eta_c$ and $\gamma^*\gamma \to \eta_b$ is under way.

2. The tools

For any pseudoscalar meson (PS), the $\gamma^*\gamma \to PS$ is expressed through $\mathcal{T}_{\mu\nu}(k_1,k_2) = T_{\mu\nu}(k_1,k_2) + T_{\mu\nu}(k_2,k_1)$, where the matrix element is:

$$T_{\mu\nu}(k_1,k_2) = \frac{e^2}{4\pi^2}\epsilon_{\mu\nu\alpha\beta}k_{1\alpha}k_{2\beta}\; G(k_1^2,k_2^2,k_1\cdot k_2)\;. \tag{1}$$

The photons' momenta are k_1 and k_2 and the meson's total momentum is $P = k_1 + k_2$ ($P^2 = -m_{PS}^2$). At leading order (rainbow-ladder truncation) in the systematic and symmetry preserving DSE truncation scheme, one can write:

$$T_{\mu\nu}(k_1, k_2) = \int \frac{d^4k}{(2\pi)^4} i\mathcal{Q}\chi_\mu(l, l_1)\Gamma_{PS}(l_1, l_2)S(l_2)i\mathcal{Q}\Gamma_\nu(l_2, l) \; . \tag{2}$$

Here $\mathcal{Q}$ is the quark charge operator (e diag$[2/3, -1/3]$ for neutral pion), $l_1 = l + k_1$, $l_2 = l - k_2$, where the kinematic constraints are: $k_1^2 = Q^2$, $k_2^2 = 0$, $k_1 \cdot k_2 = -(m_{PS}^2 + Q^2)/2$ and $P^2 = -m_{PS}^2$. The quark propagator, $S(k)$, and the Bethe-Salpeter amplitude (BSA), $\Gamma_{PS}(q_1, q_2)$, are obtained by solving the corresponding DSEs and the BSEs. On the other hand, the quark-photon vertex is constructed via the *gauge technique*, [10]. We give the details in the following subsections.

2.1. Quark propagator and Bethe-Salpeter amplitudes

Most generally, the quark propagator is written as $S(p) = -i\gamma \cdot p\sigma_v(p^2) + \sigma_s(p^2)$, while the BSA is:

$$\Gamma_{PS}(k; P) = \gamma_5[i\, E_{PS}(k; P) + \gamma \cdot P F_{PS}(k; P) + \gamma \cdot k\; k \cdot P G_{PS}(k; P) + \sigma_{\mu\nu} k_\mu P_\nu H_{PS}(k; P)] \; , \tag{3}$$

The corresponding DSE and BSE, in the rainbow-ladder truncation, are:

$$S^{-1}(p) = \mathcal{Z}_{2F}(S^0)^{-1}(p) + \mathcal{Z}_{1F} \int \frac{d^4q}{(2\pi)^4} G(p-q) D^0_{\mu\nu}(p-q, \mu) \frac{\lambda^a}{2}\gamma_\mu S(q, \mu) \frac{\lambda^a}{2}\gamma_\nu, \tag{4}$$

$$\Gamma_{PS}(p, P) = -\int \frac{d^4q}{(2\pi)^4} \frac{G((p-q)^2)}{(p-q)^2} \frac{\lambda^c}{2}\gamma_\mu S^a(q + \eta P)\Gamma_{PS}(q, P) S^b(q - (1-\eta P))\frac{\lambda^c}{2}\gamma_\nu \; , \tag{5}$$

with $G(p-q)$ being the effective coupling described in [11]. Once we obtain the solution of Eq. (4), we can parameterize it in the form of a complex conjugate pole representation:

$$S(p) = \sum_{j=1}^{N} \left(\frac{z_j}{i\gamma \cdot p + m_j} + \frac{z_j^*}{i\gamma \cdot p + m_j^*} \right) \; , \tag{6}$$

constrained to the ultraviolet conditions of free quark propagator; $N = 2$ is accurate enough for our purposes. We then solve Eq. (5), and parameterize its solutions with the perturbation theory integral representation (PTIR):

$$\begin{aligned}
\mathcal{F}(k; P) &= \mathcal{F}^i(k; P) + \mathcal{F}^u(k; P) \; , \\
\mathcal{F}^i(k; P) &= c^i_{\mathcal{F}} \int_{-1}^{1} dz\; \rho_{\nu^i_{\mathcal{F}}}(z)[a_{\mathcal{F}} \hat{\Delta}^4_{\Lambda^i_{\mathcal{F}}}(k_z^2) + \bar{a}_{\mathcal{F}} \hat{\Delta}^5_{\Lambda^i_{\mathcal{F}}}(k_z^2)] \; , \\
E^u(k; P) &= c^u_E \int_{-1}^{1} dz\; \rho_{\nu^u_E}(z) \hat{\Delta}^{1+\alpha}_{\Lambda^u_E}(k_z^2) \; , \\
F^u(k; P) &= c^u_F \int_{-1}^{1} dz\; \rho_{\nu^u_F}(z) k^2 \Lambda^u_F \Delta^{2+\alpha}_{\Lambda^u_F}(k_z^2) \; , \\
G^u(k; P) &= c^u_G \int_{-1}^{1} dz\; \rho_{\nu^u_G}(z) \Lambda^u_G \Delta^{2+\alpha}_{\Lambda^u_G}(k_z^2) \; ,
\end{aligned}$$

where $\mathcal{F}(k; P) = E, F, G$ and i, u stand for IR and UV; $\hat{\Delta}_\Lambda(s) = \Lambda^2 \Delta_\Lambda(s)$, $\Delta_\Lambda(s) = (\Lambda^2 + s)^{-1}$, $k_z^2 = k^2 + zk \cdot P$, $\bar{a}_E = 1 - a_E$, $\bar{a}_F = 1/\Lambda^i_F - a_F$, $\bar{a}_G = [1/\Lambda^i_G]^3 - a_G$. The parameters a, c, Λ, α

and ν are fitted to the numerical data; α simulates the logarithmic UV behavior, characteristic of QCD, and ν defines the spectral density

$$\rho_\nu(z) = \frac{1}{\sqrt{\pi}}\frac{\Gamma(3/2+\nu)}{\Gamma(1/2+\nu)}(1-z^2)^\nu \,. \tag{7}$$

The amplitude $H(k;P)$ has only a very small impact on the final results and can safely be neglected. Such representations have a quadratic form in the denominator, which will be practically useful in the computation of Eq. (2).

2.2. Quark-photon vertex

PTIRs are not available yet for the quark-photon vertex. Instead, we use the following *ansatz* for the unamputated vertex:

$$\begin{aligned}\chi_\mu(k_f,k_i) &= \gamma_\mu \Delta_{k^2\sigma_V} + [s\gamma\cdot k_f\gamma_\mu\gamma\cdot k_i + \bar{s}\gamma\cdot k_i\gamma_\mu\gamma\cdot k_f]\Delta_{\sigma_V} \\ &+ [s(\gamma\cdot k_f\gamma_\mu + \gamma_\mu\gamma\cdot k_i) + \bar{s}(\gamma\cdot k_i\gamma_\mu + \gamma_\mu\gamma\cdot k_f)]i\Delta_{\sigma_S}\,,\end{aligned} \tag{8}$$

where $\Delta_F = [F(k_f^2) - F(k_i^2)]/[k_f^2 - k_i^2]$, $q = k_f - k_i$ and $\bar{s} = 1 - s$. This vertex *ansatz* is obtained using the *gauge technique*. By construction, it satisfies the longitudinal Ward-Green-Takahashi identity, is free of kinematic singularities, reduces to the bare vertex in the free-field limit, and has the same Poincaré transformations properties as the bare vertex.

Up to transverse pieces associated with the scalar s, $\chi_\mu(k_f,k_i)$ is equivalent to $S(k_f)\Gamma_\mu S(k_i)$. Nothing material would be gained herein by making them identical because any difference is power-law suppressed in the ultraviolet; but computational effort would increase substantially. We define s as follows:

$$s = 1 + s_0 \,\mathrm{Exp}[-\mathcal{E}_{PS}/M_E] \tag{9}$$

where $\mathcal{E}_{PS} = \sqrt{Q^2/4 + m_{PS}^2} - m_{PS}$ is the *Breit-frame* energy of the pseudoscalar and, $M_E = \{p|p^2 = M^2(p^2),\ p^2 > 0\}$ is the Euclidian constituent-quark mass. The parameter $s \neq 1$ has to do with the value of $G(Q^2)$ in a neighborhood of $Q^2 = 0$. In the case of the pion, owing to the Abelian anomaly, it is impossible to simultaneously conserve the vector and axial vector currents associated with Eq. (2), but, with a proper choice of s_0 in Eq. (9), vector currents are conserved and the Abelian anomaly is satisfied. With everything expressed in terms the same functions, we will now see how the appropriate parameterizations of $S(p)$ and $\Gamma_{PS}(k,P)$ allow us to compute $G(Q^2)$ in the whole range of space-like momenta.

3. The Calculation

All elements of Eq. (2) are written in terms of $S(p)$ and $\Gamma_{PS}(k,P)$, expressed as complex conjugate pole representation or PTIRs, respectively. Computation of $G(Q^2)$ reduces to the task of summing a series of terms, all of which involve a single four-momentum integral.

As we saw before, our quark-photon vertex construction allow us to satisfy one of the constraints of the transition form factor, namely, the conservation of vector current (ensuring the Abelian anomaly is correctly recovered for the pion). In other words, this constraint fixes the value of $G(Q^2 = 0)$. One should also understand the connection of $G(Q^2)$ with the meson's PDA and its evolution with the factorization scale of QCD. As the same PDA also governs the Q^2 dependence of the pion electromagnetic form factor, we will have a unified prediction of both the form factors in the asymptotic limit of QCD. We detailed the algorithm below.

3.1. The algorithm
Because of the representations employed for $S(p)$ and $\Gamma_{PS}(k,P)$, the denominator in every term is a product of l-quadratic forms. We perform Feynman parameterization. After a proper change of variables, all the momentum integrals can be solved analytically. Once we solved the four-momentum integrals, one computes a finite number of simple integrals, namely, the ones over the Feynman parameters and z, the spectral integral. The complete result for $G(Q^2)$ is obtained after summing the series.

The peculiar perturbation theory like parameterizations and the procedure employed are the reason we are able to compute $G(Q^2)$, for arbitrarily large Q^2 space-like momenta, for the first time in a continuum approach directly connected to QCD. For a complete explanation and more detail, please consult our published article [9]. This procedure has also previously been employed to compute the pion's elastic electromagnetic form factor and pion's PDA [7, 8].

3.2. Pion's PDA and its evolution
Various studies indicate that the pion PDA is a broad concave function, and at the resolution scales achieved so far, it is far from being similar to the asymptotic PDA, $\phi_{asy}(x) = 6x(1-x)$. According to [7], it is defined by the expression:

$$f_\pi \phi(x;\zeta) = N_c \mathrm{tr} Z_2(\zeta,\Lambda) \int \frac{d^4k}{(2\pi)^4} \delta_n^u(k_\eta)\gamma_5 \gamma \cdot n S(k_\eta)\Gamma_\pi(k_{\eta\bar{\eta}};P)S(k_{\bar{\eta}}) \,, \tag{10}$$

where $N_c = 3$; the trace is over spinor indices; $Z_2(\zeta,\Lambda)$ is the quark wave-function renormalisation constant; $\delta_n^u(k_\eta) := \delta(n\cdot k_\eta - un\cdot P)$, with $n^2 = 0$, $n \cdot P = -m_\pi$; and $k_{\eta\bar{\eta}} = [k_\eta + k_{\bar{\eta}}]/2$, $k_\eta = k + \eta P$, $k_{\bar{\eta}} = k - (1-\eta)P$, $\eta \in [0,1]$. The way it evolves towards its asymptotic limit is described through the Efremov-Radyushkin-Brodsky-Lepage (ERBL) evolution equations [5, 12].

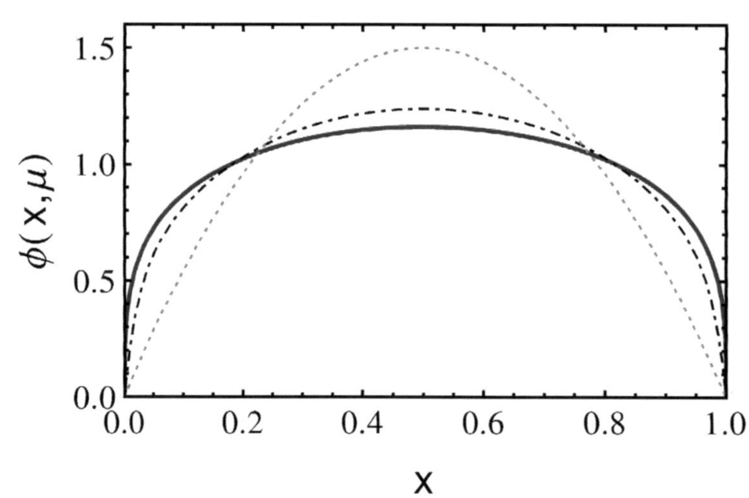

Figure 1. Pion's PDA: [Solid, blue] $\phi(x, \mu = 2$ GeV). [Dot-dashed, black] ERBL evolution to $\phi(x, \mu = \sqrt{50}$ GeV). [Green, dotted] Asymptotic PDA, $\phi_{asy}(x) = 6x(1-x)$. As we can see, the broad concave PDA, $\phi(x, \mu = 2$ GeV), slowly approaches to its asymptotic form with increasing momentum scale.

From figure 1, we see that pion's PDA at resolution scale $\mu = 2$ GeV, $\phi(x, \mu = 2 \text{ GeV})$, slowly evolves to its asymptotic form. Evolution enables the dressed-quark and antiquark degrees of freedom, in terms of which the wave function is expressed at a given scale, to split into less well-dressed partons via the addition of gluons and sea quarks in the manner prescribed by QCD dynamics. The connection of $\phi(x, \mu)$ with $G(Q^2)$ is given by the leading twist expression for the transition form factor:

$$G(Q^2) = 4\pi^2 f_\pi \int_0^1 dx T_H(x, Q^2, \alpha(\mu); \mu)\phi(x, \mu) \,, \tag{11}$$

where $T_H(\mu)$ is the photon-quark-antiquark scattering amplitude at some scale μ. One expects to reach the asymptotic limit from below, otherwise one would have to explain why $G(Q^2)$ grows bigger and then decreases towards $2f_\pi$. QCD is not known to have an additional scale to set in at a higher Q^2 to make this happen. Only logarithmic corrections have a minor role to play. We shall see that the evolution of $\phi(x, \mu)$ is crucial in understanding the asymptotic behavior of the pion transition form factor. Further details of pion's PDA can be found in Javier Cobos' contribution to this proceedings volume and in [7].

4. Results and Conclusions

4.1. Results

Our main result is depicted in figure 2. We obtain an interaction radius of $r_{\pi^0} = 0.68$ fm, which is practically identical to the one computed in [8] within the same scheme. The Abelian anomaly, namely $2f_\pi G(Q^2 = 0) = 1$, is satisfied; the asymptotic limit, $2f_\pi$, is reached from below except for a logarithmic miss-match. Our result agrees fairly well with all available data below $Q^2 < 10$ GeV^2, and with Belle data at large Q^2 scales. However, it fails to reconcile with the data reported by the Babar collaboration.

From figure 2, we see that $G(Q^2)$ exceeds (logarithmically) the asymptotic limit $2f_\pi$, although we expected such limit to be reached from below. However, as mentioned earlier, the growth is only logarithmic, and at some point, it settles onto the value $2f_\pi$. This discrepancy originates from the failure of the rainbow-ladder (RL) truncation to reproduce the complete set of gluon and quark splitting effects contained in QCD and hence its inability to fully express interferences between the anomalous dimensions of those n-point Schwinger functions which are relevant in the computation of a given scattering amplitude.

4.2. Conclusions

We describe a computation of the pion transition form factor, in which all elements employed are determined by the solutions of the QCD's DSEs, obtained in the RL truncation.

- The novel analysis techniques made it possible to compute $G(Q^2)$, on the entire domain of space-like momenta, in a continuum framework directly connected to QCD.
- Our work unifies the description and explanation of this transition with the charged pion electromagnetic form factor and its PDA.
- This enables us to demonstrate that a fully self-contained and consistent treatment can readily connect a pion PDA, that is a broad and concave function at the hadronic scale, with the perturbative QCD prediction for the transition form factor in the hard photon limit.

Full discussion and details are found in our work in [9].

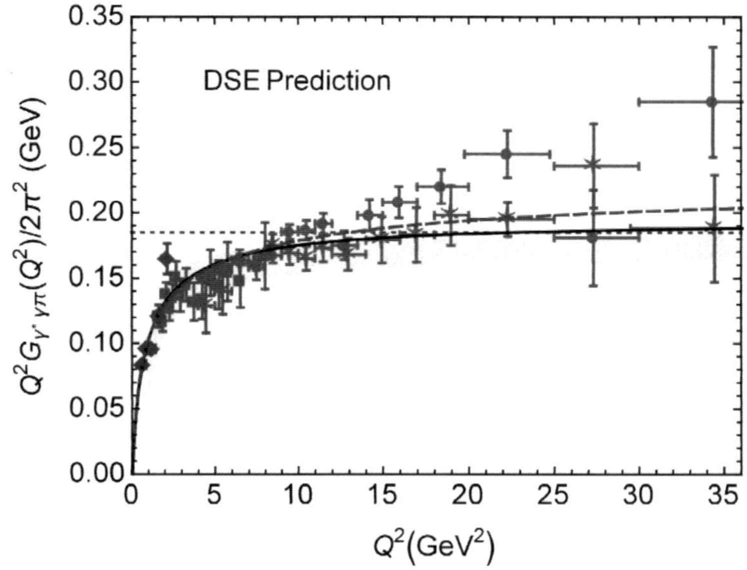

Figure 2. Pion's transition form factor: [Solid, black] DSE Prediction obtained with the ERBL evolution of pion BSA. [Dashed, blue] DSE Prediction without evolution at frozen scale $\mu = 2$ GeV. **Data:** [Circles, red] Babar [3], [Diamonds, purple] CELLO [1], [Squares, blue] CLEO [2], [Stars, green] Belle [4]. The (green) shaded band is described in [13].

5. Acknowledgements
I want to acknowledge the organizing committee for the financial support and my collaborators in [9] for fruitful discussions.

References
[1] H. J. Behrend *et al.* [CELLO Collaboration], Z. Phys. C **49**, 401 (1991). doi:10.1007/BF01549692
[2] J. Gronberg *et al.* [CLEO Collaboration], Phys. Rev. D **57**, 33 (1998) doi:10.1103/PhysRevD.57.33 [hep-ex/9707031].
[3] B. Aubert *et al.* [BaBar Collaboration], Phys. Rev. D **80**, 052002 (2009) doi:10.1103/PhysRevD.80.052002 [arXiv:0905.4778 [hep-ex]].
[4] S. Uehara *et al.* [Belle Collaboration], Phys. Rev. D **86**, 092007 (2012) doi:10.1103/PhysRevD.86.092007 [arXiv:1205.3249 [hep-ex]].
[5] G. P. Lepage and S. J. Brodsky, Phys. Rev. D **22**, 2157 (1980). doi:10.1103/PhysRevD.22.2157
[6] P. Maris and P. C. Tandy, Phys. Rev. C **65**, 045211 (2002) doi:10.1103/PhysRevC.65.045211 [nucl-th/0201017].
[7] L. Chang, I. C. Cloet, J. J. Cobos-Martinez, C. D. Roberts, S. M. Schmidt and P. C. Tandy, Phys. Rev. Lett. **110**, no. 13, 132001 (2013) doi:10.1103/PhysRevLett.110.132001 [arXiv:1301.0324 [nucl-th]].
[8] L. Chang, I. C. Clot, C. D. Roberts, S. M. Schmidt and P. C. Tandy, Phys. Rev. Lett. **111**, no. 14, 141802 (2013) doi:10.1103/PhysRevLett.111.141802 [arXiv:1307.0026 [nucl-th]].
[9] K. Raya, L. Chang, A. Bashir, J. J. Cobos-Martinez, L. X. Gutirrez-Guerrero, C. D. Roberts and P. C. Tandy, Phys. Rev. D **93**, no. 7, 074017 (2016) doi:10.1103/PhysRevD.93.074017 [arXiv:1510.02799 [nucl-th]].
[10] R. Delbourgo and P. C. West, J. Phys. A **10**, 1049 (1977). doi:10.1088/0305-4470/10/6/024
[11] S. x. Qin, L. Chang, Y. x. Liu, C. D. Roberts and D. J. Wilson, Phys. Rev. C **84**, 042202 (2011) doi:10.1103/PhysRevC.84.042202 [arXiv:1108.0603 [nucl-th]].
[12] A. V. Efremov and A. V. Radyushkin, Phys. Lett. B **94**, 245 (1980). doi:10.1016/0370-2693(80)90869-2
[13] A. P. Bakulev, S. V. Mikhailov, A. V. Pimikov and N. G. Stefanis, Phys. Rev. D **86**, 031501 (2012) doi:10.1103/PhysRevD.86.031501 [arXiv:1205.3770 [hep-ph]].

Journal of Physics: Conference Series **761** (2016) 012040 doi:10.1088/1742-6596/761/1/012040

CP-violation phases and Majorana neutrino magnetic moments in left-right models

D. Delepine[1] **and H. Novales-Sánchez**[2]

[1]División de Ciencias e Ingenierías, Universidad de Guanajuato Campus León, Loma del Bosque 103, Colonia Lomas del Campestre, 37150 León, Guanajuato, México

[2]Facultad de Ciencias Físico Matemáticas, Benemérita Universidad Autónoma de Puebla, Apartado Postal 1152 Puebla, Puebla México

E-mail: hector.novales@correo.buap.mx

Abstract. An implication of nonzero neutrino masses is the existence of neutrino magnetic moments, which arise in extensions of the Standard Model. Among the whole set of electromagnetic properties, these physical quantities have received much attention, both theoretically and experimentally. In the present paper we review the contributions to neutrino magnetic moments from new physics described by a *leff-right model*, with Majorana neutrinos, which might be as large as $10^{-11}\mu_{\rm B}$. These electromagnetic moments depend on Majorana phases. It turns out that, in presence of CP violation, specific sets of values of these phases can cancel up to two magnetic moments, while the remaining one must necessarily be nonzero and large.

1. Brief chronicle of neutrino mass

Originally proposed by W. Pauli in 1930, and called neutrons at the time, neutrinos were assumed to be electrically neutral fermions with a tiny mass [1]. It was at the famous experiments in the Savannah River Plant, leaded by Clyde Cowan and Frederick Reines, where the existence of these elusive particles was finally confirmed by the observation of antineutrinos through the process $\overline{\nu}+p\to e^{+}+n$ [1], in 1956. A year later, Bruno Pontecorvo proposed, for the first time, the phenomenon of neutrino oscillations [2], according to which a neutrino produced by a source with a definite flavor has a nonzero probability of being measured, after traveling some distance, with a different flavor. For instance, in the two-family formalism oscillations of ultrarelativistic neutrinos in vacuum are described by the transition probability [3, 4]

$$P(\nu_e\to\nu_\mu)=\sin 2\theta\,\sin^2\left(\frac{\Delta m^2 L}{4E}\right), \tag{1}$$

where E is the energy of a massless neutrino, L is the distance traveled by it, θ is a mixing angle and Δm^2 is a difference of squared neutrino masses. The presence of the mixing angle θ in Eq. (1) illustrates that neutrino mixing is a necessary ingredient for neutrino oscillations to

[1] Even so, note that neutrinos are assumed to be massless in the Standard Model.

occur. In the case of three families such mixing is described by the unitary transformation [3, 5]

$$|\nu_\alpha\rangle = \sum_{j=1}^{3} U^*_{\alpha j}|\nu_j\rangle, \qquad (\alpha = e,\, \mu,\, \tau) \tag{2}$$

that relates neutrino mass eigenstates, $|\nu_j\rangle$, to flavor states, $|\nu_\alpha\rangle$, which participate in charged currents involving charged leptons of definite flavors. Another significant element of Eq. (1) is the squared-mass difference Δm^2, because it sets the requirement that neutrino masses must be different of each other, and particularly nonzero, for neutrino oscillations to exist.

From around 1967 and for about three decades, Raymond Davis Jr. headed an experiment in the Homestake mine at Lead, South Dakota, with the purpose of measuring neutrinos produced by nuclear fusion in the Sun [6, 7]. Following the solar model, John Bahcall performed a theoretical calculation of the expected rate of solar electron-neutrinos that should be detected[2]. The result was that the experiment of Davis and collaborators only measured 1/3 of the solar neutrinos predicted by Bahcall. Such disagreement, known as the *solar neutrino problem*, was solved by the confirmation that neutrinos oscillate. The first experimental evidence of neutrino oscillations was observed in 1998 by the Super-Kamiokande Collaboration, which aimed at atmospheric neutrinos [9]. Under direction of Takaaki Kajita, this collaboration was able to establish the bounds $\sin 2\theta_{23} > 0.82$ and $4 \times 10^{-4}\,\text{eV} < \Delta m^2 < 6 \times 10^{-3}\,\text{eV}$ on the *atmospheric mixing angle* θ_{23} and the squared-mass difference Δm^2. Shortly after, Arthur B. McDonald leaded measurements, at the Sudbury Neutrino Observatory, of fluxes of neutrinos coming from the Sun [10]. It was found that the sum of all the flux components, corresponding to the different neutrino flavors, was consistent with the solar model, used in Bahcall's calculations, thus finally settling the discrepancy. In 2012 the Daya Bay and RENO Collaborations measured the mixing angle θ_{13} [11, 12], which is the last piece of the confirmation that neutrinos oscillate and thus that they mix and they are massive.

2. Dirac and Majorana neutrinos

The confirmation that neutrinos oscillate, which requires the existence of neutrino mixing and neutrino mass, is a quite relevant conclusion, since it incarnates experimental evidence of new physics, beyond the reach of the Standard Model. Moreover, it opens the question of whether neutrinos are Dirac or Majorana fermions. The Dirac equation can be split into two coupled equations for chiral spinor fields, ψ_L and ψ_R:

$$i\gamma^\mu\partial_\mu\psi_L = m\psi_R, \qquad i\gamma^\mu\partial_\mu\psi_R = m\psi_L, \tag{3}$$

that are decoupled if $m = 0$, in which case the corresponding fermion field is described by a two-component spinor. It was Ettore Majorana who realized that it is also possible to describe a massive fermion by using a two-component spinor if the chiral fermions ψ_L and ψ_R are not independent, but they fulfill the condition $\psi_R = \mathcal{C}\overline{\psi_L}^{\mathrm{T}}$ [3, 13], where $\mathcal{C}$ is the charge-conjugation matrix. This yields the equations

$$i\gamma^\mu\partial_\mu\psi_L = m\mathcal{C}\overline{\psi_L}^{\mathrm{T}}, \qquad \text{(Majorana equation)} \tag{4}$$

$$\psi = \mathcal{C}\overline{\psi}^{\mathrm{T}}, \qquad \text{(Majorana constraint)} \tag{5}$$

with Eq. (5) indicating that Majorana fermions ψ coincide with their antiparticles [13]. For the last statement to be true, note that Majorana fermions must be electrically neutral particles,

[2] A discussion can be found in Ref. [8].

because otherwise electric charge would set a clear difference among particles and antiparticles. Since neutrinos are massive and neutral particles, they might be either Dirac or Majorana fermions. While this issue is nowadays an open question and theoretical and experimental investigations on the matter are being carried out, simplicity in the sense of the number of degrees of freedom has been used as an argument favoring the scenario of Majorana neutrinos [14].

In general, a main difference between Dirac and Majorana fermions is the number of degrees of freedom characterizing them. If CPT conservation is assumed, the former have 4 degrees of freedom (2 helicities, particle-antiparticle), whereas Majorana fermions have only 2 (2 helicities) [3]. Lepton number violation is currently being used to establish the nature of neutrinos. Experiments that are nowadays looking for the elusive neutrinoless double beta decay are well motivated, since the observation of this rare process, forbidden in the Standard Model, would be evidence pointing towards a world in which neutrinos are Majorana particles. A claim of a measurement of this physical process was made some years ago [15], but up to now this discussion has not arrived at a conclusion [16]. Dirac neutrinos are also different from Majorana ones in the parametrization of the matrix describing neutrino mixing, also known as the Pontecorvo-Maki-Nakagawa-Sakata (PMNS) matrix. The parametrization of the PMNS mixing matrix for Dirac neutrinos looks the same as that used for the Cabibbo-Kobayashi-Maskawa (CKM) matrix [17], which mixes quarks in the Standard Model[3]. The PMNS matrix is conveniently written in terms of three mixing angles, θ_{12}, θ_{23}, θ_{13}, and one complex CP-violating phase, δ, which is known as the *Dirac phase* [3, 5]. In the case of Majorana neutrinos, the mixing matrix, U, can be written as [3, 5] $U = VP$, where V is a CKM-like matrix, with the aforementioned parametrization, and P is a diagonal matrix that is given by $P = \mathrm{diag}(1, e^{i\varphi_2}, e^{i\varphi_3})$, where φ_2 and φ_3 are the so-called Majorana phases, which are sources of nonconservation of CP symmetry.

It is worth emphasizing that neutrino oscillations are not sensitive to Majorana phases [3]. On the other hand, measurements of oscillations of solar, atmospheric and reactor neutrinos have been translated into estimations of the values of all other parameters of neutrino mixing [17, 18]. The current best values of so-called *atmospheric angle* θ_{23}, which is the largest mixing angle, are so far compatible with $\pi/4$. In 2015 the T2K Collaboration reported that normal hierarchy of neutrino mass is weakly favored over inverted hierarchy [19]. Then this Collaboration provided the values $\sin^2\theta_{23} = 0.528^{+0.055}_{-0.038}$ and $|\Delta m^2_{32}| = (2.51 \pm 0.11) \times 10^{-3}\,\mathrm{eV}^2$. One of the objectives of current and future experimental investigations at experimental facilities such as T2K and NOvA is the determination of the deviation, with sign, of θ_{23} from $\pi/4$ [18]. In Ref. [20] recent results on solar neutrinos from the Super-Kamiokande experiment are presented, featuring the value $\Delta m^2_{21} = (7.49^{+0.19}_{-0.17}) \times 10^{-5}\,\mathrm{eV}$ and an estimation of the *solar angle* θ_{12} given by $\sin^2\theta_{12} = 0.305 \pm 0.013$. Future improvements on the precision of these values are planned for experiments like JUNO, to start in 2020 [21]. It is considered that 2012 was quite important for neutrino physics because in this year the last mixing angle, θ_{13}, was measured [11, 12] at a relatively large value. Furthermore, the determination of a nonzero value for this angle implied that CP violation may be measured in the leptonic sector. Recently, the Daya Bay Collaboration updated its value of θ_{13}, improving it to $\sin^2 2\theta_{13} = 0.084 \pm 0.005$ [22]. In contrast with the Majorana phases, the Dirac phase does introduce effects in neutrino oscillations, which are characterized by the CP asymmetry $A^{(l,l')}_{CP} = P(\nu_l \to \nu_{l'}) - P(\overline{\nu}_l \to \overline{\nu}_{l'})$ [3]. The T2K Collaboration has excluded values of δ within the ranges $(0.19\pi, 0.80\pi)$, for normal hierarchy, and $(-\pi, -0.97\pi)$ and $(-0.04\pi, \pi)$, for inverted hierarchy, at 90% C.L. This collaboration pointed out that the value $\delta = -\pi/2$ is favored by data [23].

[3] Of course, the values of the parameters of neutrino mixing are quite different from those of the CKM matrix [17].

3. Electromagnetic properties of neutrinos

Even though neutrinos are at present known to be neutral particles, they may possess electromagnetic properties that are generated by quantum loops' effects. The electromagnetic neutrino vertex $\nu_j \nu_k \gamma$ is parametrized by [24, 25]

$$\Lambda_\mu^{jk}(q) = \left(\gamma_\mu - \frac{q_\mu \not{q}}{q^2}\right)\left[f_Q^{jk}(q^2) + f_A^{jk}(q^2) q^2 \gamma_5\right] - i\sigma_{\mu\nu} q^\nu \left[f_M^{jk}(q^2) + i f_E^{jk}(q^2)\gamma_5\right], \tag{6}$$

where q is the outgoing momentum of the external photon. The electromagnetic form factors f_Q^{jk}, f_A^{jk}, f_M^{jk} and f_E^{jk}, are respectively named *charge form factor*, *anapole form factor*, *magnetic form factor*, and *electric form factor*. Those form factors with $j \neq k$ are called transition form factors, whereas those in which $j = k$ are known as diagonal form factors. Upon taking the photon on shell, which comes along with the condition $q^2 = 0$, such form factors define the electromagnetic moments. On general grounds [14, 24] and at the phenomenological level [26] it has been shown that the only nonzero electromagnetic form factor of a Majorana neutrino is the anapole form factor. This differs from the case of Dirac neutrinos, whose diagonal electromagnetic form factors can all be nonzero. In both cases, all transition form factors are allowed to exist [24].

Magnetic moments are the most widely studied static electromagnetic properties ($q^2 = 0$) of neutrinos. They have been calculated in the *minimally extended Standard Model* [27], in which the Standard Model is endowed with sterile right-handed neutrinos, yielding nonzero Dirac neutrino masses [3]. In this context, magnetic moments are of order $10^{-19}\mu_{\rm B}$ [24], where $\mu_{\rm B}$ is the Bohr magneton. Among a variety of searches for neutrino magnetic moments, the GEMMA experiment, which uses a Germanium detector that is placed near a reactor in the Kalinin Nuclear Power Plant, has established the best bound so far [28]: $\mu_\nu < 2.9 \times 10^{-11}\mu_{\rm B}$, at 90% C.L. The huge difference among the experimental sensitivity and the theoretical prediction of the minimal Standard Model extension, not likely to improve greately in the short term [25], has been a motivation to explore different extensions of the Standard Model that generate large neutrino magnetic moments, with hope in measuring a large neutrino magnetic moment and thus evidence of new physics and its proper description. For instance, calculations of neutrino magnetic moments have been performed in the 331 model [29, 30, 31], Supersymmetry [32, 33], and in left-right extensions [34, 35, 36, 37, 38, 39].

4. Majorana neutrino magnetic moments in left-right

The defining characteristic of left-right models is the gauge group $\mathrm{SU}(2)_L \times \mathrm{SU}(2)_R \times \mathrm{U}(1)_{B-L}$ [34, 35, 36], which is an extension of the electroweak group of the Standard Model. Originally meant to describe parity violation as a remnant of a spontaneously broken high-energy symmetry, a variety of versions of left-right models have been realized and explored. In particular, models with nonmanifest left-right symmetry, in which left and right coupling constants are assumed to be different of each other ($g_L \neq g_R$), are interesting because they keep the lower bound on the mass of heavy charged gauge bosons around the relatively light value 0.3 TeV [40], in contrast with minimal versions, in which lower bounds on such masses range within the TeV scale [41, 42]. A feature that distinguishes different left-right formulations is the scalar sector [37]. While scalar sectors made of Higgs doublets yield Dirac neutrinos, much attention has been paid to scalar sectors that involve scalar triplets, since they lead to Majorana neutrinos. Tipically, in these models there are two scalar triplets, Δ_L and Δ_R, and one scalar bidoublet, Φ. In a first stage of symmetry breaking, the right triplet acquires a vaccum expectation value $\langle \Delta_R \rangle$ which is characterized by a high-energy scale v_R. This breaks the left-right symmetry group into the Standard Model electroweak symmetry, which is then spontaneously broken by the vacuum expectation values $\langle\Delta_L\rangle$ and $\langle\Phi\rangle$, of the remaining scalar structures, into the electromagnetic group. Then, a *seesaw mechanism* [36] drives the definition of light and heavy Majorana neutrino masses.

The enlargement of the Standard Model electroweak symmetry group, by the addition of an SU(2), introduces new gauge degrees of freedom. Among the new fields, an $\mathrm{SU}(2)_R$ right charged boson W_R is defined, which mixes, by an angle ζ, with the left boson W_L, corresponding to the left group $\mathrm{SU}(2)_L$ to define two charged bosons, W_1 (light) and W_2 (heavy), with definite masses:

$$W_L^{\pm\mu} = \cos\zeta\, W_1^{\pm\mu} - \sin\zeta\, W_2^{\pm\mu}, \tag{7}$$
$$W_R^{\pm\mu} = e^{i\omega}\left(\sin\zeta\, W_1^{\pm\mu} + \cos\zeta\, W_2^{\pm\mu}\right), \tag{8}$$

where $e^{i\omega}$ is a complex phase. Such W bosons couple to neutrinos and charged leptons through charged currents [39]

$$\sum_{a=1,2}\sum_{j=1,2,3}\sum_{\alpha=e,\mu,\tau}\left[W^+_{a\,\mu}\overline{\nu}_j\gamma^\mu(v_{a,j\alpha}-a_{a,j\alpha}\gamma_5)l_\alpha + W^-_{a\,\mu}\overline{l}_\alpha\gamma^\mu(v^*_{a,j\alpha}-a^*_{a,j\alpha}\gamma_5)\nu_j\right], \tag{9}$$

with the coefficients $v_{a,j\alpha}$ and $a_{a,j\alpha}$ given in terms of left and right neutrino mixing matrices[4], which are respectively denoted by $\mathcal{L}_{\alpha j}$ and $\mathcal{R}_{\alpha j}$. These factors also involve the left coupling constant, g_L, and the right one, g_R, as well, with the assumption $g_L \neq g_R$, mentioned above. Finally, the coefficients also incorporate the angles ζ and ω. In the context of a left-right model with scalar triplets, the contributions of these currents to the Majorana neutrino transition magnetic moments are generated by the diagrams shown in Fig. 1 [5], but dominant contributions

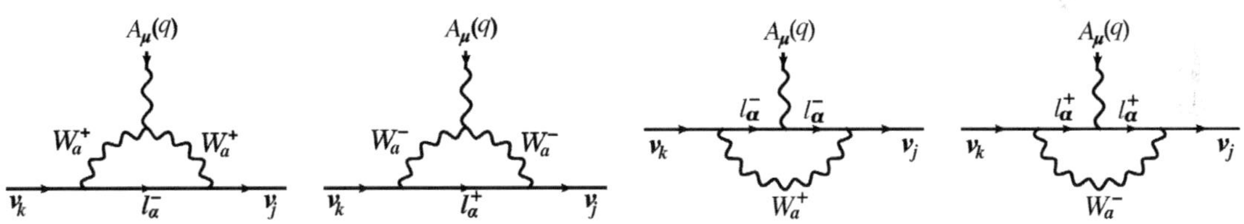

Figure 1. Diagrams contributing to anomalous magnetic moments at one loop.

come only from the first two diagrams. The contributions to the magnetic dipole form factors can be written as [39]

$$f_M^{jk}(q^2) = \frac{m_e}{8\pi^2}\mu_B\sum_{j=1}^{3}\sum_{a=1,2}\left[(a_{a,j\alpha}\,a^*_{a,k\alpha} - a_{a,k\alpha}\,a^*_{a,j\alpha})I_1 + (v_{a,j\alpha}\,v^*_{a,k\alpha} - v_{a,k\alpha}\,v^*_{a,j\alpha})I_2\right]. \tag{10}$$

In this expression the factors I_1 and I_2 have been used to denote complicated parametric integrals, arisen because of the use of the technique of Feynman parameters. From Eq. (10), it can be appreciated that diagonal magnetic form factors vanish exactly, leaving only transition form factors, which is consistent with the properties of Majorana neutrinos. Note that the magnetic dipole form factors are exactly zero if the coefficients $a_{a,j\alpha}$ and $v_{a,j\alpha}$ are real quantities.

[4] In Ref. [39], massive neutrinos were denoted by greek indices $\alpha, \beta \ldots$ and lepton flavors were labeled by lowercase indices $j, k, \ldots$. In the present paper we have swapped these labels, so $\alpha, \beta, \ldots$ denote lepton flavors and $j, k, \ldots$ denote neutrino mass eigenstates.

[5] In the case of Dirac neutrinos, only the first and the third diagrams of Fig. 1 would exist.

Such situation would correspond to charged currents that preserve CP invariance [39]. So, the existence of these neutrino magnetic moments requires violation of CP. It must be emphasized that this result is gauge dependent, since so far the photon field has been assumed to be off shell. Indeed, a full calculation of these form factors would require scalar contributions and pseudo-Goldstone bosons contributions as well. Taking the photon on shell and using the unitary gauge reduces the number of diagrams and simplifies the expressions.

5. Magnetic moments and Majorana phases in left-right

Assuming that the left and right mixings are equal, that is $\mathcal{L} = \mathcal{R}$, a scenario of *maximal right mixing* is defined [39]. Considering a heavy W mass of few TeVs, the magnetic moment is expressed as [39]

$$\begin{aligned} |\mu_{jk}| \lesssim\ & \mu_{\rm B}(4\times 10^{-11}\,{\rm GeV}^{-1})|m_{1,jk}\sin\phi_{jk} + (m_{2,jk} - m_{2,kj})\sin\delta\cos\phi_{jk} \\ & +(m_{2,jk} + m_{2,kj})\cos\delta\sin\phi_{jk}|, \end{aligned} \tag{11}$$

which includes the Majorana phase differences $\phi_{jk} = (\varphi_j - \varphi_k)/2$, with $\varphi_1 = 0$, and the Dirac phase δ. Moreover, the factors $m_{r,jk}$, whose definitions are given in Ref. [39], involve the masses of the charged leptons and the mixing angles θ_{12}, θ_{23}, and θ_{13} as well. These leading contributions are independent of small neutrino masses, which has been noticed and reported in different investigations [37, 43]. It is worth mentioning that any magnetic moment given by Eq. (11) depends on just one Majorana phase difference ϕ_{jk}. To illustrate the expression given in Eq. (11), we provide the graph in the left of Fig. 2, which shows the magnetic moment $|\mu_{21}|$ as a function of the Majorana phase difference ϕ_{12}. The region colored in light green represents all the possible values of this Majorana neutrino magnetic moment, and the solid curve in the boundary of this region is the upper bound. To plot this function, we have used the value $\delta = -\pi/2$ for the Dirac phase (there is CP violation), as suggested by the results of the T2K Collaboration [23]. As it can be appreciated from this figure, the maxima of the neutrino magnetic moments lie around $\sim 10^{-11}\mu_{\rm B}$. The horizontal red dashed line in this graph represents what the contribution would be if neutrinos were Dirac instead of Majorana. The zeros of the three transition magnetic moments correspond to the following Majorana phases [39]:

$$|\mu_{21}| \approx 0, \qquad \varphi_2 \approx -36.86^{\rm o}, \tag{12}$$

$$|\mu_{21}| \approx 0, \qquad \varphi_3 \approx -31.98^{\rm o}, \tag{13}$$

$$|\mu_{32}| \approx 0, \qquad \phi_{32} \approx -172.87^{\rm o},\ 7.13^{\rm o}. \tag{14}$$

Using Eqs. (12) and (13), note that $\phi_{32} \approx 2.44^{\rm o}$, which does not match Eq. (14). This means that a situation in which the three magnetic moments are zero cannot occur: in presence of CP violation, there is at least one large nonzero neutrino magnetic moment.

Another scenario, named the *CKM-like mixing* [39], is achieved by assuming that the mixing angles are very small, so that the neutrino right mixing resembles the pattern of the CKM matrix. Again, by taking the unitary gauge and considering a mass of the heavy charged boson that is in the range of a few TeVs, the Majorana neutrino magnetic moments can be expressed as

$$|\mu_{jk}| \lesssim \mu_{\rm B}(2\times 10^{-11}\,{\rm GeV}^{-1})(|c_{e,jk}| + |c_{\mu,jk}| + |c_{\tau,jk}|), \tag{15}$$

where the coefficients $c_{\alpha,jk}$, explicitly given in Ref. [39], depend on the charged lepton masses, on the Dirac phase and on Majorana phase differences $\phi_{jk} = (l_j - r_k)/2$, where l_j and r_k denote left and right Majorana phases, respectively. In contrast with the scenario of maximal mixing, each magnetic moment depends on two Majorana phase differences, and not just on one. In each case, one of such phase differences is dominant in the sense that they appear in separate

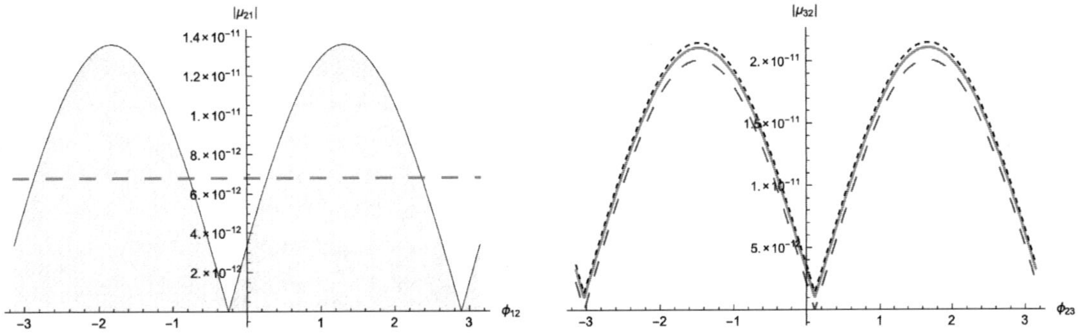

Figure 2. Magnetic moments as functions of Majorana phases: (1) maximal right mixing (plot in the left); (2) CKM-like right mixing (plot in the right).

terms and each of such terms depend multiplicatively on one charged lepton mass. The size of the mass corresponding to each phase difference then determines which one has a greater impact on the numeric result. This behavior is better noticed in the case of the magnetic moment $|\mu_{32}|$, which is illustrated by the right graph of Fig. 2. Three plots of $|\mu_{32}|$, as a function of ϕ_{23}, have been added, each one corresponding to a different phase difference ϕ_{32}: $\phi_{32} = \pi/4$ for the solid magenta curve; $\phi_{32} = \pi/2$ for the dashed black curve, with small dashes; and $\phi_{32} = 0$ for the dashed blue curve, with large dashes. Just as in the previous scenario, maxima are located around $10^{-11}\mu_{\rm B}$. For $\delta = -\pi/2$ [23], the zeros of these magnetic moments are achieved at the left and right Majorana phases that we show below:

$$|\mu_{21}| \approx 0, \qquad r_2 \approx -24^{\rm o}, \qquad l_2 = 0^{\rm o}, \tag{16}$$

$$|\mu_{31}| \approx 0, \qquad r_3 \approx 28.82^{\rm o}, \qquad l_3 = 180^{\rm o}, \tag{17}$$

$$|\mu_{32}| \approx 0, \quad \phi_{23} \approx 6.40^{\rm o}, -173.60^{\rm o}, \quad \phi_{32} = 0^{\rm o}, 180^{\rm o}. \tag{18}$$

According to Eqs. (16) and (17), we have the values $\phi_{23} = (l_2 - r_3)/2 \approx -14.41^{\rm o}$ and $\phi_{32} = (l_3 - r_2)/2 \approx 102^{\rm o}$, which are not in agreement with Eq. (18). Therefore, as in the other scenario that we discussed, it is concluded that violation of CP enforces the presence of at least one large magnetic moment, while two of such contributions are allowed to vanish.

Acknowledgments
We acknowledge financial support from SNI and CONACYT (México).

References
[1] Cowan Jr C L, Reines F, Harrison F B, Kruse H W and McGuire A D 1956 *Science* **124** 103-104
[2] Pontecorvo B 1957 *Sov. Phys. JETP* **6** 429
[3] Giunti C and Kim C W 2007 *Fundamentals of Neutrino Physics and Astrophysics* (New York: Oxford University Press)
[4] Sakurai J J and Napolitano J 2011 *Modern Quantum Mechanics* (San Francisco: Addison-Wesley)
[5] Petcov S T 2014 *International Journal of Modern Physics* A **29** 1430028
[6] Davis R, Harmer D S and Hoffman K C 1968 *Phys. Rev. Lett.* **20** 1205-9
[7] Cleveland B T, Daily T, Davis Jr , Distel J R, Lande K, Lee C K, Wildenhain P S and Ullman J 1998 *Astrophys. J.* **496** 505-26
[8] Bahcall J N, Serenelli A M and Basu S 2005 *Astrophys. J.* **621** L85-8
[9] Fukuda Y *et al* (Super-Kamiokande Collaboration) 1998 *Phys. Rev. Lett.* **81** 1562-67
[10] Ahmad Q R *et al* (SNO Collaboration) 2002 *Phys. Rev. Lett.* **89** 011301
[11] An F P *et al* 2012 *Phys. Rev. Lett.* **108** 171803
[12] Ahn J K (RENO Collaboration) 2012 *Phys. Rev. Lett.* **108** 191802

[13] Majorana E 1937 *Nuovo Cim.* **14** 171-184
[14] Schechter J and Valle J W F 1981 *Phys. Rev.* D **24** 1883
[15] Klapdor-Kleingrothaus H V *et al* 2004 *Phys. Lett.* B **586** 198
[16] Gando A *et al* (KamLAND-Zen Collaboration) 2013 *Phys. Rev. Lett.* **110** 062502
[17] Olive K A *et al* (Particle Data Group) 2015 *Chin. Phys.* C **38** 090001
[18] Stanco L 2016 *Rev. Phys.* **1** 90-100
[19] Abe K *et al* (T2K Collaboration) 2015 *Phys. Rev.* D **91** 072010
[20] Renshaw A (for the Super-Kamiokande Collaboration) 2015 *Phys. Procedia* **61** 345-54
[21] An F *et al* 2015 *J. Phys.* G **43** 300401
[22] An F P *et al* (Daya Bay Collaboration) 2015 *Phys. Rev. Lett.* **115** 111802
[23] Abe K *et al* (T2K Collaboration) 2014 *Phys. Rev. Lett.* **112** 061802
[24] Broggini C, Giunti C and Studenikin A 2012 *Adv. High Energy Phys.* **2012** 459526
[25] Giunti C, Kouzakov K A, Li Y F, Lokhov A V, Studenikin A I and Zhou S 2015 *Annalen Phys.* **528** 198-215
[26] Kayser B 1982 *Phys. Rev.* D **26** 1662-70
[27] Dvornikov M and Studenikin A 2004 *Phys. Rev.* D **69** 073001
[28] Beda A G, Brudanin V B, Egorov V G, Medvedev D V, Pogosov V S, Shirchenko M V and Starostin A S 2012 *Adv. High Energy Phys.* **2012** 350150
[29] Pisano F and Pleitez V 1992 *Phys. Rev.* D **46** 410-7
[30] Frampton P H 1992 *Phys. Rev. Lett.* **69** 2889-91
[31] Montero J C and Pleitez V 2001 *Mod. Phys. Lett.* A **16** 1605-14
[32] Góźdź M, Kamiński W A and Šimkovic 2006 *Int. J. Mod. Phys.* E **15** 441-5
[33] Góźdź M, Kamiński W A, Šimkovic and Faessler A 2006 *Phys. Rev.* D **74** 055007
[34] Pati J C and Salam A 1974 *Phys. Rev.* D **10** 2775
[35] Mohapatra R N and Pati J C 1975 *Phys. Rev.* D **11** 566
[36] Senjanović G and Mohapatra R N 1975 *Phys. Rev.* D **12** 1502
[37] Czakon M, Gluza J and Zralek M 1998 *Phys. Rev.* D **59** 013010
[38] Boyarkin O M and Boyarkina G G 2014 *Phys. Rev.* D **90** 025001
[39] Delepine D and Novales-Sánchez H 2015 *Phys. Rev.* D **92** 095016
[40] Olness F I and Ebel M E 1984 *Phys. Rev.* D **30** 1034-40
[41] Zhang Y, An H, Ji X, and Mohapatra R N 2008 *Nucl. Phys.* B **802** 247-79
[42] Nemevšek M, Nesti F, Senjanović G and Zhang Y 2011 *Phys. Rev.* D **83** 115014
[43] Bég M A B and Maciano W J 1978 *Phys. Rev.* D **17**1395-401

Matter neutrino oscillations, an approximation in a parametrization-free framework

L. J. Flores [1*] **and O. G. Miranda** [1†]

[1] Departamento de Física, Centro de Investigación y de Estudios Avanzados del IPN
Apdo. Postal 14-740 07000 Mexico, DF, Mexico

E-mail: *jflores@fis.cinvestav.mx, †omr@fis.cinvestav.mx

Abstract. Neutrino oscillations are one of the most studied and successful phenomena since the establishment of the solar neutrino problem in late 1960's. In this work we discuss the exact expressions for the probability $P_{\alpha\beta}$ in a constant density medium, in terms of the standard vacuum parameters and the medium density. Besides of being compact, these expressions are independent of any particular parametrization, which could be helpful in the application of unitary tests of the mixing matrix. In addition, we introduce a new approximation on $P_{\alpha\beta}$ and compare it with the most commonly used, discussing their main differences.

1. Introduction

Since the confirmation of the oscillation of neutrinos in 1998 by the Super-Kamiokande collaboration [1], a considerable amount of analysis has been made concerning the measurement of the parameters involved in the oscillation [2]. In the standard three families oscillation scheme, four parameters appear in the mixing matrix: three mixing angles and a CP violation phase. The former have been measured with great accuracy; however, the later had not been properly constrained. Several parameterizations for the leptonic mixing matrix exist in the literature [3, 4, 5, 6, 7]. They are suited for different purposes; for example, to investigate the nature of the neutrino (Dirac or Majorana), or to probe extra neutrino states in oscillation experiments [5, 6]. Most of the analyses have been done with the Particle Data Group parametrization [7], but if we are looking to perform a unitary test [8], it is desirable to use a formulation as independent of the parametrization as possible. Moreover, to obtain precise results, it is necessary to take into account the matter effects on the oscillation [9]. The next generation of experiments is looking to use this effect in future analysis. Therefore, the analysis gets more complicated, requesting the use of computational solutions or approximated expressions [10, 11, 12, 13, 14, 15, 16].

In the first section of this work we give exact expressions for the oscillation probability $P_{\alpha\beta}$ in a constant density medium, which are formulated without a particular parametrization; these will come in handy for future unitary tests. Additionally, in the second section we introduce new approximated and simple expressions, which can be defined to a desired degree of accuracy due to their series expansion nature. Finally, we compare this approximation with the most currently used and other two found in the literature. It is important to notice that this work is based on [17].

2. The exact case

The evolution equation for a neutrino ν_j with definite mass is given by

$$i\frac{d}{dt}\nu_j = \frac{m_j^2}{2E}\nu_j. \tag{1}$$

These mass states are related to the flavor states ν_α through the leptonic mixing matrix, U, via $\nu_\alpha = U_{\alpha j}\nu_j$, where a sum over j is understood. The mixing matrix can be parametrized depending on the context of the study. From (1) one can calculate the known vacuum oscillation probability, from ν_α to ν_β. If extra heavy neutrinos exist, they can not participate in the oscillation because the energy will not be sufficient, but the unitarity of U will be clearly affected. The 3×3 mixing matrix responsible of the oscillation is now just a block of the whole $n \times n$ matrix U. This effect changes the usual 3×3 expression into

$$P_{\nu_\alpha\to\nu_\beta} = \sum_{\ell,j}^{3} U_{\alpha\ell}^{*}U_{\beta\ell}U_{\alpha j}U_{\beta j}^{*} - 4\sum_{\ell>j}^{3}\mathrm{Re}\left[U_{\alpha\ell}^{*}U_{\beta\ell}U_{\alpha j}U_{\beta j}^{*}\right]\sin^2\left(\frac{\Delta m_{\ell j}^2 L}{4E}\right)$$
$$+2\sum_{\ell>j}^{3}\mathrm{Im}\left[U_{\alpha\ell}^{*}U_{\beta\ell}U_{\alpha j}U_{\beta j}^{*}\right]\sin\left(\frac{\Delta m_{\ell j}^2 L}{2E}\right). \tag{2}$$

Notice that now there is zero distance effect due to the non-unitary mixing matrix. Additionally, if matter effects are to be considered, we need to add the charge-current potential V_{CC} to the evolution equation in (1) in the form of $A = 2EV_{CC}$:

$$i\frac{d}{dt}\nu_j = \frac{1}{2E}\left(m_j^2\nu_j + \sum_k AU_{ej}^{*}U_{ek}\nu_k\right). \tag{3}$$

As usual, if we consider unitarity in U, the expression for the probability in matter holds the same functional form as in vacuum; with the replacement of $\Delta m_{\ell j}^2$ by the effective mass difference, $\Delta M_{\ell j}^2$, and the replacement of U by the effective mixing, V, that is related to the vacuum expression through

$$V = UW^T, \tag{4}$$

where W is the unitary matrix diagonalizing the Hamiltonian in the right-hand side of equation (3). One of the purposes of this work is to express the matter probability as a function entirely of the vacuum parameters and the potential. Therefore our problem reduces to find the matrix W. This can be done following the procedure of [11] but with the difference that we are leaving explicitly the elements $U_{\alpha i}$, in other words, without a specific parametrization. The Hamiltonian

$$H_M = \begin{pmatrix} A|U_{e1}|^2 & AU_{e1}^{*}U_{e2} & AU_{e1}^{*}U_{e3} \\ AU_{e2}^{*}U_{e1} & \Delta m_{21}^2 + A|U_{e2}|^2 & AU_{e2}^{*}U_{e3} \\ AU_{e3}^{*}U_{e1} & AU_{e3}^{*}U_{e2} & \Delta m_{31}^2 + A|U_{e3}|^2 \end{pmatrix}, \tag{5}$$

has the characteristic polynomial form

$$\lambda^3 - \alpha\lambda^2 + \beta\lambda - \gamma = 0, \tag{6}$$

with

$$\begin{aligned} \alpha &= \Delta m_{21}^2 + \Delta m_{31}^2 + A(|U_{e1}|^2 + |U_{e2}|^2 + |U_{e3}|^2) \\ \beta &= \Delta m_{31}^2\Delta m_{21}^2 + A\Delta m_{21}^2(|U_{e1}|^2 + |U_{e3}|^2) + A\Delta m_{31}^2(|U_{e1}|^2 + |U_{e2}|^2) \\ \gamma &= A\Delta m_{21}^2\Delta m_{31}^2|U_{e1}|^2 \\ \eta &= \cos\left[\frac{1}{3}\arccos\left(\frac{2\alpha^3 - 9\alpha\beta + 27\gamma}{2\sqrt{(\alpha^2 - 3\beta)^3}}\right)\right]. \end{aligned} \tag{7}$$

Here we have left indicated on purpose the expressions involving the condition of U been unitary, to show how these expressions can be easily changed in the case of a non-unitary study. Now we can write the squared effective masses in terms of the previous coefficients

$$\begin{aligned} M_1^2 \equiv \lambda_1 &= \frac{\alpha}{3} - \frac{1}{3}\sqrt{\alpha^2 - 3\beta}\eta - \frac{\sqrt{3}}{3}\sqrt{\alpha^2 - 3\beta}\sqrt{1-\eta^2}, \\ M_2^2 \equiv \lambda_2 &= \frac{\alpha}{3} - \frac{1}{3}\sqrt{\alpha^2 - 3\beta}\eta + \frac{\sqrt{3}}{3}\sqrt{\alpha^2 - 3\beta}\sqrt{1-\eta^2}, \\ M_3^2 \equiv \lambda_3 &= \frac{\alpha}{3} + \frac{2}{3}\sqrt{\alpha^2 - 3\beta}\eta. \end{aligned} \quad (8)$$

The columns that constitute the diagonalizing matrix W correspond to the eigenvectors of H_M. A suitable choice of them gives us

$$(W^T)_{kj} = \frac{\Lambda_k}{C_k}\delta_{kj} + (1-\delta_{kj})A\frac{U_{ek}U_{ej}^*\left(M_k^2 - \sum_i[\Delta m_{i1}^2 \epsilon_{ijk}^2]\right)}{C_k}, \quad (9)$$

where we have defined

$$\Lambda_j = M_j^4 - \sum_{i\neq j}\left[M_j^2\left(\Delta m_{i1}^2 + A|U_{ei}|^2\right) - A\Delta m_{i1}^2|U_{ek}|^2 - \frac{1}{2}\Delta m_{i1}^2\Delta m_{k1}^2\right], \quad \text{for} \quad k \neq i, \quad (10)$$

and a normalization constant, C_j, as

$$C_j = \sqrt{\Lambda_j^2 + A^2|U_{ej}|^2\sum_{i\neq j}|U_{ei}|^2(M_j^2 - \Delta m_{k1}^2)^2}, \quad \text{for} \quad k \neq i. \quad (11)$$

With the help of the relation (4) we are able to write down a compact expression, relating the entries of the effective matter mixing matrix V:

$$V_{\beta j} = \frac{\Lambda_j}{C_j}U_{\beta j} + A\sum_{k\neq j}\frac{U_{\beta k}U_{ei}U_{ej}^*\left(M_k^2 - \sum_i[\Delta m_{i1}^2\epsilon_{ijk}^2]\right)}{C_k}. \quad (12)$$

It is easy to note that this expression reduces to the vacuum case when $A = 0$. This relation is useful in the description of neutrino matter oscillation probabilities as a function of the vacuum parameters, without a particular choice of parametrization for the leptonic mixing matrix. The unitarity of U is not assumed. That makes it useful in case a unitary test would be implemented.

3. Approximated expressions

Now that we have presented the exact formula for the neutrino oscillation probability, free of any parametrization, we will start with the process of deducing an approximated formula. Although there exist several approximations in the literature, our expression will stick to the purpose of a formulation without parametrization. To accomplish this, we are looking to approximate the effective masses appearing in (8).

First, we notice that $\Delta m_{21}^2 = 0$ implies $\gamma = 0$, and the polynomial in (6) reduces to a quadratic form. If this is the case, we have:

$$\eta = \cos\left[\frac{1}{3}\arccos\left(\frac{2\alpha^3 - 9\alpha\beta}{2\sqrt{(\alpha^2-3\beta)^3}}\right)\right], \quad (13)$$

which reduces to

$$\eta = \cos\theta = \frac{-\frac{1}{2}\alpha}{\sqrt{\alpha^2 - 3\beta}}. \tag{14}$$

Now we go to the real case: it is well known that $\Delta m^2_{21} \ll \Delta m^2_{31}$, therefore, γ is different from zero and $\gamma \ll \alpha\beta$. We suggest a correction ε that needs to fulfill

$$\cos\theta = \frac{-\frac{1}{2}\alpha + \varepsilon}{\sqrt{\alpha^2 - 3\beta}} \tag{15}$$

and

$$\cos 3\theta = 4\cos^3\theta - 3\cos\theta \simeq \frac{2\alpha^3 - 9\alpha\beta + 27\gamma}{2\sqrt{(\alpha^2 - 3\beta)^3}}. \tag{16}$$

Working up to first-order terms in γ/β is easy to find that $\varepsilon = \frac{3\gamma}{2\beta}$ fulfills both conditions (15) and (16). The eigenvalues in (8) are then approximated to

$$\begin{aligned}
M_1^2 \equiv \lambda_3 &\simeq \frac{2}{3}\varepsilon. \\
M_2^2 \equiv \lambda_1 &\simeq \frac{1}{2}\left(\alpha - \frac{2}{3}\varepsilon\right) - \frac{1}{2}\sqrt{\left(\alpha + \frac{2}{3}\varepsilon\right)^2 - 4\left[\beta + \left(\frac{2}{3}\varepsilon\right)^2\right]}, \\
M_3^2 \equiv \lambda_2 &\simeq \frac{1}{2}\left(\alpha - \frac{2}{3}\varepsilon\right) + \frac{1}{2}\sqrt{\left(\alpha + \frac{2}{3}\varepsilon\right)^2 - 4\left[\beta + \left(\frac{2}{3}\varepsilon\right)^2\right]},
\end{aligned} \tag{17}$$

If we substitute these back in the polynomial in (6), it takes the form:

$$\lambda^3 - \alpha\lambda^2 + \beta\lambda - \left(\frac{2}{3}\beta\varepsilon - \frac{4}{9}\alpha\varepsilon^2 + \frac{8}{27}\varepsilon^3\right) = 0. \tag{18}$$

We want to go beyond this and find an expression for ε at second order in γ/β. Therefore, we propose

$$\varepsilon = \frac{3\gamma}{2\beta} + a_2\left(\frac{\gamma}{\beta}\right)^2. \tag{19}$$

Substituting this in (18), and demanding that returns to the form in (6) up to second-order terms, throws the result $a_2 = \frac{3}{2}\frac{\alpha}{\beta}$. Hence, we have

$$\varepsilon = \frac{3\gamma}{2\beta} + \frac{3\alpha}{2\beta}\frac{\gamma^2}{\beta^2}. \tag{20}$$

We can go further with this method and find the coefficients a_k in a recursive way. This states that we can write the correction ε as an infinite series in powers of γ/β and, in principle, obtain the exact case again. The correction then takes the form

$$\varepsilon = \sum_{k=1}^{\infty} a_k(\alpha,\beta)\left[\frac{\gamma}{\beta}\right]^k \tag{21}$$

with

$$a_1 = \frac{3}{2}, \qquad a_k(\alpha,\beta) = \frac{3}{2\beta}\left(\frac{4\alpha}{9}\sum_{\substack{i,j\\ i+j=k}} a_i a_j - \frac{8}{27}\sum_{\substack{i,j,l\\ i+j+l=k}} a_i a_j a_l\right);\ k \geq 2. \tag{22}$$

Now that we have defined the approximation, we would like to test its accuracy by comparing it against the exact formula and also to other known approximations: by Akhmedov [14] and Minakata & Parke [15]. To do so, we first assume unitarity in the mixing matrix and adopt the standard parametrization. The central values for the mixing parameters are the reported ones in [2] ($\sin^2\theta_{12} = 0.320$, $\sin^2\theta_{23} = 0.613$, $\sin^2\theta_{13} = 0.0246$) along with the squared mass differences ($\Delta m^2_{21} = 7.62\times 10^{-5}$ eV^2, $\Delta m^2_{31} = 2.55\times 10^{-3}$ eV^2). We have taken $\delta = 3\pi/2$ as the value of the CP phase. We computed the oscillation probabilities P_{ee} and $P_{\mu e}$ as function of the baseline at an energy of 1 GeV. In Fig. (1) we compare our result with two other approximations from [14, 15]; we can see that our result is competitive with the others, in particular for a small distance (short-baseline experiments). The precision between different orders of our approximation is shown in Fig. (2). Let us notice that, at first order approximation, our result works well, especially for a short baseline and a great development is achieved for higher orders.

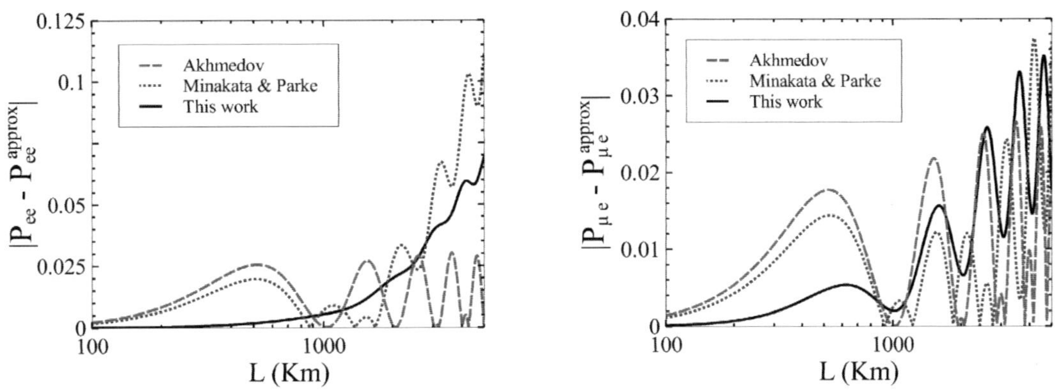

Figure 1. Comparison between the exact neutrino oscillation probability in matter and some approximations. The absolute difference for the electron neutrino survival probability P_{ee} (left panel) and the the muon to electron probability $P_{\mu e}$ (right panel) are plotted against the distance L at an energy of $E_\nu = 1$ GeV. The central values of the mixing matrix parameters and the square mass differences are taken from [2] and an electron density of $N_e = 5.92 \times 10^9$ eV^3 has been assumed. The dashed magenta curve accounts for the approximation by Akhmedov [14] as well as the dotted blue curve stands for the expression from Minakata & Parke [15]. The solid curve represents the difference with our approximation at first order.

4. Conclusions

In this work, we studied the scenario of the oscillation of three neutrino families propagating through a constant potential caused by matter. First we reconsidered the exact expressions for the oscillation probability, but without any parametrization. Additionally, we found an approximation for these expressions, using the fact that $\Delta m^2_{21} \ll \Delta m^2_{31}$ and keeping thus the parametrization-free scheme. We found that this result could be expanded to any order in a series expansion, making simple the choice of required precision. Our result is competitive with the ones found in the literature, and its value goes closer to the exact case for short baselines.

Acknowledgments

This work has been supported by the CONACyT Grant No. 166639 (Mexico).

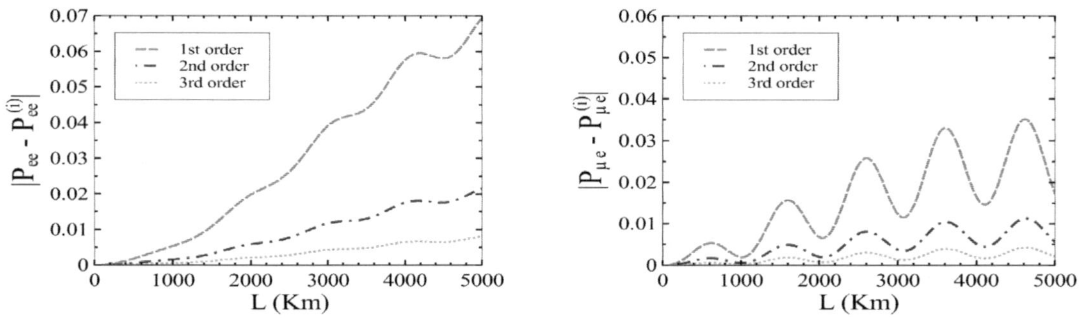

Figure 2. Comparison between the exact formula and our approximation for the neutrino oscillation probability. The absolute differences for P_{ee} (left panel) and $P_{\mu e}$ (right panel) are plotted against the distance L, at an energy of $E_\nu = 1$ GeV. The dashed, dash-dotted and dotted line correspond to the first, second and third order approximation, respectively.

References

[1] Y. Fukuda *et al.* [Super-Kamiokande Collaboration], Phys. Rev. Lett. **81**, 1562 (1998) doi:10.1103/PhysRevLett.81.1562 [hep-ex/9807003].
[2] D. V. Forero, M. Tortola and J. W. F. Valle, Phys. Rev. D **90**, 093006 (2014) [arXiv:1405.7540 [hep-ph]].
[3] J. Schechter and J. W. F. Valle, Phys. Rev. D **22**, 2227 (1980).
[4] W. Rodejohann and J. W. F. Valle, Phys. Rev. D **84**, 073011 (2011) [arXiv:1108.3484 [hep-ph]].
[5] F. J. Escrihuela, D. V. Forero, O. G. Miranda, M. Tortola and J. W. F. Valle, Phys. Rev. D **92**, 053009 (2015) doi:10.1103/PhysRevD.92.053009 arXiv:1503.08879 [hep-ph].
[6] A. de Gouva and A. Kobach, arXiv:1511.00683 [hep-ph].
[7] K. A. Olive *et al.* [Particle Data Group Collaboration], Chin. Phys. C **38**, 090001 (2014).
[8] S. Parke and M. Ross-Lonergan, arXiv:1508.05095 [hep-ph].
[9] L. Wolfenstein, Phys. Rev. D **17**, 2369 (1978). S. P. Mikheev and A. Y. .Smirnov, Nuovo Cimento Soc. Ital. Fis. C **9**, 17 (1986).
[10] V. D. Barger, K. Whisnant, S. Pakvasa and R. J. N. Phillips, Phys. Rev. D **22**, 2718 (1980).
[11] H. W. Zaglauer and K. H. Schwarzer, Z. Phys. C **40**, 273 (1988).
[12] A. Cervera, A. Donini, M. B. Gavela, J. J. Gomez Cadenas, P. Hernandez, O. Mena and S. Rigolin, Nucl. Phys. **B579**, 17 (2000) [Nucl. Phys. **B593**, 731 (2001)] doi:10.1016/S0550-3213(00)00221-2 [hep-ph/0002108].
[13] M. Freund, Phys. Rev. D **64**, 053003 (2001) [hep-ph/0103300].
[14] E. K. Akhmedov, R. Johansson, M. Lindner, T. Ohlsson and T. Schwetz, J. High Energy Phys. 04 (2004) 078. [hep-ph/0402175].
[15] H. Minakata and S. J. Parke, arXiv:1505.01826 [hep-ph].
[16] K. Asano and H. Minakata, J. High Energy Phys. 06 (2011) 022 doi:10.1007/JHEP06(2011)022 [arXiv:1103.4387 [hep-ph]].
[17] L. J. Flores and O. G. Miranda, Phys. Rev. D **93**, no. 3, 033009 (2016) doi:10.1103/PhysRevD.93.033009 [arXiv:1511.03343 [hep-ph]].

Mrk 421: The Multi-TeV emission and its astrophysical origin

Sarira Sahu[1], Luis Salvador Miranda[1] and Subhash Rajpoot[2]

[1] Instituto de Ciencias Nucleares, Universidad Nacional Autónoma de México, Circuito Exterior, C.U., A. Postal 70-543, 04510 Mexico DF, Mexico

[2] Department of Physics & Astronomy, California State University, 1250 Bellflower Boulevard, Long Beach, CA, 90840, USA

E-mail: sarira@nucleares.unam.mx, luis.miranda@correo.nucleares.unam.mx, Subhash.Rajpoot@csulb.edu

Abstract. The TeV blazar Markarian 421 underwent multi-TeV flaring during April 2004 and simultaneously observations in the X-ray and TeV energies were made. It was observed that the TeV outbursts had no counterparts in the lower energy range. One implication of this is that it might be an orphan flare. We show that Fermi-accelerated protons of energy ≤ 168 TeV can interact with the low energy tail of the background synchrotron self-Compton photons in the inner region of the blazar to produce the multi-TeV flare and our results fit very well with the observed spectrum.

1. Introduction

Active galactic nuclei (AGN) emit electromagnetic radiation, the spectrum of which stretches from radio to gamma-rays and exhibit large luminosity variations on time scales ranging from less than an hour up to several years. A super massive black hole is believed to sit at the center of the AGN surrounded by an accretion disk in the inner region and a torus of gas cloud in the outer region. Oppositely directed relativistic jets are ejected from the AGN which are perpendicular to the accretion disk and the torus. The spectral energy distribution (SED) of these AGN have a double peak structure in the $\nu - \nu F_\nu$ plane (frequency vs frequency times flux). In the context of the leptonic model, the low energy peak corresponds to the synchrotron radiation from a population of relativistic electrons in the jet. Although the general consensus is that the high energy peak corresponds to the synchrotron self Compton (SSC) scattering of the high energy electrons with their self-produced synchrotron photons, this result remains inconclusive for various reasons. However, this leptonic model is very successful in explaining the multi wavelength emission from blazars and FR I galaxies[1].

Also the inevitable outcome of the leptonic scenario is that, emission in multi-TeV energy has to be accompanied by a simultaneously enhanced emission in the synchrotron peak. Unfortunately the enhanced synchrotron emission was not observed in the flaring of 1ES 1959+650 in June 2002[2] and also probably in the flaring of Mrk 421 in April 2004[3], which implies that the SSC scenario may not be efficient enough to contribute in the multi-TeV regime.

Previously it was shown that the above hadronic processes are inefficient to explain the multi-TeV emission. However, this problem can be circumvented if one assumes that within the jet there is a dense inner jet region[4]. The multi-TeV peak can also be produced by pp interactions.

It was shown earlier that normally the pp process is inefficient to produce γ-rays in the blazar environment unless the jet-cloud interaction is taken into account[5].

2. Photohadronic Model

Sahu et al. [4, 6, 7] have employed the hadronic model to address the orphan TeV flaring of 1ES 1959+650 and the multi-TeV emission from Cen A and M87. Recently this model is also used to make estimates of the neutrinos flux from the TeV-blazars and their spatial correlation with the IceCube events[8]. Here we use the same hadronic model to explain the multi-TeV orphan flaring of Mrk 421. In this model the Fermi-accelerated high energy protons interact with the SSC photons in the core region of the jet. Subsequently the Δ-resonance decays to charged and neutral pions as follows,

$$p + \gamma \rightarrow \Delta^+ \rightarrow \left\{ \begin{array}{l} p\,\pi^0, \\ n\,\pi^+ \rightarrow n e^+ \nu_e \nu_\mu \bar{\nu}_\mu \end{array} \right. . \quad (1)$$

The decay of neutral pions to TeV photons gives the multi-TeV SED. Throughout our work we use natural units $c = \hbar = 1$. The π^0-decay TeV photon energy E_γ and the target SSC photon energy ϵ_γ in the observer frame are related as follows,

$$E_\gamma \epsilon_\gamma \simeq 0.032 \frac{\mathcal{D}^2}{(1+z)^2} \text{ GeV}^2, \quad (2)$$

where $\mathcal{D}$ is the Doppler factor and z is the redshift. The observed TeV γ-ray energy and the proton energy E_p are related through

$$E_p = \frac{10\Gamma}{\mathcal{D}} E_\gamma, \quad (3)$$

where Γ is the bulk Lorentz factor of the relativistic jet. These emissions are classified into non-flaring and flaring events where (henceforth $'$ implies the jet comoving frame) and satisfying the kinematical conditions in Eq.(2). As discussed in Ref.[4], the flaring occurs within a compact and confined volume of a smaller inner cone which is enclosed in a bigger outer cone. We can express the injected proton spectrum for both non-flaring and flaring processes in a unified manner as[4, 9]

$$\frac{dN_p}{dE_p} \propto E_p^{-\alpha} \left\{ \begin{array}{ll} 1, & non-flaring \\ e^{-E_p/E_{p,c}}, & flaring \end{array} \right. , \quad (4)$$

where $E_{p,c}$ is the break energy for the high energy protons.

The high energy protons will interact in the flaring region where the comoving photon number density is $n'_{\gamma,f}$ to produce the Δ-resonance. The photon density in the flaring region is much higher than the rest of the blob (non-flaring region with the photon density n'_γ) [10]. The optical depth of the Δ-resonance process in the inner jet region is given by

$$\tau_{p\gamma} = n'_{\gamma,f} \sigma_\Delta R'_f, \quad (5)$$

where the R'_f is the inner blob radius corresponding to the inner jet and the resonant cross section is $\sigma_\Delta \sim 5 \times 10^{-28}\, cm^2$.

In the inner region we compare the dynamical time scale $t'_d = R'_f$ with the $p\gamma$ interaction time scale $t'_{p\gamma} = (n'_{\gamma,f}\sigma_\Delta K_{p\gamma})^{-1}$ to constraint the seed photon density so that multi-TeV photons can be produced. For a moderate efficiency of this process, we can assume $t'_{p\gamma} > t'_d$

Journal of Physics: Conference Series **761** (2016) 012042 doi:10.1088/1742-6596/761/1/012042

and this gives $\tau_{p\gamma} < 2$, where the inelasticity parameter is assigned the usual value of $K_{p\gamma} = 0.5$. Also by assuming the Eddington luminosity is equally shared by the jet and the counter jet, the luminosity within the inner region for a seed photon energy ϵ'_γ will satisfy $(4\pi n'_{\gamma,f} R'_f \epsilon'_\gamma) \ll L_{Edd}/2$. This puts an upper limit on the seed photon density as

$$n'_{\gamma,f} \ll \frac{L_{Edd}}{8\pi R'^2_f \epsilon'_\gamma}. \tag{6}$$

We simply assume the scaling behavior of the photon densities in different background energies as follows:

$$\frac{n'_{\gamma,f}(\epsilon'_{\gamma1})}{n'_{\gamma,f}(\epsilon'_{\gamma2})} = \frac{n'_\gamma(\epsilon'_{\gamma1})}{n'_\gamma(\epsilon'_{\gamma2})}. \tag{7}$$

This implies that, the ratio of photon densities at two different background energies $\epsilon'_{\gamma1}$ and $\epsilon'_{\gamma2}$ in the flaring and the non-flaring states remains the same. The photon density in the outer region is calculated from the observed flux in the usual way. In the photohadronic scenario, the number of π^0-decay photons at a given energy is proportional to both the number of high energy protons and the density of the SSC background photons in the jet, i.e. $N(E_\gamma) \propto N(E_p) n'_\gamma$. For the flaring case n'_γ is replaced by the photon density in the flaring region ($n'_{\gamma,f}$). The γ-ray flux from the π^0 decay is deduced to be

$$F_\gamma(E_\gamma) \equiv E_\gamma^2 \frac{dN(E_\gamma)}{dE_\gamma} \propto E_p^2 \frac{dN(E_p)}{dE_p} n'_{\gamma,f}. \tag{8}$$

Using the scaling behavior of Eq.(7), the observed multi-TeV photon flux from π^0-decay at two different observed photon energies $E_{\gamma1}$ and $E_{\gamma2}$ can be expressed as

$$\frac{F_\gamma(E_{\gamma_1})}{F_\gamma(E_{\gamma_2})} = \frac{n'_\gamma(\epsilon_{\gamma_1})}{n'_\gamma(\epsilon_{\gamma_2})} \left(\frac{E_{\gamma_1}}{E_{\gamma_2}}\right)^{-\alpha+2} e^{-(E_{\gamma_1}-E_{\gamma_2})/E_c}, \tag{9}$$

where $E_{\gamma_{1,2}}$ correspond to the proton energy $E_{p_{1,2}}$. In this derivation we have used the relations $E_{p_1}/E_{p_2} = E_{\gamma_1}/E_{\gamma_2}$, and $E_{p_{1,2}}/E_{p,c} = E_{\gamma_{1,2}}/E_c$, where E_c is the γ-ray cut-off energy corresponding to $E_{p,c}$. By using the known flux at a particular energy in the flaring/non-flaring state, we can calculate the flux at other energies using Eq.(9). The normalization constant can also be calculated.

In terms of SSC photon energy and its luminosity, the photon number density n'_γ is expressed as

$$n'_\gamma(\epsilon_\gamma) = \eta \frac{L_{\gamma,SSC}(1+z)}{\mathcal{D}^{2+\kappa} 4\pi R'^2_b \epsilon_\gamma}, \tag{10}$$

where η is the efficiency of SSC process and κ describes whether the jet is continuous ($\kappa = 0$) or discrete ($\kappa = 1$). In this work we take $\eta = 1$ for 100% efficiency. The SSC photon luminosity is expressed in terms of the observed flux ($\Phi_{SSC}(\epsilon_\gamma) = \epsilon_\gamma^2 dN_\gamma/d\epsilon_\gamma$) and is given by

$$L_{\gamma,SSC} = \frac{4\pi d_L^2 \Phi_{SSC}(\epsilon_\gamma)}{(1+z)^2}. \tag{11}$$

Furthermore, by using Eq.(2), we can simplify the ratio of photon densities given in Eq.(7) to

$$\frac{n'_\gamma(\epsilon_{\gamma_1})}{n'_\gamma(\epsilon_{\gamma_2})} = \frac{\Phi_{SSC}(\epsilon_{\gamma1})}{\Phi_{SSC}(\epsilon_{\gamma2})} \frac{E_{\gamma_1}}{E_{\gamma_2}}. \tag{12}$$

Here we take $\kappa = 0$ as representative value for the calculation of n'_γ. We note that $\kappa = 1$ will only lead to the suppression of the photon number density for the outer enveloping jet without changing the main conclusions of our work. This follows from Eq.(12) where proportionality factors depending on κ cancel out. Eq.(9) in terms of observed SSC flux and E_γ takes the following form,

$$\frac{F_\gamma(E_{\gamma 1})}{F_\gamma(E_{\gamma 2})} = \frac{\Phi_{SSC}(\epsilon_{\gamma 1})}{\Phi_{SSC}(\epsilon_{\gamma 2})} \left(\frac{E_{\gamma 1}}{E_{\gamma 2}}\right)^{-\alpha+3} e^{-(E_{\gamma 1}-E_{\gamma 2})/E_c}, \tag{13}$$

and uses the SED of SSC photon calculated using the leptonic model. Here the multi-TeV flux is proportional to $E_\gamma^{-\alpha+3}$ and $\Phi_{SSC}(\epsilon_\gamma)$ while in Eq.(9) it is proportional to $E_\gamma^{-\alpha+2}$ and to the photon number density $n'_\gamma(\epsilon_\gamma)$. In the photohadronic process ($p\gamma$), the multi-TeV photon flux is expressed as

$$F(E_\gamma) = A_\gamma \Phi_{SSC}(\epsilon_\gamma) \left(\frac{E_\gamma}{TeV}\right)^{-\alpha+3} e^{-E_\gamma/E_c}, \tag{14}$$

Both ϵ_γ and E_γ satisfy the condition given in Eq.(2) and the dimensionless constant A_γ is given by

$$A_\gamma = \left(\frac{F(E_{\gamma 2})}{\Phi_{SSC}(\epsilon_{\gamma 2})}\right) \left(\frac{TeV}{E_{\gamma 2}}\right)^{-\alpha+3} e^{E_{\gamma 2}/E_c}. \tag{15}$$

Eq.(14) will be used to calculate the multi-TeV flux from both non-flaring (without exponential decay term) and flaring events from AGN and their subclasses if the emission is due to photohadronic process from the core region. We can calculate the Fermi accelerated high energy proton flux F_p from the TeV γ-ray flux through the relation

$$F_p(E_p) = 7.5 \times \frac{F_\gamma(E_\gamma)}{\tau_{p\gamma}(E_p)}. \tag{16}$$

The optical depth $\tau_{p\gamma}$ is given in Eq.(5). For the observed highest energy proton energy E_p, $F_p(E_p)$ will always be smaller than the Eddington flux F_{Edd}. This condition puts a lower limit on the optical depth of the process and is given by

$$\tau_{p\gamma}(E_p) > 7.5 \times \frac{F_\gamma(E_\gamma)}{F_{Edd}}. \tag{17}$$

From the comparison of different time scales and from Eq.(17) we will be able to constraint the seed photon density in the inner jet region.

3. Results

The flaring of Mrk 421 in April 2004 was observed in the energy range $0.25\,TeV(6.0 \times 10^{25}\,Hz) \leq E_\gamma \leq 16.85\,TeV(4.1 \times 10^{27}\,Hz)$ by the Whipple telescope. In the hadronic model alluded to previously, this corresponds to the Fermi accelerated proton energy in the range $2.5\,TeV \leq E_p \leq 168\,TeV$ and the corresponding background photon energy will lie in the range $23.6\,MeV(5.7 \times 10^{21}\,Hz) \geq \epsilon_\gamma \geq 0.35\,MeV(8.4 \times 10^{19}\,Hz)$. This range of ϵ_γ is the shaded region shown in Fig. 1 which is in the low energy tail of the SSC photons. For the calculation of the normalized multi-TeV flux we take into account one of the observed TeV fluxes from the flare at its corresponding energy. In this model the free parameters are the spectral index α and the TeV γ-ray cut-off energy E_c. The best fit is obtained for the values $\alpha = 2.7$ and $E_c = 6.2$ TeV. This value of E_c corresponds to the proton cut-off energy $E_{p,c} = 62$ TeV and the SSC photon energy $\epsilon_{\gamma,SSC} = 0.96\,MeV$ $(2.3 \times 10^{20}\,Hz)$ which is very close to the beginning of the SSC energy as shown in Fig. 1. This shows that for orphan flaring, the cut-off energy E_c is due to the change from the synchrotron band to the SSC band and can be calculated from their crossover energy

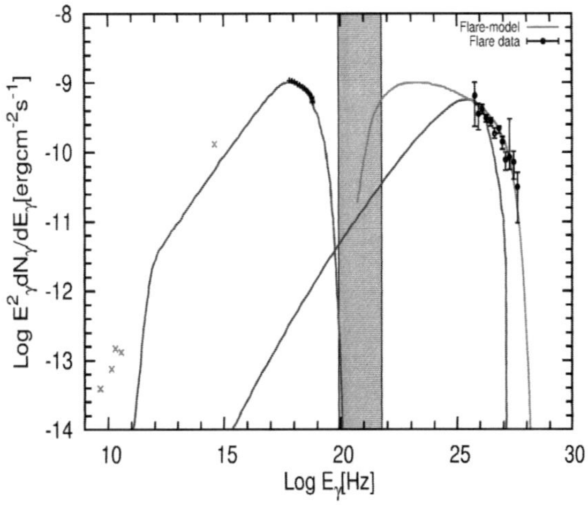

Figure 1. The SED of Mrk 421 is shown in all the energy bands (the leptonic model part in blue line) which are taken from Ref.[3]. The flare of April 2004 in multi-TeV energy is shown here. The hadronic model fit to the April 2004 data is shown as continuous line to the extreme right. The shaded region is the energy range of SSC photons where the Fermi-accelerated protons collide to produce the Δ-resonance.

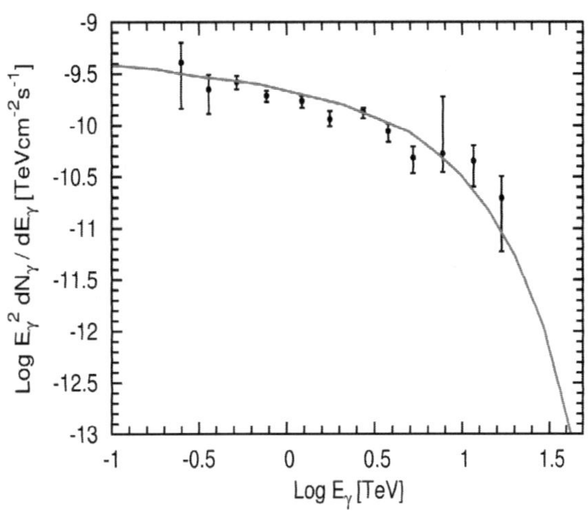

Figure 2. The continuous curve is the hadronic model fit to the multi-TeV flaring data of Mrk 421.

region. In Fig. 2 we show the observed multi-TeV SED and the predictions of our model. In our results, the presence of $\Phi_{SSC}(\epsilon_\gamma)$ in Eq.(14) modifies the power-law with the exponential fall-off scenario. From the best fit parameters we obtain the value of the dimensionless constant $A_\gamma \simeq 20$ in Eq. (15).

In the flaring state, as has been alluded to before, in general, the flux of the two opposing jets can be as high as $F_{Edd}/2$. However, the highest energy protons with $E_p = 168$ TeV must have flux $F_p < F_{Edd}/2$. Using Eq.(16) this constraint translates into $\tau_{p\gamma} > 0.02$ which corresponds to $n'_{\gamma,f} > 1.3 \times 10^{10}\, cm^{-3}$ in the inner jet. However, the hidden jet lies between R_s and R'_b. As one representative value we take $R'_f \simeq 3 \times 10^{15}$. From Eq.(6) the seed photon density for $\epsilon_\gamma = 0.35$ MeV satisfies the inequality $n'_{\gamma,f} < 8.9 \times 10^{10} cm^{-3}$ which translates to the optical depth to be constrained as $\tau_{p\gamma} < 0.13$. The upshot of all this is that while the value of $\tau_{p\gamma} < 2$ is a generic constraint, a more refined present analysis constrains the optical depth to lie in the range $0.02 < \tau_{p\gamma} < 0.13$. From this the constraint on the photon number density in the inner jet region is determined to be $1.3 \times 10^{10}\, cm^{-3} < n'_{\gamma,f} < 8.9 \times 10^{10} cm^{-3}$, which shows that the photon density in this region is high. Due to the adiabatic expansion of the inner blob, the photon density will be reduced to n'_γ and the energy will dissipate once these photons cross into the bigger outer cone. In spite of the two-zones structure, only the outer zone will be responsible for the observed synchrotron and the IC peaks. From the leptonic model fit to the SED, the magnetic field in the outer jet region is determined to be $B' = 0.26$ G. In general, higher values of the magnetic field are expected to be present in the inner jet region. The maximum proton energy in the hidden jet region will be $E_{p,max} \sim 10^{18}(B'_f/1G)$ eV and for larger magnetic fields, $E_{p,max}$ can even be higher. In comparison, in the non-flaring jet scenario case we estimate the proton flux needed to explain the observed TeV γ-rays as follows: Corresponding to the seed photon energy $\epsilon_\gamma = 0.35$ MeV, the photon density and the optical depth are respectively $n'_\gamma \sim 4 \times 10^3\, cm^{-3}$ and $\tau_{p\gamma} \sim 1.4 \times 10^{-8}$. This corresponds to the observed TeV photon energy $E_\gamma = 16.8$ TeV and the proton energy $E_p = 168$ TeV respectively. The high energy γ-ray flux for $E_\gamma = 16.8$ TeV is $F_\gamma \sim 3.16 \times 10^{-11}\, erg\, cm^{-2}\, s^{-1}$. Using Eq.(16) we obtain the high energy proton flux as $F_p \simeq 10^6 \times F_{Edd}$. This shows that the normal jet model needs super Eddington power in the protons to explain the high energy peak, whereas our inner jet scenario eliminates this extreme energy requirement.

In passing we note that, in the energy range considered, the high energy protons will be accompanied by high energy electrons. These electrons will emit synchrotron photons in the energy range $4 \times 10^{19}\, Hz$ to $2 \times 10^{23}\, Hz$ when encountering the magnetic field of the jet. These energy range photons lie in the lower part of the SSC spectrum and thus will not be observed due to their flux being low. The high energy electrons will also emit SSC photons and their energy is given by $E_{IC} \sim \gamma_e^2 \epsilon_{syn}$. By considering the electron Lorentz factor in the range $7 \times 10^2 \leq \gamma_e \leq 4 \times 10^4$ and the peak energy of the synchrotron photons to be $\epsilon_{syn} \sim 10^{18}\, Hz$, we determine the SSC process contribution to the flux in the energy range $2\, GeV \leq E_{IC} \leq 6.6\, TeV$. The details of the SSC flux depend on the breaks in SSC spectrum and the spectral index. It is observed that during the flaring of Mrk 421, the X-ray emission reached the peak days after the TeV emission, which poses a serious challenge to the SSC model[3].

The multi-TeV photons in the energy range $0.25\, TeV \leq E_\gamma \leq 16.8\, TeV$ can interact with the background photons to produce e^+e^- pairs in which the individual electrons or positrons have energy $E_\gamma/2$. To produce the e^+e^- pair, the required threshold seed photon energy of $\epsilon_\gamma \geq 2m_e^2/E_\gamma$ is needed. During the flaring, multi-TeV γ-rays will interact with soft seed photons in the energy range $0.05\, eV \leq \epsilon_\gamma \leq 3.5\, eV$ (in between the infrared and the visible range), where $\sigma_{\gamma\gamma} \sim 1.7 \times 10^{-25}\, cm^2$ is the maximum cross section. For higher ϵ_γ, $\sigma_{\gamma\gamma}$ will be smaller. We consider the sources of these soft photons to be from the synchrotron emission of 1 to 10 GeV electrons interacting with the magnetic field $B \sim 1$ Gauss and the flux of the ambient photons coming from the disk. On the other hand the multi-TeV γ-rays are produced beyond this region where the photons make the low energy tail ($0.35\, MeV - 23.5\, MeV$ range) of the IC photons. The regions $0.05\, eV - 3.5\, eV$ and $0.35\, MeV - 23.5\, MeV$ represent two distinct energy ranges. The TeV photons will mostly encounter the tail region of the SSC spectrum. The pair production cross section for $\epsilon_\gamma \geq 0.35\, MeV$ is very small $\sigma_{\gamma\gamma} \leq 10^{-30}\, cm^2$, which corresponds to a mean free path of $\lambda_{\gamma\gamma} \geq 10^{19}\, cm$. Hence, the TeV photons will not be attenuated much due to the e^+e^- pair production. Also it has been observed that, during the flaring of Mrk 421, the variation in the light curves at optical and radio wavelengths are minimal[3], which implies that the low energy photon production was suppressed. In addition to the above scenario, the positrons produced from the π^+ decay have energy $E_\gamma/2$ and will radiate synchrotron photons in the energy range $2 \times 10^{17}\, Hz$ to $9 \times 10^{20}\, Hz$. The photon flux

$F_{e^+,syn}$ from the synchrotron radiation of e^+ will be much smaller than $F_\gamma(E_\gamma = 0.25\,TeV)/8$, i.e., $F_{e^+,syn} \ll 8\times10^{-11}\,erg\,cm^{-2}\,s^{-1}$. From the above analysis our conclusion is that the photon fluxes from the synchrotron emission of the electrons and positrons are not observable during the flaring event of Mrk 421 in April 2004. This situation is very much like the one discussed in the orphan flaring of 1ES 1959+650. Hence we conclude that the flaring of Mrk 421 is orphan in nature. In principle, the flux of multi-TeV γ-rays from the extragalactic sources is attenuated due to interactions with the diffuse extragalactic background light (EBL) through the process $\gamma_{\rm TeV} + \gamma_{\rm b} \to e^+e^-$ due to the energy dependent optical depth. But for low redshifts and the energy range of our interest, the optical depth $\tau_{\gamma\gamma}$ does not vary much. Hence we can assume an almost constant optical depth[9] so that the spectral shape remains nearly unchanged.

4. Conclusions

We conclude that the orphan flaring of Mrk 421 can be explained well by the photohadronic model [11]. We observe that, in this model, the multi-TeV photon flux is proportional to $\Phi_{SSC}(\epsilon_\gamma)\,E_\gamma^{-\alpha+3}$ supplemented with an exponential decay term as shown in Eq. (14). This implies that the Fermi accelerated protons interact with the background photons (in the low energy tail) of the SSC spectrum. In the present work we have shown that the higher photon density in the internal jet scenario eliminates the requirement of the super Eddington luminosity to explain the multi-TeV emission. During the April 2004 flaring of Mrk 421, we have demonstrated that the flux from the synchrotron emission from the high energy e^+ and e^- is suppressed relative to the normal flux, implying that the flaring was orphan in nature, like the one observed in 1ES 1959+650.

References

[1] P. Roustazadeh and M. Böttcher 2011 *Astrophys. J.* **728** 134.
[2] H. Krawczynski, S. B. Hughes, D. Horan, F. Aharonian, M. F. Aller, H. Aller, P. Boltwood and J. Buckley *et al.* 2004 *Astrophys. J.* **601** 151.
[3] M. Blazejowski, G. Blaylock, I. H. Bond, S. M. Bradbury, J. H. Buckley, D. A. Carter-Lewis, O. Celik and P. Cogan *et al.* 2005 *Astrophys. J.* **630** 130.
[4] S. Sahu, A. F. O. Oliveros and J. C. Sanabria 2013 *Phys. Rev.* D **87** 103015.
[5] A. Atoyan and C. D. Dermer 2001 *Phys. Rev. Lett.* **87** 221102.
[6] S. Sahu, B. Zhang and N. Fraija 2012 *Phys. Rev.* D **85** 043012.
[7] S. Sahu and E. Palacios 2015 *Eur. Phys. J.* C **75** 52.
[8] S. Sahu and L. S. Miranda 2015 *Eur. Phys. J.* C **75** 273.
[9] F. Aharonian *et al.* [HEGRA Collaboration] 2003 *Astron. Astrophys.* **406** L9.
[10] D. Giannios, D. A. Uzdensky and M. C. Begelman 2009 *Mon. Not. Roy. Astron. Soc.* **395** L29.
[11] S. Sahu, L. S. Miranda and S. Rajpoot 2016 *Eur. Phys. J.* C **76** 127.

New limits on neutrino magnetic moments from low energy neutrino data

B C Cañas[1], O G Miranda[1], A Parada[2], M Tórtola[3] and J W F Valle[3]

[1] Departamento de Física, Centro de Investigación y de Estudios Avanzados del IPN Apdo. Postal 14-740 07000 Mexico, DF, Mexico

[2] Universidad Santiago de Cali, Campus Pampalinda, Calle 5 No. 6200, 760001, Santiago de Cali, Colombia

[3] AHEP Group, Institut de Física Corpuscular – C.S.I.C./Universitat de València, Parc Cientific de Paterna. C/Catedratico José Beltrán, 2 E-46980 Paterna (València) - SPAIN

E-mail: bcorduz@fis.cinvestav.mx

Abstract.
Here we give a brief review on the current bounds on the general Majorana transition neutrino magnetic moments (TNMM) which cover also the conventional neutrino magnetic moments (NMM). Leptonic CP phases play a key role in constraining TNMMs. While the Borexino experiment is the most sensitive to the TNMM magnitudes, one needs complementary information from reactor and accelerator experiments in order to probe the complex CP phases.

1. Introduction

The study of neutrino electromagnetic properties is of great importance, since it could open a new window to investigate physics beyond the Standard Model. Though there are various types of electromagnetic properties [1, 2], such as a neutrino charge radius [3, 4, 5, 6] or a neutrino milli-charge [7, 8], here we concentrate on the case of TNMM [9, 10, 11]. These are appealing for several reasons. For instance, like neutrinoless double beta decay [12], TNMMs can shed light on the fundamental issue of the Dirac or Majorana nature of neutrinos [9, 10, 11]. They are also expected to be calculable in a gauge theory, their finite values given in terms of fundamental neutrino properties, such as masses and mixing parameters, in addition to other genuine new physics parameters such as new messenger particle masses. The main constraints on neutrino electromagnetic properties come from reactor neutrino studies [13, 14] as well as from solar neutrino data [15, 16]. There are, moreover, many proposals aiming to improve the current bounds, one of them using a megacurie ^{51}Cr neutrino source and a large liquid Xenon detector [17]. Based on the analysis reported in [18], we summarize some of the most relevant constraints on TNMM and discuss their dependence on the CP phases.

We include different types of neutrino data samples, such as the most recent data from the TEXONO reactor experiment [14], as well as the latest results from the Borexino experiment [19]. Data from the reactor experiments Krasnoyarsk [20], Rovno [21] and MUNU [22] along with the accelerator experiments LAMPF [23] and LSND [24] are also included. In addition, we take into account the updated values of the neutrino mixing parameters as determined in global oscillation fits [25], including the value of θ_{13} implied by Daya-Bay [26] and RENO [27] reactor data, as well as accelerator data [28]. Besides, we stress on the role played by the, yet unknown, leptonic CP violating phases.

2. Neutrino magnetic moments

The interaction between Majorana neutrinos and the electromagnetic field is described by the general effective Hamiltonian [9]

$$H_{em}^{M} = -\frac{1}{4}\nu_L^T C^{-1}\lambda\sigma^{\alpha\beta}\nu_L F_{\alpha\beta} + h.c., \tag{1}$$

with $\lambda = \mu - id$ a complex antisymmetric matrix in generation space, implying that $\mu^T = -\mu$ and $d^T = -d$ are pure imaginary. Therefore, we need six real parameters to describe the Majorana NMM. The Majorana NMM matrix can be written in the flavor (mass) basis, λ ($\tilde{\lambda}$), as follows

$$\lambda = \begin{pmatrix} 0 & \Lambda_\tau & -\Lambda_\mu \\ -\Lambda_\tau & 0 & \Lambda_e \\ \Lambda_\mu & -\Lambda_e & 0 \end{pmatrix}, \qquad \tilde{\lambda} = \begin{pmatrix} 0 & \Lambda_3 & -\Lambda_2 \\ -\Lambda_3 & 0 & \Lambda_1 \\ \Lambda_2 & -\Lambda_1 & 0 \end{pmatrix}. \tag{2}$$

Here we have defined $\lambda_{\alpha\beta} = \varepsilon_{\alpha\beta\gamma}\Lambda_\gamma$, through the complex parameters: $\Lambda_\alpha = |\Lambda_\alpha|e^{i\zeta_\alpha}$, $\Lambda_i = |\Lambda_i|e^{i\zeta_i}$. Having described our theoretical framework, we now discuss the relation of these observables with the parameters measured in current neutrino experiments. For neutrino-electron scattering, the differential cross section for the NMM contribution will be given by

$$\left(\frac{d\sigma}{dT}\right)_{em} = \frac{\pi\alpha^2}{m_e^2\mu_B^2}\left(\frac{1}{T} - \frac{1}{E_\nu}\right){\mu_\nu}^2, \tag{3}$$

where μ_ν is an effective magnetic moment accounting for the NMM contribution to the scattering process. It is defined in terms of the components of the NMM matrix in Eq. (2). In the flavor basis, this parameter can be written as [16]

$$(\mu_\nu^F)^2 = a_-^\dagger\lambda^\dagger\lambda a_- + a_+^\dagger\lambda\lambda^\dagger a_+, \tag{4}$$

where we have denoted the negative and positive helicity neutrino amplitudes by a_- and a_+, respectively. The flavour and mass neutrino basis are connected through the neutrino mixing matrix U

$$\tilde{a}_- = U^\dagger a_-, \qquad \tilde{a}_+ = U^T a_+, \qquad \tilde{\lambda} = U^T\lambda U, \tag{5}$$

such that the effective NMM in the mass basis is given by

$$(\mu_\nu^M)^2 = \tilde{a}_-^\dagger\tilde{\lambda}^\dagger\tilde{\lambda}\tilde{a}_- + \tilde{a}_+^\dagger\tilde{\lambda}\tilde{\lambda}^\dagger\tilde{a}_+. \tag{6}$$

Notice that there are six complex phases in the effective NMM parameter: ζ_1, ζ_2 and ζ_3 from the NMM matrix; δ and two-Majorana phases from the leptonic mixing matrix. However, as noticed in Ref. [29], it is clear that only three of these six complex phases are independent. To carry out our analysis [18], we choose the Dirac CP phase δ, and the two relative phases, $\xi_2 = \zeta_3 - \zeta_1$ and $\xi_3 = \zeta_2 - \zeta_1$.

In reactor experiments, we initially have only an electron antineutrino flux and, therefore, $a_+^1 = 1$ will be the only nonzero entry. With this initial condition, we get from Eq. (4) the following expression for the effective Majorana NMM for reactor antineutrino experiments in the flavor basis:

$$(\mu_R^F)^2 = |\Lambda_\mu|^2 + |\Lambda_\tau|^2. \tag{7}$$

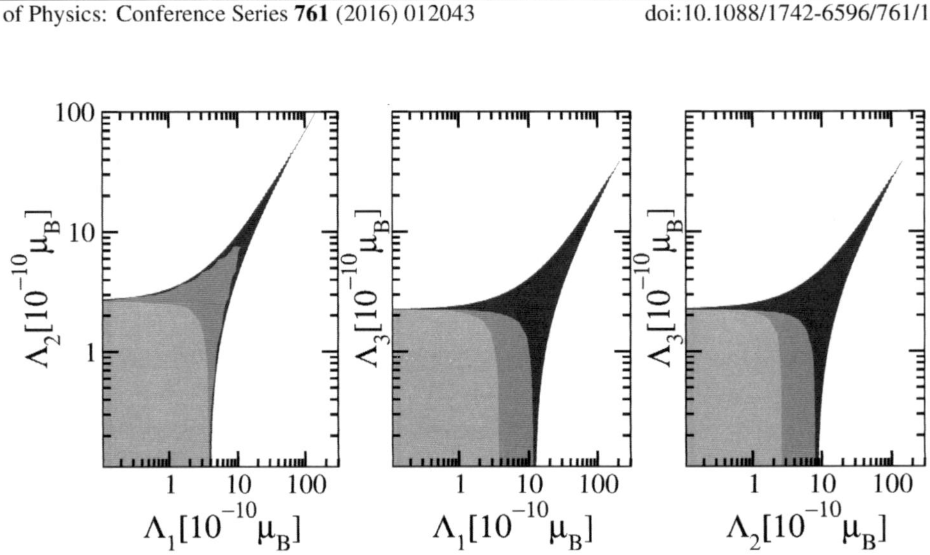

Figure 1. Allowed regions for the TNMM, at 90 % CL, obtained from TEXONO data. The two-dimensional regions $(|\Lambda_i|, |\Lambda_j|)$ have been computed by marginalizing over the third component, $|\Lambda_k|$. For the purple (outer) region we have fixed the phases to be $\delta = 0$ and $\xi_2 = \xi_3 = 0$ (see Eq. (9)), while in the magenta (middle) region we have set $\delta = 3\pi/2$ and $\xi_2 = \xi_3 = 0$. The orange (inner) region corresponds to $\delta = 3\pi/2$, $\xi_2 = 0$ and $\xi_3 = \pi/2$.

On the other hand, for the mass basis, we have the expression

$$\begin{aligned}(\mu_R^M)^2 &= |\mathbf{\Lambda}|^2 - s_{12}^2 c_{13}^2|\Lambda_2|^2 - c_{12}^2 c_{13}^2|\Lambda_1|^2 - s_{13}^2|\Lambda_3|^2 \\ &- 2s_{12}c_{12}c_{13}^2|\Lambda_1||\Lambda_2|\cos\delta_{12} - 2c_{12}c_{13}s_{13}|\Lambda_1||\Lambda_3|\cos\delta_{13} \\ &- 2s_{12}c_{13}s_{13}|\Lambda_2||\Lambda_3|\cos\delta_{23},\end{aligned} \tag{8}$$

with $c_{ij} = \cos\theta_{ij}$, $s_{ij} = \sin\theta_{ij}$. The phases in this equation depend on the three independent CP phases already mentioned: $\delta_{12} = \xi_3$, $\delta_{23} = \xi_2 - \delta$, and $\delta_{13} = \delta_{12} - \delta_{23}$. The dependence on these CP phases is very interesting and adds an extra complexity to the interpretation of the experimental constraints. For instance, in the particular case where all the independent phases vanish, i.e, $\delta_{12} = \delta_{23} = \delta_{13} = 0$, the effective Majorana NMM in Eq. (8) is given by

$$(\mu_R^M)^2 = |\mathbf{\Lambda}|^2 - (c_{12}c_{13}|\Lambda_1| + s_{12}c_{12}c_{13}|\Lambda_2| + s_{13}|\Lambda_3|)^2. \tag{9}$$

It is easy to notice that if we also impose the conditions

$$|\Lambda_1| = c_{12}c_{13}|\mathbf{\Lambda}|, \quad |\Lambda_2| = s_{12}c_{13}|\mathbf{\Lambda}|, \quad |\Lambda_3| = s_{13}|\mathbf{\Lambda}|, \tag{10}$$

$(\mu_R^M)^2$ cancels exactly and, hence, in this case reactor experiments would not be sensitive to this parameter. This is illustrated in Fig. 1.

A similar situation will appear in accelerator experiments such as the LAMPF [23] and LSND [24], where a dependence on the CP phases will appear [18].

On the other hand, in solar neutrino experiments like Borexino [19], the effective NMM parameter in the mass basis is given by

$$(\mu_{\rm sol}^M)^2 = |\mathbf{\Lambda}|^2 - c_{13}^2|\Lambda_2|^2 + (c_{13}^2 - 1)|\Lambda_3|^2 + c_{13}^2 P_{e1}^{2\nu}(|\Lambda_2|^2 - |\Lambda_1|^2), \tag{11}$$

where we have defined the effective two-neutrino oscillation probabilities as $P_{e1,e2}^{2\nu}$. Due to the loss of coherence, the effective NMM measured from the solar neutrino flux is independent of any phase, a fact already noticed in [16]. Notice also that the analysis presented here takes into account the non-zero value of the reactor angle θ_{13} [18].

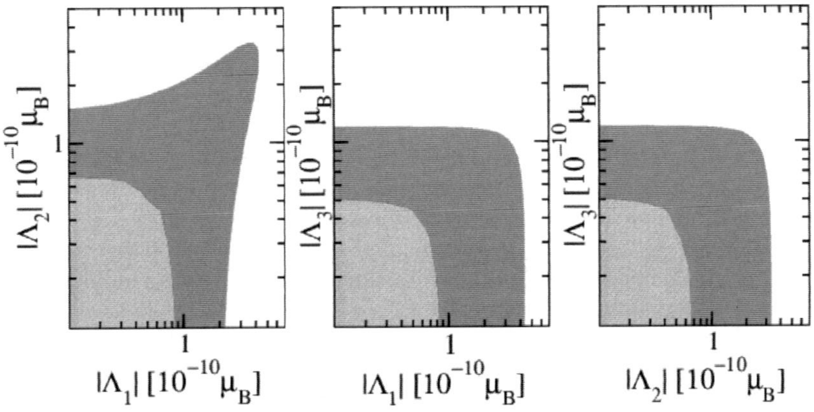

Figure 2. Allowed regions, at 90 % CL, for the TNMM [18]. The magenta region shows the constraints from reactor and accelerator data when all the phases vanish, except for $\delta = 3\pi/2$. The turquoise zone shows the corresponding bounds from the Borexino data that are phase-independent.

3. Analysis of the neutrino data

We perform a combined analysis of the experimental data in order to get constraints for the three different TNMM components Λ_i. In order to carry out the statistical analysis we use the following χ^2 function:

$$\chi^2 = \sum_{i=1}^{N_{bin}} \left(\frac{O^i - N^i(\mu)}{\Delta_i} \right)^2 , \tag{12}$$

where O^i and N^i are the observed number of events and the predicted number of events in the presence of an effective NMM, μ, at the i-th bin, respectively. Here Δ_i is the statistical error for each bin. In our analysis, we have included the experimental results reported by Krasnoyarsk [20], Rovno [21], MUNU [22], and TEXONO [14] reactor experiments. We have also included the experimental data reported by the LAMPF [23] and LSND [24] collaborations, as well as the most recent measurements of the Beryllium solar neutrino flux reported in Ref. [19] by Borexino.

We perform a complete analysis taking into account the role of the CP phases in the reactor and accelerator data. For the particular case of reactor neutrinos, we have carried out a statistical analysis of TEXONO data [14] taking different values of the complex phases of Λ_i, ζ_i, and taking all nonzero TNMM amplitudes. The result of this analysis is shown in Fig. 1. Notice that the resulting restrictions on the TNMM $|\Lambda_1|$ and $|\Lambda_2|$ depend on the chosen CP phase combinations.

Finally, we carried out a combined analysis [18] of all the reactor and accelerator data for a particular choice of phases ($\delta = 3\pi/2$ and $\xi_i = 0$) and compared it with the corresponding χ^2 analysis obtained from the Borexino data. The results, shown in Fig. 2, illustrate how Borexino [19] is more sensitive in constraining the magnitude of the TNMM. We stress that the Borexino effective NMM depends only on the square magnitudes of these TNMM and hence, its constraints are almost the same as those in the one-parameter-at-a-time analysis. On the other hand, future reactor and accelerator experiments are the only ones that could give information on individual TNMMs as well as on the Majorana phases discussed here, an information inaccessible to the Borexino experiment.

4. Conclusions

In this short review, we have discussed the current status of the bounds on the transition neutrino magnetic moments. These parameters are very important, because they encode the Majorana CP phases present both in the mixing matrix and in the NMM matrix. The conventional neutrino magnetic moment emerges as a particular effective case. The Borexino solar experiment plays a key role in constraining the electromagnetic neutrino properties due to the low energies (below 1 MeV) which are probed as well as its robust statistics. Indeed, it provides the most stringent constraints on the absolute magnitude of the transition magnetic moments. However, the Borexino experiment can not probe the Majorana phases, due to the incoherent nature of the solar neutrino flux. Although less sensitive to the absolute value of the transition magnetic moment strengths, reactor and accelerator experiments provide the only chance to obtain a hint of the complex CP phases, as illustrated in Fig. 1.

Acknowledgements
Work supported by MINECO grants FPA2014-58183-P, Multidark CSD2009- 00064 and SEV-2014-0398 (MINECO); by EPLANET, by the CONACyT grant 166639 (Mexico) and the PROMETEOII/2014/084 grant from Generalitat Valenciana. M. Tórtola is also supported by a Ramon y Cajal contract of the Spanish MINECO. The work of A.Parada was partially supported by Universidad Santiago de Cali under grant 935-621114-031.

References

[1] Giunti C *et al.* 2016 *Annalen Phys.* **528** 198
[2] Giunti C and Studenikin A 2015 *Rev. Mod. Phys* **87** 531
[3] Bardeen W A Gastmans R and Lautrup B E 1972 *Nucl. Phys. B* **46** 319
[4] Lee S Y 1972 *Phys. Rev. D* **6** 1701
[5] Barranco J Miranda O G and Rashba T I 2008 *Phys. Lett. B* **662** 431
[6] Kosmas T S *et al.* 2015 *Phys. Rev. D* **92** 013011
[7] Marciano W J and Sanda A I 1977 *Phys. Lett. B* **67** 303
[8] Papoulias D K and Kosmas T S 2015 *Phys. Lett. B* **747** 454
[9] Schechter J and Valle J W F 1981 *Phys. Rev. D* **24** 1883 Erratum: [Phys. Rev. D **25**, 283 (1982)]
[10] Kayser B 1982 *Phys. Rev. D* **26** 1662
[11] Bell N F 2007 *Int. J. Mod. Phys. A* **22** 4891
[12] J. Schechter and J. W. F. Valle, Phys. Rev. D **25** (1982) 2951. doi:10.1103/PhysRevD.25.2951
[13] Li H B *et al.* 2003 *Phys. Rev. Lett.* **90** 131802
[14] Deniz M *et al.* [TEXONO Collaboration] 2010 *Phys. Rev.* D **81** 072001
[15] Beacom J F and Vogel P 1999 *Phys. Rev. Lett.* **83** 5222
[16] Grimus W *et al.* 2003 *Nucl. Phys.* B **648** 376
[17] Coloma P Huber P and Link J M 2014 *JHEP* **1411** 042
[18] Canas B C *et al.* 2016 *Phys. Lett.* B **753** 191 Addendum: *Phys. Lett.* B **757** 568
[19] Bellini G *et al.* 2011 *Phys. Rev. Lett.* **107** 141302
[20] Vidyakin G S *et al.* 1992 *JETP Lett.* **55** 206
[21] Derbin A I *et al.* 1993 *JETP Lett.* **57** 768
[22] Daraktchieva Z *et al.* [MUNU Collaboration] 2005 *Phys. Lett.* B **615** 153
[23] Allen R C *et al.* 1993 *Phys. Rev.* D **47** 11
[24] Auerbach L B *et al.* [LSND Collaboration] 2001 *Phys. Rev.* D **63** 112001
[25] Forero D V Tortola M and Valle J W F 2014 *Phys. Rev.* D **90** 093006
[26] An F P *et al.* [Daya Bay Collaboration] 2014 *Phys. Rev. Lett.* **112** 061801
[27] Ahn J K *et al.* [RENO Collaboration] 2012 *Phys. Rev. Lett.* **108** 191802
[28] Abe K *et al.* [T2K Collaboration] 2014 *Phys. Rev. Lett.* **112** 061802
[29] Grimus W and Schwetz T 2000 *Nucl. Phys.* B **587** 45

CP violation in $\mu - \tau$ symmetric models

D.C. Rivera-Agudelo [a,1] **and A. Pérez-Lorenzana** [a,2]

[a] Departamento de Física, Centro de Investigación y de Estudios Avanzados del I.P.N. Apdo. Post. 14-740, 07000, México, Distrito Federal, México

E-mail: [1] `drivera@fis.cinvestav.mx`, [2] `aplorenz@fis.cinvestav.mx`

Abstract. The observed masses and mixings in the neutrino sector suggest an approximate $\mu - \tau$ symmetry in the neutrino mass matrix. We explore the corresponding parameter space linked to a small breaking of this symmetry to show that in such a scenario neutrino masses must be quasi degenerated, whereas Majorana CP phases shall be non zero and strongly correlated to the Dirac CP phase.

1. Introduction

Global fits of the available data of different neutrino oscillation experiments obtained in the three flavor scheme indicate that [1], within the 3σ range, the mixing angles are given by solar $\sin^2\theta_{12} \equiv \sin^2\theta_\odot \approx 0.308^{+0.064}_{-0.055}$, atmospheric $\sin^2\theta_{23} \equiv \sin^2\theta_{ATM} \approx 0.437^{+0.189}_{-0.063}$ $(0.455^{+0.186}_{-0.075})$, and reactor $\sin^2\theta_{13} \approx 0.0234^{+0.0061}_{-0.0058}$ $(0.0240^{+0.0058}_{-0.0062})$, for the normal (inverted) hierarchy. The squared mass differences are given by the so-called solar scale $\Delta m^2_{21} = m_2^2 - m_1^2 \equiv \Delta m^2_{sol} \approx 7.54 \times 10^{-5} eV^2$ and the atmospheric scale $|\Delta m^2_{31}| \sim \Delta m^2_{ATM} \sim 2.4 \times 10^{-3} eV^2$. However, the sign in $\Delta m^2_{31} = m_3^2 - m_1^2$, and thus the neutrino mass hierarchy pattern, is still unknown. A very recent global analysis of neutrino oscillation data was provided by [2].

The mixing angles measured in oscillation experiments are large, except for θ_{13}, which was found to be rather small but certainly non-zero. In the standard parametrization, the mixings are given by [3, 4],

$$U_{mix} = \begin{pmatrix} c_{12}c_{13} & s_{12}c_{13} & s_{13}e^{-i\delta_{CP}} \\ -s_{12}c_{23} - c_{12}s_{23}s_{13}e^{i\delta_{CP}} & c_{12}c_{23} - s_{12}s_{23}s_{13}e^{i\delta_{CP}} & s_{23}c_{13} \\ s_{12}s_{23} - c_{12}c_{23}s_{13}e^{i\delta_{CP}} & -c_{12}s_{23} - c_{23}s_{12}s_{13}e^{i\delta_{CP}} & c_{23}c_{13} \end{pmatrix} \cdot K\ , \tag{1}$$

where c_{ij} and s_{ij} denote $\cos\theta_{ij}$ and $\sin\theta_{ij}$, respectively, and δ_{CP} is the Dirac CP-violating phase, whereas K =Diag$(e^{-i\beta_1/2}, e^{-i\beta_2/2}, 1)$ is a diagonal matrix containing two Majorana CP-violating phases.

The neutrino mass terms are given by the effective operator $\bar{\nu}_{\alpha L}(M_\nu)_{\alpha\beta}\nu^c_{\beta L} + h.c.$. Let us consider a basis where the weak flavor eigenstates of the charged lepton are identified with their masses eigenstates, then the mixing matrix also becomes that which diagonalizes the mass terms, and $U_{mix} = U_\nu \cdot K$. Therefore, the neutrino mass matrix can be written in terms of a diagonal matrix, as

$$M_\nu = U_\nu^* \cdot M_{diag} \cdot U_\nu^\dagger, \tag{2}$$

where M_{diag} = Diag(m_1, m_2, m_3), and we have denoted $m_j \equiv |m_j|e^{-i\beta_j}$, for $j = 1, 2$.

When θ_{13} and θ_{23} take the critical values, zero and $\pi/4$, respectively, the U_{mix} matrix in Eq. (1) becomes the one that holds the equality $|U_{\tau i}| = |U_{\mu i}|$, and gives rise to the so called $\mu - \tau$ symmetry [5]. As a consequence $|m_{e\mu}| = |m_{e\tau}|$ and $|m_{\mu\mu}| = |m_{\tau\tau}|$ in Eq. (2). Nowadays, we know that the value $\theta_{13} = 0$ is disfavored, but $\theta_{23} = \pi/4$ is still allowed at the 1σ level. Then, the observed values of these angles could be understood as a result of the breaking of the $\mu - \tau$ symmetry. That is why we claim that there must be an approximate $\mu - \tau$ symmetry behind the observed pattern of the mixing matrix.

In this study, we provide a detailed analysis of the magnitude of the $\mu - \tau$ symmetry breaking using the current data for neutrino mixing parameters, and we identify the mass spectrum and *CPVP* (CP violating phases) required to obtain a small symmetry breaking.

2. $\mu - \tau$ symmetry breaking parameters

Without loss of generality, any neutrino mass matrix can be written out as $M_\nu = M_{\mu-\tau} + \delta M$, where $M_{\mu-\tau}$ posses the $\mu - \tau$ exchange symmetry, and δM is defined by only two non-zero elements,

$$\delta M = \begin{pmatrix} 0 & 0 & \delta \\ 0 & 0 & 0 \\ \delta & 0 & \epsilon \end{pmatrix} , \tag{3}$$

where $\delta = m_{e\tau} - m_{e\mu}$ and $\epsilon = m_{\tau\tau} - m_{\mu\mu}$ are considered the breaking parameters. Two dimensionless parameters can be defined as $\hat{\delta} \equiv \frac{\delta}{m_{e\mu}}$ and $\hat{\epsilon} \equiv \frac{\epsilon}{m_{\mu\mu}}$, and using Eq. (2) we get

$$\begin{aligned} \hat{\delta} &= \frac{\sum_i (U^*_{ei} U^*_{\tau i} - U^*_{ei} U^*_{\mu i}) m_i}{\sum_i U^*_{ei} U^*_{\mu i} m_i}, \\ \hat{\epsilon} &= \frac{\sum_i (U^*_{\tau i} U^*_{\tau i} - U^*_{\mu i} U^*_{\mu i}) m_i}{\sum_i U^*_{\mu i} U^*_{\mu i} m_i} , \end{aligned} \tag{4}$$

where $i = 1, 2, 3$. The absolute masses can be rewritten as

$$\begin{aligned} |m_2| = \sqrt{m_0^2 + \Delta m^2_{sol}} \quad &\textbf{and} \quad |m_3| = \sqrt{m_0^2 + |\Delta m^2_{ATM}|} \ \textbf{ for NH}. \\ |m_1| = \sqrt{m_0^2 + |\Delta m^2_{ATM}|} \quad &\textbf{and} \quad |m_2| = \sqrt{m_0^2 + |\Delta m^2_{ATM}| + \Delta m^2_{sol}} \ \textbf{ for IH}. \end{aligned} \tag{5}$$

In the equations above, m_0 is the lightest neutrino mass that becomes $|m_1|$ for the normal mass hierarchy (NH) and m_3 for the inverted mass hierarchy (IH). Thus, $\hat{\delta}$ and $\hat{\epsilon}$, depend on nine observables: three mixing angles, two mass squared differences, the lightest neutrino mass, and three *CPVP*. The neutrino oscillation experiments have already provided accurate values for the first five. Hence, the breaking parameter space $\hat{\delta} - \hat{\epsilon}$ that can accommodate neutrino mixings and oscillation mass scales within experimental uncertainties should, in principle, remain undetermined due to the arbitrariness of the m_0 and the *CPVP*. However, we suggest that specific predictions for the lightest neutrino mass and CP phases can be extracted by assuming small values of $|\hat{\delta}|$ and $|\hat{\epsilon}|$. We infer that the resulting bounded parameter space will not be consistent with any arbitrary values of the neutrino observables. In order to obtain some insights into the expected conditions required for small symmetry breaking we write the Eq. (4) in the analytical form

$$\hat{\delta} = \frac{y_- f s_{13} - y_+}{1 + f\ s_{13} \tan\ \theta_{23}} , \quad \hat{\epsilon} = \frac{g \cos\ 2\theta_{23} - s_{13} h}{1 + g s_{23}^2 + s_{13} h/2} , \tag{6}$$

where

$$y_\pm = \frac{c_{23} \pm s_{23}}{c_{23}},$$

$$\begin{aligned} f &= \frac{(c_{12}^2|m_1|e^{-i\beta_1}+s_{12}^2|m_2|e^{-i\beta_2})e^{-i\delta_{CP}}-m_3e^{i\delta_{CP}}}{c_{12}s_{12}(|m_1|e^{-i\beta_1}-|m_2|e^{-i\beta_2})}\,, \\ g &= \frac{(c_{12}^2s_{13}^2-s_{12}^2)|m_1|e^{-i\beta_1}+(s_{12}^2s_{13}^2-c_{12}^2)|m_2|e^{-i\beta_2}+m_3c_{13}^2}{s_{12}^2|m_1|e^{-i\beta_1}+c_{12}^2|m_2|e^{-i\beta_2}}\,, \\ h &= \frac{(|m_1|e^{-i\beta_1}-|m_2|e^{-i\beta_2})\sin\,2\theta_{23}\,\sin\,2\theta_{12}e^{-i\delta_{CP}}}{s_{12}^2|m_1|e^{-i\beta_1}+c_{12}^2|m_2|e^{-i\beta_2}}\,. \end{aligned} \tag{7}$$

Using the expressions above, let us analyse the conditions for $|\hat{\delta}|,|\hat{\epsilon}| \ll 1$, in the three possible approaches given by the hierarchies and using the central values for the mixing parameters [1].

For NH, where $|m_1| \ll |m_2| \approx \sqrt{\Delta m^2_{sol}} \ll m_3 \approx \sqrt{\Delta m^2_{ATM}}$, we have

$$f \approx \frac{e^{i(\delta_{CP}+\beta_2)}}{s_{12}\;c_{12}}\sqrt{\frac{\Delta m^2_{ATM}}{\Delta m^2_{sol}}}\left(1-\mathcal{O}\left(\frac{\Delta m^2_{sol}}{\Delta m^2_{ATM}}\right)\right)\,,\;|f|\sim 12.5\,, \tag{8}$$

that implies that $|\hat{\delta}| \sim 3.26$. Thus, in NH, the breaking of $\mu-\tau$ symmetry by $\hat{\delta}$ is always large.

For IH, we have $|m_1| \approx \sqrt{\Delta m^2_{ATM}}\,,\; |m_2| \approx \sqrt{\Delta m^2_{sol}+\Delta m^2_{ATM}} \gg m_3$, which gives

$$f \approx \frac{-e^{-i\delta_{CP}}(c_{12}^2e^{-i(\beta_1-\beta_2)}+s_{12}^2+\frac{s_{12}^2}{2}\frac{\Delta m^2_{sol}}{\Delta m^2_{ATM}})}{s_{12}\;c_{12}\left(1-e^{-i(\beta_1-\beta_2)}+\frac{1}{2}\frac{\Delta m^2_{sol}}{\Delta m^2_{ATM}}\right)}\,. \tag{9}$$

We can see that $|f|$ is very large when $\beta_1-\beta_2=0$, whereas, if $\beta_1-\beta_2=\pm\pi$ then $|f|\sim 1$, which gives $|\hat{\delta}|\sim 0.1$. Also, from Eq. (6)

$$g \approx \frac{(c_{12}^2s_{13}^2-s_{12}^2)e^{-i(\beta_1-\beta_2)}+(s_{12}^2s_{13}^2-c_{12}^2)}{s_{12}^2e^{-i(\beta_1-\beta_2)}+c_{12}^2}\,,\quad h\approx\frac{(e^{-i(\beta_1-\beta_2)}-1)\sin\,2\theta_{23}\,\sin\,2\theta_{12}e^{-i\delta_{CP}}}{s_{12}^2e^{-i(\beta_1-\beta_2)}+c_{12}^2}\,. \tag{10}$$

Thus, $\beta_1-\beta_2=\pm\pi$ gives $g\approx -1$ and $h\approx 4e^{-i\delta_{CP}}$, then $|\hat{\epsilon}|\sim 0.6$. Hence, the largest contribution to $\mu-\tau$ breaking in this case comes from $|\hat{\epsilon}|$ rather than $|\hat{\delta}|$.

For the degenerate hierarchy (DH), where $|m_1|\approx|m_2|\approx m_3$, we obtain

$$f \approx \frac{-e^{-i\delta_{CP}}(c_{12}^2e^{-i(\beta_1-\beta_2)}+s_{12}^2)-e^{i\delta_{CP}}e^{i\beta_1}}{s_{12}\;c_{12}(1-e^{-i(\beta_1-\beta_2)})}\,, \tag{11}$$

$$g \approx \frac{(c_{12}^2s_{13}^2-s_{12}^2)e^{-i(\beta_1-\beta_2)}+(s_{12}^2s_{13}^2-c_{12}^2)+c_{13}^2e^{-i\beta_1}}{s_{12}^2e^{-i(\beta_1-\beta_2)}+c_{12}^2}\,, \tag{12}$$

$$h \approx \frac{(e^{-i(\beta_1-\beta_2)}-1)\sin\,2\theta_{23}\,\sin\,2\theta_{12}e^{-i\delta_{CP}}}{s_{12}^2e^{-i(\beta_1-\beta_2)}+c_{12}^2}\,. \tag{13}$$

We again note that $|f|$ is strongly enhanced for $\beta_1-\beta_2=0$, which implies that $|\hat{\delta}|>1$, e.g., for $\beta_1=\beta_2=\pi$ and $\delta_{CP}=-\pi/2$, we obtain $|f|\sim 68$, which leads to $|\hat{\delta}|\sim 2$. However, other values of $CPVP$ may lead to small breaking parameters. As an example, for $\beta_1=\pi$, $\beta_2=\pi/2$ and $\delta_{CP}=-\pi/2$, we obtain $|f|\sim 0.6$, $|g|\sim 2$, and $|h|\sim 1.7$, such that $|\hat{\delta}|\sim 0.2$ and $|\hat{\epsilon}|\sim 0.1$, which is desirable.

An interesting conclusion of the last analytical approach is that the case of equal Majorana phases is strongly disfavored in any hierarchy. It is also consistent with the numerical analysis presented in the following.

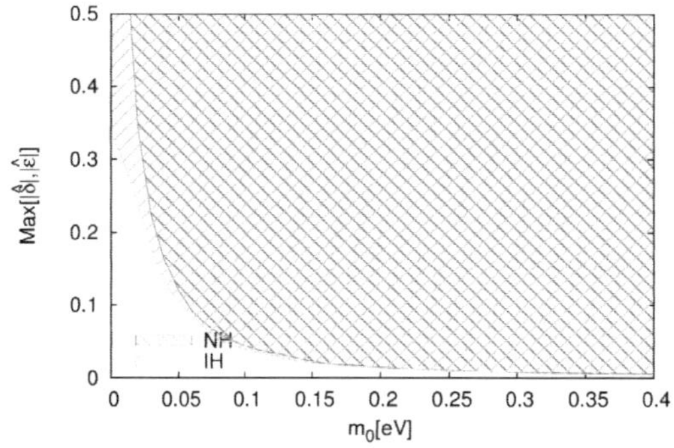

Figure 1. Allowed region (hashed area) based on the experimental data for Max[$|\hat{\delta}|$, $|\hat{\epsilon}|$], the maximum between $|\hat{\delta}|$ and $|\hat{\epsilon}|$, as a function of the lightest neutrino mass, m_0, for NH and IH. The regions are obtained by varying $CPVP$ within $[0, 2\pi]$ interval.

In Fig. 1, we show the allowed region for the symmetry breaking parameters obtained by varying $CPVP$ in the $(0 - 2\pi)$ interval, and taking the mixing parameters within 3σ level. We depict the Max[$|\hat{\delta}|$,$|\hat{\epsilon}|$], as a function of m_0. According to this figure, we can see that for $m_0 \sim 0$ eV in the NH, the breaking parameters are very large regardless of the $CPVP$ values, as found in the previous analysis. In contrast, in the IH the breaking is at least of 30%. Finally, we can see that for $m_0 \sim 0.05$ eV, which corresponds to the DH, the breaking can be less than 10%. This indicates that DH is the best hierarchy for regarding $\mu - \tau$ as a good approximate symmetry, which is consistent with the results obtained in [6] using a different approach.

If we impose the phenomenological requirement of Max[$|\hat{\delta}|, |\hat{\epsilon}|$] < 0.1, or any other selected small value, we can examine the $CPVP$ that match with this condition. We will study this point of view in the following.

3. Majorana phases and $\mu - \tau$ symmetry

We have shown that the requirement for small symmetry breaking favors the DH. Accordingly, we pick up $m_0 = 0.1$ eV in the following and use the exact expressions in Eqs. (6) and (7) for our numerical analysis. Once we have selected m_0, $\hat{\delta}$ and $\hat{\epsilon}$ only depend on 3 $CPVP$. Thus, by selecting a suitable value of the Dirac phase, δ_{CP}, any two given values for the Majorana phases will provide a unique set of absolute breaking parameters, and vice versa. Therefore, only a well-defined allowed region in the Majorana CP phases space will be consistent with our symmetry breaking requirement. This is shown in Fig. 2, where the left-hand side shows the allowed region for the two Majorana phases which match with the requisite that Max[$|\hat{\delta}|$, $|\hat{\epsilon}|$]≤ 0.25. We select $\delta_{CP} = -\pi/2$, which is near to the best fit value reported by [1, 2]. The other mixing parameters are varied up to 1 σ and 3 σ from their central values, respectively. We can see that there is a strong correlation between the two Majorana phases, which can be inferred from the expressions in Eqs. (11 -13), where the relative phase $\beta_1 - \beta_2$ appears. This figure shows that the values of $\beta_1 = \pi$ and $\beta_2 = \pi/2$ analysed in the previous section, which lead to Max[$|\hat{\delta}|$, $|\hat{\epsilon}|$]≈ 0.2, are contained well within the allowed regions.

A similar study with different values for the Dirac CP phase shows a strong correlation of the allowed Majorana phases with the former.

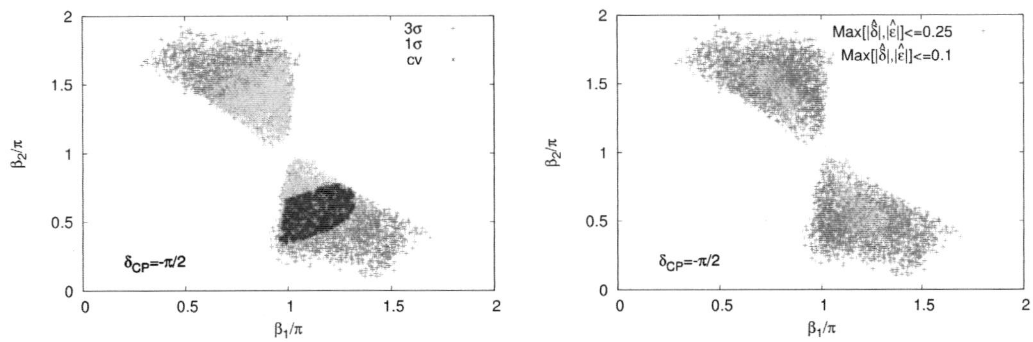

Figure 2. In the left-hand plot, the green color (cross signs) corresponds to the mixing parameters varied up to 1σ, the red color (plus signs) corresponds to the 3σ, and the blue color (stars) corresponds to the central values (cv) of the mixing parameters, which match with Max[$|\hat{\delta}|$, $|\hat{\epsilon}|$]≤ 0.25. In the right-hand plot, the green color (cross signs) corresponds to Max[$|\hat{\delta}|$, $|\hat{\epsilon}|$]≤ 0.1 and the red color (plus signs) corresponds to Max[$|\hat{\delta}|$, $|\hat{\epsilon}|$]≤ 0.25, where the experimental values are varied within 3σ.

4. Concluding remarks

As it is shown in Fig. 2, null Majorana phases are completely excluded at the 3σ level, which is consistent with the analysis presented in sec. 2. The allowed region is not modified greatly when we move the experimental values from 1σ to 3σ deviation limits, as we can see from the left plot in Fig. 2. Thus, the correlation is only slightly sensitive to small mixing angle variations. By contrast, in the right hand side of the same plot, when we restrict the upper bound condition even more on Max[$|\hat{\delta}|$, $|\hat{\epsilon}|$], the allowed area is reduced greatly. Therefore, we can conclude that the size of the breaking of $\mu - \tau$ symmetry is highly sensitive to the Majorana phases and vice-versa. Finally, we can make two important conclusions. First, if we regard $\mu - \tau$ as a good approximate symmetry, then the Majorana phases will definitely be non-zero. Second, a future determination of the Dirac phase can restrict the Majorana phases that are compatible with the condition of a weak breaking of $\mu - \tau$. It is interesting that a recent global analysis provided by [2] confirmed the previous preference for δ_{CP} near $-\pi/2$, which brings our attention to the robust prediction of the Majorana phases obtained for this value. More detailed information about this work can be found in [7].

Acknowledgments

DCRA thanks the organizing committee of the XXX Annual Meeting of the Division of Particles and Fields of the Mexican Physical Society, for the economical support to attend the event. This work was partially supported by CONACyT, México, under grant No.237004.

References

[1] Capozzi F *et al.* 2014 *Phys. Rev.* **D89** 093018.
[2] Capozzi F *et al.* 2016 *Nucl. Phys.* **B908** 218.
[3] Pontecorvo B 1957 *J. Exptl. Theoret. Phys.* **33** 549.
[4] Maki Z, Nakagawa M, and Sakata S 1962 *Prog. Theor. Phys.* **28** 870.
[5] Xing Z and Zhao Z 2016 *Rep. Prog. Phys.* **79** 076201.
[6] Gupta S, Joshipura A, and Patel K 2013 *J. High Energy Phys.* **09** 035.
[7] Rivera-Agudelo D and Pérez-Lorenzana A 2016 *Phys. Lett.* **B760** 153.

Fermion masses as mixing parameters in the SM

U J Saldaña-Salazar

Facultad de Ciencias Físico-Matemáticas, Benemérita Universidad Autónoma de Puebla, C.P. 72570, Puebla, Pue., México.

E-mail: ulisesjesus@protonmail.ch

Abstract. Flavor transitions via the charged current interactions are parametrized by a three dimensional and unitary transformation. This so called mixing matrix requires of four mixing parameters. Here we show that under the phenomenological observation of hierarchical fermion masses, $m_3 \gg m_2 \gg m_1$, a mixing parametrization can be built with its mixing parameters being the corresponding four independent mass ratios of each fermion sector, i.e., m_u/m_c, m_c/m_t, m_d/m_s, and m_s/m_b and m_e/m_μ, m_μ/m_τ, $m_{\nu 1}/m_{\nu 2}$, and $m_{\nu 2}/m_{\nu 3}$, for the quark and lepton sector, respectively.

1. Introduction

Physical theories with a less number of independent parameters are not only more predictive when compared to those with a larger number, but do also offer a better and more comprehensible description of Nature. Here we show that the Standard Model (SM) of the electroweak interactions can have its set of independent parameters significantly reduced [1].

The SM Lagrangian, written in the interaction basis, has the fermion mass matrices as three by three arbitrary and complex matrices. It is by virtue of a biunitary transformation that the mass matrices can be expressed in its diagonal form with all its entries being now real and positive. This set of biunitary transformations are nothing more than the needed independent rotations which must be, in general, simultaneously applied to the left- and right-handed fields of each fermion type. Generically, we have the following picture

(i) Weak or Interaction basis:

$$\mathbf{M} = \mathbf{L}^\dagger \mathbf{\Sigma} \mathbf{R} = \begin{pmatrix} m_{11} & m_{12} & m_{13} \\ m_{21} & m_{22} & m_{23} \\ m_{31} & m_{32} & m_{33} \end{pmatrix}, \tag{1}$$

where $\mathbf{L}$ and $\mathbf{R}$ are unitary transformations acting, respectively, in the left- and right-handed fields of the same fermion type, $\mathbf{\Sigma} \equiv \mathrm{Diag}(m_1, m_2, m_3)$ with $m_j \geq 0$, and m_{ij} are complex numbers.

(ii) Mass basis:

$$\mathbf{L}\mathbf{M}\mathbf{R}^\dagger = \mathbf{\Sigma} = \begin{pmatrix} m_1 & 0 & 0 \\ 0 & m_2 & 0 \\ 0 & 0 & m_3 \end{pmatrix}, \tag{2}$$

notice that $\mathbf{L}\mathbf{M}\mathbf{M}^\dagger\mathbf{L}^\dagger = \mathbf{\Sigma}^2$ and $\mathbf{R}\mathbf{M}^\dagger\mathbf{M}\mathbf{R}^\dagger = \mathbf{\Sigma}^2$.

Up to now, nothing has been said about the charged currents

$$\mathcal{J}^{\mu}_{q,\mathrm{IB}} \propto \bar{u}_{L,i}\gamma^{\mu}\delta_{ij}d_{L,j} \qquad \text{and} \qquad \mathcal{J}^{\mu}_{\ell,\mathrm{IB}} \propto \bar{e}_{L,i}\gamma^{\mu}\delta_{ij}\nu_{L,j}, \tag{3}$$

which in the interaction basis (IB) are diagonal and thus conserve flavor. After moving to the mass basis (MB), the charged currents suffer the following change

$$\mathcal{J}^{\mu}_{q,\mathrm{MB}} \propto \bar{u}'_{L,i}\gamma^{\mu}\mathbf{V}_{ij}d'_{L,j} \qquad \text{and} \qquad \mathcal{J}^{\mu}_{\ell,\mathrm{MB}} \propto \bar{e}'_{L,i}\gamma^{\mu}\mathbf{U}_{ij}\nu'_{L,j}, \tag{4}$$

where $\mathbf{V} = \mathbf{L}_u\mathbf{L}_d^{\dagger}$ and $\mathbf{U} = \mathbf{L}_e\mathbf{L}_{\nu}^{\dagger}$ are the quark and lepton mixing matrix, respectively. In fact, they are named after their inventors as Cabibbo–Kobayashi–Maskawa (CKM) and Pontecorvo–Maki–Nakagawa–Sakata (PMNS) for the quark and lepton sector, respectively. From this point onwards, they will be denoted, following tradition, by $\mathbf{V}_{\mathrm{CKM}} \equiv \mathbf{V}$ and $\mathbf{U}_{\mathrm{PMNS}} \equiv \mathbf{U}$.

A three dimensional unitary transformation requires nine parameters to be fully described; three of them are real while the remaining six are complex phases. Freedom to redefine the fields by a global phase implies five independent transformations per fermion sector. This then translates into a mixing matrix whose description requires only four independent mixing parameters (i.e., three real and one complex phase). There is no unique way to parametrize the mixing among fermions. In fact, different alternatives can be found in the literature [1–5]. And it is precisely a mixing parametrization in terms of fermion masses what we are trying to propose here.

A mixing parametrization can also have all of its mixing parameters purely real. The only requirement it should fulfill though is that after rephasing all the corresponding fields the unitary transformation cannot be made real. A good example of this is the well known Wolfenstein parametrization which is used in the description of quark mixing in its unitary form.

A basis-independent measure of CP violation can be achieved through the rephasing invariant quantity called Jarlskog invariant [6]

$$J_f \sim \mathrm{Im}\left[\mathrm{Det}\left([\mathbf{M}_a\mathbf{M}_a^{\dagger}, \mathbf{M}_b\mathbf{M}_b^{\dagger}]\right)\right], \tag{5}$$

where $f = q, \ell$, $a = u, \nu$, and $b = d, e$.

Current knowledge about fermion mixing data shows the following values [7]

$$|\mathbf{V}_{\mathrm{CKM}}| = \begin{pmatrix} 0.97427 \pm 0.00014 & 0.22536 \pm 0.00061 & 0.00355 \pm 0.00015 \\ 0.22522 \pm 0.00061 & 0.97343 \pm 0.00015 & 0.0414 \pm 0.0012 \\ 0.00886^{+0.00033}_{-0.00032} & 0.0405^{+0.0011}_{-0.0012} & 0.99914 \pm 0.00005 \end{pmatrix}, \tag{6}$$

with the Jarlskog invariant equal to $J_q = (3.06^{+0.21}_{-0.20}) \times 10^{-5}$. In the standard parametrization by the Particle Data Group (PDG), the central values give the following mixing angles,

$$\theta^q_{12} \approx 13.3^{\circ}, \qquad \theta^q_{13} \approx 0.2^{\circ}, \qquad \theta^q_{23} \approx 2.4^{\circ}. \tag{7}$$

Whereas the most recent update on the 3σ allowed ranges of the elements of the PMNS mixing matrix gives [8]

$$|\mathbf{U}_{\mathrm{PMNS}}| = \begin{pmatrix} 0.801 \to 0.845 & 0.514 \to 0.580 & 0.137 \to 0.158 \\ 0.225 \to 0.517 & 0.441 \to 0.699 & 0.614 \to 0.793 \\ 0.246 \to 0.529 & 0.464 \to 0.713 & 0.590 \to 0.776 \end{pmatrix}, \tag{8}$$

where the best fit points for the mixing angles are

$$\theta^{\ell}_{12} \approx 33.48^{\circ}, \qquad \theta^{\ell}_{13} \approx 8.50^{\circ}, \qquad \theta^{\ell}_{23} \approx 42.3^{\circ}. \tag{9}$$

Some remarks now follow. The quark mixing matrix is close to the unit matrix with $|V_{12}|$ being its largest off-diagonal element. In the PDG parametrization this is translated into two very small angles ($\sim 1^\circ$) and a third one with order 10°. On the other hand, the leptonic mixing matrix is by far quite different from this picture. Its smallest matrix element is of the same order as the largest one in the quark sector. In the PDG parametrization this gets translated into two large angles (one of them close to maximal) and one small angle. So we have two mixing matrices that from their structure really seem to come from different origins. This aspect is one of many which are commonly known as the *flavour puzzle.* So how can we understand this observed discrepancy? What we want to show here is that we do not need to go beyond the standard model to understand the observed values in mixing phenomena.

We take the Gatto–Sartori–Tonin (GST) relation [9], which coincides surprisingly well with the Cabibbo angle,

$$\tan\theta_{12}^{q} \approx \sqrt{\frac{m_d}{m_s}} \approx 0.22, \tag{10}$$

as the strongest hint to understand this aspect of the flavour puzzle. This relation is pointing out to a possibility: *a mixing parametrization with fermion mass ratios as its mixing parameters.* But, does it make any sense to think on this possibility? We have six masses per fermion sector and only four mass ratios are independent: $m_{a,1}/m_{a,2}$, $m_{a,2}/m_{a,3}$, $m_{b,1}/m_{b,2}$, and $m_{b,2}/m_{b,3}$. In fact, only in the two and three family cases we have sufficient (more or an equal number of) mass ratios as mixing parameters [1]. Beyond three families we cannot think of building such a mixing parametrization with mass ratios.

In the following, we try to guide the reader through a list of questions and answers.

2. From where we should start?

Recall the generic mixing matrix definition

$$\mathbf{V} = \mathbf{L}_a \mathbf{L}_b^{\dagger}, \tag{11}$$

where $a = u, e$ and $b = d, \nu$ and $\mathbf{V}$ equals the CKM or PMNS mixing matrix, respectively. As we want to find

$$\mathbf{V} = \mathbf{V}\left(\frac{m_{a,1}}{m_{a,2}}, \frac{m_{a,2}}{m_{a,3}}, \frac{m_{b,1}}{m_{b,2}}, \frac{m_{b,2}}{m_{b,3}}\right), \tag{12}$$

this implies we should look out for the possibility of expressing $\mathbf{L}_f = \mathbf{L}_f(m_{f,1}/m_{f,2}, m_{f,2}/m_{f,3})$. How can we connect this unitary transformations acting on the left-handed fields to the corresponding fermion masses? We need to consider that they diagonalize the left Hermitian product of the mass matrices,

$$\mathbf{L}_f \mathbf{M}_f \mathbf{M}_f^{\dagger} \mathbf{L}_f^{\dagger} = \mathrm{Diag}\left[m_{f,1}^2, m_{f,2}^2, m_{f,3}^2\right]. \tag{13}$$

Any existing relation between mass ratios and the unitary transformations should emerge from here. The use of the three matrix invariants gives,

$$\mathrm{Det}\left[\mathbf{M}_f \mathbf{M}_f^{\dagger}\right] = x_1 x_2 x_3 - x_1|y_3|^2 - x_2|y_2|^2 - x_3|y_1|^2 + 2\,\mathrm{Re}(y_1 y_2^* y_3) = m_{f,1}^2 m_{f,2}^2 m_{f,3}^2, \tag{14}$$

$$\mathrm{Tr}\left[(\mathbf{M}_f \mathbf{M}_f^{\dagger})^2\right] = x_1^2 + x_2^2 + x_3^2 + 2(|y_1|^2 + |y_2|^2 + |y_3|^2) = m_{f,1}^4 + m_{f,2}^4 + m_{f,3}^4, \tag{15}$$

$$\mathrm{Tr}\left[\mathbf{M}_f \mathbf{M}_f^{\dagger}\right] = x_1 + x_2 + x_3 = m_{f,1}^2 + m_{f,2}^2 + m_{f,3}^2. \tag{16}$$

where we have denoted by $x_i = [\mathbf{M}_f\mathbf{M}_f^\dagger]_{ii}$ and $y_1 = [\mathbf{M}_f\mathbf{M}_f^\dagger]_{12}$, $y_2 = [\mathbf{M}_f\mathbf{M}_f^\dagger]_{13}$, and $y_3 = [\mathbf{M}_f\mathbf{M}_f^\dagger]_{23}$, the matrix elements of the Hermitian product

$$\mathbf{M}_f\mathbf{M}_f^\dagger = \begin{pmatrix} |m_{11}|^2+|m_{12}|^2+|m_{13}|^2 & m_{11}m_{21}^*+m_{12}m_{22}^*+m_{13}m_{23}^* & m_{11}m_{31}^*+m_{12}m_{32}^*+m_{13}m_{33}^* \\ & |m_{21}|^2+|m_{22}|^2+|m_{23}|^2 & m_{21}m_{31}^*+m_{22}m_{32}^*+m_{23}m_{33}^* \\ & & |m_{31}|^2+|m_{32}|^2+|m_{33}|^2 \end{pmatrix}. \tag{17}$$

No unique and exact solution exists within the SM for this non-linear system of equations.

3. What about an approximated solution?

An important feature masses of the charged fermion species share is their observed hierarchical structure

$$m_{f,1} \ll m_{f,2} \ll m_{f,3}, \tag{18}$$

which in terms of orders of magnitude becomes

$$\begin{gathered} m_u : m_c : m_t \approx 10^{-5} : 10^{-3} : 1, \qquad m_d : m_s : m_b \approx 10^{-4} : 10^{-2} : 1, \\ m_e : m_\mu : m_\tau \approx 10^{-4} : 10^{-2} : 1. \end{gathered} \tag{19}$$

Even for neutrinos, the two squared mass differences measured from neutrino oscillations obey a hierarchy although much weaker,

$$\Delta m_{21}^2 : \Delta m_{31(32)}^2 \approx 10^{-2} : 1. \tag{20}$$

Could we use this hierarchical nature of fermion masses to find out our desired approximated solution? The answer is yes.

3.1. A first step towards an approximated solution

Two questions must be answered before considering an approximated solution. Which conditions should be met by the fermion masses in order to guarantee that we have a unique solution? Which theorems will support the uniqueness of our approximations?

The answer for the former question follows from the Singular Value Decomposition (SVD) of a complex matrix. The two unitary transformations diagonalizing it are unique up to some global complex phases whenever the singular values (fermions masses) have non-degeneracy and can thus be ordered from the smallest to the largest one.

On the other hand, the answer to the second question comes from the following. When considering some of the singular values as zero (because when compared to the rest are rather small) then we need to consider a theorem named Schmidt–Mirsky approximation theorem that tells us we are safe to use such lower rank approximations and still guarantee the uniqueness of the computed unitary transformations.

3.2. A generic treatment

We choose to work in the interaction basis which has all mass matrices still non-diagonal. From now on we will denote them generically by $\mathbf{M}$ and its singular values (fermion masses) by m_i $(i = 1, 2, 3)$.

The SVD provides the relation,

$$\mathbf{M} = \sum_i \ell_i m_i r_i^\dagger = \left[\left(l_1 \frac{m_1}{m_2} r_1^\dagger + l_2 r_2^\dagger\right)\frac{m_2}{m_3} + l_3 r_3^\dagger\right] m_3, \tag{21}$$

where ℓ_i and r_i are the singular vectors of each singular value. This expression points to the fact that the fermion mass ratios m_1/m_2 and m_2/m_3 play the dominant rôle in determining the structure of the mass matrix whereas m_3 sets the overall mass scale.

The lower rank approximations can be obtained from either $m_3 \gg m_2, m_1$

$$\mathbf{M}_{r=1} = m_3 l_3 r_3^\dagger, \tag{22}$$

or $m_3, m_2 \gg m_1$

$$\mathbf{M}_{r=2} = \left(\frac{m_2}{m_3} l_2 r_2^\dagger + l_3 r_3^\dagger\right) m_3. \tag{23}$$

3.3. The rank one approximation

In general, the SVD tells us that the rank one approximation is given by

$$\mathbf{M}_{r=1} = m_3 \begin{pmatrix} \ell_{31} r_{31}^* & \ell_{31} r_{32}^* & \ell_{31} r_{33}^* \\ \ell_{32} r_{31}^* & \ell_{32} r_{32}^* & \ell_{32} r_{33}^* \\ \ell_{33} r_{31}^* & \ell_{33} r_{32}^* & \ell_{33} r_{33}^* \end{pmatrix}. \tag{24}$$

However, if the first and second families are massless we should expect no mixing among the fermions $\mathbf{L}(m_1 = 0, m_2 = 0) = \mathbf{1}_{3\times 3}$. Therefore, we actually know the left Hermitian product

$$\mathbf{M}_{r=1}\mathbf{M}_{r=1}{}^\dagger = m_3^2 \ell_3 \ell_3^\dagger = m_3^2 \begin{pmatrix} 0 & 0 & 0 \\ 0 & 0 & 0 \\ 0 & 0 & 1 \end{pmatrix}. \tag{25}$$

In order to find $\mathbf{M}_{r=1}$ and not only the left Hermitian product we need to remember that it should be possible to consider an electroweak basis which explicitly shows no coupling between the massless families and the Higgs field. In that basis

$$|\mathbf{M}_{r=1}| = m_3 \begin{pmatrix} 0 & 0 & 0 \\ 0 & 0 & 0 \\ 0 & 0 & 1 \end{pmatrix}. \tag{26}$$

3.4. The rank two approximation

Setting the mass of the second family different from zero can produce mixing between the second and third fermion generations,

$$\mathbf{L}(m_1 = 0, m_2) = \mathbf{L}_{23}\left(\frac{m_2}{m_3}\right) = \begin{pmatrix} 1 & 0 & 0 \\ 0 & c_{23} & s_{23} e^{-i\delta_{23}} \\ 0 & -s_{23} e^{i\delta_{23}} & c_{23} \end{pmatrix}, \tag{27}$$

At this point we still do not know what the relation between the rotation angle θ_{23} and the mass ratio m_2/m_3 is. But what we do know is the hierarchy which means a small angle $\theta_{23} \ll 1$. This allows us to take the Taylor expansion to first order

$$\mathbf{L}(m_1 = 0, m_2) = \mathbf{L}_{23}\left(\frac{m_2}{m_3}\right) = \begin{pmatrix} 1 & 0 & 0 \\ 0 & 1 & \theta_{23} e^{-i\delta_{23}} \\ 0 & -\theta_{23} e^{i\delta_{23}} & 1 \end{pmatrix}, \tag{28}$$

which then gives to second order,

$$|\mathbf{M}_{r=2}\mathbf{M}_{r=2}{}^\dagger| \sim m_3^2 \begin{pmatrix} 0 & 0 & 0 \\ 0 & |\theta_{23}|^2 & |\theta_{23}| \\ 0 & |\theta_{23}| & 1 + |\theta_{23}|^2 \end{pmatrix}. \tag{29}$$

To know the matrix form of $|\mathbf{M}_{r=2}|$ we need to assume that the right Hermitian product $\mathbf{M}^\dagger\mathbf{M}$ is equal to its left counterpart $\mathbf{M}\mathbf{M}^\dagger$. This is equivalent to assuming the mass matrices to be normal. A normal matrix $\mathbf{A}$ satisfies the relation $\mathbf{A}\mathbf{A}^\dagger = \mathbf{A}^\dagger\mathbf{A}$. We find to second order a similar expansion as before

$$|\mathbf{M}_{r=2}| \sim m_3 \begin{pmatrix} 0 & 0 & 0 \\ 0 & |\theta_{23}|^2 & |\theta_{23}| \\ 0 & |\theta_{23}| & 1+|\theta_{23}|^2 \end{pmatrix} . \tag{30}$$

We take this hierarchical structure and translate it into the following matrix form

$$\mathbf{M}_{r=2} = \begin{pmatrix} 0 & 0 & 0 \\ 0 & m_{22} & m_{23} \\ 0 & m_{32} & m_{33} \end{pmatrix} \tag{31}$$

with the two conditions $|m_{23}| = |m_{32}|$ and $|m_{22}| = |m_{23}|^2$. Complex phases are constrained by a single trascendental equation with tangent functions. Because we expect $|m_{22}| \sim \theta_{23}^2 \ll 1$ to be much smaller than one, then, when we study the left Hermitian product of the mass matrices $\mathbf{M}\mathbf{M}^\dagger$ its appearance can be neglected as it only provides contributions of order three and four, $\mathcal{O}(\theta^3)$ and $\mathcal{O}(\theta^4)$. Thus we choose to work with the approximation

$$\mathbf{M}_{r=2} \simeq \begin{pmatrix} 0 & 0 & 0 \\ 0 & 0 & m_{23} \\ 0 & m_{32} & m_{33} \end{pmatrix} . \tag{32}$$

In fact, this assumption needs a more careful treatment and its details can be found in [10].

It is straightforward to show that $|m_{23}| = \sqrt{m_2 m_3}$ and $|m_{33}| = m_3 - m_2$. Moreover, we find that the diagonalization of such a matrix requires the angle of rotation to satisfy a GST-like relation

$$\tan^2\theta_{23} = \frac{m_2}{m_3}. \tag{33}$$

3.5. Ansatz on the complex phases

The special unitary transformation diagonalizing $\mathbf{M}_{r=2}$ has one phase and one angle

$$\mathbf{L}_{23} = \mathbf{L}_{23}(\theta_{23}, \delta_{23}) \tag{34}$$

with the angle satisfying the GST relation. However, we cannot expect any new dependence on more free parameters. If the four mass ratios can really behave as mixing parameters no significant effect should come from the complex phases apart from determining if either we have orthogonal transformations in its two varieties: clockwise ($\delta_{23} = 0$) or counter-clockwise ($\delta_{23} = \pi$); or special unitary transformations having an i factor in the off diagonal elements also in its two possible varieties ($\delta_{23} = \pi/2,\ 3\pi/2$). Therefore, we consider this set of four possible values as an *ansatz* that complex phases should follow [11, 12].

4. What is the effect produced by each complex phase value?

One can show that complex phases will always appear as phase differences,

$$\delta_{ij}^a - \delta_{ij}^b. \tag{35}$$

Thus, from now on we will consider that only one of them takes a value while the other is zero,

$$\delta_{ij}^a = 0 \qquad \text{and} \qquad \delta_{ij}^b = 0,\ \frac{\pi}{2},\ \pi,\ \frac{3\pi}{2}. \tag{36}$$

We will not go here into more details but one can also see that when studying mixing between two families three different cases arise: a minimal ($\delta^b_{ij} = 0$) or maximal ($\delta^b_{ij} = \pi$) value of mixing without CP violation or a medium value of mixing ($\delta^b_{ij} = \pi/2, 3\pi/2$) with CP violation.

Then, when studying mixing in the full rank picture, which basically means within one fermion type

$$L^f = L^f_{12} L^f_{13} L^f_{23}, \tag{37}$$

that is, three successive rotations in the three different two-family planes, what we could, on general grounds, anticipate is the fact that we will be allowed to consider $4^3 = 64$ possible combinations. However, mixing in the 2-3 sector (two families) should not introduce any CP violation. Therefore, we can only consider it to have either minimal or maximal mixing. Then this reduces our combinations to $4^2 \times 2 = 32$. Moreover, a symmetry argument pointing to the fact that we should only consider a complex phase implying CP violation in the 1-2 sector now follows. In general, our mass matrices can be considered as rank one rather than rank two. That is, the contribution added from the rank one to the rank two is negligible. Therefore, we should expect our mass matrices to approximately conserved the global and accidental $U(2)^3$ flavour symmetry. Hence, we will constrain our combinations to:

$$L^f = L^f_{12}\left(\frac{m_1}{m_2}, \frac{\pi}{2} \text{ or } \frac{3\pi}{2}\right) L^f_{13}\left(\frac{m_1}{m_3}, 0 \text{ or } \pi\right) L^f_{23}\left(\frac{m_2}{m_3}, 0 \text{ or } \pi\right). \tag{38}$$

Implying we will have $2^3 = 8$ possible combinations. We lack a principle that could explain us which case we should expect. But what we know is that if the four mass ratios parametrization really works we should find one possible case from the eight agreeing with what has been already observed.

5. How are the approximations improved?

In the full rank picture, passing from two massive families to three, we need to set m_1 different from zero. But, this in return, as expected from the lower rank approximation theorem, should provide contributions proportional to it in all the matrix elements. Even in the 2-3 sector which we had already diagonalized.

Thus, recall we have already diagonalized the 2-3 sector. However, after moving to the full rank picture we should consider diagonalizing again this sector by some angle proportional to m_1. We treat this by introducing the ansatz:

$$L_{23} = L^{(2)}_{23}\left(\frac{m_1 m_2}{m_3^2}\right) L^{(1)}_{23}\left(\frac{m_1}{m_3}\right) L^{(0)}_{23}\left(\frac{m_2}{m_3}\right). \tag{39}$$

Similar, for the 1-3 sector we have the ansatz

$$L_{13} = L^{(2)}_{13}\left(\frac{m_1 m_2}{m_3^2}\right) L^{(1)}_{13}\left(\frac{m_2^2}{m_3^2}\right) L^{(0)}_{13}\left(\frac{m_1}{m_3}\right). \tag{40}$$

The 1-2 sector has no approximations as rotating this sector already includes the two corresponding masses. Then, we only require a single rotation

$$L_{12} = L^{(0)}_{12}\left(\frac{m_1}{m_2}\right). \tag{41}$$

Note that all the new phases appearing in the added two more rotations in the 2-3 and 1-3 sector should be taken such that they either minimize or maximize the initial value if the initial rotation was meant to produce minimal or maximal mixing, respectively.

The complete expression to diagonalize the mass matrix of one fermion type is thus

$$L_f = L_{12} L_{13} L_{23}. \tag{42}$$

6. Discussions and Conclusions

The quark sector is found to be well described by minimal mixing in both the 2-3 and 1-3 sector, while medium mixing in the 1-2 sector with $\pi/2$. The agreement to experiment is simply great

$$|V_{\rm CKM}^{\rm th}| = \begin{pmatrix} 0.974^{+0.004}_{-0.003} & 0.225^{+0.016}_{-0.011} & 0.0031^{+0.0018}_{-0.0015} \\ 0.225^{+0.016}_{-0.011} & 0.974^{+0.004}_{-0.003} & 0.039^{+0.005}_{-0.004} \\ 0.0087^{+0.0010}_{-0.0008} & 0.038^{+0.004}_{-0.004} & 0.9992^{+0.0002}_{-0.0001} \end{pmatrix} \tag{43}$$

with the Jarlskog invariant being $J_q = (2.6^{+1.3}_{-1.0}) \times 10^{-5}$.

This approach can be extended to the lepton sector. Knowledge of the electron and muon mass plus the value of $|[\mathbf{U}_{\rm PMNS}]_{12}| \simeq 0.54$ allows the computation of the neutrino masses, which are found to be

$$m_{\nu 1} = (0.0041 \pm 0.0015)\,{\rm eV}, \quad m_{\nu 2} = (0.0096 \pm 0.0005)\,{\rm eV}, \quad m_{\nu 3} = (0.050 \pm 0.001)\,{\rm eV}. \tag{44}$$

These values in return are then used to calculate the full leptonic mixing matrix, which in the PDG parametrization results into

$$\sin^2\theta_{23}^{\rm th} = 0.54^{+0.03}_{-0.03} \qquad \sin^2\theta_{12}^{\rm th} = 0.30^{+0.07}_{-0.09} \qquad \sin^2\theta_{13}^{\rm th} = 0.019^{+0.009}_{-0.007}, \tag{45}$$

with the Jarlskog invariant being $J_\ell = -0.031^{+0.006}_{-0.007}$. Leptonic mixing requires maximal mixing in the 2-3 sector, minimal in the 1-3 sector, and medium in the 1-2 sector with $3\pi/2$.

Hierarchical masses within the quark sector led us to the computed mixing formulas. However, the fact that neutrinos also give an excellent agreement without having such a strong hierarchy leaves the sensation that some mechanism independent of the hierarchy in the masses is being responsible for the way mixing happens. That is, the formulas we found are applicable for all possible values of fermion masses and are not a consequence of hierarchical masses. In this regard, what we called the Flavor-Blind Principle gives a reason on why such Yukawas appear and why the GST relation is not related to hierarchical masses but rather to the symmetrical origin of the Yukawa matrices.

A last remark about the usefulness of this approach is that it has been seen to play an important role in solving the strong CP problem [13].

Acknowledgments

I am grateful to Wolfgang G. Hollik for a careful reading of this manuscript. This work has been supported by CONACyT-Mexico under Contract No. 220498.

References

[1] W. G. Hollik and U. J. Saldana Salazar, Nucl. Phys. B **892**, 364 (2015) doi:10.1016/j.nuclphysb.2015.01.019 [arXiv:1411.3549 [hep-ph]].
[2] M. Kobayashi and T. Maskawa, Prog. Theor. Phys. **49**, 652 (1973). doi:10.1143/PTP.49.652
[3] J. Schechter and J. W. F. Valle, Phys. Rev. D **22**, 2227 (1980). doi:10.1103/PhysRevD.22.2227
[4] L. Wolfenstein, Phys. Rev. Lett. **51**, 1945 (1983). doi:10.1103/PhysRevLett.51.1945
[5] L. L. Chau and W. Y. Keung, Phys. Rev. Lett. **53**, 1802 (1984). doi:10.1103/PhysRevLett.53.1802
[6] C. Jarlskog, Phys. Rev. Lett. **55**, 1039 (1985). doi:10.1103/PhysRevLett.55.1039
[7] K. A. Olive *et al.* [Particle Data Group Collaboration], Chin. Phys. C **38**, 090001 (2014). doi:10.1088/1674-1137/38/9/090001
[8] M. C. Gonzalez-Garcia, M. Maltoni and T. Schwetz, JHEP **1411**, 052 (2014) doi:10.1007/JHEP11(2014)052 [arXiv:1409.5439 [hep-ph]].
[9] R. Gatto, G. Sartori and M. Tonin, Phys. Lett. B **28**, 128 (1968). doi:10.1016/0370-2693(68)90150-0
[10] U. J. Saldana-Salazar, Phys. Rev. D **93**, no. 1, 013002 (2016) doi:10.1103/PhysRevD.93.013002 [arXiv:1509.08877 [hep-ph]].
[11] I. Masina, Phys. Lett. B **633**, 134 (2006) doi:10.1016/j.physletb.2005.10.097 [hep-ph/0508031].
[12] I. Masina and C. A. Savoy, Nucl. Phys. B **755**, 1 (2006) doi:10.1016/j.nuclphysb.2006.07.021 [hep-ph/0603101].
[13] J. L. Diaz-Cruz, W. G. Hollik and U. J. Saldana-Salazar, arXiv:1605.03860 [hep-ph].

Journal of Physics: Conference Series **761** (2016) 012046 doi:10.1088/1742-6596/761/1/012046

The remnant CP transformation and its implications

Felix Gonzalez-Canales

Departamento de Física, Centro de Investigación y de Estudios Avanzados del Instituto Politécnico Nacional, Apartado Postal 14-740, CDMX 07000, México.

E-mail: `fgonzalez@fis.cinvestav.mx`

Abstract. In the context of remnant CP transformations, I briefly discuss a generalized $\mu - \tau$ reflection symmetry, where the "Majorana" phases have CP conserving values, which are directly related with the CP parities of neutrino states. Also, one finds that the "Dirac-like" CP violation phase is correlated with the atmospheric mixing angle, giving important phenomenological implications for current and future long baseline oscillation neutrino experiments.

1. Introduction

Nowadays it is well known that the flavour oscillations phenomenon is possible only if neutrinos have mass. This result contradicts the prediction of the Standard Model and is one of the strongest evidence that there must be a more complete theory. One of the main aims of theoretical physicists, is to identify a theory that responds to the question of existence of very small neutrino masses and, at the same time, explain the measured values of the parameters that govern its oscillations. At present, we are in the precision era for the determination of lepton mixing angles. However, it is not the same situation for the CP violation phase factors.

A non-zero reactor mixing angle, θ_{13}, implies the existence of CP violation in lepton sector or vice versa. Hence, one of the main goals of long baseline neutrino oscillation experiments (T2K [1], NOνA [2] and DUNE [3]) is the determination of the "Dirac-like" CP violation phase. Also, $\theta_{13} \neq 0$ reduces the viability of several symmetry groups to be considered as flavour symmetries in Standard Model extensions.

In agreement with the last results of global fits of neutrino oscillation data [4], the elements of second and third row of lepton mixing matrix, satisfy the following approximate relation $|U_{\mu i}| \simeq |U_{\tau i}|$ with $i = 1, 2, 3$. Then, so called $\mu - \tau$ symmetry is obtained if $|U_{\mu i}| = |U_{\tau i}|$ exactly holds. We obtain this equality if and only if one of following two sets of conditions is satisfied: $\theta_{23} = \frac{\pi}{4}$ and $\theta_{13} = 0$, or $\theta_{23} = \frac{\pi}{4}$ and $\delta_{\rm CP} = \pm\frac{\pi}{2}$. From the first set of conditions, one can see that neutrino mass term is invariant under the transformations: $\nu_e \to \nu_e$, $\nu_\mu \to \nu_\tau$ and $\nu_\tau \to \nu_\mu$, which are called $\mu - \tau$ *permutation symmetry* [5]. However, this kind of transformations is disfavoured by neutrino oscillation data, since θ_{13} is non-zero [4]. The last set of conditions implies that neutrino mass term is invariant under the transformations: $\nu_e \to \nu_e^c$, $\nu_\mu \to \nu_\tau^c$ and $\nu_\tau \to \nu_\mu^c$, where c denotes the charge conjugation of neutrino field. This kind of transformations are called $\mu - \tau$ *reflection symmetry* [5]. Also, the values $\theta_{23} = \frac{\pi}{4}$ and $\delta_{\rm CP} = \pm\frac{\pi}{2}$ are allowed at 3σ level, while $\theta_{23} = \frac{\pi}{4}$ and $\delta_{\rm CP} = -\frac{\pi}{2}$ are allowed at 1σ or 2σ level [4].

In this work, the residual CP transformations are used for redefining the flavour lepton mixing matrix. Thus, in the theoretical framework of a generalization of $\mu - \tau$ reflection symmetry, the

Journal of Physics: Conference Series **761** (2016) 012046 doi:10.1088/1742-6596/761/1/012046

possible values of CP phases are constrained, and its phenomenological implication for current and future long baseline neutrino oscillation experiments is shown.

2. The hidden symmetries in the fermion mass matrices

In order to construct the mass term of known neutrinos at low energy, here it is considered that neutrinos are Majorana particles. The leptonic flavour mixing implied by the neutrino oscillation phenomena can be described with help of the Lagrangian

$$\mathcal{L} = -\frac{g}{\sqrt{2}}\,\bar{\ell}_L \gamma^{\mu} \nu_L\, W_{\mu} - \frac{1}{2} \nu_L^{\top} \mathbf{C}^{-1} \mathbf{M}_{\nu}\, \nu_L - \bar{\ell}_R\, \mathbf{M}_{\ell}\, \ell_L + \text{h. c.}\,, \tag{1}$$

where the charged lepton and Majorana neutrino mass terms are included. In the above expression the L and R subscripts denote the left- and right-handed field respectively, while $\ell_L = (e_L, \mu_L, \tau_L)^{\top}$ and $\ell_R = (e_R, \mu_R, \tau_R)^{\top}$ are the charged lepton fields, and finally $\nu_L = (\nu_{eL}, \nu_{\mu L}, \nu_{\tau L})^{\top}$ is the left-handed neutrino field. Note that the mass matrix $\mathbf{M}_{\nu}$ must be symmetric, since neutrinos are Majorana particles, while $\mathbf{M}_{\ell}$ does not have any special feature so it must be represented through a 3×3 complex matrix. Respectively, the neutrino and charged lepton mass matrices can be brought to its diagonal form through the following unitary transformations

$$\mathbf{M}_{\nu} = \mathbf{U}_{\nu}^{*}\, \mathbf{M}_{\nu}^{\text{diag}}\, \mathbf{U}_{\nu}^{\dagger} \quad \text{and} \quad \mathbf{M}_{\ell} = \mathbf{V}_{\ell}\, \mathbf{M}_{\ell}^{\text{diag}}\, \mathbf{U}_{\ell}^{\dagger}\,, \tag{2}$$

where $\mathbf{M}_{\nu}^{\text{diag}} = \text{diag}\,(m_{\nu_1}, m_{\nu_2}, m_{\nu_3})$ and $\mathbf{M}_{\ell}^{\text{diag}} = \text{diag}\,(m_e, m_{\mu}, m_{\tau})$. The neutrino masses m_{ν_i} are real and non-negative. In agreement with the singular value decomposition theorem, $\mathbf{U}_{\ell}$ and $\mathbf{V}_{\ell}$ are the unitary matrices through which the Hermitian matrices $\mathbf{M}_{\ell}\mathbf{M}_{\ell}^{\dagger}$ and $\mathbf{M}_{\ell}^{\dagger}\mathbf{M}_{\ell}$ can be brought to its diagonal form, respectively. The PMNS lepton mixing matrix has the shape $\mathbf{U}_{\text{PMNS}} = \mathbf{U}_{\ell}^{\dagger}\mathbf{U}_{\nu}$ [6]. Now, since one is working in a independent model framework and without loss of generality, the charged lepton mass matrix will be considered with a diagonal shape. In the charged lepton diagonal basis the $\mathbf{U}_{\ell}$ unitary matrix is reduced to unit matrix, consequently the PMNS matrix takes the form $\mathbf{U}_{\text{PMNS}} = \mathbf{U}_{\nu}$, which means that all information about the leptonic flavour mixing comes from the neutrino oscillations. In order to find the flavour symmetry hidden in the neutrino mass matrix is necessary to apply the following CP transformation of the left-handed neutrino fields to the lepton mass terms [7]

$$\nu_L\,(x) \to i\mathbf{X}_{\nu}\, \gamma^0\, \mathbf{C}\, \bar{\nu}_L^{\top}\,(x_p)\,, \tag{3}$$

where $\mathbf{X}_{\nu}$ is a unitary matrix acting on family space, $\mathbf{C}$ is the charge conjugation matrix, and $x_p = (t, -x)$. If the neutrino mass matrix satisfies the relation

$$\mathbf{M}_{\nu}^{*} = \mathbf{X}_{\nu}^{\top} \mathbf{M}_{\nu} \mathbf{X}_{\nu} \quad \Longrightarrow \quad \mathbf{M}_{\nu} = \mathbf{X}_{\nu}^{\dagger} \mathbf{X}_{\nu}^{\top} \mathbf{M}_{\nu} \mathbf{X}_{\nu} \mathbf{X}_{\nu}^{*}, \tag{4}$$

the corresponding mass term in the Lagrangian, eq. (1), is invariant. The second expression in the last equation means that one flavour symmetry transformation is equivalent to successive application of two CP-transformations [7]. From eqs. (2) and (4) we have

$$\mathbf{U}_{\nu}^{\dagger}\, \mathbf{X}_{\nu}\, \mathbf{U}_{\nu}^{*} = e^{in_1\pi}\text{diag}\left(1, e^{in_{21}\pi}, e^{in_{31}\pi}\right), \tag{5}$$

where $n_{21} = n_2 - n_1$ and $n_{31} = n_3 - n_1$ with n_i elements of natural numbers. The phase factor $e^{in_1\pi}$ can be absorbed by neutrino fields, so there exist four possible combinations of signs, which can be expressed as

$$(\mathbf{X}_{\nu})_i = \mathbf{U}_{\nu}\, \mathbf{d}_i\, \mathbf{U}_{\nu}^{\top}, \tag{6}$$

where $\mathbf{d}_1 = \text{diag}\,(1, -1, -1)$, $\mathbf{d}_2 = \text{diag}\,(1, -1, 1)$, $\mathbf{d}_3 = \text{diag}\,(1, 1, -1)$, and $\mathbf{d}_4 = \text{diag}\,(1, 1, 1)$. From eq. (5) or (6) it is easy to conclude that the residual flavour symmetry of Majorana neutrinos mass matrix is the Klein group, namely $Z_2 \otimes Z_2$. A direct consequence of eq. (6) is that $\mathbf{X}_\nu = \mathbf{X}_\nu^\top$. Therefore, the $\mathbf{X}_\nu$ is a unitary symmetric matrix. Applying the Takagi factorization to $\mathbf{X}_\nu$ matrix, the eq. (6) can be written as $\mathbf{O}_\nu = \mathbf{O}_\nu^*$ or $\mathbf{O}_\nu^\top = \mathbf{O}_\nu^\dagger$, where $\mathbf{O}_\nu = \mathbf{\Sigma}^\top \mathbf{U}_\nu \text{diag}\left(e^{-i\frac{n_1}{2}\pi}, e^{-i\frac{n_2}{2}\pi}, e^{-i\frac{n_3}{2}\pi}\right)$ [8, 9]. Also, $\mathbf{O}_\nu \mathbf{O}_\nu^\dagger = \mathbf{O}_\nu \mathbf{O}_\nu^\top = \mathbf{I}$, where $\mathbf{I}$ is the unit matrix. So it is very easy conclude that $\mathbf{O}_\nu$ matrix is a real orthogonal matrix. In this context the lepton flavour mixing matrix takes the shape [8, 9]

$$\mathbf{U}_{\text{PMNS}} = \mathbf{U}_\nu = \mathbf{\Sigma}_\nu \, \mathbf{O}_\nu \, \mathbf{Q}_\nu, \tag{7}$$

where

$$\mathbf{O}_\nu = \begin{pmatrix} c_2 c_3 & c_2 s_3 & s_2 \\ -c_3 s_1 s_2 - c_1 s_3 & c_1 c_3 - s_1 s_2 s_3 & c_2 s_1 \\ s_1 s_3 - c_1 c_3 s_2 & -c_3 s_1 - c_1 s_2 s_3 & c_1 c_2 \end{pmatrix} \tag{8}$$

and $\mathbf{Q}_\nu = \text{diag}\left(e^{-i\frac{n_1}{2}\pi}, e^{-i\frac{n_2}{2}\pi}, e^{-i\frac{n_3}{2}\pi}\right)$. Here, $c_i = \cos\theta_i$ and $s_i = \sin\theta_i$.

3. Generalized $\mu - \tau$ symmetry

Here a generalized $\mu - \tau$ reflection symmetry is proposed, defined as [8, 9]:

$$\mathbf{X}_\nu = \begin{pmatrix} e^{i\alpha} & 0 & 0 \\ 0 & e^{i\beta}\cos\Theta & ie^{i\frac{(\beta+\gamma)}{2}}\sin\Theta \\ 0 & ie^{i\frac{(\beta+\gamma)}{2}}\sin\Theta & e^{i\gamma}\cos\Theta \end{pmatrix} \xrightarrow[\Theta=\pm\frac{\pi}{2}]{\alpha=0,\,\beta=0,\,\gamma=0} \begin{pmatrix} 1 & 0 & 0 \\ 0 & 0 & i\kappa \\ 0 & i\kappa & 0 \end{pmatrix}, \tag{9}$$

where $\kappa = \pm 1$, the phase factors α, β, γ, and the angle Θ are real parameters. In the particular case of having $\alpha = 0$, $\beta = 0$, $\gamma = 0$, and $\Theta = \pm\frac{\pi}{2}$, the $\mathbf{X}_\nu$ matrix is reduced to standard $\mu - \tau$ reflection symmetry [5]. The Takagi factorization of $\mathbf{X}_\nu$ matrix has the shape

$$\mathbf{\Sigma}_\nu = \begin{pmatrix} e^{i\frac{\alpha}{2}} & 0 & 0 \\ 0 & e^{i\frac{\beta}{2}}\cos\Theta & ie^{i\frac{\beta}{2}}\sin\Theta \\ 0 & ie^{i\frac{\gamma}{2}}\sin\Theta & e^{i\frac{\gamma}{2}}\cos\Theta \end{pmatrix}. \tag{10}$$

If the $\mathbf{\Sigma}_\nu$ matrix has the form given in the above expression, the relation between flavour mixing angles and the entries of the lepton mixing matrix, eq. (7), are:

$$\begin{aligned} \sin^2\theta_{12} &= \tfrac{|U_{e2}|^2}{1-|U_{e3}|^2} = \sin^2\theta_3, \quad \sin^2\theta_{13} = |U_{e3}|^2 = \sin^2\theta_2, \\ \sin^2\theta_{23} &= \tfrac{|U_{\mu 3}|^2}{1-|U_{e3}|^2} = \tfrac{1}{2}\left(1 - \cos\Theta\cos 2\theta_1\right). \end{aligned} \tag{11}$$

The phase factors associated to CP violation are determined through the Jarlskog invariant $\mathcal{J}_{\text{CP}} = \mathcal{I}m\left\{U_{e1}^* U_{\mu 3}^* U_{e3} U_{\mu 1}\right\}$ and the invariants $\text{I}_1 = \mathcal{I}m\left\{U_{e2}^2 U_{e1}^{*2}\right\}$ and $\text{I}_2 = \mathcal{I}m\left\{U_{e2}^2 U_{e1}^{*2}\right\}$. The explicit form of these phases are [8]:

$$\sin\delta_{\text{CP}} = \tfrac{\sin\Theta\,\text{sign}[\sin\theta_2\sin 2\theta_3]}{\sqrt{1-\cos^2\Theta\cos^2 2\theta_1}}, \qquad \phi_{12} = \tfrac{n_{21}}{2}\pi \quad \text{and} \quad \phi_{13} = \tfrac{n_{31}}{2}\pi. \tag{12}$$

Here, δ_{CP} is the Dirac-like phase which is involved in the neutrino oscillation. Therefore, this phase can be measured in the long baseline neutrino oscillation experiments. On the other hand, ϕ_{12} and ϕ_{13} are the Majorana phases which are involved in the neutrinoless double beta decay. The PMNS mixing matrix in the standard parametrization has six free parameters, three angles and three phase factors, while in the context of generalized $\mu - \tau$ reflection symmetry this matrix has just four free parameters, which are four angles. In our theoretical framework, we have no authentic prediction for the lepton mixing angles and Dirac-like CP violation phase. However, we have an important prediction for the Majorana phases. The ϕ_{12} and ϕ_{13} phases are CP conserving and have a direct relation with the CP parities of the neutrino states [8].

4. Neutrino oscillation in matter

A viable way of observing CP violation in the leptonic sector is through the differences in the oscillation probabilities that involve neutrinos and antineutrinos. This difference in the vacuum has the form [10]: $\Delta P_{\alpha\beta} \equiv P\left(\nu_\alpha \to \nu_\beta\right) - P\left(\bar{\nu}_\alpha \to \bar{\nu}_\beta\right) = -16\, J_{\alpha\beta} \sin\Delta_{21} \sin\Delta_{23} \sin\Delta_{31}$. Here, $\Delta_{kj} = \Delta m^2_{kj} L/(4E)$ and $\Delta m^2_{kj} = m^2_{\nu_k} - m^2_{\nu_j}$, L is the baseline, E is the energy of neutrino beam, and $\mathcal{J}_{\alpha\beta}$ is the Jarlskog invariant in its leptonic version, whose definition is $\mathcal{J}_{\alpha\beta} = \mathcal{I}m\left(U_{\alpha 1} U_{\beta 2} U^*_{\alpha 2} U^*_{\beta 1}\right) = \pm J_{CP}$. The positive sign correspond to a cyclic permutation of the flavour indices e, μ and τ, while the negative sign is for an anti-cyclic permutation.

The present long baseline experiments like T2K and NOνA as well as the proposed experiment DUNE, are interested in the measurements of the parameters that govern the $\nu_\mu \to \nu_e$ oscillation. In the vacuum, the transition probabilities $P\left(\nu_\mu \to \nu_e\right)$ and $P\left(\bar{\nu}_\mu \to \bar{\nu}_e\right)$ have the form [10]

$$\begin{aligned} P\left(\nu_\mu \to \nu_e\right) &\simeq P_{\rm atm} + 2\sqrt{P_{\rm atm}}\sqrt{P_{\rm sol}} \cos\left(\Delta_{32} + \delta_{\rm CP}\right) + P_{\rm sol}\,, \\ P\left(\bar{\nu}_\mu \to \bar{\nu}_e\right) &\simeq P_{\rm atm} + 2\sqrt{P_{\rm atm}}\sqrt{P_{\rm sol}} \cos\left(\Delta_{32} - \delta_{\rm CP}\right) + P_{\rm sol}\,, \end{aligned} \tag{13}$$

where $\sqrt{P_{\rm atm}} = \sin\theta_{23} \sin 2\theta_{13} \sin\Delta_{31}$ and $\sqrt{P_{\rm sol}} = \cos\theta_{23} \cos\theta_{13} \sin 2\theta_{12} \sin\Delta_{21}$. Then, the asymmetry between transition probabilities $P\left(\nu_\mu \to \nu_e\right)$ and $P\left(\bar{\nu}_\mu \to \bar{\nu}_e\right)$ is given by

$$A_{\mu e} = \frac{P\left(\nu_\mu \to \nu_e\right) - P\left(\bar{\nu}_\mu \to \bar{\nu}_e\right)}{P\left(\nu_\mu \to \nu_e\right) + P\left(\bar{\nu}_\mu \to \bar{\nu}_e\right)} = -\frac{2\sqrt{P_{\rm atm}}\sqrt{P_{\rm sol}} \sin\Delta_{32} \sin\delta_{\rm CP}}{P_{\rm atm} + 2\sqrt{P_{\rm atm}}\sqrt{P_{\rm sol}} \cos\Delta_{32} \cos\delta_{\rm CP} + P_{\rm sol}}\,. \tag{14}$$

However, to make a realistic description of $\nu_\mu \to \nu_e$ oscillation in the long baseline experiments, it is necessary to include the matter effects associated with neutrino propagation inside the Earth. Then, the expressions for $\sqrt{P_{\rm atm}}$ and $\sqrt{P_{\rm sol}}$ in matter are [10]:

$$\sqrt{P_{\rm atm}} = \sin\theta_{23} \sin 2\theta_{13} \frac{\sin\left(\Delta_{31} - aL\right)}{\Delta_{31} - aL} \Delta_{31}\,, \quad \sqrt{P_{\rm sol}} = \cos\theta_{23} \sin 2\theta_{12} \frac{\sin(aL)}{aL} \Delta_{21}\,. \tag{15}$$

Here, $a = G_F N_e/\sqrt{2}$, where G_F is the Fermi constant and N_e is the density of electrons. For the earth crust the parameter a is approximately $a \approx (3500{\rm km})^{-1}$ [9]. The neutrino transition probability, $P\left(\nu_\mu \to \nu_e\right)$, is obtained by replacing the eq. (15) into first expression of eq. (13). However, for transition probability $P\left(\bar{\nu}_\mu \to \bar{\nu}_e\right)$ we can not only make the previous substitution, because the involved particles are the antineutrinos and we need make the change $a \to -a$. Therefore, we make the redefinitions $\sqrt{P_{\rm atm}} \to \sqrt{\mathcal{P}_{\rm atm}}$ and $\sqrt{P_{\rm sol}} \to \sqrt{\mathcal{P}_{\rm sol}}$, where

$$\sqrt{\mathcal{P}_{\rm atm}} = \sin\theta_{23} \sin 2\theta_{13} \frac{\sin\left(\Delta_{31} + aL\right)}{\Delta_{31} + aL} \Delta_{31}\,, \quad \sqrt{\mathcal{P}_{\rm sol}} = \cos\theta_{23} \sin 2\theta_{12} \frac{\sin(aL)}{aL} \Delta_{21}\,. \tag{16}$$

The antineutrino oscillation probability $P(\bar{\nu}_\mu \to \bar{\nu}_e)$ is obtained replacing the eq. (16) in second expression of eq. (13). Finally, the asymmetry between transition probabilities $P\left(\nu_\mu \to \nu_e\right)$ and $P\left(\bar{\nu}_\mu \to \bar{\nu}_e\right)$ including matter effects takes the form

$$\mathcal{A}_{\mu e} = \frac{\left(P_{\rm atm} - \mathcal{P}_{\rm atm}\right) + 2\sqrt{P_{\rm sol}}\left[\left(\sqrt{P_{\rm atm}} - \sqrt{\mathcal{P}_{\rm atm}}\right) \cos\Delta_{32} \cos\delta_{\rm CP} - \left(\sqrt{P_{\rm atm}} + \sqrt{\mathcal{P}_{\rm atm}}\right) \sin\Delta_{32} \sin\delta_{\rm CP}\right]}{2P_{\rm sol} + \left(P_{\rm atm} + \mathcal{P}_{\rm atm}\right) + 2\sqrt{P_{\rm sol}}\left[\left(\sqrt{P_{\rm atm}} + \sqrt{\mathcal{P}_{\rm atm}}\right) \cos\Delta_{32} \cos\delta_{\rm CP} - \left(\sqrt{P_{\rm atm}} - \sqrt{\mathcal{P}_{\rm atm}}\right) \sin\Delta_{32} \sin\delta_{\rm CP}\right]}\,. \tag{17}$$

In the context of generalized $\mu - \tau$ reflection symmetry, the Dirac-like phase, atmospheric mixing angle and Θ parameter that characterizes the $\mathbf{X}_\nu$ CP transformation matrix, are related each other through the expression

$$\sin^2\delta_{\rm CP} \sin^2 2\theta_{23} = \sin^2\Theta \quad \Rightarrow \quad \delta_{\rm CP} = k_1\,\pi + (-1)^{k_2} \arcsin\left(\frac{\sin\Theta}{\sin 2\theta_{23}}\right), \tag{18}$$

Journal of Physics: Conference Series **761** (2016) 012046 doi:10.1088/1742-6596/761/1/012046

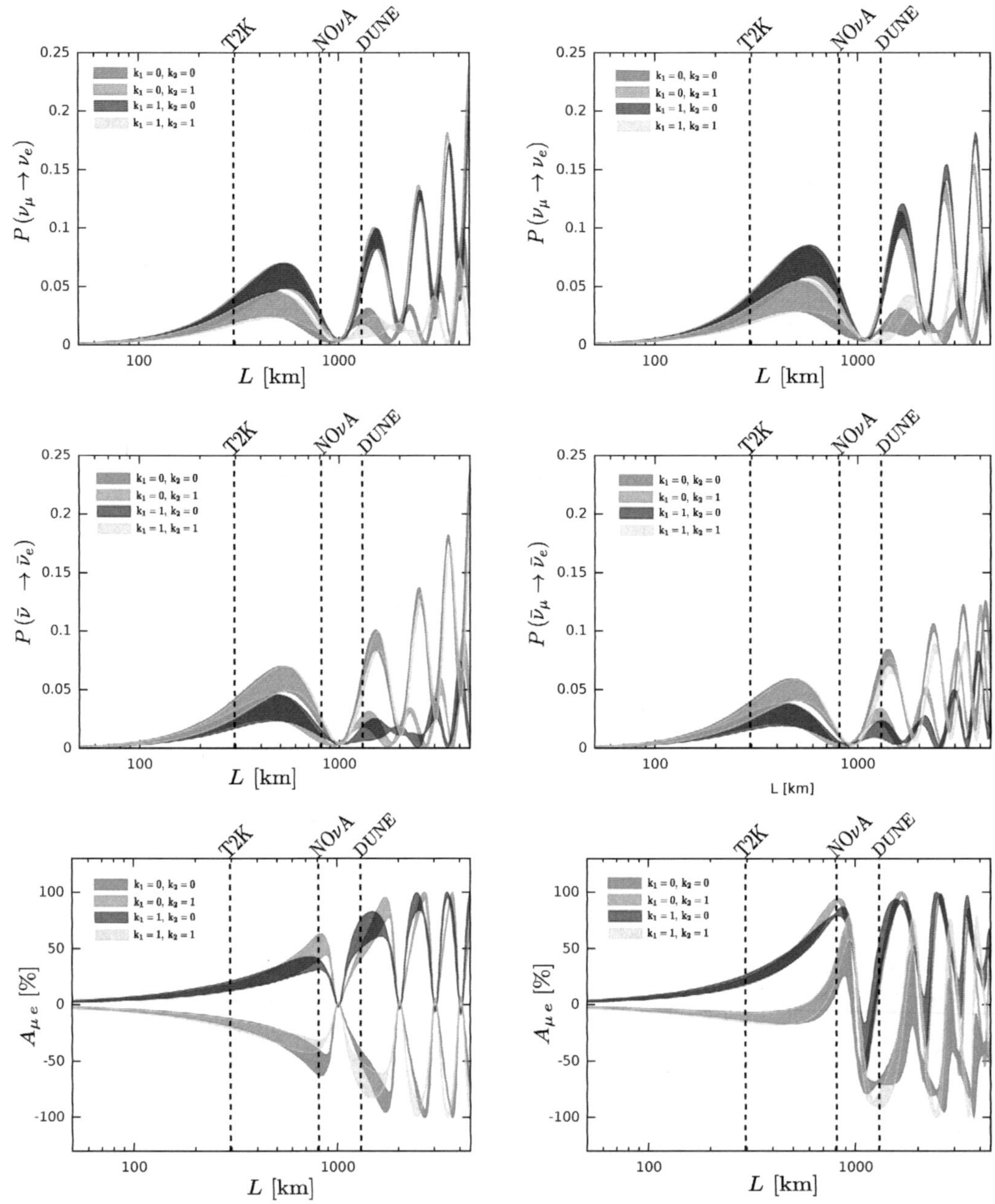

Figure 1. The transition probabilities $\nu_\mu \to \nu_e$ and $\bar{\nu}_\mu \to \bar{\nu}_e$, as well as the $A_{\mu e}$ and $\mathcal{A}_{\mu e}$ asymmetries for a energy of neutrino beam $E = 1$ GeV. Respectively, the left and right panels correspond to neutrino oscillation in vacuum and matter. For a normal hierarchy in the neutrino mass spectrum, the atmospheric mixing angle is taken within 3σ range $0.393 \leq sin^2\theta_{23} \leq 0.643$ [4]. The other neutrino oscillation parameters are taken at their best fit point values: $\Delta m^2_{21} = 7.60 \times 10^{-5}$ eV^2, $|\Delta m^2_{31}| = 2.48 \times 10^{-3}$ eV^2, $\sin^2\theta_{12} = 0.323$ and $\sin^2\theta_{13} = 0.0226$, while Θ parameter is $3\pi/8$.

where $k_{1,2} = 0, 1$. This correlation allow us to obtain the value range of the Dirac-like CP violation phase. In the particular case when the Θ angle is $\pm\pi/4$, the other two parameters, θ_{23} and δ_{CP}, involved in the above expression are maximal. In agreement with the results of global fits of neutrino oscillation data [4], a maximal atmospheric angle is disfavoured. Consequently, the called standard $\mu - \tau$ reflection symmetry is disfavoured. So, a non-maximal atmospheric mixing angle makes that our generalized $\mu - \tau$ reflection symmetry scenario be a good alternative for the CP violation study. The explicit form of $\nu_\mu \to \nu_e$ and $\bar{\nu}_\mu \to \bar{\nu}_e$ transition probabilities, as well as the asymmetry between these transition probabilities, are obtained by substituting the second expression of eq. (18) in eqs. (13), (14) and (17). For instance, the $P(\nu_\mu \to \nu_e)$ and $P(\bar{\nu}_\mu \to \bar{\nu}_e)$ transition probabilities in matter have the following explicit shape

$$\begin{aligned} P(\nu_\mu \to \nu_e) &\simeq P_{\rm atm} + 2\sqrt{P_{\rm atm}}\sqrt{P_{\rm sol}}\cos(k_1\pi)\cos\left(\Delta_{32} + (-1)^{k_2}\arcsin\left(\tfrac{\sin\Theta}{\sin 2\theta_{23}}\right)\right) + P_{\rm sol}\,, \\ P(\bar{\nu}_\mu \to \bar{\nu}_e) &\simeq \mathcal{P}_{\rm atm} + 2\sqrt{\mathcal{P}_{\rm atm}}\sqrt{P_{\rm sol}}\cos(k_1\pi)\cos\left(\Delta_{32} + (-1)^{k_2+1}\arcsin\left(\tfrac{\sin\Theta}{\sin 2\theta_{23}}\right)\right) + P_{\rm sol}\,. \end{aligned} \tag{19}$$

In the Fig. 1 is shown the behaviour of $\nu_\mu \to \nu_e$ and $\bar{\nu}_\mu \to \bar{\nu}_e$ transition probabilities, as well as $A_{\mu e}$ and $\mathcal{A}_{\mu e}$ asymmetries.

5. Conclusions

We proposed a generalized $\mu - \tau$ reflection symmetry where the PMNS mixing matrix has four free parameters only, in contrast with the six parameters in the standard parametrization. We obtained that the "Majorana" phases have CP conserving values which are directly related with the CP parities of neutrino states. On the other hand, the "Dirac-like" CP violation phase is correlated with the θ_{23} atmospheric mixing angle and Θ angle, which characterizes the CP transformation $\mathbf{X}_\nu$ matrix. This correlation has important implications for the long baseline oscillation neutrino experiments T2K, NOνA and DUNE. For a review of the phenomenological implications of our generalized $\mu - \tau$ reflection symmetry for the neutrinoless double beta decay and the leptogenesis see [8, 9].

Acknowledgments

This talk was based on work in collaboration with P. Chen, G. J. Ding and J. W. F. Valle, and has been partially supported by *CONACYT-SNI (México)* and *CONACYT* under grant 236394. The author thanks the support provided by the Laboratorio Nacional de Supercómputo del Sureste de México through the grant number O-2016/039.

References

[1] K. Abe *et al.* [T2K Collaboration], Phys. Rev. D **91** (2015) no.7, 072010 doi:10.1103/PhysRevD.91.072010 [arXiv:1502.01550 [hep-ex]].
[2] J. Bian [NOvA Collaboration], arXiv:1510.05708 [hep-ex].
[3] R. Acciarri *et al.* [DUNE Collaboration], arXiv:1512.06148 [physics.ins-det].
[4] D. V. Forero, M. Tortola and J. W. F. Valle, Phys. Rev. D **90** (2014) no.9, 093006 doi:10.1103/PhysRevD.90.093006 [arXiv:1405.7540 [hep-ph]].
[5] Z. z. Xing and Z. h. Zhao, Rept. Prog. Phys. **79** (2016) no.7, 076201 doi:10.1088/0034-4885/79/7/076201 [arXiv:1512.04207 [hep-ph]].
[6] K. A. Hochmuth, S. T. Petcov and W. Rodejohann, Phys. Lett. B **654**, 177 (2007) doi:10.1016/j.physletb.2007.08.072 [arXiv:0706.2975 [hep-ph]].
[7] P. Chen, C. C. Li and G. J. Ding, Phys. Rev. D **91** (2015) 033003 doi:10.1103/PhysRevD.91.033003 [arXiv:1412.8352 [hep-ph]].
[8] P. Chen, G. J. Ding, F. Gonzalez-Canales and J. W. F. Valle, Phys. Lett. B **753** (2016) 644 doi:10.1016/j.physletb.2015.12.069 [arXiv:1512.01551 [hep-ph]].
[9] P. Chen, G. J. Ding, F. Gonzalez-Canales and J. W. F. Valle, Phys. Rev. D **94** (2016) no.3, 033002 doi:10.1103/PhysRevD.94.033002 [arXiv:1604.03510 [hep-ph]].
[10] H. Nunokawa, S. J. Parke and J. W. F. Valle, Prog. Part. Nucl. Phys. **60** (2008) 338 doi:10.1016/j.ppnp.2007.10.001 [arXiv:0710.0554 [hep-ph]].

Journal of Physics: Conference Series **761** (2016) 012047 doi:10.1088/1742-6596/761/1/012047

Properties of vector mesons in four pseudoscalars electroproduction

G. Toledo Sánchez

Instituto de Física, Universidad Nacional Autónoma de México. México D. F. C.P. 04510

E-mail: toledo@fisica.unam.mx

Abstract.
Vector mesons and the W gauge boson share some common features, such as the spin and a very short lifetime. The electromagnetic properties of the W are linked to its gauge nature and thus a big effort has been devoted to measure them, while accounting for its instability. In this work, we elaborate on how these ideas can be translated to study vector mesons. We focus in the unstable property of such states and the restrictions from electromagnetic gauge invariance, describing the differences and similarities. Then, we describe the four pseudoscalars electroproduction as an analog to the $e^+e^- \to 4fermions$ process, used to study the electromagnetic properties of the W boson. We point out that the current experimental capabilities are reaching the possibility to measure the magnetic dipole moment of light vector mesons.

1. Introduction

The trilinear gauge boson interactions $WW\gamma$ and WWZ come from the kinetic term of the SU(2) gauge field, upon rewriting it in terms of the mass eigenstates. Thus, it probes the non-abelian nature of the standard model. The corresponding $\gamma(q,\lambda) \to W^-(k_-,\nu)W^+(k_+,\mu)$ vertex, takes the form:

$$ie\left[g^{\mu\nu}(k_- - k_+)^\lambda + g^{\nu\lambda}(-q-k_-)^\mu + g^{\lambda\mu}(q+k_+)^\nu\right], \quad (1)$$

which can be set, using momentum conservation and taking the W particles on-shell as:

$$\Gamma_0^{\lambda\mu\nu} = g^{\mu\nu}(k_- - k_+)^\lambda + 2\left(q^\nu g^{\lambda\mu} - q^\mu g^{\nu\lambda}\right), \quad (2)$$

where we omit the ie factor and the subindex 0 denotes this tree level result.

A general parametrization of the vertex fullfiling the C, P and CP symmetries is given by[1, 2]:

$$\Gamma_{eff}^{\lambda\mu\nu} = g_1(k_- - k_+)^\lambda g^{\mu\nu} + (g_1+\kappa+\lambda)\left(q^\nu g^{\lambda\mu} - q^\mu g^{\nu\lambda}\right) - \frac{\lambda}{m_W^2}(k_- - k_+)^\lambda q^\mu q^\nu. \quad (3)$$

Comparing the above expressions we identify that for the SM the effective parameters are: $g_1 = 1$, $1+\kappa+\lambda = 2$ and $\lambda = 0$. In the static limit, such parameters are related to the electromagnetic multipoles as follows: $\mathcal{Q} = g_1 = 1$ is the electric charge (in e units), $\mu_W = 1+\kappa+\lambda = 2$ is the magnetic dipole moment (in $e/2M_W$ units) and the electric quadrupole is $X_E = -(\kappa-\lambda) = -1$ (in e/M_W^2 units). These values are usually taken as a reference for vector mesons. We have

Journal of Physics: Conference Series **761** (2016) 012047 doi:10.1088/1742-6596/761/1/012047

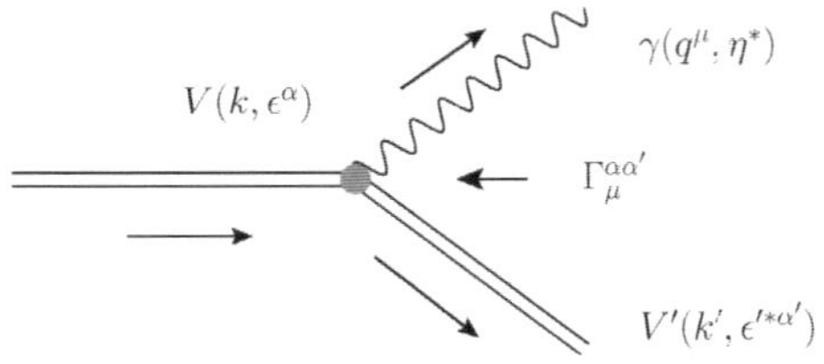

Figure 1. Vector meson electromagnetic vertex, in a generic convention. Within parenthesis are the corresponding momentum and polarization tensor.

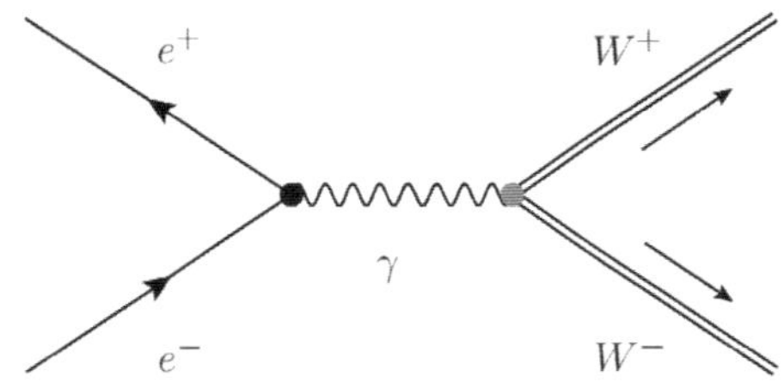

Figure 2. W gauge boson pair production. The full $e^+e^- \to 4fermions$ process has the analog $e^+e^- \to P^+P^-P^0P^0$ process for vector mesons.

also shown that the analog of Burnett and Kroll theorem [3] for polarized radiative interferences favors the $|\vec{\mu}| = 2$ value [4]. In Figure 1, we show the generic vertex of a vector particle and the photon.

The measurement performed, for example, by The DELPHI Collaboration [5] at LEP2 considered the $e^+e^- \to 4fermions$ process, where the four fermions were associated to a W^+W^- production (Figure 2), which then decay to pairs of fermions, identified as jets and/or leptons. The observations were found to be consistent with the SM values.

In this proceedings, we describe a similar approach for vector mesons which mainly decay into two pseudoscalars (P). Thus, processes of the form $e^+e^- \to P^+P^-P^0P^0$ can be useful to explore the electromagnetic structure of the vector mesons. We devote most of the work to discuss the general features of the elements relevant for the description of the process and we close with particular cases for light vector mesons, namely for the ρ and K^* mesons.

2. The elements in the $e^+e^- \to P^+P^-P^0P^0$ process

Let us discuss each element relevant for the description of the probability amplitude of this process. The electron-positron annihilation into a photon is properly accounted by a QED description. The description of the hadronic part relevant for our purposes considers that the photon becomes a neutral vector meson, based on the vector meson dominance model (VMD), which then produces two charged vector mesons, each decaying into two pseudoscalar mesons, as depicted in Figure 2. The following elements are important in order to account for the probability

- The vector meson propagator and the unstable feature
- The three vector mesons vertex
- Bose-Einstein and Charge conjugation symmetries
- Gauge invariance of the amplitude

Let us address each point:

2.1. The vector mesons propagator and the unstable feature

The finite decay width (Γ) of the W became relevant with the advent of colliders capable of producing this gauge boson on-shell. In order to avoid the propagator pole, several prescriptions were introduced on phenomenological grounds. The so-called complex-mass scheme [6, 7] succeeded in this task while fulfilling the electromagnetic gauge invariance condition. The

fermion-loop scheme [8, 9, 10] provided a proper quantum field theory ground by accounting for the absorptive contributions of fermions to the electromagnetic vertex and the propagator, linked by the Ward-Takahashi identity.
These schemas are consistent with each other upon renormalization of the gauge field. Thus, it is common to consider the W propagator as given in the complex-mass scheme:

$$D^{\mu\nu}[q^2, W] = i\left(\frac{-g^{\mu\nu} + \frac{q^\mu q^\nu}{M_W - iM_W\Gamma}}{q^2 - M_W^2 + iM_W\Gamma}\right), \tag{4}$$

where the width can be taken as either constant or momentum dependent.

In the case of vector mesons, the same prescription can be translated to account for their finite width by noticing that the decay width is driven by the decay into two pseudoscalars. The boson-loop scheme [11] considered the absorptive contributions of such particles. Splitting the absorptive correction into transverse ($\mathrm{Im}\Pi^T$) and longitudinal ($\mathrm{Im}\Pi^L$) ones, the propagator in general becomes [11]:

$$D^{\mu\nu}[q^2, V] = \frac{i}{q^2 - M_V^2 + i\mathrm{Im}\Pi^T}\left\{-g^{\mu\nu} + q^\mu q^\nu \frac{q^2 + i\left(\mathrm{Im}\Pi^T - \mathrm{Im}\Pi^L\right)}{q^2\left(M_V^2 - i\mathrm{Im}\Pi^L\right)}\right\}, \tag{5}$$

where $\mathrm{Im}\Pi^T$ is directly proportional to the width and $\mathrm{Im}\Pi^L$ to the squared mass difference of the bosons in the loop. Notice that for the ρ the particles in the loop are the π^+ and π^0, thus the longitudinal contribution is nearly zero. However, for the K^* they correspond to $K^+\pi^0$ and $K^0\pi^+$ and important corrections from $SU(3)_F$ symmetry breaking are expected [12].

2.2. The three vector mesons vertex

The fact that there is no resonant state coupled to the photon producing the W pair allows to directly identify the couplings with the electromagnetic properties of the W. However, the electromagnetic structure of the vector meson as a function of the momentum can be accomplished by the inclusion of resonances coupled to the photon. The structure of the triple vector meson vertex is expected to follow the same form as the electromagnetic vertex previously discussed but for a set of different couplings, such that the electric charge form factor is written as:

$$(k_- - k_+)^\lambda g^{\mu\nu} \to \sum_{j=V'} \frac{-1}{g_j}\frac{g_{jVV}M_j^2}{q^2 - M_j^2 + iM_j\Gamma_j}(k_- - k_+)^\lambda g^{\mu\nu}, \tag{6}$$

where j runs over all the possible resonances, and the g_j and g_{jVV} are related to the photon and vector meson couplings of such resonances respectively. The condition that at $q^2 \to 0$ both expressions must be the same, corresponding to the electric charge of the particle, imposes a relation between the couplings.

We have computed the correction to the multipoles of the W, ρ and K^* particles, exclusively from the fact they are unstable and found them to be relatively small [13]. Thus, such correction is usually not included.

2.3. Bose-Einstein and Charge conjugation symmetries

Now, we get into the properties of the amplitude of the process. We take as a definite case the production of the four pseudoscalars to illustrate the conditions to be fulfilled due to the nature of the particles produced and the electromagnetic origin. This is particularly useful for effective theories. In the case of the four fermions, the SM construction takes care of many of the details.

Let us set our notation as: $e^+(k_1)e^-(k_2) \to P^+(p_1)P^0(p_2)P^-(p_3)P^0(p_4)$, in parenthesis are the corresponding 4-momenta. The total amplitude can be written as:

$$\mathcal{M} = \frac{-ie}{(k_1+k_2)^2} l^\mu h_\mu(p_1,p_2,p_3,p_4), \tag{7}$$

where the leptonic current is given by $l^\mu \equiv \bar{v}(k_2)\gamma^\mu u(k_1)$ and h_μ represents the four pseudoscalars electromagnetic current. In general, this is not limited to the channel where the pair of intermediate vector mesons is produced. The fact that it involves identical neutral particles imposes the requirement of Bose-Einstein symmetry. That is, the amplitude must be symmetric under the exchange of the identical neutral particles. This is represented by:

$$h_\mu(p_1,p_2,p_3,p_4) = h_\mu(p_1,p_4,p_3,p_2). \tag{8}$$

In addition, the current must be symmetric under charge conjugation (C invariance). In the case that the neutral particles are their own anti-particles, and the charged ones are the antiparticle of each other, it corresponds to the exchange of them:

$$h_\mu(p_1,p_2,p_3,p_4) = -h_\mu(p_3,p_2,p_1,p_4). \tag{9}$$

Thus, in order to account for such requirements, the total contribution can be written as the sum of the four possible momenta configurations, represented by a reduced amplitude $\mathcal{M}_{r\mu}$ no longer constrained by such symmetries [14]:

$$\begin{aligned} h_\mu(p_1,p_2,p_3,p_4) &= \mathcal{M}_{r\mu}(p_1,p_2,p_3,p_4) + \mathcal{M}_{r\mu}(p_1,p_4,p_3,p_2) \\ &- \mathcal{M}_{r\mu}(p_3,p_2,p_1,p_4) - \mathcal{M}_{r\mu}(p_3,p_4,p_1,p_2). \end{aligned} \tag{10}$$

Note that, since the different channels add up to build the total amplitude, we can impose this construction channel by channel. An example of the reduced amplitude is the one involving the intermediate pair of vector mesons:

$$\mathcal{M}_r^\mu(p_1,p_2,p_3,p_4) = (p_2-p_1)_\lambda D^{\delta\lambda}[s_{12},V^+]\Gamma_{\gamma\delta\epsilon}D^{\epsilon\nu}[s_{34},V^-](p_4-p_3)_\nu D^{\mu\gamma}[q^2,V'], \tag{11}$$

where we have defined $s_{ij} \equiv (P_i+P_j)^2$ and the derivative coupling of the vector meson to the pseudoscalars was used.

2.4. *Gauge invariance of the amplitude*

In general, not all the diagrams contributing to the amplitude are independent, gauge invariance condition requires some of them to be related in such a way that the condition is fulfilled. The leptonic current is certainly gauge invariant. Thus, the condition is translated to the hadronic part, which can be stated as follows:

$$q^\mu h_\mu(p_1,p_2,p_3,p_4) = 0. \tag{12}$$

In a general case, this condition can be broken and the current can be split into two parts, one which satisfies the condition (GI) and other which fails to do so (NGI):

$$q^\mu h_\mu(p_1,p_2,p_3,p_4) = q^\mu h_\mu^{GI}(p_1,p_2,p_3,p_4) + q^\mu h_\mu^{NGI}(p_1,p_2,p_3,p_4)0 \neq 0, \tag{13}$$

Thus, in order to have a fully gauge invariant amplitude, a counter term must be added to the original amplitude, built out from the NGI contraction:

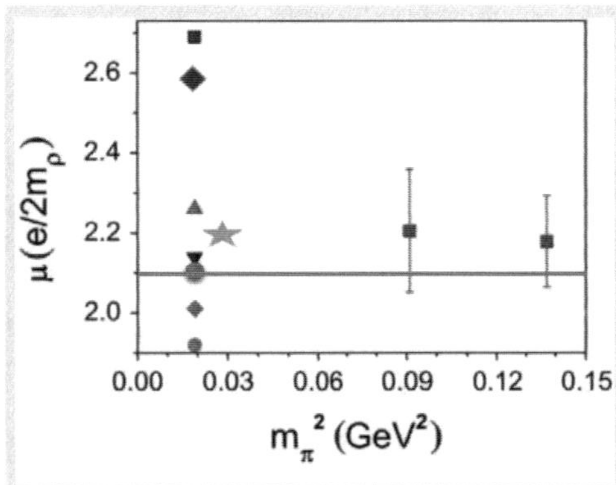

Figure 3. ρ Magnetic dipole moment value determined from BaBar data (solid line) compared with the predictions from effective models of QCD and Lattice results at several pion mass values (symbols).

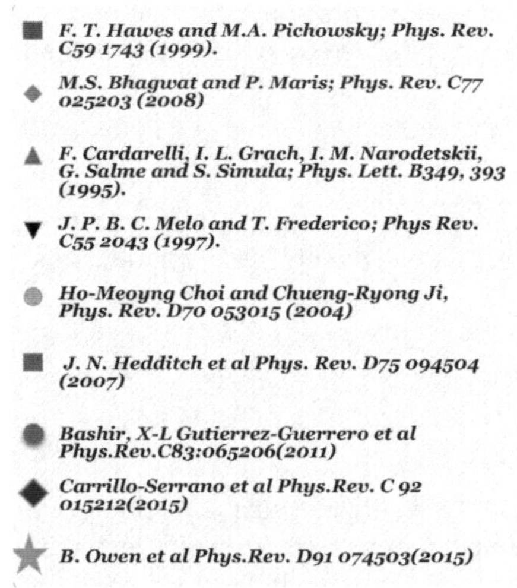

Figure 4. References for the symbols in Figure 3.

$$h_\mu^{Total}(p_1,p_2,p_3,p_4) = h_\mu(p_1,p_2,p_3,p_4) + h_\mu^{CT}(p_1,p_2,p_3,p_4), \tag{14}$$

such that now Eqn. (12) is fulfilled.

Due to the charge conservation relation of the gauge invariance, contributions coming from charge couplings from different diagrams are related to this level by gauge invariance. Higher order terms are usually introduced as gauge invariant by themselves. An explicit example of a higher order gauge invariant term is the magnetic dipole moment structure in the electromagnetic vertex Eqn. (3). Certainly, these can be worked out at the reduced amplitude level.

3. Light vector mesons

Experimental information exist on the $e^+e^- \to \pi^+\pi^-2\pi^0$ process in the low energy regime [15, 16, 17, 18, 19, 20, 21]. The leading channel contribution to the total cross section involves the $\omega - \pi$ mesons intermediate state. Measurements at increasing energies made clear that this process was also sensitive to the channel where the pions are produced in pairs through the $\rho - \rho$ meson resonant states [22]. This particularity opens the door to explore the ρ electromagnetic vertex and by consequence the corresponding electromagnetic multipoles. The general procedure discussed above, has been used to determine the magnetic dipole moment of the ρ meson, by studying its effect in the $e^+e^- \to \pi^+\pi^-2\pi^0$ cross section. The hadronic part was modeled in the VMD approach, which involves several channels. Using preliminary BaBar data [22], while using other observables to fix all the remaining parameters, we have found [23, 24] that the best fit implies a value of $\mu_\rho = 2.1 \pm 0.5\,[\frac{e}{2m_\rho}]$. The quoted error bar takes into account the uncertainties coming from the couplings of the different channels and model assumptions, dominated by the lack of information on the *rho* properties. In Figure 3 we compare the central value of this result with the predictions from effective models of QCD and the Lattice result at several pion mass values [25, 26, 27, 28, 29, 30, 31, 32, 33].

Another example is the $e^+e^- \to K^+K^-2\pi^0$ process, where an intermediate pair of K^* vector mesons are produced. Experimental information on this process still has large uncertainties [34], but the experimental capabilities are certainly increasing. In this particular case, the $SU(3)_F$ symmetry breaking for the vector meson is large and might have an appreciable effect.

4. Conclusions

We have described the general procedure to study the $e^+e^- \to P^+P^-P^0P^0$ in order to gain insight into the properties of vector mesons. Using the W gauge boson analog, as a baseline, we discussed each element relevant for the description of the probability amplitude of this process, paying particular attention to the vector meson electromagnetic vertex and its unstable feature, where not only the finite decay width is important but also isospin symmetry breaking corrections might be explored. An application of the formalism to determine the magnetic dipole moment of the ρ meson has proven the sensitivity of this process to the electromagnetic structure of the vector mesons and open the door to explore similar processes like that involving the K^* meson. Finally, we would like to stress the importance of having fine data on these kind of processes to improve our knowledge of the vector mesons.

Acknowledgments

We congratulate the Mexican Division of Particles and Fields for its 30th annual meeting, and thank the organizers for a very pleasant gathering.

References

[1] K. Hagiwara, R. D. Peccei, D. Zeppenfeld and K. Hikasa, Nucl. Phys. B282, 253 (1987)
[2] J. F. Nieves and P. B. Pal, Phys. Rev. D **55** 3118 (1997)
[3] T.H. Burnett and N.M. Kroll, Phys. Rev. Lett. **20**, 86(1968)
[4] G. Toledo Sánchez, Phys. Rev. D **66** 097301(2002)
[5] Abdallah,J. et al (DELPHI Coll.), Eur. Phys. J. C **66** 35(2010)
[6] D. Yu. Bardin, A. Leike, T. Riemann, and M. Sachwitz, Phys. Lett. B **206** 546(1988)
[7] G. López Castro, J. L. M. Lucio, and J. Pestieau, Mod. Phys. Lett. A **6** 3679(1991)
[8] U. Baur and D. Zeppenfeld, Phys. Rev. Lett. **75** 1002(1995)
[9] E. N. Argyres et al., Phys. Lett. B **358** 339(1995)
[10] M. Beuthe, R. Gonzales Felipe, G. López Castro, and J. Pestieau, Nucl. Phys. B **498** 55(1997)
[11] G. Lopez Castro and G. Toledo Sanchez, Phys. Rev. D **61** 033007(2000)
[12] L. A. Jiménez-Pérez, G. Toledo Sánchez and P. Roig (In preparation)
[13] D. García Gudiño and G. Toledo Sánchez, Phys. Rev. D **81** 073006(2010)
[14] G. Ecker and R. Unterdorfer, Eur. Phys. J. C **24** 535(2002)
[15] M. N. Achasov et al. (SND Collaboration), J. Exp. Theor. Phys. **96** 789(2003)
[16] Cosme et al. (M3N Collaboration), Nucl. Phys. B **152** 215(1979)
[17] C. Bacci et al., Nucl. Phys. B **184** 31(1981)
[18] L. M. Kurdadze et al., JETP Lett. **43** 643(1986)
[19] D. Bisello et al., Nucl. Phys. B, Proc. Suppl. **21** 111(1991)
[20] S. I. Dolinsky et al., Phys. Rep. **202** 99(1991)
[21] R. R. Akhmetshin et al., Phys. Lett. B **466** 392(1999)
[22] V. P. Druzhinin, arXiv: 0710.3455 [hep-ex]
[23] G. Toledo Sánchez and D. García Gudiño, Int. J. Mod. Phys. Conf. Series **35** 1460463(2014)
[24] D. García Gudiño and G. Toledo Sánchez, Int. J. Mod. Phys. A **30** 1550114(2015)
[25] F. T. Hawes and M.A. Pichowsky, Phys. Rev. C **59** 1743(1999)
[26] M.S. Bhagwat and P. Maris, Phys. Rev. C **77** 025203(2008)
[27] Ho-Meoyng Choi and Chueng-Ryong Ji, Phys. Rev. D **70** 053015(2004)
[28] F. Cardarelli, et al, Phys. Lett. B **349** 393(1995)
[29] J. P. B. C. Melo and T. Frederico, Phys Rev. C **55** 2043(1997)
[30] J. N. Hedditch et al, Phys. Rev. D **75** 094504(2007)
[31] H. L. L. Roberts et al, Phys Rev. C **83** 065206(2011)
[32] B. Owen et al, Phys Rev. C **91** 074503(2015)
[33] Carrillo Serrano et al, Phys Rev. C **92** 015212(2015)
[34] J. P. Lees et al, (BABAR Coll.), Phys. Rev. D **86** 012008(2012)

Spontaneous Symmetry Breaking in Hyperbolic Field Theory

Oscar Meza-Aldama

Facultad de Ciencias Físico-Matemáticas, Benemérita Universidad Autónoma de Puebla. Av. San Claudio y 18 Sur, C.U. 72570 Puebla, México

E-mail: `oscar.meza@email.ucr.edu`

Abstract. We study a non-compact analog to the U(1) symmetry group. The Higgs potential is obtained as a transversal section of the $\lambda\phi^4$ potential possessing symmetry under the action of this group. Both the spontaneous symmetry breaking and the uniform exact symmetry scenarios are obtained as particular cases. We then study an extension of the U(1) group by taking the direct product of it with this non-compact generator. In particular we obtain an expression for the mass terms of the potential after spontaneous symmetry breaking.

1. Introduction

Hyperbolic numbers appeared in the literature for the first time in a James Cockle article in the year of 1848 [1]. Despite having appeared since so long, applications of them in physical contexts haven't been developed until recent years. The progress done however has shown (see [2]) that a lot of structures used in physics, usually built in terms of complex numbers, admit definitions in terms of hyperbolic numbers. In particular, studies of relativistic quantum mechanics and relativistic wave equations [3], [4], and even a formulation of a gravito-electromagnetism theory [5], have been done systematically. The present work is based on the more detailed study [6].

Hyperbolic numbers can be introduced rigorously as one of the three possible (up to isomorphism) real commutative unital algebras [2]. More general hypercomplex systems are defined as finite-dimensional (not necessarily commutative) real unital algebras [7]. In this work we do not take this approach; we make practical definitions and present only the properties that will be needed. In section 2 we introduce the hyperbolic number system, the usual arithmetic operations are presented and the hyperbolic phases, or "versors", are defined. In section 3 we study the potential of the $\lambda\phi^4$ theory over the hyperbolic numbers, and its respective spontaneous symmetry breaking (SSB) realization. In section 4 we extend the number system of hyperbolic numbers by combining them with the usual complex numbers, thus obtaining the bicomplex number system. Finally, in section 5 we present the case of the bicomplex $\lambda\phi^4$ and one of its possible SSB scenarios. Conclusions are presented in section 6.

2. Hyperbolic numbers

Complex numbers can be constructed heuristically by taking all the formal combinations $x + iy$ with i, called the imaginary unit, is such that $i^2 = -1$, and $x, y \in \mathcal{R}$, but obviously $i \notin \mathcal{R}$. In a similar way we define hyperbolic numbers as the set of all numbers of the form $x + jy$, with $x, y \in \mathcal{R}$, and j is a new quantity, called *hyperbolic* imaginary unit, with the property $j^2 = +1$,

Journal of Physics: Conference Series **761** (2016) 012048 doi:10.1088/1742-6596/761/1/012048

but $j \notin \mathcal{R}$, in particular $j \neq \pm 1$. We will denote this set as $\mathcal{D}$, i.e. $\mathcal{D} \equiv \{x + jy | x, y \in \mathcal{R}\}$. We define sum, subtraction and multiplication in the natural way: for any $z_1, z_2 \in \mathcal{D}$, with $z_1 = x_1 + jy_1$ and $z_2 = x_2 + jy_2$,

$$z_1 \pm z_2 \equiv (x_1 \pm x_2) + j(y_1 \pm y_2) \tag{1}$$

$$z_1 z_2 \equiv (x_1 x_2 + y_1 y_2) + j(x_1 y_2 + y_1 x_2). \tag{2}$$

Analogously to the usual complex case, we define a *hyperbolic* conjugation for any $z = x + jy \in \mathcal{D}$, $\bar{z} \equiv x - jy$. The modulus or norm is $|z|^2 \equiv z\bar{z} = x^2 - y^2$. Note that this "norm" is not positive definite; in particular $|z|^2 = 0$ whenever $y = x$; these numbers of the form $z = x + jx$ are called *null* hyperbolic numbers. The (multiplicative) inverse of z is denoted by z^{-1} and defined as

$$z^{-1} \equiv \frac{\bar{z}}{|z|^2}. \tag{3}$$

We can see that null numbers do not possess an inverse. Finally we define hyperbolic phases or "versors" in the hyperbolic analogous to the Euler formula:

$$e^{j\alpha} \equiv \cosh\alpha + j\sinh\alpha. \tag{4}$$

3. The hyperbolic $\lambda\varphi^4$ model

In this section we analyze the potential of a massive self-interacting hyperbolic scalar field. We propose the following lagrangian

$$\mathcal{L} = \partial^\mu \bar{\varphi} \partial_\mu \varphi - V(\varphi, \bar{\varphi}), \tag{5}$$

$$V = m^2 \bar{\varphi}\varphi + \frac{\lambda}{2}(\bar{\varphi}\varphi)^2, \tag{6}$$

where $\varphi(x) = \varphi_1(x) + j\varphi_2(x) \in \mathcal{D}$. This potential is invariant under rescaling by hyperbolic phases, i.e. under the transformations

$$\varphi \to e^{j\alpha} \quad , \quad \bar{\varphi} \to e^{-j\alpha}\bar{\varphi}. \tag{7}$$

The graph of the potential (as a function of the real and hyperbolic parts of φ, denoted by $\mathrm{Re}(\varphi)$ and $\mathrm{Hy}(\varphi)$, respectively) is shown in figure 1. The solid black lines correspond to the SSB (for $\mathrm{Hy}(\varphi) = 0$) and non-SSB (for $\mathrm{Re}(\varphi) = 0$) scenarios of a *real* scalar field.

One interesting thing to notice is that the full hyperbolic potential *always* presents SSB, independent of the sign of the (squared) mass parameter. This can be seen in figure 2, where the potential is plotted for both positive (figure 3) and negative (figure 3) mass term. Both figures are qualitatively equal, the only difference is a 90 degree rotation in field space, i.e., the change from a real to a tachyonic mass amounts to make the discrete transformation $(\varphi_1, \varphi_2) \to (\varphi_2, -\varphi_1)$.

We now proceed to make explicit the SSB of this potential. First we write it in terms of φ_1 and φ_2,

$$V(\varphi_1, \varphi_2) = m^2(\varphi_1^2 - \varphi_2^2) + \frac{\lambda}{2}(\varphi_1^2 - \varphi_2^2)^2. \tag{8}$$

The extrema of this function are located at $\varphi_1 = \varphi_2 = 0$ and at the hyperbola defined by $m^2 + \lambda(\varphi_1^2 - \varphi_2^2)$. The former is a saddle point, which can be seen from the figure or by evaluating the trace of the Hessian matrix of V at $\varphi_1 = \varphi_2 = 0$, while the latter are the true minima of the potential. We have then

$$|\varphi|^2_{min} = \varphi^2_{1_{min}} - \varphi^2_{2_{min}} = -\frac{m^2}{\lambda} \equiv K^2. \tag{9}$$

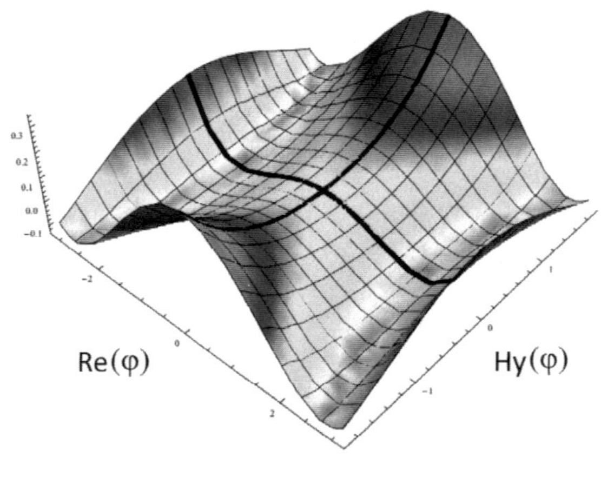

Figure 1. Potential for a massive self-interacting hyperbolic field.

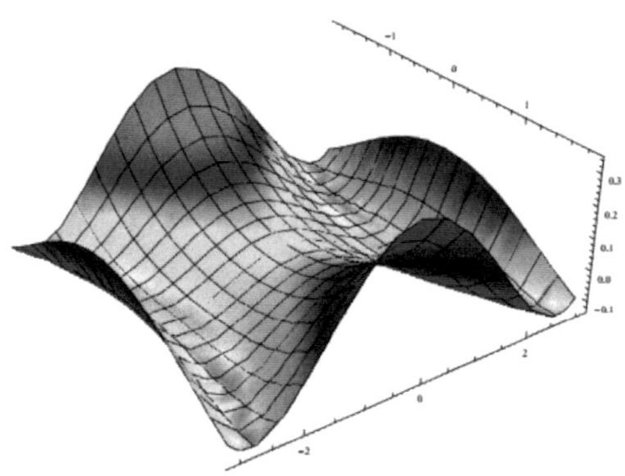

(a) Potential for $m^2 > 0$.

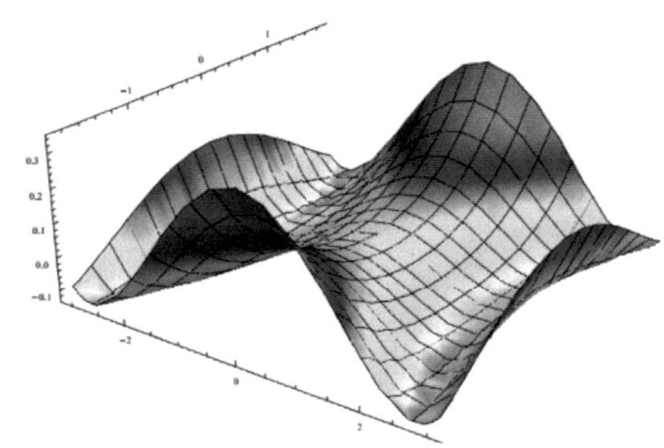

(b) Potential for $m^2 < 0$.

Figure 2. Potentials for real and imaginary mass.

For $m^2 < 0$ the simplest choice is $\varphi 1_{min} = K$ and $\varphi 2_{min} = 0$, so we define the shifted fields χ_1 and χ_2 by

$$\varphi_1 \equiv \chi_1 + K \quad , \quad \varphi_2 \equiv \chi_2. \tag{10}$$

In terms of these the potential reads

$$V = 2m^2\chi_1^2 + \frac{\lambda}{2}\chi_1^4 + \frac{\lambda}{2}\chi_2^4 - \lambda\chi_1^2\chi_2^2 + 2\lambda K\chi_1(\chi_1^2 - \chi_2^2). \tag{11}$$

We can see that χ_2 is now a Goldstone boson which has lost its mass, while χ_1 doubled the value of its mass term. Both have quartic self-interactions and cubic and quartic interactions with each other. This is not very different from the usual complex case.

4. Bicomplex numbers

We now introduce the bicomplex number system, denoted as $\mathcal{H}$, which is basically the direct product of complex and hyperbolic numbers. We define $z \in \mathcal{H}$ as a number of the form

$$z = x + iy + jv + ijw, \tag{12}$$

with $x, y, v, w \in \mathcal{R}$. Addition, subtraction and multiplication of bicomplex numbers are defined in the natural way. Complex conjugation now acts both on i and j: $\bar{z} \equiv x - iy - jv + ijw$, so the first and last terms do not change sign. The "norm" of z is

$$|z|^2 = \bar{z}z = x^2 + y^2 - v^2 - w^2 + 2ij(xw - yv). \tag{13}$$

Now note that $|z|^2$ is now, in general, not even a real number, but it is *hermitian* in the sense that it is invariant under bicomplex conjugation; we call this type of numbers (the ones of the form $a + ijb$) *hybrid* numbers. The set of all hybrid numbers has a couple of nice features: it is closed under sum and multiplication, and any number $z = a + ijb$ always has a multiplicative inverse (as long as $a^2 + b^2 \neq 0$). We can also define the bicomplex phases:

$$e^{i\alpha+j\beta} \equiv e^{i\alpha}e^{j\beta} = \cos\alpha\cosh\beta + i\sin\alpha\cosh\beta + j\cos\alpha\sinh\beta + ij\sin\alpha\sinh\beta. \tag{14}$$

In general a bicomplex number has 4 "degrees of freedom", however in the following sections we will be interested in the simplest generalization of a complex field, which has only two. To reduce the number of components of the bicomplex field, then, we will assume the following relations of proportionality:

$$x = \beta w \quad , \quad y = \beta v, \tag{15}$$

where $\beta \in \mathcal{R}$ is a constant which in principle could take any value; we will nevertheless see below that in some cases we will have to restrict its range of values to a certain interval. Using (15) we can write

$$z = (\beta + ij)w + (i\beta + j)v \quad , \quad \bar{z} = (\beta + ij)w - (i\beta + j)v. \tag{16}$$

The norm is now given by

$$|z|^2 = (\beta^2 - 1)(v^2 + w^2) + 2ij\beta(w^2 - v^2). \tag{17}$$

The relations (15) are the only ones consistent with both (circular and hyperbolic) invariances of this norm. For instance, the identifications $x = \beta v$ and $y = \beta w$ would be inconsistent with the circular symmetry.

5. The bicomplex $\lambda\psi^4$ model

We now study the lagrangian

$$\mathcal{L} = \partial^\mu\bar{\psi}\partial_\mu\psi - V(\psi, \bar{\psi}), \tag{18}$$

$$V = a\frac{m^2}{2}\bar{\psi}\psi + \frac{\lambda}{4!}(\bar{\psi}\psi)^2, \tag{19}$$

where $a = \pm 1$ is just a constant that allows us to control the sign of the mass term, and ψ is a bicomplex field with the restrictions described in the previous section, i.e. it has the form (16). Also, to have a bounded potential (here with "bounded" we mean that both the real and the ij components of the potential are bounded), we have to assume that the mass and self-coupling parameters are hybrid:

$$m^2 = m_R^2 + ijm_H^2, \quad \lambda = \lambda_R + ij\lambda_H. \tag{20}$$

The potential (19) only depends on the combination $\bar{\psi}\psi$, so it takes only hybrid values. Then we can write it as $V = V_R + ijV_H$, where

$$\begin{aligned} V_R \;=\; & a\left(\frac{\beta^2-1}{2}m_R^2 + \beta m_H^2\right)v^2 + a\left(\frac{\beta^2-1}{2}m_R^2 - \beta m_H^2\right)w^2 \\ & + \frac{\lambda_R}{6}\left[\frac{(\beta^2-1)^2}{4}(v^2+w^2)^2 - \beta^2(v^2-w^2)^2\right] \\ & - \frac{\lambda_H}{6}\beta(\beta^2-1)(w^4 - v^4), \end{aligned} \tag{21}$$

$$\begin{aligned} V_H &= a\left(\frac{\beta^2-1}{2}m_H^2-\beta m_R^2\right)v^2+a\left(\frac{\beta^2-1}{2}m_H^2+\beta m_R^2\right)w^2 \\ &+\frac{\lambda_R}{6}\beta(\beta^2-1)(w^4-v^4) \\ &+\frac{\lambda_H}{6}\left[\frac{(\beta^2-1)^2}{4}(v^2+w^2)^2-\beta^2(v^2-w^2)^2\right]. \end{aligned} \tag{22}$$

The vacuum conditions are the usual ones:

$$\psi_0=\bar{\psi}_0=0 \tag{23}$$

or

$$am^2+\frac{\lambda}{6}\bar{\psi}_0\psi_0=0. \tag{24}$$

The first option corresponds to the origin of field space and is not very interesting to us. The second can be rewritten in terms of the components of ψ as

$$(1-\beta^2)(\lambda_R+\lambda_H)(v_0^2+w_0^2)=6a(\lambda_R m_R^2+\lambda_H m_H^2), \tag{25}$$

$$\beta(\lambda_R^2+\lambda_H^2)(v_0^2-w_0^2)=3a(\lambda_R m_H^2-\lambda_H m_R^2). \tag{26}$$

We can solve these two equations for v_0 and w_0, however from the analysis of the hessian matrix of the potentials V_R and V_H we see that the points (v_0, w_0) determined by (25) and (26) are saddle points unless one of the expectation values v_0 or w_0 vanishes. We assume then that $w_0=0$ (the other choice, $v_0=0$, is completely analogous), which implies

$$\frac{m_H^2}{m_R^2}=\frac{(1-\beta^2)\lambda_H+2\beta\lambda_R}{(1-\beta^2)\lambda_R-2\beta\lambda_H},\quad v_0^2=\frac{6am_R^2}{(1-\beta^2)\lambda_R-2\beta\lambda_H}. \tag{27}$$

In this work we will only analyze the case with the simplifying condition $\lambda_R=\lambda_H\equiv\lambda$. We also assume $\beta^2\neq 1$. With these assumptions (27) simplifies to

$$\frac{m_H^2}{m_R^2}=\frac{\beta^2-2\beta-1}{\beta^2+2\beta-1},\quad v_0^2=\frac{6m_R^2}{a\lambda(1-2\beta-\beta^2)}. \tag{28}$$

The latter equation implies that

$$a\lambda(1-2\beta-\beta^2)>0. \tag{29}$$

In the following we assume $a=+1$, $\lambda>0$, and $-1-\sqrt{2}<\beta<-1+\sqrt{2}$ so the above inequality holds. (Actually, from the analysis of the hessian, the allowed interval for producing a stable vacuum is $\beta\in\sim(-0.2, 0.2)$, which is within the interval on which the condition (29) holds.)

We can now rewrite the potentials as

$$V_R=\frac{a}{2}P_R^v m_R^2 v^2+\frac{a}{2}P_R^w m_R^2 w^2+\frac{\lambda}{6}\left[\frac{(\beta^2-1)^2}{4}(v^2+w^2)^2-\beta^2(v^2-w^2)^2-\beta(\beta^2-1)(w^4-v^4)\right],$$

$$V_H=\frac{a}{2}P_H^v m_R^2 v^2+\frac{a}{2}P_H^w m_R^2 w^2+\frac{\lambda}{6}\left[\frac{(\beta^2-1)^2}{4}(v^2+w^2)^2-\beta^2(v^2-w^2)^2+\beta(\beta^2-1)(w^4-v^4)\right],$$

where

$$P_R^v=\frac{\beta^4+4\beta^3-6\beta^2-4\beta+1}{\beta^2+2\beta-1},\quad P_R^w=\frac{(\beta^2+1)^2}{\beta^2+2\beta-1}, \tag{30}$$

$$P_H^v=\frac{\beta^4-4\beta^3-6\beta^2+4\beta+1}{\beta^2+2\beta-1},\quad P_H^w=P_R^w. \tag{31}$$

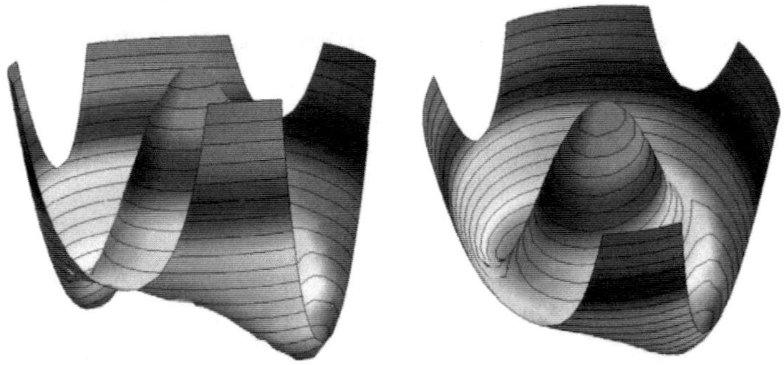

Figure 3. The plots shown here correspond to V_R/m_R^2 and V_H/m_H^2.

V_R and V_H are qualitatively equal; their plots are shown in figure 3. We can see that they have more or less the shape of the traditional mexican hat potential, however now the minima are just two points, located at the bottom of the red valleys.

Once we choose a specific value for the vacuum and expand the lagrangian around that point, we obtain that the quadratic terms (i.e. the mass terms) reduce to

$$-ijP_H^v m_R^2 v^2. \tag{32}$$

So the field w has lost both of its mass components, while v still has a mass in the hybrid sense, though its real part vanishes.

6. Conclusions

We have developed the $\lambda\phi^4$ theory for both hyperbolic and a bicomplex fields. We have shown that the hyperbolic potential contains both the SSB and non-SSB scenarios of a single real scalar field. Also, it is qualitatively insensitive to the sign of the mass term, leading to a hyperbolic SSB scenario whether mass is real or tachyonic. In the more general bicomplex case, we saw that in some cases the incorporation of the new imaginary unit j leads to a deformation of the mexican hat potential, reducing its vacuum manifold to a set of only two points.

Acknowledgements

This work was economically supported by CONACYT. I would like to thank Dr. R. Cartas-Fuentevilla for useful discussions.

References

[1] Cockle J 1848 *London-Edinburgh-Dublin Philosophical Magazine* **33** 37
[2] Hucks J 1993 *J. Math. Phys.* **34** 5986
[3] Ulrych S 2005 *Phys. Lett.* B **625** 313
[4] Ulrych S 2005 *Phys. Lett.* B **632** 417
[5] Ulrych S 2006 *Phys. Lett.* B **633** 631
[6] Cartas-Fuentevilla R and Meza-Aldama O 2016 *Eur. Phys. J.* C **76** 98
[7] Kantor I L and Solodovnikov A S 1989 *Hypercomplex Numbers. An Elementary Introduction to Algebras.* (Springer-Verlag)

MWPF IOP Publishing
Journal of Physics: Conference Series **761** (2016) 012049 doi:10.1088/1742-6596/761/1/012049

Available on CMS information server **CMS CR -2016/187**

The Compact Muon Solenoid Experiment

Conference Report

Mailing address: CMS CERN, CH-1211 GENEVA 23, Switzerland

19 August 2016

Measurement of the J/ψ cross section in pp collisions at $\sqrt{s} = 13$ TeV

Cristian Heber Zepeda Fernández for the CMS Collaboration

Abstract

The double-differential cross sections of the promptly-produced J/ψ mesons, in transverse momentum, p_T, and absolute rapidity, $|y|$, are measured in pp collisions at $\sqrt{s} = 13$ TeV, using a dimuon sample collected by CMS in 2015, corresponding to an integrated luminosity of 2.4 fb $^{-1}$. The results are based on a two−dimensional analysis of the dimuon invariant mass and decay length, and cover a broad p_T range, extending from 20 to 120 GeV, in the $|y| < 1.2$ region.

Presented at *XXX Reunion Anual de la Division de Particulas y Campos de la SMF*

Measurement of the J/ψ cross section in pp collisions at $\sqrt{s} = 13$ TeV

Heber Zepeda for the CMS Collaboration

Centro de Investigación y de Estudios Avanzados del Instituto Politécnico Nacional. Av. Instituto Politécnico Nacional No. 2508 Col. San Pedro Zacatenco 07360, Ciudad de México.

E-mail: hzepeda@fis.cinvestav.mx

Abstract. The double-differential cross-sections of the promptly-produced J/ψ mesons, in transverse momentum, p_T, and absolute rapidity, $|y|$, are measured in pp collisions at $\sqrt{s} =$ 13 TeV, using a dimuon sample collected by CMS in 2015, corresponding to an integrated luminosity of 2.4 fb^{-1}. The results are based on a two-dimensional analysis of the dimuon invariant mass and decay length, and cover a broad p_T range, extending from 20 to 120 GeV, in the $|y| < 1.2$ region.

1. Introduction

The study of quarkonium production in hadronic collisions has been the subject of a large number of experiments since the discovery of heavy-quark bound states. The theoretical treatment of the subject is nowadays based on an effective field theory, Non-Relativistic QCD (NRQCD) [1-3] which assumes factorization of the production process into two phases. The perturbative part of the process generates the heavy $Q\bar{Q}$ pair in a given spin and orbital angular momentum state in either a color singlet or color octet configuration. The subsequent hadronization producing the bound state is controlled by so-called Long Distance Matrix Elements (LDMEs) that are process-independent. Run-1 experiments at the LHC provided precise measurements for J/ψ. These measurements include the differential cross-sections [5] and polarizations [6] at $\sqrt{s} = 7$ TeV calculations by several groups [6-10]. The most recent theory calculations have shown improved agreement with the experimental observation that the polarization at production for all the quarkonium states remains small for transverse momenta up to 70 GeV. Comparison of the pp collision results with Run-2 data at 13 TeV to the corresponding 7 TeV measurements provides a good opportunity to test the factorization hypotheses of NRQCD by using the same LDMEs and perturbative calculations appropriate to the two center-of-mass energies. In addition, the extended p_T reach obtained with the higher center-of-mass energy and production cross section, together with the improved statistical precision, can provide further comparisons with theoretical calculations.

We used the data taken by CMS [4] in the so-called RUN-2 in 2015. This analysis uses the same techniques previously employed by CMS [5] cross-sections of promptly produced quarkonium states, which are calculated in several dimuon p_T and $|y|$ bins, using the formula:

Journal of Physics: Conference Series **761** (2016) 012049 doi:10.1088/1742-6596/761/1/012049

$$Br(J/\psi \to \mu^+\mu^-)\frac{d^2\sigma^{c\bar{c}}}{dp_T dy} = \frac{N^{c\bar{c}}(p_T, y)}{\mathcal{L}\Delta y \Delta p_T}\left\langle\frac{1}{\mathcal{A}(p_T, y)\epsilon(p_T, y)}\right\rangle, \tag{1}$$

where $Br(J/\psi \to \mu^+\mu^-)$ is the branching fraction of the J/ψ into two muons, $N^{c\bar{c}}(p_T, y)$ is the number of signal events from likelihood fit in the (p_T, y) bin under consideration. $\mathcal{L}$ is the integrated luminosity 2.4 fb^{-1}, Δy and Δp_T are the widths of the bin, and $\left\langle\frac{1}{\mathcal{A}(p_T,y)\epsilon(p_T,y)}\right\rangle$ is the average of the inverse acceptance times efficiency for all the events in the bin. The acceptance is calculated using simulation, while the efficiency is derived using a data-driven approach. The non-prompt components for J/ψ, due to decays of b-hadrons, are separated using the pseudo-proper decay length (PPDL). The analysis does not distinguish feed-down from decays of heavier quarkonium states and direct production of quarkonia. The results obtained by CMS at $\sqrt{s} = 7$ TeV [12] are shown in Fig. 1. The polarization of J/ψ has been measured in a previous CMS work and it is found consistent with zero. In this analysis the acceptance is calculated in the hypothesis of null polarization, presently theorical. Models are being improved to describe simultaneously J/ψ differential cross section and polarization measurements.

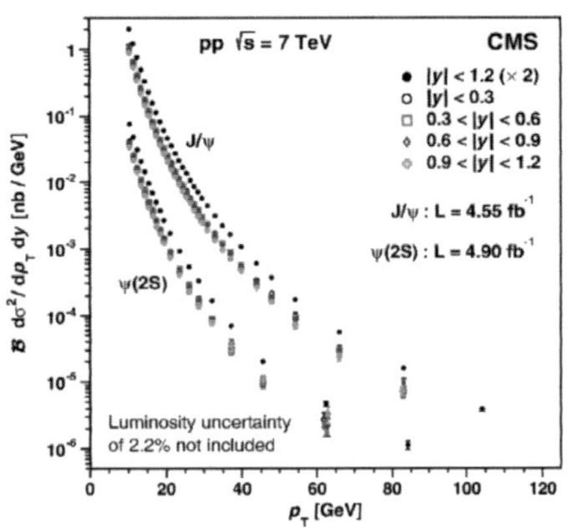

Figure 1. The J/ψ and $\psi(2S)$ differential p_T cross-sections times the dimuon branching fractions for four rapidity bins and integrated over the range $|y| < 1.2$. Taken from **Phys. Rev. Lett. 114, 191802**.

2. Dataset and event selection

J/ψ candidates are formed by pairing opposite-sign muons. In this analysis, to ease the calculation of efficiency, we considered only the case in which the two muons bend away from each other in the CMS magnetic field. This measurement was performed in the kinematical region $20 < p_T(J/\psi) < 120$ GeV and $|y(J/\psi)| < 1.2$. A fit of the two muon tracks to a common vertex is performed, and the fit χ^2 probability required to be above 0.5%. Both muons have to be identified as reconstructed tracks with at least five measurements in the silicon tracker, and at least one in the pixel detector. The track is required to match at least one muon segment identified by a station in the muon detector. Loose cuts are applied on the longitudinal and transverse impact parameter in order to reject cosmic rays and in-flight decays. Opposite-sign muons are then paired and the dimuon vertex-χ^2 probability requires to be greater than 1%.

3. Determination of the yields

The yields are measured using an extended unbinned maximum likelihood fit to the mass distribution. The J/ψ mass is modelled with a Crystall Ball and a Gaussian function. An exponential is used to describe the dimuon mass background.

An additional non-prompt component originanting from the decay of b-hadrons must be taken into account. The prompt and non-prompt yields are measured by simultaneously fitting the mass and the PPDL [13] distribution. The PPDL distribution is modelled by a prompt signal component represented by a resolution function, which uses the event-by-event uncertainty information, a non-prompt term given by an exponential function convoluted with the resolution function, and the continuum background term represented by the empirical sum of two Gaussian functions centered at zero plus an exponential decay function to take into account prompt and non-prompt background components. The resolution function was modelled by two Gaussians, with gaussianly-constrained parameters derived from fits to p_T-integrated data. Fig. 2 shows the maximum likelihood fit projected onto the mass and PPDL dimensions, for a particular (p_T, y) bin as an example.

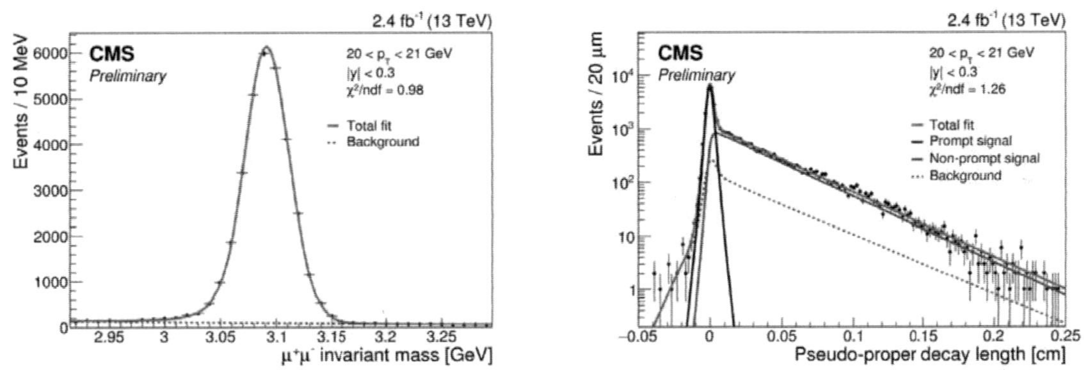

Figure 2. Fits of J/ψ mass (right) and lifetime (left) in a particular (p_T, y) bin. **Taken from CMS-PAS-BPH-15-005.**

4. Acceptance

The acceptance is calculated using simulated events produced by Monte Carlo. The simulation takes into account final-state photon radiation. The acceptance for events in a given (p_T, y) range is defined as:

$$\mathcal{A} = \frac{N^{gen}_{kin}(p_T, y)}{N^{gen}(p_T, y)}, \tag{2}$$

where $N^{gen}_{kin}(p_T, y)$ is the number of generated events which pass the kinematic selections described above for data events and $N^{gen}(p_T, y)$ is the total of the simulated events in that p_T and y range.

5. Efficiency

The muon reconstruction, identification and trigger efficiencies were measured individually from data as a function of muon p_T and pseudorapidity,

$$\epsilon_{\mu\mu}(p_T, y) = \epsilon(p_{T1}, \eta_1) \cdot \epsilon(p_{T2}, \eta_2) \cdot \rho(p_T, y) \cdot \epsilon^2_{tk}. \tag{3}$$

Efficiencies are measurement from data. $\epsilon(p_{T1}, \eta_1)$ and $\epsilon(p_{T2}, \eta_2)$ are obtained from Tag-and-Probe [14] technique that takes advantage of dedicated efficiency triggers, $\rho(p_T, y)$ is the correlation factor and ϵ^2_{tk} is the tracking efficiency. The dimuon efficiency is, similarly to the acceptance, calculated on a event-by-event basis and stored in (p_T, y) histograms.

6. Systematic Uncertainties

Systematic uncertainties include uncertainties on dimuon acceptance and efficiencies, and the determination of the signal yields. A 2.7% uncertainty on the luminosity measurement is assigned. We considered the systematics uncertainties:

- Uncertainties in the estimation of the yield.
- Statistical.
- Non-prompt fraction statistical.
- Acceptance statistical.
- ρ statistical.
- Conditional single efficiencie.
- L3 scale factor.
- Single efficiency.
- Non-prompt fraction systematical.
- Yield systematic.
- ρ Systematical.

As an example, we explain two of them.

6.1. Uncertainties in the measurement of the yields

Uncertainties on the estimation of yields were evaluated by changing the signal and background models used in the maximum likelihood fits. To assess the systematic uncertainty in the signal modeling, n (a parameters of the Crystall Ball function) was varied by $\pm 5\sigma$. The systematic uncertainty originating from a possibly imperfect description of the background was evaluated by changing the background model from an exponential to a linear function. The observed differences from the nominal were taken as systematic uncertainty. The uncertainty on the estimation of the non-prompt fraction arises from the choice of primary vertex and from the modeling of the signal and background in the transverse decay length distribution. Similarly, we assess the impact of modeling the non-prompt signal by fixing the parameterization of the exponential. The systematic uncertainty contribution from the primary vertex choice was added in quadrature with half of the difference between the maximum deviation observed from the nominal fit to the other contributions.

6.2. Uncertainties in the acceptance and efficiency correction

To take into account statistical uncertainties in the measurement of the single-muon efficiencies, we varied the three parameters of the parameterization function within their uncertainties and repeated the efficiency calculation.

7. Results

The cross-sections for the J/ψ is computed using Eq. (1). Results are presented in Fig. 3 in for rapidity range. Fig. 4 shows the comparison of cross section measured by CMS at 7 TeV [5] with the respective analysis. Both 7 TeV and 13 TeV data are presented for the rapidity-integrated range $|y| < 1.2$, which was obtained as the weighted average of the individual rapidity ranges at $\sqrt{s} = 13$ TeV. As shown in the bottom panel of Fig. 4, the 13 TeV cross-sections of the meson is factor of 2 to 3 larger than the corresponding 7 TeV cross-sections, changing slowly as a function of dimuon p_T. A detailed comparison with theory awaits an updated NRQCD calculation for 13 TeV.

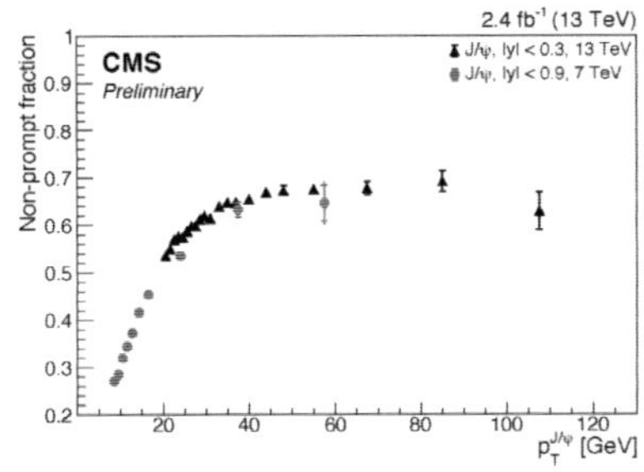

Figure 3. Comparison of the non-prompt fraction of J/ψ as function of dimuon p_T for 13 TeV and 7 TeV. The inner error bars represent the statistical uncertainty while the total errors show the statistical and systematic uncertainties. **Taken from CMS-PAS-BPH-15-005.**

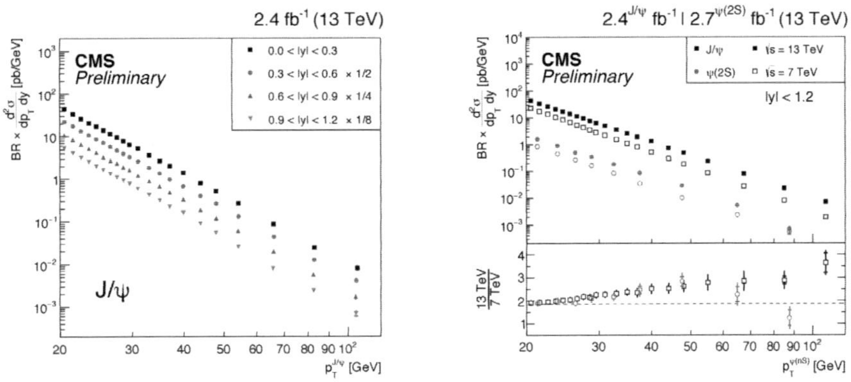

Figure 4. Prompt cross-sections times branching ratios for the J/ψ (left). Comparison of 7 TeV and 13 TeV cross-sections.

8. Conclusions

The double differential production cross-sections of J/ψ has been measured in pp collisions at $\sqrt{s} = 13$ TeV with CMS detector at LHC. The measurement has been performed in the central rapidity region ($|y| < 1.2$) as a function of p_T in several rapidity ranges, extending up to 120 GeV. Ratios of cross-sections measured at 13 TeV and 7 TeV are included in this report. These results shall contribute to consolidate the underlying hypotheses of NRQCD and provide further input to constrain the theory parameters.

9. Acknowledgments

I want to thank Dr. Alberto Sánchez for letting me participate in this work, which also, is the analysis for my doctoral thesis. And finally thank Dr. Pablo Roig for the invitation to contribute to the Proceedings commemorating the XXXth anniversary of the Mexican Division of Particles and Fields.

References

[1] Bodwin G T, Braaten E, and Lepage G P 1997 *Phys. Rev.* D **51** 1125
[2] Cho P and Leibovich A K 1996 *Phys. ReV.* D **53** 150
[3] Cho P and Leibovich A K 1996 *Phys. ReV.* D **53** 6203
[4] CMS Collaboration *JINST 3 (2008) S08004*
[5] CMS Collaboration 2015 *Phys. Rev. Lett.* **114** 191802
[6] CMS Collaboration 2013 *Phys. Lett.* B **727** 381
[7] Kang Z B, Ma Y Q, Qiu J W and Sterman G 2015 *Phys. Rev.* D **91** 014030
[8] Butenschön M and Kniehl B A 2012 *Phys. Rev. Lett.* **108** 172002
[9] Chao K T et al. 2012 *Phys. Rev. Lett.* **108** 242004

[10] Faccioli P et al. 2014 *Phys. Lett.* B **736** 98
[11] Bodwin G T, Chung H S, Kim U R and Lee J 2014 *Phys. Rev. Lett.* **113** 022001
[12] Khachatryan V et al. 2015 *Phys. Rev. Lett.* **114** 191802
[13] CMS Collaboration 2011 *Eur. Phys. J.* C **71** 1575
[14] CMS Collaboration 2011 *JHEP* **01** 080

Top quark electric and magnetic color dipole moments in a Two Higgs Doublet Model with CP violation

E A Garcés[1], R Gaitan [2], R Martinez [3] and J H Montes de Oca [2]

[1] Instituto de Física, Universidad Nacional Autónoma de México, Apdo. Postal 20-364, 01000 Ciudad de México, México.
[2] Departamento de Fśica, FES-Cuautitlán, UNAM, C.P. 54770, Estado de México, México
[3] Departamento de Fśica, Universidad Nacional de Colombia, Bogotá D.C., Colombia

E-mail: egarces@fisica.unam.mx

Abstract.
In this work we study the anomalous top quark-gluon couplings Chromoelectric Dipole Moment and Chromomagnetic Dipole Moment in a general THDM with CP violation also known as THDM Type III. We include the contribution to the color electric and magnetic couplings arising from the Y_{tt} Yukawa coupling. We provide phenomenological restrictions to chromoelectric and chromomagnetic anomalous couplings within the model.

1. Introduction

The top quark is one of the most interesting fermions in the Standard Model (SM), it is the heaviest one with a mass close to the electroweak symmetry breaking scale. There is now in operation the largest machine ever built, fortunately the Large Hadron Collider (LHC) happens to be the place where most top quark pairs are produced allowing the study of top quark properties and providing an optimal site to measure its couplings to the gauge fields and to perform searches for new physics. Top quark Chromomagnetic dipole moment (CMDM) is induced at one loop level in the SM and chromoelectric dipole moment (CEDM) appears from the complex CKM phase only at three loop level [1]. In extended models with new sources of CP violation the study of new physics from anomalous couplings becomes important. CMDM and CEDM are defined through the effective Lagrangian

$$\mathcal{L} = \bar{u}(t)\frac{-g_s}{2m_t}\sigma_{\mu\nu}G^{\mu\nu a}T^a\left(\Delta\tilde{\kappa} + i\gamma_5\Delta\tilde{d}\right)u(t), \tag{1}$$

the real part of the anomalous coupling $\Delta\tilde{\kappa}$ is the CMDM and in the imaginary part appears $\Delta\tilde{d}$ the CEDM, $G^{\mu\nu a}$ is the gluon field strength, and T^a are the QCD fundamental generators of $SU(3)_c$ [2] .

Recently, the CMS collaboration reported the following bound $Re(\Delta\tilde{\kappa}) = 0.037 \pm 0.041$, at 95%CL [3]. Also from Tevatron and Atlas results on the cross section $m_{t\bar{t}}$ there are reported limits to CMDM and CEDM, $|\Delta\tilde{\kappa}| < 0.05$ and $\Delta\tilde{d} < 0.16$ at 95%C.L. [4]. Within uncertainties still in agreement with the SM prediction to the CMDM, wich is $\Delta\tilde{\kappa} \sim 5.6 \times 10^{-2}$ [5]. In the near future with increased integrated luminosity and energy the experimental bound to CEDM expected to be improved up to $\Delta\tilde{d} < 0.05$ [4].

In this work we study the THDM Type-III [1] contribution in the quark top CMDM and CEDM [2]. This kind of model explicitly violates the CP symmetry in the scalar potential, and the top quark CMDM is generated at one loop level.

2. The general two Higgs doublet model

The scalar potential for the general two Higgs doublet model is

$$\begin{aligned} V &= -\mu_1^2\Phi_1^\dagger\Phi_1 - \mu_2^2\Phi_2^\dagger\Phi_2 - \left[\mu_{12}^2\Phi_1^\dagger\Phi_2 + h.c.\right] \\ &+ \frac{1}{2}\lambda_1\left(\Phi_1^\dagger\Phi_1\right)^2 + \frac{1}{2}\lambda_2\left(\Phi_2^\dagger\Phi_2\right)^2 + \lambda_3\left(\Phi_1^\dagger\Phi_1\right)\left(\Phi_2^\dagger\Phi_2\right) + \lambda_4\left(\Phi_1^\dagger\Phi_2\right)\left(\Phi_2^\dagger\Phi_1\right) \\ &+ \left[\frac{1}{2}\lambda_5\left(\Phi_1^\dagger\Phi_2\right)^2 + \lambda_6\left(\Phi_1^\dagger\Phi_1\right)\left(\Phi_1^\dagger\Phi_2\right) + \lambda_7\left(\Phi_2^\dagger\Phi_2\right)\left(\Phi_1^\dagger\Phi_2\right) + h.c.\right], \end{aligned} \tag{2}$$

where μ_1^2, μ_2^2, λ_1, λ_2, λ_3 and λ_4 are real parameters and μ_{12}^2, λ_5, λ_6, and λ_7 could have complex values allowing explicit CP violation. The neutral components in the fields are defined as $\frac{1}{\sqrt{2}}(v_1 + \eta_1 + i\chi_1)$ and $\frac{1}{\sqrt{2}}(v_2 + \eta_2 + i\chi_2)$. The vacuum expectation values (VEV) can be taken real because complex phases can be reabsorbed by the complex parameters of scalar potential.

Due to the explicit CP symmetry breaking, there is mixing among the CP-odd and CP-even scalar sectors. Defining $\tan\beta = \frac{v_1}{v_2}$, we take the scalar field ($\eta_3 = -\chi_1 s_\beta + \chi_2 c_\beta$) orthogonal to the Would-be Goldstone component corresponding to the Z gauge boson. After symmetry breaking, the mass eigenstates of the neutral Higgs bosons are related to the CP mixed η_j states as

$$h_i = \sum_{j=1}^{3} R_{ij}\eta_j, \tag{3}$$

where $i = 1, 2, 3$ and the R matrix is given by [6, 7]:

$$R = \begin{pmatrix} c_1c_2 & s_1c_2 & s_2 \\ -(c_1s_2s_3 + s_1c_3) & c_1c_3 - s_1s_2s_3 & c_2s_3 \\ -c_1s_2c_3 + s_1s_3 & -(c_1s_3 + s_1s_2c_3) & c_2c_3 \end{pmatrix}, \tag{4}$$

with $c_i = \cos\alpha_i$ and $s_i = \sin\alpha_i$, with $i = 1, 2, 3$. The eigenstates h_i are not CP pure states and their mixing is given by the R matrix. Note that in the limit case $s_2 = s_3 = 0$ the case of THDM without CP violation is recovered.

The Yukawa Lagrangian for the quark sector has the general form

$$-\mathcal{L}_{Yukawa} = \sum_{i,j=1}^{3}\sum_{a=1}^{2}\left(\bar{q}_{Li}^0 Y_{aij}^{0u}\tilde{\Phi}_a u_{Rj}^0 + \bar{q}_{Li}^0 Y_{aij}^{0d}\Phi_a d_{Rj}^{0'} + h.c.\right). \tag{5}$$

In the above equation, $Y_a^{u,d,l}$ are the 3×3 Yukawa matrices. q_L denotes the left handed quark doublets and u_R, d_R, represent the right handed quark singlets under $SU(2)_L$. The mass matrix after spontaneous symmetry breaking is

$$M^{u,d} = \sum_{a=1}^{2}\frac{v_a}{\sqrt{2}}Y_a^{u,d}, \tag{6}$$

[1] The Mexican high energy physics community has been very productive in the study of models beyond the SM with an extended scalar sector. See for instance talk contributions by J. Barranco, A. Bolaños, L. Díaz-Cruz, Montes de Oca, J. Orduz and D. Rojas, just to mention a few.

$Y_a^f = V_L^f Y_a^{0f} \left(V_R^f\right)^\dagger$, for $f = u, d$, and $V_{L,R}^f$ are the rotation matrices that diagonalize the mass matrix. The Yukawa matrix Y_2^u as a function of M^u and Y_1^a gives THDM-II Lagrangian, with tree level flavor changing. For the up sector the Yukawa Lagrangian can be written as

$$\begin{aligned} -\mathcal{L}_Y &= \frac{1}{v\sin\beta}\sum_{ijk} \bar{u}_i M_{ij}^u \left(A_k^u P_L + A_k^{*u} P_R\right) u_j h_k \\ &+ \frac{1}{\sin\beta}\sum_{ijk} \bar{u}_i Y_{ij}^u \left(B_k^u P_L + B_k^{*u} P_R\right) u_j h_k, \end{aligned} \tag{7}$$

where

$$\begin{aligned} A_k^u &= R_{k2} - iR_{k3}\cos\beta, \\ B_k^u &= R_{k1}\sin\beta - R_{k2}\cos\beta + iR_{k3}. \end{aligned} \tag{8}$$

The Y_{ij} also gives a contribution to the anomalous couplings CEDM and CMDM of the same order of the one in the THDM-II coming from $(\bar{u}_i M_{ij}^u \left(A_k^u P_L + A_k^{*u} P_R\right) u_j h_k)$.

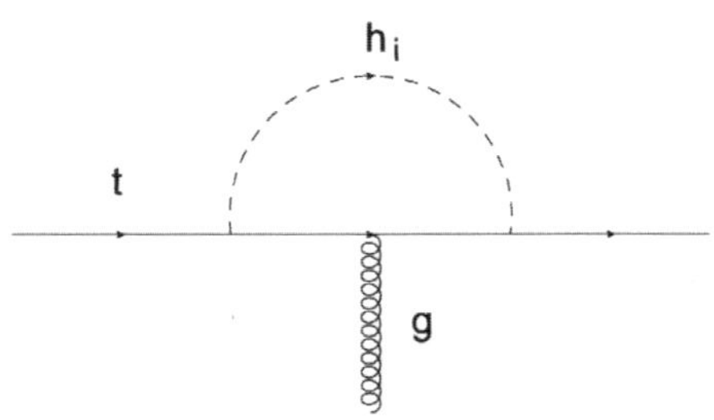

Figure 1. Feynman Diagram for the anomalous quark-gluon couplings in the general THDM.

3. CMDM and CEDM in the general THDM

We will take into account three contributions in the model to the total CMDM and CEDM. (1) The anomalous couplings contributions arising from the diagram in Figure (1) from now on will be denoted as $\Delta\tilde{\kappa}$ and $\Delta\tilde{d}$, (2) the contribution from the Yukawa coupling Y_{ij} denoted as $\Delta\tilde{\kappa_{tt}}$ and $\Delta\tilde{d_{tt}}$, and finally (3) contributions from the coupling proportional to M^u in one vertex and Y^u in the other vertex also known as an interference term $\Delta\tilde{\kappa_{in}}$ and $\Delta\tilde{d_{int}}$.

To obtain the analytic expressions for the above mentioned loop integrals we apply the method presented in [8, 9, 10]. Anomalous couplings will be dependent on A_k^u and B_k^u in Eq. 8, linear combinations of the $\alpha_{1,2,3}$ mixing angles and $\tan\beta$, they will also be dependent on the masses of the Higgses h_1, h_2, and h_3. For simplicity we will take the hierarchical arrangement where h_1 is the SM Higgs, and $m_{h_2} = m_{h_3} = m_{H^+}$. In this partially degenerate case there is CP violation for $\alpha_2 \neq 0$ [11]. We denote the three contributions as $\Delta\tilde{\kappa}_t = \Delta\tilde{\kappa} + \Delta\tilde{\kappa}_{tt} + \Delta\tilde{\kappa}_{int}$ and $\Delta\tilde{d}_t = \Delta\tilde{d} + \Delta\tilde{d}_{tt} + \Delta\tilde{d}_{int}$. Explicit expressions can be found in [2].

In Figure 2 we show the scatter plot in the $\Delta\tilde{\kappa}_t$ and $\Delta\tilde{d}_t$ plane with random values of the mixing angles α_1, α_2 and α_3 for a set of fixed values of $\tan\beta$. In Figure 3 and Figure 4 also the mixing angle α_3 is fixed to $\alpha_3 = 0$ and $\alpha_3 = \pi/4$ respectively. The masses of the heavier Higgses in this case are $m_{h_2} = m_{h_3} = m_{H^+} = 500\ GeV$.

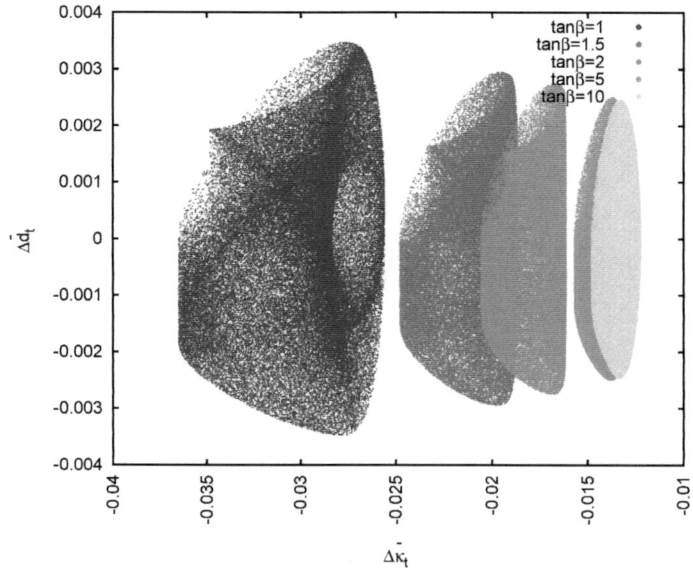

Figure 2. Scatter plot in the $\Delta\tilde{\kappa}_t$ and $\Delta\tilde{d}_t$ plane with random values of the mixing angles α_1, α_2 and α_3, for the specified values of $\tan\beta = 1,\ 1.5,\ 2,\ 5,\ 10$.

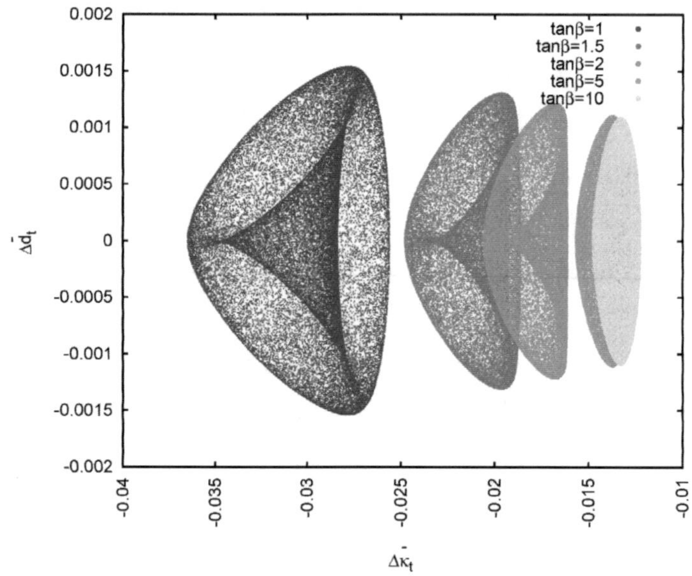

Figure 3. Scatter plot in the $\Delta\tilde{\kappa}_t$ and $\Delta\tilde{d}_t$ plane with random values of the mixing angles α_1, α_2 and $\alpha_3 = 0.$, for the specified values of $\tan\beta = 1,\ 1.5,\ 2,\ 5,\ 10$.

4. Results in the restricted parameter space

We will study nine regions of interest in the α_1-α_2 parameter space, this approximate regions are described in Table 1 and are those already under consideration in [12]. The allowed regions

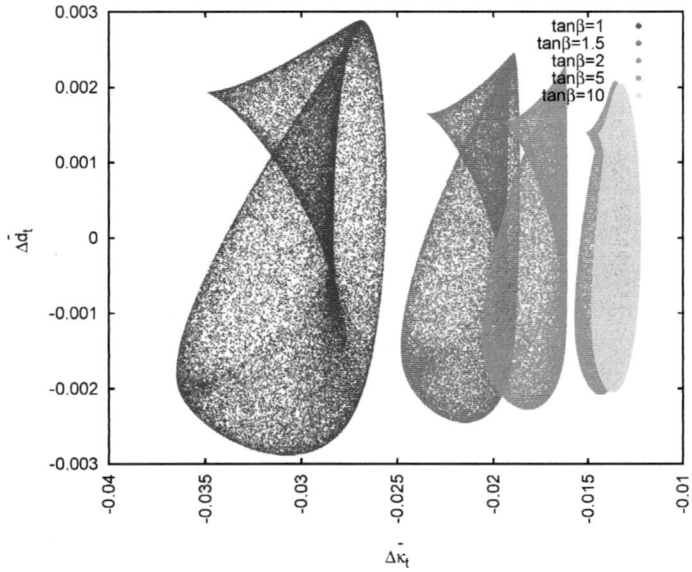

Figure 4. Scatter plot in the $\Delta\tilde{\kappa}_t$ and $\Delta\tilde{d}_t$ plane with random values of the mixing angles α_1, α_2 and $\alpha_3 = \pi/4.$, for the specified values of $\tan\beta = 1,\ 1.5,\ 2,\ 5,\ 10$.

Table 1. Definition of the phenomenologically allowed regions under study, we specify the values of $M_{H^+}$and $\tan\beta$ in each region [2]. In all cases $\alpha_3 = 0$.

	α_1	α_2	M_{H^+}(GeV)	$\tan\beta$
$R1$	$0.67 \leq \alpha_1 \leq 0.8$	$0 \leq \alpha_2 \leq 0.23$	300	1
$R2$	$0.8 \leq \alpha_1 \leq 1.14$	$-0.25 \leq \alpha_2 \leq 0.$	300	1
$R3$	$1.18 \leq \alpha_1 \leq 1.55$	$-0.51 \leq \alpha_2 \leq 0$	500	1
$R4$	$-1.57 \leq \alpha_1 \leq -1.3$	$-0.46 \leq \alpha_2 \leq 0.$	350	1.5
$R5$	$0.93 \leq \alpha_1 \leq 1.57$	$-0.61 \leq \alpha_2 \leq 0.$	350	1.5
$R6$	$-1.57 \leq \alpha_1 \leq -1.28$	$-0.38 \leq \alpha_2 \leq 0.$	350	2
$R7$	$1.08 \leq \alpha_1 \leq 1.57$	$-0.46 \leq \alpha_2 \leq 0.$	350	2
$R8$	$-1.39 \leq \alpha_1 \leq -1.3$	$-0.13 \leq \alpha_2 \leq 0.$	350	2.5
$R9$	$1.16 \leq \alpha_1 \leq 1.5$	$-0.43 \leq \alpha_2 \leq -0.1$	350	2.5

$R_{1..9}$ in the $\alpha_1 - \alpha_2$ plane, are defined from experimental bounds in $R_{\gamma\gamma}$ [13].

The radiative process ($B \to X_s\gamma$) strongly restricts the parameter space $M_{H^\pm}$ vs. $\tan\beta$ [14, 15]. The charged Higgs mass is also restricted to be of around 300 GeV [16] for small $\tan\beta$. A global analysis of B meson decays restricts $M_{H^\pm} < 400$ GeV and $\tan\beta < 10$ [17, 18, 19]. Table 1 resumes the R_i regions for the given values of M_{H^+} and $\tan\beta$. In each case we set the masses of the neutral Higgses m_{h_2} and m_{h_3} equal to the mass of the charged Higgs $m_{h_2} = m_{h_3} = M_{H^+}$.

Once all contributions to CMDM and CEDM are included, we find that for the phenomenologically allowed parameter space the anomalous moments are restricted to be at most of $\Delta\tilde{\kappa} \approx 10^{-2}$ and $\Delta\tilde{d} \approx 10^{-4}$.

5. Conclusions

In this work we have studied regions of interest in the $\alpha_1 - \alpha_2$ parameter space, we calculate the contribution to the top anomalous couplings CMDM and CEDM in the context of a general THDM with CP violation. We find that contributions arising from the Y_{tt} Yukawa coupling and from the interference between M^u and Y_u need to be taken into account.

New physics properties of the top quark such as CMDM and CEDM are an excellent probe of different SM extensions. It will be very exciting if any SM deviation is measured. In the future years with more LHC experimental results we will be able to tell more about the top quark structure.

6. Acknowledgments

This work is supported in part by Sistema Nacional de Investigadores (SNI) in México. R. M. thanks to COLCIENCIAS for the financial support. E.A.G. thanks postdoctoral CONACYT grant and UNAM PAPIIT IN111115.

References

[1] I. B. Khriplovich, Phys. Lett. B **173**, 193 (1986) [Sov. J. Nucl. Phys. **44**, 659 (1986)] [Yad. Fiz. **44**, 1019 (1986)].
[2] R. Gaitan, E. A. Garces, J. H. M. de Oca and R. Martinez, Phys. Rev. D **92**, no. 9, 094025 (2015)
[3] CMS Collaboration [CMS Collaboration], *Search for Anomalous Top Chromomagnetic Dipole Moments from angular distributions in $t\bar{t}$ Dileptonic events at $\sqrt{s} = 7\,TeV$ with the CMS detector*, CMS-PAS-TOP-14-005 (2014).
[4] J. F. Kamenik, M. Papucci and A. Weiler, Phys. Rev. D **85**, 071501 (2012) [Erratum-ibid. D **88**, no. 3, 039903 (2013)] [arXiv:1107.3143 [hep-ph]].
[5] R. Martinez, M. A. Perez and N. Poveda, Eur. Phys. J. C **53** (2008) 221 [hep-ph/0701098].
[6] G. C. Branco, P. M. Ferreira, L. Lavoura, M. N. Rebelo, M. Sher and J. P. Silva, Phys. Rept. **516**, 1 (2012) [arXiv:1106.0034 [hep-ph]].
[7] I. F. Ginzburg and M. Krawczyk, Phys. Rev. D **72**, 115013 (2005) [hep-ph/0408011].
[8] R. Martinez and J. A. Rodriguez, Phys. Rev. D **55**, 3212 (1997) [hep-ph/9612438].
[9] R. Martinez and J. A. Rodriguez, Phys. Rev. D **65**, 057301 (2002) [hep-ph/0109109].
[10] R. Martinez, M. A. Perez and O. A. Sampayo, Int. J. Mod. Phys. A **25**, 1061 (2010) [arXiv:0805.0371 [hep-ph]].
[11] A. W. El Kaffas, W. Khater, O. M. Ogreid and P. Osland, Nucl. Phys. B **775**, 45 (2007) [hep-ph/0605142].
[12] R. Gaitán, R. Martinez, J. H. Montes de Oca and S. R. Romo, Eur. Phys. J. C **74**, 2788 (2014) [arXiv:1312.0044 [hep-ph]].
[13] L. Basso, A. Lipniacka, F. Mahmoudi, S. Moretti, P. Osland, G. M. Pruna and M. Purmohammadi, JHEP **1211** (2012) 011 [arXiv:1205.6569 [hep-ph]].
[14] J. Walsh [BaBar and Belle Collaborations], arXiv:1412.4946 [hep-ex].
[15] M. Misiak *et al.*, Phys. Rev. Lett. **114**, no. 22, 221801 (2015) doi:10.1103/PhysRevLett.114.221801 [arXiv:1503.01789 [hep-ph]].
[16] [CMS Collaboration], CMS-PAS-HIG-12-052.
[17] A. Arbey, M. Battaglia, F. Mahmoudi and D. Martínez Santos, Phys. Rev. D **87**, no. 3, 035026 (2013) [arXiv:1212.4887 [hep-ph]].
[18] U. Haisch and F. Mahmoudi, JHEP **1301**, 061 (2013) [arXiv:1210.7806 [hep-ph]].
[19] CMS Collaboration [CMS Collaboration], CMS-PAS-HIG-12-011.

The decay $h \to \mu\tau$ in the Littlest Higgs Model with T-parity

S. Chamorro-Solano, A. Moyotl and M. A. Perez.

Departamento de Física, Centro de Investigación y de Estudios Avanzados del Instituto Politecnico Nacional, Apdo. Postal 14-740, 07000 México D.F., México.

E-mail: mchamorro@fis.cinvestav.mx
amoyotl@fis.cinvestav.mx
mperez@fis.cinvestav.mx

Abstract. We calculate the one loop contributions to $h \to \mu\tau$ decay in the littlest Higgs model with T-parity. We include mirror fermions and odd gauge bosons loops. We find that for a special choice of the mixing matrix of mirror fermion $k_{ii} = 0.2$ and $k_{ii} = 0.45$, the branching fraction is the order $10^{-3} - 10^{-2}$ and $10^{-2} - 10^{-1}$, respectively, with an energy scale $f = 0.5 - 2$TeV. However, our results are a function of the square of the product of mixing matrices $|V^{\dagger}_{H\mu} V_{H\tau}|^2$.

1. Introduction

Some months ago, the CMS experiment [1] reported a 2σ excess in a search for $h \to \mu\tau$ decay, corresponding to an integrated luminosity of 19.7 fb^{-1} at $\sqrt{s}$ =8 TeV. The corresponding branching ratio reported is $Br(h \to \tau\mu) = 0.84^{+0.39}_{-0.37}\%$, while ATLAS [2] observed only a small excess in one of the signal regions and reported the upper limit $Br(h \to \tau\mu) = 0.77 \pm 0.62\%$ (1.2σ). These results provide great motivation for the search of physics beyond the Standard Model (SM). In the SM the neutral flavour changing processes (NFCP) are not induced at tree level. However, in some SM extensions NFCP are associated with heavy particles, thus the $h \to \mu\tau$ decay could be induced at one loop level. In this regard, in the effective lagrangian approach $Br(h \to \tau\mu) = 10^{-1} - 10^{-2}$ was reported in [3], whereas in the generic supersymmetric SM without R parity [4] the branching ratio can exceed 10^{-5} [4]. This last result is in agreement with [5], where the off-diagonal $\ell\ell h$ boson vertex was induced by one-loop electroweak corrections in the SM framework, in which one right-handed neutrino for each family has been introduced. Moreover, the study of this process in the Simplest little Higgs Model has also been reported in [6] they found that $Br(H \to \mu e) \sim 10^{-12}$ as low as and $Br(H \to \tau\ell)$ is enhanced by a factor of order 300. In this work, we present the results of $h \to \tau\mu$ decay in the littlest Higgs model (LHM) with T-parity [7], induced only by a heavy $W^{\pm}_{H}$ boson. The LHM with T parity is a minimal version of a type of models known as *Little Higgs Model* [8], which are characterized by a global $SU(5)$ and gauge $[SU(2)_1 \times U(1)_1] \times [SU(2)_2 \times U(1)_2]$ symmetries and they are spontaneously broken at scale $f \sim \mathcal{O}(TeV)$. In this model, the Higgs boson is considered as pseudo-Goldstone boson; an important aspect of the model is that it remains protected from quadratic divergences and thus this model offers an alternative to the hierarchy problem. Nevertheless, the LHM with

T parity is affected by electroweak precision tests [9], which led to the inclusion of the so called T-parity and in consequence not only the number of parameters are reduced, but also the scale energy f can be relaxed considerably.
This paper is organized as follows. In section 2, we briefly review the LHM with T parity and show the relevant couplings of the Higgs boson that are necessary for our work. In section 3, we present our calculations for $h \to \mu\tau$ decay at one loop level. Numerical results and analysis are given in Section 4 and finally, our conclusions are given in Section 5.

2. Littlest Higgs Model with T-Parity Framework

The Littlest Higgs Model [10] is based on a non-linear sigma model with global $SU(5)$ symmetry, which is spontaneously broken down to $SO(5)$ via the vacuum expectation value of the scalar field Σ. On other hand, the sub group $[SU(2)\otimes U(1)]^2$ is contained in $SU(5)$, but Σ_0 at the scale f breaks the gauge symmetry to the diagonal odd-T group $SU(2)\otimes U(1)$, which is identified with the SM gauge group; in other words $[SU(2)\otimes U(1)]^2 \to SU(2)\otimes U(1)$.

The gauge-kinetic Lagrangian of LHM is

$$\mathcal{L} = \frac{1}{2}\frac{f^2}{4} Tr|D_\mu \Sigma|^2 \tag{1}$$

were $f = \frac{\Lambda}{4\pi}$ is the collective symmetry breaking scale, Λ the cut off of the model, and the covariant derivative is given by

$$D_\mu\Sigma = \partial_\mu\Sigma - \sum_{j=1}^{2}\left\{ig_j W_j^a\left(Q_j^a\Sigma + \Sigma Q_j^{aT}\right) + ig'_j B_j\left(Y_j\Sigma + \Sigma Y_j^T\right)\right\} \tag{2}$$

Q_j^a and Y_j are generators of the gauged symmetries, g_j and g'_j are coupling constants. The breaking $SU(5) \to SO(5)$ generates 14 Goldstone bosons, 4 of which are eaten to give mass to 4 linear combinations of the gauge fields (see next subsection).

One way to implement T-parity is by the following requirement on the fields, $W_1^a \leftrightarrow W_2^a$, $B_1 \leftrightarrow B_2$. In this way, the gauge kinetic lagrangian (1) is invariant under T-parity and the heavy gauge bosons are odd while the SM ones are even, obtaining a reduction in the total number of model parameters; in particular, for the gauge sector $g_1 = g_2 = \sqrt{2}g$ and $g'_1 = g'_2 = \sqrt{2}g'$, where g and g' would shortly be identified with the SM $SU(2)$ and $U(1)$ gauge couplings, respectively.

2.1. Gauge Boson Sector

The T-odd bosons acquire mass at order $\mathcal{O}(v^2/f^2)$ in the expansion of the Eq. (1) and the mass spectrum is given by

$$M_{W_H} = fg\left(1-\frac{v^2}{f^2}\right), \qquad M_{W_H} = M_{Z_H}, \qquad M_{A_H} = \frac{fg'}{\sqrt{5}}\left(1-\frac{5v^2}{8f^2}\right), \tag{3}$$

and v can be given by $v \simeq v_{SM}(1+\frac{1}{12}\frac{v_{SM}^2}{f^2})$, where $v_{SM} = 246$ GeV.

The mass of the even-T bosons is generated by the second breaking of the symmetry $SU(2)_L\otimes U(1)_Y \to U(1)_{EM}$, which takes place via the usual Higgs mechanism. In this case the mass of the bosons is given as a small correction; however, the relations of SM for the masses of gauge bosons are satisfied at tree level.

2.2. Fermion Sector

For the implementation of parity-T in the fermion sector we introduce mirror fermions. This is done by including fermion doublets in $SU(5)$ and a right handed multiplet Ψ_R of $SO(5)$ is introduced,

$$\mathcal{L}_{mirror} = -\kappa_{ij} f \left(\bar{\Psi}^i_2 \xi + \bar{\Psi}^i_1 \Sigma_0 \Omega \xi^\dagger \Omega\right) \Psi^j_R, \tag{4}$$

where the sum runs over the generation indices $i, j = 1, 2, 3$, κ_{ij} is a mixing matrix and Ψ are the fermion multiplets. After expansion of Eq. (4), we obtain expressions for the masses of mirrors leptons and neutrinos, and neutrino couplings with Higgs boson are given according to

$$m^\nu_H = \sqrt{2}k_{ii} f, \qquad m^\ell_H = \sqrt{2}k_{ii} f \left(1 - \frac{v^2}{8f^2}\right), \qquad h\bar{\nu}^i_H \nu^i_{HR} \sim \frac{i\kappa_{ii}}{2\sqrt{2}}\left(\frac{v}{f}\right). \tag{5}$$

On the other hand, in Ref. [11] it is discussed in detail the existence of two unitary mixing matrices $V_{H\ell}$ and $V_{H\nu}$; this is one of the principal ingredients in the mirror fermionic sector. These matrices are associate with the heavy bosons and satisfy the relation $V^\dagger_{H\nu} V_{H\ell} = V_{PMNS}$, were V_{PMNS} is the Pontecorvo-Maki-Nakagata-Saki (PMNS) matrix, which contains information on the mixture (flavour violating) of the quantum heavy states with SM fermions when they interact through heavy gauge bosons W_H, Z_H or A_H.

As we saw, the heavy charged leptons are predicted by the LHM with T parity; however, they do not have direct couplings with the Higgs boson, and we do not consider these couplings in our calculations. The coupling of leptons and mirror neutrinos with heavy bosons W_H [12] and the coupling between Higgs boson and heavy bosons W_H's are given respectively by

$$\bar{\nu}^i_H W^{+\mu}_H \ell^j \sim \frac{ig}{\sqrt{2}} (V_{H\ell})_{ij} \gamma^\mu P_L, \qquad h W^{+\mu}_H W^\nu_H \sim -\frac{ig^2 v}{2} g^{\mu\nu}, \tag{6}$$

were $P_L = \frac{1}{2}(1 - \gamma_5)$.

3. $h \to \bar{\tau}\mu$ Decay

In general, there are four diagrams contributing at this order to the process $h \to \bar{\tau}\mu$, see Figure 1. In this paper we present the results for the contributions with W_H bosons and a mirror neutrino in the loop. The couplings with heavy particles and SM particles used to perform this calculation are given in the equations (5) and (6).

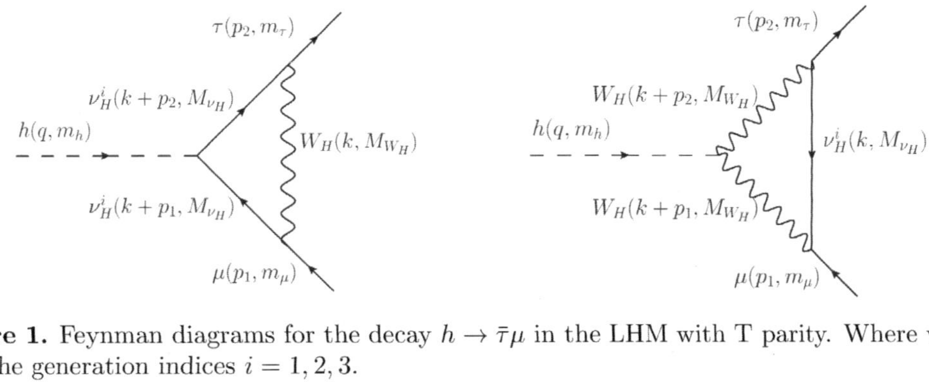

Figure 1. Feynman diagrams for the decay $h \to \bar{\tau}\mu$ in the LHM with T parity. Where we sum over the generation indices $i = 1, 2, 3$.

We consider two complementary diagrams for each diagram considered in Figure (1), In order to remove the divergences. In complementary diagrams, we note that the largest contribution will come from that in which the matrix element is proportional to the mass of the tau.
To solve the integrals we use the method called *Feynman parametrization.*

4. Results and Analysis

In general, the transition amplitude is expressed as follows

$$i\mathcal{M} = i\left(\mathcal{M}_1 + \mathcal{M}_2 + \mathcal{M}_3\right) = i\overline{u}(p_2, m_\tau)\left(A + iB\gamma_5\right)u(p_1, m_\mu), \tag{7}$$

where A and B are structure functions which depend on the scale energy f; we plot their behavior in Figure 2, for two different value of κ_{ii} [1]. A kinematical study is done in [13] for the permitted limits and it was found that $\kappa_{ii} \leq 0.45$ it is a good limit, and excludes $\kappa_{ii} \leq 0.1$, since in this case the quark partners become lighter than the heavy photon A_H and therefore stable. On the other hand, the numerical value of the masses in Eq. (5) are dependent on the energy scale considered and on the parameter κ_{ii}. In our calculations we choose 0.5 TeV $\leq f \leq$ 2 TeV, because when we include the symmetry T parity in LHM the constraints are mitigated with electroweak precision data, starting at a scale of 500GeV [14]. Although the most recent data cover up to 25 fb^{-1} analyzed [15] for Higgs searches in CMS and ATLAS, they excluded parameter space regions at 95% and 99% CL respectly; the exclusion limit for f is $f \geq 607$ GeV.

In the following analyses we have used $\kappa_{ii} = 0.2$ and $\kappa_{ii} = 0.45$. In Figure 2 we present our results as a function of the product $V^*_{H\mu}V_{H\tau}$ mixing matrices.

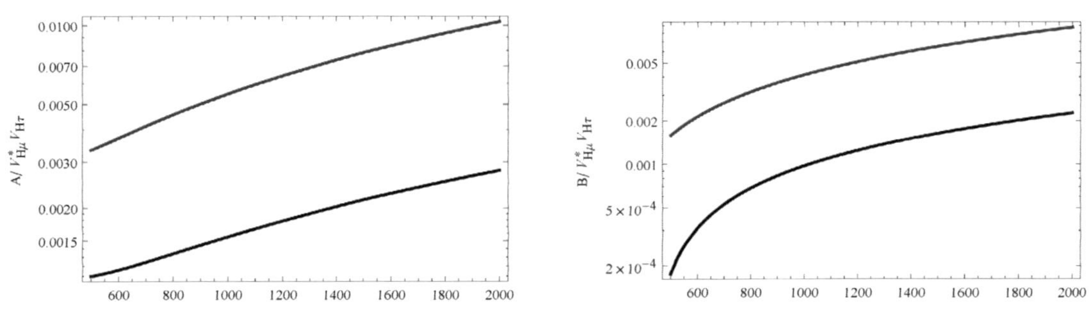

Figure 2. Structure functions A and B as a function of the symmetry breaking scale f. We have used $\kappa_{ii} = 0.2$ (black line) and $\kappa_{ii} = 0.45$ (blue line). The mixing matrices are in general arbitrary.

In our case the decay width has the form,

$$\Gamma = \frac{m_h}{8\pi}\sqrt{\lambda\left(1, \frac{m_\tau^2}{m_h^2}, \frac{m_\mu^2}{m_h^2}\right)}\left\{A^2\left(1 - \frac{(m_\mu + m_\tau)^2}{m_h^2}\right) + B^2\left(1 - \frac{(m_\tau - m_\mu)^2}{m_h^2}\right)\right\}, \tag{8}$$

where $\lambda(x, y, z) = x^2 + y^2 + z^2 - 2xy - 2zy - 2xz$. We will calculate the fraction of the total decay using the experimental results shown in Table 1.

[1] We have seen that κ_{ij} is a free parameter of the model, making the fermion spectrum dependent on this value besides the scale f.

Table 1. Experimental values of the decay width of the Higgs boson in different channels [16].

$X\bar{X}$	$b\bar{b}$	$\tau^-\tau^+$	WW^*	ZZ^*	gg	$\gamma\gamma$
$\Gamma(h \to X\bar{X})$MeV	$4.2^{+7.3}_{-2.6}$	$0.3^{+1.0}_{-0.7}$	$1.2^{+0.5}_{-0.4}$	$0.14^{+0.06}_{-0.04}$	$0.20^{+0.16}_{-0.09}$	$0.04^{+0.05}_{-0.02}$

Finally, we plot the branching ratio in Figure 3. We can see that the branching ratio of $h \to \mu\tau$ decay can reach an order of 10^{-1} for $\kappa_{ii} = 0.45$ (blue line) in agreement with the results of CMS and ATLAS experiments; for $\kappa_{ii} = 0.2$ (black line) we get branching ratio up 10^{-2}, these results are in the energy range allowed by the LHM with T parity.

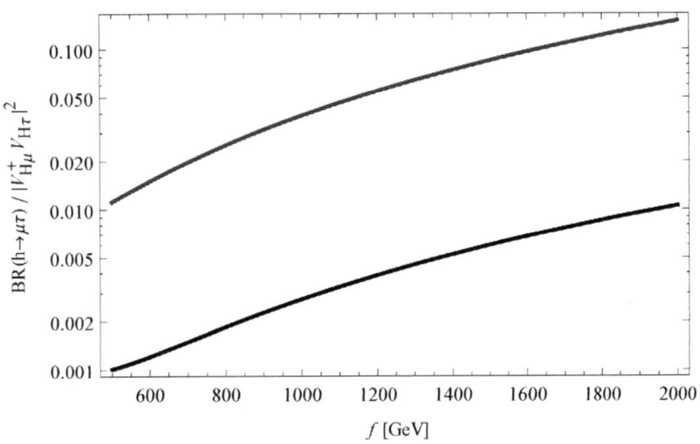

Figure 3. Branching ratio for decay $h \to \mu\tau$ as a function of the symmetry breaking scale f. We have used $\kappa_{ii} = 0.2$ (black line) and $\kappa_{ii} = 0.45$ (blue line).

5. Conclusions and Prospects

We have studied the flavor violating Higgs boson decay to $\mu\tau$, which is induced at one loop level by charged heavy gauge bosons and mirror heavy neutrinos, in the Little Higgs Model with T-parity. For this purpose, we consider symmetry breaking scale f between 500 and 2000 GeV. Moreover, we have used values of $\kappa_{ii} = 0.2$ and $\kappa_{ii} = 0.45$, which was obtained to consider that the mirror quarks are heavier than all the heavy gauge bosons of the LHM with T parity. Our result shows that the branching fraction is of the order $(10^{-3} - 10^{-2})|V^\dagger_{H\mu}V_{H\tau}|^2$ and $(10^{-2} - 10^{-1})|V^\dagger_{H\mu}V_{H\tau}|^2$, but our result decreases when it considers any particular scenario for the mixing matrices. However, there are additional contributions arising from neutral heavy gauge bosons and charged mirror leptons, but it is showed that the charged heavy gauge bosons is the main contribution. The details of these analyses will be reported soon.

6. Acknowledgements

We acknowledge support from CONACYT (Mexico).

References

[1] V. Khachatryan et al. [CMS Collaboration], "Search for Lepton-Flavour-Violating Decays of the Higgs Boson", Phys. Lett. 749, 337 (2015).

[2] G. Aad et al. [ATLAS Collaboration], " Search for Lepton-Flavour-Violating $H \to \mu\tau$ Decays of the Higgs Boson with the ATLAS Detector", arXiv:1508.03372 [hep-ex].

[3] J. L. Diaz-Cruz and J. Toscano, "Probing Lepton Flavour Violation with Higgs Boson Decays $H \to \ell_i + \ell_j$", Phys.Rev. D62, 116005 (2000).

[4] A. Arhrib , Y. Cheng, O. Kong, "$H \to \mu\tau$ Decay in Supersymmetry without R-Parity", Europhys. Lett. 101 (2013) 31003.

[5] A. Pilaftsis, "Lepton Flavor nonconservation in h^0 Decays", Phys. Lett. B285 (1992).

[6] A. Lami and P. Roig, "$H \to \ell\ell'$ in the Simplest Little Higgs Model", arXiv:1603.09663v1 [hep-ph] 31 Mar 2016.

[7] N. Arkani, A. G. Cohen, E. Katz and A. E. Nelson, " The Littlest Higgs" JHEP 0207, 034 (2002).

[8] N. Arkani, A. Cohen, and H. Georgi, "Electroweak Symmetry Breaking from Dimensional Deconstruction", Phys.Lett. B513 (2001).

[9] C. Csaki, J. Hubisz, G. D. Kribs, P. Meade, and J. Terning, "Big Corrections from a Little Higgs", Phys.Rev. D67 (2003) 115002.
J. Hewett, F. Petriello, and T. Rizzo, "Constraining the littlest Higgs", JHEP 0310 (2003) 062.

[10] H. Cheng, I. Low, " TeV Symmetry and the Little Hierarchy Problem", JHEP 0309, 051 (2003).

[11] J. Hubisz, S.J. Lee and G. Paz,"The flavor of a little Higgs with T-parity", JHEP 06, 041 (2006), hep-ph/0512169.

[12] M. Blanke et al. "Rare and CP-Violating K and B Decays in the Littlest Higgs Model with T-Parity", JHEP01 (2007) 066.

[13] Jrgen Reuter, Marco Tonini, Maikel de Vries, "Littlest Higgs with T-parity: Status and Prospects", arXiv:1310.2918v2[hep-ph] 20 Feb, 2014.

[14] J. Hubisz, P. Meade, A. Noble, and M. Perelstein, "Electroweak Precision Constraints on the Littlest Higgs Model with T Parity", J. High Energy Phys. 01 (2006) 135.

[15] J. Reuter, M. Tonini and M. de Vries, "Littlest Higgs with T-Parity: Status and Prospects", JHEP02 (2014) 053.

[16] CMS collaboration, arXiv:1202.1487v1[hep-ex]; arXiv:1202.1488v1.
ATLAS collaboration, ATLAS-CONF-2012-019, March 7, 2012.
The TEVNPH Working Group for the CDF and D0 Collaboration, FERMILAB-CONF-12-065-E.

Journal of Physics: Conference Series **761** (2016) 012052 doi:10.1088/1742-6596/761/1/012052

The diphoton resonance and Higgs results from the CMS Detector of the LHC

M.I.Pedraza-Morales on behalf of the CMS Collaboration

Facultad de Ciencias Físico-Matemáticas, Benemérita Universidad Autónoma de Puebla

E-mail: mpedraza@fcfm.buap.mx

Abstract. Despite the discovery of the Higgs boson, which contributed to the success of the Standard Model, there are at least 60 different analyses carried out in parallel with the CMS detector looking for new physics. In this talk the current status of the Higgs analysis and the search for physics beyond the Standard Model in the two photons final state are presented.

1. Introduction

During the Run I of the CERN LHC, a new particle was discovered by both ATLAS [1] and CMS [2] collaborations. Subsequent studies using the full LHC Run I data set in various decay channels and production modes and combined measurements from ATLAS and CMS [[3]-[19]] showed that the properties are so far consistent with expectations from the Standard Model Higgs boson.

In this paper, the results with the Run II of the most sensitive channels with a mass around 125 GeV: $H \to ZZ \to 4l$ [6] and $H \to \gamma\gamma$ [7] are presented, including the search for resonant production of high mass photon pairs [8]. The analyses followed the strategy described in [9], [10] and [11] respectively. The data collected by CMS in 2016 so far corresponds to 12.9 fb^{-1} of integrated luminosity of pp collisions at $\sqrt{s} = 13\ TeV$.

2. CMS Detector

The Compact Muon Solenoid (CMS) [12] is a multipurpose detector operating at the Large Hadron Collider (LHC) at CERN, which has been successfully collecting data since 2009. The central feature is a superconducting solenoid, 13m in length and 6m in diameter, which provides an axial magnetic field of 3.8 T. CMS detector subsystems relevant for these searches are the tracker (silicon pixel and strip tracker), the electromagnetic calorimeter (ECAL), the hadronic calorimeter, and the muon spectrometer. The physical objects used in these analyses are built as follows: Muons are reconstructed independently in the tracker and the muon spectrometer, and are combined using statistical combination; electrons are reconstructed from clusters of ECAL cells with a matched track in the tracker; and photons are directly obtained from the ECAL measurement. A particle-flow event algorithm [[13], [14]] reconstructs and identifies each individual particle with an optimized combination of information from the various elements of the CMS detector.

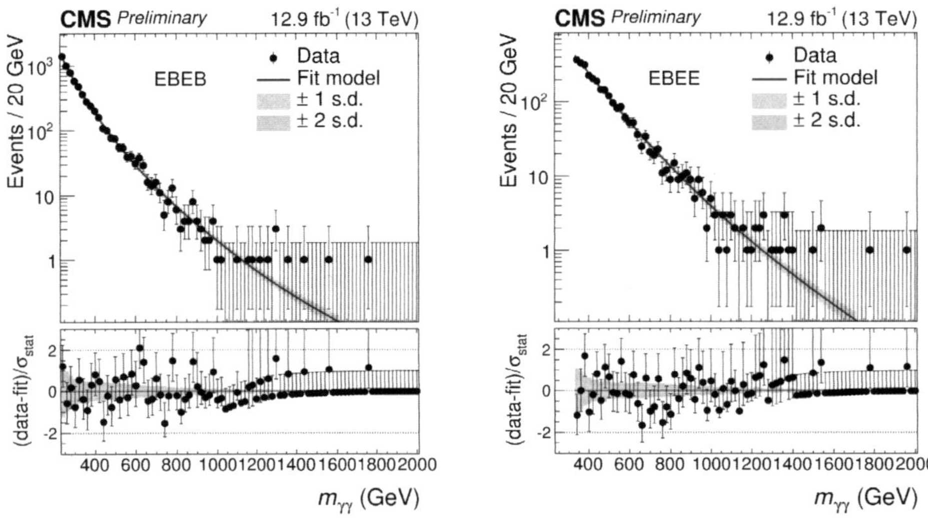

Figure 1. Observed invariant mass spectra for the EBEB (Left, both γ's in the barrel) and EBEE (Right,only one γ in the barrel, one in the endcap). No event with $m_{\gamma\gamma} > 2000$ GeV is selected in the analysis. The results of a likelihood fit to the background-only hypothesis are also shown. The shaded regions show the 1 and 2 standard deviation uncertainty bands associated with the fit, and reflect the statistical uncertainty of the data. The lower panels show the difference between the data and fit, divided by the statistical uncertainty in the data points.

3. Data Sets

These analyses makes use of an integrated luminosity of 12.9 fb^{-1} of pp collisions collected by the CMS experiment at $\sqrt{s} = 13$ TeV in 2016. The main triggers for the $H \to ZZ \to 4l$ analysis select either a pair of electrons or muons, or an electron and a muon. While a diphoton trigger with asymmetric transverse energy (ET) thresholds is used for the $H \to \gamma\gamma$ analysis, and the high mass photon pairs search. The trigger selection is fully efficient for resonance masses above 500 GeV.

All signal and background generators are interfaced with PYTHIA 8 [15] to simulate the multiparton interaction and hadronization effects. The generated events go through a detailed simulation of the CMS detector based on GEANT4 [16] and are reconstructed with the same algorithms that are used for data. The simulated events include pp interactions and have been reweighted so that the distribution of the number of interactions per LHC bunch crossing matches that observed in data.

4. Diphoton resonances

The diphoton channel provides fully reconstructed resonances with a good mass resolution, an opportunity to discover new particles. It is an experimentally robust channel with small systematic uncertainties and relatively low background at hadron colliders. The event selection looks for final states with two high-p_T isolated photons passing the selection of $p_T > 75$ GeV, $|\eta| < 2.5$, excluding the transition barrel-encap region $1.44 < |\eta| < 1.57$. Two event categories are defined EBEB, and EBEE. The background model is a parametric fit to data with an empirical function $f(m_(\gamma\gamma)) = m_{\gamma\gamma}^{a+b*log(m_{\gamma\gamma})}$. The mass spectra is shown in Figure 1 for both categories.

The signal shape is a convolution of intrinsic line-shape of the resonance, derived using

Journal of Physics: Conference Series **761** (2016) 012052 doi:10.1088/1742-6596/761/1/012052

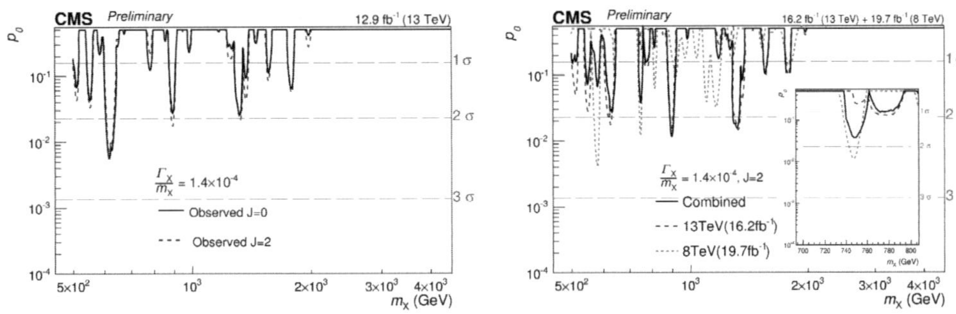

Figure 2. (Left) Observed background-only p-value for resonances with $\Gamma_X/m_X = 1.4\times10^{-4}$ as a function of the resonance mass m_X, from the analysis of the 12.9 fb^{-1} data collected at $\sqrt{s} = 13$ TeV in 2016. (Left) Observed background-only p-value for resonances with $\Gamma_X/m_X = 1.4\times10^{-4}$ and spin-2 as a function of the resonance mass m_X, from the combined analysis of the 8 and 13 TeV data [8].

PYTHIA, and detector resolution. The signal mass resolution, quantified through the ratio of the full width at half maximum of the distribution, divided by 2.35, to the peak position, is roughly 1% and 1.5% for the EBEB and EBEE categories respectively. Nine mass values in the range 500-4500 GeV with spin-0 and spin-2 with the value of the width $\Gamma_X/m_X = 1.4\times10^{-4}$, 1.4×10^{-2}, 5.6×10^{-2} were tested. The observed p-value obtained for selected width is shown in Figure 2. No significant excess in proximity of 750 GeV.

5. Higgs boson studies

5.1. Diphoton

The event selection requires two photon candidates with $p_T^1 > 30$ GeV and $p_T^2 > 20$ GeV. Both photons must satisfy the requirement $|\eta| < 2.5$, excluding the transition region between the barrel and the endcap $1.44 < |\eta| < 1.57$. Additionally the energy sum of 3X3 crystals centered on the most energetic crystal in the candidate electromagnetic cluster divided by the energy of the candidate (R_9 variable) should be larger than 0.8, or the charged hadron isolation < 20 GeV, or the charged hadron isolation relative to $p_T < 0.3$.

In order to increase the signal background ratio and improve the analysis sensitivity the events are divided into eight categories, labelled as: ttH Leptonic Tag, ttH Hadronic Tag, VBF Tag 0, VBF Tag 1, Untagged 0, Untagged 1, Untagged 2 and Untagged 3. Figure 3 shows the multivariate discriminator, Boosted Decison Tree (BDT) as described in [17], used to define the Untagged categories, labelled from best resolution to worst (left) and the shows the invariant mass for all categories (right). The dominant background to $H\to\gamma\gamma$ consists of the irreducible prompt diphoton production, and the reducible backgrounds from $\gamma + jet$ and QCD multijet, where the jets are misidentified as isolated photons. The model used to describe the background is extracted from data with the discrete profiling method [18].

A significance of 5.6 σ is observed (6.2 σ expected) at 125.09 GeV. The best fit for the signal is strength $\hat{\mu} = \sigma/\sigma_{SM} = 0.95^{+0.21}_{-0.18} = 0.95\pm0.17(stat.)^{+0.08}_{-0.05}(theo.)^{+0.10}_{-0.07}(syst.)$. The signal strength is also measured for the bosonic and fermioninc components $\hat{\mu}_{VBF,VH} = 1.59^{+0.73}_{-0.45}$ and $\hat{\mu}_{ggH,t\bar{t}H} = 0.80^{+0.14}_{-0.18}$. Figure 4 shows the signal strength per production processes. The production mechanism signal strengths are compatible with the SM.

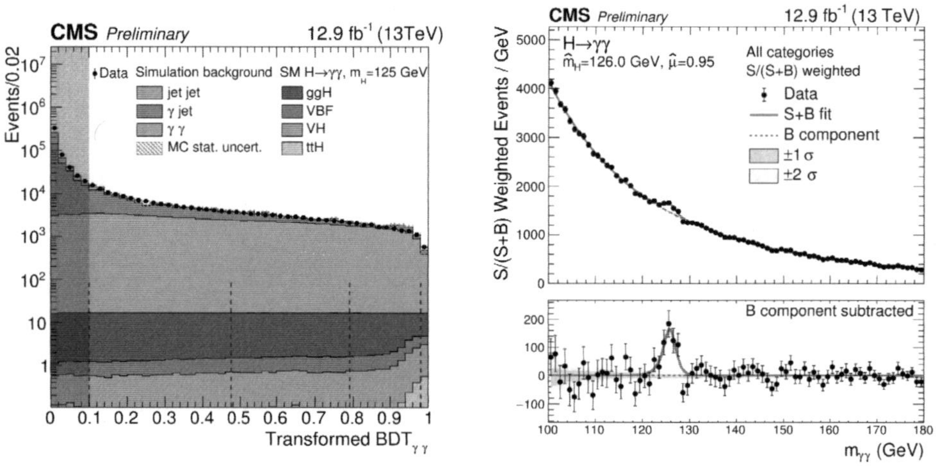

Figure 3. (Left) Transformed $BDT_{\gamma\gamma}$ classifier score in data (black points) and simulation (stacked histograms) for events in the region $100 < m_{\gamma\gamma} < 180\ GeV$. Dashed vertical lines separate the Untagged categories from left (Untagged 0) to right (Untagged 3). The gray band represents events rejected in the analysis. (Right) Data points (black) and signal plus background model fits for all categories summed weighted by their sensitivity. The 1 standard deviation (green) and 2 standard deviation bands (yellow) include the uncertainties of the fit. The bottom plot shows the residuals after background subtraction.

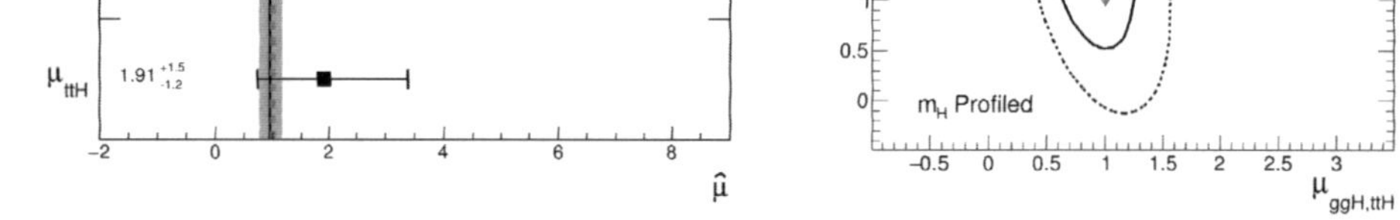

Figure 4. (Left) Signal strength measured for each process (black points) for profiled m_H, compared to the overall signal strength (green band) and to the SM expectation (dashed red line). Since this analysis does not include any categories targeting the VH process, we impose $\mu_{VH} = 1$. (Right) The two-dimensional best-fit (black cross) of the signal strengths for fermionic (ggH, $t\bar{t}H$) and bosonic (VBF, ZH, WH) production modes compared to the SM expectations (red diamond). The Higgs boson mass is profiled in the fit. The solid (dashed) line represents the 1 standard deviation (2 standard deviation) confidence region.

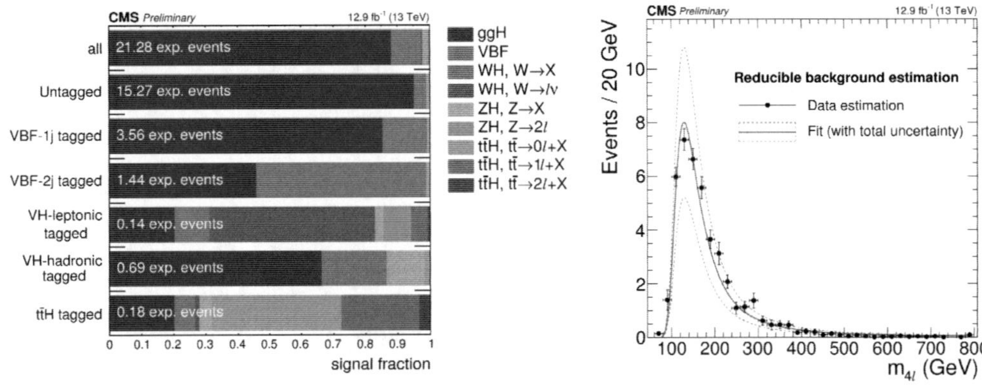

Figure 5. (Left) Signal relative purity of the six event categories in terms of the 5 main production mechanisms of the H(125) boson in a $118 < m_{4l} < 130$ GeV window. The WH, ZH and $t\bar{t}H$ processes are split according to the decay of associated objects, whereby X denotes anything else than a lepton. (Right) Combination of the OS and SS method predictions for the reducible background in the signal region and the parametrized m_{4l} shape. The yellow band shows the total uncertainty on the prediction.

5.2. *Four leptons*

This analysis relies on the efficiency to select leptons, the large signal to background ratio, excellent resolution, and the complete reconstruction of the final state. It is called *the golden channel* for discovery and property measurements. The event selection requires at least 4 leptons in the event, Z candidates are then built with pairs of leptons of the same flavor and opposite-charge (e^+e^-, $\mu^+\mu^-$) and required to have $12 < m_{l^+l^-} < 120$ GeV. Then they are combined into ZZ candidates, we denote as Z_1 the Z candidate with an invariant mass closest to the nominal Z boson mass. The flavors involved define three subchannels: $4e$, 4μ and $2e2\mu$.

The events are separated in 6 mutually exclusive categories in order to increase sensitivity, relying on the number of jets, b-tags, additional leptons, and cuts on the 4 production discriminants. The Figure 5 shows the signal fraction for each category. The main and irreducible background is the ZZ production, the reducible background coming from $Z + X$ is data driven estimated. Two independent methods are applied : Same-sign (SS) and Opposite-sign (OS), using orthogonal control regions. The Figure 5 shows the reducible background estimation using both methods.

The results of the event selection show a good agreement over the whole range of m_{4l} in the three final states ($4e$, 4μ and $2e2\mu$). The highest-mass candidate has a mass of 802 GeV. Figure 6 shows the invariant mass combined. The observed significance from simultaneous fit of the 2D likelyhood in 3 final states by 6 categories is 6.2 σ (6.5σ expected) at $125.09GeV$, the mass value obtained in the combination of Run1 [19]. Figure 6 shows the minimum of p-value gotten at $m_H = 124.3$ GeV, a significance of 6.4 σ is observed (6.3 σ expected).

The combined signal strength at $m = 125.09$ is $\hat{\mu} = 0.99^{+0.33}_{-0.26}$. The signal strength is also measured for the bosonic and fermionic components $\hat{\mu}_{VBF,VH} = 0.91^{+1.56}_{-0.91}$ and $\hat{\mu}_{ggH,t\bar{t}H} = 1^{+0.39}_{-0.32}$. Figure 7 shows the signal strength per production processes. The production mechanism signal strengths are compatible with the SM.

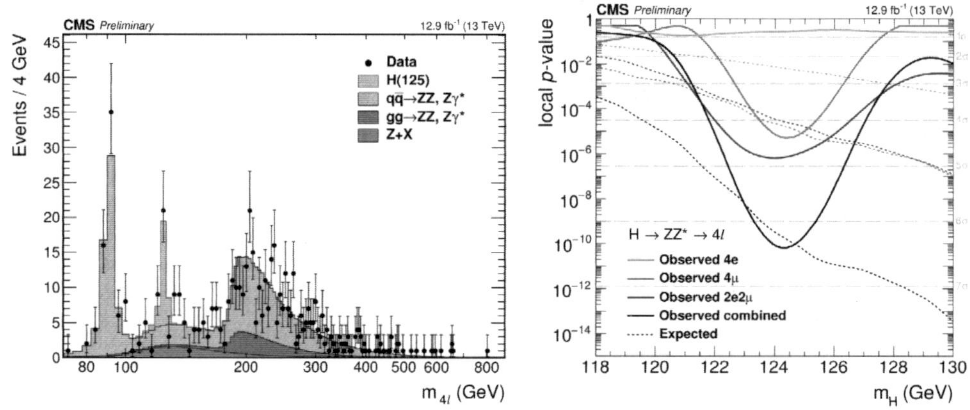

Figure 6. (Left) Distribution of the four-lepton reconstructed invariant mass m_{4l} in the full mass range. Points with error bars represent the data and the stacked histograms represent expected distributions from the 125 GeV Higgs boson signal and the ZZ backgrounds, both normalized to the SM expectation. (Right) Significance of the local fluctuation with respect to the SM expectation as a function of the Higgs boson mass. Dashed lines show the mean expected significance of the SM Higgs boson for a given mass hypothesis

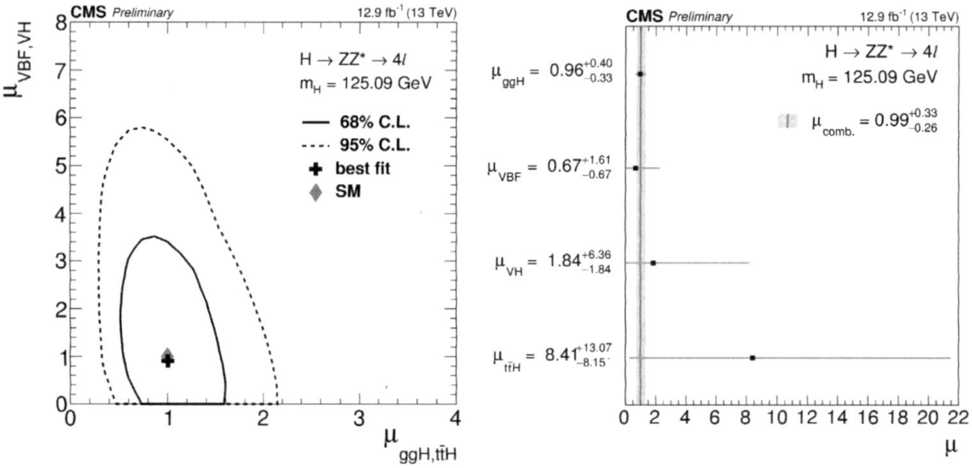

Figure 7. (Left) Result of the 2D likelihood scan for the $\mu_{ggH,t\bar{t}H}$ and $\mu_{VBF,VH}$ signal strength. The solid and dashed contours show the 68% and 95% CL regions, respectively.The cross indicates the best-fit values, and the diamond represents the expected values for the SM Higgs boson. (Right) Results of likelihood scans for the signal strength corresponding to 4 main Higgs boson production modes, compared to the combined μ shown as a vertical line. The horizontal bars and the filled band indicate the $\pm 1\sigma$ uncertainties. The uncertainties include both statistical and systematic sources.

Summary
The observation of the Higgs boson decaying in the four lepton and the diphoton channel and the measurement of some of its properties is presented, as well as the search for diphoton resonances in the high mass region. The analysis used 12.9 fb^{-1} of data collected by the CMS experiment in 2016. The analyses follow closely the strategies used for Run 1. A clear signal, of the Higgs boson in both channels, is observed. The observed significance at $m_H = 125.09$ in the four lepton channel is 6.2 σ, and in the diphoton channel is 5.6 σ. No significant excess of events was observed for high mass diphoton resonances, with 2016 data, above the standard model predictions. Using leading order calculations from PYTHIA 8.2, RS gravitons with masses below 3.85 and 4.45 TeV are excluded for $k = 0.1$ and 0.2 respectively. For $k = 0.01$, graviton masses below 1.95 TeV are excluded, except for the region between 1.75 TeV and 1.85 TeV.

References

[1] G. Aad *et al.* [ATLAS Collaboration], "Observation of a new particle in the search for the Standard Model Higgs boson with the ATLAS detector at the LHC," Phys. Lett. B **716**, 1 (2012) doi:10.1016/j.physletb.2012.08.020 [arXiv:1207.7214 [hep-ex]].

[2] S. Chatrchyan *et al.* [CMS Collaboration], "Observation of a new boson at a mass of 125 GeV with the CMS experiment at the LHC," Phys. Lett. B **716**, 30 (2012) doi:10.1016/j.physletb.2012.08.021 [arXiv:1207.7235 [hep-ex]].

[3] V. Khachatryan *et al.* [CMS Collaboration], "Precise determination of the mass of the Higgs boson and tests of compatibility of its couplings with the standard model predictions using proton collisions at 7 and 8 TeV," Eur. Phys. J. C **75**, no. 5, 212 (2015) doi:10.1140/epjc/s10052-015-3351-7 [arXiv:1412.8662 [hep-ex]].

[4] G. Aad *et al.* [ATLAS Collaboration], "Measurements of the Higgs boson production and decay rates and coupling strengths using pp collision data at $\sqrt{s} = 7$ and 8 TeV in the ATLAS experiment," Eur. Phys. J. C **76**, no. 1, 6 (2016) doi:10.1140/epjc/s10052-015-3769-y [arXiv:1507.04548 [hep-ex]].

[5] G. Aad *et al.* [ATLAS and CMS Collaborations], "Combined Measurement of the Higgs Boson Mass in pp Collisions at $\sqrt{s} = 7$ and 8 TeV with the ATLAS and CMS Experiments," Phys. Rev. Lett. **114**, 191803 (2015) doi:10.1103/PhysRevLett.114.191803 [arXiv:1503.07589 [hep-ex]].

[6] CMS Collaboration [CMS Collaboration], "Measurements of properties of the Higgs boson and search for an additional resonance in the four-lepton final state at sqrt(s) = 13 TeV," CMS-PAS-HIG-16-033.

[7] CMS Collaboration [CMS Collaboration], "Updated measurements of Higgs boson production in the diphoton decay channel at $\sqrt{s} = 13$ TeV in pp collisions at CMS.," CMS-PAS-HIG-16-020.

[8] CMS Collaboration [CMS Collaboration], "Search for resonant production of high mass photon pairs using 12.9 fb^{-1} of proton-proton collisions at $\sqrt{s} = 13$ TeV and combined interpretation of searches at 8 and 13 TeV," CMS-PAS-EXO-16-027.

[9] V. Khachatryan *et al.* [CMS Collaboration], "Observation of the diphoton decay of the Higgs boson and measurement of its properties," Eur. Phys. J. C **74**, no. 10, 3076 (2014) doi:10.1140/epjc/s10052-014-3076-z [arXiv:1407.0558 [hep-ex]].

[10] CMS Collaboration [CMS Collaboration], "Studies of Higgs boson production in the four-lepton final state at $\sqrt{s} = 13$ TeV," CMS-PAS-HIG-15-004.

[11] V. Khachatryan *et al.* [CMS Collaboration], "Search for Resonant Production of High-Mass Photon Pairs in Proton-Proton Collisions at $\sqrt{s}$ =8 and 13 TeV," Phys. Rev. Lett. **117**, no. 5, 051802 (2016) doi:10.1103/PhysRevLett.117.051802 [arXiv:1606.04093 [hep-ex]].

[12] S. Chatrchyan *et al.* [CMS Collaboration], "The CMS experiment at the CERN LHC," JINST **3**, S08004 (2008). doi:10.1088/1748-0221/3/08/S08004

[13] CMS Collaboration [CMS Collaboration], "Particle-Flow Event Reconstruction in CMS and Performance for Jets, Taus, and MET," CMS-PAS-PFT-09-001

[14] "Commissioning of the particle-flow event with the first LHC collisions recorded in the CMS detector," CMS-PAS-PFT-10-001

[15] T. Sjstrand *et al.*, "An Introduction to PYTHIA 8.2," Comput. Phys. Commun. **191**, 159 (2015) doi:10.1016/j.cpc.2015.01.024 [arXiv:1410.3012 [hep-ph]].

[16] J. Allison *et al.*, "Geant4 developments and applications," IEEE Trans. Nucl. Sci. **53**, 270 (2006). doi:10.1109/TNS.2006.869826

[17] V. Khachatryan *et al.* [CMS Collaboration], "Performance of Photon Reconstruction and Identification with the CMS Detector in Proton-Proton Collisions at sqrt(s) = 8 TeV," JINST **10**, no. 08, P08010 (2015) doi:10.1088/1748-0221/10/08/P08010 [arXiv:1502.02702 [physics.ins-det]].

[18] P. D. Dauncey, M. Kenzie, N. Wardle and G. J. Davies, "Handling uncertainties in background shapes

: the discrete profiling method," JINST **10**, no. 04, P04015 (2015) doi:10.1088/1748-0221/10/04/P04015 [arXiv:1408.6865 [physics.data-an]].

[19] G. Aad *et al.* [ATLAS and CMS Collaborations], "Combined Measurement of the Higgs Boson Mass in *pp* Collisions at $\sqrt{s} = 7$ and 8 TeV with the ATLAS and CMS Experiments," Phys. Rev. Lett. **114**, 191803 (2015) doi:10.1103/PhysRevLett.114.191803 [arXiv:1503.07589 [hep-ex]].

Has a Flavon with $M = 750$ GeV been detected at the LHC13 ?

Azucena Bolaños

Facultad de Física
Universidad Veracruzana, 91000, Xalapa, Veracruz, México

E-mail: abolanos@uv.mx

Abstract. Higgs-flavon fields appear as a part of the Froggatt-Nielsen (FN) mechanism, which attempts to explain the hierarchy of Yukawa couplings. We explore the possibility that the 750 GeV diphoton resonance recently reported at the LHC13, could be identified with a low-scale Higgs-flavon field H_F and find the region of the parameter space consistent with CMS and ATLAS data. It is found that the extra vector-like fermions of the ultraviolet completion of the FN mechanism are necessary in order to reproduce the observed signal. We consider a standard model (SM) extension that contains two Higgs doublets (a standard one and an inert one) and one complex FN singlet. The inert doublet includes a stable neutral boson, which provides a viable dark matter candidate, while the mixing of the standard doublet and the FN singlet induces flavor violation in the Higgs sector at the tree-level. Constraints on the parameters of the model are derived from the LHC Higgs data, which include the search for the lepton flavor violating decay of the SM Higgs boson $h \to \bar{\mu}\tau$.

1. Introduction

Several ideas have been proposed to address the flavor problem [1]. For instance, textures and GUT-inspired relations, flavor symmetries and radiative generation, etc. The flavor symmetry approach can be supplemented with the Froggatt-Nielsen (FN) mechanism, which assumes that above some scale M_F there is a symmetry that forbids the appearance of Yukawa couplings; SM fermions are charged under this symmetry [which could be of Abelian type $U(1)_F$]. However, the Yukawa matrices can arise through non-renormalizable operators. The Higgs spectrum of these models could include light Higgs-flavons, which could mix with the scalar bosons. In these models, the diagonal flavor conserving (FC) couplings of the SM-like Higgs boson could deviate from the SM, while FV couplings could be induced at small rates too, but still produce detectable signals. On the other hand, extending the Higgs sector of the SM opens up the possibility of including a scalar dark matter (DM) candidate, such as occurs with the well studied inert doublet model (IDM). There are important motivations to supplement this model with a complex singlet, for instance to have extra sources of CP violation, as in the IDM with a complex singlet (IDMS) recently studied [7].

2. The IDMS-FN model

We shall consider a model with a multi-Higgs model that includes one standard-like Higgs doublet, denoted as Φ_s, as well as an inert-type doublet Φ_n. Furthermore, we shall also include Froggart-Nielsen scalar field (SM singlet S). The possibility of having light flavon fields was studied in ref. [3], and more recently in [4, 5]. Thus, besides breaking the EW symmetry, and thereby generating the masses of the W and Z gauge bosons, the field Φ_s gives masses to up-type quarks, d-type quarks and leptons. We shall impose a discrete symmetry in such a way that the doublet (Φ_n) is of the inert-type, and therefore contains a Dark matter candidate [6]. The FN mechanism could be UV completed through the introduction of heavy mirror fermions, as we shall discuss later these fermions could play a role in order to help reproducing the 750 GeV signal hinted at LHC.

In our model only the Z_2-even fields Φ_s and S_F acquire vacuum expectation values v and u, respectively. We shall use the following field decomposition around the vacuum state:

$$\Phi_s = \begin{pmatrix} G^+ \\ \frac{1}{\sqrt{2}}\left(v + \phi^0 + iG_z\right) \end{pmatrix}, \qquad \Phi_n = \begin{pmatrix} H^+ \\ \frac{1}{\sqrt{2}}\left(H + iA\right) \end{pmatrix}, \tag{1}$$

$$S_F = \frac{1}{\sqrt{2}}(u + s_1 + ip_1). \tag{2}$$

2.1. The Higgs potential

The Higgs potential of our model resembles the one of the IDMS model, which was studied in [7]. The IDMS model includes a $U(1)$ symmetry to simplify the number of free parameters of the model. Here such symmetry has a physical motivation, namely it is identified as the flavor symmetry, whose breaking helps to address the hierarchy of the Yukawa couplings associated with the broad spectrum of quark and lepton masses. Furthermore, our scalar potential is invariant under CP symmetry, and it takes the following form:

$$V = -\frac{1}{2}\left[m_1^2\Phi_s^\dagger\Phi_s + m_2^2\Phi_n^\dagger\Phi_n\right] + \frac{1}{2}\left[\lambda_1\left(\Phi_s^\dagger\Phi_s\right)^2 + \lambda_2\left(\Phi_n^\dagger\Phi_n\right)^2\right]$$

$$+\lambda_3\left(\Phi_s^\dagger\Phi_s\right)\left(\Phi_n^\dagger\Phi_n\right) + \lambda_4\left(\Phi_s^\dagger\Phi_n\right)\left(\Phi_n^\dagger\Phi_s\right) + \frac{\lambda_5}{2}\left[\left(\Phi_s^\dagger\Phi_n\right)^2 + \left(\Phi_n^\dagger\Phi_s\right)^2\right]$$

$$-\frac{m_3^2}{2}S_F^*S_F - \frac{m_4^2}{2}(S_F^{*2} + S_F^2) + \lambda_{s1}(S_F^*S_F)^2 + \lambda_{ss}(\Phi_s^\dagger\Phi_s)(S_F^*S_F) + \lambda_{sn}(\Phi_n^\dagger\Phi_n)(S_F^*S_F)$$

2.2. The Yukawa Lagrangian

On the other hand, the FN lagrangian of the model includes the terms that would become the Yukawa couplings after the $U(1)$ flavor symmetry is spontaneously broken. The FN lagrangian is given by:

$$\mathcal{L}_Y = \rho^u_{ij}(\frac{S_F}{\Lambda_F})^{n_{ij}}\bar{Q}_i d_j\tilde{\Phi} + \rho^d_{ij}(\frac{S_F}{\Lambda_F})^{p_{ij}}\bar{Q}_i u_j\Phi + \rho^l_{ij}(\frac{S_F}{\Lambda_F})^{q_{ij}}\bar{L}_i l_j\Phi + h.c. \tag{3}$$

where n_{ij}, p_{ij}, q_{ij} denote the combination of abelian charges, for each fermion type. The flavon field S_F is assumed to have flavor charge equal to -1, such that $\mathcal{L}_Y$ is $U(1)_F$-invariant. Then, Yukawa couplings arise after the spontaneous breaking of the flavor symmetry, i.e. $\lambda_x = (\frac{<S_F>}{\Lambda_F})^{n_x}$, where $< S >$ denotes the flavon vacuum expectation value, while M_F denotes the heavy mass scale, which represents the mass of heavy fields that transmit such symmetry breaking to the quarks and leptons. For examples of specific structures for the Yukawa matrices of the fermion types, see [6].

After substituting the Higgs-Flavon mass eigenstates, one gets finally the following interaction lagrangian for the Higgs-fermion couplings,

$$\begin{aligned}\mathcal{L}_Y &= \frac{1}{v}[\bar{U}M_uU + \bar{D}M_dD + \bar{L}M_lL](c_\alpha h + s_\alpha H_F) \\ &+\frac{v}{\sqrt{2}u}[\bar{U}_i\tilde{Z}^uU_j + \bar{D}_i\tilde{Z}^dD_j + \bar{L}_i\tilde{Z}^lL_j](-s_\alpha h + c_\alpha H_F + iA_F)\end{aligned} \tag{4}$$

Here, the information about the size of FV Higgs couplings is contained in the matrices Z. Thus, the (diagonal and non-diagonal) interactions of the scalars (h, H_F, A_F) with any fermion f are:

$$
\begin{aligned}
(\bar{f}_i f_i h) &= \frac{c_\alpha}{v}\bar{M}^f_{ii} - \frac{s_\alpha v}{\sqrt{2}u}\tilde{Z}^f_{ii} \\
(\bar{f}_i f_j h) &= -\frac{s_\alpha v}{\sqrt{2}u}\tilde{Z}^f_{ij} \\
(\bar{f}_i f_i H_F) &= \frac{s_\alpha}{v}\bar{M}^f{}_{ii} + \frac{c_\alpha v}{\sqrt{2}u}\tilde{Z}^f_{ii} \\
(\bar{f}_i f_j H_F) &= \frac{c_\alpha v}{\sqrt{2}u}\tilde{Z}^f_{ij} \\
(\bar{f}_i f_j A_F) &= \frac{iv}{\sqrt{2}u}\tilde{Z}^f_{ij}
\end{aligned}
\qquad (5)
$$

Besides the Yukawa couplings, we also need to specify the Higgs couplings with vector bosons, which is written as $g_{hVV} = \chi^{h_i}_V g^{sm}_{hVV}$, with the factor $\chi^{h_i}_V$ given as: $\chi^h_V = \cos\alpha$ and $\chi^{H_F}_V = \sin\alpha$.

Moreover, since the Higgs couplings to first generation quarks and leptons is highly suppressed, in order to study the FV Higgs coupling, which depends on the matrices $\tilde{Z}^f$, we shall consider the 2-3 family sub-system. Namely, for up quarks the Z-matrix (in mass eigenstate basis), is given by:

$$
\tilde{Z}^u = \begin{pmatrix} Y^u_{22} & Y^u_{23} \\ Y^u_{23} & 2s_u Y^u_{23} \end{pmatrix} \qquad (6)
$$

and similarly for d-quarks and leptons. We find a relation among the parameters, such that we can express the $\rho^{u,d}_{ij}$'s in terms of the ratios of masses and the CKM angle $V_{cb} \simeq s_{23}$. Namely, we define: $r_u = m_c/m_t$, $r_d = m_s/m_b$, and $r^u_1 = Y^u_{22}/Y^u_{33}$, $r^u_2 = Y^u_{23}/Y^u_{33}$. Similarly: $r^d_1 = Y^d_{22}/Y^d_{33}$, $r^d_2 = Y^d_{23}/Y^d_{33}$. Within this approximation we have: $\tilde{Y}^f_{33} \simeq Y^f_{33}$ for $f = u, d$. Then, $r^f_1 = r_f + r^f_2$, and the ratios of Yukawas must satisfy the following relation:

$$
r^u_2 = r^d_2 \frac{1 + r_d}{1 + r_u} - \frac{s_{23}}{1 + r_u} \qquad (7)
$$

3. Scenarios for Higgs couplings in the IDMS-FN model

Here we shall discuss the formulae needed to express all relevant Higgs couplings, which will permit us to define some benchmark points.

(i) **FC Higgs couplings.** First, we shall use LHC data to derive bounds on the Higgs couplings, following the analysis presented in ref. [8]; Here, the deviation from the SM Higgs couplings are assumed to be small and are expressed as: $g_{hXX} = g^{sm}_{hXX}(1 + \epsilon_X)$, where $\eta^X = 1 + \epsilon_X$. The results obtained in [8] give the following allowed ranges at 95 % C.L.: $\epsilon_t = -0.21 \pm 0.23$, $\epsilon_b = -0.19 \pm 0.30$, $\epsilon_\tau = 0.00 \pm 0.18$; while for W (Z) bosons they find: $\epsilon_W = -0.15 \pm 0.14$, $\epsilon_Z = -0.01 \pm 0.13$.
We shall use the strongest constraints, which come from a combination of ϵ_Z and ϵ_t, in such a way that the resulting constraints on the mixing angles is $0.86 < \cos\alpha < 1.0$.

(ii) **FV Higgs couplings with up-type quarks.** Then, regarding the couplings with up-type quarks, we consider the following sample values: $r^2_d = 0.05, 0.1, 0.3$, then table 1 shows the values of the entries for the $\tilde{Z}^u$ matrix for the 2nd-3rd family case for up-type quarks. We choose to focus on the up-quark sector, because we want to get an estimate for the most relevant predictions of the model.

Journal of Physics: Conference Series **761** (2016) 012053 doi:10.1088/1742-6596/761/1/012053

Scenario	$\tilde{Z}^u_{33}$	$\tilde{Z}^u_{23}$	$\tilde{Z}^u_{22}$
X1	4×10^{-4}	2×10^{-2}	2×10^{-4}
X2	1.4×10^{-2}	1.2×10^{-1}	7.2×10^{-3}
X3	0.27	0.52	0.14

Table 1. Relevant elements of the matrix $\tilde{Z}^u_{ij}$ for up-type quarks

(iii) **LFV Higgs couplings.** These couplings are written in terms of the parameters ρ_{ij}, which appear in the charged lepton mass matrix, and are of $O(1)$. Namely,

$$\begin{aligned} \tilde{Z}^l_{33} &= 2\sqrt{2}\frac{m_\tau}{v} \simeq 1.95 \times 10^{-2} \\ \tilde{Z}^l_{23} &= 4\lambda^4 \rho^l_{23} \simeq 10^{-2}\rho^l_{23} \\ \tilde{Z}^l_{22} &= 4.1 \times 10^{-2}\rho^l_{23} + 2.41 \times 10^{-3} \end{aligned} \tag{8}$$

We shall consider values of $\rho^l_{23} = 0.25, 0.75$.

An interesting probe of FV Higgs couplings is provided by the decay $h \to \tau\mu$, which was initially studied in refs. [9, 10]. Subsequent studies on detectability of the signal appeared soon after [11, 12, 13]. Precise loop calculations with massive neutrinos, SUSY and other models appeared in [14, 15, 16, 17]. The recent search for this decay at LHC [18], have resulted in a bound for the corresponding branching ratio of order $B.r.(h \to \tau\mu) < 1.51 \times 10^{-2}$ at 95% c.l.. Furthermore, given that the best fit to the data gives $B.r.(h \to \tau\mu) = 0.84^{+0.39}_{-0.37} \times 10^{-2}$, many more papers have appeared recently, trying to explain this result [19]. The search for this LFV Higgs decay could be one great opportunity to find new physics at the LHC RunII.

4. Branching ratios

Given the interactions of the model, we can evaluate the decays of the heavy flavon state H_F, with parameters satisfying the constraints from LHC Higgs searches. They are shown in figure 1,2. In figure 1, we show the branching ratios for the tree-level modes with the vev $u = 500$ Gev, including $H_F \to t\bar{t}, b\bar{b}, ZZ, WW, hh$. We notice that the modes ZZ, WW, are the dominant ones, with B.R. of order 0.7 and 0.35, respectively, even after the threshold for the $t\bar{t}$ decay, which also reaches at most a $B.R. \simeq 0.2$. We also notice that the decay $H \to hh$ could reach a B.R. of order 0.1, which should be searched at LHC13.

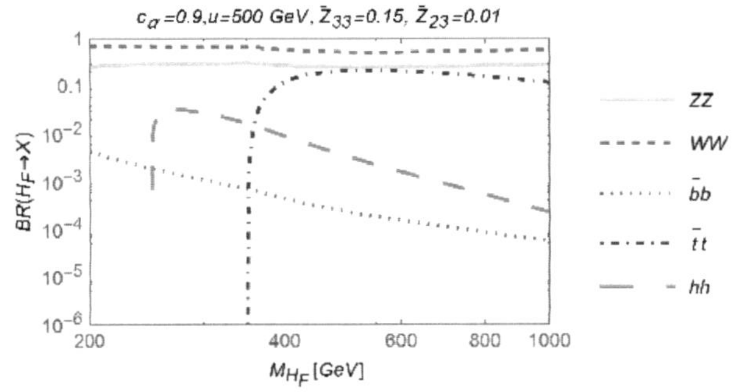

Figure 1. The BR for the tree-levl decays of the heavy flavon state.

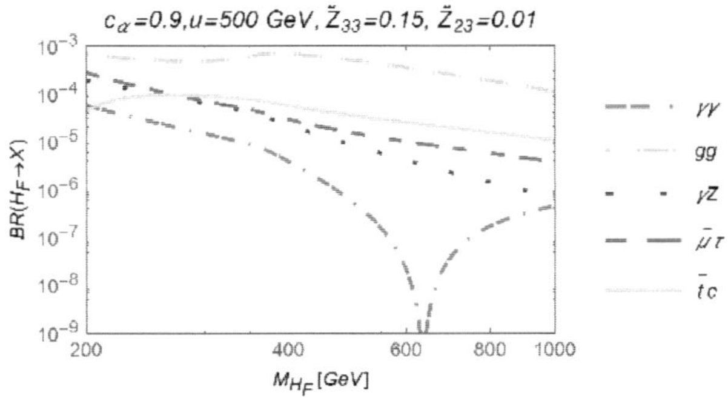

Figure 2. The BR for the rare decays of the heavy flavon state.

To explore further the interesting mode $H_F \to hh$, we show in figure 3 the contour regions for the Branching Ratio in the plane $u - M_{H_F}$. We can notice that for the full range $500 < u < 3000$ GeV and $250 < M_{H_F} < 1500$ GeV, the corresponding B.R. is larger than about 0.01, which seems amenable to be searched at LHC13.

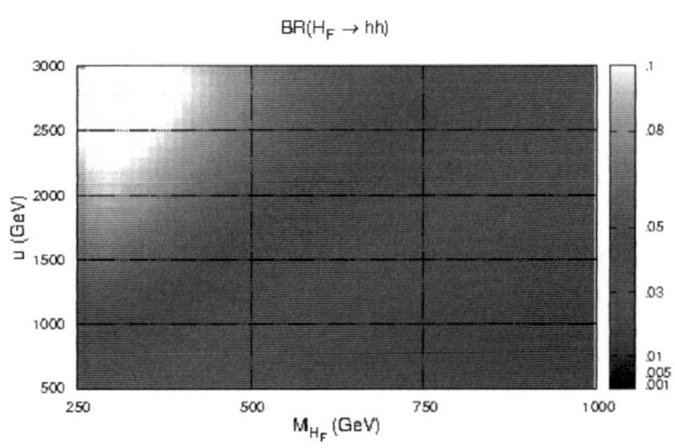

Figure 3. The BR of the heavy flavon state $H_F \to hh$.

5. Conclusions and outlook

We have studied the Higgs couplings, within a model within the IDMS-FN model, where one Higgs doublet participates in SSB while the second one is of inert-type and contains a dark matter candidate. This model also includes mixing of the Higgs doublets with a Flavon field, which generates the Yukawa hierarchies and induces Flavor-violating Higgs couplings at acceptable rates.

Constraints on these couplings, derived from Higgs search at LHC, and their implications for Higgs anomalies observed at LHC. We find that within this model it is possible to explain both the decay $h \to \tau\mu$, as well as the new 750 GeV resonance observed at LHC13 in the two-photon final state.

Support from CONACYT-SNI (Mexico) and VIEP(BUAP) are acknowledged.

References

[1] G. Isidori, Y. Nir and G. Perez, Ann. Rev. Nucl. Part. Sci. **60**, 355 (2010) [arXiv:1002.0900 [hep-ph]].
[2] C. Bonilla, D. Sokolowska, J. L. Diaz-Cruz, M. Krawczyk and N. Darvishi, arXiv:1412.8730 [hep-ph].
[3] I. Dorsner and S. M. Barr, Phys. Rev. D **65**, 095004 (2002) [hep-ph/0201207].
[4] K. Tsumura and L. Velasco-Sevilla, Phys. Rev. D **81**, 036012 (2010) [arXiv:0911.2149 [hep-ph]].
[5] E. L. Berger, S. B. Giddings, H. Wang and H. Zhang, Phys. Rev. D **90**, no. 7, 076004 (2014) [arXiv:1406.6054 [hep-ph]].
[6] J. L. Diaz-Cruz, arXiv:1405.0990 [hep-ph].
[7] C. Bonilla, D. Sokolowska, J. L. Diaz-Cruz, M. Krawczyk and N. Darvishi, arXiv:1412.8730 [hep-ph].
[8] P. P. Giardino, K. Kannike, I. Masina, M. Raidal and A. Strumia, arXiv:1303.3570 [hep-ph].
[9] A. Pilaftsis, Phys. Lett. B **285**, 68 (1992).
[10] J. L. Diaz-Cruz and J. J. Toscano, Phys. Rev. D **62**, 116005 (2000) [hep-ph/9910233].
[11] T. Han and D. Marfatia, Phys. Rev. Lett. **86**, 1442 (2001) [hep-ph/0008141].
[12] K. A. Assamagan, A. Deandrea and P. A. Delsart, Phys. Rev. D **67**, 035001 (2003) [hep-ph/0207302].
[13] S. Kanemura, T. Ota and K. Tsumura, Phys. Rev. D **73**, 016006 (2006) [hep-ph/0505191].
[14] E. Arganda, A. M. Curiel, M. J. Herrero and D. Temes, Phys. Rev. D **71**, 035011 (2005) [hep-ph/0407302].
[15] J. L. Diaz-Cruz, JHEP **0305**, 036 (2003) [hep-ph/0207030];
[16] A. Brignole and A. Rossi, Nucl. Phys. B **701**, 3 (2004) [hep-ph/0404211].
[17] J. L. Diaz-Cruz, D. K. Ghosh and S. Moretti, Phys. Lett. B **679**, 376 (2009) [arXiv:0809.5158 [hep-ph]].
[18] V. Khachatryan *et al.* [CMS Collaboration], arXiv:1502.07400 [hep-ex].
[19] For a recent review see: A. Vicente, arXiv:1503.08622 [hep-ph].

MWPF IOP Publishing
Journal of Physics: Conference Series **761** (2016) 012054 doi:10.1088/1742-6596/761/1/012054

Available on CMS information server **CMS CR -2016/171**

The Compact Muon Solenoid Experiment

Conference Report

Mailing address: CMS CERN, CH-1211 GENEVA 23, Switzerland

27 July 2016

Offline Data Quality Monitoring for the RPC of the CMS Detector

Brenda Fabela and Isabel Pedraza for the CMS Collaboration

Abstract

Data Quality Monitoring is an important activity that guarantees the proper operation of the detectors in high energy experiments. Besides of the online environment for data quality monitoring there is an offline environment which main goal is the reconstruction and validation of the calibration results, software releases and simulated data. The offline monitoring system of the Resistive Plate Chambers at the Muon Detector of the CMS consists on the data analysis of each run in order to verify that all the subsystems are working correctly, at an efficiency greater than 95pct. This work presents the latest upgrade done to the offline monitoring software based on the restyling of the efficiency code for the 2016 data taking that is about to start. This update allows to do the general analysis in less time in a more complete way and makes the code more flexible to do specific analysis.

Presented at *DPC-SMF2016 XXX Reunin Anual de la Divisin de Partculas y Campos de la SMF*

Journal of Physics: Conference Series **761** (2016) 012054 doi:10.1088/1742-6596/761/1/012054

Offline Data Quality Monitoring for the RPC of the CMS Detector

B. Fabela[1] and I. Pedraza[1] on behalf of the CMS Collaboration

1 Benemérita Universidad Autónoma de Puebla, Puebla, México

E-mail: brenda.fabela.enriquez@cern.ch

Abstract. Data Quality Monitoring (DQM) is an important activity that guarantees the proper operation of the detectors in high energy experiments. Besides of the online environment for data quality monitoring there is an offline environment which main goal is the reconstruction and validation of the calibration results, software releases and simulated data. The offline monitoring system of the Resistive Plate Chambers at the Muon Detector of the CMS consists on the data analysis of each run in order to verify that all the subsystems are working correctly, at an efficiency greater than 95%. This work presents the latest upgrade done to the offline monitoring software based on the restyling of the efficiency code for the 2016 data taking that is about to start. This update allows to do the general analysis in less time in a more complete way and makes the code more flexible to do specific analysis.

1. Introduction

The primary goal of the CMS Data Quality Monitoring (DQM) system is to guarantee high quality physics data. Its functions are monitoring the detector and trigger performance, debugging hardware and certifying recorded data [1]. For this system there is an *online environment* and an *offline environment.*

The **offline environment** accumulates monitoring data from several workflows [2], [3]

- Tier-0 prompt reconstruction,
- Tier-1s re-reconstruction,
- Validation of the alignment and calibration results, the software releases and all the simulated data.

Offline monitoring is unique and formaly divided in two steps:

(i) **Data analysis**. It analizes the RAW (RPC Monitor) or RECO or FEVT (Express) data. It produces a ROOT file with the result of the analysis.

(ii) **Efficiency calculation**. Histograms and monitoring information are extracted and merged to yield full statistics. Efficiencies are calculated, summary plots are produced and quality tests are performed.

There are several parameters that the Resistive Plate Chamber (RPC) DQM group monitors, including occupancy, multiplicity, cluster size, syncronization, noise and detection efficiency [1].

2. RPC Efficiency: Code Restyling

The latest modifications to the efficiency offline monitoring code were mainly done to improve the summary of the efficiency as a last step in the generation of the plots. The old code produced output files in different formats (ROOT, TXT, folders with PNG, etc). The funcionality of the code is preserved. The first step of the analysis is unmodified and the second one has been upgraded. Currently, the code preserves its funcionality but now it is shorter and updated with one new module and with some additional macros. The output files are only in ROOT format allowing to the user an easy browsing through the plots and the information needed. The efficiency calculation is done in less time.

3. Results

In this conference part of the results of the analyzed runs up to May 4th, 2016 were presented. The first stable collision runs analyzed with $B = 0$ T and a duration greater than 30 minutes were 271336, 271646, and the Cosmic Runs Analyzed At Full Tesla (CRAFT) with a duration greater than 40 minutes were 272133, 272134, 272137, 272138, 272139, 272143, 272148. The following efficiency maps show the performance of each subdetector allowing to monitor the operation of chambers (endcap) or *sectors* (barrel) [4] during the taking of collision data and cosmic data.

3.1. Collision runs

In this case, it is observed that the most majority of the subsystems are above 95% of efficiency.

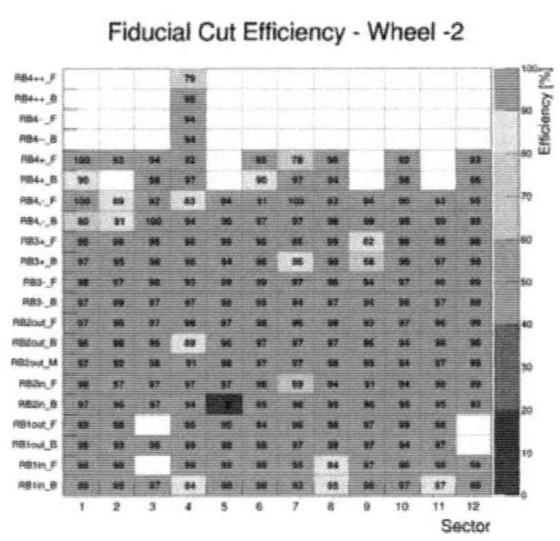

Figure 1. Efficiency map for the wheel -2 of the barrel.

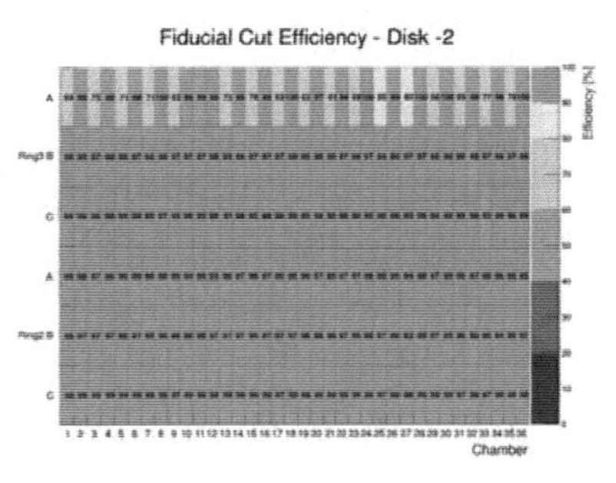

Figure 2. Efficiency map for the disk -2 of the endcap.

3.2. Cosmic runs

The ideal performance of each subdetector must be above 95%. However, it can be seen that for the barrel there is an important quantity of chambers that are not working at the required efficiency, but these results are extrapolated from the data. These results are a consequence of the way the cosmic rays go across the detector, and therefore, they do not cross the endcap completely.

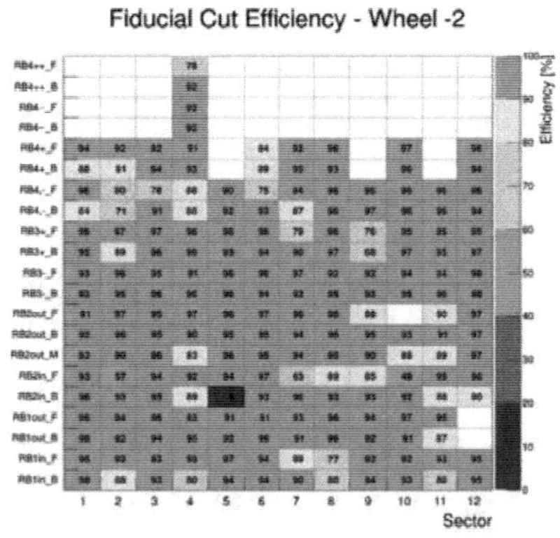

Figure 3. Efficiency map for the wheel -2 of the barrel.

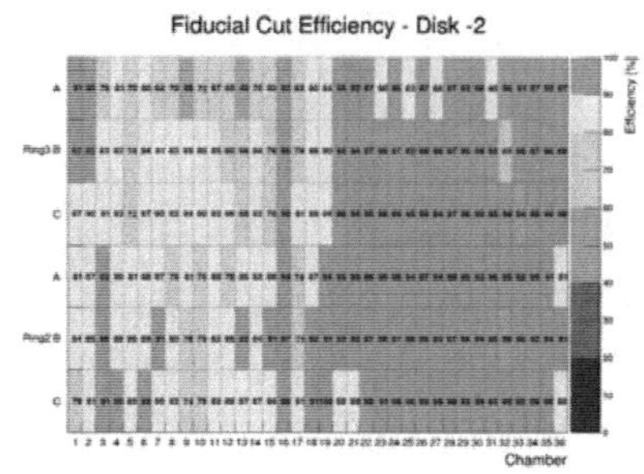

Figure 4. Efficiency map for the disk -2 of the endcap.

4. Conclusion

The most recent implementation to the RPC offline analysis code has been focused on the efficiency plots, making the management of all the information easier and stored in an unique format ROOT file. After this upgrade, the second step of the offline analysis takes only few minutes to be completed, saving time and concentrating the monitoring information in a more complete way.

Acknowledgments

We congratulate our colleagues in the CERN accelerator departments for the excellent performance of the LHC and thank the technical and administrative staffs at CERN and at other CMS institutes for their contributions to the success of the CMS effort. In addition, we gratefully acknowledge the computing centers and personnel of the Worldwide LHC Computing Grid for delivering so effectively the computing infrastructure essential to our analyses. Finally, we acknowledge the enduring support for the construction and operation of the LHC and the CMS detector provided by the following funding agencies: BMWFW and FWF (Austria); FNRS and FWO (Belgium); CNPq, CAPES, FAPERJ, and FAPESP (Brazil); MES (Bulgaria); CERN; CAS, MoST, and NSFC (China); COLCIENCIAS (Colombia); MSES and CSF (Croatia); RPF (Cyprus); MoER, ERC IUT and ERDF (Estonia); Academy of Finland, MEC, and HIP (Finland); CEA and CNRS/IN2P3 (France); BMBF, DFG, and HGF (Germany); GSRT (Greece); OTKA and NIH (Hungary); DAE and DST (India); IPM (Iran); SFI (Ireland); INFN (Italy); MSIP and NRF (Republic of Korea); LAS (Lithuania); MOE and UM (Malaysia); BUAP, LNS, CINVESTAV, CONACYT, SEP, and UASLP-FAI (Mexico); MBIE (New Zealand); PAEC (Pakistan); MSHE and NSC (Poland); FCT (Portugal); JINR (Dubna); MON, RosAtom, RAS and RFBR (Russia); MESTD (Serbia); SEIDI and CPAN (Spain); Swiss Funding Agencies (Switzerland); MST (Taipei); ThEPCenter, IPST, STAR and NSTDA (Thailand); TUBITAK and TAEK (Turkey); NASU and SFFR (Ukraine); STFC (United Kingdom); DOE and NSF (USA).

References

[1] Cimmino A 2010 *Data quality monitoring and performance studies of the resistive plate chamber detector at the CMS experiment at LHC* Ph.D. thesis Naples U. URL `http://web2.infn.it/cms_napoli/index.php/theses`

[2] Cimmino A, Paolucci P and Lomidze D (CMS) 2010

Journal of Physics: Conference Series **761** (2016) 012054 doi:10.1088/1742-6596/761/1/012054

[3] Chatrchyan S *et al.* (CMS) 2010 *JINST* **5** T03017 (*Preprint* 0911.4045)
[4] 2014 URL https://cds.cern.ch/record/1667879

The search for Physics Beyond the SM: Wins and Losses

J. Lorenzo Díaz-Cruz
Facultad de Ciencias Físico - Matemáticas, BUAP
Apdo. Postal 1364, C.P. 72000, Puebla, Pue. México
E-mail: ldiaz@ifuap.buap.mx

Abstract.
In this paper we review briefly the motivations for physics Beyond the Standard Model, which generate many possible signals that can be searched at the LHC. To get some perspective on the current affairs, we discuss some landmark success in the sub-fields of theoretical particle physics: formal QFT developments, Model Building and Phenomenology. By studying how these developments were done, we can learn some lessons about how to be cautious and optimistic at the same time, when looking at potential signals of new physics. We conclude by considering the case of the 750 GeV state and what we have learned from the theoretical activity that tried to reproduce this signal.

1. Introduction

The Standard Model (SM) of fundamental particles and interactions has been extremely successful to explain most data in the field. In particular, the discovery of a Higgs-like particle with $m_h = 125$ GeV at the LHC [1, 2], has tested the mechanism of electroweak symmetry breaking [3]. The agreement of the measured Higgs mass with the range predicted by electroweak precision tests [4], confirms the success of the SM. Similarly, the measurements of spin, parity, and couplings, seem also consistent with the expectations from the SM.

Despite this success, there are several arguments that lead us to expect that the SM could not be the final theory and that some form of new physics will come after the SM. Some scenarios for physics beyond the SM have been conjectured in order to address some of the SM open problems, such as hierarchy, flavor, unification, etc [5]. What is not know is the energy/mass scale (Λ) at which effects of such new physics could appear. It could be that $\Lambda = O(\text{TeV})$ and it is possible that such effects could be seen at LHC or near future colliders. Here we could include the supersymmetric extensions of the SM [6], but many more models could be included too.

However, the LHC has not detected, so far, any sign of new physics, but instead has provided bounds on the scale Λ. For instance, the LHC bounds on the mass of superpartners (gluinos and squarks) are entering into the Multi-TeV range. These results suggest that SUSY breaking involves a heavy mass scale, and put some questioning into the original motivation to solve the hierarchy/naturalness problem, as the resulting constraints are difficult to satisfy in the most constrained versions of the MSSM, namely for the cMSSM or minimal SUGRA. More recently, the possibility to have superpartners with masses beyond the reach of LHC, or large CPV phases, has been considered [7], and this has opened the discussion about the construction of future accelerators. In particular, a recent interesting proposal for future collider machine working at 100 TeV has been discussed.

It could also happen that the scale of new physics Λ is near the GUT or Planck scales, and then its effects are far from conventional colliders, but still in such case the Intensity and Cosmology frontiers provide tools that may allow us to test such effects; for instance by looking at proton decay, neutrino oscillations or LFV interactions, it could be possible to probe very high energy scales. Test of the fundamental principles, such as Lorentz invariance and CPT symmetry have been considered too.

Recently CERN LHC found hints of a diphoton resonance with $\Lambda = 750$ GeV , with statistical level of 2.7 σ-2.9 σ (after considering LEE). The data from 2015 [8, 9], could not allow to conclude whether the signal was just a fluctuation or a real effect of a possible major discovery. However weak the signal could had been, it attracted lots of attention from the theoretical community, with more than 500 new papers trying to explain such effect. The new 2016 results [10] did not bring what some of us were expecting or dreamed. Essentially the signal is now gone to the land of less than 2 σ fluctuations.

As part of the activities of the Annual Meeting of the Mexican Division of Particles and fields, held in May 2016, in the city of Puebla, I acted as coordinator of a round table devoted to discuss the 750 resonance, together with my colleagues Azucena Bolaños, Isabel Pedraza and Saúl Ramos. Then, I was asked to write a short note in this topic, but given the speed of recent events, with a signal being present one semester and gone next one, I decided to write the coming section first, to look in perspective these events and the lessons we can learn for the future. In blogs of different ideology, we could find some harsh comments about the particle theory community. But really? Should the theorist be taken to the flames for their actions? Care and some knowledge on how the field has evolved and achieved some of its greatest success, could help to keep a cool mind. Furthermore, at some stages the known principles do not allow to discriminate among several models, and to make things worse, in some epochs the experiments find full agreement between theory (SM) and experiment. Thus, model building has some lottery side and it is nature who chooses the right one.

2. Theoretical Particle Physics

One could say that theoretical particle physics started with the discovery of the electron, proton and neutron. Then came beta decay and the neutrino, followed by the diversity of hadrons found in the 40-50's. The completion of QED was the first great theoretical construction, which allowed to calculate cross sections and magnetic moments of leptons from first principles, with techniques being developed to renormalize the infinities. The generalization to Yang-Mills theories with its non-abelian symmetries provided another turn of the screw. The discovery of SSB, DIS, Renormalization of gauge theories, the Electroweak model and QCD provided the next theoretical tools that shaped our current understanding of the fundamental constituents of matter and its interactions; namely, quarks and leptons interacting through vector gauge bosons, with masses arising from SSB.

How come that the HEP community was able to build such great intellectual machinery? Through lots and lots of success and failures, with efforts being aligned into three main avenues; namely,

2.1. Formal Development

Formal developments are perhaps the area that has the far reaching goals. The current theoretical framework is provided by Quantum Field Theory. When progress is made in this area, the results could have some influence even other areas of physics. Here we can list a few examples where success is obtained

(a) SSB: After the works of Englert and Brout and P. Higgs, particle physics acquired a general method to provide masses to gauge vector bosons. From these techniques, it was a matter of time, experiments and model building, to find out the model that was chosen by nature to describe the fundamental interactions.

(b) Renormalization of Gauge Theories: This is another example where the product of research provided a general method to build renormalizable gauge theories with massive vector bosons. Not only one model, but a whole class of models were at the disposal of model builders to describe the subatomic world.

(c) Renormalization Group Methods: Here it was the genius of K. Wilson, who developed new methods to treat the evolution of parameters, which has such influence that it was used in statistical physics.

2.2. Model Building

In this area one tries to apply the ideas an methods obtained in formal developments, either to describe phenomena already observed or to predict new phenomena from the model. When some phenomenon is already observed, for example with β-decay and weak interactions, it could take a long time, but finally the correct model is found. This happened for the electroweak unified model, but also for the construction of quark model and QCD, which provide another of such examples.

This process has some lottery side, as sometimes many models could do the job. For instance, Weinberg proposed a model of leptons with gauge group $SU(2)\times U(1)$, that used SSB to generate the masses. However, the rules of the game and the lack of experimental data, allowed other possibilities, and so he proposed the $SU(3)_L \times U(1)_Y$ model too. But then the models are useful at least as generators of signals, which motivate the experimental searches.

In a few occasions, a model is created with the greatest goals, and here I included as examples, the GUT $SU(5)$ model of Georgi and Glashow, as well as the Pati-Salam model, with lepton number as the 4th color. But is nature that has the final word, and as beautiful as the model could look, the proton decay was not observed at the rates predicted by the model.

2.3. Phenomenology

Finally, we have a third area of research, phenomenology, which at some times it is seen as less glamorous, but it plays a crucial role as it allows to test the predictions of models of particle physics, as well as the validity of the principles that support them. Here we can include great work that requires gigantic calculations, programming and careful treatment of statistics. For example, the area of perturbative QCD provides the radiative corrections that are of vital importance, for instance to study Higgs production at hadron colliders, The detection of the Higgs particle at LHC is a great output of this area.

Very solid results could be obtained when one uses axiomatic properties of QFT to analyze some signal. In general, it requires a lot of ingenuity to extract the SM parameters from experimental data, ranging from masses of light quarks, CKM angles and phases, gauge couplings up to the Higgs mass and its couplings. I ask for an applause for the practitioners of such artcraft.

3. The 750 GeV resonance is gone

The 2015 results of the LHC run with c.m. energy of 13 TeV has shown surprising hints of a new resonance in the di-photon channel with invariant mass of 750 GeV, which clearly represents a signal of Physics beyond the SM. ATLAS [8] collected 3.2 fb^{-1} of data and reported a signal with significance of 3.6 (local), which becomes 2.3 (after LEE), while CMS [9] collected an integrated luminosity of 2.6 fb^{-1} and reported a significance 2.6 (local) that became 2.0 (after LEE).

For our contribution to this subject, we explored the possibility that the 750 GeV di-photon resonance, could be identified with a low-scale flavon field [11]. We worked within an SM extension of the IDMS-type that contains two Higgs doublets and one complex FN singlet; one of the doublets is of the inert type, and thus provides a viable DM candidate. Mixing of the doublet and singlet induces flavor violation in the Higgs sector, which predicts a branching ratio for LFV Higgs decay $h \to \tau\mu$ of order 10^{-5}, which could be searched at LHC too. We find that within this model it was possible to explain both the decay $h \to \mu\tau$, as well as the new 750 GeV resonance observed at LHC13 in the two-photon final state. We also studied the Higgs couplings, within a model of the IDMS-FN type, where fermion masses for each fermion type (up, down and leptons) are generated by one Higgs doublet each, while a fourth doublet is of inert-type and contains a dark matter candidate [12]. Constraints on these couplings, are derived from the Higgs search at LHC, and their implications for Higgs anomalies observed at LHC is also studied.

Another contribution that studied the properties of the 750 resonance is ref. [13], also made in Mexico, focused on the large width of the resonance. This is very difficult to accommodate within weakly interacting BSM theories, hinting a composite scenario. By means of forward sum-rules for $\gamma\gamma$ and gg scattering, the authors show that a spin-0 resonance with mass of the order of the TeV and a sizable partial width -of the order of a few GeV- must be accompanied by higher spin resonances with $JR \geq 2$ with similar properties, as expected in strongly coupled extensions of the Standard Model or, alternatively, in higher dimensional deconstructed duals. Furthermore, independently of whether the 750 GeV candidate is a scalar or a tensor, the large contribution to the forward sum-rules in the referred scenario implies the presence of states in the spectrum with $JR \geq 2$, being these high spin particles a manifestation of new extra-dimensions or composite states of a new strong sector.

However, the 2016 data delivered by LHC [10] indicates that essentially the signal was gone to the territory of less than 2 σ fluctuations. The implications for our model is that flavons with generic couplings can not be as light as 750 GeV, unless we somehow fine-tune the parameters of the model. LHC and future colliders should keep looking for low-scale flavons, either a signal is found or we will keep increasing the bounds on the flavon mass. On the other hand, the conclusions of ref. [13], which used more general QFT methods, remain valid for hypothetical or real particles with large width.

4. Conclusions

In this paper we have reviewed the motivations for physics Beyond the Standard Model, which generate many possible signals that can be searched at the LHC. After the vanishing of the signal for the 750 GeV resonance in the 2016 LHC data, which was reported first by LHC in 2015, some pessimism has invaded the field. However, in order to get some perspective on the current affairs, we review briefly some of the landmark success in each of the subfields of theoretical particle physics, namely: formal QFT developments, Model Building and Phenomenology. I think that we can learn some lessons about to judge potential signals of new physics by looking at these issues from a historical perspective. In the first place, it is not easy that nature will deliver any surprises. What we know about particles respects QFT, and it is very likely that it will continue to do so for quite some time. Overall, we have learned something from the theoretical activity on this hypothetical signal. I do not think that it was a waste of time.

We ended this work by summarizing the contribution from the Mexican HEP community to the 750 GeV state. In our own work [11], we tried to interpret the signal as coming from Higgs-Flavon mixing, while the work of ref. [13] focused on the implications of the large width of the resonance.

References

[1] G. Aad *et al.* [ATLAS Collaboration], Phys. Lett. B **716**, 1 (2012) [arXiv:1207.7214 [hep-ex]].
[2] S. Chatrchyan *et al.* [CMS Collaboration], Phys. Lett. B **716**, 30 (2012) [arXiv:1207.7235 [hep-ex]].
[3] J. F. Gunion, H. E. Haber, G. L. Kane and S. Dawson, Front. Phys. **80**, 1 (2000).
[4] J. Erler, AIP Conf. Proc. **917**, 244 (2007) [hep-ph/0701261].
[5] A. Pomarol, CERN Yellow Report CERN-2012-001, 115-151 [arXiv:1202.1391 [hep-ph]].
[6] S. P. Martin, In *Kane, G.L. (ed.): Perspectives on supersymmetry II* 1-153 [hep-ph/9709356].
[7] E. Arganda, J. L. Díaz-Cruz and A. Szynkman, Phys. Lett. B **722**, 100 (2013) [arXiv:1301.0708 [hep-ph]]; E. Arganda, J. L. Díaz-Cruz and A. Szynkman, Eur. Phys. J. C **73**, 2384 (2013) [arXiv:1211.0163 [hep-ph]]; A. Chakraborty, B. Das, J. L. Díaz-Cruz, D. K. Ghosh, S. Moretti and P. Poulose, arXiv:1301.2745 [hep-ph].
[8] Search for resonances decaying to photon pairs in 3.2 fb^{1} of pp collisions at s = 13 TeV with the ATLAS detector, Technical Report ATLAS-CONF-2015- 081, CERN, Geneva (2015), URL http://cds.cern.ch/record/2114853.
[9] CMS Collaboration, Search for new physics in high mass diphoton events in proton-proton collisions at s = 13 TeV (2015) CMS-PAS-EXO-15-004, URL https://cds.cern.ch/record/2114808.
[10] See the presentations of CMS and ATLAS at ICHEP16, Chicago (USA).
[11] A. Bolaños, J. L. Díaz-Cruz, G. Hernández-Tomé and G. Tavares-Velasco, arXiv:1604.04822 [hep-ph].
[12] J. L. Díaz-Cruz and U. Saldaña-Salazar, arXiv:1405.0990 [hep-ph].
[13] P. Roig and J. J. Sanz-Cillero, arXiv:1605.03831 [hep-ph].

Possible violations of spacetime symmetries

Luis Urrutia

Instituto de Ciencias Nucleares, Universidad Nacional Autónoma de México, 04510 Ciudad de México, México

E-mail: `urrutia@nucleares.unam.mx`

Abstract. The identification of symmetries has played a fundamental role in our understanding of physical phenomena. Nevertheless, in most cases such symmetries constitute only a zeroth-order approximation and they need to be broken so that the predictions of the theory are consistent with experimental observation. In particular, the almost sacred CPT and Lorentz symmetries, which are certainly part of the fundamental ideas of modern physics, need to be probed experimentally. Recently, such efforts have been intensified because different theoretical approaches aiming to understand the microstructure of space-time suggest the possibility that such symmetries could present minute violations. Up to now, and with increasing experimental sensitivities, no signs of violation have been found. Nevertheless, we observe that even the persistence of such negative results will have a profound impact. On one hand, they will provide those symmetries with a firm experimental basis. On the other, they will set stringent experimental bounds to be compared with the possible emergence of such violations in quantum gravity models based upon a discrete structure of space. We present a very general perspective of the research on Lorentz symmetry breaking, together with a review of a few specific topics.

1. Introduction

The intuitive idea we have about some object exhibiting symmetry is that of some organized structure which shows some kind of regularity. From the point of view of physics we require a more precise statement of what a symmetry is. The basic idea is that some configuration is symmetrical if we perform operations (motions, changes or transformations) on it and we recover the original pattern afterward. In mathematics, the idea of a transformation (e.g. translations, reflections and rotations, for example) is encoded in what we call a group of motions or simply a group. The elements of the group are a set of transformations that satisfy some well known rules. We say that a system is symmetric under such group if its time evolution does not change under the application of any operation of the group. This requirement puts severe restrictions upon the form of the Hamiltonian of a symmetric system and usually one can use group-theoretical methods to find the spectrum.

A more general approach to describe a symmetry is in terms of the action S of the system, which contains all the information about its time evolution, both at the classical and the quantum level. The action is a functional of the fields $\Phi^M(x^\mu)$ which describe the system, and it is given by

$$S[\Phi^M] = \int_R d^4x\, \mathcal{L}(\Phi^M, \Phi^M_{,\mu}), \qquad \Phi^M_{,\mu} = \partial_\mu \Phi^M, \qquad \partial_\mu \mathcal{L} = 0, \qquad \partial_\mu \equiv \frac{\partial}{\partial x^\mu}, \tag{1}$$

where $\mathcal{L}$ is the Lagrangian density (LD), $\mu = 0, 1, 2, 3$ are Minkoswki spacetime indices (with metric $\eta_{\mu\nu} = \text{diag}(+, -, -, -)$) and M are generalized labels that run over internal, as well as over spinor and tensor indices which specify the type of field we are dealing with. Let us emphasize that we are only considering LDs which do not depend explicitly upon the spacetime coordinates. The Euler-Lagrange equations of motion, obtained by requiring the action to be an extreme keeping the fields fixed at the initial and final time slices, together with appropriate boundary conditions at spacial infinite in the region R, are

$$0 = \frac{\partial \mathcal{L}}{\partial \Phi^M} - \partial_\mu \left(\frac{\partial \mathcal{L}}{\partial \Phi^M_{,\mu}} \right) \equiv \mathcal{E}_M. \tag{2}$$

1.1. *Global continuous symmetries*

Suppose now that we perform some global transformations $\delta_{\epsilon_{k,B}} x^\mu$, $\delta_{\epsilon_{k,B}} \Phi^M$, $\delta_{\epsilon_{k,B}} \Phi^M_{,\mu} = \partial_\mu \left(\delta_{\epsilon_{k,B}} \Phi^M \right)$, depending upon some independent infinitesimal constant parameters $\epsilon_{k,B}$, with $k = 1, \cdots, p$, for each type of field B. We say that these transformations are a symmetry of the system if we can write

$$\delta_{\epsilon_{k,B}} \mathcal{L} = \partial_\mu \Omega^\mu_{\epsilon_{k,B}} (\Phi^M, \Phi^M_{,\nu}), \tag{3}$$

without using the equations of motion. Here the function $\Omega^\mu_{\epsilon_{k,B}}(\Phi^M, \Phi^M_{,\nu})$ depends linearly on the parameters $\epsilon_{k,B}$. The transformations of the fields Φ^M, together with those of their derivatives, are dictated by the spin content of the indices M. Calculating the explicit variation in the left hand side of Eq. (3), and now using the equations of motion (2), we obtain

$$0 = \partial_\mu \left(\frac{\partial \mathcal{L}}{\partial \Phi^M_{,\mu}} \delta_{\epsilon_{k,B}} \Phi^M - \Omega^\mu_{\epsilon_{k,B}} \right) \equiv \epsilon_{k,B}\, \partial_\mu J^\mu_{k,B}, \tag{4}$$

from where we can read the conserved currents $J^\mu_{k,B}$. This is the celebrated first Noether's theorem, which states that each independent global transformation of coordinates and fields which is a symmetry of the LD as defined in Eq. (3), induces a conserved current given by Eq. (4) [1]. With appropriate boundary conditions, the corresponding charges

$$Q_{k,B}(t) \equiv \int d^3\mathbf{x}\, J^0_{k,B}(t, \mathbf{x}) \tag{5}$$

satisfy $dQ_{k,B}(t)/dt = 0$ and generate the Lie algebra of the corresponding symmetry group, after rewriting them in terms of the canonical variables of the system. The most typical examples of the application of these ideas are: (1) energy-momentum conservation, arising from the translation symmetry $\delta_{\epsilon_\alpha} x^\mu = \epsilon^\mu$ of the LDs which do not depend explicitly upon the coordinates and (2) angular momentum (spin plus orbital) conservation arising from Lorentz invariance expressed in the infinitesimal form $\delta_{\omega_{\alpha\beta}} x^\mu = \omega^{\mu\alpha} x_\alpha$ with $\omega_{\alpha\beta} = -\omega_{\beta\alpha}$. As usual, indices in Minkowski space are raised and lowered by the metric $\eta_{\mu\nu}$ and its inverse $\eta^{\mu\nu}$.

1.2. *Local continuous symmetries*

They are also known as gauge symmetries and are characterized by spacetime-dependent infinitesimal parameters $\theta_k(x)$ such that the field variations can be written as $\delta\Phi^M = P_{M,k}(\Phi, \Phi_{,\mu})\theta_k(x) + Q^\mu_{M,k}(\Phi, \Phi_{,\mu})\partial_\mu \theta_k(x)$. Here we further restrict ourselves to internal transformations where $\delta x^\mu = 0$. Imposing that such local transformations leave the action invariant and calculating the variation of the Lagrangian density under them, for arbitrary fields Φ and without using the equations of motion, leads to the relation

$$\int d^4x \left[\mathcal{E}_M \delta\Phi^M + \partial_\mu \left(\frac{\partial \mathcal{L}}{\partial \Phi^M_{,\mu}} \delta\Phi^M \right) \right] = 0, \tag{6}$$

where $\mathcal{E}_M$ was defined in the Eq. (2). After imposing the conditions that $\theta_k(x)$ and $\partial_\mu \theta_k(x)$ are zero on the boundary, but arbitrary in the bulk region, we obtain the second Noether's theorem

$$\mathcal{E}_M P_{M,k} - \partial_\mu \left(\mathcal{E}_M Q^\mu_{M,k} \right) = 0, \tag{7}$$

which provides p relations among the equations of motion. This is an alternative route to see that not all the equations in a gauge theory are independent, in such a way that arbitrary functions appear in their solutions, which is a consequence of the constraints that arise in the Hamiltonian formulation. In the well known case of electrodynamics coupled to an external current J_μ, we have $\Phi^M \to A^\mu$ together with the gauge transformation $\delta A_\mu = \partial_\mu \theta$ (only one parameter), which leads to $P_{\mu,1} = 0$ and $Q^\nu_{\mu,1} = \delta^\nu_\mu$. Then Eq. (7) reduces to $0 = \partial^\mu \mathcal{E}_\mu = \partial^\mu \left(\partial^\nu F_{\nu\mu} - J_\mu \right) = \partial^\mu J_\mu$, thus yielding the conservation of the external current. Here $F_{\nu\mu} = \partial_\nu A_\mu - \partial_\mu A_\nu = -F_{\mu\nu}$. The standard point of view is that gauge invariance forbids the term $m^2 A_\mu A^\mu$ in the LD of electrodynamics, which states that the photon has zero mass. The same argument follows for the additional gauge bosons: the gluons and the graviton. In subsection (4.3) we present an alternative way of looking at gauge bosons in terms of spontaneous Lorentz symmetry breaking.

2. Spacetime symmetries

They include the discrete symmetries C (charge conjugation), P (parity) and T (time reversal) together with the continuous ones associated to translational invariance and Lorentz transformations. Even in the broken symmetry cases, we will always respect the translational invariance of the LD in order to have energy-momentum conservation. It is also interesting to observe that P and CP are not exact symmetries in Nature, since they are violated in the weak interactions. In fact, in 1956 Lee and Yang proposed tests to probe parity symmetry [2] and in the next year C.S. Wu et al. found P violation in the beta decay of Co^{60} [3]. In the same year, this violation was independently confirmed by Garwin et al. [4]. Later in 1964, Cronin and Fitch found CP violation in the decay of neutral kaons [5].

This contribution will focus on the discussion of some specific topics in Lorentz symmetry breaking (LSB) and does not do any justice to the huge literature in this field. Before going into the details it is worth recalling two basic theorems relating spacetime symmetries. The first one is the CPT theorem [6], which states that any Lorentz invariant, local, quantum field theory with an hermitian Hamiltonian must exhibit CPT invariance. The second one is called the anti-CPT theorem [7], which proves that CPT violation implies Lorentz violation, but not the converse. That is to say, in LSB theories we can have both CPT even and CPT odd contributions in the Lagrangian.

3. Why testing Lorentz symmetry

3.1. Motivation

There are at least two important reasons for probing this symmetry. The most obvious one is that, since physics is an experimental science, it is of fundamental importance to test, with the utmost precision, the validity of such a fundamental symmetry as Lorentz invariance. In order to do this one must have models that include and parameterize a violation of Lorentz symmetry, in order to design experiments that are able to set up bounds upon these parameters, determining the range of its validity. The second reason is related to the description of space at very short distances, where one assumes that quantum gravity becomes relevant. Most of the contending theories in this field start from the idea that space is no longer a continuum in that limit, but rather that is has a discrete or granular structure. Naively, one can entertain the possibility that Lorentz symmetry, basically rooted on a continuum space, is no longer valid under such circumstances and that very small violations will permeate to standard model energies and distances. A definite answer to this question will come only after the semiclassical limit of such quantum gravities theories is achieved, that is to say after the reconstruction of standard Minkowski space is attained. Then it will be possible to decide upon the impact of such discrete structure in the continuum limit associated with the Lorentz symmetry. The construction of such semiclassical approximation is still an open problem in quantum gravity theories [8]. In this process, the strict bounds for LSB already determined will serve as a set of experimental benchmarks for contending quantum gravity theories, providing an experimental basis to decide upon the correct one. For a review of the present day status in testing Lorentz invariance see for example Ref. [9].

3.2. The Standard model extension

To test Lorentz symmetry we require models that break this invariance, in order to introduce parameters that quantify the violations, which subsequently are bounded by experiments. The most general of such models, which has been extensively used by experimentalists to present their results, is the Standard Model Extension (SME) [10]. This is an effective field theory which extends the LD of the Standard Model (SM) of particle physics by adding all possible dimension three and four Lorentz violating operators that can be constructed with the fields appearing in the SM, thus being consistent with the particle content and interactions of the SM. The SME has been also generalized to incorporate the gravitational interaction [11] and also to include higher order Lorentz violating operators [12]. The full range of applications of the SME, both at the theoretical and experimental levels, can be appreciated in Refs. [13]. The basic idea is that Lorentz symmetry breaking (LSB) arises from non-zero vacuum expectation values (VEVs) of tensorial fields in a spontaneous symmetry breaking of a more fundamental theory. Such VEVs fill all spacetime, define privileged directions and couple to the fields of the SM thus yielding the LSB operators. Since LSB is expected to be suppressed by some huge quantum gravity scale, in a first approximation one considers only linear modifications arising from such VEVs. When the gravitational interaction is not included, the dynamics of such VEVs is not relevant and they behave as non-dynamical fields entering in the action. In this case LSB is equivalent to an explicit breaking, in contrast with a spontaneous breaking which is mandatory in the presence of gravity [11, 14]. It is important to emphasize that we are not violating what is called observer (passive) Lorentz invariance, which is the freedom we have to select an arbitrary coordinate frame to describe the system. When passing from one reference frame to another, both the dynamical fields as well as the tensorial VEVs transform accordingly. Here one must be careful with the effects of boosts, which may kinematically amplify the values of the VEVs, giving rise to enormous effects in the LSB which are not consistent with observation. The natural reference frames in which we describe observations in the laboratory are such that LSB effects are much suppressed and we want to restrict ourselves only to these set of frames, which are called concordant frames. What we violate is particle (active) Lorentz transformations which, in a given reference frame, are applied to the dynamical fields but not to the VEVs. In a pictorial representation we can imagine that the earth is immersed in the flux of constant tensor fields (VEVs) defining fixed directions in space. As the earth moves through this aether, the relation among the relevant fields of our experiment in the laboratory changes with respect to such directions, thus providing sidereal or daily variations in the signals produced by the coupling of matter and gauge fields to the VEVs. This type of variations is what most of the experiments searching for LSB look for. A more vivid analogy is that of a physicist confined to a laboratory immersed in a constant magnetic field and not having access to the exterior. For him, a privileged direction appears and rotational symmetry is violated.

Just to give an example, the LD corresponding to the photon sector of the SME is

$$L = -\frac{1}{4}F_{\mu\nu}F^{\mu\nu} - \frac{1}{4}(k_F)^{\mu\nu\alpha\beta}F_{\mu\nu}F^{\alpha\beta} - J^{\mu}A_{\mu}, \tag{8}$$

where $F_{\mu\nu}$ is the Faraday tensor. We have the standard Maxwell term plus corrections that incorporate LSB via the constant dimensionless VEV $(k_F)^{\mu\nu\alpha\beta}$. This tensor has zero epsilon-trace plus Riemann-like symmetries, yielding 19 independent components which have been bounded by different experiments or observations. The system described in Eq. (8) can be cast in the standard form of electrodynamics in a medium by introducing the corresponding constitutive relations in terms of 3×3 permittivity, permeability and mixing matrices. Further linear combinations of these matrix elements $((\tilde{\kappa}_{e+})_{ij}, (\tilde{\kappa}_{o-})_{ij}, ((\tilde{\kappa}_{e-})_{ij}, (\tilde{\kappa}_{o+})_{ij}, (\tilde{\kappa}_{\rm tr}))$, not to be specified here, are subjected to bounds from different sources. For example, astrophysical tests produce the bounds $(\tilde{\kappa}_{e+})_{ij} \leq 10^{-32}$, $(\tilde{\kappa}_{o-})_{ij} \leq 10^{-32}$ and $(\tilde{\kappa}_{\rm tr}) \leq 10^{-14}$, while resonator tests lead to $(\tilde{\kappa}_{e-})_{ij} \leq 10^{-17}$ together with $(\tilde{\kappa}_{o+})_{ij} \leq 10^{-14}$. The situation for the remaining sectors of the SME is completely similar.

The amount of experimental work dedicated to search for LSB has been enormous, ranging from table top experiments in atomic physics and data analysis in high energy physics to astrophysical observations. No signal of LSB has been detected so far, but a huge list of bounds obtained for the VEVs codifying LSB in the SME has been obtained. These bounds, together with their sources are reported in the Data Tables for Lorentz and CPT Violation [15], which is revised every year in order to include the improved results obtained from recent experiments.

The results obtained in the study of LSB can be considered an important contribution to the subject of Quantum Gravity Phenomenology, which attempts to provide experimental guidance to the problem of constructing a viable quantum gravity theory [16]. Other approaches to study the impact of space granularity upon the propagation of particles at standard model energies have been considered. Some of them are: Deformed (Double) Special Relativity [17], Minimal Length Scenarios [18], Finsler Geometries [19], Horava-Lifshitz Gravity [20] and Very Special Relativity [21], for example.

4. Some particular topics in LSB

4.1. Modified dispersion relations and photon propagation

One of the simplest kinematical consequences of LSB is the modification of the energy-momentum relations for propagating particles, as well as the changes introduced in the threshold conditions that allow some specific reaction. Of course, a detailed description of a particular process requires the knowledge of the dynamics, via the cross section for example, which can be calculated in a given LSB field theory, as it is the case of the SME, for example. Much attention has been devoted to the information that extragalactic photons could provide by observing gamma ray bursts (GRB), pulsars and active galactic nuclei (AGN), among others. The modified dispersion relations (MDR) for photons propagating with momentum $\mathbf{k}$ and energy E can be presented as

$$c^2\mathbf{k}^2 = E^2\left(1 \pm \xi\left(\frac{E}{E_{\rm QG}}\right) + O(E^2) + \cdots\right), \quad \longrightarrow \quad |\mathbf{v}| = |\nabla_{\mathbf{k}}E| = c\left(1 \mp \xi\left(\frac{E}{E_{\rm QG}}\right) + \cdots\right), \tag{9}$$

where $E_{\rm QG}$ is the quantum gravity scale and ξ is a parameter of order one. One must keep an open mind regarding $E_{\rm QG}$, which is usually taken as the Planck mass $M_P = 10^{19}$GeV. Such MDR predicts an energy dependent velocity, which produces a time delay Δt in the arrival of photons having an energy difference ΔE, which were simultaneously emitted at a distance L from the detector. A first approximation gives

$$\Delta t = \xi\,\frac{L}{c}\frac{\Delta E}{E_{\rm QG}}. \tag{10}$$

As emphasized in the seminal proposal [22], the large suppression arising from $1/E_{\rm QG}$ requires a huge amplifying factor in such a way that the effect becomes observable. This is provided by taking photon sources located at cosmological distances. Some typical numbers like $\Delta E = 20$ MeV, $E_{QG}/\xi = 10^{19}$ GeV, together with the distance $L = 10^{10}$ l.y. produce $\Delta t = 10^{-3}$ s, which is within the precision of cosmological gamma ray detection. Numerous observations of this kind have been carried on and they have set lower bounds for E_{QG}/ξ, which range from 1.8×10^{15} GeV [23] to $(1.4 - 122) \times 10^{19}$ GeV [24]. The latter bound either indicates that modified linear dispersion relations are not consistent with observation, or that the quantum gravity scale needs to be revised. Soon after the proposal of Ref. [22], some specific models of quantum gravity corrections to standard particle dynamics incorporating LSB appeared [25, 26, 27, 28]. The first heuristic derivation of a quantum gravity induced modified Hamiltonian for Maxwell equations in a Loop Quantum Gravity inspired description of Einstein-Maxwell theory was reported in [25]. This sparked a great deal of interest and it was subsequently followed by generalizations to the fermionic case [27], further elaborations in the Maxwell case [28] and the first steps towards the Yang-Mills case [29]. The idea behind all these heuristic derivations starts from the corresponding well defined operators in the quantum theory [30] and relies in the construction of quantum gravity semiclassical states which approximate a flat geometry at macroscopic distances while retaining their quantum granular structure at scales of the order of the Planck length. Since those states were not known, their expected properties were postulated and estimations of the required expectation values were performed. Accordingly, this procedure left overall dimensionless coefficients undetermined. Furthermore, extracting the dynamics is a process less clear to develop due to the known difficulties with time evolution. [31]. The construction of such semiclassical states is still an open problem in canonical quantum gravity [8] and the main difficulty resides in finding the appropriate quantum gravity ground state which yields the Minkoswki space.

4.2. Radiative corrections

Most of the quantum gravity inspired corrections to the dynamics at standard model energies arise through powers of the factor (E/E_{QG}), which are directly relevant at inaccessible energies $E \approx E_{QG}$. An alternative possibility of probing such high energies is through the inclusion of radiative corrections, because the internal momenta are integrated up to the maximum allowed in a given reference frame. The standard folklore with respect to any new physics entering at high energy scales (here Planck scale) via an effective low energy field theory is that such scales should produce negligible effects upon the leading-order low-energy physics (here standard model energies). Contrary to this belief, the first indications that this is not necessarily true in the LIV case appeared in Ref. [32], where a specific form of the coupling was proposed in order to avoid such contributions. Subsequently it was shown that the proposed cure did not survive higher orders in perturbation theory [33]. The fact that LIV induced by modeling space granularity via the introduction of a physical cutoff, defined in a preferred reference frame, lead to unsuppressed dimension four LIV contributions, was finally recognized in Ref. [34]. To this end, the calculation of one-loop radiative corrections in the Yukawa model

$$\mathcal{L} = \frac{1}{2}(\partial\Phi)^2 - \frac{1}{2}\mu^2\Phi^2 + \bar{\Psi}(i\gamma^\mu\partial_\mu - m)\Psi + g\Phi\bar{\Psi}\Psi + (\text{LVT}) \tag{11}$$

was considered. Here LVT refers to the highly suppressed zeroth order Lorentz violating terms that take into account the previously discussed free particle dynamical modifications. The space granularity was modeled by introducing a physical cut-off Λ, in such a way that the magnitude of the three-momentum in any loop is bounded by this quantity. The parameter Λ defines the onset of the scale at which the granularity of space becomes manifest. A convenient way of incorporating this requirement is to introduce the physical cutoff function $f(|\mathbf{k}|/\Lambda)$ with $f(0) = 1$, $f(\infty) = 0$, which suppresses the internal momenta having $|\mathbf{k}| \geq \Lambda$. The one-loop

Journal of Physics: Conference Series **761** (2016) 012056 doi:10.1088/1742-6596/761/1/012056

calculation of the boson self-energy $\Pi(E, \mathbf{p}, \Lambda)$ produces a finite contribution

$$\delta\Pi(E, \mathbf{p}, \Lambda) = \eta\, p_\mu p_\nu W^\mu W^\nu, \qquad \eta = \frac{g^2}{12\pi^2}\left[1 + 2\int_0^\infty dx x(f'(x))^2\right], \tag{12}$$

where W^μ is the four-velocity of the frame in which the momentum cut-off Λ is defined. Using standard model couplings, the coefficient is estimated in the range $\eta \geq 10^{-3}$. Similar results are found in the calculation of the fermion self-energy. These huge corrections would make LSB easily observable, which contradicts the tight bounds found so far. This result also poses serious problems for the preferred frame interpretation of the explicit dynamical modifications possibly induced by quantum gravity effects [34]. Nevertheless, spontaneous Lorentz symmetry breaking still remains an open possibility.

4.3. Gauge particles as Goldstone bosons

Since a definite signal for LSB has not being found so far, it is interesting to consider a somewhat opposite point of view regarding Lorentz violation which, having a long history in physics, has been recently revisited in the literature. The main goal in this approach is to determine the conditions under which spontaneous Lorentz symmetry breaking (SLSB) in a more fundamental theory becomes unobservable, thus providing a dynamical origin to the massless gauge bosons. The motivation comes from the property that spontaneous symmetry breaking of a global symmetry produces zero-mass excitations (Nambu-Goldstone bosons (NGB)). In particular, the global breaking of continuous spacetime symmetries including tensorial VEVs, can be adjusted to produce massless spin one and spin two fields describing the corresponding NGB. Then, the question whether or not it is possible to identify such NGB with the corresponding gauge particles (photons and gravitons, for example) arises. An affirmative answer would provide a dynamical basis for the gauge principle, which has played a fundamental role in the construction of modern physics. This idea goes back to Dirac [35], Bjorken [36] and Nambu [37], among others, and has been recently revisited in the context of Electrodynamics [38], Yang-Mills theories [39] and gravity [40]. In analogy with the nonlinear sigma model describing pion physics, a very efficient way to deal with the low energy sector of a theory exhibiting SLSB, which includes only the corresponding NGB, is by considering the so called Nambu models. Their LD is given by that of the gauge theory which gauge bosons one tries to interpret as NGB, plus a nonlinear constraint that encodes the SLSB. The simplest case, from which electrodynamics arises, is the abelian Nambu model coupled to a conserved current [37]

$$\mathcal{L} = -\frac{1}{4}F^{\mu\nu}F_{\mu\nu} - J^\mu A_\mu, \quad A^\mu A_\mu = (n_\beta n^\beta)\, M^2, \qquad \partial_\mu J^\mu = 0. \tag{13}$$

Here n_μ denotes the direction of the vacuum which breaks the symmetry, M is the energy scale at which such breaking occurs and the nonlinear constraint (NLC) indicates that we are taking the VEV $\langle A_\mu\rangle = Mn_\mu$. This constraint is to be solved and substituted into the LD. An adequate parametrization to solve the constraint in Eq. (13) is by introducing the field a_μ as

$$A_\mu = a_\mu + n_\mu\left(M^2 - n^2 a^2\right)^{1/2}/n^2, \quad n^2 \neq 0, \quad n^\mu a_\mu = 0. \tag{14}$$

Since a_μ is orthogonal to the vacuum it includes only the NGB of the model. The simplest realization of the ANM arises from taking standard QED plus the nonlinear constraint in Eq. (13), which is subsequently solved in terms of a_μ according to Eq. (14), in such a way that

$$\mathcal{L}_{\rm ANM}(\bar\psi, \psi, a_\mu) = \mathcal{L}_{\rm QED}(\bar\psi, \psi, A_\mu(a_\alpha)). \tag{15}$$

Journal of Physics: Conference Series **761** (2016) 012056 doi:10.1088/1742-6596/761/1/012056

The substitution $A_\mu(a_\alpha)$ has to be rewritten as an expansion in powers of a^μ/M and yields a highly nonlinear LD, which provides additional LSB contributions to standard QED processes, where photons are now described by the NGB field a_μ. Perturbative calculations in the tree and one loop approximation have been reported for some specific processes, with the surprising result that all LSB contributions cancel, yielding the standard QED amplitudes [37, 41]. Motivated by these unexpected conclusions and as a first step to understand this question, we have examined in a non-perturbative way, the conditions under which the equivalence between a non-abelian Nambu model (NANM) and its corresponding Yang-Mills theory is achieved. The NANM is the direct generalization of the ANM in Eq. (13) where $A_\mu \to A^a_\mu$, $F_{\mu\nu} \to F^a_{\mu\nu}$, $J_\mu \to J^a_\mu$, $n_\mu \to n^a_\mu$, still with only one nonlinear constraint. Here a labels the corresponding group index. For the sake of simplicity we next illustrate the main steps of the procedure in terms of the ANM, given by Eq. (13), and refer the reader to Refs. [42] for the generalization to the NANM. The strategy is to start from the canonical formulation of the ANM and arrive to the standard canonical formulation of ED in terms of the canonically conjugated fields A_i, E^j, $i,j = 1,2,3$, with Hamiltonian density

$$\mathcal{H}_{\rm ED} = \frac{1}{2}\left(\mathbf{E}^2 + \mathbf{B}^2\right) + J^i A_i - \left(\partial_i E_i - J^0\right) A_0, \quad E_i = -\partial_0 A_i - \partial_i A_0, \quad (\partial_i E_i - J^0) \approx 0. \quad (16)$$

To begin we introduce the fields Φ_k, where we split the indices $k = 1,2,3$ into $\bar{k} = 1,2$ and 3, together with the alternative parametrization

$$A_0 = \Phi_3\left[1 + \frac{N}{4\,\Phi_3\Phi_3}\right], \qquad A_3 = \Phi_3\left[1 - \frac{N}{4\,\Phi_3\Phi_3}\right], \qquad A_{\bar{k}} = \Phi_{\bar{k}}, \quad (17)$$

where $N = \Phi_{\bar{k}}\Phi_{\bar{k}} + n^2M^2$. In terms of the new coordinates Φ_k the NLC in Eq. (13) is identically satisfied. Let us now focus on the invertible change of variables $A_i = A_i(\Phi_j)$, which yields $\dot{A}_i = \dot{A}_i(\Phi,\dot{\Phi})$ and $\dot{A}_0 = \dot{A}_0(\Phi,\dot{\Phi})$, recalling that we also know $A_0 = A_0(\Phi)$. The first relation gives

$$\dot{A}_i = \dot{\Phi}_k \frac{\partial A_i}{\partial \Phi_k}, \quad \to \quad \frac{\partial \dot{A}_i}{\partial \dot{\Phi}_k} = \frac{\partial A_i}{\partial \Phi_k}. \quad (18)$$

We observe that we will not require the explicit form of the transformations $A_i = A_i(\Phi_j)$ in the following. In this way, the LD of the ANM can be written as $\mathcal{L}_{\rm ANM}(\Phi,\dot{\Phi})$. A convenient way to make later contact with ED is to rewrite the LD in Eq. (13) as

$$\mathcal{L}_{\rm ANM}(\Phi,\dot{\Phi}) = \frac{1}{2}\left(\mathbf{E}^2 - \mathbf{B}^2\right) - J^\mu A_\mu, \quad (19)$$

where $E_i, B_i = \epsilon_{ijk}\partial_j A_k$ and A_i are just labels for the corresponding functions of Φ and $\dot{\Phi}$, according to previous definitions. The coordinates Φ_i together with the ANM canonically conjugated momenta Π^j satisfy the standard canonical algebra and define a regular system. The usual definition of the electric field in Eq.(16), yields the relations

$$\Pi_i = \frac{\partial \mathcal{L}}{\partial \dot{\Phi}_i} = \frac{\partial A_j}{\partial \Phi_i} E_j, \qquad E_i = \frac{\partial \Phi_j}{\partial A_i}\Pi_j, \qquad \Pi_i\dot{\Phi}_i = E_j\dot{A}_j, \quad (20)$$

which follow because the dependence on $\dot{\Phi}$ arises only in E_i, together with the relations in Eq. (18). Next we calculate the ANM Hamiltonian $\mathcal{H}_{\rm ANM}(\Pi,\Phi) = \Pi\dot{\Phi} - \mathcal{L}_{\rm ANM}(\Phi,\dot{\Phi})$ and explicitly verify that it can be written as

$$\mathcal{H}_{\rm ANM} = \frac{1}{2}\left(E^a_i E^a_i + B^a_i B^a_i\right) + J^{ai}A^a_i - \left(D_i E^b_i - J^{0b}\right) A^b_0, \quad (21)$$

in terms of the corresponding labels. Also, the relations (20) would allow the explicit calculation of the algebra among A_i^a, E^{bj}, in terms of the ANM canonical algebra. Let us observe that in the Hamiltonian form of the action for the ANM, the last relation in Eq. (20) identifies E^j as the canonically conjugated momenta to the coordinates A_i of ED. Moreover, since the transformation among the variables (E, A) and (Π, Φ) is generated only by the coordinate transformations $A_i = A_i(\Phi)$, we can assert that this is indeed a canonical transformation [42]. In this way we recover the canonical algebra for the ED phase-space variables A_i, E^j without any further calculation. Summarizing, up to now we have proved that the ANM canonical algebra yields the canonical algebra of ED. Nevertheless, both theories are still not equivalent because Eq. (21) is not the Hamiltonian of ED, in spite of its almost identical form. In fact, A_0 is not an arbitrary function, but only a label for some combinations of the coordinates in the ANM. Also, the Gauss function $G \equiv (\partial_i E_i - J^0)$ is not a constraint, as it should be in a gauge theory. To proceed, we calculate the time evolution of the Gauss function under the ANM dynamics, obtaining

$$\dot{G} = -\partial_\mu J^\mu + MG + \partial_i(N^i G), \tag{22}$$

where M, N^i are known functions of ANM phase space. Since we have coupled the ANM to a conserved current, it is enough to impose the Gauss constraint $G = 0$ at some initial time slice t_0 in order to recover gauge invariance. In fact, under this requirement Eq. (22) guarantees that $G = 0$ for all time. Then, such condition can be added to the Hamiltonian Eq. (21) as an additional constraint via a Lagrange multiplier R which now replace A_0 by $S = A_0 + R$. The canonical algebra of ED, together with current conservation, guarantees that no additional constraints appear. Summarizing, the equivalence between ED and the ANM coupled to a conserved current is established once the Gauss law $G = 0$ is imposed as an initial condition upon the ANM dynamics.

Acknowledgments

LFU acknowledges support from the projects UNAM (Dirección General de Asuntos del Personal Académico) # IN104815 and CONACyT # 237503.

References

[1] Noether E 1918 *Nachr. v. d. Ges. d. Wiss. zu Gttingen, Mathphys. Klasse* 235
Bañados M and Reyes I A 2016 A short review on Noether's theorems, gauge symmetries and boundary terms *Preprint* arXiv:1601.03616 [hep-th]
Brading K and Brown H R 2000 Noether's Theorems and Gauge Symmetries *Preprint* arXiv: hep-th/0009058
[2] Lee T D and Yang C N 1956 *Phys. Rev.* **104** 254
[3] Wu C S, Ambler E, Hayward R W, Hoppes D D and Hudson R P 1957 *Phys. Rev.* **104** 1413
[4] Garwin R L, Lederman L M and Weinrich M 1957 *Phys. Rev.* **105** 1415
[5] Cronin J and Fitch V 1964 *Phys. Rev. Lett.* **13** 138
[6] Schwinger J 1951 *Phys. Rev.* **82** 914
Luders G 1954 *Kgl. Dansk. Vidensk. Selsk. Mat.-Fys Medd.* **28** # 5
Pauli W 1955 *Niels Bohr and the Development of Physics* (McGraw-Hill, New York and Pergamon, London)
[7] Greenberg O W 2002 *Phys. Rev. Lett.* **89** 231602
[8] Date G and Hossain G M 2012 *SIGMA* **8** 010
[9] Mattingly D 2005 *Living Rev. Relativity* **8** 5
[10] Colladay D and Kostelecký V A 1997 *Phys. Rev.* D **55** 6760
Colladay D and Kostelecký V A 1998 *Phys. Rev.* D **58** 116002
[11] Kostelecký V A 2004 *Phys. Rev.* D **69** 105009
[12] Kostelecký V A and Mewes M 2009 *Phys. Rev.* D **80** 015020
Kostelecký V A and Mewes M 2013 *Phys. Rev.* D **88** 096006
[13] 1999 *Proceedings of the Meeting on CPT and Lorentz Symmetry*, ed V A Kostelecký (World Scientific, Singapore)
2001 *Proceedings of 2nd Meeting on CPT and Lorentz Symmetry*, ed V A Kostelecký (World Scientific, Singapore)

2004 *Proceedings of 3rd on CPT and Lorentz Symmetry*, ed V A Kostelecký (World Scientific, Singapore)
2007 *Proceedings of 4th Meeting on CPT and Lorentz Symmetry*, ed V A Kostelecký (World Scientific, Singapore)
2010 *Proceedings of 5th Meeting on CPT and Lorentz Symmetry*, ed V A Kostelecký (World Scientific, Singapore)
2014 *Proceedings of 6th Meeting on CPT and Lorentz Symmetry*, ed V A Kostelecký (World Scientific, Singapore)

[14] Bluhm R 2015 *Phys. Rev.* D **91** 065034
[15] Kostelecký V A and Russell N 2011 *Rev. Mod. Phys.* **83** 11
[16] Amelino-Camelia G 2013 *Living Rev. Relativity* **16** 5
[17] Amelino-Camelia G 2010 *Symmetry* **2** 230
[18] Hossenfelder S 2013 *Living Rev. Relativity* **16** 2
[19] Vacaru S I 2012 *Int. J. Mod. Phys.* D **21** 1250072
Girelli F, Liberati S and Sindoni L 2007 *Phys. Rev.* D **75** 064015
[20] Horava P 2009 *Phys. Rev.* D **79** 084008
[21] Cohen A G and Glashow S L 2006 *Phys. Rev. Lett.* **97** 021601
Alfaro J, González P and Avila R 2015 *Phys. Rev.* D **91** 105007
[22] Amelino-Camelia G, Ellis J R, Mavromatos N E, Nanopoulos D V, and Sarkar S, 1998 *Nature (London)* **393** 763
[23] Kaaret P 1999 *Astron. Astrophys.* **345** L2
[24] Abdo A A et al. 2009 *Nature (London)* **462** 331
[25] Gambini R and Pullin J 1999 *Phys. Rev.* D **59** 124021
[26] Ellis J R, Mavromatos N E and Nanopoulos D V 2000 *Gen. Rel. and Grav.* D **59** 124021
[27] Alfaro J, Morales-Técotl H A and Urrutia L F 2000 *Phys. Rev. Lett.* **84** 2318
Alfaro J, Morales-Técotl H A and Urrutia L F 2002 *Phys. Rev.* D **66** 124006
[28] Alfaro J, Morales-Técotl H A and Urrutia L F 2002 *Phys. Rev.* D **65** 103509
[29] Alfaro J, Morales-Técotl H A, Reyes M and Urrutia L F 2003 *J. Phys.* A **36** 12097
[30] Thiemann T 1996 *Phys. Lett.* B **380** 257
[31] Bojowald M, Morales-Técotl H A and Sahlmann H 2005 *Phys. Rev.* D **71** 084012
[32] Myers R C and Pospelov M 2003 *Phys. Rev. Lett.* **90** 211601
[33] Pérez A and Sudarsky D 2003 *Phys. Rev. Lett.* **91** 179101
[34] Collins J, Pérez A, Sudarsky D, Urrutia L and Vucetich H 2004 *Phys. Rev. Lett.* **93** 191301
[35] Dirac P A M 1951 *Nature (London)* **168** 906
[36] Bjorken J D 1963 *Ann. Phys. (N.Y.)* **24** 174
[37] Nambu Y 1968 *Prog. Theor. Phys. Supplement* **E68** 190
[38] Righi R and Venturi G 1982 *Int. J. Theor. Phys.* **21** 63
[39] Righi R, Venturi G and Zamiralov V 1978 *Nuovo Cimento* A **47** 518
Chkareuli J L, Froggatt C D and Nielsen H B 2009 *Nucl. Phys.* B **821** 65
[40] Chkareuli J L, Jejelava J G and Tatishvili G 2011 *Phys. Lett.* B **696** 124
[41] A.T. Azatov and J.L. Chkareuli 2006 *Phys. Rev.* D **73** 065026
[42] Escobar C A and Urrutia L F 2015 *Phys. Rev.* D **92** 025013
Escobar C A and Urrutia L F 2016 Emergence of Yang Mills theory from the Non-Abelian Nambu Model *Preprint* arXiv:1607.08587 [hep-th], these Proceedings

MWPF IOP Publishing
Journal of Physics: Conference Series **761** (2016) 012057 doi:10.1088/1742-6596/761/1/012057

The CONNIE experiment

A. Aguilar-Arevalo[1], X. Bertou[2], C. Bonifazi[3], M. Butner[4], G. Cancelo[4], A. Castaneda Vazquez[1], B. Cervantes Vergara[1], C.R. Chavez[5], H. Da Motta[6], J.C. D'Olivo[1], J. Dos Anjos[6], J. Estrada[4], G. Fernandez Moroni[7,8], R. Ford[4], A. Foguel[3,6], K.P. Hernandez Torres[1], F. Izraelevitch[4], A. Kavner[9], B. Kilminster[10], K. Kuk[4], H.P. Lima Jr.[6], M. Makler[6], J. Molina[5], G. Moreno-Granados[1], J.M. Moro[11], E.E. Paolini[7,12], M. Sofo Haro[2], J. Tiffenberg[4], F. Trillaud[1], and S. Wagner[6,13]

[1]Universidad Nacional Autónoma de México, Ciudad de México, México
[2]Centro Atómico Bariloche - Instituto Balseiro, CNEA/CONICET, Argentina
[3]Universidade Federal do Rio de Janeiro, Instituto de Física, Rio de Janeiro, Brazil
[4]Fermi National Accelerator Laboratory, Batavia, IL, U.S.A.
[5]Facultad de Ingeniería - Universidad Nacional de Asunción, Paraguay
[6]Centro Brasileiro de Pesquisas Fisicas, Rio de Janeiro, Brazil
[7]Departamento de Ingeniera Eléctrica y de Computadores, Universidad Nacional del Sur, Bahía Blanca, Argentina
[8]Instituto de Investigaciones en Ingeniería Eléctrica "Alfredo Desages", CONICET - Universidad Nacional del Sur, Bahía Blanca, Argentina
[9]University of Michigan, Ann Arbor, MI, U.S.A.
[10]Universität Zürich Physik Institut, Zurich, Switzerland
[11]Depto. de Ingeniería, Universidad Nacional del Sur, Bahía Blanca, Argentina
[12]Comisión de Investigaciones Científicas Provincia Buenos Aires, La Plata, Argentina
[13]Pontificia Universidade Católica, Rio de Janeiro, Brazil

E-mail: `alexis@nucleares.unam.mx`

Abstract. The CONNIE experiment uses fully depleted, high resistivity CCDs as particle detectors in an attempt to measure for the first time the Coherent Neutrino-Nucleus Elastic Scattering of antineutrinos from a nuclear reactor with silicon nuclei. This talk, given at the XV Mexican Workshop on Particles and Fields (MWPF), discussed the potential of CONNIE to perform this measurement, the installation progress at the Angra dos Reis nuclear power plant, as well as the plans for future upgrades.

1. Introduction

The Coherent Elastic Neutrino-Nucleus Scattering (CEνNS), is a Standard Model (SM) process where a neutrino, or antineutrino, interacts with a nucleus as a whole entity [1]. It arises from the coherent enhancement of the interaction cross-section with the constituent nucleons, when the 4-momentum transfer is small compared to the reciprocal of the nuclear size: $|q^2| < 1/R^2$; in the laboratory frame, this corresponds roughly to incident neutrino energies $E_\nu < 50$ MeV. Its differential cross-section, to lowest order in T/E_ν, is [2]

$$\frac{d\sigma}{dT}(E_\nu, T) = \frac{G_F^2}{8\pi}\left[Z(4\sin^2\theta_W - 1) + N\right]^2 M\left(2 - \frac{MT}{E_\nu^2}\right)|f(q^2)|^2\,, \quad (1)$$

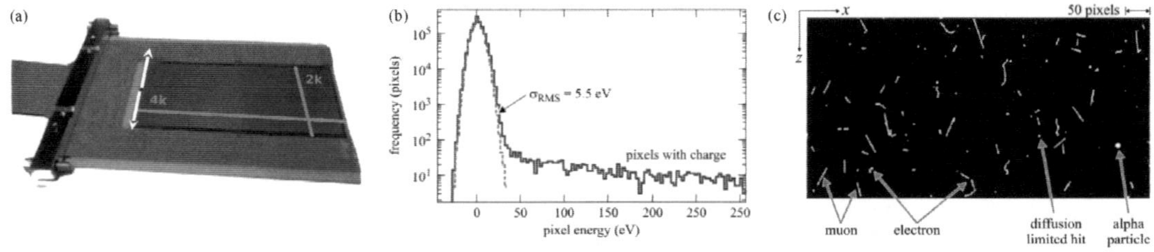

Figure 1. (a) A packaged 8 Mpix CONNIE CCD. (b) Single pixel value distribution at -140°C with and without a ^{60}Co gamma source. (c) Sea level exposure showing various particle tracks.

where T is the nuclear recoil energy, G_F is the Fermi constant, M is the mass of the nucleus, Z, and N are the number of protons and neutrons in the nucleus respectively, θ_W is the weak mixing angle, and $f(q^2)$ is the nuclear form factor, which is a function only of the 4-momentum transfer $q^2 = 2E_\nu TM/(T - E_\nu)$. Integration of Eq. (1) over T with the approximations $f(q^2) \approx 1$ (valid for $E_\nu < 20$ MeV), and $\sin^2\theta_W \approx 1/4$, yields a total cross section as a function of neutrino energy: $\sigma_T(E_\nu) \approx 4.22 \times 10^{-45} N^2 (E_\nu/1\ \mathrm{MeV})^2\ \mathrm{cm}^2$.

This process has never been observed experimentaly, primarily because, in addition to its small cross-section ($\mathcal{O} \sim 10^{-41}$ cm^2), the nuclear recoil energies involved are very small ($<\sim$15 keV), and escape the capabilities of most sufficiently massive detectors. Many detector technologies used in direct WIMP dark matter (DM) searches, combining low thresholds with large target masses, have been considered for the detection of CEνNS. Such DM experiments are expected to eventually reach the sensitivity to the "neutrino floor": an irreducible background from the coherent scattering of solar, atmospheric and difuse supernova neutrinos [3].

Recent developments in the technology of Charge Coupled Devices (CCD) have made it possible to build and operate detectors with very low thresholds (of order $\sim$10 eV) and relatively large sensitive mass ($\sim$5 g per CCD); these have been successfully used in the DAMIC experiment [4] at SNOLAB to perform direct DM searches in the low mass region < 10 GeV$/c^2$, demostrating the unique capabilities of CCDs for the detection of low-energy recoils.

The Coherent Neutrino Nucleus Interaction Experiment (CONNIE) is a detector of solid-state technology installed at a nuclear reactor with an array of CCDs at its core, designed to observe the CEνNS reaction for the first time. In 2014 a prototype detector was installed at one of the reactors at the Angra dos Reis nuclear power plant in Brazil, where the characterization of the detectors response to the backgrounds is being studied.

2. The CONNIE CCDs

The CCDs are developed by the LBNL Microsystems Laboratory. They are fabricated on high-resistivity (10-20 kΩ cm) silicon, which allows their operation at full depletion with moderate substrate voltages. The sensitive area is divided into several million square (15 μm$\times$15 μm) pixels, reaching a few tens of square centimeters, and thicknesses up to $\sim$ 650 μm, providing detector masses of a few grams per CCD. They exhibit very low charge transfer inefficiencies and dark current when operated at low temperatures ($\sim$ 140 K). The energy response to electronic and nuclear recoils has been extensively studied [5, 6, 7, 8]. The CCDs are packaged in supporting structures and bonded to flex readout cables, as shown in Figure 1 (a), at the Fermilab Silicon Detector Facility. An optimal readout noise of $\sigma_{\mathrm{RMS}} = 1.5\mathrm{e}^-$ (equivalent to $\sim$5.8 eV) on the pixel values can be achieved by appropriate choice of the output amplifier integration time window. Figure 1 (b) shows the distribution of pixel values in a typical exposure with a ^{60}Co source. A typical image from a CCD exposure at sea level in which different particle tracks are visible is shown in Figure 1 (c). Events depositing energy within the volume of a single pixel

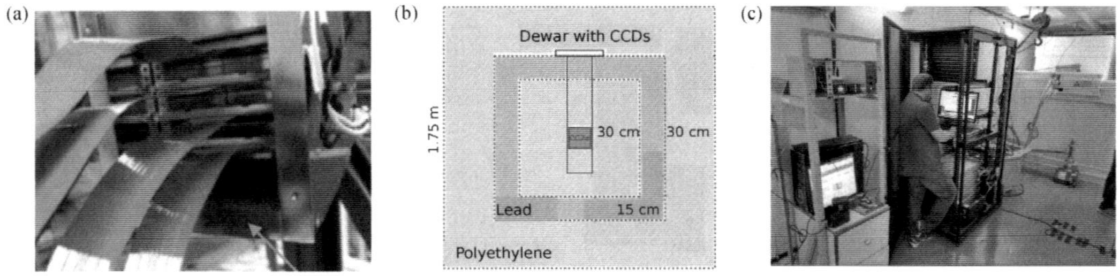

Figure 2. (a) CCD copper box. (b) Full shield concept. (c) Full installation inside container.

produce point-like events where ionized charge diffuses over one or a few neighboring pixels as it is drifted towards the CCD gate electrodes. These "diffusion limited hits" are expected from nuclear recoils such as those produced by the coherent neutrino-nucleus scattering.

3. Prototype installation at Angra-II

The prototype CCD array in the initial installation (2014 to mid 2016) has four 4k×2k pixel (8 Mpix) CCDs, each with a thickness of 250 μm, an area of 6 cm×3 cm, and a mass of $\sim$ 1 g. These are "engineering grade" sensors having several defects and significant contamination from U and Th in the AlN substrate used in the package design; they were deployed with the purpose of testing the system operation at the reactor site, and have preliminary estimates of the environmental background. Only two of the sensors were of sufficient quality for the background analysis.

The CCDs are stacked in a high purity copper box, see Figure 2 (a) kept at a temperature of $-140°$C inside a vacuum vessel (10^{-7} torr). A 15 cm-long lead cylinder inside the vessel shields the CCD array from gammas coming from radioactive contaminants in the readout electronics. A radiation shield was installed in two phases: Phase-I consisting of an inner 30 cm layer of polyethylene followed by 5 cm of lead, and Phase-II consisting of an additional 10 cm of lead (totaling 15 cm of Pb) and an outer 30 cm layer of polyethylene. The concept of the full shield is shown in Figure 2 (b). A picture of the inside of the container with the final installation of the detector, electronics, and support systems is shown in Figure 2 (c).

The detector is located at a distance of 30 m, see Figure 3 (a) and (b), from the core of the Angra-II reactor of the Almirante Alvaro Alberto nuclear power plant in Angra dos Reis, near Rio de Janeiro, Brasil. It is housed in a standard shipping container outside the main reactor building. The Angra-II reactor is a 3.95 GW$_{\text{th}}$ Presurized Water Reactor, and generates $\sim 8.7 \times 10^{20}$ $\bar{\nu}_e$s^{-1}. At the detector location this corresponds to $\sim 7.8 \times 10^{20}$ $\bar{\nu}_e$cm^{-2}s^{-1}, coming mainly from β decay chains of the products of the fragmentation of fissile isotopes ^{235}U, ^{238}U, ^{239}Pu, and ^{241}Pu, and n-capture processes in the reactor core (avg. fission rate of $\sim 1.2 \times 10^{20}$fis/s). The antineutrino flux expected from the reactor is shown in Figure 3 (c).

4. Expected event rates and forecast

Convolution of the antineutrino energy spectrum in Figure 3 (c), with the differential cross section in Eq. (1) yields the nuclear recoil energy spectrum. This is converted into the measured spectrum by means of the quenching factor $Q(T)$, relating the "measured" energy, T_m, to the recoil energy, T, via $T_m = T\,Q(T)$. Event rates are then calculated as a function of the detection threshold, T_{thr}, by integration of the measured energy spectrum weighted by the detection efficiency. Details of the event rate calculations and the efficiency estimate assumed here are given elsewhere [9]. The left-hand side plot in Figure 4 compares the expected recoil energy

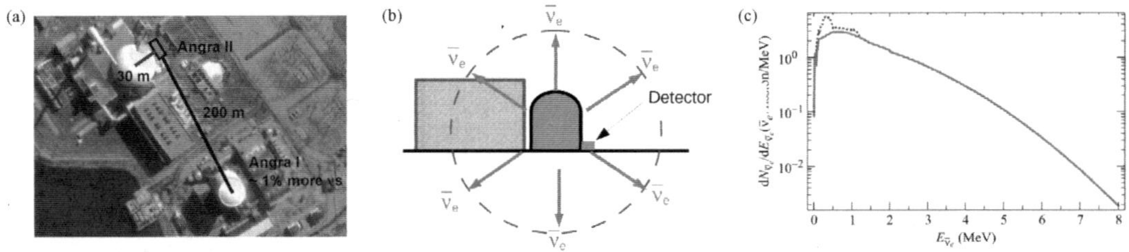

Figure 3. (a) Detector location in reactor complex. (b) Detector scketch. (c) Neutrino flux at detector.

spectrum and the spectrum of measured energies assuming the Lindhard [10] quenching factor with (green dashed line) and without (red dashed line) the effect of the detector efficiency. The right-hand side plot shows the total event rate as a function of the threshold, T_{thr}, for different choices of Q showing a weak dependence on the details of the quenching factor. From these calculations a signal rate of 16.15 evt/kg/day is extracted, using the Lindhard quenching factor and the estimated efficiency. As shown in the right-hand side plot, the contribution to the total number of signal events above 300 eV is negligible, hence defining a signal region from $\sim 28 - 300$ eV. Earlier studies have shown that with passive shielding at shallow depths, a background rate at sea level of ~ 600 evt/day/keV/kg can be achieved [11]; over the expected signal region this amounts to a background contribution of 163.2 evt/day.

4.1. Forecast for a 52 g detector

For the purpose of evaluating the potential of CONNIE, in what follows we assume a 52 g detector array (ten 8 Mpix CCDs with 650 μm and fiducial mass of 5.2 g each) running for T days. As shown in [9], in such a detector we expect a signal rate of $\sim 0.84T$ events, and a background rate of $\sim 8.5T$ events. Adding a rate of false positive events of $\sim$3.8 evt/day due to readout noise values above the threshold gives a total of $\sim 11.68T$ background events. The significance of a given signal can be calculated as a function of the running time by the expression $N_\sigma = 0.84\,T/\sqrt{11.68\,T} = 0.25\sqrt{T}$. Table 1 shows the running time in days required to reach a given level of significance with this setup. A $\sim 3\sigma$ significance can be achieved with 150 live days under the above assumptions.

Table 1. Days to reach a given confidence level (C.L.).

C.L. [%]	80	90	95	98	99
T (days)	12	28	45	70	150

5. Timeline

Plans to deploy a CCD-based detector making use of the available space already aproved for the Angra-Neutrinos project at the Angra-II reactor site began in 2013. The detectors were tested and packaged at Fermilab during 2014. The fully assembled vacuum vessel (containing the CCDs), the cooling and vacuum systems, and the readout electronics, were all shipped from Fermilab to Rio de Janeiro on August 2014, arriving (by sea) on September. The installation of the system took place from October through November; producing the first images of events for background studies as early as December. The full (lead+polyethylene) shield was assembled during July and August of 2015. From August through September, 2015 more than a month of reactor-on data were acquired with the full shield in place. From September through October,

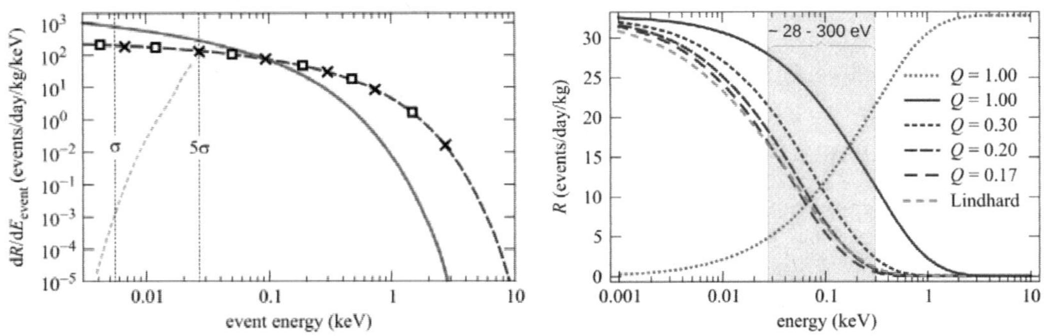

Figure 4. Left: The expected recoil energy spectrum (blue dashed line); the visible energy spectrum assuming a Lindhard Q factor (red solid line); the visible energy spectrum with Lindhard Q factor and the effect of the detector efficiency (green dashed line). Right: Total event rate as a function of the energy threshold for various quenching factors. The shaded area shows the expected signal region for a threshold of $5\sigma_{RMS} \simeq 28$ eV, and realistic quenching factors. Adapted from [9].

2015, one full month of reactor-off data were collected during a scheduled refueling outage. These data sets constitute the first CONNIE engineering run, whose results have been published since the time of this conference [12].

6. Summary and outlook

The low noise and low threshold of CCDs make them an ideal detector technology for the measurement of low energy depositions such as those expected from the coherent scattering of neutrinos off nuclei. The CONNIE Collaboration has been operating a prototype CCD detector at the Angra-II reactor in Brazil since December of 2014, and has taken data with different shield configurations, and with the reactor on and off. Results have been published since the XV MWPF [12] demonstrating the feasibility of the operations and background stability. The current setup is not expected to distinguish the coherent neutrino scattering events, only to demonstrate operations at the reactor site and measure environmental backgrounds and stability. Plans for upgrade to a CONNIE100 are ongoing.

6.1. Update since the XV MWPF

The collaboration plans to conduct an upgrade to a 100 g detector (CONNIE100) during 2016. The plan includes an upgrade of the packaging design that will reduce the intrinsic background of the detector (*e.g.* removal of the AlN substrate). It will incorporate a total of 18 engineering grade sensors of superior quality, with 675 μm thickness and 5.7 g of active mass per CCD. The upgraded detector should achieve sensitivity to the SM neutrino-nucleus coherent scattering as discussed in [9].

Acknowledgements

We thank *Central Nuclear Almirante Alvaro Alberto Electronuclear* for access to the reactor and their technical support, in particular to Ilson Soares. We thank the Silicon Detector Facility team at Fermilab for being the host lab for the assembly and testing of the detector components. We acknowledge support from Ministério da Ciencia, Tecnologia e Inovação (MCTI) and the Brazilian funding agencies FAPERJ (grant E-26/110.145/2013), CNPQ, and FINEP; México's CONACYT (grant No. 240666), and DGAPA-UNAM (PAPIIT grants IB100413 and IN112213); Argentina's Agencia Nacional de Promoción Científica y Tecnológica (PICT-2014-1225, PICT-2013-2128). We thank Ricardo Galvão and Ronald Shellard for supporting the experiment.

Journal of Physics: Conference Series **761** (2016) 012057 doi:10.1088/1742-6596/761/1/012057

References

[1] D. Z. Freedman, Phys. Rev. D9, 1389 (1974).
[2] Barranco J, Miranda O and Rashba TI, 2005, *J. High Energy Phys.* JHEP12 (2005) 21.
[3] Cushman P., *et al.* Snowmass CF1 Summary: WIMP Dark Matter Direct Detection (2013) [arXiv:1310.8327].
[4] J.R.T. de Mello Neto *et al.*, [DAMIC Collaboration], [arXiv:1510.02126].
[5] J.S. Tiffenberg, Proceedings of the ICRC 2013 [arXiv:1310.6688].
[6] A. Aguilar-Arevalo *et al.* [DAMIC Collaboration], JINST 10 (2015), no. 8, P08014 [arXiv:1506.02562].
[7] J. Tiffenberg, https://kicp-workshops.uchicago.edu/2015-lowecal/depot/talk-tiffenberg-javier.pdf (2015).
[8] A. Chavarria, https://kicp-workshops.uchicago.edu/2015-lowecal/depot/talk-chavarria-alvaro.pdf (2015).
[9] G. Fernandez Moroni, *et al.*, Phys. Rev. D **91**, 072001 (2015).
[10] J. Lindhard, V. Nielsen, M. Scharf, and P.V. Thomsen, Mat. Fys. Medd. K. Dan. Vidensk. Selsk **33**, 10 (1963).
[11] G. Heusser, Annu. Rev. Nucl. Part. Sci. 45, 543 (1995).
[12] A. Aguilar-Arevalo *et al.* [CONNIE Collaboration], JINST 11 (2016) no. 07, P07024 [arXiv:1604.01343].

Journal of Physics: Conference Series **761** (2016) 012058 doi:10.1088/1742-6596/761/1/012058

Emergence of Yang Mills theory from the Non-Abelian Nambu Model

C. A. Escobar[1] **and L. F. Urrutia**[1]

[1] Instituto de Ciencias Nucleares, Universidad Nacional Autónoma de México, A. Postal 70-543, 04510 México D.F., México

E-mail: `carlos.escobar@correo.nucleares.unam.mx`

Abstract.
The equivalence between the Non-Abelian Nambu model (NANM) and Yang Mills theory is proved, after demanding the Gauss laws at some initial time to the first one. Thereby, the Lorentz violation encoded into the constraint that defines the NANM is physically unobservable. As result, the Goldstone bosons in the NANM arising from the spontaneous symmetry breaking can be identified as the standard gauge fields.

Lorentz invariance violation (LIV) has attracted considerable attention since some theories, such as Quantum Gravity [1] and Strings [2], provide some mechanisms where the fundamental Lorentz symmetry can be broken. Some schemas seek for deviations of Lorentz symmetry experimentally measurable [3], while others attempt to incorporate the concept of LIV in such a way that the physical effects are unobservable. As an example of the last idea, the Nambu model [4] is motivated by means of Lorentz violation arising from a spontaneous symmetry breaking. Perturbative and nonperturbative calculations show that, under some conditions, the Nambu model is equivalent to the standard Electrodynamics [5]. In this way, the Nambu model provides a dynamical interpretation of the corresponding massless gauge bosons in terms of the Goldstone bosons (GB). Additional previous works in Yang Mills theories and gravitation can be found in Refs. [6] and Refs. [7], respectively.

In the present work, following a Hamiltonian analysis we show a nonperturbative equivalence between the Non-Abelian Nambu model and Yang Mills (YM) theory. To prove the foregoing equivalence we have to deal with the issues: i) the number of degrees of freedom (DOF) in the NANM is larger than that of the Yang Mills theory, ii) the NANM is not gauge invariant, iii) the equations of motion of the NANM and Yang Mills theory do not match and iv) current conservation does not hold in the NANM. The equivalence will be proved showing that by means of a suitable canonical transformation of the NANM variables plus the Gauss laws as initial conditions, both the Hamiltonian and canonical algebra of the NANM are the same as those describing the Yang Mills theory. The appendix includes a review of the Hamiltonian analysis of Yang Mills theory, which we use as a benchmark to state the proposed equivalence.

1. The Non-Abelian Nambu model

The Non-Abelian Nambu model is defined by the Lagrange density

$$\mathcal{L}(A^a_\mu) = -\frac{1}{4}F^a_{\mu\nu}F^{a\mu\nu} - J^{a\mu}A^a_\mu, \tag{1}$$

plus the condition

$$A^a_\mu A^{a\mu} = n^2 M^2, \qquad M^2 > 0, \quad \mu = 0,1,2,3, \quad a = 1,...,N. \tag{2}$$

Here, n_μ is a properly oriented constant vector, such that $n^2 = \pm 1, 0$, while M is the proposed scale associated with the spontaneous Lorentz symmetry breaking (SLSB) and $\mu, \nu = 0,1,2,3$ and $a = 1,2,...,N$ indices in the Lorentz and gauge group with N generators, respectively. The constraint (2) can be understood as providing a nonzero vacuum expectation value $\langle A_\mu \rangle = n_\mu M$, which produces spontaneous symmetry breaking of Lorentz invariance and the appearing of Goldstone bosons via the Goldstone theorem.

A convenient way to deal with the different cases of the NANM is by means of the parametrization

$$A^a_0 = B^a \left(1 + \frac{N}{4B^2}\right), \qquad A^a_3 = B^a \left(1 - \frac{N}{4B^2}\right), \tag{3}$$

with

$$N = \left(A^b_{\bar{\imath}} A^b_{\bar{\imath}} + n^2 M^2\right), \quad 4B^2 \pm N \neq 0, \qquad \bar{\imath} = 1,2, \tag{4}$$

which certainly satisfies the condition (2) and it is written in terms of the $3N$ independent GB $B^a, A^b_{\bar{\imath}}$. Notice that n^2 can take the values ± 1 and 0.

After the substitution of (3) in the Lagrangian density (1), the variation of the corresponding action with respect to $B^a, A^b_{\bar{\imath}}$ yields the equations of motion

$$\delta A^a_{\bar{\imath}} : \quad \mathcal{E}^{\bar{\imath}a} + \frac{B^b}{2B^2}\left[\mathcal{E}^{0b} - \mathcal{E}^{3b}\right] A^a_{\bar{\imath}} = 0, \tag{5}$$

$$\delta B^a : \quad \left(\left(1 + \frac{N}{4B^2}\right)\delta^{ab} - \frac{N}{4B^2}\frac{2B^b B^a}{B^2}\right)\mathcal{E}^{0b} + \left(\left(1 - \frac{N}{4B^2}\right)\delta^{ab} + \frac{N}{4B^2}\frac{2B^b B^a}{B^2}\right)\mathcal{E}^{3b} = 0, \tag{6}$$

with the notation

$$\mathcal{E}^{\nu a} = \left(D_\mu F^{\mu\nu} - J^\nu\right)^a. \tag{7}$$

Let us recall that in the case of the $SO(N)$ Yang Mills theory the equations of motion are just given by $\mathcal{E}^{\nu a} = 0$. The equations of motion (5) and (6) do not imply current conservation $D_\nu J^{\nu a} = 0$, basically because the condition (2) breaks the gauge invariance. A way to recover the Yang Mills equations of motion together with gauge invariance is to impose the Gauss laws $\mathcal{E}^{0a} = 0$. In this way, under the conditions $4B^2 \pm N \neq 0$, Eq. (6) yields the solution $\mathcal{E}^{3b} = 0$. These two conditions in (5) provide the final set $\mathcal{E}^{\bar{\imath}a} = 0$.

In order to unify the notation when going to the Hamiltonian formulation we introduce the $3N$ DOF Φ^a_A, $A = 1,2,3$,

$$\Phi^a_1 = A^a_1, \quad \Phi^a_2 = A^a_2, \quad \Phi^a_3 = B^a, \tag{8}$$

in such a way that the coordinate transformation

$$A^a_i = A^a_i(\Phi^b_A), \tag{9}$$

arising from (3) is invertible. The relevant property of the transformation (9) is that

$$\dot{A}^a_i = \frac{\partial A^a_i}{\partial \Phi^b_B}\dot{\Phi}^b_B, \quad \rightarrow \quad \frac{\partial \dot{A}^a_i}{\partial \dot{\Phi}^b_B} = \frac{\partial A^a_i}{\partial \Phi^b_B}, \tag{10}$$

together with the invertibility of the velocities

$$\dot{\Phi}^a_A = \frac{\partial \Phi^a_A}{\partial A^b_i}\dot{A}^b_i. \tag{11}$$

In the following we will not require the explicit form of the transformations (3), but only its generic form (9), together with the property that this transformation can be inverted.

Next we proceed to calculate the Hamiltonian density of the NANM in terms of the canonically conjugated variables Φ^b_A, Π^b_A and employing a procedure that allows to make direct contact with both the YM Hamiltonian density (A.8) and the YM canonical algebra (A.9). After the substitutions (9) together with $A^a_0 = A^a_0(\Phi^b_A)$ are made, the Lagrangian density (1) can be splitted as

$$\mathcal{L}_{\rm NANM}(\Phi, \dot{\Phi}) = \frac{1}{2}E^a_i E^a_i - \frac{1}{2}B^a_i B^a_i - J^{a\mu} A^a_\mu, \tag{12}$$

where

$$E^a_i = \dot{A}^a_i - D_i A^a_0, \qquad B^a_i = \frac{1}{2}\epsilon_{ijk} F^a_{jk}, \tag{13}$$

with $E^a_i = E^a_i(\Phi, \dot{\Phi})$, $B^a_i = B^a_i(\Phi)$. The canonically conjugated momenta are calculated as

$$\Pi^a_A = \frac{\partial \mathcal{L}_{\rm NANM}(\Phi, \dot{\Phi})}{\partial \dot{\Phi}^a_A} = E^b_i \frac{\partial \dot{A}^b_i}{\partial \dot{\Phi}^a_A} = E^b_i \frac{\partial A^b_i}{\partial \Phi^a_A},$$

according to (10). The inverse of Eqs. (9) allows us to write the colored electric fields E^a_i as functions of the momenta Π^b_A of the NANM

$$E^b_i(\Phi, \Pi) = \frac{\partial \Phi^a_A}{\partial A^b_i} \Pi^a_A. \tag{14}$$

It can be proved that the Wronskian of the system is

$$\det\left(\frac{\partial^2 \mathcal{L}_{\rm NANM}(\Phi, \dot{\Phi})}{\partial \dot{\Phi}^a_A \partial \dot{\Phi}^b_B}\right) = \det\left(\frac{\partial \Pi^a_A}{\partial \dot{\Phi}^b_B}\right) = \det\left(\frac{\partial \dot{A}^c_i}{\partial \dot{\Phi}^b_B}\frac{\partial A^c_i}{\partial \Phi^a_A}\right) = \det\left(\frac{\partial A^c_i}{\partial \Phi^b_B}\frac{\partial A^c_i}{\partial \Phi^a_A}\right) \neq 0. \tag{15}$$

In this way, the NANM is exhibited as a regular system in the parameterization (3), so that no constraints are present.

The NANM Hamiltonian density is

$$\mathcal{H}_{\rm NANM} = \Pi^a_A \dot{\Phi}^a_A - \left(\frac{1}{2}E^a_i E^a_i - \frac{1}{2}B^a_i B^a_i - J^{a\mu} A^a_\mu\right), \tag{16}$$

which we rewrite in successive steps as

$$\mathcal{H}_{\rm NANM}(\Phi, \Pi) = \frac{1}{2}E^a_i E^a_i + \frac{1}{2}B^a_i B^a_i - \left(D_i E^b_i - J^{b0}\right) A^b_0 + J^{ai} A^a_i, \tag{17}$$

where we have used the relations (11), (13) and (14), together with an integration by parts in the term containing the covariant derivative. The dependence of $\mathcal{H}_{\rm NANM}$ upon the canonical variables Φ, Π is clearly established by the change of variables (9) and (14). The NANM canonical variables satisfy the standard PB algebra

$$\left\{\Phi^a_A(\mathbf{x}), \Phi^b_B(\mathbf{y})\right\} = 0, \quad \left\{\Pi^a_A(\mathbf{x}), \Pi^b_B(\mathbf{y})\right\} = 0, \quad \left\{\Phi^a_A(\mathbf{x}), \Pi^b_B(\mathbf{y})\right\} = \delta^{ab}\delta_{AB}\delta^3(\mathbf{x}-\mathbf{y}). \tag{18}$$

Now we can consider the NANM Hamiltonian density (17) from the perspective of the fields A^a_i, E^a_i. The following relation arising from the velocity dependent term of NANM Hamiltonian action

$$\int d^4x\, \Pi^a_A \dot{\Phi}^a_A = \int d^4x\, E^a_i \dot{A}^a_i = \int d^4x\, (-E^{ai}) \dot{A}^a_i, \tag{19}$$

establishes $(-E^{ai})$ as the canonically conjugated momenta of A_i^a. In this way (17) can be read as a Hamiltonian density $\mathcal{H}(A,E)$ obtained from $\mathcal{H}_{NANM}(\Phi,\Pi)$ via the substitution of the phase space transformations

$$(\Phi,\Pi)\rightarrow(A,E), \tag{20}$$

which follow from the inverses of Eqs.(9) and (14) plus the relation

$$A_0^a=\frac{A_3^a}{\sqrt{A_3^bA_3^b}}\left(\sqrt{A_i^bA_i^b+n^2M^2}\right), \tag{21}$$

in terms of the new variables. But, since the transformations (14) are generated by the change of variables (9) in the coordinate space, we know from classical mechanics that the full transformation in the phase space is a canonical transformation. In this way we automatically recover the PB algebra

$$\left\{A_i^a(\mathbf{x}),A_j^b(\mathbf{y})\right\}=0,\quad \left\{E^{ai}(\mathbf{x}),E^{bj}(\mathbf{y})\right\}=0,\quad \left\{A_i^a(\mathbf{x}),E^{bj}(\mathbf{y})\right\}=-\delta^{ab}\delta_i^j\delta^3(\mathbf{x}-\mathbf{y}) \tag{22}$$

from Eq. (18). Summarizing, from each Hamiltonian version of the NANM, defined by the different values of n^2, we can regain, via a canonical transformation, the Hamiltonian density (17) together with the canonical algebra (A.9). The Hamiltonian density (17) differs from the YM Hamiltonian density (A.8) only in the fact that the Gauss laws $\Omega^b=\left(D_iE_i^b-J^{b0}\right)=0$ do not appear as first class constraints, because A_0^a are not arbitrary Lagrange multipliers, but functions of the coordinates as shown in Eq. (21). To deal with this issue we study the time evolution of the Gauss functions under the dynamics of the NANM, starting from the Hamiltonian density (17). After some calculations, in the NANM we find [8]

$$\dot{\Omega}^a=-gC^{abc}A_0^b\Omega^c-D_\mu J^{\mu a}+D_3\left(\frac{A_0^1}{A_3^1}\Omega^a\right)-D_3\left(\frac{A_3^a}{A_3^cA_3^c}A_0^b\Omega^b\right)+D_i\left(\frac{A_i^a}{N}A_0^b\Omega^b\right), \tag{23}$$

where A_0^b is given by Eq. (21). It follows that, imposing the Gauss constraints as initial conditions $(\Omega^a(t=t_0)=0)$ upon Eqs. (5)-(6), the standard Yang-Mills equations of motion $(\mathcal{E}^{\nu a}=0)$ are recovered and they are valid at $t=t_0$. As a consequence of the antisymmetry of Maxwell tensor, the relation

$$0=D_\nu\mathcal{E}^{\nu a}=D_\nu(D_\mu F^{\mu\nu}-J^\nu)^a=-D_\nu J^{\nu a}=0,\qquad (\text{at } t=t_0), \tag{24}$$

holds, yielding current conservation at $t=t_0$. Using $\Omega^a(t=t_0)=0$ and $D_\nu J^{\nu a}|_{t=t_0}=0$ in Eq. (23), we obtain $\dot{\Omega}^a(t=t_0)=0$. Since the relations $\Omega^a(t=t_0)=0$ and $\dot{\Omega}^a(t=t_0)=0$ are fulfilled, we obtain

$$\begin{aligned}\Omega^a(t=t_0+\delta t_1)&=\Omega^a(t=t_0)+\dot{\Omega}^a(t=t_0)\delta t_1+\cdots,\\&=0.\end{aligned} \tag{25}$$

Using that $\Omega^a(t=t_0+\delta t_1)=0$, we observe that the Yang Mills equations are now valid at $t=t_0+\delta t_1$, which again, due to antisymmetry of the Maxwell tensor, imply current conservation at $t=t_0+\delta t_1$ and $\dot{\Omega}^a(t=t_0+\delta t_1)=0$ via Eq. (23). The relations $\Omega^a(t=t_0+\delta t_1)=0$ and $\dot{\Omega}^a(t=t_0+\delta t_1)=0$ imply $\Omega^a(t=t_0+\delta t_1+\delta t_2)=0$. Iterating the previous process, it follows that the Gauss laws and current conservation are now valid for all time. Summarizing, the above analysis shows that in the case of the Non-Abelian Nambu model, the specific gauge structure of the theory allows us to impose only the Gauss constraints as initial conditions, which necessarily yield current conservation for all time. In this way we can recover the $SO(N)$ Yang-Mills theory by imposing the Gauss laws as Hamiltonian constraints, with arbitrary functions N^a adding $-N^a\Omega^a$ to $\mathcal{H}_{\rm NANM}$ and redefining $A_0^a + N^a = \Theta^a$. This leads to

$$\mathcal{H}_{\rm NANM} = \frac{1}{2}(\mathbf{E}^2 + \mathbf{B}^2) - \Theta^a\Omega^a + J_i^a A^{ia}, \tag{26}$$

where Θ^a are now arbitrary functions, thus getting back to the YM Hamiltonian density (A.8) plus the canonical algebra (A.9). It is worth to mention that the imposition of the N Gauss laws as first class constraints in the NANM Hamiltonian reduces the $3N$ degrees of freedom of the NANM to $2N$, which are the number of degrees of freedom of the Yang Mills theory.

2. Summary

A nonperturbative equivalence between the Yang Mills theory and the Non-Abelian Nambu model (NANM) has been established, after the Gauss laws are imposed as initial conditions for the latter. The possible interpretation of gauge particles (i.e. photons and gravitons, for example) as the Goldstone bosons (GB) modes arising from some spontaneous symmetry breaking is an interesting hypothesis which would provide a dynamical setting to the gauge principle. In this work we have taken the Nambu approach in which the spontaneous Lorentz symmetry breaking is incorporated in an effective way in the model, by means of a non-linear constraint. The challenge posed by this setting is to show under which conditions, the violations of Lorentz symmetry and gauge invariance, introduced by the non-linear constraint, are unobservable in such a way that the appearing Goldstone bosons can be interpreted as the gauge particles of an unbroken gauge theory. In the case of the NANM we have proved that such conditions are the Gauss laws just as initial conditions. The strategy employed to prove the equivalence between both theories was to show that, by means of a suitable transformation of the NANM variables plus the Gauss laws as initial conditions, the Hamiltonian and canonical algebra of the NANM are reduced to their corresponding in the Yang Mills theory, in such a way that both theories are undistinguishable.

Acknowledgments

L. F. U has been partially supported by the project CONACyT # 237503. C. A. E. and L. F. U acknowledge support from the project UNAM (Dirección General de Asuntos del Personal Académico) # IN104815.

Appendix A: The SO(N) Yang-Mills (YM) theory

We present a brief review of the Hamiltonian formulation of the standard $SO(N)$ Yang-Mills theory. The Yang- Mills Lagrangian density is given by

$$\mathcal{L} = Tr\left[-\frac{1}{4}\mathbf{F}_{\mu\nu}\mathbf{F}^{\mu\nu} - \mathbf{J}^\mu\mathbf{A}_\mu\right], \tag{A.1}$$

where boldfaced quantities denote matrices in the Lie algebra of the internal symmetry group $SO(N)$ with $N(N-1)/2$ generators t^a; i.e. $\mathbf{M} = M^a t^a$. This algebra is generated by $\left[t^a, t^b\right] = C^{abc}t^c$, where the structure constants C^{abc} are completely antisymmetric.

The canonical Hamiltonian density arising from the Lagrange density (A.1) is given by

$$\mathcal{H} = \Pi_a^i \dot{A}_i^a - \mathcal{L} = \frac{1}{2}(\mathbf{E}^2 + \mathbf{B}^2) - A_0^a(D_iE_i - J^0)^a - J_i^a A_i^a. \tag{A.2}$$

We employ the Dirac's method to construct the canonical theory due to the fact that primary constraints

$$\Sigma^a = \Pi_0^a \simeq 0, \tag{A.3}$$

are present. The extended Hamiltonian density is

$$\mathcal{H}_E = \frac{1}{2}(\mathbf{E}^2 + \mathbf{B}^2) - A_0^a(D_iE_i - J^0)_a + J_i^a A_a^i + \lambda^a \Sigma_a, \tag{A.4}$$

where λ^a are arbitrary functions. The evolution condition of the primary constraints $\dot{\Sigma}^a(\mathbf{x}) = \{\Sigma^a, \mathcal{H}_E\}$ leads to the Gauss laws

$$\Omega^a = (D_iE_i - J^0)^a \simeq 0. \tag{A.5}$$

It is not difficult to prove that (A.3) and (A.5) are the only constraints present and that they constitute a first class set. In fact, calculating the time evolution of Ω^a yields

$$\dot{\Omega}^a = -\partial_0 J_0^a - C^{abc}A_0^b J_0^c - C^{abc}A_0^b\Omega^c - D_k J^{ak} = -C^{abc}A_0^b\Omega^c - D_\mu J^{a\mu}, \tag{A.6}$$

which is zero, modulo the constraints and using current conservation.

Normally one fixes

$$\Pi_0^a \simeq 0, \qquad A_0^a \simeq \Theta^a, \tag{A.7}$$

with Θ^a being arbitrary functions to be consistently determined after the remaining first class constraints Ω^a are fixed.

The final Hamiltonian density is

$$\mathcal{H}_E = \frac{1}{2}(\mathbf{E}^2 + \mathbf{B}^2) - \Theta^a(D_iE_i - J^0)_a + J_i^a A_a^i. \tag{A.8}$$

Once Π_0^a and A_0^a are fixed strongly, the Dirac brackets among the remaining variables are

$$\{A_i^a(\mathbf{x},t), A_j^b(\mathbf{y},t)\}^* = 0, \qquad \{E^{ai}(\mathbf{x},t), E^{bj}(\mathbf{y},t)\}^* = 0, \qquad \{A_i^a(\mathbf{x},t), E^{bj}(\mathbf{y},t)\}^* = -\delta_i^j\delta^{ab}\delta(\mathbf{x}-\mathbf{y}). \tag{A.9}$$

The final count of degrees of freedom (DOF) per point in coordinate space yields

$$\#DOF = \frac{1}{2}(2\times 4N - 2\times 2N) = 2N. \tag{A.10}$$

At this point, the dynamics of Yangs Mills is determined by the Hamiltonian (A.8) together with the canonical algebra (A.9). The goal to obtain the proposed equivalence, between YM theory and the NANM, will be to obtain the same Hamiltonian and canonical algebra from the NANM, after the Gauss laws are imposed as initial conditions.

References

[1] Gambini R and Pullin J 1999 *Phys. Rev* D **59** 124021; Alfaro J, Morales-Técotl H A and Urrutia L F 2002 *Phys. Rev* D **65** 103509; Alfaro J, Morales-Técotl H A and Urrutia L F 2000 *Phys. Rev. Lett.* **84** 2318; Carroll S M, Harvey J A, Kostelecký V A, Lane C D and Okamoto T 2001 *Phys. Rev. Lett.* **87** 141601; Amelino-Camelia G, Ellis J R, Mavromatos N E, Nanopoulos D V and Sarkar S 1998 *Nature* **393** 763; Magueijo J and Smolin L 2003 *Phys. Rev* D **67** 044017

[2] Kostelecký V A and Samuel S 1989 *Phys. Rev* D **39** 683; Kostelecký V A and Potting R 1991 *Nucl. Phys.* B **359** 545

[3] Colladay D and Kostelecký V A 1997 *Phys. Rev* D **55** 6760; Colladay D and Kostelecký V A 1998 *Phys. Rev* D **58** 116002; Kostelecký V A 2004 *Phys. Rev* D **69** 105009

[4] Nambu Y 1968 *Suppl. Prog. Theor. Phys.* **E68** 190 (1968)

[5] Escobar C A and Urrutia L F 2015 *Phys. Rev* D **92** 025042; Franca O J, Montemayor R and Urrutia L F 2012 *Phys. Rev* D **85** 085008; Azatov A T and Chkareuli J L 2006 *Phys. Rev* D **73** 065026

[6] Chkareuli J L and Kepuladze Z R 2007 *Phys. Lett.* B **644** 212; Chkareuli J L and Jejelava J G 2008 *Phys. Lett.* B **659** 754; Chkareuli J L, Froggatt C D and Nielsen H B 2009 *Nucl. Phys.* B **821** 65; Chkareuli J L, Froggatt C D, Jejelava J G and Nielsen H B 2008 *Nucl. Phys.* B **796** 211

[7] Chkareuli J L, Jejelava J G, and Tatishvili G 2011 *Phys. Lett.* B **696** 124

[8] Escobar C A and Urrutia L F 2015 *Phys. Rev* D **92** 025013

Journal of Physics: Conference Series **761** (2016) 012059 doi:10.1088/1742-6596/761/1/012059

Study of the semileptonic $\tau^- \to \pi^-\pi^0\nu_\tau$ and $\tau^- \to \pi^-\pi^0\ell^+\ell^-\nu_\tau$ ($\ell = e$, μ) decays.

J. L. Gutiérrez Santiago.

Departamento de Física, Cinvestav-IPN, Apdo. Postal 14-740, 07000 Ciudad de México, México.

E-mail: jlgutierrez@fis.cinvestav.mx

Abstract. We analyze the decay modes $\tau^- \to \pi^-\pi^0\nu_\tau$ and $\tau^- \to \pi^-\pi^0\ell^+\ell^-\nu_\tau$, $\ell = e$, μ. The first one is studied in order to show that the three resonances $\rho(770)$, $\rho(1450)$ and $\rho(1700)$ are necessary to fix the model vector dominance parameters on the two pion system dynamics. We have obtained $\mathcal{BR}(\tau^- \to \pi^-\pi^0\nu_\tau)$=(25.40±0.22)% which is in good agreement with the world average reported in the PDG. The second decay, which is the target of our study, is important because it can participate as a background in processes of lepton flavour and lepton number violation and thus its accurate description is essential in searches for such processes. It is also important to test the model dependence of the most important ($\pi\pi$) contribution to the hadron vacuum polarization (HVP) entering the muon anomalous magnetic moment, if evaluated using hadronic tau decay data. In these preliminary results, we have only taken into account the total Inner Bremmsstrahlung (IB) contribution to the $\tau^- \to \pi^-\pi^0\ell^+\ell^-\nu_\tau$ decays obtaining $\mathcal{BR}(\tau^- \to \pi^-\pi^0 e^+e^-\nu_\tau)$=(8.09 ±0.26) $\times 10^{-5}$ and $\mathcal{BR}(\tau^- \to \pi^-\pi^0\mu^+\mu^-\nu_\tau)$= (2.094 ±0.004) $\times 10^{-6}$. These results have been obtained using the VEGAS routine to integrate numerically the phase space.

1. Introduction

We report on a preliminary study of the $\tau^- \to \pi^-\pi^0\ell^+\ell^-\nu_\tau$ ($\ell = e$, μ) decays [1] where we have considered only the IB contribution. The inclusion of structure dependent effects is in progress. $\tau^- \to \pi^-\pi^0\nu_\tau$ decays involve the interplay of strong and weak interactions. Since the latter are under well theoretical control, these processes offer a clean environment to study QCD currents in the GeV region. These decays proceed through the hadronization of the vector piece of the quark current coupling to the W boson. This is dominated by the creation of intermediate particles (resonances) whose masses corresponds to the $\rho(770)$, $\rho(1450)$ and $\rho(1700)$ states. There must be conservation of certain quantum numbers (like isospin, G-parity and C-parity) in the process. In the $\tau^- \to \pi^-\pi^0\nu_\tau$ decays, the isospin and G-parity have the values $I = 1$, $G = +$, corresponding to the properties of the weak vector current that produces. We note that while the branching ratio can be reproduced just with the $\rho(770)$ resonance contribution, the invariant mass spectrum requires at least three resonances for its proper description, as shown in Figure 2. This decay mode dominates the τ lepton width, with a branching ratio (25.52 $\pm$ 0.09)% [2]. As a result, if using tau data to evaluate the HVP to a_μ, this decay mode becomes essential in the precise determination demanded by accuracy of the measurement of a_μ. According to the vector dominance model the form factor can be parametrized in terms of Breit-Wigner functions which correspond to the three mentioned resonances. Model parameters are fixed using data on

Journal of Physics: Conference Series **761** (2016) 012059 doi:10.1088/1742-6596/761/1/012059

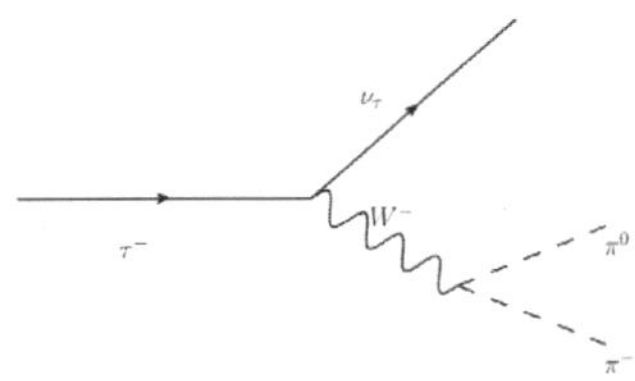

Figure 1. Feynman diagram of $\tau^- \to \nu_\tau \pi^- \pi^0$ decay.

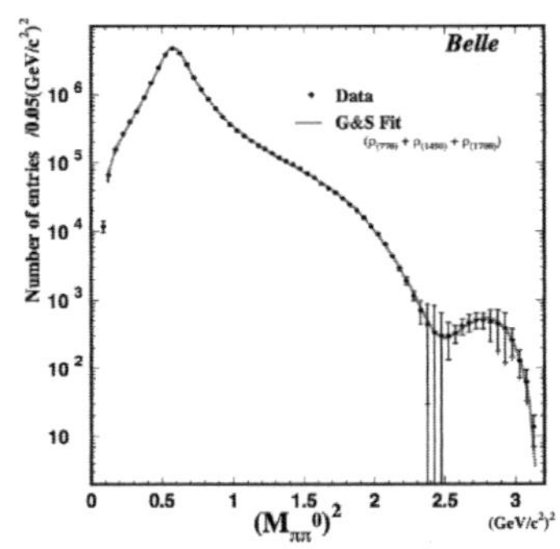

Figure 2. Invariant mass spectrum of the $\pi^-\pi^0$ system. The peaks of the $\rho(770)$, $\rho'(1450)$ and $\rho''(1700)$ resonances are shown [3].

the invariant mass spectrum of $\pi^-\pi^0$ system [3].

As our main result we predict the $\mathcal{BR}(\tau^- \to \nu_\tau \pi^- \pi^0 \ell^+ \ell^-)$ using the same simplified form factor tested in the $\tau^- \to \nu_\tau \pi^- \pi^0$ channel. The evaluation of structure-dependent contributions to these decays is in progress [1].

2. The $\tau^- \to \pi^- \pi^0 \nu_\tau$ decay.

The leading Feynman Diagram contributing to this decay is shown in Figure 1. We use the following momenta convention in this decay:

$$\tau^-(P) \to \nu_\tau(q) \pi^-(P_-) \pi^0(P_0) \tag{1}$$

The effective Hamiltonian is given by

$$\mathcal{H}_{eff} = \frac{G_F}{\sqrt{2}} \left[V_{ud} \bar{d} \gamma_\mu (1-\gamma_5) u \right] \left[\bar{\nu} \gamma^\mu (1-\gamma_5) \tau \right] + h.c., \tag{2}$$

where the Cabibbo-Kobayashi-Maskawa matrix element is $|V_{ud}| = 0.97425 \pm 0.00022$ and $G_F = 1.1663787(6)$ x 10^{-5}GeV^{-2} is the Fermi constant [2]. Thus the invariant amplitude is given by

$$\mathcal{M}(\tau^- \to \nu_\tau \pi^- \pi^0) = \frac{G_F V_{ud}^*}{\sqrt{2}} \langle \nu_\tau | \bar{\nu} \gamma_\mu (1-\gamma_5) \tau | \tau^- \rangle \left\langle \pi^- \pi^0 \left| \bar{d} \gamma^\mu (1-\gamma_5) u \right| 0 \right\rangle.$$

The axial part of the hadronic current (γ_5) does not contribute because this term does not conserve G-parity, thus $\left\langle \pi^- \pi^0 \left| \bar{d} \gamma^\mu \gamma_5 u \right| 0 \right\rangle = 0$. A general parametrization of the hadronic current is $\left\langle \pi^- \pi^0 \left| \bar{d} \gamma_\mu u \right| 0 \right\rangle = f_+(t)(P_- - P_0)_\mu + f_-(t)(P_- + P_0)_\mu$. However, in this work, we use the alternative form given by

$$\langle \pi^- \pi^0 \left| \bar{d} \gamma_\mu u \right| 0 \rangle = f_+(t) \left[(P_- - P_0)_\mu - \frac{\Delta^2}{t}(P_- + P_0)_\mu \right] + f_0(t) \frac{\Delta^2}{t} (P_- + P_0)_\mu, \tag{3}$$

where $\Delta^2 = m_{\pi^-}^2 - m_{\pi^0}^2$. The reason for using this basis is that this is orthogonal since f_+ and f_0 are related with the $J^P = 1^-, 0^+$ configurations, respectively, of the pionic system (while this does not happen with f_+ and f_-). The invariant mass spectrum distribution of the pionic system is

$$\frac{d\Gamma(\tau^- \to \pi^-\pi^0\nu)}{dt} = \frac{G_F^2 \left|V_{ud}\right|^2 m_\tau^3 S_{EW}}{6(4\pi)^3}\beta^3(t)\left[1-\frac{t}{m_\tau^2}\right]^2 \left|f_+(t)\right|^2 \left(1+\frac{2t}{m_\tau^2}\right), \tag{4}$$

where $t=(P_-+P_0)^2=(P-q)^2$, $\beta=[\lambda^{1/2}(t,m_{\pi^-}^2,m_{\pi^0}^2)]/t$, $\lambda(t,m_{\pi^-}^2,m_{\pi^0}^2)=t^2+m_{\pi^-}^4+m_{\pi^0}^4-2tm_{\pi^-}^2-2tm_{\pi^0}^2-2m_{\pi^-}^2m_{\pi^0}^2$. $S_{EW}=1.0201\pm0.0003$ is electroweak radiative correction factor at short distances [4]. For the form factor we use a parametrization in agreement with the vector dominance model [5]

$$f_+(t) = \frac{1}{1+\beta e^{i\phi}+\gamma}\left[BW_\rho + \beta e^{i\phi}BW_{\rho'}(t) + \gamma BW_{\rho''}(t)\right], \tag{5}$$

here BW_ρ, $BW_{\rho'}$ and $BW_{\rho''}$ are Breit-Wigner functions for the resonances of the ρ meson and its excitations. β, γ and ϕ are real parameters. For $BW_{\rho'}$ and $BW_{\rho''}$ we have used $\Gamma_{\rho'}$ and $\Gamma_{\rho''}$ as constants. The masses m_τ, m_{π^-}, m_{π^0} and mean life of τ lepton are given in PDG [2] while the parameters m_ρ, Γ_ρ, $m_{\rho'}$, $\Gamma_{\rho'}$, $m_{\rho''}$, $\Gamma_{\rho''}$, β, γ and ϕ are given in reference [6] for the fit to the Belle Spectrum. With this parameters we have obtained the branching ratio

$$\mathcal{B}(\tau \to \pi^-\pi^0\nu_\tau) = (25.40\pm0.22)\%, \tag{6}$$

Comparing this result with the PDG value [2], $\mathcal{B}(\tau \to \pi^-\pi^0\nu_\tau) = (25.52\pm0.09)\%$, there is a difference of 0.13 %, which indicates that the result is a very good approximation as expected because the parameters were obtained fitting Belle data. In latter stages of this study, a dispersive representation of this form factor [7] (consistent with elastic unitarity and analiticity) will be used.

3. The $\tau^- \to \pi^-\pi^0\ell^+\ell^-\nu_\tau$ decay.

This decay is interesting for at least three reasons: it is a prominent background in searches of Lepton Flavour Violation in the charged sector, and of Lepton Number Violation ($\tau^- \to \pi^-\pi^0\ell^+\ell^-\nu_\tau$ can emulate the signal of processes of the kind $L^- \to \ell^-\ell'^+\ell'^-$). It is also very relevant to verify the radiative corrections used in the computation of the HVP contribution to a_μ using τ hadronic decay data. We use the following convention for the involved momenta

$$\tau(P) \to \pi^-(p_-) + \pi^0(p_0) + \ell^+(p_{\ell^+}) + \ell^-(p_{\ell^-}) + \nu_\tau(q), \tag{7}$$

where $\ell = e,\ \mu$. The matrix element is given by

$$\begin{aligned} T &= eG_FV_{ud}\frac{l^\mu}{k^2}\{F_\nu\overline{u}(q)\gamma^\nu(1-\gamma_5)(M_\tau+\not{P}-\not{k})\gamma_\mu u(P)+ \\ &\quad (V_{\mu\nu}-A_{\mu\nu})\overline{u}(q)\gamma^\nu(1-\gamma_5)u(P)\}, \end{aligned} \tag{8}$$

with $l^\mu = \overline{u}_{\ell^-}(p_{\ell^-})\gamma^\mu v(p_{\ell^+})$ the leptonic current, $k = p_{\ell^-}+p_{\ell^+}$ the momentum of the virtual photon. The first term describes the Bremmsstrahlung off the initial τ^-. In this part $F_\nu = [(p_- - p_0)_\nu f_+(t)]/(k^2-2P\cdot k)$ and $t=(p_-+p_0)^2$. The form factor $f_+(t)$ rules the non-radiative decay (its parameters have been fixed in the preceding section). The second part of (8) describes the vector and axial-vector model-dependent contributions to the process

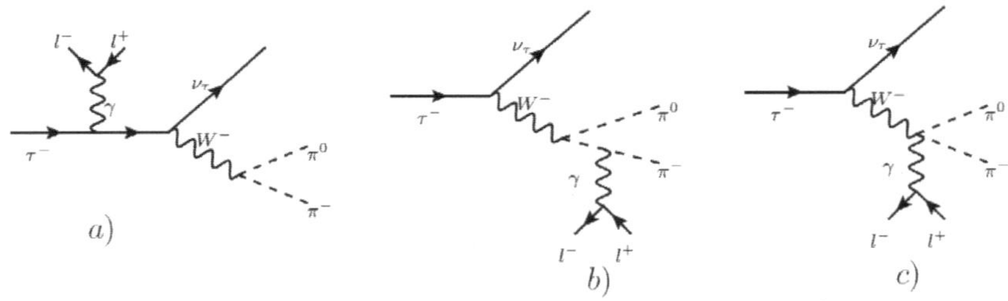

Figure 3. Leading diagrams contributing to the $\tau^- \to \nu_\tau \pi^- \pi^0 \ell^+ \ell^-$ decays. The a) part corresponds to Inner Bremmsstrahlung off the τ^- lepton. The b) diagram correspond to Inner Bremmsstrahlung off the π^- meson. The c) diagram contains both and inner Bremmsstrahlung and a structure-dependent part.

$W^-(P-q) \to \pi^-(p_-)\pi^0(p_0)\gamma(k)$. The hadronic tensor $V_{\mu\nu}$ contains a part that corresponds to Bremmsstrahlung off the π^- in the final state. With respect to the influence of the model-dependent part (V_μ and A_μ) on the branching ratio, it is expected to be minimal for $\ell = e$, but dominant for $\ell = \mu$ [8]. The evaluation of this structure-dependent part is work in progress. Leading Feynman diagrams of the process are shown in Figure 3. The total Inner Bremmsstrahlung amplitude is given by [9]

$$\begin{aligned}\mathcal{M}_{IB} &= \frac{e^2}{k^2} G_F V_{ud}^* l^\mu \left\{ \frac{(p_- - p_0)_\nu}{k^2 - 2P\cdot k} f_+(t) \overline{u}_\nu(q)\gamma^\nu(1-\gamma_5)(\not{P} - \not{k} + M_\tau)\gamma_\mu u_\tau(P) \right. \\ &+ \left[f_+(t')\frac{2p_{-\mu}}{2p_-\cdot k + k^2}(p_- + k - p_0)_\nu - f_+(t')g_{\mu\nu} + \right. \\ & \left.\left. 2\left(\frac{f_+(t') - f_+(t)}{2(p_0+p_-)\cdot k + k^2}\right)(p_- + p_0)_\mu (p_0 - p_-)_\nu \right] L^\nu \right\}, \end{aligned} \tag{9}$$

here $t' = (P-q)^2$ and $L^\nu = \overline{u}_\nu(q)\gamma^\nu(1-\gamma_5)u_\tau(P)$. In order to get an easier calculation we have divided the square amplitude summed over polarizations as $\overline{|\mathcal{M}_{IB}|^2} = \overline{|\mathcal{M}_{IB\tau}|^2} + \overline{|\mathcal{M}_{IB\pi}|^2} + \overline{2Re[\mathcal{M}_{IB\tau}\mathcal{M}_{IB\pi}^*]}$. Using these amplitudes we have obtained in $\overline{|\mathcal{M}_{IB\pi}|^2}$ and $\overline{2Re|\mathcal{M}_{IB\tau}\mathcal{M}_{IB\pi}^*|^2}$ terms of the form $\epsilon_{\mu\nu\rho\sigma}q^\mu p_-^\nu p_0^\rho p_{\ell^-}^\sigma$, which we have expressed in terms of Lorentz invariants. For the kinematics of this five body decay we have used the reference [10]. Following this reference we can write all the scalar products $p_i \cdot p_j$ in terms of eight independent variables which we have defined as $s_1 = (P-q)^2$, $s_2 = (P-q-p_-)^2$, $s_3 = (P-q-p_- - p_0)^2$, $u_1 = (P-p_-)^2$, $u_2 = (P-p_0)^2$, $u_3 = (P-p_{\ell^-})^2$, $t_2 = (P-p_- - p_0)^2$, $t_3 = (P-p_- - p_0 - p_{\ell^-})^2$, and the auxiliary variables $s_0 = M^2$, $s_4 = m_{\ell^+}^2$, $u_0 = s_1$ and $t_1 = u_1$.

Using the PDG data [2] for the masses and considering the isospin limit $m_{\pi^-} = m_{\pi^0} = m_\pi$ (since we are considering a radiative process, their inequality is a next-to-leading order effect in $SU(2)$ breaking). Therefore, we take $m_\pi = \left[(2m_{\pi^-}^2 + m_{\pi^0}^2)/3\right]^{1/2}$. As the main result of this analysis, we have obtained the branching fractions shown in Table 1. As it is seen, the IB off the τ yields in bobth cases the smallest (but non-negligible) contribution to the branching ratios. Interference effects between IB off the τ and off the π are very relevant in both decay channels.

4. Conclusions

In this work, we have checked a simple description of the vector form factor of the pion against data. In this way, we have shown that the invariant mass spectrum of the $\pi^-\pi^0$ system is well

Table 1. Results for $\tau^- \to \nu_\tau \pi^- \pi^0 \ell^+ \ell^-$ decay.

Contribution	$\tau^- \to \nu_\tau \pi^- \pi^0 e^+ e^-$	$\tau^- \to \nu_\tau \pi^- \pi^0 \mu^+ \mu^-$
$\overline{\vert\mathcal{M}_{IB\tau}\vert^2}$	(5.960 ± 1.39) x 10^{-6}	(4.947 ± 0.021)x 10^{-7}
$\overline{\vert\mathcal{M}_{IB\pi}\vert^2}$	(3.416 ± 0.211) x 10^{-5}	(7.243 ± 0.139) x 10^{-7}
$2\mathrm{Re}\overline{\vert\mathcal{M}_{IB\tau}\mathcal{M}^*_{IB\pi}\vert}$	(4.075 ± 0.723) x 10^{-5}	(8.756 ± 0.220) x 10^{-7}
Total	(8.087± 0.260) x 10^{-5}	(2.094 ± 0.004) x 10^{-6}

described by three resonances $\rho(770)$, $\rho(1450)$ and $\rho(1700)$ and we have fixed their parameters in $\tau^- \to \pi^- \pi^0 \nu_\tau$decays. The obtained branching ratio agrees with the world average PDG value.

For the $\tau^- \to \pi^- \pi^0 \ell^+ \ell^- \nu_\tau$ decays, the IB contributions to the branching fractions are of the order of magnitude expected. Consequently, these branching ratios are at a measurable level in Belle-II, and therefore they constitute an important background in the search for lepton flavor and lepton number violation in related processes. The considered decays can mimic the $\tau^- \to \mu^- \mu^+ \mu^-$, $\tau^- \to \mu^- e^+ e^-$, $\tau^- \to e^- \mu^+ \mu^-$, $\tau^- \to e^- e^+ e^-$, $\tau^- \to e^- \pi^+ \pi^-$, $\tau^- \to \mu^- \pi^+ \pi^-$, $\tau^- \to e^+ \pi^- \pi^-$, $\tau^- \to \mu^+ \pi^- \pi^-$ decays, with current limits of order 10^{-8}. The results using the dispersive pion form factor and including the structure-dependent contributions will be presented elsewhere.

Acknowledgments
We thank to Cinvestav-IPN, the organizing committee and Conacyt for financial support.

References
[1] J. L. Gutiérrez-Santiago, G. López-Castro and P. Roig, work in progress.
[2] K.A. Olive et al. (Particle Data Group), Chin. Phys. C38, 090001 (2014).
[3] M. Fujikawa et al. (Belle Collab.), Phys. Rev. D78, 072006 (2008).
[4] J. Erler, Rev. Mex. Fis. **50** (2004) 200.
[5] A. Flores-Tlalpa, Ph. D. Thesis, *Modelo de dominancia de mesones para decaimientos semileptónicos de sabores pesados*, Cinvestav (2008).
[6] G. López Castro, P. Roig and G. Toledo Sánchez, Nucl. Part. Phys. Proc. 260, 70 (2015).
[7] D. Gómez Dumm and P. Roig, Eur. Phys. J. C **73** (2013) no.8, 2528.
[8] P. Roig, A. Guevara, and G. López Castro, Phys. Rev. D 88, 033007 (2013).
[9] V. Cirigliano, G. Ecker and H. Neufeld, JHEP **0208** (2002) 002.
[10] R. Kumar, Phys. Rev. Vol. 185, num. 5, 1865 (1969).
[11] G. P. Lepage, J. Comput. Phys. 27, 192 (1978).

Journal of Physics: Conference Series **761** (2016) 012060 doi:10.1088/1742-6596/761/1/012060

Detection of supernova neutrinos via coherent ν-nucleus scattering

Catalina Espinoza[1] and Sergio Palomares-Ruiz[2]

[1]Instituto de Física, UNAM. Apartado Postal 20-364, 01000 México D.F.

[2]Instituto de Física Corpuscular (IFIC); CSIC-Universitat de València; Apartado de Correos 22085; E-46071 Valencia; Spain.

E-mail: `m.catalina@fisica.unam.mx`

Abstract. We consider the possible detection of Supernova neutrinos via Coherent ν-Nucleus Scattering in a scintillation detector. The envisaged applications include the separate determination of the energy and temperature of the neutrinos and antineutrinos of the second and third family by exploiting the spectral information encoded in the recoiling nuclei and inherited from the impinging neutrinos. Our results suggest that despite a significant quenching factor and a high detector threshold, determination of one of the neutrino observables is achievable for a wide range of the parameter space.

1. Introduction

Supernova events are extremely powerful explosions exhibitive of the last stages of the evolution of stars with masses above the Chandrasekhar limit, roughly 1.44 solar masses. Of particular interest to neutrino phenomenology are *core collapse* supernovae, in which a large massive star burns out all its fuel and its interior nucleus rapidly shrinks by gravitational pressure until it reaches nuclear densities. Then, the nucleon degeneracy pressure abruptly stops the collapse and the falling outer parts of the dying star literally bounce back after crashing with the wall of dense inner matter, initiating a series of events, in which neutrinos are suspected to play a prominent role, that trigger the explosion [1].

When a large massive star becomes a core collapse supernova, several solar masses of material end up ejected in the explosion into interstellar space with kinetic energies of the order of 10^{51} erg, which amounts to approximately 10^{30} Hiroshima nuclear bombs. The fingerprints of the ejected material are easily identified even long afterwards from the distance as beautiful shapes known as nebulae. From the perspective of human history, the most prominent feature of a supernova is the accompanying luminous explosion when the star increases its luminosity and becomes, for several days, as bright as its hosting galaxy. Despite the copious amounts of energy involved in these processes, it is estimated that only 0.01% and 1% of the total energy released in a supernova is accounted for, respectively, by the luminous explosion and the ejected material kinetic energy.

Surprisingly, the 99% of the energy in a core collapse supernova event, around 3×10^{53} erg, is released through neutrinos of all flavours. Moreover, the neutrino pulse accompanying such an event only lasts a few seconds and precedes the luminous explosion. In the short time interval of the neutrino pulse, an estimate of 10^{58} neutrinos are emitted from the supernova with an average energy of 10 MeV. This means that the luminosity in neutrinos of a core collapse supernova is comparable to the combined luminosity in photons of all the stars in the visible universe [2].

From a global perspective supernovae occur quite often in the whole universe, approximately 1 event per second. Of course, we are forced due to technological impediments to focus our attention to the immediate local environments. In particular, when we are interested in the detection of supernova neutrinos only galactic supernovae offer viable scenarios for detection of these particles. An exception occured in 1987 when a neutrino signal was obtained in several neutrino detectors operating at the time, correlated with the discovery of the supernova explosion in the Large Magellanic satellite galaxy orbiting the Milky Way. The estimated core-collapse supernova rate in the Galaxy is about 3 times per century, though external effects such as interstellar dust blocking the light of an event might explain the lack of reported observations in the last four centuries, since the last galactic supernova observation in 1604. Present neutrino detectors can easily observe a supernova anywhere in the Galaxy or its immediate satellite companions but do not have large enough volumes to observe one in even the nearest galaxy, some 700 kpc away.

In a supernova event, because different flavours of neutrinos have different interactions with matter, and because the nascent neutron star's temperature is decreasing with increasing radius, the neutrino decoupling temperatures are different. Theoretical and numerical estimates are model dependent, in the usual model the ν_μ and ν_τ neutrinos and their antiparticles have a temperature of about 8 MeV, the $\bar{\nu}_e$ neutrinos about 5 MeV, and the ν_e neutrinos about 3.5 MeV. In the following we shall assume these values in our calculations.

At present most neutrino detectors operate in the charged current detection channel, so it is to be expected that in the event of an actual supernova neutrino signal the properties of the electron neutrino flavour and its antiparticle will be probed with good statistics. It was pointed out some time ago [3] that the neutral current detection channel also offered the possibility of determining separately the energy and temperature of the supernova neutrinos of the second and third generation, for example through the elastic neutrino-proton dispersion channel. In this complementary channel the information on the recoiling proton energy is crucial, which is non-relativistic and therefore invisible on Cerenkov detectors but detectable in scintillation ones. Soon afterwards [4] the same idea was explored with the detection via neutrino-nucleus elastic scattering taking advantage of the very large coherent cross section and the sensitivity to all flavours of neutrinos and antineutrinos.

Backgrounds for supernova neutrinos can be handled relatively easy, mostly because all of the events are concentrated in a few second interval. On the other hand an important obstacle present in these neutral current detection channels is the phenomenon of quenching, which refers to energy losses by ionization. The low energy recoiling particles are highly ionizing in such a way that their scintillating light is reduced (quenched) relative to a minimum ionizing particle like an electron. Because of these energy losses the recoiled particle spectrum is affected significantly. This feature is accounted for by introducing the quenching factor Q_f. As an illustration of the effect of Q_f on the yield of recoiling nuclei in coherent neutrino-nucleus scattering on a Xenon detector we show in Figure 1 how the number of events per ton above an illustrative detector threshold cut is notoriously reduced by the effects of ionization.

This quenching effect is particularly damaging for the electron (anti)neutrino contribution to the neutrino-nucleus elastic scattering channel signal. While in principle this channel is sensitive to all types of neutrinos, the electron neutrinos and antineutrinos cannot be detected in real situations due to their relative low decoupling temperatures. For the same reason, the corrections due to quenching are notoriously strong for the electron neutrino and its antiparticle. That being said, we shall ignore possible *spectral split* effects [5] where oscillation between neutrino types might produce changes that would be perceived as different decoupling temperatures than the ones assumed here.

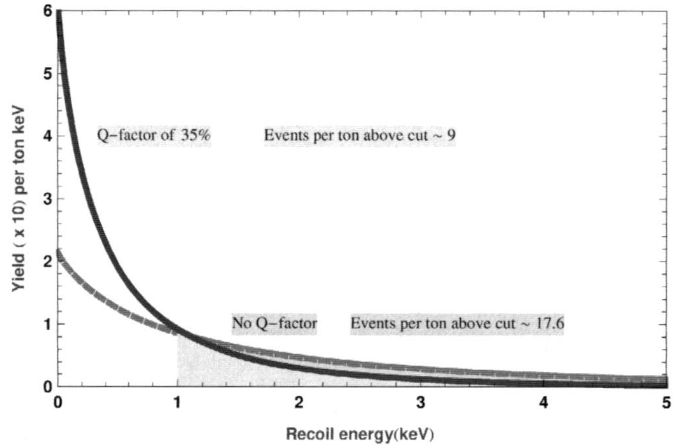

Figure 1. Calculated yield for the elastic neutrino-nucleus scattering channel. Because of energy losses by ionization, the spectrum of recoiling nuclei is affected significantly, this is accounted for by the quenching factor Q_f. The solid (blue) line illustrates the effect of Q_f on the recoiled nuclei spectrum, showing a decrease in the number of events per ton relative to the ideal case illustrated by the dashed (red) line, for a Xe detector with an illustrative detector threshold of 1 keV.

Multi-ton liquid scintillation detectors are attractive proposals in the experimental race to detect the elusive dark matter particles, though their use also include for example measurement of neutrinoless double beta decay of ^{136}Xe, the pp solar neutrino spectrum and in particular, the neutrino flux and temperature from a Galactic supernova [6]. In this letter we perform simulations of detectors sensitivities for the determination of the temperature and energy of galactic core collapse supernova neutrinos of the second and third generation based on neutrino-nucleus elastic scattering in scintillation materials. We take into account realistic quenching factors and detector thresholds and show the viability of this complementary channel for future multi-ton scintillation detectors.

2. Cross section, yield and results

Neutrino-nucleus elastic cross section can be calculated accurately with very little theoretical uncertainty,

$$\frac{d\sigma}{d\Omega} = \frac{G_F^2}{4\pi^2} k^2 (1 + \cos\theta) \frac{Q_w^2}{4} F(Q^2)^2$$

for a neutrino of energy k scattering at angle θ, G_F being the Fermi constant. This coherent cross section depends on the square of the weak charge $Q_w = N - (1 - 4\sin^2\Theta_W)Z$ of a nucleus with N neutrons and Z protons, $\sin^2\Theta_W \approx 0.231$ being the weak mixing angle and F the Helm [7] nuclear form factor at momentum transfer Q, $Q^2 = 2k^2(1 - \cos\theta)$. Assuming a Boltzmann spectra for the neutrinos allows us to obtain an analytic expression for the yield of recoiling nuclei with energy E and mass M:

$$Y(E) = \frac{G_F^2 Q_w^2}{4\pi} M F^2(2ME) \left(\frac{N_t}{4\pi d^2}\right) \sum_{i=\nu_e,\bar{\nu}_e,\nu_x} N_i \left(1 + \left(\frac{ME}{2T_i^2}\right)^{1/2}\right) e^{-\left(\frac{ME}{2T_i^2}\right)^{1/2}}$$

where T_{ν_x} and N_{ν_x} are the temperature and number of (anti)neutrino x, $x = \mu, \tau$ and similarly for the electron (anti)neutrino, the sum being over all neutrino and anti-neutrinos flavours. Here

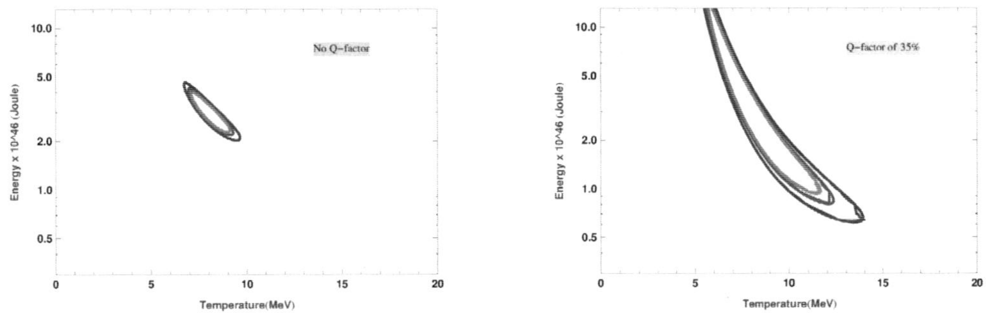

(a) (b)

Figure 2. These plots show the effect of Q_f on the detector sensitivity, the confident level (CL) contours in the right panel expand for a much larger area than in the left panel where no energy losses are assumed. Contours shown correspond to 90%, 95% and 99% of C.L. for a Xenon detector with 10 tons and a detector threshold of 8 keVs.

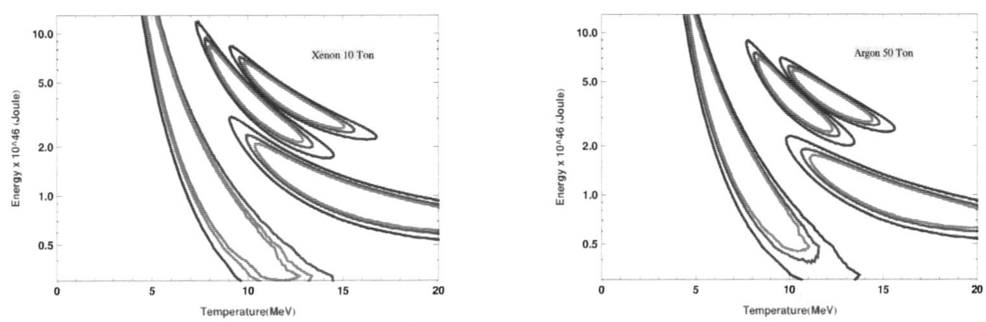

(a) (b)

Figure 3. Detector sensitivity plots for various values of $(T_{\nu_x}, E_{\nu_x}) = (7,2), (10,4), (12,4)$ and $(15,1)$ in the units shown. In the left panel we present results for a detector based on 10 tons of Xenon with a threshold of 8 keVs; the right panel is for 50 tons of Argon with a detector threshold of 20 keVs. In both cases a conservative value of 0.35 is assumed for the quenching factor. Contours shown correspond to 90%, 95% and 99% of C.L.

N_t is the total number of target atoms in the detector, and d is the distance to the supernova which we assume to be 10 kpc. We note that the yield is dominated by the contributions of the neutrinos of the second and third generation due to the relative low values of the temperature of the electron (anti)neutrinos, specially for large recoil energy E. This effect is enhanced by the correction due to quenching effects which practically filters out the contribution of the electron (anti)neutrinos.

As mentioned before, access to the spectrum of the recoiling nuclei is of great importance and to take advantage of this information it is necessary to bin the simulated data in energy. By performing chi-square fits on the simulated data we obtain best fit points for the parameters of interest, determining the regions of sensitivity of the assumed detector configurations, the χ^2 function being defined as

$$\chi^2 = \sum_j 2\left(n_j^t - n_j^f - n_j^f \ln \frac{n_j^t}{n_j^f}\right)$$

where the sum runs over energy bins. Here, a superscript t is used to denote the predicted number of events n_j^t in the j-th bin, while a superscript f is employed to denote the number of events obtained from the fitting to the simulated data, n_j^f.

In Figure 2 we show sensitivity regions for a 10 ton Xenon detector with (right panel) and without (left panel) energy losses taking into account. The importance of ionizing effects are reflected in the detector sensitivity which is found to worsen significantly for the realistic case of energy losses. Finally in Figure 3 [8] we present our results for two types of scintillating detectors and several assumed (true) values of the temperature and energy of the supernova neutrinos.

3. Conclusions

We have performed simulations of detection of neutrinos from core collapse supernova events in scintillating detectors in the coherent neutrino-nucleus scattering channel with realistic detector thresholds and quenching factor. We have presented sensitivity regions for two types of scintillating materials currently studied for multi-ton detector proposals mainly in the experimental search for weakly interacting dark matter. Our results allow us to conclude that with these set-ups a good determination of the neutrino ν_μ and ν_τ parameters, their temperature and energy, could be achieved for energies above $2 \times 10^{46} J$ and temperatures above 8 MeV.

Acknowledgments

C. E. acknowledges the support of CONACyT-México, and in part of CONACyT project no. 132059.

References

[1] Janka H T, Langanke K, Marek A, Martinez-Pinedo G and Mueller B 2007 *Phys. Rept.* **442** 38–74 (*Preprint* astro-ph/0612072)
[2] Raffelt G 2012 *ArXiv e-prints* (*Preprint* 1201.1637)
[3] Beacom J F, Farr W M and Vogel P 2002 *Phys. Rev.* **D66** 033001 (*Preprint* hep-ph/0205220)
[4] Horowitz C J, Coakley K J and McKinsey D N 2003 *Phys. Rev.* **D68** 023005 (*Preprint* astro-ph/0302071)
[5] Duan H, Fuller G M, Carlson J and Qian Y Z 2006 *Phys. Rev.* **D74** 105014 (*Preprint* astro-ph/0606616)
[6] Arisaka K, Beltrame P, Lam C W, Smith P F, Ghag C, Cline D B, Lung K, Meng Y, Pantic E, Scovell P R and Teymourian A 2012 *Astroparticle Physics* **36** 93–122 (*Preprint* 1107.1295)
[7] Helm R H 1956 *Phys. Rev.* **104** 1466
[8] Espinoza C and Palomares-Ruiz S in preparation

New scalar contributions to the weak dipole moments of charged leptons

M.A. Arroyo-Ureña, G. Hernández-Tomé, and G. Tavares-Velasco
Facultad de Ciencias Físico Matemáticas, Benemérita Universidad Autónoma de Puebla, Apartado postal 1152, 72001 Puebla, Pue., México
E-mail: `gerardoht@alumnos.fcfm.buap.mx`

Abstract. We calculate the contributions of new scalar particles to the weak anomalous dipoles moments in alternative electroweak gauge theories.

1. Introduction

It is well known that the study of the static electromagnetic properties of fermions provides a unique opportunity to look for new physics effects. The theoretical study of both the magnetic dipole moment (MDM) and the electric dipole moment (EDM) of fermions has long received considerable attention, which has been boosted in recent years due to the significant progress in the experimental area. Along with the interest in the study of the electromagnetic properties of a fermion, there has also been interest in the static weak properties, which are associated with its interaction with the Z gauge boson. The analogues of the MDM and the EDM are the weak magnetic dipole moment (WMDM) a_f^W and the weak electric dipole moment (WEDM) d_f^W, respectively, which are defined at the Z-pole via the dipole terms of the $Z\bar{f}f$ vertex function $ie\bar{u}(p)\Gamma^{\mu}_{Z\bar{f}f}u(p')$

$$\Gamma^{\mu}_{Z\bar{f}f}\left(q^2\right) = F_2\left(q^2\right) i\sigma^{\mu\nu}q_\nu + F_3\left(q^2\right)\sigma^{\mu\nu}\gamma_5 q_\nu, \tag{1}$$

with $q = p - p'$ the Z transfer momentum. The WMDM is defined as $a_f^W = -2m_f F_2\left(m_Z^2\right)$ and the WEDM is given by $d_f^W = -eF_3\left(m_Z^2\right)$. Only the weak dipole moments (WDMs) of heavy fermions are worth studying as those of lighter fermions would be beyond the reach of experimental detection. In the SM, $a_\tau^W = -(2.10 + 0.61i) \times 10^{-6}$ arises at the one-loop level [3], and $d_\tau^W < 8 \times 10^{-34}$ is induced up to the three-loop level [4]. Although the sensitivity reached at the LEP was beyond such precision level, potentially large contributions from SM extensions can be at the reach of future experiments. In this work we are interested in analyzing new potential contributions arising from models with an extended scalar sector. We will thus calculate the one-loop contributions to the WDMs induced by neutral, singly and doubly charged scalar bosons, which can arise in several models with extended scalar sectors. Our calculation will thus be somewhat general: instead of working out the WDMs within a specific model, we will consider the scenario of a theory with various nondegenerate neutral, singly and doubly charged scalar bosons with the most general renormalizable couplings to leptons and the Z gauge boson that can induce the WDMs at the one-loop level. Because of space limitations, in this report we only show the contributions to the WMDM, as for the contributions to the WEDM they can be found in [1].

2. New scalar contributions to the WMDM of charged leptons

We first consider lepton number conserving (LNC) interactions mediated by scalar bosons (lepton number violating interactions (LNV) can be induced by the doubly charged scalar bosons). For the couplings of a lepton-antilepton pair with a neutral or singly charged scalar particle (denoted ϕ_i or ϕ_j from now on) we will consider the following renormalizable interaction

$$\mathcal{L} = g\,\bar{\ell}_l\left(S_{ilm} + P_{ilm}\gamma_5\right)\ell_m\phi_i + \text{H.c.}, \tag{2}$$

and a similar expression for any other scalar boson. Here ℓ_l is a charged lepton and ℓ_m is a lepton whose charge depends on that of the scalar boson: if ϕ_i is a neutral (charged) scalar boson, ℓ_m is a charged (neutral) lepton. Note that we are considering the most general case where the neutral scalar bosons are a mixture of CP-even and CP-odd states, which can arise for instance in general THDM with CP violation.

As for the interactions of Z gauge boson with two nondegenerate neutral or charged scalar bosons ϕ_i and ϕ_j, it will be written as follows

$$\mathcal{L} = ig\,g_{Z\phi_i\phi_j}Z^\mu\phi_i^\dagger\overleftrightarrow{\partial_\mu}\phi_j + \text{H.c}, \tag{3}$$

whereas the couplings of the type $ZV\phi_i$, with V a neutral (charged) gauge boson and ϕ_i a neutral (charged) scalar boson, are given by

$$\mathcal{L} = g\,g_{\phi_i VZ}m_Z Z^\mu V_\mu\phi_i + \text{H.c.}, \tag{4}$$

where V stands for a SM gauge boson or another one predicted by a SM extension. Such coupling can be for instance the $\phi_i ZZ$ and the $\phi^\pm ZW^\mp$ ones. The latter can arise at the tree-level in Higgs triplet models. We also need the interaction between a lepton-antilepton pair with a neutral or charged gauge boson V, which we write as

$$\mathcal{L} = g\,\bar{\ell}_l\gamma_\mu\left(g_V^{Vlm} - g_A^{Vlm}\gamma_5\right)V^\mu\ell_m + \text{H.c.} \tag{5}$$

Once a specific model is considered, the couplings S_{ilm}, P_{ilm}, $g_{Z\phi_i\phi_i}$, etc. will acquire a particular form, which will depend on the parameters of such model. Since we are primarily interested on any contributions arising from new hypothetical scalar bosons, we will not consider those contributions that have already been discussed in the context of the SM . At the one-loop level these couplings lead to contributions to the WDMs of a charged lepton via the Feynman diagrams depicted in Fig. 1. We denote the external charged lepton by ℓ_l and the internal lepton (neutral or charged) by ℓ_m, whereas ϕ_i and ϕ_j represent scalar bosons, and V is a gauge boson. Evidently once the electric charge of the scalar bosons are fixed, the charges of the internal lepton ℓ_m and the gauge boson V will also become fixed by charge conservation in each vertex. For instance, if ϕ_i and ϕ_j are neutral scalar bosons, ℓ_m is a charged lepton e, μ, τ. Thus, the contributions of new neutral scalar bosons require the vertices $\phi_i\ell_m\ell_l$, $\phi_i ZZ$, $Z\phi_i\phi_j$ and $Z\bar{\ell}_m\ell_m$. On the other hand, when ϕ_i and ϕ_j are charged scalar bosons, the internal lepton is a neutrino $\ell_m = \nu_m$. This class of contributions requires the vertices $\phi_i^+\bar{\nu}_m\ell_l$, $\phi_i^- W^+ Z$, $Z\phi_i^-\phi_j^-$ and $Z\bar{\nu}_m\nu_m$.

We have obtained the following expressions for the contributions of each type of Feynman diagram to the WMDM of a charged lepton via the Feynman parameter technique.

The contributions to the WMDM can be written as follows

$$a_l^{W-I} = \frac{\alpha\sqrt{x_l}}{4\pi s_W^3}\sum_{i,j,m} 16\left(1-\delta_{ij}\right)\,\text{Re}\left[S_{ilm}S_{jlm}^* g_{Z\phi_i\phi_j}^*\right]A_I^{m\phi_i\phi_j} + \begin{pmatrix} \sqrt{x_m} \to -\sqrt{x_m} \\ S_{ilm} \to P_{ilm} \\ S_{jlm} \to P_{jlm} \end{pmatrix}, \tag{6}$$

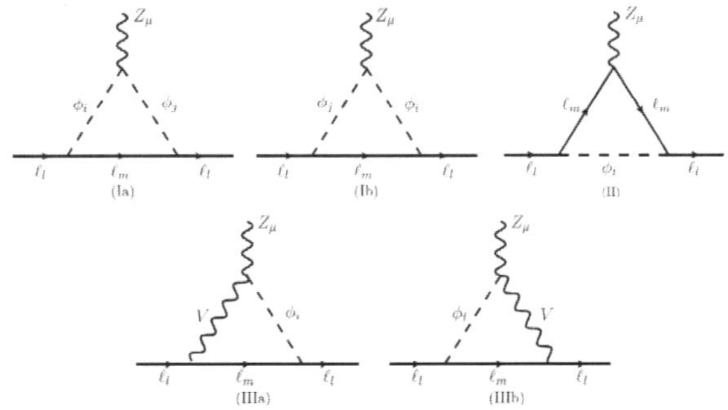

Figure 1. Generic Feynman diagrams for the new scalar contributions to the WDMs of a charged lepton. Here ℓ_l stands for a charged lepton, whereas ℓ_m is a lepton whose charge depends on that of the ϕ_i and ϕ_j scalar bosons (diagrams I and II) and that of the V gauge boson and the ϕ_i scalar boson (diagrams III).

$$
\begin{aligned}
a_l^{W-II} &= \frac{\alpha\sqrt{x_l}}{4\pi s_W^3}\sum_{i,m} 16\left(g_V^{Zmm}\|S_{imm}\|^2 A_{II_1}^{\phi_i mm} + g_A^{Zmm}\sqrt{x_l}\,\mathrm{Re}\left[S_{imm}P_{imm}^*\right]A_{II_2}^{\phi_i mm}\right)\\
&+ \begin{pmatrix} \sqrt{x_m}\to-\sqrt{x_m} \\ S_{imm}\leftrightarrow P_{imm}\end{pmatrix},
\end{aligned}
\tag{7}
$$

and

$$
a_l^{W-III} = \frac{\alpha\sqrt{x_l}}{4\pi s_W^3}\sum_{i,m,V}\frac{2g_{\phi_i VZ}}{x_V}\mathrm{Re}\left[S_{ilm}\,g_V^{Vlm*}\right]A_{III}^{m\phi_i V} - \begin{pmatrix}\sqrt{x_m}\to-\sqrt{x_m}\\ S_{ilm}\to P_{ilm}\\ g_V^{Vlm}\to g_A^{Vlm}\end{pmatrix}, \tag{8}
$$

where $x_a = m_a^2/m_Z^2$ and δ_{ij} is the Kronecker delta. It is understood that these sums run over all the possible combinations of internal particles predicted by a particular theory. Furthermore, the term between parenthesis means that the previous term but with the respective substitutions is to be added. As far as the A_i^{ABC} functions are concerned, they depend on the masses of the particles circulating into each triangular loop and that of the external lepton (the superscript letters stand for the distinct particles in the loop), which can be found in [1]. It is also important to mention that our expressions are complementary to those reported in Ref. [2], where the WMDM of a fermion was obtained using the Feynman-'t Hooft gauge along with the Passarino-Veltman reduction scheme.

In addition to the above results, we also need to consider the $\Delta L = 2$ LNV contributions, which can be mediated by a doubly charged scalar boson. This class of interactions can be written as

$$
\mathcal{L}^{\Delta L=2} = g\,\ell_l^T C\left(S'_{ilm} + P'_{ilm}\gamma^5\right)\ell_m\phi^i + \mathrm{H.c} \tag{9}
$$

where C is the charge conjugation matrix. Doubly charged scalar bosons contribute to the WDMs of charged leptons via the Feynman diagrams shown in Fig. 2, where the fermion-flow arrows either clash or emerge from LNV vertices as opposed to LNC vertices where the fermion flow follows the same direction. Note that we have considered type-III′ Feynman diagrams, where Y stands for a doubly charged gauge boson. Some models with Higgs triplets, for instance 331

Journal of Physics: Conference Series **761** (2016) 012061 doi:10.1088/1742-6596/761/1/012061

models, predict such particles. We will consider the following interaction of a doubly charged gauge boson with charged leptons

$$\mathcal{L}_Y^{\Delta L=2} = g\,\ell_l^T C\left(g_V^{Ylm} - g_A^{Ylm}\gamma^5\right)\gamma^\mu \ell_m Y_\mu^{++} + \text{H.c.}. \tag{10}$$

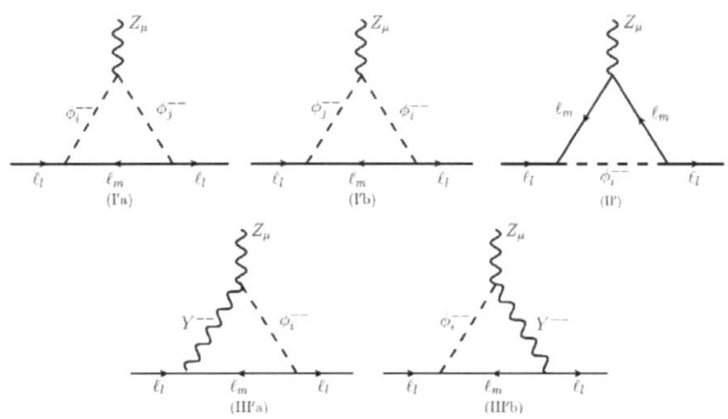

Figure 2. Contributions to the WDMs of charged leptons mediated by doubly charged scalar bosons. Here ℓ_l and ℓ_m are both charged leptons. For completeness we also include the contributions from a doubly charged gauge bosons Y.

Due to the presence of the charge conjugation matrix and the transposed spinors, the application of the Feynman rules to obtain the amplitudes of these Feynman diagrams must be done carefully. We have followed the approach presented in Ref. [5, 6] for calculating amplitudes with fermion number violating vertices. After some algebra we have found that the results arising from the Feynman diagrams of Fig. 1 [i.e. (6)- (8)] hold true for the contributions of a doubly charged scalar bosons except that a multiplicative factor of two for each LNV vertex must be inserted when the leptons are identical ($l = m$):

$$a_l^{W-I'} \rightarrow (1+\delta_{lm})^2 a_l^{W-I} \tag{11}$$

$$a_l^{W-II'} \rightarrow (1+\delta_{lm})^2 a_l^{W-II} \tag{12}$$

$$a_l^{W-III'} \rightarrow (1+\delta_{lm})^2 a_l^{W-III} \tag{13}$$

where it is understood that one must replace the appropriate couplings and masses involved in each contribution. This situation was also noted in the calculation of the anomalous magnetic moment of a lepton [5].

References

[1] M.A. Arroyo-Ureña, G. Hernández-Tomé, and G. Tavares-Velasco, Weak dipole moments of the tau lepton in models with extended scalar sectors, Work in progress.

[2] W. Hollik, J. I. Illana, S. Rigolin, C. Schappacher and D. Stockinger, Nucl. Phys. B **551**, 3 (1999) Erratum: [Nucl. Phys. B **557**, 407 (1999)]

[3] J. Bernabeu, G. A. Gonzalez-Sprinberg, M. Tung and J. Vidal, Nucl. Phys. B **436**, 474 (1995)

[4] W. Bernreuther, U. Low, J. P. Ma and O. Nachtmann, Z. Phys. C **43**, 117 (1989).

[5] S. R. Moore, K. Whisnant and B. L. Young, Phys. Rev. D **31**, 105 (1985).

[6] A. Denner, H. Eck, O. Hahn and J. Kublbeck, Nucl. Phys. B **387**, 467 (1992).

Lepton flavor violation in an extended MSSM

R Espinosa-Castañeda[1], F Flores-Baez[2], M Gomez Bock[3] and M Mondragon[4]

[1] CERN
[2] Facultad de Ciencias Físico-Matemáticas, Universidad Autónoma de Nuevo León Ciudad Universitaria, San Nicolás de los Garza, Nuevo León, 64450, México.
[3] DAFM, Universidad de las Américas Puebla, Ex. Hda. Sta. Catarina Mártir s/n, San Andrés Cholula, Puebla. México
[4] Instituto de Física, Universidad Nacional Autónoma de México, Apdo. Postal 20-364, Cd. México 01000, México.

E-mail: melina.gomez@udlap.mx, fflores@fcfm.uanl.mx

Abstract. In this work we explore a lepton flavor violation effect induced at one loop for a flavor structure in an extended minimal standard supersymmetric model, considering an ansatz for the trilinear term. In particular we find a finite expression which will show the impact of this phenomena in the $h \to \mu\tau$ decay, produced by a mixing in the trilinear coupling of the soft supersymmetric Lagrangian.

1. Introduction

In the Standard Model (SM) lepton flavor violation processes were forbidden by the lepton number conservation, which is not associated with a gauge symmetry. In the SM, the spontaneous breaking of the electroweak symmetry produces eigenstates of the remaining gauge group that are not in general eigenstates of the mass matrix [1, 2, 3, 4]. But after diagonalization of the mass matrix, the electroweak coupling matrix is also diagonal in the mass basis, therefore there is no possibility for lepton flavor violation. Nevertheless, this symmetry is lost in neutrino oscillation found in experiments [5, 6, 7, 8], which forces the model structure to go beyond the SM.

This lost symmetry on leptons observed in the neutrino mixing evidence opens also the possibility of lepton flavor violation in the charged sector. Experimental data taken from CMS at $8\ TeV$ with $19.7 fb^{-1}$ had shown an excess for $BR(h^0 \to \tau\mu)$ of 2.4σ with best fit branching fraction of $0.84^{+0.39}_{-0.37}\%$ [9]. Nevertheless data from ATLAS has shown only a 1σ significance for the same process [10]. Moreover, also recent measurements at $13\ TeV$, although with only $2.3 fb^{-1}$ of data, has shown no evidence of excess. It was even reported [11] a best fit branching fraction of -0.56%. The sign change may imply a statistical error of the data. Even for those latest reports, if they are confirmed, they will indicate a very low range for this lepton flavor violation processes to occur at these energies. These will set stringent bounds to any model beyond the SM. One of the works toward this direction is done in [12], where it has been explored it in a Two Higgs Doublet Model with flavor violation in the Yukawa couplings, a model known as THDM-III [13]. In order to fit these restrictions bounds, the model we proposed in this work of a scalar flavor extended MSSM, should show exclusions regions in the parameter space.

The most recent data by the LHC at 13 TeV have not shown evidence for supersymmetry for different channels and observables as events with one lepton as final state [11], jets and leptons or three leptons [14], missing energy [15]. The experimental search is mainly for the Lightest Supersymmetric Particle (LSP) as missing energy, these reports analyses the data for simplified supersymmetric models at this specific energy. The results of these analysis have reduced the parameter space for Minimal Supersymmetric Standard Model (MSSM). Nevertheless we claim that is important to fully explore the possible parameter space for different non-simple supersymmetric low energy models knowing that Supersymmetry still solves many phenomenological issues [16, 17] and also is needed for many GUT models. A review on flavor violation processes in the MSSM considering neutrino mixing can be found in [18].

2. Flavor mixed sleptons

As the evidence on flavor violation in charged lepton is not yet conclusive but gives low values for these branching ratios, one possibility is to have a mixed flavor structure in an unseen sector as the sfermions, and then this mixing will induce flavor violation through radiative corrections. The mixing of slepton states will be given by a flavor structure in the mass matrix, specifically we explore the possibility of having mixing flavor terms in the trilinear couplings of the soft supersymmetric Lagrangian.

$$\mathcal{L}^f_{soft} = -\sum_{\tilde{f}_i} \tilde{M}^2_{\tilde{f}} \tilde{\bar{f}}_i \tilde{f}_i - (A_{\tilde{f},i} \tilde{\bar{f}}^i_L H_1 \tilde{f}_R{}^i + h.c), \tag{1}$$

where $\tilde{f}$ are the scalar fields in the supermultiplet. In the case of sfermions, as they are scalar particles, the L, R are just labels which point out to the fermionic SM partners, but they no longer have left and right $SU(2)$ properties. In general they may mix in two physical states. Once the EW symmetry breaking is considered, the trilinear term of the soft SUSY Lagrangian for the sleptonic sector takes the following form

$$\mathcal{L}_{H\tilde{f}_i\tilde{f}_j} = \frac{A^{ij}_l}{\sqrt{2}} \left[(\phi^0_1 - i\chi^0_1)\tilde{l}^*_{iR}\tilde{l}_{jL} - \sqrt{2}\phi^-_1 \tilde{l}^*_{iR}\tilde{\nu}_{jL} + v_1 \tilde{l}^*_R \tilde{l}_L \right] + h.c.$$

The contribution to the elements of the sfermion mass matrix come from the interaction of the Higgs scalars with the sfermions, which appear in different terms of the superpotential and soft-SUSY breaking terms as is fully explained in [19, 20]. In the case of the slepton mass matrix, as we said before, the contributions coming from *mass soft terms* are $\tilde{M}^2_{\tilde{l},LL}$, $\tilde{M}^2_{\tilde{l},RR}$, from trilinear couplings after EW symmetry breaking A^l_{ij} and from the $F, D-$terms. We arrange them in a block mass matrix as follows

$$\tilde{M}^2_{\tilde{l}} = \begin{pmatrix} m^2_{LL,l} & m^2_{LR,l} \\ m^2_{RL,l} & m^2_{RR,l} \end{pmatrix} . \tag{2}$$

The elements of the sleptons mass matrix eq. (2), for the different flavors given by $i, j = e, \mu, \tau$ are

$$m^2_{LL,l} = \tilde{M}^2_{\tilde{L},l} + m^2_{l_L} + \frac{1}{2}\cos 2\beta(2M^2_W - M^2_Z), \tag{3}$$

$$m^2_{RR,l} = \tilde{M}^2_{\tilde{E},l} + m^2_{l_R} - \cos 2\beta \sin^2\theta_W M^2_Z, \tag{4}$$

$$m^2_{LR,l} = \frac{A_l v \cos\beta}{\sqrt{2}} - m_l \mu_{susy} \tan\beta. \tag{5}$$

The soft terms are not the only contributions to the sfermion mass elements, the supersymmetric auxiliary fields F and D coming from the superpotential also contribute to this mass matrix. In the present work, we assume the mass terms coming from SUSY breaking terms dominate over the EW terms coming from auxiliary F and D fields so we can safely approximate the diagonal elements to a soft susy scale m_S as $m_{RR}^2 \approx m_{LL}^2 = m_S^2 \mathbf{1}_{2\times 2}$.
In order to analyze the consequences of a flavor structure we construct an ansatz for the trilinear terms A_t. Our procedure is similar to the work done in Ref. [21] for FCNC's in the quark sector through an ansatz of soft-susy terms. In our case we consider the whole two families contributions and of the same order of magnitude, having the following form for the trilinear term

$$A_l = \begin{pmatrix} 0 & 0 & 0 \\ 0 & w & y \\ 0 & y & 1 \end{pmatrix} A_0. \tag{6}$$

In this case, one could have at tree level the selectrons in a singlet irreducible representation decoupled from the other two families of sleptons. This would give rise to a 4×4 matrix, diagonalizable through a unitary matrix $Z_{\tilde{l}}$, such that $Z_{\tilde{l}}^\dagger \tilde{M}_l^2 Z_{\tilde{l}} = \tilde{M}_{diag}^2$.
We will have then physical non-degenerate slepton masses[1]

$$\begin{aligned} m_{\tilde{\tau}_{1,2}}^2 &= \frac{1}{2}(2\tilde{m}_S^2 - X_m - X_t \pm R) \ , \\ m_{\tilde{\mu}_{1,2}}^2 &= \frac{1}{2}(2\tilde{m}_S^2 + X_m + X_t \pm R) \ , \end{aligned} \tag{7}$$

where $R = \sqrt{4A_y^2 + (X_t - X_m)^2}$ with $A_y = \frac{1}{\sqrt{2}} y A_0 v \cos\beta$, $X_m = \frac{1}{\sqrt{2}} w A_0 v \cos\beta - \mu_{susy} m_\mu \tan\beta$ and $X_t = \frac{1}{\sqrt{2}} A_0 v \cos\beta - \mu m_\tau \tan\beta$.
We may write the transformation which diagonalizes the mass matrix as a 4×4 rotation matrix for sleptons $Z_{\tilde{l}}$, which is in turn a 2×2 block matrix $Z_{\tilde{l}}^\dagger \tilde{M}_{\mu-\tau}^2 Z_{\tilde{l}} = \tilde{M}_{l,diag}^2$, explicitly having the form

$$Z_{\tilde{l}} = \frac{1}{\sqrt{2}} \begin{pmatrix} \Phi & -\Phi \\ \Phi\sigma^3 & \Phi\sigma^3 \end{pmatrix} , \tag{8}$$

where σ_3 is the Pauli matrix and

$$\Phi = \begin{pmatrix} -\sin\frac{\varphi}{2} & -\cos\frac{\varphi}{2} \\ \cos\frac{\varphi}{2} & -\sin\frac{\varphi}{2} \end{pmatrix}, \quad \tan\varphi = \frac{2A_y}{X_m - X_t}, \tag{9}$$

The non-physical states are transformed to the physical eigenstates as

$$\begin{pmatrix} \tilde{\mu}_L \\ \tilde{\tau}_L \\ \tilde{\mu}_R \\ \tilde{\tau}_R \end{pmatrix} = Z_{\tilde{l}} \begin{pmatrix} \tilde{l}_1 \\ \tilde{l}_2 \\ \tilde{l}_3 \\ \tilde{l}_4 \end{pmatrix} \tag{10}$$

[1] We assign the label $\tilde{\tau}, \tilde{\mu}$ to the masses to show the relation to the non-FV sleptons.

Table 1. Expressions of the respective interactions of the Higgs boson h^0 with two sleptons

$g_{h^0\tilde{f}\tilde{f}}$	$\tilde{\mu}_1$	$\tilde{\mu}_2$	$\tilde{\tau}_1$	$\tilde{\tau}_2$
$\tilde{\mu}_1$	$s_\varphi^2 T_{\tilde{\tau}}^+ + c_\varphi^2 T_{\tilde{\mu}}^+ - \frac{1}{4}G$	$\frac{1}{4}G(1-4s_w^2)$	0	$s_\varphi c_\varphi (T_{\tilde{\tau}}^+ - T_{\tilde{\mu}}^+)$
$\tilde{\mu}_2$	*	$s_\varphi^2 T_{\tilde{\tau}}^- + c_\varphi^2 T_{\tilde{\mu}}^- - \frac{1}{4}G$	$s_\varphi c_\varphi (T_{\tilde{\tau}}^- - T_{\tilde{\mu}}^-)$	0
$\tilde{\tau}_1$	0	*	$s_\varphi^2 T_{\tilde{\mu}}^- + c_\varphi^2 T_{\tilde{\tau}}^- - \frac{1}{4}G$	$\frac{1}{4}G(1-4s_w^2)$
$\tilde{\tau}_2$	*	0	*	$s_\varphi^2 T_{\tilde{\mu}}^+ + c_\varphi^2 T_{\tilde{\tau}}^+ - \frac{1}{4}G$

2.1. Higgs flavor violation coupling with sleptons

The Lagrangian which gives the interaction of scalar neutral light Higgs h^0-slepton-slepton is given as

$$\mathcal{L}_{h^0\tilde{l}\tilde{l}} = Q_{\tilde{l}}\left[\tilde{l}_L^*\tilde{l}_L + \tilde{l}_R^*\tilde{l}_R\right]h^0 + G\left[(-\frac{1}{2}+s_w^2)\tilde{l}_L^*\tilde{l}_L - s_w^2\tilde{l}_R^*\tilde{l}_R\right]h^0 + \chi_{\tilde{l}}\left[\tilde{l}_L^*\tilde{l}_R + \tilde{l}_R^*\tilde{l}_L\right] \tag{11}$$

where

$$Q_{\tilde{\mu},\tilde{\tau}} = \frac{gm_{\tilde{\mu},\tilde{\tau}}^2 \sin\alpha}{M_w \cos\beta}\ , \tag{12}$$

$$G = g_z M_z \sin(\alpha+\beta)\ , \tag{13}$$

$$X_{\tilde{\mu},\tilde{\tau}} = \frac{gm_{\tilde{\mu},\tilde{\tau}} \sin\alpha}{2M_w cos\beta}(A_{\tilde{\mu},\tilde{\tau}} - \mu\cot\alpha)\ . \tag{14}$$

Then, in the slepton physical states, as rotated by (10), the couplings to the light Higgs boson are given as in table 1, where $T_{\tilde{l}}^{\pm} = Q_{\tilde{l}} \pm X_{\tilde{l}}$. We have simplified the notation using $s_\varphi = \sin\frac{\varphi}{2}$ and $c_\varphi = \cos\frac{\varphi}{2}$.

3. Radiative Higgs flavour violation decay

In this section we show the construction of the radiative induced Higgs flavor violation decay. We assume the following convention: the slepton which interacts with the muon (tau) is labeled with the index i (j), see Fig 1. The notation used for the coupling between the bino $\tilde{B}$, the

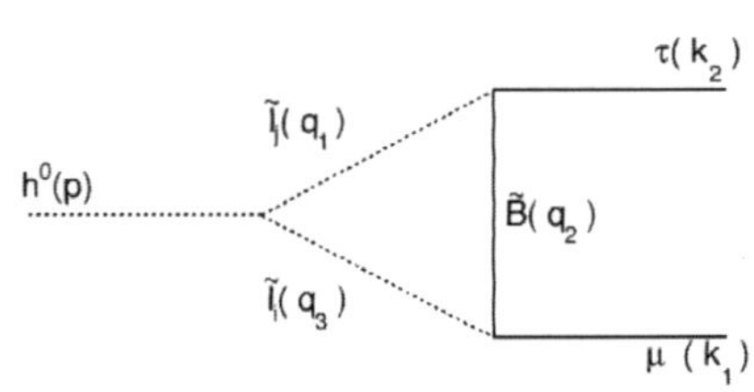

Figure 1. *1-loop SUSY slepton flavor mixing contribution to $h^0 \to \mu\tau$.*

slepton $\tilde{l}$ and the lepton l, for $l = \mu, \tau$, which is denoted as $\tilde{B}l\tilde{l}$, can be written in terms of three types of coefficients for each lepton. We use a_k, for the ones in the coupling with the muon and

Table 2. Coefficients for $a_{2,3}$ and $b_{2,3}$ of Bino-slepton-lepton couplings $g_{\tilde{B}\tilde{l}l}$ given in eq. (15)

	$i,j=1$	2	3	4		1	2	3	4
a_2^i	1	3	3	1	a_3^i	3	1	1	3
b_2^j	1	3	3	1	b_3^j	3	1	1	3

b_k for coefficients in the coupling with the tau; with $k = 1,2,3$; then we have this couplings written as:

$$g_{\tilde{B}\tilde{l}_i\mu} = a_1^i[a_2^i + a_3^i\gamma^5] \ ,$$
$$g_{\tilde{B}\tilde{l}_j\tau} = b_1^j[b_2^j + b_3^j\gamma^5] \ , \tag{15}$$

where the first type of coefficients are given as

$$a_1^i = \frac{g}{4}\tan\theta_W a_{\tilde{l}_i} \ , \tag{16}$$
$$b_1^j = -\frac{g}{4}\tan\theta_W b_{\tilde{l}_j} \ , \tag{17}$$

being $a_{\tilde{l}_i} = s_\varphi(-c_\varphi)$ for $\tilde{\mu}_{1,2}(\tilde{\tau}_{1,2})$, while $b_{\tilde{l}_j} = c_\varphi, (s_\varphi)$ for $\tilde{\mu}_{1,2}(\tilde{\tau}_{1,2})$. And the rest of the coefficients, type 2 and 3 are numbers shown in table 2.

Now we write the invariant amplitude for the $h^0 \to \mu\tau$ decay with slepton $\tilde{l}_i, \tilde{l}_j$ running inside the loop as indicated in lines above, in such a way that it is possible to identify every vertex, propagator and external fermion legs, see figure 3

$$\mathcal{M}_{ij} = \int \frac{d^4q}{(2\pi)^4}\bar{v}(k_1)[\imath g_{\tilde{B}\tilde{l}_i\mu}]\left[\frac{\imath(\not{q} - \not{k}_2 + m_B)}{(q-k_2)^2 - m_B^2}\right][\imath g_{\tilde{B}\tilde{l}_j\tau}]\left[\frac{\imath}{(q-p)^2 - m_i^2}\right]\left[\frac{\imath}{q^2 - m_j^2}\right][\imath g_{h\tilde{l}_i\tilde{l}_j}]u(k_2) \tag{18}$$

After the loop momentum integration over $q = q_1$, the amplitude can be written in the following form

$$\mathcal{M}_{ij} = \bar{v}(k_1)[A^{ij} + B^{ij}\gamma^5]u(k_2) \ , \tag{19}$$

where

$$A^{ij} = a_1^i b_1^j\big[(a_2^i b_2^j + a_3^i b_3^j)m_B F^{ij} + (-a_2^i b_2^j + a_3^i b_3^j)m_\tau F^{ij}$$
$$+ (a_2^i b_2^j - a_3^i b_3^j)m_\tau F_{II}^{ij} - (a_2^i b_2^j - a_3^i b_3^j)m_\mu F_{III}^{ij}\big] \ , \tag{20}$$
$$B^{ij} = a_1^i b_1^j\big[(a_2^i b_3^j + a_3^i b_2^j)m_B F^{ij} - (a_3^i b_2^j - a_2^i b_3^j)m_\tau F^{ij}$$
$$+ (a_3^i b_2^j - a_2^i b_3^j)m_\tau F_{II}^{ij} - (a_2^i b_3^j - a_3^i b_2^j)m_\mu F_{III}^{ij}\big] \ . \tag{21}$$

The F functions are given as

$$F^{ij} = -g_{h\tilde{l}_i\tilde{l}_j}\frac{\imath}{16\pi^2}C0[m_H^2, m_\mu^2, m_\tau^2, m_j^2, m_i^2, m_B^2] \ , \tag{22}$$

$$\begin{aligned} F_{II}^{ij} &= -g_{h\tilde{l}_i\tilde{l}_j}\frac{\imath}{(16\pi^2)[m_H^2-(m_\mu+m_\tau)^2][m_H^2-(m_\mu-m_\tau)^2]} \\ &\times \big[- C0(m_H^2, m_\mu^2, m_\tau^2, m_j^2, m_i^2, m_B^2)\big[m_B^2(m_H^2+m_\mu^2-m_\tau^2) \\ &+ m_H^4 + m_\mu^2(m_i^2-2m_j^2) - m_H^2(m_i^2+m_\mu^2+2m_\tau^2) + m_\tau^2(m_i^2-m_\mu^2) + m_\tau^4\big] \\ &- 2m_\mu^2 B0(m_\mu^2, m_B^2, m_i^2) + B0(m_\tau^2, m_B^2, m_j^2)(m_\mu^2+m_\tau^2-m_H^2) \\ &- B0(m_H^2, m_i^2, m_j^2)(m_\tau^2-m_\mu^2-m_H^2)\big] \ , \end{aligned} \tag{23}$$

$$\begin{aligned} F_{III}^{ij} &= -g_{h\tilde{l}_i\tilde{l}_j}\frac{\imath}{(16\pi^2)[m_H^2-(m_\mu+m_\tau)^2][m_H^2-(m_\mu-m_\tau)^2]} \\ &\times \big[- C0(m_H^2, m_\mu^2, m_\tau^2, m_j^2, m_i^2, m_B^2)\big[m_\tau^2(m_B^2+m_H^2-2m_i^2+m_j^2+m_\mu^2) \\ &- m_\tau^4 + (m_B^2-m_j^2)(m_H^2-m_\mu^2)\big] + 2m_\tau^2 B0(m_\tau^2, m_B^2, m_j^2) \\ &+ B0(m_\mu^2, m_B^2, m_i^2)(m_H^2-m_\mu^2-m_\tau^2) - B0(m_H^2, m_i^2, m_j^2)(m_H^2-m_\mu^2+m_\tau^2)\big] \ . \end{aligned} \tag{24}$$

Then the differential width for the process $h^0 \to \tau\mu$ can be calculated using the amplitude transition matrix as

$$d\Gamma = \frac{1}{32\pi^2}\sum_{spin}|\mathcal{M}_T|^2\frac{|p|}{m_H^2}d\Omega, \tag{25}$$

$$|p| = \sqrt{(m_H^2-(m_\tau-m_\mu)^2)(m_H^2-(m_\tau+m_\mu)^2)} \ . \tag{26}$$

Notice that the total amplitude includes all possible combinations of sleptons in the internal lines, then

$$\mathcal{M}_T = \sum_{i,j}\mathcal{M}_{ij}. \tag{27}$$

4. Conclusions

In this work we consider a flavor structure on trilinear soft terms, assuming a two family mixing in the sleptons we explore the consequences of this structure in a particular process involving lepton flavor violation for the Higgs boson. We obtain non-degenerate slepton masses for four of the sleptons which are decoupled from the first family and mixed in flavor. We also found that in physical basis two specific couplings of the Higgs boson with sleptons are zero, *i.e.* $g_{h^0\tilde{\mu}_1\tilde{\tau}_1} = g_{h^0\tilde{\mu}_2\tilde{\tau}_2} = 0$.
We obtain the expression for the one-loop radiative correction of the specific process $h^0 \to \tau\mu$. The expression we obtain is found to be UV-finite and can be used to restrict the parameter space of the supersymmetric model applied to this process, as it is very restricted by the experimental data. This kind of structure also gives extra contribution to $BR(\tau \to \mu\gamma)$ and to the muon anomalous magnetic moment $g-2$. So a complete exploration of the parameter for all these processes will be a goal for a further work.

Acknowledgments

This work was partially supported by a Consejo Nacional de Ciencia y Tecnología (Conacyt), Posdoctoral Fellowship and SNI México. R. E-C wants to thank CONACyT RED-Física de Altas

Enerías. F. F-B was partially supported by Universidad Autónoma de Nuevo León through the grant PAICYT-2015. M. G-B acknowledges partial support from Universidad de las Américas Puebla. This work was also partially supported by grants UNAM PAPIIT IN111115 and Conacyt 132059.

References

[1] Weinberg S 1967 *Phys.Rev.Lett.* **19** 1264–1266
[2] Weinberg S 1972 *Phys.Rev.* **D5** 1412–1417
[3] Glashow S 1961 *Nucl.Phys.* **22** 579–588
[4] Glashow S, Iliopoulos J and Maiani L 1970 *Phys.Rev.* **D2** 1285–1292
[5] Cleveland B, Daily T, Davis Raymond J, Distel J R, Lande K *et al.* 1998 *Astrophys.J.* **496** 505–526
[6] Fukuda Y *et al.* (Super-Kamiokande Collaboration) 1998 *Phys.Rev.Lett.* **81** 1562–1567 (*Preprint* hep-ex/9807003)
[7] Ahmad Q *et al.* (SNO Collaboration) 2002 *Phys.Rev.Lett.* **89** 011301 (*Preprint* nucl-ex/0204008)
[8] Ahn M *et al.* (K2K Collaboration) 2003 *Phys.Rev.Lett.* **90** 041801 (*Preprint* hep-ex/0212007)
[9] Khachatryan V *et al.* (CMS) 2015 *Phys. Lett.* **B749** 337–362 (*Preprint* 1502.07400)
[10] Aad G *et al.* (ATLAS) 2016 (*Preprint* 1604.07730)
[11] 2016 Search for supersymmetry in events with one lepton in proton-proton collisions at sqrt(s)=13 TeV with the CMS experiment Tech. Rep. CMS-PAS-SUS-15-006 CERN Geneva URL https://cds.cern.ch/record/2140638
[12] Aristizabal Sierra D and Vicente A 2014 *Phys. Rev.* **D90** 115004 (*Preprint* 1409.7690)
[13] Gomez-Bock M and Noriega-Papaqui R 2006 *J. Phys.* **G32** 761–776 (*Preprint* hep-ph/0509353)
[14] Aad G *et al.* (ATLAS) 2016 *Eur. Phys. J.* **C76** 259 (*Preprint* 1602.09058)
[15] Bradmiller-Feld J (CMS) 2016 Search for supersymmetry in the multijet and missing transverse momentum final state *51st Rencontres de Moriond on EW Interactions and Unified Theories La Thuile, Italy, March 12-19, 2016* (*Preprint* 1605.05762) URL https://inspirehep.net/record/1462256/files/arXiv:1605.05762.pdf
[16] Haber H E 2001 *Nucl.Phys.Proc.Suppl.* **101** 217–236 (*Preprint* hep-ph/0103095)
[17] Heinemeyer S, Stal O and Weiglein G 2012 *Phys.Lett.* **B710** 201–206 (*Preprint* 1112.3026)
[18] Vicente A 2015 *Adv. High Energy Phys.* **2015** 686572 (*Preprint* 1503.08622)
[19] Kuroda M 1999 (*Preprint* hep-ph/9902340)
[20] Okumura K i and Roszkowski L 2003 *JHEP* **10** 024 (*Preprint* hep-ph/0308102)
[21] Diaz-Cruz J L, He H J and Yuan C P 2002 *Phys. Lett.* **B530** 179–187 (*Preprint* hep-ph/0103178)

Journal of Physics: Conference Series **761** (2016) 012063 doi:10.1088/1742-6596/761/1/012063

The scalar-photon 3-point vertex in massless quenched scalar QED

Y Concha-Sánchez[1], **L X Gutiérrez-Guerrero**[2] **and L A Fernández-Rangel**[3]

[1] Facultad de Ingeniería Civil, Universidad Michoacana de San Nicolás de Hidalgo, Apartado Postal 2-82, Morelia, Michoacán 58040, México.
[2] Mesoamerican Centre for Theoretical Physics, Universidad Autonoma de Chiapas, Carretera Zapata Km. 4, Real del Bosque (Terán), Tuxtla Gutiérrez, Chiapas 29040, México.
[3] Instituto de Física y Matemáticas, Universidad Michoacana de San Nicolás de Hidalgo, Apartado Postal 2-82, Morelia, Michoacán 58040, México.

E-mail: yconcha@umich.mx

Abstract. Non perturbative studies of Schwinger-Dyson equations (SDEs) require their infinite, coupled tower to be truncated in order to reduce them to a practically solvable set. In this connection, a physically acceptable *ansatz* for the three point vertex is the most favorite choice. Scalar quantum electrodynamics (sQED) provides a simple and neat platform to address this problem. The most general form of the scalar-photon three point vertex can be expressed in terms of only two independent form factors, longitudinal and transverse. Ball and Chiu have demonstrated that the longitudinal vertex is fixed by requiring the Ward-Fradkin-Green-Takahashi identity (WFGTI), while the transverse vertex remains undetermined. In massless quenched sQED, we propose the transverse part of the non perturbative scalar-photon vertex.

1. Introduction

Gauge theories of fundamental interactions have been the cornerstone of describing the physical world at the most basic level. Their enormous success primarily lies in the region where the coupling strength is small enough and the tools of perturbation theory are reliable. However, not all interesting phenomena can be accessed in this approximation scheme. In the non perturbative sector of quantum chromodynamics (QCD), two major phenomena emerge: 1) color confinement, and 2) dynamical chiral symmetry breaking (DCSB). For studying strongly interacting bound states, a reliable understanding of these phenomena is essential. However, it can be achieved solely through non perturbative techniques such as lattice QCD, SDEs, [1, 2], chiral perturbation theory and effective quark models. Keeping this in mind, our interest is focussed on the study and physically acceptable truncations of SDEs beyond perturbation theory.

SDEs are the fundamental equations of motion of any quantum field theory (QFT). They form an infinite set of coupled integral equations that relate the n-point Green function to the $(n+1)$-point function. As their derivation requires no assumption regarding the strength of the interaction, they are ideally suited for studying interactions like QCD, where one single theory has diametrically opposed perturbative and non perturbative facets in the ultraviolet and infrared regimes of momenta, respectively. Unfortunately, being an infinite set of coupled equations, they are intractable without some simplifying assumptions. Typically, in the non

Journal of Physics: Conference Series **761** (2016) 012063 doi:10.1088/1742-6596/761/1/012063

perturbative region, SDEs are truncated at the level of two-point Green functions (propagators). We must then use an *ansatz* for the full three point vertex. This has to be done carefully.

- It must satisfy the **Ward-Fradkin-Green-Takahashi identity** (WFGTI), [3, 4, 5].
- It must satisfy the **local gauge covariance** properties of the theory.
- It must ensure the **multiplicative renormalizability** (MR) of the two point propagator.
- It should reduce to its **perturbation theory** Feynman expansion in the limit of weak coupling.
- It must have the same **symmetry properties** as the bare vertex under charge conjugation, parity and time reversal.
- One loop perturbation suggests that it should be free of any **kinematic singularities**.

The scalar-photon three point vertex $\Gamma^\mu(k,p)$ must be symmetric under the exchange of momenta k and p.

2. The SDE for the Scalar Propagator

The SDE for the scalar propagator $S(k)$ in sQED, in the quenched approximation, is shown in Fig. 1:

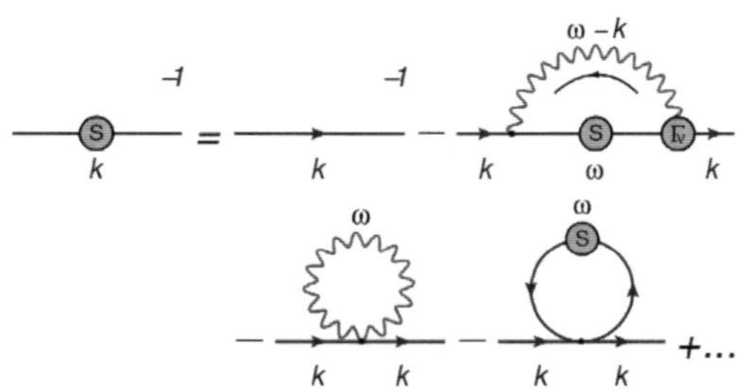

Figure 1. The SDE for the scalar propagator. The color-filled solid blobs labelled with S and Γ_ν stand for the full scalar propagator and the full scalar-photon vertex, respectively. The dots $(\cdots)$ represent all the diagrams whose contribution begins at the two-loop level.

Mathematically, this is written as:

$$\begin{aligned} -iS^{-1}(k) &= -iS_0^{-1}(k) + e^2 \int_M \frac{d^4\omega}{(2\pi)^4}(\omega+k)^\mu S(\omega)\Gamma^\nu(\omega,k)\Delta_{\mu\nu}(q) \\ &\quad -e^2 \int_M \frac{d^4\omega}{(2\pi)^4}\Gamma_0^{\mu\nu}(k,-\omega,k,\omega)\Delta_{\mu\nu}(\omega) - \int_M \frac{d^4\omega}{(2\pi)^4} S(\omega)\Gamma_0(k,\omega) + \cdots, \end{aligned} \quad (1)$$

where e is the electromagnetic coupling, $q = \omega - k$, and the subscript M indicates integration over the entire Minkowski space.

$\Delta_{\mu\nu}(\omega)$ and $S_0(k)$ are the bare photon and scalar propagators. $S(k)$ is the full scalar propagator. For massless scalars, $S(k)$ can be expressed in terms of the so-called wavefunction renormalization $F(k^2, \Lambda^2)$, so that

$$S(k) = \frac{F(k^2,\Lambda^2)}{k^2}, \quad (2)$$

where Λ is the ultraviolet cut-off used to regularize the divergent integrals involved. The bare scalar propagator is given by $S_0(k) = 1/k^2$. The bare photon propagator is

$$\Delta_{\mu\nu}(q) = \frac{1}{q^2}\left[g_{\mu\nu} + (\xi - 1)\frac{q_\mu q_\nu}{q^2}\right] , \tag{3}$$

and it remains unrenormalized in the quenched approximation. $\Gamma_0^{\mu\nu}(k, -\omega, k, \omega) = 2ie^2 g^{\mu\nu}$ and $\Gamma_0(k, \omega) = -i\lambda$ are the bare four point scalar-scalar-photon-photon and the four-scalar vertices, respectively. The last two diagrams of the gap equation, Eq. (1), will be referred to as the photon and the scalar bubble diagrams, in that order. $\Gamma^\nu(\omega, k)$ is the full three point scalar-photon vertex, for which we must make an *ansatz* in order to solve Eq. (1). The WFGTI for this vertex, i.e.,

$$q_\mu \Gamma^\mu(\omega, k) = S^{-1}(\omega) - S^{-1}(k) , \tag{4}$$

allows us to decompose it as a sum of longitudinal and transverse components, as suggested by Ball and Chiu, [6]:

$$\Gamma^\mu(\omega, k) = \Gamma_L^\mu(\omega, k) + \Gamma_T^\mu(\omega, k) . \tag{5}$$

The *longitudinal* part $\Gamma_L^\mu(\omega, k)$ satisfies the WFGTI, Eq. (4), by itself, and the *transverse* part $\Gamma_T^\mu(\omega, k)$, which remains completely undetermined, is naturally constrained by

$$q_\mu \Gamma_T^\mu(\omega, k) = 0 . \tag{6}$$

Moreover,

$$\Gamma_T^\mu(k, k) = 0 . \tag{7}$$

In order to satisfy the WFGTI in a manner free of kinematic singularities, we follow Ball and Chiu and write

$$\Gamma_L^\mu(\omega, k) = \frac{S^{-1}(\omega) - S^{-1}(k)}{\omega^2 - k^2}(\omega + k)^\mu . \tag{8}$$

This construction implies that the ultraviolet divergences solely reside in the longitudinal part. Moreover, recall the following relations between the renormalized and bare quantities:

$$S^R(p) = \mathcal{Z}_2^{-1} S(p) , \; \Gamma_R^\mu(k, p) = \mathcal{Z}_1 \Gamma^\mu(k, p) . \tag{9}$$

Thus, the form of the longitudinal vertex in Eq. (8) guarantees the relation $\mathcal{Z}_1 = \mathcal{Z}_2$. Consequently, the running of the coupling is dictated by the corrections to the photon propagator alone. In the approximation of quenched sQED, the coupling does not run. If we unquench the theory, it is easy to calculate $\mathcal{Z}_3$ and the running coupling constant with the well known expression:

$$\alpha(Q^2) = \frac{\alpha(Q_0^2)}{1 - \left(\alpha(Q_0^2)/12\pi\right)\ln(Q^2/Q_0^2)} . \tag{10}$$

The ultraviolet finite transverse vertex can be expanded out in terms of one unknown function $\tau(\omega^2, k^2, q^2)$, [6]:

$$\Gamma_T^\mu(\omega, k) = \tau(\omega^2, k^2, q^2) T^\mu(\omega, k) , \tag{11}$$

where

$$T^\mu(\omega, k) = (\omega \cdot q)\, k^\mu - (k \cdot q)\, \omega^\mu \tag{12}$$

is the transverse basis vector in the Minkowski space and fulfils Eqs. (6,7). To begin with, the form factor $\tau(\omega^2, k^2, q^2)$ is an unconstrained scalar function (representing an 8-fold simplification of the spinor QED/QCD case).

Journal of Physics: Conference Series **761** (2016) 012063 doi:10.1088/1742-6596/761/1/012063

3. The Transverse Vertex and Perturbation Theory Constraints

The three point vertex at one-loop level in perturbation theory has been calculated by Concha-Sánchez *et al.*, [7], using dimensional regularization, in arbitrary gauge ξ and dimensions d.

For the massless case, in dimension $d = 4$, they report

$$\begin{aligned}\tau_{BCD}(k^2,p^2,q^2) &= \frac{\alpha}{8\pi\Delta^2}\Bigg\{(k^2+p^2-4k\cdot p)\left(k\cdot pJ_0+\ln\left(\frac{q^4}{p^2k^2}\right)\right)\\ &+\frac{(k^2+p^2)q^2-8p^2k^2}{p^2-k^2}\ln\left(\frac{k^2}{p^2}\right)\\ &+(\xi-1)\left[k^2p^2J_0+\frac{2[k^2p^2+k\cdot p(k^2+p^2)]}{k^2-p^2}\right]\ln\left(\frac{p^2}{k^2}\right)\\ &+\frac{2k\cdot p}{k^2-p^2}\left[k^2\ln\left(\frac{q^2}{p^2}\right)-p^2\ln\left(\frac{q^2}{k^2}\right)\right]\Bigg\}, \end{aligned}\tag{13}$$

where

$$J_0^{4,0} = \frac{2}{i\pi^2}\int_M d^4\omega\frac{1}{\omega^2\,(p-\omega)^2\,(k-\omega)^2}, \tag{14}$$

with $q = k - p$. We propose the following vertex *ansatz* in Minkowski space [8]

$$\begin{aligned}\tau(k^2,p^2) &= -\frac{3}{2}\frac{1}{(k^2-p^2)}\left[\frac{1}{F(k^2)}-\frac{1}{F(p^2)}\right]-\frac{\xi}{2}\frac{1}{(k^2-p^2)}\frac{F(k^2)+F(p^2)}{s(k^2,p^2)}\\ &\left[\frac{1}{F(k^2)}-\frac{1}{F(p^2)}\right]+\frac{1}{4}\frac{1}{(k^2-p^2)}\frac{1}{s(k^2,p^2)}\left[W\left(\frac{k^2}{p^2}\right)-W\left(\frac{p^2}{k^2}\right)\right], \end{aligned}\tag{15}$$

where we have introduced $F(k^2) \equiv F(k^2,\Lambda^2)$ as a simplifying notation. We also introduce the definition

$$s(k^2,p^2) = F(k^2)\frac{k^2}{p^2}+F(p^2)\frac{p^2}{k^2} \tag{16}$$

and for the sake of simplicity we can choose the trivial solution $W(x) = 0$ for any dimensionless ratio x of momenta.

$$\begin{aligned}\tau(k^2,p^2) &= -\frac{3}{2}\frac{1}{(k^2-p^2)}\left[\frac{1}{F(k^2)}-\frac{1}{F(p^2)}\right]-\frac{\xi}{2}\frac{1}{(k^2-p^2)}\frac{F(k^2)+F(p^2)}{s(k^2,p^2)}\\ &\left[\frac{1}{F(k^2)}-\frac{1}{F(p^2)}\right]. \end{aligned}\tag{17}$$

In order to compare the vertex *ansatz*, Eq. (17), based upon multiplicative renormalizability, against its one loop perturbative form, Eq. (13), it is convenient to take the asymptotic limit $k^2 \gg p^2$ of external momenta in the latter vertex. The resulting $\tau_{\rm BCD}$ in the leading logarithms approximation (LLA) is

$$\tau_{\rm BCD}^{\rm asym}(k^2,p^2) \overset{k^2\gg p^2}{=} -3\frac{\alpha}{4\pi}\frac{1}{k^2}\ln\left(\frac{k^2}{p^2}\right). \tag{18}$$

Expectedly, it is independent of q^2 and hence we drop this dependence from its argument. Note that this expression is also independent of the covariant gauge parameter ξ. It is unlike spinor QED where the leading asymptotic vertex is proportional to ξ. For a numerical check, we define

$$\tilde{\tau}(x) = -\frac{k^2\,\tau(k^2,xk^2)}{\alpha\ln x}, \tag{19}$$

where $x = p^2/k^2$ and we have suppressed the q^2 dependence for notational simplification. Thus:

$$\tilde{\tau}^{\rm asym}_{\rm BCD}(x) = -\frac{3}{4\pi}\,. \tag{20}$$

In Fig. (2), we plot $\tilde{\tau}^{\rm asym}_{\rm BCD}(x)$ and $\tilde{\tau}_{\rm BCD}(x)$ as a function of x, the latter for different values of the gauge parameter ξ and for a fixed value of q^2, chosen arbitrarily. In the asymptotic limit, all curves converge to a single value, as expected.

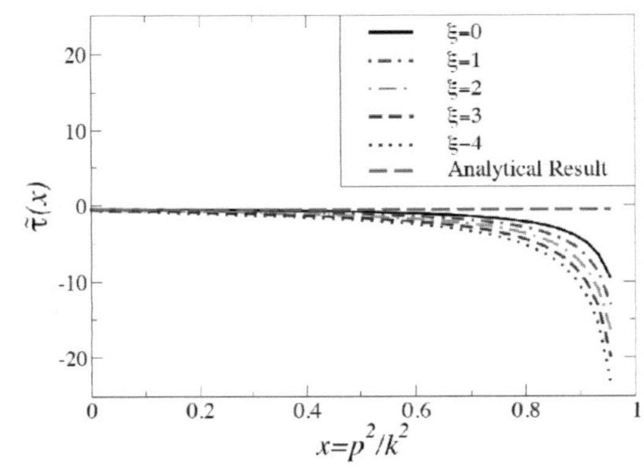

Figure 2. The analytical result, long dashed lines representing a constant value given in Eq. (20) for the asymptotic transverse form factor $\tilde{\tau}^{\rm asym}_{\rm BCD}(x)$, agrees with the numerical plot of $\tilde{\tau}_{\rm BCD}(x)$ obtained from Eq. (13) in the limit $x \to 0$ for different gauges and an arbitrarily chosen value of $q^2 = -0.7GeV^2$.

Using the perturbative expression for $F(k^2)$ in Eq. (17), and taking the asymptotic limit $k^2 \gg p^2$, we have

$$\tau^{\rm asym}(k^2,p^2) \overset{k^2 \gg p^2}{=} \frac{3}{2}\frac{\alpha}{4\pi}\frac{(\xi-3)}{k^2}\ln\left(\frac{k^2}{p^2}\right) \tag{21}$$

in the LLA. Note that the transverse form factors, Eqs. (18) and (21) have the functional form $(1/k^2)\ln(k^2/p^2)$. Furthermore, they are the same in the Feynman gauge ($\xi = 1$). In order for them to be the same in an arbitrary gauge ξ, we must seek a non-trivial W-function in Eq. (15). Perhaps the simplest such choice for W is

$$W\left(\frac{k^2}{p^2}\right) = \lambda\frac{k^2}{p^2}\ln\left(\frac{k^2}{p^2}\right) + \frac{\lambda}{2}\frac{k^2}{p^2}\,, \tag{22}$$

with $\lambda = -3\alpha(\xi-1)/2\pi$. In the Feynman gauge ($\xi = 1$) $W = 0$, i.e., there is no necessity of a non-trivial W-function since both perturbative vertices, Eqs. (18) and (21) are already the same. Using the choice in Eq. (22) for W in the vertex, Eq. (15), we can define the transverse form factor as:

$$\begin{aligned}\tau(k^2,p^2) &= -\frac{3}{2}\frac{1}{(k^2-p^2)}\left[\frac{1}{F(k^2)}-\frac{1}{F(p^2)}\right] - \frac{\xi}{2}\frac{1}{(k^2-p^2)}\frac{F(k^2)+F(p^2)}{s(k^2,p^2)}\\ &\times\left[\frac{1}{F(k^2)}-\frac{1}{F(p^2)}\right] - \frac{(\xi-1)}{(k^2-p^2)\,s(k^2,p^2)}\frac{3\alpha}{8\pi}\left[\frac{k^2}{p^2}+\frac{p^2}{k^2}\right]\ln\left(\frac{k^2}{p^2}\right)\,. \end{aligned}\tag{23}$$

Note that the Eqs. (5,8,11,23) define our full vertex *ansatz*. It ensures the following key features of sQED:

- It satisfies the WFGTI by construction, [3, 4, 5].
- It guarantees the LKFT property of the scalar propagator and can be checked by employing it in its SDE. In other words, it ensures the multiplicative renormalizability (MR) of the two point scalar propagator.
- It reduces to its one loop perturbation theory Feynman expansion in the limit of small coupling and asymptotic values of momenta $k^2 \gg p^2$.
- It has the same symmetry properties as the bare vertex under charge conjugation, parity and time reversal, which imply symmetry under $k \leftrightarrow p$.
- It is free of any kinematic singularities when $k^2 \Rightarrow p^2$, i.e.,

$$\lim_{k^2 \Rightarrow p^2} (k^2 - p^2)\, \tau(k^2, p^2) = 0\,. \tag{24}$$

Note that the kinematic dependence of the vertex on q^2 plays no role asymptotically and the standard analysis proceeds without reference to it. On the infrared domain, however, the kinematic dependence on q^2 may be important. Our vertex has this pitfall but its simplicity is reason enough for us to ignore this dependence.

Acknowledgments

We are grateful to Adnan Bashir for helpful discussions. This work was partly supported by CIC, CONACyT and PRODEP grants.

References

[1] Dyson F 1949 *Phys. Rev.* **75** 1736
[2] Schwinger J S 1951 *Proc. Nat. Acad. Sci.* **37** 452
[3] Ward J C 1950 *Phys. Rev.* **78** 182
[4] Green H 1953 *Proc. Phys. Soc.* **A66** 873
[5] Takahashi Y 1957 *Nuovo Cim.* **6** 371
[6] Ball J S and Chiu T W 1980 *Phys. Rev.* D **22** 2542
[7] Bashir A, Concha-Sánchez Y and Delbourgo R 2007 *Phys. Rev.* D **76** 065009
[8] Fernández-Rangel L A, Bashir A, Gutiérrez-Guerrero L X and Concha-Sánchez Y 2016 *Phys. Rev.* D **93** 065022

Journal of Physics: Conference Series **761** (2016) 012064 doi:10.1088/1742-6596/761/1/012064

A principle for the Yukawa couplings

U J Saldaña-Salazar

Facultad de Ciencias Físico-Matemáticas, Benemérita Universidad Autónoma de Puebla, C.P. 72570, Puebla, Pue., México.

E-mail: ulisesjesus@protonmail.ch

Abstract. Yukawa couplings in the Standard Model are introduced in its most general form, that is, completely arbitrary complex numbers. Here we show that their origin could not be general but dictated by a principle.

1. Introduction

There is a huge gap between knowing and not knowing. There are times though when their separation becomes null and not knowing is knowing or vice-versa. Curiously, this is the current situation on the Yukawa couplings, which are introduced in order to parametrize the interactions between the two handedness of fermions and the Higgs field, with their inclusion being made by following the principle of generality. Therewith, we know Yukawa couplings are, in general, complex and arbitrary numbers. But that is all we know about them, i.e., that we do not know them. Here we try to briefly summarize the essence of a recently proposed principle called the Flavor-Blind Principle [1] that gives an origin to the Yukawa couplings in agreement to the emerged structures appearing under the phenomenological application of hierarchical fermion masses [2].

The same scalar field required to break spontaneously the electroweak symmetry can be employed to give masses in a gauge invariant and renormalizable way to all the fermions. This kind of interactions between the two handedness of fermions with the Higgs field are named Yukawa interactions. The Yukawa Lagrangian is normally written by demanding its most general form

$$-\mathcal{L}_Y = \sum_{i,j=1}^{3} \left(\sum_{a=u,\nu} \mathbf{Y}^a_{ij} \bar{\psi}^a_{L,i} (i\sigma_2 \Phi^*) \psi^a_{R,j} + \sum_{b=d,e} \mathbf{Y}^b_{ij} \bar{\psi}^b_{L,i} \Phi \psi^b_{R,j} \right) + h.c., \tag{1}$$

where $\mathbf{Y}^{a(b)}_{ij}$ are the Yukawa couplings which, due to generality, are arbitrary complex numbers. With only one scalar field, Φ, the Yukawa matrices are proportional to the mass matrices. The constant producing such distinction appears after the neutral component of the Higgs field acquires a non-zero vacuum expectation value, $v = 246$ GeV.

It has become a well established fact, that matter in the subatomic world replicates itself thrice. That is, each of the four types of fermions, whose distinction is given by their quantum numbers, appears three-fold; it is their mass the only one aspect allowing their differentiation. Each group of four is called family or generation, whereas, each one of the three members of a given fermion type is called flavor.

Journal of Physics: Conference Series **761** (2016) 012064 doi:10.1088/1742-6596/761/1/012064

Phenomenologically, one finds that masses among the three flavors and on each fermion type obey the hierarchical structure

$$m_1(M_Z) \ll m_2(M_Z) \ll m_3(M_Z). \tag{2}$$

Neutrinos, however, may be the exception; their squared mass differences still satisfy a hierarchy

$$\Delta m_{21}^2 \ll \Delta m_{31}^2, \tag{3}$$

but this is not necessarily the case for their individual masses.

Recently, by virtue of the aforementioned hierarchy, a mixing parametrization was built [2] wherein its mixing parameters correspond to the four independent fermion mass ratios of the quark or lepton sector, respectively. The developed procedure found that fermion mixing originates from a very particular sequence of Yukawa matrices

$$\mathbf{M}_f = v\mathbf{Y}_f = v\left(\mathbf{Y}_{f,3} + \mathbf{Y}_{f,2} + \mathbf{Y}_{f,1}\right), \tag{4}$$

where

$$\mathbf{Y}_{f,3} = \begin{pmatrix} 0 & 0 & 0 \\ 0 & 0 & 0 \\ 0 & 0 & y_3 \end{pmatrix}, \qquad \mathbf{Y}_{f,2} = \begin{pmatrix} 0 & 0 & 0 \\ 0 & 0 & y_2 \\ 0 & \pm y_2 & 0 \end{pmatrix}, \quad \text{and} \quad \mathbf{Y}_{f,1} = \begin{pmatrix} 0 & y_1 & y_1' \\ \pm y_1 & 0 & 0 \\ \pm y_1' & 0 & 0 \end{pmatrix}, \tag{5}$$

with y_3 being real while y_2, y_1, and y_1' being either real or purely imaginary. Also, an unexplained feature was that between these three matrices there must exist a hierarchy in correspondance to $m_{f,3} \gg m_{f,2} \gg m_{f,1}$. It is our goal to find such an origin that gives precisely these Yukawa structures.

This paper is organized with only two sections: in Section 2 we give account of the Flavor-Blind principle and briefly comment about all its implications while in Section 3 we conclude.

2. The Flavor-Blind Principle

Consider the Standard Model (SM) Lagrangian[1] with its Yukawa part turned off. That is, we have not introduced yet the interactions between the Higgs field with the fermions. Also, assume we have an arbitrary number of families. Universality in the gauge interactions for n families will imply an accidental global symmetry group given as

$$\mathcal{G}_F = U_L^Q(n) \times U_R^u(n) \times U_R^d(n) \times U_L^\ell(n) \times U_R^e(n) \times U_R^\nu(n). \tag{6}$$

Basically, we cannot distinguish between any of the n fermion families. The Flavor-Blind principle (FBP) then states that Yukawa interactions should be introduced as to conserve flavor-blindness. That is, the n fermion fields should equally couple to the scalar field

$$\mathbf{Y}^{1\leftrightarrow 2\leftrightarrow 3\leftrightarrow\cdots\leftrightarrow(n-1)\leftrightarrow n} = y_n \begin{pmatrix} 1 & 1 & 1 & \cdots & 1 \\ 1 & 1 & 1 & \cdots & 1 \\ 1 & 1 & 1 & \cdots & 1 \\ \vdots & \vdots & \vdots & \vdots & \vdots \\ 1 & 1 & 1 & \cdots & 1 \end{pmatrix}. \tag{7}$$

The symmetry properties of this matrix are described by the permutational symmetry group $S_{nL} \otimes S_{nR}$. As this matrix is rank one then a change of basis will show that $(n-1)$ families are

[1] From this point onwards, we refer by SM a theory including massive neutrinos. Also, for the moment and due to simplicity, we take the massive nature of neutrinos to be of the Dirac type.

Journal of Physics: Conference Series **761** (2016) 012064 doi:10.1088/1742-6596/761/1/012064

still massless while the remaining one is massive, $m_n = ny_nv/\sqrt{2}$. Notice that the same weak basis transformation applies to all fermions and it has the form

$$\mathbf{O}_n = \begin{pmatrix} \frac{1}{\sqrt{2}} & \frac{1}{\sqrt{3\cdot2}} & \frac{1}{\sqrt{4\cdot3}} & \cdots & \frac{1}{\sqrt{n(n-1)}} & \frac{1}{\sqrt{n}} \\ -\frac{1}{\sqrt{2}} & \frac{1}{\sqrt{3\cdot2}} & \frac{1}{\sqrt{4\cdot3}} & \cdots & \frac{1}{\sqrt{n(n-1)}} & \frac{1}{\sqrt{n}} \\ 0 & -\frac{2}{\sqrt{3\cdot2}} & \frac{1}{\sqrt{4\cdot3}} & \cdots & \frac{1}{\sqrt{n(n-1)}} & \frac{1}{\sqrt{n}} \\ 0 & 0 & -\frac{3}{\sqrt{4\cdot3}} & \cdots & \frac{1}{\sqrt{n(n-1)}} & \frac{1}{\sqrt{n}} \\ \vdots & \vdots & \vdots & \cdots & \vdots & \vdots \\ 0 & 0 & 0 & \cdots & \frac{1}{\sqrt{n(n-1)}} & \frac{1}{\sqrt{n}} \\ 0 & 0 & 0 & \cdots & -\frac{n-1}{\sqrt{n(n-1)}} & \frac{1}{\sqrt{n}} \end{pmatrix}. \tag{8}$$

Interestingly, due to the generic form the Yukawa matrices acquire in the new basis

$$\widetilde{\mathbf{Y}}^{1\leftrightarrow2\leftrightarrow3\leftrightarrow\cdots\leftrightarrow(n-1)\leftrightarrow n} = ny_n \begin{pmatrix} 0 & 0 & 0 & \cdots & 0 \\ 0 & 0 & 0 & \cdots & 0 \\ 0 & 0 & 0 & \cdots & 0 \\ \vdots & \vdots & \vdots & \vdots & \vdots \\ 0 & 0 & 0 & \cdots & 1 \end{pmatrix}, \tag{9}$$

we can easily see that our initial accidental and global symmetry group has broken into

$$\mathcal{G}_F \to U_L^Q(n-1) \times U_R^u(n-1) \times U_R^d(n-1) \times U_L^\ell(n-1) \times U_R^e(n-1) \times U_R^\nu(n-1), \tag{10}$$

plus some additional $U(1)$ factors.

A new Yukawa matrix is now required in order to give mass to the still massless $(n-1)$ fermion fields. Application of the FBP means

$$\mathbf{Y}^{1\leftrightarrow2\leftrightarrow3\leftrightarrow\cdots\leftrightarrow(n-1)} = \begin{pmatrix} \beta_{n-1} & \beta_{n-1} & \cdots & \beta_{n-1} & \alpha_{n-1} \\ \beta_{n-1} & \beta_{n-1} & \cdots & \beta_{n-1} & \alpha_{n-1} \\ \vdots & \vdots & \cdots & \vdots & \vdots \\ \beta_{n-1} & \beta_{n-1} & \cdots & \beta_{n-1} & \alpha_{n-1} \\ \alpha'_{n-1} & \alpha'_{n-1} & \cdots & \alpha'_{n-1} & \gamma_{n-1} \end{pmatrix}. \tag{11}$$

The arbitrary parameters α_{n-1}, β_{n-1}, α'_{n-1}, and γ_{n-1} are, in general, complex parameters. The symmetry properties of this matrix are described by the permutational symmetry group $S_{(n-1)L} \otimes S_{(n-1)R}$. It has rank two and using the same weak basis transformation of Eq. (8) brings it into the form

$$\widetilde{\mathbf{Y}}^{1\leftrightarrow2\leftrightarrow3\leftrightarrow\cdots\leftrightarrow(n-1)} = \begin{pmatrix} 0 & 0 & 0 & \cdots & 0 & 0 & 0 \\ 0 & 0 & 0 & \cdots & 0 & 0 & 0 \\ 0 & 0 & 0 & \cdots & 0 & 0 & 0 \\ \vdots & \vdots & \vdots & \vdots & \vdots & \vdots & \vdots \\ 0 & 0 & 0 & \cdots & 0 & 0 & 0 \\ 0 & 0 & 0 & \cdots & 0 & \widetilde{y}_{n-1,n-1} & \widetilde{y}_{n-1,n} \\ 0 & 0 & 0 & \cdots & 0 & \widetilde{y}_{n,n-1} & \widetilde{y}_{n,n} \end{pmatrix}, \tag{12}$$

where

$$\widetilde{\mathcal{Y}}_{n-1,n-1} = \frac{n-1}{n}[\beta_{n-1} + \gamma_{n-1} - \alpha_{n-1} - \alpha'_{n-1}], \tag{13}$$

$$\widetilde{\mathcal{Y}}_{n-1,n} = \frac{\sqrt{n-1}}{n}[(n-1)(\beta_{n-1} - \alpha'_{n-1}) + \alpha_{n-1} - \gamma_{n-1}], \tag{14}$$

$$\widetilde{\mathcal{Y}}_{n,n-1} = \frac{\sqrt{n-1}}{n}[(n-1)(\beta_{n-1} - \alpha_{n-1}) + \alpha'_{n-1} - \gamma_{n-1}], \tag{15}$$

$$\widetilde{\mathcal{Y}}_{n,n} = \frac{1}{n}\{\gamma_{n-1} + (n-1)[(n-1)\beta_{n-1} + \alpha_{n-1} + \alpha'_{n-1}]\}. \tag{16}$$

We can now notice that besides the n-th family being massive the $(n-1)$-th one can also become massive after diagonalizing the complete 2×2 submatrix. We can see that our previous global symmetry group $U(n-1)^6$ has now broken into $U(n-2)^6$ plus some additional $U(1)$ factors.

In this way, by each step, we are providing mass to only one fermion family. When reaching the $U(1)^6$ last step, i.e., with only one massless family, we need to consider that the properties of the needed Yukawa matrix should be S_{2A} as shown in [1].

The repetition of this procedure leads to some important observations:

(i) The accidental global symmetry group $\mathcal{G}_F$ needs to has the sequential breaking

$$U(n)^6 \to U(n-1)^6 \to U(n-2)^6 \to \cdots \to U(2)^6 \to U(1)^6 \to U(1)_B \times U(1)_L, \tag{17}$$

with B and L being baryon and lepton number, respectively. When $n=3$ this reminds us of minimal flavor violation which requires a similar breaking [3–5].

(ii) The introduced set of Yukawa matrices also has a sequential breaking of its symmetrical properties

$$S_{nL} \otimes S_{nR} \to S_{(n-1)L} \otimes S_{(n-1)R} \to \cdots \to S_{2L} \otimes S_{2R} \to S_{2A}. \tag{18}$$

(iii) Each step, after the weak basis transformation of Eq. (8), brings the following Yukawa structures

$$\begin{pmatrix} & & \\ & & \\ & & \times \end{pmatrix} \to \begin{pmatrix} & & \\ & \times & \times \\ & \times & \times \end{pmatrix} \to \begin{pmatrix} & & & \\ & \times & \times & \times \\ & \times & \times & \times \\ & \times & \times & \times \end{pmatrix} \to \cdots \to \begin{pmatrix} \times & \times & \times & \times & \cdots & \times \\ \times & \times & \times & \times & \cdots & \times \\ \times & \times & \times & \times & \cdots & \times \\ \vdots & \vdots & \vdots & \vdots & \vdots & \vdots \\ \times & \times & \times & \times & \times & \times \end{pmatrix} \tag{19}$$

which basically are telling us how mixing phenomena occurrs. First, mixing in the $(n-1)$-n sector with its initial individual mixing angle proportional to m_{n-1}/m_n. After it, nevertheless, we still require to introduce mixing in the same sector proportional to each of the new mass terms divided by m_n. Next, mixing in the $(n-2)$-n sector happens. Then, mixing in $(n-3)$-n and so on and so forth until reaching mixing between the first family with the last one. After all these rotations, the mass matrices become diagonal by parts with one $(n-1) \times (n-1)$ submatrix plus another 1×1 one. Similarly, mixing between the $(n-1)$-th family with the rest happens and so on and so forth. This same observation was found when applying the mass hierarchy $m_3 \gg m_2 \gg m_1$ to the study of fermion mixing [2].

(iv) Application of the FBP to the minimal case, that is, two families, naturally implies that individual mixing should happen satisfying a Gatto–Sartori–Tonin-like relation[2] [1]

$$\tan\theta_{12} = \sqrt{\frac{m_1}{m_2}}. \tag{21}$$

(v) Application of the FBP to the three family case gives a solvable system of linear equations with which we can easily reproduce the matrix structures of Eq. (5) [1].

3. Conclusions

We have seen how by demanding that massless fermion families equally couple to the Higgs field, therefore respecting their initial flavor-blindness in the gauge interactions, gives us a series of phenomenological implications which are in agreement to the ones already found in the study of fermion mixing by virtue of hierarchical fermion masses [2] and minimal flavor violation [3–5]. This petition is called the Flavor-Blind Principle [1].

Also, a very interesting remark about the usefulness of this approach is that it has recently been seen to play an important role in solving the strong CP problem [7].

Acknowledgments

I am grateful to Wolfgang G. Hollik for a careful reading of this manuscript. This work has been supported by CONACyT-Mexico under Contract No. 220498.

References

[1] U. J. Saldana-Salazar, Phys. Rev. D **93**, no. 1, 013002 (2016) doi:10.1103/PhysRevD.93.013002 [arXiv:1509.08877 [hep-ph]].
[2] W. G. Hollik and U. J. Saldana-Salazar, Nucl. Phys. B **892**, 364 (2015) doi:10.1016/j.nuclphysb.2015.01.019 [arXiv:1411.3549 [hep-ph]].
[3] R. Barbieri, G. R. Dvali and L. J. Hall, Phys. Lett. B **377**, 76 (1996) doi:10.1016/0370-2693(96)00318-8 [hep-ph/9512388].
[4] R. Barbieri, L. J. Hall, S. Raby and A. Romanino, Nucl. Phys. B **493**, 3 (1997) doi:10.1016/S0550-3213(97)00134-X [hep-ph/9610449].
[5] R. Barbieri, L. J. Hall and A. Romanino, Phys. Lett. B **401**, 47 (1997) doi:10.1016/S0370-2693(97)00372-9 [hep-ph/9702315].
[6] R. Gatto, G. Sartori and M. Tonin, Phys. Lett. B **28**, 128 (1968). doi:10.1016/0370-2693(68)90150-0
[7] J. L. Diaz-Cruz, W. G. Hollik and U. J. Saldana-Salazar, arXiv:1605.03860 [hep-ph].

[2] The Gatto–Sartori–Tonin relation [6] was the first expression to relate a mixing angle with a mass ratio as

$$\tan\theta_C \approx \sqrt{\frac{m_d}{m_s}} \approx 0.22, \tag{20}$$

which gives a very good approximation to the Cabibbo angle.

MWPF IOP Publishing
Journal of Physics: Conference Series **761** (2016) 012065 doi:10.1088/1742-6596/761/1/012065

Neutrino-electron scattering with a new source of CP violation

J Barranco*, D Delepine†, M Napsuciale‡ and A Yebra§

Departamento de Física, División de Ciencias e Ingenierías, Campus León, Universidad de Guanajuato, León 37150, México

E-mail: *jbarranc@fisica.ugto.mx, †delepine@fisica.ugto.mx, ‡mauro@fisica.ugto.mx §azarael@fisica.ugto.mx,

Abstract. According to previous works, there is a possibility for increasing the difference between Dirac and Majorana cross section for a neutrino-electron scattering process if we take into account that the neutrino longitudinal polarization can be different from minus one. In this work, we study the difference between Dirac and Majorana scattering process but we introduce an additional effective interaction that depends on complex coupling constants for the neutrinos. Thus, in this more general case, we have two additional parameters, the phase of the neutrino couplings and one parameter ϵ related to the effective coupling of the new interaction.

1. Introduction

One of the open questions in particle physics about neutrinos is referred to the nature of this particle, that is, if the neutrino is a Dirac or a Majorana particle. We can remember now that in Dirac case the particle is different than its associated antiparticle, otherwise, if we have a Majorana particle then it is identical to its own antiparticle [1]. That is one crucial difference between them. It is important to establish a way to determine to which of the both categories the neutrino belongs, although we know that there is a clear difference in both cases, it is very difficult to determine via a neutrino-electron scattering process if the neutrino belongs to one or other kind, because Dirac process differs from Majorana by terms that are proportional to the neutrino mass, which is usually small. Then, in order to increase such difference other authors have explored in the past the possibility of including the neutrino parallel polarization term in the calculation of the cross section, that is, they take this value different from minus one [2].

In this work, additionally to this approach we propose and consider a new effective interaction similar in its functional form to Standard Model neutral current interaction when the polarization term is considered again. With all these ingredients the cross section for the neutrino-electron scattering process is calculated but not only we explore the possibilities to increase the difference between Dirac and Majorana processes but actually we are able to define a region in which a Dirac neutrino is indistinguishable from a Majorana neutrino, for a certain neutrino energy and polarization values.

Journal of Physics: Conference Series **761** (2016) 012065 doi:10.1088/1742-6596/761/1/012065

2. Neutrino-electron scattering in the Standard Model

In the Standard Model the amplitudes for the neutrino-electron scattering process are given by

$$\mathcal{M}_D = -i\frac{G_F}{\sqrt{2}}\left[\bar{u}^f_{\nu_\ell}\gamma^\mu\left(1-\gamma^5\right)u^i_{\nu_\ell}\right]\left[\bar{u}^f_e\gamma_\mu\left(g^\ell_V - g^\ell_A\gamma^5\right)u^i_e\right], \quad (1)$$

in the Dirac case, and by

$$\mathcal{M}_M = -i\frac{G_F}{\sqrt{2}}\left[\bar{u}^f_e\gamma_\mu\left(g^\ell_V - g^\ell_A\gamma^5\right)u^i_e\right]\left[\bar{u}^f_{\nu_\ell}\gamma^\mu\left(1-\gamma^5\right)u^i_{\nu_\ell} - \bar{v}^f_{\nu_\ell}\gamma^\mu\left(1-\gamma^5\right)v^i_{\nu_\ell}\right], \quad (2)$$

for the Majorana case. The effective coupling constants are defined as follows,

$$g^\ell_V = -\frac{1}{2} + 2\sin^2\theta_W + \delta_{\ell e}, \quad g^\ell_A = -\frac{1}{2} + \delta_{\ell e}, \quad (3)$$

with $\ell = e,\, \mu,\, \tau$.

In the last expression the second term, the neutrino block, includes the contribution from the anti-neutrino-electron scattering because in Majorana case the particle is identical to its own anti-particle, then if we pretend to obtain the completed amplitude for the scattering process this contribution cannot be despised. Related to the last equation, if we consider that the neutrino is a Majorana particle, then the following identity is valid [3]

$$\bar{v}^f_{\nu_\ell}\gamma_\mu\left(1-\gamma^5\right)v^i_{\nu_\ell} = \bar{u}^f_{\nu_\ell}\gamma_\mu\left(1+\gamma^5\right)u^i_{\nu_\ell}, \quad (4)$$

and the amplitude that appears in equation (2) can be rewritten as

$$\mathcal{M}_M = i\frac{2G_F}{\sqrt{2}}\left[\bar{u}^f_e\gamma^\mu\left(g^\ell_V - g^\ell_A\gamma^5\right)u^i_e\right]\left[\bar{u}^f_{\nu_\ell}\gamma_\mu\gamma^5 u^i_{\nu_\ell}\right]. \quad (5)$$

Now it is possible to see that Majorana amplitude, as opposed to Dirac amplitude, does not have a vector term in the neutrino block, thus identity (4) allows cancellation of the vectorial part in the amplitude.

As we can see from equations (5) and (1), their functional forms are different, for this reason it is reasonable to expect that after making all calculations needed to find the cross section for the scattering process, these cross sections do not look identical therefore we would expect to see a difference, but experimentally it does not happen because of the very small neutrino mass.

Now, considering that incident neutrino has a longitudinal polarization different than minus one, that is, they are not completely left-handed particles, it is possible to gain a certain degree of dissimilarity between Dirac and Majorana processes. To quantify how dissimilar the cross sections can be, a function is defined [4]

$$D_{DM}\left(E^{Lab}_{\nu_\ell}, s_\parallel\right) = \frac{\left|\frac{d\sigma^D}{dE^{Lab}_{\nu_\ell}} - \frac{d\sigma^M}{dE^{Lab}_{\nu_\ell}}\right|}{\frac{d\sigma^D}{dE^{Lab}_{\nu_\ell}}}, \quad (6)$$

which measures the percentage of difference between Dirac and Majorana as a function of neutrino energy and polarization. Of course, all variables involved are expressed in the laboratory reference frame. Figure 1 shows the difference between Dirac and Majorana for a set of distinct polarization values.

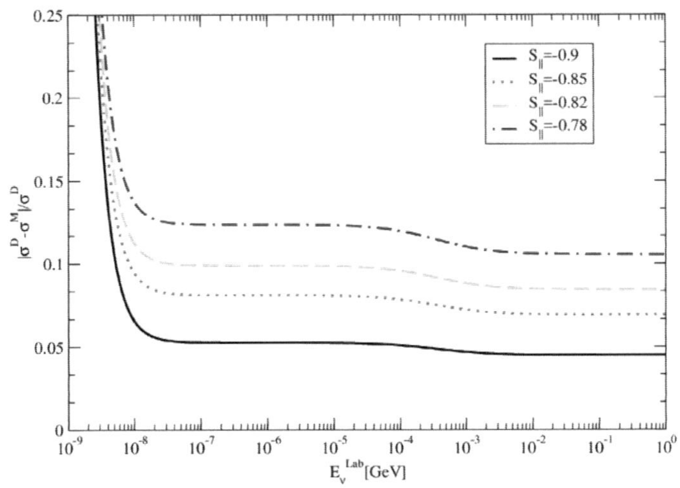

Figure 1. Difference between the Majorana and Dirac neutrino-electron elastic scattering cross section as a function of the incident neutrino energy and $m_\nu = 1eV$. The black solid line corresponds to a polarization $s_{\|} = -0.9$, dotted red line for $s_{\|} = -0.85$, broken green line for $s_{\|} = -0.82$ and chain blue line for $s_{\|} = -0.78$.

3. A new effective interaction

A straightforward way to introduce a more general interaction is to suppose an additional neutral current interaction with some special properties. Thus, we consider that its functional form remains equal than Standard Model neutral current interaction, but neutrino coupling constants for this case are the complex numbers $g_A^{\prime\nu_\ell} = e^{i\alpha}$ and $g_V^{\prime\nu_\ell} = e^{-i\beta}$, and not the typical values for this scattering process [5]. Thus, in the general case of complex coupling constants there are additional parameters, the phases of these constants just as reference [6] reports. Moreover, the electron axial and vector coupling constants do not change and have the same ordinary values.

3.1. Dirac case

We are interested again in the total amplitude for the neutrino-electron scattering process, but considering now that there is an additional contribution to amplitude (1) that comes from the new interaction described above. In Dirac case we have

$$\mathcal{M}_{D'} = -i\frac{G_F}{\sqrt{2}}\left[\bar{u}^f_{\nu_\ell}\gamma^\mu\left(\tilde{g}_V^{\nu_\ell} - \tilde{g}_A^{\nu_\ell}\gamma^5\right)u^i_{\nu_\ell}\right]\left[\bar{u}^f_e\gamma_\mu\left(g_V^\ell - g_A^\ell\gamma^5\right)u^i_e\right], \tag{7}$$

where

$$\tilde{g}_V^{\nu_\ell} = 1 + \epsilon g_V^{\prime\nu_\ell} \quad\text{and}\quad \tilde{g}_A^{\nu_\ell} = 1 + \epsilon g_A^{\prime\nu_\ell}. \tag{8}$$

The small dimensionless parameter ϵ is defined as follows

$$\epsilon = \frac{G'_F}{G_F}. \tag{9}$$

Here, G'_F corresponds to the effective coupling for the new physics interaction and it is according with $G'_F \ll G_F$ and consequently implies that $\epsilon \ll 1$. When $\epsilon = 0$, the new interaction is totally suppressed and we will recover the standard amplitudes in both cases, Majorana and Dirac.

3.2. Majorana case

On the other hand, if the neutrino is a Majorana particle, then it is necessary to incorporate the contribution that comes from new interaction for the neutrino and anti-neutrino to the total amplitude in (2), that is

$$\mathcal{M}_{M'} = -i\frac{G_F}{\sqrt{2}}\left[\bar{u}_e^f \gamma_\mu \left(g_V^\ell - g_A^\ell \gamma^5\right) u_e^i\right]\left[\bar{u}_{\nu_\ell}^f \gamma^\mu \left(\tilde{g}_V^{\nu_\ell} - \tilde{g}_A^{\nu_\ell}\gamma^5\right) u_{\nu_\ell}^i - \bar{v}_{\nu_\ell}^f \gamma^\mu \left(\tilde{g}_V^{\nu_\ell^*} - \tilde{g}_A^{\nu_\ell^*}\gamma^5\right) v_{\nu_\ell}^i\right]. \quad (10)$$

In the same way in which the identity (4) was obtained by applying the fact that the neutrino is a Majorana particle, we can construct an equivalent expression for the anti-neutrino term in (10) but considering that neutrino coupling constants are complex numbers, so the following identity is valid

$$\bar{v}_{\nu_\ell}^f \gamma^\mu \left(\tilde{g}_V^{\nu_\ell^*} - \tilde{g}_A^{\nu_\ell^*}\gamma^5\right) v_{\nu_\ell}^i = \bar{u}_{\nu_\ell}^f \gamma^\mu \left(\tilde{g}_V^{\nu_\ell^*} + \tilde{g}_A^{\nu_\ell^*}\gamma^5\right) u_{\nu_\ell}^i. \quad (11)$$

Using last equation, the Majorana amplitude in (10) is now given by

$$\mathcal{M}_{M'} = -i\frac{G_F}{\sqrt{2}}\left[\bar{u}_{\nu_\ell}^f \gamma_\mu \left(\xi_V - \xi_A \gamma^5\right) u_{\nu_\ell}^i\right]\left[\bar{u}_e^f \gamma^\mu \left(g_V^\ell - g_A^\ell \gamma^5\right) u_e^i\right], \quad (12)$$

where the new effective coupling constants are $\xi_V = 2i\mathrm{Im}\left\{\tilde{g}_V^{\nu_\ell}\right\}$ and $\xi_A = 2\mathrm{Re}\left\{\tilde{g}_A^{\nu_\ell}\right\}$, or equivalently

$$\xi_V = -2i\epsilon\sin(\beta) \quad \text{and} \quad \xi_A = 2\left[1 + \epsilon\cos(\alpha)\right]. \quad (13)$$

On the one hand, equation (7) looks similar to (1), actually we only rewrote that equation in a more general expression, where axial and vector coupling constants are complex and they depend on ϵ and phases α and β, but on the other hand the amplitude for a scattering process in Majorana case with that additional interaction, does not look like (5), because of their functional forms are disparate, that is, while in the Dirac case the amplitudes have axial and vector terms in the neutrino block, but different values, in Majorana case when a new interaction is introduced in the model, with the properties described above, the resulting amplitude has, not only a vector term as in (5), but an axial and a vector terms. Thus, the use of identity (11) does not cancel the vector part in the Majorana amplitude for the scattering process with a new interaction.

We will assume for simplicity that the value of the phases of neutrino complex coupling constants are identical, that is $\alpha = \beta = \delta$. Thus, for a given value of incoming neutrino energy $E_{\nu_\ell}^{Lab}$ the total cross sections will be three-parametric families that depend on the dimensionless parameter ϵ, the phase of vector and axial couplings δ and the neutrino longitudinal polarization $s_\parallel$. Figure 2 shows the function defined in (6) but changing $d\sigma^M/dE_{\nu_\ell}^{Lab}$ by $d\sigma^{M'}/dE_{\nu_\ell}^{Lab}$, where $d\sigma^{M'}/dE_{\nu_\ell}^{Lab}$ is the differential cross section referred to the lab frame for a neutrino-electron scattering process with the new interaction, and $\ell \neq e$ (left) and $\ell = e$ (right) in (5). The dotted blue lines and the dashed red lines in the plots correspond to $D_{DM'}\left(E_{\nu_\ell}^{Lab}, s_\parallel = -0.9, \epsilon = 0.01, \delta\right)$ for $\delta = 0$ (up) and $\delta = \pi$ (down), respectively, while green dash-dotted line is the difference between $d\sigma^D/dE_{\nu_\ell}^{Lab}$ and $d\sigma^{D'}/dE_{\nu_\ell}^{Lab}$ when $\delta = \pi$. The solid line corresponds to the difference function $D_{DM'}\left(E_{\nu_\ell}^{Lab}, s_\parallel = -0.9, \epsilon = 0\right)$.

4. Conclusion

For some appropriate values for parameters ϵ, $s_\parallel$ and δ it is possible to obtain an overlap region, just as we can see from plots in Figure 2. Once we take all contributions for neutrino-electron scattering process into account, this overlap region suggests that for a certain energy $E_{\nu_e}^{Lab}$ of the incident neutrino there is confusion between Dirac and Majorana neutrinos if we include an additional interaction with vector and axial complex coupling constants for the neutrino, then it is not possible to distinguish with this model if the neutrino is a Dirac or Majorana particle taken specific values for the involved parameters, even if the longitudinal polarization $s_\parallel$ is different from minus one.

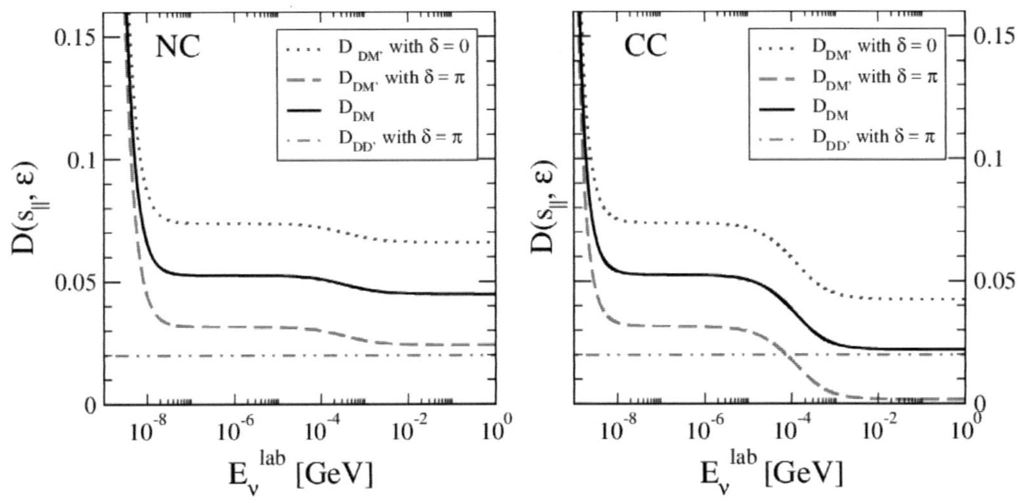

Figure 2. The blue dotted line and the red dashed line are the differences between $d\sigma^D/dE^{Lab}_{\nu_\ell}$ and $d\sigma^{M'}/dE^{Lab}_{\nu_\ell}$ for $\delta = 0$ and $\delta = \pi$, respectively; dash-dotted line is the plot for $D_{DD'}$ with $\delta = \pi$; the solid line corresponds to equation (6). The parameter values for these plots are $s_{||} = -0.9$, $\epsilon = 0.01$ and a neutrino mass $m_\nu = 1$ eV.

Acknowledgments
We thank CONACyT and Division of Particles and Fields of the Mexican Physical Society for the financial support for the XV Mexican Workshop on Particles and Fields.

References
[1] Majorana E 1937 *Nuovo Cimento* 5 171-84
[2] Kayser B 1982 *Phys. Lett.* B **112** 137-42
[3] Campagne J E 1995 *Effects of the nature of Dirac neutrino or of Majorana neutrino as well as the effects of the mass, on the behavior of neutrinos (in french)* LAL-95-23
[4] Barranco J, Delepine D, Gonzalez-Macias V, Lujan-Peschard C and Napsuciale M 2014 *Phys. Lett.* B **739** 343-7
[5] 't Hooft G 1971 *Phys. Lett.* B **37** 195-6
[6] Ciechanowicz S and Sobków W 2005 *Phys. Rev.* D **71**

Journal of Physics: Conference Series **761** (2016) 012066 doi:10.1088/1742-6596/761/1/012066

Hadronic matter at the edge: A survey of some theoretical approaches to the physics of the QCD phase diagram

Alejandro Ayala

Instituto de Ciencias Nucleares, Universidad Nacional Autónoma de México, Apartado Postal 70-543, México Distrito Federal 04510, Mexico
and
Centre for Theoretical and Mathematical Physics, and Department of Physics, University of Cape Town, Rondebosch 7700, South Africa.

Abstract.
In the past few years a wealth of high quality data has made possible to test current theoretical ideas about the properties of hadrons subject to extreme conditions of density and temperature. The relativistic heavy-ion program carried out at the CERN-SPS and under development at the BNL-RHIC and CERN-LHC has provided results that probe the evolution of collisions of hadronic matter at high energies from the initially large density to the late dilute stages. In addition, QCD on the lattice has produced results complementing these findings with first principles calculations for observables in a regime where perturbative techniques cannot describe the nature of strongly coupled systems. This work aims to review some recent developments that make use of field theoretical methods to describe the physics of hadrons at finite temperature and density. I concentrate on two of the main topics that have been explored in the last few years: (1) The search for the structure of the phase diagram and (2) analytical signals linked to the chiral symmetry restoration/deconfinement.

1. The QCD phase diagram

The different phases in which matter made up of quarks and gluons arranges itself depends, as for any other substance, on temperature and density or equivalently, on the chemical potentials coupled to the corresponding quark quantum numbers. Our experimental knowledge on the structure of the phase diagram comes basically from information collected in relativistic heavy-ion collisions. A working theoretical assumption is that local equilibrium is a good description after the early stages of the collision. This assumption, together with the large amount of particles produced in the reactions, emphasizes the need to account for the density variable. Since u, d and s are the relevant quark species for temperatures and densities close to the phase boundaries, a complete description should in principle be given in terms of possibly distinct chemical potentials describing the density associated to these quarks. Nevertheless, in equilibrium, these chemical potentials are not independent from each other. There are two basic requirements that they need to satisfy: beta equilibrium and charge neutrality. This means that out of the three chemical potentials, we can choose one of them as independent. In this manner, the phase diagram becomes two dimensional.

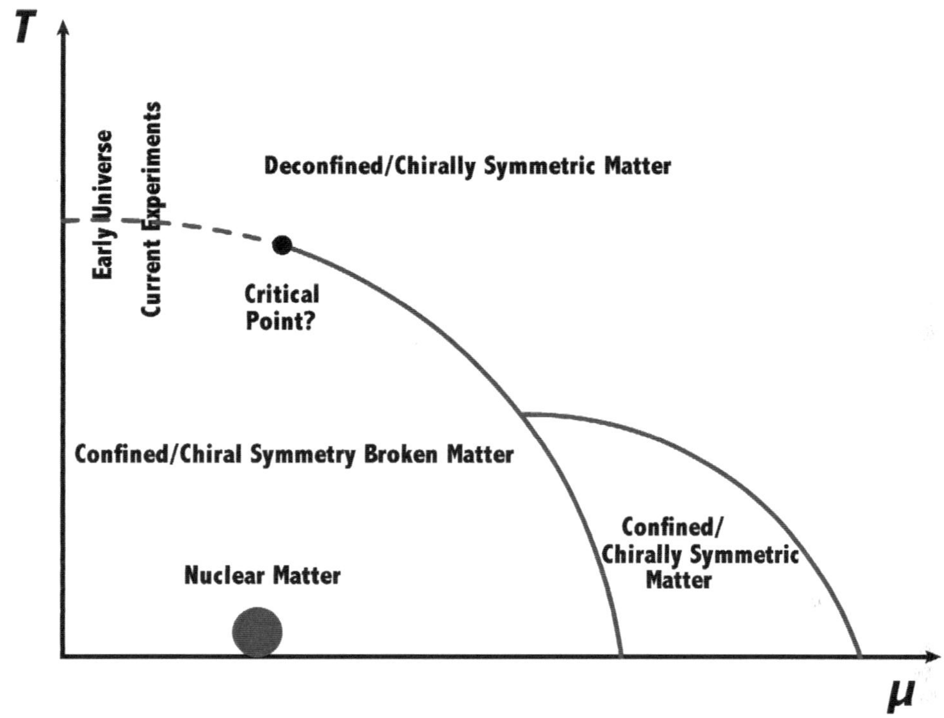

Figure 1. Schematic representation fo the QCD phase diagram in the $T - \mu$ plane.

A popular representation of this two-dimensional phase diagram, where the independent chemical potential chosen is μ, the light-quark chemical potential, is given in Fig. 1. μ is related to the baryon chemical potential μ_B by $\mu_B = 3\mu$. It should be noted that most of our knowledge of the phase diagram is restricted to the $T \geq 0$, $\mu = 0$ axis. The phase diagram is by and large unknown. For physical quark masses and $\mu = 0$ lattice calculations have shown [1] that the change of phase, from where the dominant degrees of freedom are quarks and gluons (QGP), to where the dominant ones are hadrons, is an analytic crossover and thus it is not a real phase transition. This means that the change is smooth, as opposed to presenting a jump in some of the thermodynamic variables. This result was first obtained analytically in Ref. [2] from a model of phase transitions for statistical systems that can possess different bound states, described as compact clusters (see also Ref. [3]). Given that there is no true phase transition to the QGP as the temperature increases, the definition of the (pseudo) critical temperature, T_c, is to some extent a matter of choice. Since for systems where there is a true phase transition, the transition lines are defined in terms of the equality of the free energies in one phase and the other, one can resort to analyzing observables sensitive to the free-energy to define T_c.

To describe chiral symmetry restoration, one of these observables is the quark condensate $\langle\bar{\psi}\psi\rangle$ which, from the Hellmann-Feynman theorem generalized to statistical systems, can be

Journal of Physics: Conference Series **761** (2016) 012066 doi:10.1088/1742-6596/761/1/012066

written as

$$\langle\bar{\psi}\psi\rangle = \frac{d}{dm}\Omega, \tag{1}$$

where Ω is the grand potential or free-energy density and m is the (light) quark mass (hereby I make no distinction between the mass of u and d-quarks).

In the chiral limit ($m = 0$) the quark condensate is a true order parameter since it is finite in the chirally broken (Nambu-Goldstone) phase and vanishes in the chirally symmetric (Wigner-Weyl) phase. In this limit the (true) critical temperature can be obtained from the *chiral susceptibility*

$$\chi_m = \frac{\partial}{\partial m}\langle\bar{\psi}\psi\rangle, \tag{2}$$

which has the critical behavior $\chi_m \sim m^{1/\delta - 1}$. For $O(N)$ scaling $\delta > 1$. Therefore χ_m diverges at the critical temperature, where $m = 0$.

For physical quark masses neither $\langle\bar{\psi}\psi\rangle$ vanishes nor χ_m diverges at the pseudocritical temperature. Nevertheless these quantities retain a reminiscent behavior of the corresponding one in the chiral limit. In particular χ_m has a peaked structure as a function of T and it is customary to define T_c as the temperature for which χ_m reaches its peak. Other susceptibilities, such as $\chi_T = \frac{\partial}{\partial T}\langle\bar{\psi}\psi\rangle$, can also be used to define T_c. It has been shown [4] that the critical temperatures thus obtained coincide within a narrow band and therefore using any of these susceptibilities yields basically the same T_c.

Lattice calculations have provided values for T_c extracted from the peak of χ_m for 2+1 flavors using different types of improved staggered fermions [5]. These values show some discrepancies. The MILC collaboration [6] obtained $T_c = 169(12)(4)$ MeV. The BNL-RBC-Bielefeld collaboration [7] reported $T_c = 193(7)(4)$ MeV. The Wuppertal-Budapest collaboration [8] has consistently obtained smaller values, the latest being $T_c = 147(2)(3)$ MeV. The HotQCD collaboration [9] has reported $T_c = 154(9)$ MeV. The differences could perhaps be attributed to different lattice spacings.

To describe deconfinement a variable that is often used is the Polyakov loop L, which is regarded as an order parameter in such a way that the thermal average of its trace, $\langle \mathrm{Tr}\ L\rangle$, vanishes in the confined phase and is finite in the deconfined phase. The Polyakov loop is a particular kind of Wilson loop where the integration path C is taken along the periodic time direction, namely

$$L = \mathcal{P}e^{ig\oint_C dx^\mu A_\mu}, \tag{3}$$

where $\mathcal{P}$ is the path order operator, g the QCD coupling constant and A_μ is the gauge field. It can be shown [10] that $\langle \mathrm{Tr}\ L\rangle$ satisfies

$$\langle \mathrm{Tr}\ L\rangle \propto e^{-\Delta F_q/T}, \tag{4}$$

where ΔF_q is the difference between the free-energy of the gauge theory containing an isolated static quark and the free-energy of a pure gauge system. Equation (4) means that $\langle \mathrm{Tr}\ L\rangle = 0$ if and only if, for a given T, ΔF_q becomes infinite. This is interpreted as the impossibility to have an isolated quark and therefore this situation must correspond to the confined case. However, when $\langle \mathrm{Tr}\ L\rangle \neq 0$ for a given T, ΔF_q must be finite. This means that quarks are deconfined, in the sense that they exist as isolated, finite energy states bearing their corresponding quantum numbers.

An isolated static quark can be regarded as an infinitely heavy one. Therefore, for physical quark masses the Polyakov loop is no longer an order parameter and the extraction of T_c is based on less firm soil. Nevertheless, lattice calculations have also turned their attention to the computation of the Polyakov loop. The HotQCD collaboration has made systematic studies showing that, once discretization effects are accounted for, the Polyakov loop yields consistent estimates of the thermal properties of the theory with a small quark mass dependence in the transition region. The Polyakov loop however rises faster for heavier quarks at higher temperatures [9].

The unambiguous picture presented by lattice QCD for $T \geq 0$, $\mu = 0$ cannot be easily extended to the case $\mu \neq 0$, the reason being that standard Monte Carlo simulations can only be applied to the case where μ is either 0 or purely imaginary. Simulations with $\mu \neq 0$ are hindered by the *sign problem* [11]. Recall that in the computation of the QCD partition function with finite μ, integration over each fermion field produces a *fermion determinant*, i.e. a factor $\mathrm{Det}M \equiv \mathrm{Det}(\not{D} + m + \mu\gamma_0)$, where M is the fermion matrix. Let us consider in general a complex μ. Taking the determinant on both sides of the identity

$$\gamma_5(\not{D} + m + \mu\gamma_0)\gamma_5 = (\not{D} + m - \mu^*\gamma_0)^\dagger, \tag{5}$$

we get

$$\mathrm{Det}(\not{D} + m + \mu\gamma_0) = [\mathrm{Det}(\not{D} + m - \mu^*\gamma_0)]^*, \tag{6}$$

which shows that the fermion determinant is not real unless $\mu = 0$ or purely imaginary. For real μ the direct sampling of a finite density ensemble by Monte Carlo methods is not possible, since the sampling requires a real non-negative measure. This problem is referred to as the sign problem, although a more appropriate name would seem to be the *phase problem*.

That the integrand of the partition function becomes complex would seem to be only a minor inconvenience. A naive approach to deal with the sign problem would be to write [12]

$$\mathrm{Det}M = |\mathrm{Det}M|e^{i\theta}. \tag{7}$$

To compute the thermal average of an observable O in QCD one writes

$$\langle O\rangle = \frac{\int DUe^{S_{YM}}\mathrm{Det}M\ O}{\int DUe^{S_{YM}}\mathrm{Det}M} = \frac{\int DUe^{S_{YM}}|\mathrm{Det}M|e^{i\theta}\ O}{\int DUe^{S_{YM}}|\mathrm{Det}M|e^{i\theta}}, \tag{8}$$

where S_{YM} is the Yang-Mills action. Notice that written in this manner, the simulations could be implemented in terms of the phase-quenched (pq) theory where the measure involves $|\mathrm{Det}M|$ and the thermal average could be written as

$$\langle O\rangle = \frac{\langle Oe^{i\theta}\rangle_{\mathrm{pq}}}{\langle e^{i\theta}\rangle_{\mathrm{pq}}}. \tag{9}$$

The average phase factor (also referred to as the average sign) in the phase-quenched theory can be written as

$$\langle e^{i\theta}\rangle_{\mathrm{pq}} = e^{-V(f-f_{pq})/T}, \tag{10}$$

where f and f_{pq} are the free-energy densities in the full and phase-quenched theories, respectively and V is the three-dimensional volume. If $f - f_{pq} \neq 0$, the average phase factor decreases exponentially when V goes to infinity (the thermodynamic limit) and/or T goes to zero. Under these circumstances the signal to noise ratio worsens. This is referred to as the *severe sign problem*.

To circumvent the sign problem, a possibility is to determine the first Taylor coefficients in the expansion of a given observable in powers of μ_B/T. The coefficients of the expansion can be expressed as expectation values of traces of matrix polynomials taken in the $\mu_B = 0$ ensemble. Although care has to be taken regarding the growth of the statistical errors with the order of the expansion, this strategy has provided a very important result: the curvature of the transition line at $\mu_B = 0$. The curvature κ is defined as the dimensionless coefficient of the quadratic term in the Taylor expansion of the pseudocritical line

$$T_c(\mu_B) = T_c\left(1 - \kappa\frac{\mu_B^2}{T_c^2}\right), \qquad \kappa \equiv -\left(T_c\frac{dT_c(\mu_B)}{d\mu_B^2}\right)\Bigg|_{\mu_B=0}. \tag{11}$$

Values for κ between 0.006 – 0.02 have been reported [13]. It should be noted that since the phases are separated by a crossover, the curvature should depend in principle on the observable considered. Nevertheless, these curvatures give considerable smaller values than that of the chemical freeze-out curve obtained from statistical models [14]. This observation could be of potential importance for if the pseudo critical line is flatter than the chemical freeze-out line, the distance between the possible QCD critical end point (CEP) and the freeze-out curve increases. If this happens then possible experimental evidences for criticality may be washed out by the moment when particle abundances are established after a heavy-ion collision.

To distinguish between chiral and deconfinement transitions it has been suggested that one can look at the effects of a magnetic background, which enters as a new control parameter for the thermodynamics. Peripheral heavy-ion collisions might be used as a physically realizable situation to explore this possibility since they generate very intense magnetic fields, even stronger than magnetars. The possible splitting of the deconfinement and chiral transitions in an external magnetic field has been studied in Ref. [15] within the linear sigma model coupled to quarks (with two flavors) and to Polyakov loops. The authors found that the vacuum correction from quarks on the phase structure was dramatic. When ignoring this correction, the confinement and chiral phase transition lines coincide. However, inclusion of the correction led to an splitting of the confinement and chiral transition lines, and both chiral and deconfining critical temperatures became increasing functions of the magnetic field. The vacuum contribution from the quarks drastically affected the chiral sector as well. Since these conclusions were drawn from a model that does not reproduce the behavior of the critical temperature as a function of the magnetic field, found by lattice QCD calculations, they are nowadays regarded as incorrect. Furthermore, a recent lattice result shows that no significant difference between chiral and deconfinement transition temperatures exists up to fields as intense as 3.25 GeV^2 [16]. On the contrary, the different definitions of T_c tend to approach each other and they cluster between 109 - 112 MeV at this high value of the magnetic field.

The behavior of strongly interacting matter in the presence of magnetic fields has become a subject of increasing interest over the past few years. This interest has been sparked by recent lattice results showing that the transition temperature with 2 + 1 quark flavors decreases with increasing magnetic field and that the strength of the condensate decreases for temperatures close and above the phase transition [17, 18, 19]. This behavior has been dubbed *inverse magnetic catalysis*, and it has been the center of attention in a large number of model-dependent analyses [20] – [31]. For recent reviews see Refs. [32, 33]. In general terms, it seems that inverse magnetic catalysis is not obtained in mean field approaches describing the thermal environment [15], [34] – [38], nor when calculations beyond mean field do not include magnetic effects on the coupling constants [39].

The novel feature implemented in effective models, able to account self-consistently for inverse magnetic catalysis, is the decrease of the coupling constants with increasing field strength obtained from the model itself [40] without resorting to *ad hoc* parametrizations. This has been achieved within the Abelian Higgs model and the linear sigma model coupled to quarks (LSMq). This behavior is made possible by accounting for the screening properties of the plasma, which have been recently formulated consistently for theories with spontaneous symmetry breaking. This results in a formalism beyond the mean field approximation. Screening is also important to obtain a decrease of the QCD coupling constant with the magnetic field in the Hard Thermal Loop approximation [41]. In the opposite limit, it has also been shown by means of a QCD sum rules analysis at zero temperature, that both, the threshold energy for the onset of the continuum in the quark vector current spectral density (a phenomenological parameter that signals the onset of deconfinement) and the gluon condensate increase with increasing magnetic field, as expected [42].

The LSMq has been also used to explore the phase diagram without magnetic fields [43]. It was found that there are values for the model couplings that allow locating a critical end point (CEP) in the region where lattice inspired calculations find it [44]. Since the LSMq does not exhibit confinement, this behavior is attributed to the proper treatment of plasma screening, instead of to the existence of a given confinement length scale [45]. In addition, recent LQCD calculations [16] show that for very strong magnetic fields, inverse magnetic catalysis prevails and the phase transition becomes first order at asymptotically large values of the magnetic field for vanishing quark chemical potential μ. A similar behavior is obtained in the Nambu Jona-Lasinio model if one includes a magnetic field dependence of the critical temperature in agreement with lattice QCD results [46].

In Ref. [47] the LSMq was used to explore the consequences of a proper handling of the plasma screening properties in the description of the magnetized effective QCD phase diagram. It was shown that when including self-consistently magnetic field effects in the calculation of both the effective potential as well as on the thermo-magnetic dependence of the coupling constants, the CEP moves toward smaller values of the critical quark chemical potential and larger values of the critical temperature. In addition, above a certain value of the field strength the CEP moves toward the T-axis. This behavior can be understood on general grounds, as the magnetic field produces a dimensional reduction, whereby virtual charged particles from the vacuum are effectively constrained to occupy Landau levels, thus restricting their motion to a plane. This makes these particles to lay closer to each other on average, thus reducing the interaction strength for strongly coupled theories. This situation takes place regardless of how weak the external field is.

The search for the CEP is one of the main motivations for the RHIC beam energy scan (BES) program. The two RHIC experiments, PHENIX and STAR, have reported their results for the BEC phase I (see for example Refs. [48] and [49]) in Au + Au collisions at $\sqrt{s_{NN}} = 7.7, 11.5, 19.6, 27, 39,$ and 62.4 GeV. As the collision energy decreases, the stopping power becomes larger and therefore μ_B becomes also larger which in turn allows to experimentally probe deeper into the phase diagram. The experimental findings reveal that for central collisions, as the collision energy lowers, several signals of the QGP fade away. Among them we can mention the disappearance of the elliptic flow scaling with the number of constituent quarks, the growth in the difference of the elliptic flow between particles and anti-particles [50] and the smaller strength of the pion R_{AA} [51].

Of particular relevance for identifying and locating the CEP is the search for event-by-event fluctuations of conserved charges. The idea is that these fluctuations are proportional to powers of the relevant correlation length which in turn must diverge, or at least present a rapid growth, near criticality. The analogue is the liquid-gas transition where the *critical opalescence* phenomenon takes place as the critical point is approached. In this region of the phase diagram, the sizes of the gas and liquid domains begin to fluctuate over larger length scales. As the size of the density fluctuations (the correlation length) becomes comparable to the wavelength of light, this last is scattered in all directions causing the normally transparent liquid to appear cloudy, or opal-like, and hence the name of the phenomenon.

It has been recently argued that in order to look for critical behavior it is important to target moments of higher order than quadratic of the event-by-event conserved charge's distribution, since these are proportional to higher powers of the correlation length and thus in principle easier to detect [52]. It is even better to look for ratios of these moments (or of the cumulants) [53] since parameters such as the fireball volume, which are poorly known, cancel and the ratios are better comparable to lattice predictions [54]. The shapes of the distributions are also expected to be sensitive to the presence of the critical point [55]. PHENIX has measured the skewness and the kurtosis of net charge distributions for four different collision energies $\sqrt{s_{NN}} = 200, 62.4, 39$, and 7.7 GeV finding no excess above the values from simulations that do not include effects of critical behavior. In contrast, the CERN-NA49 collaboration has reported a significant intermittency signal in the transverse momentum correlations of opposite sign pions in Si + Si collisions at top SPS energies [56] approaching in size some predictions for critical behavior in QCD [57].

The experimental search for the structure of the phase diagram for hot and dense matter will continue with the RHIC-BES phase II scheduled to start in 2018 and for cold nuclear matter with the starting of the EIC program sometime after 2024. The NA61/SHINE beam energy and system size scan, started in 2009, will continue pursuing evidence for the existence of the critical point hinted by NA49 fluctuation data [58]. In addition new planned facilities like the Facility for Antiprotons and Ion Research (FAIR) in GSI and the Heavy-Ion Collider Facility NICA at JINR-Dubna are under construction and promise to deepen the exploration of the QCD phase diagram in the future.

2. Analytical signals of chiral symmetry restoration/deconfinement

The connection between parameters describing chiral symmetry restoration/deconfinement and observables such as the hadron spectral density is provided by the QCD sum rules (QCDSRs) approach. This analytical method is, to a large extent, complementary to the numerical lattice simulations of hadrons. In this section we concentrate on summarizing recent progress employing the QCDSRs approach for the computation of the hadronic spectral density.

The QCDSRs method was developed more than thirty years ago [59] and ever since it has become a prime working tool in hadron phenomenlogy. QCDSRs rest on two pillars [60], (i) the operator product expansion (OPE) of current correlators at short distances beyond perturbation theory, and (ii) Cauchy's theorem in the complex squared energy s-plane (a dispersion integral), which relates the (hadronic) discontinuity across the cut on the real positive semi-axis with the integral around a contour of radius $|s_0|$ (which could be taken to infinity) where the OPE is expected to be valid. The latter is usually referred to as quark-hadron duality.

An important object in QCD is the amplitude for the quark-pair creation and annihilation, the so called current correlator. The formal expression for this amplitude at $T, \mu = 0$ can be written as

$$\Pi_{\mu\nu}(q) = i \int d^4x e^{iq\cdot x} \langle 0|\mathcal{T}[j_\mu(x) j_\nu^\dagger(0)]|0\rangle, \tag{12}$$

where $q^\mu = (q_0, \mathbf{q})$ is the four-momentum carried by the current, $j_\mu = \bar{\psi}\gamma_\mu\psi$ is the colorless quark-current of a given flavor (for our purposes $\psi = u, d, s, c \dots$ or a linear combination of these), $\langle 0| \dots |0\rangle$ is the vacuum expectation value and $\mathcal{T}$ is the time order product operator. Current conservation dictates that the Lorentz structure of the correlator is

$$\Pi_{\mu\nu}(q) = -q^2 \left(g_{\mu\nu} - \frac{q_\mu q_\nu}{q^2} \right) \Pi(q^2). \tag{13}$$

$\Pi(q^2 \equiv s)$ contains all the dynamical effects.

At finite temperature and/or density the correlator can be written as

$$\begin{aligned}\Pi_{\mu\nu}(q_0^2, \mathbf{q}^2) &= i\int d^4x e^{iq\cdot x}\langle \mathcal{T}[j_\mu(x)j_\nu^\dagger(0)]\rangle \\ &= -q^2\left[\Pi_0(q_0^2, \mathbf{q}^2)P_{\mu\nu}^T + \Pi_1(q_0^2, \mathbf{q}^2)P_{\mu\nu}^L\right], \end{aligned} \tag{14}$$

where now $\langle\cdots\rangle$ refers to a thermal average and $P_{\mu\nu}^T$ and $P_{\mu\nu}^L$ are the three-dimensional transverse and longitudinal projectors (both are four dimensionally transverse), respectively. At finite temperature and/or density it is customary to work in the *static* limit $\mathbf{q} \to 0$. The rationale is that in equilibrium and for temperatures around the (pseudo)critical temperature, the average three-momentum is small compared to the binding energy (mass) of the relevant hadrons and therefore these last can be considered as slowly moving. In this limit $\Pi_{\mu\nu}$ contains only spatial components and therefore the only remaining coefficient in Eq. (14) is $\Pi_0(q_0^2 \equiv s)$ which now carries the information on the dynamical effects. The temperature behavior of the light-quark vector current correlator in the framework of thermal QCDSRs was first discussed in Ref. [61] and later reanalyzed by others [62].

Let us argue how the spectral density is related to the physics of hadrons. $\Pi(s)$ and $\Pi_0(s)$ can be thought of as analytical functions of s defined for negative (space-like), positive (time-like) or even complex values of s. Also, for large positive s, the integral in Eqs. (12) and (14) are dominated by large values of x^2 [60]. This means that the long distance QCD interactions become important, forcing the quarks in the current to eventually form hadrons. In particular, a quark-antiquark pair created by the current j_μ with $J^P = 1^-$ becomes a neutral vector meson. The ground-state vector mesons with the isospin $I = 1$ and $I = 0$ are ρ and ω having the quark content $(\bar{u}u - \bar{d}d)/\sqrt{2}$ and $(\bar{u}u + \bar{d}d)/\sqrt{2}$, respectively. The lightest vector meson created by the current $\bar{s}\gamma_\mu s$ is ϕ. For large negative values of s, the integral in Eqs. (12) and (14) are dominated by small values of x^2. Therefore quarks propagate predominantly at short distances and during short time intervals. The QCD interactions are suppressed due to asymptotic freedom. Therefore, as a first approximation, one may calculate the current correlator using perturbation theory. At finite temperature and/or density, one needs to restrict this perturbative piece to the leading, one-loop level, as the treatment of the appearance of two scales in $\alpha_s(q^2, T)$, i.e., Λ_{QCD} and T_c, remains an open problem. Corrections to this approximation should account for the fact that the QCD vacuum influences this propagation. This influence is parametrized in terms of the condensates and expressed in terms of the OPE.

A direct measurement of the in-medium hadronic spectral density in the vector channel, Π_V, is provided by the invariant-mass spectrum of low mass dileptons. Dileptons and photons constitute what is generically called *electromagnetic probes* in heavy-ion collisions. These probes are emitted continuously from the early quark-gluon plasma phase through to the late hadronic phase, therefore revealing the entire thermal evolution of the reaction. Since their mean free path is larger than the size of the fireball, these probes escape from the medium without further interaction. For invariant masses below 1 GeV, the spectrum is dominated by the light vector mesons ρ, ω and ϕ. The temperature and density dependence of the spectral function for these particles reflects their mass distribution and thus opens a window to experimentally study the dynamical mass generation mechanism in QCD. Among these vectors, ρ is an ideal test particle to sample in-medium changes of parameters such as mass, width and leptonic decay constant, given its short lifetime and large coupling to pions and muons. There is no similar experimental knowledge of the in-medium axial vector spectral density, Π_A.

Data from the CERN-NA60 Collaboration [63, 64] has allowed to settle a long standing controversy regarding the origin of the dilepton excess [65] below the ρ peak at SPS energies. The data shows that when the known sources are subtracted from the spectral function, a clear peak that broadens for the most central collisions but remains centered at its vacuum mass value

Journal of Physics: Conference Series **761** (2016) 012066 doi:10.1088/1742-6596/761/1/012066

is observed at all centralities. The total dimuon yield also increases with centrality. This result is in line with current ideas about the nature of the deconfinement/chiral symmetry restoration transition; a width which increases with increasing T and/or μ indicates deconfinement, with the spectral function becoming smooth and eventually accounted for by perturbative QCD, irrespective of the changes in the mass [66]. One could picture the extreme situation of the mass decreasing and vanishing at some critical temperature. If this behavior is not accompanied by an increasing width then the hadron would still be present in the spectral function at that temperature.

The leading theoretical descriptions of the ρ broadening at SPS energies are based on hadronic many-body or transport calculations, invoking the idea that the ρ scatters and thus melts not only at finite temperature, but also in a baryon-rich environment [67] – [71] (see also Ref. [72]). These rather successful descriptions have allowed the interpretation of a wealth of fixed target data. In principle, this approach could also be expected to describe data at RHIC and LHC, where the net density remains comparable to SPS energies although the baryon density is smaller. However, the PHENIX collaboration has found a large enhancement in central Au + Au at $\sqrt{s_{NN}} = 200$ GeV [73] which cannot be described by the in-medium hadronic effects invoked at SPS energies, albeit this description works well for lower centralities and lower energies. This enhancement is not reported by STAR [74]. Recently, PHENIX has reported new measurements with a better signal to background ratio performed with their Hadron Blind Detector which come closer to STAR results [75].

One way to connect the spectral densities Π_V and Π_A with a signal of chiral symmetry restoration is through a possible mixing of their vacuum values Π_V^0 and Π_A^0. This idea was first pursued in Ref. [76] and led to the deduction of the expression

$$\Pi_{V,A}(q^2;T) = [1-\epsilon(T)]\Pi^0_{V,A}(q^2) + \epsilon(T)\Pi^0_{A,V}(q^2). \tag{15}$$

Equation (15) is valid for low temperatures and when considering only pions in the sum over excited states in the Gibbs averages. It represents the leading T-dependence of the spectral densities which are related by the mixing parameter $\epsilon(T)$ which in the chiral limit, $m_\pi = 0$, becomes $\epsilon(T) = T^2/6f_\pi^2$, where f_π is the pion decay constant. The mixing parameter is proportional to the scalar pion density. Equation (15) satisfies Weinberg-type sum rules [77, 78] (WSR) provided $\Pi^0_{V,A}$ also satisfy them. The physical process responsible for the mixing is the interaction of ρ with pions to form a_1. In the chiral limit, when $\epsilon(T) \simeq 0.5$, Eq. (15) shows that Π_V and Π_A become degenerate, which is interpreted as signaling chiral symmetry restoration. Notice however that even in the chiral limit, this happens for $T \simeq 160$ MeV, temperature for which other than only pion degrees of freedom are expected to be present and the above expression is no longer reliable.

When the mixing scenario is pursued, it has been argued that it is possible to use Eq. (15) to extract information on the relation between mixing of spectral densities and chiral symmetry restoration in a more sophisticated setup. This has been done in Ref. [79] where use is made of vacuum spectral densities containing a postulated universal separation of nonperturbative and perturbative regimes. The former is represented by excited states and the latter by a degenerate continuum. The description includes also a finite pion mass. However, it is found that deviations from the QCDSRs set in at temperatures of the order of the pion mass suggesting the need of additional physics, beyond low-energy chiral pion dynamics, that is, the inclusion of resonances. This result may also indicate that the encoding of the finite temperature effects in such a simple mixing scenario may not be a good approximation for not too low temperatures.

More recently, another connection between chiral symmetry restoration/deconfinement and hadron spectral densities has been explored using the Finite Energy QCDSRs (FESR) approach. The method involves relating the moments of the hadronic spectral densities to condensates. However, the integrals are computed up to a *finite squared-energy* s_0, and thus the name of the method. s_0 corresponds to the square of the energy threshold for the onset of the continuum. This is a key parameter signaling deconfinement above which the hadronic spectral function is well approximated by perturbative QCD. The importance of this parameter was first pointed out in Ref. [61]. Explicit determinations of s_0 in several light and heavy-light quark systems [62], [80] – [83] find this parameter to decrease with increasing temperature, vanishing at a critical value interpreted as the deconfinement temperature.

Integrating the function $(s^N/\pi)/\Pi_0(s)$ on Eq. (14) in the complex s-plane along a contour with a fixed radius $|s| = s_0$, deforming the contour to make it pass slightly below and above the the positive real axis and assuming that $\Pi_0(s)$ is analytical within the integration contour we obtain the FESR

$$(-1)^{N-1}C_{2N}\langle O_{2N}\rangle = 8\pi^2 \left[\int_0^{s_0} ds s^{N-1}\frac{1}{\pi}\mathrm{Im}\Pi^{\mathrm{HAD}}(s) - \int_0^{s_0} ds s^{N-1}\frac{1}{\pi}\mathrm{Im}\Pi^{\mathrm{pQCD}}(s)\right], \quad (16)$$

where $N = 1, 2, \ldots$, and the leading order vacuum condensates in the chiral limit are the dimension $d = 4$ gluon condensate $C_4\langle O_4\rangle = (\pi/3)\langle\alpha_s G^2\rangle$, and the dimension $d = 6$ four-quark condensate, $C_6\langle O_6\rangle$. The implicit assumption is that $s_0(T,\mu)$ is large enough as to preserve the validity of the OPE with varying temperature and/or chemical potential.

An interesting connection between $s_0(T,\mu)$ and $\langle\bar\psi\psi\rangle(T,\mu)$ has been found in Ref. [84] in the analysis of the FESR in the axial channel. Results indicate that the critical temperatures for deconfinement and chiral symmetry restoration are basically the same within a few percent accuracy up to about $\mu = 100$ MeV where the method is reliable. FESR in the vector channel have also been used to determine the parameters describing the ρ-meson resonance. [85]. Assuming ρ saturation of the spectral function, and a Breit-Wigner resonance form

$$\frac{1}{\pi}\mathrm{Im}^{\mathrm{HAD}}(s) = \frac{1}{\pi}\frac{1}{f_\rho^2}\frac{M_\rho^3\Gamma_\rho}{(s-M_\rho^2)^2 + M_\rho^2\Gamma_\rho^2}, \quad (17)$$

it is found that a consistent solution for the three leading FESR are obtained when the ρ-width, Γ_ρ, increases and even diverges at T_c whereas the ρ-mass, M_ρ, and the leptonic decay constant, f_ρ, decrease only slightly near T_c.

This solution has recently been extended to finite μ and used to describe the dilepton spectrum around the ρ-peak [86]. The extension to finite chemical potential is made by first including the μ-dependence into the quark loop of the pQCD piece in Eq. (16) which splits the Fermi-Dirac distribution into particle–antiparticle contributions and then by incorporating the μ-dependence of the critical temperature T_c. This dependence is obtained from the results of Ref. [87] that make use of a Schwinger-Dyson approach, to find a parametrization for the crossover transition line between chiral symmetry restored and broken phases, valid for small values of μ,

$$T_c(\mu) = T_c(\mu = 0) - 0.218\mu - 0.139\mu^2. \quad (18)$$

Assuming a simple, Bjorken-like expansion cooling law, the results are in good agreement with NA60 dimuon data [64]. FESR have also been used in the axial-vector channel to explore the relation between the (pseudo)critical temperature for deconfinement transition –signaled by the vanishing of s_0– and that for chiral symmetry restoration –signaled by the vanishing of the light-quark condensate [88]. The result is that both temperatures coincide within the accuracy of the method.

The mixing scenario has been challenged in Ref. [89]. The argument is based on the dramatic difference between the vector and the axial-vector spectral functions already at $T = 0$ and on their very different evolution with increasing temperature, with the exception that s_0 is the same in both channels. With increasing T, the parameters of each channel evolve independently, thus

Journal of Physics: Conference Series **761** (2016) 012066 doi:10.1088/1742-6596/761/1/012066

keeping both spectral functions distinct. Eventually, this asymmetry is expected to vanish at deconfinement, when the ρ and the a_1 mesons disappear from the spectrum. This implies no chiral-mixing at any temperature, except obviously at $T \simeq T_c$. In addition to these differences there is an asymmetry due to the hadronic (pionic) scattering term, present in the vector channel at the leading, one-loop level, and strongly two-loop suppressed in the axial-vector case [90]. To be more specific let us consider the vector and axial-vector correlators. In a thermal bath, and in the hadronic representation one has (schematically)

$$\begin{aligned}\Pi_{\mu\nu}|_V &= \langle\pi|V_\mu(0)V_\nu(x)|\pi\rangle = \langle\pi|V_\mu(0)|\pi\rangle\langle\pi|V_\nu(x)|\pi\rangle + \langle\pi\pi|V_\mu(0)|\pi\pi\rangle\langle\pi\pi|V_\nu(x)|\pi\pi\rangle + \ldots \\ \Pi_{\mu\nu}|_A &= \langle\pi|A_\mu(0)A_\nu(x)|\pi\rangle = \langle\pi|A_\mu(0)|0\rangle\langle 0|A_\nu(x)|\pi\rangle + \langle\pi\pi\pi|A_\mu(0)|0\rangle\langle 0|A_\nu(x)|\pi\pi\pi\rangle + \ldots\end{aligned} \tag{19}$$

Notice that in Eq. (19) the chiral asymmetry is manifest at finite T, since Isospin and G-parity remain valid symmetries also at finite T. The leading term in the vector channel is the two-pion loop whereas in the axial-vector channel the leading term is the tree-level pion to-vacuum term (f_π) (pion pole) followed by a highly phase-space suppressed three-pion loop. In other words, the matrix element $\langle\pi|A_\mu(0)|0\rangle$, needed for chiral-mixing, vanishes identically at leading order. The correct matrix element, beyond the pion pole, is the phase-space suppressed second term on the second of Eqs. (19). In principle, this term could have a one-loop resonant sub-channel contribution from the matrix element $\langle\rho\pi|A_\mu(0)|0\rangle$. However, this contribution is also phase space suppressed due to the finite (albeit not too large) ρ-width.

Thermal Hilbert moment QCD sum rules have also been used to determine the behavior of hadronic parameters of charmonium in the scalar, pseudoscalar and vector channels at finite temperature [91]. From the results of the temperature dependence of the width and the leptonic coupling, the charmonium channels show signs of survival beyond the deconfinement pseudo critical temperature T_c. It should be noted in contrast that the light-light and even the heavy-light quark mesons disappear at that T_c. The difference is driven by two main effects: (1) the T-dependence of the light-light and heavy-light quarks systems in the pQCD sector is dominated by the spectral function in the time-like region, the so-called annihilation term, which is anyway relatively unimportant in relation to the light quark condensate contribution. The pQCD spectral function in the space-like region (scattering term) is highly suppressed. For heavy-heavy quark systems the scattering term becomes increasingly important with increasing temperature, while the annihilation term only contributes near threshold. (2) in the nonperturbative QCD sector of light-light and heavy-light quark correlators, the driving term in the OPE is the light quark condensate. This term is responsible for the behavior of $s_0(T)$. In contrast, for heavy-heavy quark correlators the leading power correction in the OPE is that of the gluon condensate, which has a very different temperature behavior.

The distinct signatures of the survival of charmonium states beyond T_c are the behavior of the width and the leptonic coupling. These parameters are almost independent of T up to $T \simeq 0.8T_c$ where the width begins to increase substantially. However, above $T \simeq 1.04T_c$ it starts to decrease sharply and the coupling increases also sharply. This behavior, which can mostly be traced back to that of the pQCD spectral function in the space-like region, points to the survival of the J/ψ, η_c and χ_c resonances above the deconfinement temperature. However, the QCD sum rules have no longer solutions for the hadronic parameters once the continuum threshold reaches the value $s_0|_{\min} \simeq M^2_{J/\psi}$, as there is no longer any support for the integrals. Hence, the temperature range explored with this technique does not extend beyond $T \simeq 1.22T_c$. A similar behavior is obtained for bottomonium ground states in the vector (Υ) and pseudo scalar (η_b) channels [92]. The width increases with temperature, but for $T/T_c \simeq 0.9$ it drops approaching its zero temperature value. The leptonic decay constant is basically a monotonically increasing function of T. The results are also interpreted as the survival of these bottonium states above T_c.

The QCDSRs method is powerful and can be implemented in a wide range of forms. Although it has been applied to the study of physical systems at finite temperature and density ever since only a few years after its original formulation, the heavy-ion program has provided the motivation to continue exploring this tool and to relate it to other analytical and numerical methods such as SDEs and lattice studies. The method promises to continue presenting insightful ways to look at the physical systems under extreme conditions of temperature and density.

Acknowledgments

This work has been supported in part by UNAM-DGAPA-PAPIIT grant number IN101515 and by Consejo Nacional de Ciencia y Tecnología grant number 256494.

References

[1] Y. Aoki, G. Endrödi, Z. Fodor, S. K. Katz and K. K. Szabó, Nature **443**, 675 (2006).
[2] V. I. Yukalov, E. P. Yukalova, Physica A **243**, 382-414 (1997).
[3] V. I. Yukalov, E. P. Yukalova, Phys. Part. Nucl. **28**, 37-65 (1997); PoS Baldin-ISHEPP-XXI, 046 (2012).
[4] X. Shu-Sheng, S. Yuan-Mei, Y. You-Chang , C. Zhu-Fang and Z. Hong-Shi, Chin. Phys. Lett. **32**, 121203 (2015).
[5] For a recent review of these results see L. Levkova, PoS Lattice **2011**, 011 (2011).
[6] C. Bernard *et al.* (MILC Collaboration), Phys. Rev. D **71**, 034504 (2005).
[7] M. Cheng *et al.*, Phys. Rev. D **74**, 054507 (2006).
[8] S. Borsányi *et al.*, J. High Energy Phys. 1009.073 (2010); Y. Aoki *et al.*, J. High Energy Phys. 0906.088 (2009); Y. Aoki *et al.*, Phys. Lett. B **643**, 46 (2006).
[9] A. Bazavov, PoS Lattice **2011**, 182 (2011); A. Bazavov *et al.*, Phys. Rev. D **85**, 054503 (2012).
[10] J. Greensite, *An Introduction to the Confinement Problem*, Lecture Notes in Physics 821, (Springer, Heildelberg, 2011).
[11] For a recent review see P. de Forcrand, PoS Lattice **2009**, 010 (2009).
[12] G. Aarts, PoS Lattice **2012**, 017 (2012).
[13] Z. Fodor, C. Guse, S. D. Katz, K. K. Szabo, PoS Lattice **2007**, 189 (2007); O. Kaczmarek, F. Karsch, E. Laermann, C. Miao, S. Mukherjee, P. Petreczky, C. Schmidt, W. Soeldner, W. Unger, Phys. Rev. D **83**, 014504 (2011); G. Endrödi, Z. Fodor, S. D. Katz, K. K. Szabo, J. High Energy Phys. **1104**, 001 (2011); E. Laermann, F. Meyer, M. P. Lombardo, J. Phys. Conf. Ser. **432**, 012016 (2013); P. Cea, L. Cosmai, A. Papa, Phys. Rev. D **89**, 074512 (2014); C. Bonati, M. D'Elia, M. Mariti, M. Mesiti, F. Negro, F. Sanfilippo, Phys. Rev. D **90** , 114025 (2014); C. Bonati, M. D'Elia, M. Mariti, M. Mesiti, F. Negro, F. Sanfilippo, Phys. Rev. D **92**, 054503 (2015); R. Bellwied, S. Borsanyi, Z. Fodor, J. Günther, S. D. Katz, C. Ratti, K. K. Szabo, Phys. Lett. B **751**, 559-564 (2015); P. Cea, L. Cosmai, A. Papa, Phys. Rev. D **93**, 014507 (2016); P. Hegde and H.-T. Ding (for the Bielefeld-BNL-CCNU Collaboration), PoS Lattice **2015**, 141 (2016).
[14] J. Cleymans, H. Oeschler, R. Redlich, S. Weaton, J. Phys. G **32**, S165 (2006).
[15] A. J. Mizher, M. N. Chernodub, E. S. Fraga, Phys. Rev. D **82**, 105016 (2010).
[16] G. Endrödi, J. High Energy Phys. **1507**, 173 (2015).
[17] G. S. Bali, F. Bruckmann, G. Endrödi, Z. Fodor, S. D. Katz, S. Krieg, A. Schafer, and K. K. Szabo, J. High Energy Phys. **1202**, 044 (2012).
[18] G. S. Bali, F. Bruckmann, G. Endrödi, S. D. Katz and A. Schafer, J. High Energy Phys. **1408**, 177 (2014).
[19] G. S. Bali, F. Bruckmann, G. Endrödi, Z. Fodor, S. D. Katz, and A. Schäfer, Phys. Rev. D **86**, 071502 (2012).
[20] E. S. Fraga, J. Noronha and L. F. Palhares, Phys. Rev. D **87**, 114014 (2013).
[21] F. Bruckmann, G. Endrodi and T. G. Kovacs, J. High Energy Phys. **1304**, 112 (2013).
[22] M. Ferreira, P. Costa, D. P. Menezes, C. Providencia and N. N. Scoccola, Phys. Rev. D **89**, 016002 (2014).
[23] K. Fukushima and Y. Hidaka, Phys. Rev. Lett. **110**, 031601 (2013).
[24] N. Mueller and J. M. Pawlowski, Phys. Rev. D **91**, 11, 116010 (2015).
[25] J. O. Andersen and A. Tranberg, J. High Energy Phys. **1208**, 002 (2012); J. O. Andersen, W. R. Naylor, and A. Tranberg, J. High Energy Phys. **1404**, 187 (2014).
[26] J. Braun, W. A. Mian, S. Rechenberger, arXiv:1412.6025 [hep-ph].
[27] G. S. Bali, F. Bruckmann, G. Endrodi, F. Gruber, A. Schaefer, J. High Energy Phys. **1304**, 130 (2013).
[28] E. J. Ferrer, V. de la Incera, X. J. Wen, Phys. Rev. D **91**, 054006 (2015).
[29] Sh. Fayazbakhsh and N. Sadooghi, Phys. Rev. D **90**, 105030 (2014).
[30] R. L. S. Farias, K. P. Gomes, G. Krein and M. B. Pinto, Phys. Rev. C **90**, 025203 (2014).

[31] M. Ferreira, P. Costa, O. Lourenço, T. Frederico, C. Providência, Phys. Rev. D **89**, 116011 (2014).
[32] J. O. Andersen, W. R. Naylor, A. Tranberg, arXiv:1411.7176 [hep-ph].
[33] V. A. Miransky, I. A. Shovkovy, Phys. Rept. **576**, 1-209 (2015).
[34] J. O. Andersen, W. R. Naylor, A. Tranberg, J. High Energy Phys. **1502**, 042 (2015).
[35] E. S. Fraga and A. J. Mizher, Phys. Rev. D **78**, 025016 (2008).
[36] M. Loewe, C. Villavicencio, R. Zamora, Phys. Rev. D **89**, 016004 (2014).
[37] N. O. Agasian and S. M. Fedorov, Phys. Lett. B **663**, 445 (2008).
[38] E. S. Fraga, B. W. Mintz, J. Schaffner-Bielich, Phys. Lett. B **731**, 154 (2014).
[39] A. Ayala, L. A. Hernández, A. J. Mizher, J. C. Rojas, C. Villavicencio, Phys. Rev. D **89**, 116017 (2014).
[40] A. Ayala, M. Loewe, A. J. Mizher, R. Zamora, Phys. Rev. D **90**, 036001 (2014); A. Ayala, M. Loewe and R. Zamora, Phys. Rev. D **91**, 016002 (2015).
[41] A. Ayala, J. J. Cobos-Martinez, M. Loewe, M. E. Tejeda-Yeomans, and R. Zamora, Phys. Rev. D **91**, 016007 (2015).
[42] A. Ayala, C. A. Dominguez, L.A. Hernandez, M. Loewe, J. C. Rojas, C. Villavicencio, Phys. Rev. D **92**, 1, 016006 (2015).
[43] A. Ayala, A. Bashir, J. J. Cobos-Martínez, S. Hernández, A. Raya, Nucl. Phys. B **897**, 77-86 (2015).
[44] S. Sharma, Adv. High Energy Phys. **2013**, 452978 (2013).
[45] S.-x. Qin, L. Chang, H. Chen, Y.-x. Liu and C. D. Roberts, Phys. Rev. Lett. **106**, 172301 (2011).
[46] P. Costa, M. Ferreira, D. P. Menezes, J. Moreira and C. Providência, Phys. Rev. D **92**, 036012 (2015); M. Ferreira, P. Costa, C. Providência, arXiv:1509.01181 [hep].
[47] A. Ayala, C. A. Dominguez, L. A. Hernandez, M. Loewe, R. Zamora, Phys. Rev. D **92** 096011 (2015).
[48] J. T. Mitchell (for the PHENIX Collaboration), PoS CPOD **2013** (2013), arXiv:1308.2185 [nucl-exp].
[49] H. Wang (for the STAR collaboration), J. Phys. Conf. Ser. **458**, 012001 (2013).
[50] L. Adamczyk *et al.*, (STAR Collaboration), Phys. Rev. Lett. **110**, 142301 (2013).
[51] A. Adare *et al.*, (PHENIX Collaboration), Phys. Rev. Lett. **109**, 152301 (2012).
[52] M. Stephanov, Phys. Rev. Lett. **102**, 032301 (2009).
[53] S. Gupta, PoS CPOD **2009**, 025 (2009).
[54] A. Bazavov *et al.*, Phys. Rev. Lett. **109**, 192302 (2012).
[55] R. V. Gavai, S. Gupta, Phys. Lett. B **696**, 459 (2011).
[56] N. Davis, N. G. Antoniou, F. K. Diakonos (and the NA49 Collaboration), Phys. Atom. Nucl. **75**, 661 (2012).
[57] N. G. Antoniou, Y. F. Contoyiannis, F. K. Diakonos, and G. Mavromanolakis, Nucl. Phys. A **761**, 149 (2005).
[58] M. Gazdzicki (for the NA49 NA61/SHINE Collaborations), J. Phys. G **38**, 124024 (2011).
[59] M. A. Shifman, A. I. Vainshtein and V. I. Zakharov, Nucl. Phys. B **147**, 385 (1979).
[60] For a review see e.g. P. Colangelo and A. Khodjamirian, in *At the Frontier of Particle Physics*, M. Shifman, (World Scientific, Singapore, 2001), Vol. 3, pp. 1495-1576, hep-ph/0010175.
[61] A. I. Bochkarev and M. E. Shaposnikov, Nucl. Phys. B **268**, 220 (1986).
[62] R. J. Furnstahl, T. Hatsuda, and S. H. Lee, Phys. Rev. D **42**, 1744 (1990); C. Adami, T. Hatsuda, and I. Zahed, Phys. Rev. D **43**, 921 (1991); C. Adami and I. Zahed, Phys. Rev. D **45**, 4312 (1992); T. Hatsuda, Y. Koike, and S.-H. Lee, Phys. Rev. D **47**, 1225 (1993); Nucl. Phys. B **394**, 221 (1993); Y. Koike, Phys. Rev. D **48**, 2313 (1993).
[63] R. Arnaldi *et al.* (NA60 Collaboration), Phys. Rev. Lett. **96**, 162302 (2006); Phys. Rev. Lett. **100**, 022302 (2008); S. Damjanovic *et al.* (NA60 Collaboration), J. Phys. G **35**, 104036 (2008); R. Arnaldi *et al.* (NA60 Collaboration), Eur. Phys. J. C **61**, 711 (2009).
[64] R. Arnaldi *et al.* (NA60 Collaboration), Eur. Phys. J. C **59**, 607 (2009).
[65] G. Agakichiev *et al.* (CERES Collaboration), Phys. Rev. Lett. **75**, 1272 (1995); Phys. Lett. B **422**, 405 (1998).
[66] C. A. Dominguez and M. Loewe, Phys. Lett. B **233**, 201 (1989).
[67] R. Rapp, J. Wambach, Eur. Phys. J. A **6**, 415 (1999).
[68] H. van Hees and R. Rapp, Phys. Rev. Lett. **97**, 102301 (2006); Nucl. Phys. A **806**, 339 (2008).
[69] E. L. Bratkovskaya, W. Cassing and O. Linnyk, Phys. Lett. B **670**, 428 (2009).
[70] J. Weil, K. Gallmeister and U. Mosel, PoS BORMIO **2011**, 053 (2011).
[71] For a recent review see, R. Rapp, Adv. High Energy Phys. **2013**, 148253 (2013) .
[72] S. Sarkar and S. Ghosh, J. Phys. Conf. Ser. **374**, 012010 (2012); S. Ghosh, S. Sarkar, J.-e Alam, Eur. Phys. J. C **71**, 1760 (2011).
[73] A. Adare *et al.* (PHENIX Collaboration), Phys. Rev. C **81**, 034911 (2010).
[74] F. Geurts (for the STAR Collaboration), J. Phys. Conf. Ser. **458**, 012016 (2013).
[75] E. T. Atomssa (for the PHENIX Collaboration), Nucl. Phys. A **904-905**, 561c (2013).
[76] M. Dey, V. L. Eletsky, and B. L. Ioffe, Phys. Lett. B **252**, 620 (1990).

[77] S. Weinberg, Phys. Rev. Lett. **18**, 507 (1967).
[78] T. Das, V. S. Mathur, and S. Okubo, Phys. Rev. Lett. **19**, 859 (1967).
[79] N. P. M. Holt, P. M. Hohler and R. Rapp, Phys. Rev. D **87**, 076010 (2013).
[80] C. A. Dominguez and M. Loewe, Z. Phys. C **51**, 69 (1991).
[81] C. A. Dominguez, AIP Conf. Proc. **342**, 383 (1995).
[82] C. A. Dominguez and M. Loewe, Z. Phys. C **58**, 273 (1993); C. A. Dominguez, M. S. Fetea, and M. Loewe, Phys. Lett. B **387**, 151 (1996).
[83] C. A. Dominguez and M. Loewe, Phys. Lett. B **481**, 295 (2000); C. A. Dominguez, M. Loewe, and J. S. Rozowsky, Phys. Lett. B **335**, 506 (1994); C. A. Dominguez, M. S. Fetea, and M. Loewe, Phys. Lett. B **406**, 149 (1997); C. A. Dominguez, M. Loewe, and C. van Gend, Phys. Lett. B **429**, 64 (1998); *ibid* **460**, 442 (1999); C. A. Dominguez, M. Loewe, and J. C. Rojas, J. High Energy Phys. **08** , 040 (2007); E. V. Veliv and G. Kaya, Eur. Phys. J. C **63**, 87 (2009).
[84] A. Ayala, A. Bashir, C. A. Dominguez, E. Gutierrez, M. Loewe, A. Raya, Phys. Rev. D **84**, 056004 (2011).
[85] A. Ayala, C. A. Dominguez, M. Loewe, Y. Zhang, Phys. Rev. D **86**, 114036 (2012).
[86] A. Ayala, C. A. Dominguez, L. A. Hernandez, M. Loewe, A. J. Mizher, Phys. Rev. D **88**, 114028 (2013).
[87] E. Gutierrez, A. Ahmad, A. Ayala, A. Bashir, and A. Raya, J. Phys. G **41**, 075002 (2014).
[88] C. A. Dominguez, M. Loewe, Y. Zhang, Phys. Rev. D **86**, 034030 (2012); Nucl. Phys. B Proc. Suppl. **234**, 305 (2013).
[89] A. Ayala, C. A. Dominguez, M. Loewe, Y. Zhang, Phys. Rev. D **90**, 034012 (2014).
[90] A. Ayala, C. A. Dominguez, M. Loewe, arXiv:1608.04284 [hep-ph].
[91] C. A. Dominguez, M. Loewe, J. C. Rojas, Y. Zhang, Phys. Rev. D **81**, 014007 (2010); **83**, 034033 (2011).
[92] C. A. Dominguez, M. Loewe, Y. Zhang, Phys. Rev. D **88**, 054015 (2013).

Towards the (Mexican) discovery of second class currents at Belle-II

Pablo Roig

Departamento de Física, Centro de Investigación y de Estudios Avanzados del Instituto Politécnico Nacional, Apartado Postal 14-740, 07000 México D.F., México

E-mail: proig@fis.cinvestav.mx

Abstract. Within the SM, the yet unmeasured $\tau^- \to \pi^-\eta^{(\prime)}\nu_\tau$ decays are predicted as a suppressed, isospin-violating effect with branching ratios $\lesssim \mathcal{O}(10^{-5})$. However, they can also proceed through other mechanisms (such as charged Higgs exchange) at comparable rates. This has motivated several studies of the corresponding QCD predictions for these second class current processes. In this contribution we discuss the prospects for the discovery of these decays at Belle-II emphasizing the Mexican involvement in this project. Our branching ratio prediction $\sim 1.7 \cdot 10^{-5}$ (decay channel with an η meson) is well within the reach of Belle-II. The branching fraction for the decay channel with an η' meson is expected to be between one and two orders of magnitude more suppressed.

1. Introduction

Non-strange weak left-handed hadron currents [1] are split according to their spin (J), parity (P) and G-parity (G) [1] as:
i) first class currents, with $J^{PG} = 0^{++}, 0^{--}, 1^{+-}, 1^{-+}$;
ii) second class currents (SCC), which have opposite G-parity to those considered above.
Within the SM, $\tau^- \to \pi^-\eta^{(\prime)}\nu_\tau$ decays can only arise through SCC because the $S(P)$-waves have $J^P = 0^+(1^-)$ but the $\pi\eta^{(\prime)}$ system has $G = -$. However, they are *induced* by a small isospin violating parameter $\varepsilon_{\pi\eta^{(\prime)}} \sim \frac{m_u - m_d}{m_s} \sim 10^{-2}$ at the amplitude level, which is fixed by chiral symmetry alone for the $SU(3)$ flavor octet η_8 state [2]. This isospin (equivalently G-parity) breaking can be understood as the allowed production of the $\pi^-\pi^0$ system [2] with a subsequent $\pi^0 - \eta^{(\prime)}$ mixing regulated by $\varepsilon_{\pi\eta^{(\prime)}}$. On the contrary, new physics contributions can yield enhanced rates if *genuine* SCC do exist.
Shortly after the τ lepton discovery, it was pointed out [3] that the search for $\tau^- \to \pi^-\eta\nu_\tau$ decays (a prominent scalar contribution from $a_0(980)$ meson intermediate exchange was already noticed) could lead to the discovery of SCC.
Interest in SCC was renewed ten years later, when the HRS Coll. [4] announced $BR(\tau^- \to \pi^-\eta\nu_\tau) \sim \mathcal{O}(5\%)$, against the expected suppression given by the overall $\varepsilon_{\pi\eta}^2$ factor, which would yield a branching fraction at the level of 10^{-4}, at most. The situation settled quickly, with the SM prediction of τ decays to final states including η mesons [5] evidencing some issue in the HRS

[1] We note that this classification is only determined by the quantum numbers of the final-state hadrons, being therefore independent of possible (different) intermediate resonance exchange mechanisms.
[2] In fact it is the dominant decay mode of the τ lepton with a branching fraction of $\sim 25\%$.

data, in agreement with other analyses of the $\tau^- \to \pi^-\eta\nu_\tau$ decays [6, 7, 8], including Mexican contributions already [9] [3]. Branching ratios in the range $[1.2, 1.8]\cdot 10^{-5}$ were predicted by that time [7, 10, 11] for this decay mode.

Current bounds on these decays were set by the BaBar Coll.: $BR(\tau^- \to \pi^-\eta\nu_\tau) < 9.9\cdot 10^{-5}$ [12] and $BR(\tau^- \to \pi^-\eta'\nu_\tau) < 4.0\cdot 10^{-6}$ [13], at 95 and 90% C.L., respectively. Belle-II is expected to accumulate up to two orders of magnitude more data than BaBar and Belle [14], which should make possible the discovery of SCC. Elucidating whether these are of the expected *induced* type within the SM, or of *genuine* kind (new physics contributions) motivates reconsidering the QCD prediction for these decays.

In section 2 we introduce the two form factors (scalar and vector) that encode the underlying dynamics of the considered SCC. While we find that the vector one can be extracted from data in a model-independent way, this is not possible for the scalar one, for which a fully analytic and unitary treatment is needed, as we discuss. Based on this hadronic input, we give our predictions in section 3 and state our conclusions and give an outlook in section 4.

2. Vector and scalar form factors

From one point of view, semileptonic τ decays allow for a clean study of the hadronization of QCD currents in the GeV region [15], since the electroweak half of the process is well known theoretically. On the other hand, this hadronic input limits the accuracy of the SM predictions on exclusive meson τ decays, like those considered here. The hadronic matrix element depends on two form factors, $F_0^{\pi^-\eta^{(\prime)}}(s)$ $\left(F_+^{\pi^-\eta^{(\prime)}}(s)\right)$, carrying $J^P = 0^+(1^-)$, respectively:

$$\langle \pi^-\eta^{(\prime)}|\bar{d}\gamma^\mu u|0\rangle = \left[(p_{\eta^{(\prime)}} - p_\pi)^\mu + \frac{\Delta_{\pi^-\eta^{(\prime)}}}{s}q^\mu\right] c^V_{\pi\eta^{(\prime)}} F_+^{\pi^-\eta^{(\prime)}}(s) + \frac{\Delta^{QCD}_{K^0K^+}}{s} q^\mu c^S_{\pi^-\eta^{(\prime)}} F_0^{\pi^-\eta^{(\prime)}}(s)\,, \quad (1)$$

where [16] $c^V_{\pi^-\eta^{(\prime)}} = \sqrt{2}$, $s = q^2 = (p_{\eta^{(\prime)}} + p_{\pi^-})^2$, $c^S_{\pi^-\eta} = \sqrt{\frac{2}{3}}$, $c^S_{\pi^-\eta'} = \frac{2}{\sqrt{3}}$, $\Delta_{PQ} = m_P^2 - m_Q^2$ and the superscript QCD indicates that the K^0K^+ electromagnetic mass difference does not contribute to eq. (1). It is also worth noting that the finiteness of the matrix element (1) relates the two form factors at the origin:

$$F_+^{\pi^-\eta^{(\prime)}}(0) = -\frac{c^S_{\pi^-\eta^{(\prime)}}}{c^V_{\pi^-\eta^{(\prime)}}}\frac{\Delta^{QCD}_{K^0K^+}}{\Delta_{\pi^-\eta^{(\prime)}}} F_0^{\pi^-\eta^{(\prime)}}(0)\,. \quad (2)$$

Then, the overall suppression of these decays is given by $F_+^{\pi^-\eta^{(\prime)}}(0) \sim \mathcal{O}(m_d - m_u)$ [4]. In order to display this neatly, the reduced form factors

$$\widetilde{F}_{+,0}^{\pi^-\eta^{(\prime)}}(s) = \frac{F_{+,0}^{\pi^-\eta^{(\prime)}}(s)}{F_{+,0}^{\pi^-\eta^{(\prime)}}(0)} \quad (3)$$

will be used. In fact, as first put forward in Ref. [16], the relation

$$\widetilde{F}_+^{\pi^-\eta}(s) = \widetilde{F}_+^{\pi^-\eta'}(s) = \widetilde{F}_+^{\pi^-\pi^0}(s) \quad (4)$$

allows a data-driven extraction of $\widetilde{F}_+^{\pi^-\eta^{(\prime)}}(s)$ using the charged pion vector form factor, which has been measured precisely by the Belle Coll. [17], as shown in figure 1. The values of the form

[3] Incidentally, one of the authors of Ref. [3] (J. P.) was the Ph. D. thesis advisor of one of the authors of Ref. [9] (G. L. C.). The other author of the last reference (J. L. D. C.) was a postdoc with one of the authors of Ref. [7] (A. B.). These precedents probably gave rise to the early Mexican interest in SCC.

[4] In fact $F_+^{\pi^-\eta}(0) \sim \mathcal{O}(\varepsilon_{\pi\eta})$, while an accidental cancellation makes $F_+^{\pi^-\eta'}(0) \sim \mathcal{O}(\varepsilon^2_{\pi\eta})$, see eq. (5).

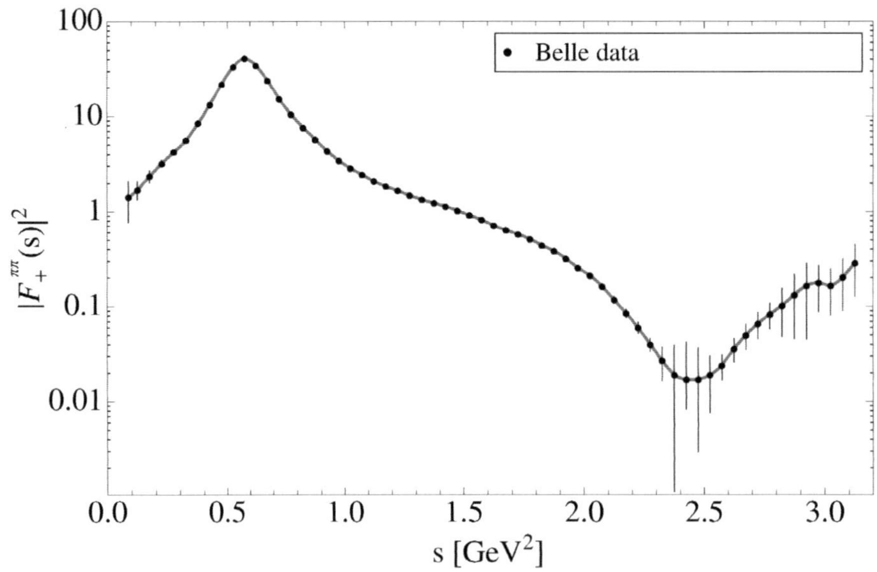

Figure 1. Vector form factor of the $\pi^-\pi^0$ as obtained in Ref. [17] (black circles). The red solid curve corresponds to our data interpolation used in this work.

factors at the origin are fixed in terms of the mixing parameters of the $\pi^0-\eta-\eta'$ system ($\varepsilon_{\pi\eta}$, $\varepsilon_{\pi\eta'}$ and $\theta_{\eta\eta'}$). At next-to-leading order in Chiral Perturbation Theory [18] in the large-N_C limit [19, 20], this reads [16]

$$F_+^{\pi^-\eta}(0)=\varepsilon_{\pi\eta}\,,\quad F_+^{\pi^-\eta'}(0)=\varepsilon_{\pi\eta'}\,, \tag{5}$$

and $\theta_{\eta\eta'}=-(13.3\pm0.5)^\circ$ [21], $\varepsilon_{\pi\eta}=(9.8\pm0.3)\cdot10^{-3}$, $\varepsilon_{\pi\eta'}=(2.5\pm1.5)\cdot10^{-4}$ [22].

According to the previous discussion, the vector form factor contribution to $\tau^-\to\pi^-\eta^{(\prime)}\nu_\tau$ decays is fixed with only a small theory error coming basically from the uncertainty on $\varepsilon_{\pi\eta^{(\prime)}}$. Unfortunately, the situation is opposite for the scalar form factor contribution. In this case there is no model-independent method for extracting $\widetilde{F}_0^{\pi^-\eta^{(\prime)}}(s)$. In Ref.[16] several approaches have been considered to obtain it. These start from the most basic Breit-Wigner-like construction (which violates analyticity maximally) and goes on considering different approximations with increasing fulfillment of unitarity and analyticity constraints (see also [23, 24, 25] for a thorough discussion of these issues in the case of strangeness-changing scalar form factors). Here we will only discuss the most satisfactory approach, which considers the three-coupled channel problem ($\pi^-\eta$, K^-K^0, $\pi^-\eta'$) in a dispersive treatment (thus fulfilling analyticity by construction) which unitarizes the relevant meson-meson scattering amplitudes obtained within the framework of $U(3)$ Chiral Perturbation Theory including resonances [26].

In the elastic region, a two-meson form factor is related to the corresponding meson-meson scattering amplitude via the optical theorem. For our particular case, this reads

$$\mathrm{Im}F_0^{\pi^-\eta^{(\prime)}}(s)=\sigma_{\pi^-\eta^{(\prime)}}(s)F_0^{\pi^-\eta^{(\prime)}}(s)t_{1,0}^{\pi^-\eta^{(\prime)}*}(s)\,, \tag{6}$$

where $t_{1,0}^{\pi^-\eta^{(\prime)}}(s)$ is the unitarized elastic $\pi^-\eta^{(\prime)}$ partial wave of the scattering amplitude of $I=1$ and $J=0$ and $\sigma_{\pi^-\eta^{(\prime)}}(s)$ is the two-particle phase space function. In Ref. [26] the relevant meson-meson scattering amplitudes have been properly unitarized through the N/D method [27, 28], allowing for the simplified perturbative solution

$$t_{1,0}^{\pi^-\eta^{(\prime)}}(s)=\frac{\sigma_{\pi^-\eta^{(\prime)}}(s)N_{1,0}^{\pi^-\eta^{(\prime)}}(s)}{\left(1+g^{\pi^-\eta^{(\prime)}}(s)N_{1,0}^{\pi^-\eta^{(\prime)}}(s)\right)}, \tag{7}$$

where $g^{\pi^-\eta^{(\prime)}}(s)$ are the $\pi^-\eta^{(\prime)}$ one-loop scalar functions defined in Eq. (33) of Ref. [26] and $N_{1,0}^{\pi^-\eta^{(\prime)}}(s)$ contain the expressions of the partial wave amplitudes up to next-to-next-to-leading order in the chiral counting in the large-N_C limit. The central values of the low-energy constants and resonance parameters are obtained from Ref. [29]. However, our ignorance on the correlations between the errors of the different parameters forbids us to give precise estimates of our uncertainties in section 3.
In the elastic region the form factor can, of course, be built using the Omnès dispersive representation [30]. However, from the computational point of view, it is more convenient to construct it employing the closed-form solution [23, 31]

$$\widetilde{F}_0^{\pi^-\eta^{(\prime)}}(s)=\prod_{i=1}\frac{1}{\left(s-s_p^i\right)\left(1+g^{\pi^-\eta^{(\prime)}}(s)N_{1,0}^{\pi^-\eta^{(\prime)}}(s)\right)}, \tag{8}$$

where the pre-factors $(s-s_p^i)^{-1}$ cancel possible poles of eq. (8), which have to be removed in $\widetilde{F}_0^{\pi^-\eta^{(\prime)}}(s)$. In our specific case, the pole resides at $s_p=1.9516$ GeV^2 corresponding to the bare (squared) mass of the scalar octet [29].

A closed-form solution becomes even more convenient in the inelastic region, where unitarity demands a coupled-channel treatment of the problem. In Ref. [16] the two coupled-channel $\pi\eta-\pi\eta'$ problem is considered as a starting point. However we find that the K^-K^0 contribution needs to be accounted for in order to obtain accurate $\widetilde{F}_0^{\pi^-\eta^{(\prime)}}(s)$ form factors [5]. In this case,

$$g_{1,0}(s)=\begin{pmatrix} g_{\pi^-\eta}(s) & 0 & 0\\ 0 & g_{K^-K^0}(s) & 0\\ 0 & 0 & g_{\pi^-\eta'}(s)\end{pmatrix} \tag{9}$$

and

$$N_{1,0}(s)=\begin{pmatrix} N_{\pi^-\eta\to\pi^-\eta}(s) & N_{\pi^-\eta\to K^-K^0}(s) & N_{\pi^-\eta\to\pi^-\eta'}(s)\\ N_{K^-K^0\to\pi^-\eta}(s) & N_{K^-K^0\to K^-K^0}(s) & N_{K^-K^0\to\pi^-\eta'}(s)\\ N_{\pi^-\eta'\to\pi^-\eta}(s) & N_{\pi^-\eta'\to K^-K^0}(s) & N_{\pi^-\eta'\to\pi^-\eta'}(s)\end{pmatrix} \tag{10}$$

encode the corresponding scalar loop functions and partial-wave amplitudes, which enter the master formula for the determination of the participant scalar form factors

$$\begin{pmatrix} F_0^{\pi^-\eta}(s)\\ F_0^{K^-K^0}(s)\\ F_0^{\pi^-\eta'}(s)\end{pmatrix}=\frac{1}{Det[D(s)]}D^{-1}(s)\begin{pmatrix} F_0^{\pi^-\eta}(0)\\ F_0^{K^-K^0}(0)\\ F_0^{\pi^-\eta'}(0)\end{pmatrix}, \tag{11}$$

where

$$D(s)=\begin{pmatrix} 1+g_{\pi^-\eta}(s)T_{\pi^-\eta\to\pi^-\eta}(s) & g_{\pi^-\eta}(s)T_{\pi^-\eta\to K^-K^0}(s) & g_{\pi^-\eta}(s)T_{\pi^-\eta\to\pi^-\eta'}(s)\\ g_{K^-K^0}(s)T_{K^-K^0\to\pi^-\eta}(s) & 1+g_{K^-K^0}(s)T_{K^-K^0\to K^-K^0}(s) & g_{K^-K^0}(s)T_{K^-K^0\to\pi^-\eta'}(s)\\ g_{\pi^-\eta'}(s)T_{\pi^-\eta'\to\pi^-\eta}(s) & g_{\pi^-\eta'}(s)T_{\pi^-\eta'\to K^-K^0}(s) & 1+g_{\pi^-\eta'}(s)T_{\pi^-\eta'\to\pi^-\eta'}(s)\end{pmatrix}. \tag{12}$$

[5] On the contrary, the $\pi^-\pi^0$ contribution is negligible in both cases. This happens because resonances cannot contribute in this channel at first order in isospin breaking.

Again, possible poles of eq. (11) need to be removed as explained before.

Our results [16], obtained by means of this unitarization procedure, are shown in figure 2. While for $\widetilde{F}_0^{\pi^-\eta}(s)$ the coupling of the $\pi^-\eta - K^-K^0$ channels gives already a good approximation to the solution of the three-coupled channels problem, this is not the case for the $\widetilde{F}_0^{\pi^-\eta'}(s)$, where no simplified solution should be used. The complicated multi-peak-and-dip structure of these form factors is a characteristic feature of the unitarization of the coupled channels problem. It must be evident that a parametrization of these form factors which does not fulfill unitarity and analyticity (for instance a sum of Breit-Wigner-like functions) will have very little (if anything) to do with this QCD prediction.

3. Predictions

Our predicted spectra for the $\tau^- \to \pi^-\eta\nu_\tau$ decays are plotted in Fig. 3. As discussed previously, our reference curve is the green dot-dashed line, corresponding to the scalar form factor obtained solving the three-coupled channels problem. Comparison with the other curves in the figure makes clear, again, that simplified approaches should not be pursued in the study of these decays.

In table 3 we compare our predictions for the $\tau^- \to \pi^-\eta\nu_\tau$ decays to current experimental bounds and to other results in the literature. Earlier analyses show a spread in the vector form factor contribution that is much larger than the uncertainty resulting from its data-driven extraction that we advocate. According to our previous discussion it seems natural that approaches that do not require unitarity to the scalar form factor tend to underestimate its contribution to the decay width.

$\mathrm{BR}_V \cdot 10^5$	$\mathrm{BR}_S \cdot 10^5$	$\mathrm{BR}\cdot 10^5$	Reference
0.25	1.60	1.85	Tisserant, Truong [10]
0.12	1.38	1.50	Bramón, Narison, Pich [7]
0.15	1.06	1.21	Neufeld, Rupertsberger [11]
0.36	1.00	1.36	Nussinov, Soffer [32]
$[0.2, 0.6]$	$[0.2, 2.3]$	$[0.4, 2.9]$	Paver, Riazuddin [33]
0.44	0.04	0.48	Volkov, Kostunin [34]
0.13	0.20	0.33	Descotes-Genon, Moussallam [35]
0.26 ± 0.02	1.41 ± 0.09	1.67 ± 0.09	**Our result** [16]
		$\mathrm{BR}\cdot 10^5$	Experimental collaboration
		< 14 (95% CL)	CLEO [36]
		< 7.3 (90% CL)	Belle [37]
		< 9.9 (95% CL)	BaBar [12]

Table 1. Our branching ratio predictions for the $\tau^- \to \pi^-\eta\nu_\tau$ decays are compared to current experimental data and other authors' estimates. In the first and second columns the contributions from the vector and scalar form factors are displayed. The source of uncertainty arises from the errors on $\varepsilon_{\pi\eta}$ and $\varepsilon_{\pi\eta'}$ and from the (uncorrelated) errors on the input values. The total branching ratio is obtained after symmetrizing and adding in quadrature all uncertainties.

Similarly, we plot our predictions for the $\tau^- \to \pi^-\eta'\nu_\tau$ decays in fig. 4, where the reference line is the green dot-dashed curve corresponding to the three-coupled channels determination of the scalar form factor. It can hardly be more evident that the unitarized solution to the three-coupled channels problem for the relevant scalar form factor is the only meaningful description for these decays.

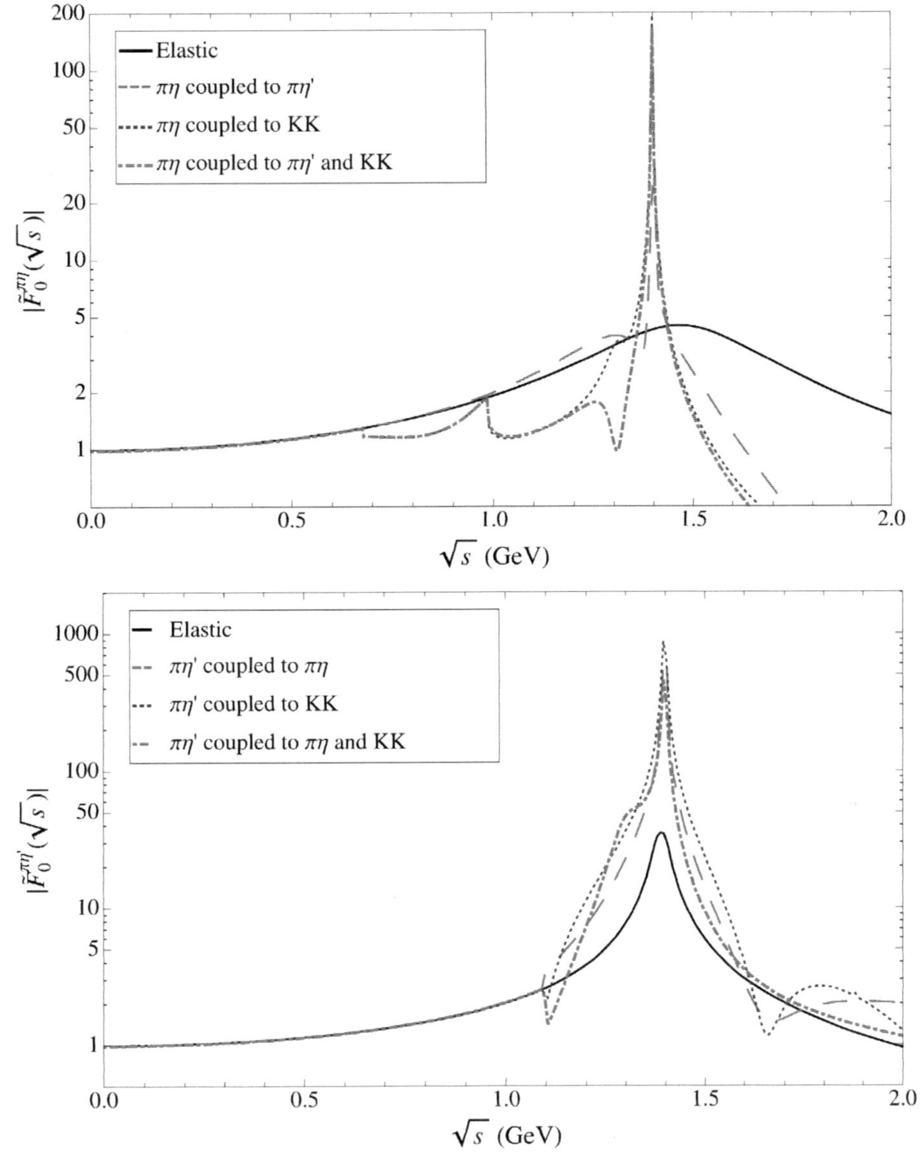

Figure 2. $\pi^-\eta$ SFF coupled to K^-K^0 and $\pi^-\eta'$ (up) and $\pi^-\eta'$ SFF coupled to $\pi^-\eta$ and K^-K^0 (down) as calculated from eq. (11) (green dot-dashed curve) compared to both the elastic case (black solid curve) and to the two coupled-channels cases (red dashed and blue dotted curves). All the expressions are normalized to unity at the origin.

In table 3 we compare our predictions for the $\tau^- \to \pi^-\eta'\nu_\tau$ decays to current experimental bounds and to other results in the literature. Again, it is obvious that simplified descriptions of the scalar form factor yield underestimated predictions for these decay rates.

In Ref. [16] we also consider the partner $\eta^{(\prime)} \to \pi^+\ell^-\nu_\tau \quad (\ell = e, \mu)$ decays. Our predictions for them indicate that their discovery is out of reach at present and next-generation colliders.

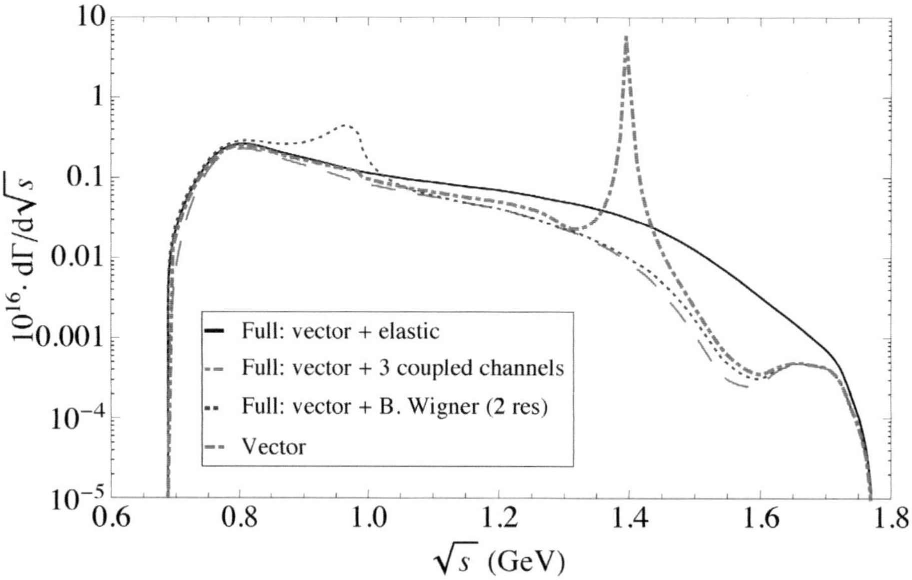

Figure 3. Decay distribution for $\tau^- \to \pi^-\eta\nu_\tau$ decays. Red-dashed curve describes the contribution corresponding to the vector form factor while the other three curves represent the full distribution by employing the scalar form factor in its elastic version (black solid curve), in the three coupled-channels analysis (green dot-dashed curve) and using a Breit-Wigner formula with two resonances (blue dotted curve).

BR_V	BR_S	BR	Reference
$< 10^{-7}$ $[0.14, 3.4]\cdot 10^{-8}$ $1.11\cdot 10^{-8}$	$[0.2, 1.3]\cdot 10^{-6}$ $[0.6, 1.8]\cdot 10^{-7}$ $2.63\cdot 10^{-8}$	$[0.2, 1.4]\cdot 10^{-6}$ $[0.61, 2.1]\cdot 10^{-7}$ $3.74\cdot 10^{-8}$	Nussinov, Soffer [38] Paver, Riazuddin [39] Volkov, Kostunin [34]
$[0.3, 5.7]\cdot 10^{-10}$	$[1\cdot 10^{-7}, 1\cdot 10^{-6}]$	$[1\cdot 10^{-7}, 1\cdot 10^{-6}]$	**Our result** [16]
		BR	Experimental collaboration
		$< 4\cdot 10^{-6}$ (90% CL) $< 7.2\cdot 10^{-6}$ (90% CL)	BaBar [13] BaBar [40]

Table 2. Our branching ratio predictions for the $\tau^- \to \pi^-\eta'\nu_\tau$ decays are compared to current experimental bounds and other authors' estimates as it was done in table 3.

We obtain $BR(\eta \to \pi^+e^-\nu_e + c.c.) = 0.6\cdot 10^{-13}$, $BR(\eta \to \pi^+\mu^-\nu_\mu + c.c.) = 0.4\cdot 10^{-13}$ and $1.7\cdot 10^{-17}$ for both η' decays. The first bound on any of these decays has just been set by BESIII: $\mathcal{B}(\eta \to \pi^+e^-\bar{\nu}_e + c.c.) < 1.7\cdot 10^{-4}$ and $\mathcal{B}(\eta' \to \pi^+e^-\bar{\nu}_e + c.c.) < 2.2\cdot 10^{-4}$, both at the 90% C.L., which are still extremely far from our predictions. The spectra of these decays can be found in Ref. [16].

4. Conclusions and outlook

According to our results, SCC should finally be discovered at Belle-II through the measurement of the $\tau^- \to \pi^-\eta\nu_\tau$ decays. A precision of $\sim$ 20% on both their measurement and the

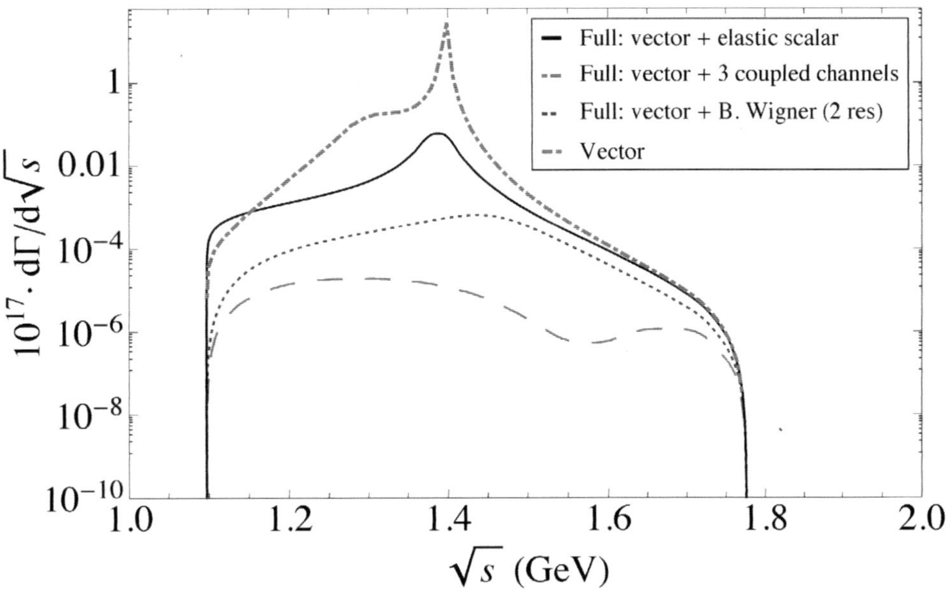

Figure 4. Decay distribution for the $\tau^- \to \pi^-\eta'\nu_\tau$ decays. Red-dashed curve describes the contribution corresponding to the vector form factor while the other three curves represent the full distribution by employing the scalar form factor in its elastic version (black solid curve), in the three coupled-channels analysis (green dot-dashed curve) and finally using the Breit-Wigner formula with two resonances (blue dotted curve).

corresponding theory prediction would allow to improve the bounds on a charged Higgs [35] (currently obtained from $B \to \tau\nu$). Our study aims at providing a QCD-based parametrization of this decay that can be useful in Belle-II searches. Upon eventual discovery, a joint analysis of meson-meson scattering amplitudes together with the Belle-II signal on $\tau^- \to \pi^-\eta^{(\prime)}\nu_\tau$ decays would yield an accurate determination of the correlations between all relevant low-energy constants and resonance parameters that would enable a consistency test of both sets of data in the search for non-SM contributions, i. e. *genuine* SCC. In these combined fits, the usefulness of our closed-form solution of the multi-coupled channels problem emerges plainly.

Several things remain to be done. In BaBar and Belle-I the background from $\tau^- \to \pi^-\pi^0\eta\nu_\tau$ decays [42] limited the searches for SCC. Within a Mexican-Polish Collaboration [43] we plan to include the Resonance Chiral Lagrangians prediction for these decays [44] in the corresponding TAUOLA currents [45] in order to ease the Belle-II searches. It turns out [46] that some resonance contributions to the $\tau^- \to \pi^-\eta^{(\prime)}\gamma\nu_\tau$ decays yield sizable backgrounds to the searches for SCC in the corresponding non-radiative decays, and therefore need to be properly modeled in TAUOLA too, as we plan to do in the future.

Acknowledgments
I enjoyed very much collaborating with R. Escribano and S. González-Solís on the research summarized here. I thank S. G. S. for a careful revision of the draft of this contribution. Mexican funding from Conacyt and SNI (Mexico) supporting this research is acknowledged, as well as from RED-FAE to attend the XV Mexican Workshop on Particles and Fields. I congratulate the organizing Committee for the fruitful conference.

Journal of Physics: Conference Series **761** (2016) 012067 doi:10.1088/1742-6596/761/1/012067

References

[1] S. Weinberg, Phys. Rev. **112**, 1375 (1958).
[2] J. Gasser and H. Leutwyler, Nucl. Phys. B **250** (1985) 465.
[3] C. Leroy and J. Pestieau, Phys. Lett. B **72** (1978) 398.
[4] M. Derrick *et al.*, Phys. Lett. B **189** (1987) 260.
[5] A. Pich, Phys. Lett. B **196** (1987) 561.
[6] E. L. Berger and H. J. Lipkin, Phys. Lett. B **189** (1987) 226.
[7] A. Bramón, S. Narison and A. Pich, Phys. Lett. B **196** (1987) 543.
[8] E. Braaten, R. J. Oakes and S. M. Tse, Int. J. Mod. Phys. A **5** (1990) 2737.
[9] J. L. Díaz-Cruz and G. López Castro, Mod. Phys. Lett. A **6** (1991) 1605.
[10] S. Tisserant and T. N. Truong, Phys. Lett. B **115** (1982) 264.
[11] H. Neufeld and H. Rupertsberger, Z. Phys. C **68** (1995) 91.
[12] P. del Amo Sánchez *et al.* [BaBar Collaboration], Phys. Rev. D **83** (2011) 032002.
[13] J. P. Lees *et al.* [BaBar Collaboration], Phys. Rev. D **86** (2012) 092010.
[14] T. Abe *et al.* [Belle-II Collaboration], arXiv:1011.0352 [physics.ins-det].
[15] A. Pich, Prog. Part. Nucl. Phys. **75** (2014) 41.
[16] R. Escribano, S. González-Solís and P. Roig, Phys. Rev. D **94** (2016) 034008.
[17] M. Fujikawa *et al.* [Belle Collaboration], Phys. Rev. D **78**, 072006 (2008).
[18] S. Weinberg, Physica A **96** (1979) 327. J. Gasser and H. Leutwyler, Annals Phys. **158** (1984) 142.
[19] G. 't Hooft, Nucl. Phys. B **72** (1974) 461.
[20] R. Kaiser and H. Leutwyler, Eur. Phys. J. C **17**, 623 (2000), also in 'Adelaide 1998, Nonperturbative methods in quantum field theory', pags. 15-29.
[21] F. Ambrosino *et al.* [KLOE Collaboration], Phys. Lett. B **648**, 267 (2007).
[22] P. Kroll, Mod. Phys. Lett. A **20**, 2667 (2005).
[23] M. Jamin, J. A. Oller and A. Pich, Nucl. Phys. B **622**, 279 (2002).
[24] R. Escribano, S. González-Solís and P. Roig, JHEP **1310** (2013) 039.
[25] R. Escribano, S. González-Solís, M. Jamin and P. Roig, JHEP **1409** (2014) 042.
[26] Z. H. Guo and J. A. Oller, Phys. Rev. D **84**, 034005 (2011).
[27] J. A. Oller and E. Oset, Phys. Rev. D **60**, 074023 (1999).
[28] J. A. Oller, E. Oset and A. Ramos, Prog. Part. Nucl. Phys. **45**, 157 (2000).
[29] Z. H. Guo, J. A. Oller and J. Ruiz de Elvira, Phys. Rev. D **86** (2012) 054006.
[30] R. Omnès, Nuovo Cim. **8**, 316 (1958).
[31] O. Babelon, J.-L. Basdevant, D. Caillerie and G. Mennessier, Nucl. Phys. B **113**, 445 (1976).
[32] S. Nussinov and A. Soffer, Phys. Rev. D **78**, 033006 (2008).
[33] N. Paver and Riazuddin, Phys. Rev. D **82**, 057301 (2010).
[34] M. K. Volkov and D. G. Kostunin, Phys. Rev. D **86**, 013005 (2012).
[35] S. Descotes-Genon and B. Moussallam, Eur. Phys. J. C **74** (2014) 2946.
[36] J. E. Bartelt *et al.* [CLEO Collaboration], Phys. Rev. Lett. **76** (1996) 4119.
[37] K. Hayasaka [Belle Collaboration], PoS EPS **-HEP2009**, 374 (2009).
[38] S. Nussinov and A. Soffer, Phys. Rev. D **80**, 033010 (2009).
[39] N. Paver and Riazuddin, Phys. Rev. D **84**, 017302 (2011).
[40] B. Aubert *et al.* [BaBar Collaboration], Phys. Rev. D **77**, 112002 (2008).
[41] M. Ablikim *et al.* [BESIII Collaboration], Phys. Rev. D **87**, 032006 (2013).
[42] K. Inami *et al.* [Belle Collaboration], Phys. Lett. B **672** (2009) 209.
[43] Michel Hernández, a Ph. D. student co-advised by Eduard de la Cruz Burelo and Gabriel López Castro, and Jorge Martínez, a posdoc of the project "Fronteras de la Ciencia 296", are involved in this project with us.
[44] D. Gómez Dumm and P. Roig, Phys. Rev. D **86** (2012) 076009.
[45] O. Shekhovtsova, T. Przedzinski, P. Roig and Z. Was, Phys. Rev. D **86** (2012) 113008.
[46] Adolfo Guevara, a Ph. D. student co-advised by Gabriel López Castro and P. R., is involved in this project with us.

Journal of Physics: Conference Series **761** (2016) 012068 doi:10.1088/1742-6596/761/1/012068

Phenomenology of Dihadron Fragmentation Function.

A. Courtoy

Departamento de Física, Centro de Investigación y de Estudios Avanzados, Apartado Postal 14-740, 07000 Ciudad de México, México
CONACyT

E-mail: acourtoy@fis.cinvestav.mx

Abstract. We report on the phenomenological results obtained through Dihadron Fragmentation Functions related processes. In 2015, an update on the fitting techniques for the Dihadron Fragmentation Functions has led to an improved extraction of the transversity PDF and, as a consequence, the nucleon tensor charge. We discuss the impact of the determination of the latter on search for physics Beyond the Standard Model, focusing on the error treatment. We also comment on the future of the extraction of the subleading-twist PDF $e(x)$ from JLab soon-to-be-released Beam Spin Asymmetry data.

1. Introduction

Dihadron Fragmentation Functions (DiFF) encode information about the fragmentation process of a quark into a hadron pair (plus something else, undetected). Fragmentation has been hardly studied through models for pion pairs, the main knowledge about those DiFFs being for now fits from data.

The rich phenomenology associated to dihadron fragmentation compares to the well-known Semi-Inclusive Deep Inelastic Scattering (SIDIS), in which a single hadron is produced in the current fragmentation region. In these proceedings, we will review the expectations rather than the acknowledged successes of the subset of Dihadron data, that can be found elsewhere [2, 3]. These proceedings will focus on the error treatment.

2. The future of Dihadron FF from data

The unpolarized DiFF have been extracted by fitting the pair distribution simulated by a Monte Carlo event generator [1, 2], to compensate for the absence of data on multiplicities. The resulting error on the parametrization of the unpolarized DiFF is obviously small. On the other hand, once the unpolarized DiFF parameterized, the chiral-odd DiFF can be extracted from the Artru–Collins asymmetry at Belle. The factorization of the process $e^+e^- \to (\pi^+\pi^-)_{\rm jet}(\pi^+\pi^-)_{\overline{\rm jet}}X$ is valid in the kinematical regime $P_h^2 = M_h^2 \ll Q^2 = -q^2 \geq 0$ and $q = k - k'$ the space-like momentum transferred, with the total momentum $P_h = P_1 + P_2$.

Belle data for the Artru–Collins asymmetry led to the extraction —and, hence,

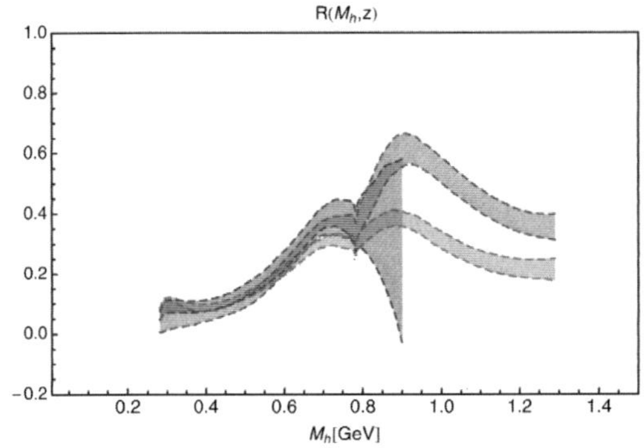

Figure 1. The ratio $R(z, M_h)$ as a function of M_h at $Q_0^2 = 1$ GeV^2 for three different $z = 0.25$ (shortest band), $z = 0.45$ (lower band at $M_h \sim 1.2$ GeV), and $z = 0.65$ (upper band at $M_h \sim 1.2$ GeV), with the value $\alpha_s(M_Z^2) = 0.139$ used in the QCD evolution equations.

parameterization— of the leading-twist DiFFs [1, 2], illustrated in Fig. 1 through their ratio

$$R(z, M_h) = \frac{|\mathbf{R}|}{M_h} \frac{H_1^{\sphericalangle\, u}(z, M_h; Q_0^2)}{D_1^u(z, M_h; Q_0^2)} , \tag{1}$$

the relevant variables being the pion pair invariant mass, M_h, and the sum of the fractional energies carried by the two final hadrons, z.

While the first fitting approach was based on the usual Hessian statistics with a variation of the chisquare equal to unity [1], the updated version incorporates a Neural Network preparation, *i.e.* the normal generation of data replicas within the —1σ— experimental errors [2]. This second technique insures the proper treatment of errors outside the data range, thus avoiding the vanishing of the errors at the endpoints, where they should be maximal. We have noticed that both fitting techniques, still based on adhoc functional forms, agree with each other as far as the chiral-odd DiFF is concerned.

In principle, the unpolarized D_1 should be extracted by global fits of the unpolarized cross section, in the same way as it is done for single-hadron fragmentation [4]. The first analysis of di-hadron multiplicities is ongoing at COMPASS [5] and an analysis of Dihadron SIDIS multiplicities has been proposed at CLAS12 [6]. The expression for the multiplicities can be written in terms of PDFs and DiFFs as

$$M^h(z, M_h, x; Q^2) \;=\; \frac{\sum_q e_q^2 f_1^q(x; Q^2)\, D_1^q(z, M_h; Q^2)}{\sum_q e_q^2 f_1^q(x; Q^2)} , \tag{2}$$

where f_1 is the well-known collinear unpolarized PDF and D_1 is the unpolarized DiFF. It corresponds to the ratio of the number $N^{\pi^+\pi^-}$ of produced pion pairs in SIDIS and the number of DIS events $N^{\rm DIS}$ for the same kinematics. With the future CLAS12 data, we will analyze the 4-dimensional binning spanning a wide kinematical range in (z, M_h, x, Q^2), for both proton and deuteron targets, allowing for a flavor separation of the valence distributions. The development expected from such a measurement and statistical analysis will qualitatively improve the present estimates for a number of processes. As most of the Distribution/Fragmentation Functions, DiFF are believed to be universal. This characteristic has recently been studied and reinforced through a confirmation on the role of transversity in proton-proton collisions [7].

3. Completing the collinear twist-2 picture

Combining the parametrization obtained for DiFF to HERMES and COMPASS data on dihadron SIDIS allowed for the extraction of the last leading-twist PDF, *i.e. the transversity PDF* [2, 3]. While the fit of the transversity is an achievement by itself, though affected of a large uncertainty outside the data kinematical range, its first Mellin moment is also of great interest. The tensor charge[1] is obtained by integrating the transversity PDF over the physical support in x,

$$\delta q_v(Q^2) \quad = \quad \int_0^1 dx\, h_1^{q_v}(x, Q^2)\,. \tag{3}$$

The error we mentioned above, consequence of the extrapolation of the PDF outside the data range, is the main source of uncertainty on the determination of the tensor charge, as shown in Table 3 of Ref. [2]. Another source of error consists in the choice of the functional form for the transversity PDF. This problem has been taken care of, in a first step, considering three different functional forms, each with a growing number of free parameters. The value for the isovector tensor charge $g_T = \delta u_v - \delta d_v$ for the functional form related to the so-called *flexible scenario* with $\alpha_s(M_Z^2) = 0.125$ is, at 1σ,

$$g_T = 0.81 \pm 0.44 \quad \text{at} \quad Q^2 = 4\,\text{GeV}^2 \quad . \tag{4}$$

It is in agreement with lattice determinations as well as with the other extractions from hadronic phenomenology [8, 9, 10], though the absolute value is slightly smaller than the lattice's. Due to its non-perturbative nature, the nucleon structure can only be unveiled using complementary methods such as effecive field theories, lattice calculations, models for the nucleon structure, Schwinger-Dyson based techniques and phenomenological extraction of observables from data. As such, the lattice field theory and the data based determinations play a similar, yet uncorrelated, role.

The role of the tensor aspect of the nucleon becomes clear in beta decay where the hadronic characterization of the process is encoded through, among others, the tensor form factor — which value at zero momentum transfer is nothing else than the tensor charge. While there is no leptonic counterpart of the tensor structure in the Standard Model, all quark bilinear Lorentz structures can be introduced through an effective Lagrangian which is relevant for beta decay observables. The beta decay observables, *e.g.* b_0^+ and the Fierz term b, involve products of the Beyond the Standard Model (BSM) couplings, ϵ_i, and the corresponding hadronic charges, g_i. The scalar (S) and tensor (T) operators, in particular, contribute linearly to the beta decay parameters through their interference with the SM amplitude, and they are, therefore, more easily detectable. We focus here on the following low-energy effective interaction (see Ref. [11] for a review)

$$\Delta\mathcal{L}_{\text{eff}} \quad = \quad -C_S \bar{p} n\, \bar{e}(\mathbb{I} - \gamma_5)\nu_e - C_T \bar{p}\sigma_{\mu\nu} n\, \bar{e}\sigma^{\mu\nu}(\mathbb{I} - \gamma_5)\nu_e \quad , \tag{5}$$

where $C_S = G_F V_{ud}\sqrt{2}\epsilon_S g_S$, and $C_T = 4G_F V_{ud}\sqrt{2}\epsilon_T g_T$ and $G_F \equiv \sqrt{2}g^2/(8M_W^2)$ is the tree-level definition of the Fermi constant. In the SM, the ϵ_S and ϵ_T coefficients vanish leaving the well-known $(V - A) \times (V - A)$ structure generated by the exchange of a W boson.

Precise measurements in beta decay set strong bounds on the combination $g_T\epsilon_T$ obtained through global fits [12, 13],

$$|g_T\epsilon_T| < 6 \cdot 10^{-4} \qquad (90\%\ \text{C.L.})\ , \tag{6}$$

[1] Note that the word charge here is a misuse of language.

which is expected to be improved by the next generation of experiments.

In order to extract a bound on the Wilson coefficient ϵ_T from Eq. (6) it is necessary to know the value of the tensor charge g_T. Hence, the sensitivity of beta decay measurements to an exotic tensor interaction depends on our knowledge of the tensor charge. The impact of data based determinations of the tensor charge has been studied in Ref. [14]. Given the sensitivity bounds on the Fierz term b, an error of about $10-15\%$ on the tensor charge is estimated to be sufficient. The values for g_T mentioned above are still far from that precision.

Originally, in Ref. [14], an alternative to the standard Hessian evaluation was considered. The argument that both the lattice QCD and experimental extractions of the couplings from β-decay are affected by systematic or theoretical uncertainty invalidates the assumption of a gaussian distribution of the error around the central value. In order to deal with this situation we followed Ref. [15] and we calculated the confidence interval on ϵ_T using the so-called R-Fit method [16]. In this scheme the theoretical likelihoods do not contribute to the χ^2 of the fit and the corresponding QCD parameters take values within certain "allowed ranges". In our case, this means that g_T is restricted to remain inside a given interval, e.g. $0.37 \leq g_T \leq 1.25$ for the current determination from di-hadron SIDIS. The chisquare function is then given by

$$\chi^2(\epsilon_T) = \min_{g_T} \left(\frac{[g_T\epsilon_T]^{\exp} - g_T\epsilon_T}{\delta\, [g_T\epsilon_T]^{\exp}} \right)^2 \tag{7}$$

where the minimization is performed varying g_T within its allowed range. In this approach, the bound on ϵ_T depends only on the lower limit of the tensor charge, as long as the experimental determination of $g_T\epsilon_T$ is compatible with zero at 1σ. In particular, the tensor charge given by Eq. (4) leads to a bound $|\epsilon_T| < 0.00162$. This bound is the largest found through hadron phenomenology (see Fig. 2 of Ref. [14]) but is expected to decrease of about 10% with the new JLab data. For comparison, the bound obtained from the analysis of LHC data carried out in Ref. [17] is $|\epsilon_T| < 0.0013$.

This error analysis presents shortcomings, *e.g.* the fact that Eq. (7) only depends on the lower bound on the tensor charge. The principal criticism would be that the R-fit method is a global method for data fitting, used here for the determination of a single parameter based on a fit output, Eq. (6).

Another alternative analysis could be implemented taking advantage on the replica technique of the tensor charge determination in DiFF related processes. The bound given by Eq. (4) is nothing else then the 64 most central values, output of the 100 fitted data replicas. The 90% C.L. is easily obtained by discarding only ten outer values (5 upper and 5 lower). Generating normally 100 replicated values for the product $\epsilon_T\, g_T$ around zero with a variance of 3×10^{-4}, we obtain a Monte Carlo-like estimate of the error, fully at 90% C.L.: $|\epsilon_T| < 0.00139$.

Extending the previous analysis to two dimensions, we can use the bound on the relevant observables obtained from the decay rate measurement, *e.g.*, b_0^+ from super-allowed Fermi transition as well as the Fierz term b. The latter can be expressed in terms of the factorized coupling [15]

$$\begin{aligned} b &= \frac{2}{1+3\lambda^2}\left[g_S\epsilon_S - 12 g_T\epsilon_T\lambda\right] < 10^{-3} \quad , \\ b_0^+ &= -2g_S\epsilon_S < 0.0022(26) \quad , \end{aligned} \tag{8}$$

with $\lambda = g_A/g_V$. The bound on b_0^+ is given in Ref. [18] and the precision limit on the Fierz term corresponds to future experiments expectations.

A simple excercise to compare various techniques for the error treatment consists in calculating the propagation of error for the (ϵ_T, ϵ_S) plane based on Eqs. (8). We propose a

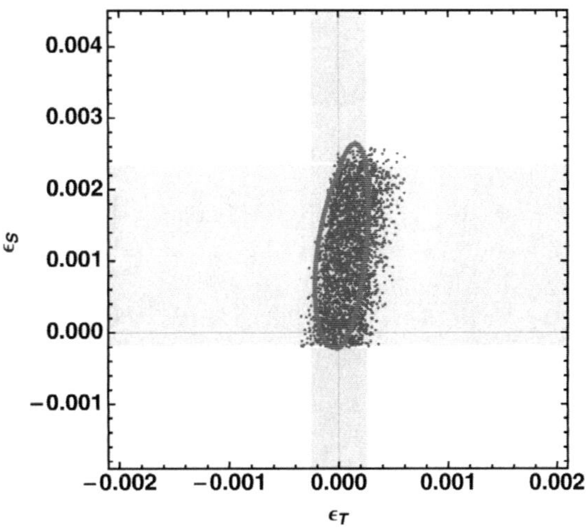

Figure 2. The (ϵ_T, ϵ_S) plane. Allowed values at 1σ evaluated in three different statistical methods. Scatter plot (blue dots), contour plot (red curve) and Hessian method for ϵ_T and ϵ_S in orange and light blue, respectively.

simplified version[2] for the sake of illustration, in which we also assume that there is no error on λ. In Fig. 2, we show the allowed region for the plane (ϵ_T, ϵ_S) using three different statistical techniques, namely the Hessian method, the R-fit method and the space scan of the permitted values of parameters. We have used $g_S = 1.02 \pm 0.11$ for the scalar charge [19].

The first technique is the most commonly used error propagation. The error on each effective coupling is given by

$$\begin{aligned}
\Delta^2\epsilon_S &= \frac{1}{4}\left[\left(\frac{\Delta b_0^+}{g_S}\right)^2 + \left(\frac{\Delta g_S}{g_S^2}\right)^2 b_0^{+\,2}\right] \quad , \\
\Delta^2\epsilon_T &= \frac{1}{(12\,\lambda)^2}\left[\Delta^2 b\left(\frac{1+3\lambda^2}{2\,g_T}\right)^2 + \left(\frac{\Delta b_0^+}{2\,g_T}\right)^2 + \left(\frac{\Delta g_T}{g_T^2}\right)^2\left[\left(\frac{b(1+3\lambda^2)}{2}\right)^2 + \left(\frac{b_0^+}{2}\right)^2\right]\right] \quad .
\end{aligned} \tag{9}$$

Notice that, here, $b = 0$. The result is given by the light orange and blue bands in Fig. 2. There are no correlation taken into account in this technique.

The second approach consists in using the R-Fit method again. In this scheme, $g_{S/T}$ is restricted to remain inside the interval defined by its error at 1σ. Notice that all values inside this range are treated on an equal footing. The chi-squared function is then given by

$$\begin{aligned}
&\chi^2(\epsilon_S, \epsilon_T) \\
&= \min_{g_{S,T}}\left[\left(\frac{(b_0^+)^{\text{exp}} - (-2\,g_S\epsilon_S)^{\text{theo}}}{(\Delta b_0^+)^{\text{exp}}}\right)^2 + \left(\frac{b^{\text{exp}} - 2/(1+3\lambda^2)\times[g_S\epsilon_S - 12 g_T\epsilon_T\lambda]^{\text{theo}}}{(\Delta b)^{\text{exp}}}\right)^2\right] \quad .
\end{aligned} \tag{10}$$

Again, $b^{\text{exp}} = 0$. The 1σ contour is found equating Eq. (10) to 1. It is depicted by the red contour in Fig. 2.

[2] Simplified as compared to accurate global fit analyses.

Journal of Physics: Conference Series **761** (2016) 012068 doi:10.1088/1742-6596/761/1/012068

Finally, we show a scan of the allowed (ϵ_S, ϵ_T) plane solving the equations for b and b_0^+ varying normally $g_{S,T}$ at 1σ. It corresponds to the area of blue points in Fig. 2. The three techniques seem to be compatible for the present case, with the available precision.

4. Twist-3 from DiFF

In the previous Section, we have presented an analysis based on the extracted tensor charge [2] together with the scalar charge from isospin breaking analysis [19]. Ideally, we would like to determine the scalar charge from phenomenology as well.

Though in a more elusive way, the scalar charge is related to a twist-3 PDF, $e(x)$. QCD equations of motion allow to decompose the chiral-odd twist-3 distributions into three terms

$$e^q(x) \quad = \quad e^q_{\text{loc}}(x) + e^q_{\text{tw-3}}(x) + e^q_{\text{mass}}(x) \quad . \tag{11}$$

The first term comes from the local operator and is exactly the contribution related to scalar charge,

$$\begin{aligned} e^q_{\text{loc}}(x) &= \frac{1}{2M}\int \frac{d\lambda}{2\pi}\, e^{i\lambda x}\langle P|\bar{\psi}_q(0)\psi_q(0)|P\rangle \quad , \\ &= \frac{\delta(x)}{2M}\langle P|\bar{\psi}_q(0)\psi_q(0)|P\rangle \quad ; \end{aligned} \tag{12}$$

the second term is a genuine twist-3 contribution, *i.e.* pure quark-gluon interaction term ; while the last term is related to the current quark mass.

The sum rule related the chiral-odd twist-3 PDF to the scalar charge is similar to Eq. (3). Only the *local* term contributes to the first Mellin moment, *i.e.* the scalar charge is nothing else than $e(x=0)$. That delta-function singularity follows from chiral symmetry and the existence of non-vanishing quark condensate [20].

Being a subleading contribution, this twist-3 PDF is hardly known. It is however accessible in single- [21] and di-hadron SIDIS. In the latter case, the chiral-odd partner of $e(x)$ is the chiral-odd dihadron fragmentation function, *i.e.* $H_1^{\sphericalangle}$. An extraction of the twist-3 PDF, $e(x)$, through the analysis of the preliminary data [22] for the $\sin\phi_R$-moment of the beam-spin asymmetry for dihadron Semi-Inclusive DIS at CLAS at 6 GeV was proposed in Ref. [23]. It is illustrated in Fig. 3 in a specific scenario —in which the asymmetry is dominated by the term containing the twist-3 PDF. By relaxing that hypothesis, we can define other scenarios: the function $e(x)$ always results non-zero.

In the next few years, future data at CLAS12 [6] will be analyzed that bring us more statistics and information on that particular function. The determination of the scalar charge will be a delicate task as the value of the function at $x=0$ is experimentally unaccessible. The solution will be to appeal to other properties of the subleading-twist function together with advanced fitting techniques.

The determination of the scalar charge represents a challenge for the hadronic physics community but an opportunity to experimentally explore the consequence of chiral symmetry on the one hand. On the other hand, the search for a new fundamental scalar interaction, Beyond the Standard Model, would benefit from the outcome of this line of research, complementing the results obtained by the lattice community.

5. Conclusions

The phenomenology of Dihadron Fragmentation Functions integrates the idea that more structures and information become accessible when incorporating a dependence on the transverse momentum. The elegance of DiFF results in less intricated and more easily taken care of

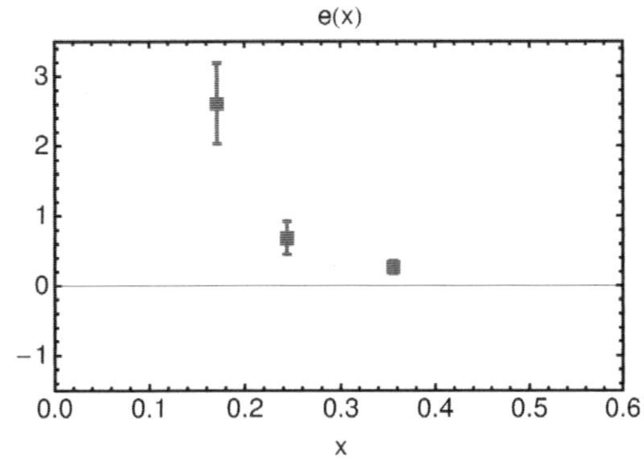

Figure 3. The valence combination $e(x) \equiv 4e^{u_V}(x_i, Q_i^2)/9 - e^{d_V}(x_i, Q_i^2)/9$ (see text). The error bars correspond to the propagation of the experimental and DiFF errors.

observables, namely collinear objects and simple products, as opposed to convolutions. This particular feature allows for more flexibility in their statistical analysis. The parameterization of the third leading-twist PDF, *the transversity*, marked the first stage of an alternative approach to transverse momentum dependent functions. Then followed the tensor charge determination, made possible thanks to the flavor structure of the analyzed observables. The role of dihadron in proton-proton collision has been studied as well. Subleading processes involving DiFF are being analyzed and will be extended to JLab@12.

Twist-3 functions encode invaluable information on the low energy dynamics of quarks and gluons, so far hardly explored. The nonperturbative nature of nucleons influences numerous observables that involve them. Chiral symmetry, the existence of a delta-function singularity, the pion-nucleon sigma-term, all pieces of a same puzzle that could further contribute to searches of New Physics, *e.g.* [24]. The possibility of obtaining the scalar and tensor charges directly from experiment with sufficient precision, not having to deal with purely theoretical uncertainties of these quantities, gives a different weight to searches for BSM involving nucleons.

The results presented here concerning the determination of the tensor charge through Dihadron Fragmentation Functions are complemented by the Deeply Virtual Meson Production analysis [9] and the Single-hadron SIDIS approach [8].

To conclude, we would like to restate that dihadron SIDIS offers a unique way to access collinear functions, subleading functions being of particular interest. Though there are few data now, future analyses —*e.g.* dihadron multiplicities at COMPASS— and experiments — especially in CLAS12 and SoLID at JLab— will increase the data set, allowing for an improved knowledge on both DiFFs at leading and subleading-twist and collinear leading and subleading PDFs.

Acknowledgments

The author acknowledges her co-authors from Pavia, A. Bacchetta, M. Guagnelli and M. Radici, discussions and broad collaboration with S. Liuti and S. Pisano. A.C. would like to thank the collaborations with SoLID and CLAS12 in general. This work is supported by a Cátedra CONACyT contract.

References

[1] A. Courtoy, A. Bacchetta, M. Radici and A. Bianconi, Phys. Rev. D **85** (2012) 114023
[2] M. Radici, A. Courtoy, A. Bacchetta and M. Guagnelli, JHEP **1505** (2015) 123
[3] A. Bacchetta, A. Courtoy and M. Radici, JHEP **1303** (2013) 119

[4] D. de Florian, R. Sassot and M. Stratmann, Phys. Rev. D **75** (2007) 114010
[5] N. Makke [COMPASS Collaboration], Phys. Part. Nucl. **45** (2014) 138
[6] S. Pisano, A. Courtoy et al.. JLab. Exp. E12-06-112B/E12-09-008B. `http://www.jlab.org/exp_prog/proposals/14/E12-06-112B_E12-09-008B.pdf`
[7] M. Radici, A. M. Ricci, A. Bacchetta and A. Mukherjee, Phys. Rev. D **94** (2016) no.3, 034012
[8] M. Anselmino, M. Boglione, U. D'Alesio, S. Melis, F. Murgia and A. Prokudin, Phys. Rev. D **87** (2013) 094019
[9] G. R. Goldstein, J. O. G. Hernandez and S. Liuti, arXiv:1401.0438 [hep-ph].
[10] Z. B. Kang, A. Prokudin, P. Sun and F. Yuan, Phys. Rev. D **93** (2016) no.1, 014009
[11] V. Cirigliano, S. Gardner and B. Holstein, Prog. Part. Nucl. Phys. **71** (2013) 93 doi:10.1016/j.ppnp.2013.03.005 [arXiv:1303.6953 [hep-ph]].
[12] R. W. Pattie, K. P. Hickerson and A. R. Young, Phys. Rev. C **88** (2013) 048501 Erratum: [Phys. Rev. C **92** (2015) no.6, 069902]
[13] F. Wauters, A. García and R. Hong, Phys. Rev. C **89** (2014) no.2, 025501 Erratum: [Phys. Rev. C **91** (2015) no.4, 049904]
[14] A. Courtoy, S. Baeler, M. González-Alonso and S. Liuti, Phys. Rev. Lett. **115** (2015) 162001
[15] T. Bhattacharya, V. Cirigliano, S. D. Cohen, A. Filipuzzi, M. Gonzalez-Alonso, M. L. Graesser, R. Gupta and H. W. Lin, Phys. Rev. D **85** (2012) 054512
[16] A. Hocker, H. Lacker, S. Laplace and F. Le Diberder, Eur. Phys. J. C **21** (2001) 225
[17] O. Naviliat-Cuncic and M. Gonzlez-Alonso, Annalen Phys. **525** (2013) 600
[18] J. C. Hardy and I. S. Towner, Phys. Rev. C **79** (2009) 055502
[19] M. González-Alonso and J. Martin Camalich, Phys. Rev. Lett. **112** (2014) no.4, 042501
[20] M. Wakamatsu and Y. Ohnishi, Phys. Rev. D **67** (2003) 114011 doi:10.1103/PhysRevD.67.114011 [hep-ph/0303007].
[21] A. V. Efremov, K. Goeke and P. Schweitzer, Phys. Rev. D **67** (2003) 114014
[22] S. Pisano [CLAS Collaboration], EPJ Web Conf. **73** (2014) 02008.
[23] A. Courtoy, arXiv:1405.7659 [hep-ph].
[24] J. R. Ellis, K. A. Olive and C. Savage, Phys. Rev. D **77** (2008) 065026

Journal of Physics: Conference Series **761** (2016) 012069 doi:10.1088/1742-6596/761/1/012069

Top Quark Properties Measurements with the ATLAS Experiment

J A Murillo Quijada[1] **on behalf of the ATLAS collaboration**

[1] Department of Physics at CINVESTAV-IPN, Av. Instituto Politécnico Nacional 2508, Col. San Pedro Zacatenco, 07360, México D.F. (+52) (01) 55 5747 3800

E-mail: `javier.murillo@cern.ch`

Abstract.
Results on recent measurements of top quark properties with the ATLAS experiment at the European Laboratory, CERN, are shown. The measurements are performed using the full data set recorded during the LHC Run-I. The full data set consists of a collected integrated luminosities $\int$Ldt of 4.6 fb^{-1} recorded at a proton-proton collision energy of $\sqrt{s}$ = 7 TeV and 20.3 fb^{-1} collected at 8 TeV. The mentioned top quark properties include: spin correlation, charge asymmetry, W-boson polarization, color flow, top mass and top width in events with a top and anti-top quark pair ($t\bar{t}$). An introduction to the LHC and the ATLAS detector is included and latest main results from this experiment. The contents include the current world benchmark results for the different properties and plans for future measurements during the ongoing LHC Run-II.

1. Performance of LHC and the ATLAS detector during Run-I

The successful operation of the Large Hadron Collider (LHC), and the ATLAS experiment is the product of the hard work of thousands of people involved in their design, construction and careful usage. The studies and results presented in this contribution could not have been possible without the support of this collaboration. The CERN laboratory directs the operation of the LHC which has a 27 km circumference that extends to both sides of the border between France and Switzerland. This massive accelerator runs at unprecedented collisions energies, being able to operate at a design luminosity of 10^{34} cm^{-2} s^{-1} [1]. ATLAS is the largest detector located around the LHC ring. It produces two independent magnetic fields around the collision point, in the inner section a superconducting Central Solenoid (CS), is used to bend all charged particles in a plane perpendicular to the beam before they reach the electromagnetic and hadronic calorimeters. In the outer section a superconducting air cored toroid surrounding the calorimeters, provides with a set of particle detectors known as Muon Spectrometer (MS), located before, inside and after the toroidal field, an excellent muon identification. The LHC has already run with proton-proton (pp), collision energies of $\sqrt{s}$ = 7 and 8 TeV during its Run-I, larger by more than a factor of four with respect Fermilab's Tevatron accelerator beam energies. It delivered an integrated luminosity of 5.46 fb^{-1} and 22.8 fb^{-1} at 7 and 8 TeV respectively. From these integrated luminosities, about 4.57 fb^{-1} and 20.3 fb^{-1}, subsequently, are categorized as good quality Data for physics anaysis [2].

2. Predictive power of the standard model, Higgs mass and super-symmetry

Rate of production of several specific processes has been measured with the ATLAS detector at 7 and 8 TeV. The results of these measurements are compared with their corresponding theoretical expectation as given by the standard model of particle physics (SM). All the production measurements performed so far are in agreement with what is predicted by the SM [3]. During the first run of the LHC a particle consistent with the so-called Higgs boson (H) was disvovered by the CMS and ATLAS collaborations. These two experiments used for this purpose mainly the H $\rightarrow \gamma\gamma$ channel (two energetic photons are produced in the final state) and the H $\rightarrow 4\ell$ channel (4 leptons are produced in the final state) to discover this particle and measure its mass value, whose value is not predicted by the SM. After combining their individual measurements with the two mentioned channels, the final reported Higgs mass by the ATLAS and CMS collaborations corresponding to the LHC Run-I is m_H = 125.09 ± 0.24 (± 0.21 ± 0.11) GeV [4].

The ATLAS collaboration has performed a wide range of direct searches for particles consistent with the theory of Super-Symmetry (SUSY), during Run-I. So far no significant excess of events over the SM expectation has been observed. Exclusion limits have been set for the masses of the supersymmetric particles. So far only 'low mass regions' in the scale of SM particles have been excluded for the different SUSY particles, indicating that if they exist they should be more massive than any particle we know [5].

3. Introduction to top quark physics

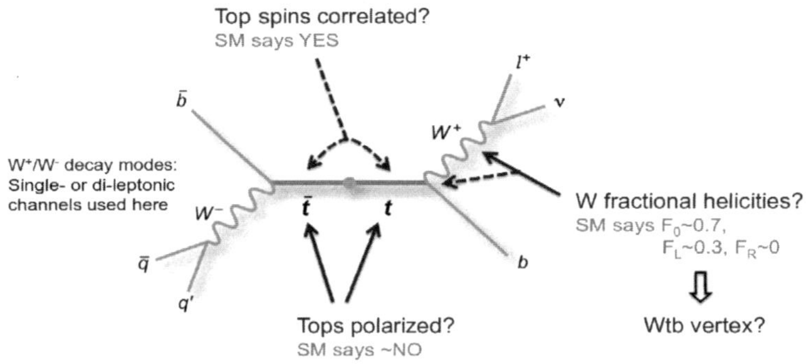

Figure 1. Diagram representing a $t\bar{t}$ pair semi-leptonic decay. From [6].

The top quark[1] is the heaviest known fundamental particle with a measured mass value equal to ∼ 172.5 GeV. The heavier particles following the top quark are the recently discovered Higgs boson with mass ∼ 125 GeV, the Z^0 boson with a mass ∼ 90 GeV and the $W^{\pm}$ boson with a mass ∼ 80 GeV. These particles, due to their large mass, compared with the rest of the fundamental particles, are the source of a huge variety of decays under study at the LHC.

The SM states that the top quark decays via the electroweak interaction to an on-shell W-boson plus one of the down-type quarks, decaying with a rate over 99 % in the form $t \rightarrow$ W-boson + b-quark [7]. A deviation from these predicted rate could indicate beyond the SM processes involving contributions from unobserved heavier quarks. Some extensions of the SM allow the mentioned rate to be smaller allowing the production of charged Higgs particles via top decays like $t \rightarrow b + H^+$. However recent searches for charged Higgs bosons show no deviations from the SM predictions [8]. At the LHC, top quarks are produced mainly as top and anti-top quark pairs ($t\bar{t}$) via the strong interaction. The nature of the event relies on the decays of the two produced W-bosons. In the selected events for the measurements discussed in this article, one of the W-bosons decays to a couple of leptons while the other to a couple of quarks, which is known as *semi-leptonic channel.* Also events where the two W-bosons decay to a couple of leptons, known as *dilepton channel,* are considered. As indicated in Figure 1, several properties associated with the top quark can be measured from these events. Study of the properties of the top quark is an ideal strategy to test the SM, as the latter delivers well defined predictions for their values.

4. Measurement of top quark properties

Brief description and results over the most relevant top quark properties measured by the ATLAS detector are shown in this section.

[1] Unless explicitly stated, top quark refers to both the top and the anti-top quark

4.1. Spin correlation in $t\bar{t}$ events

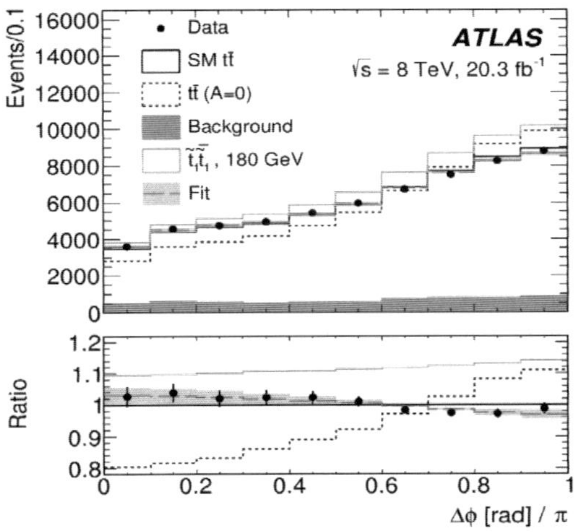

Figure 2. Distribution of the observable $\Delta\phi$ between two event leptons. Plot is drawn with the predicted distribution from the SM and the actual Data. From [9].

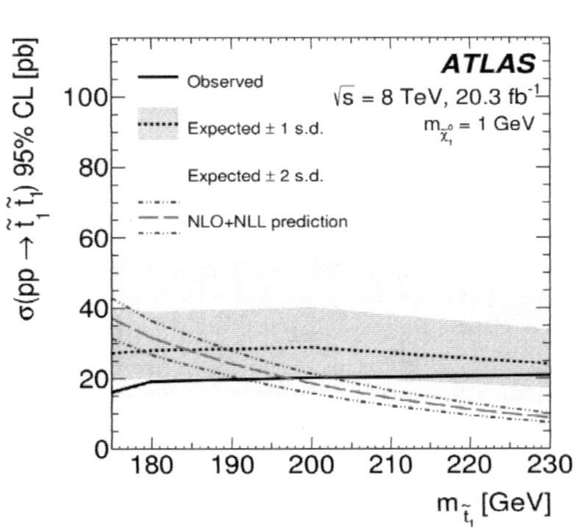

Figure 3. Measured lower limits for the production of a couple of stop quarks $\tilde{t}_1\tilde{t}_1$ compared with the SUSY prediction. From [9].

The top quark decay products give us direct access to its spin, as it decays before hadronisation ($< 10^{-24}$ s), before QCD interactions can affect this quantity. The top quark spin determines the angular distribution of its daughters when it decays. For this analysis a spin correlation coefficient $A_{helicity}$ is defined as:

$$A_{helicity} = \frac{N_{like} - N_{unlike}}{N_{like} + N_{unlike}} \tag{1}$$

where N_{like} are the number of selected events where the top and antitop quarks have their spins pointing in the same direction and N_{unlike} stands for events where the pair have their spins pointing in opposite directions. For this study only dileptonic events are selected, where the two W-bosons decay leptonically. Additional events allowing the production of super-symmetric neutralino ($\tilde{\chi}_1^0$), and stop ($\tilde{t}_1$), particles are also searched. Monte Carlo (MC) samples with different underlying values of $A_{helicity}$ are produced. These templates are compared with actual Data using the distribution of the variable $\Delta\phi$ as in Figure 2. This variable stands for the azymuthal angle between the two charged leptons in the event. In this figure it is indicated the MC template that fits better with the observed data, yielding a measured value of spin correlation ($A^{meas}_{helicity}$), divided over the SM expected value of $f_{SM} = \frac{A^{meas}_{helicity}}{A^{SM}_{helicity}} = 1.20 \pm 0.05$ (stat.) $\pm$ 0.13 (syst.) [9]. The errors correspond to 68% confidence level (1-σ errors), so the result is in agreement with the SM prediction, and is compatible with previous measurements by DØ, CDF and CMS collaborations, but more precise. The Plot in Figure 3 is constructed to set limits over the mass of the stop quark, for this purpose the rate of production of $pp \to \tilde{t}_1\tilde{t}_1$ events is measured assuming various masses for the the stop quark. The black line corresponding to observed values goes below the red line of predicted values by SUSY at the point where $m_{\tilde{t}_1} \sim 191$ GeV. It is then possible to exclude stop masses below this value.

4.2. Measurement of charge asymmetry in $t\bar{t}$ decays

The rapidity quantity y is defined as:

$$y = \frac{1}{2} ln\left(\frac{E + p_z}{E - p_z}\right) \tag{2}$$

the largest the value of this variable the closest the particle is traveling to the beam line. For this analysis the rapidities of the top and anti-top quarks are quantified in each of the events to obtain the observable $\Delta|y| = |y_t| - |y_{\bar{t}}|$, which is the difference between absolute value of top quark rapidity and the absolute value of the anti-top quark rapidity. The measured parameter A_C, regarded as charge asymmetry between top and anti-top quarks, is defined as:

$$A_C = \frac{N(\Delta|y| > 0) - N(\Delta|y| < 0)}{N(\Delta|y| > 0) + N(\Delta|y| < 0)} \tag{3}$$

where $N(\Delta|y| > 0)$ stands for the number of selected events satisfying $\Delta|y| > 0$, and $N(\Delta|y| < 0)$ for events where $\Delta|y| < 0$. For this analysis the semi-leptonic channel is used, and as for the case of the spin correlation study, several MC templates are produced with different underlying values of A_C. A statistical fit is done in this case to compare the Data and the different MC models using the distribution of the $\Delta|y|$ observable. The result of the measurement is $A_C = 0.009 \pm 0.005$, where the error includes statistical and systematic uncertainties. Independently the CMS collaboration measured the value $A_C = 0.0010 \pm 0.0068 \pm 0.0037$ [10], being then the result from ATLAS more precise. The predicted value for this parameter by the SM is $A_C = 0.0101 \pm 0.0005$ [11], both results from ATLAS and CMS are in agreement with the SM. Previously the DØ and the CDF collaborations measured another asymmetry regarded as forward/backward asymmetry (A_{FB}) [12], Figure 4 shows in a 2-dimensional plot the region allowed for the SM for the two asymmetries A_C and A_{FB} (small rectangle in purple). The figure shows that the current measurements of these two asymmetries are in agreement with the SM prediction [11].

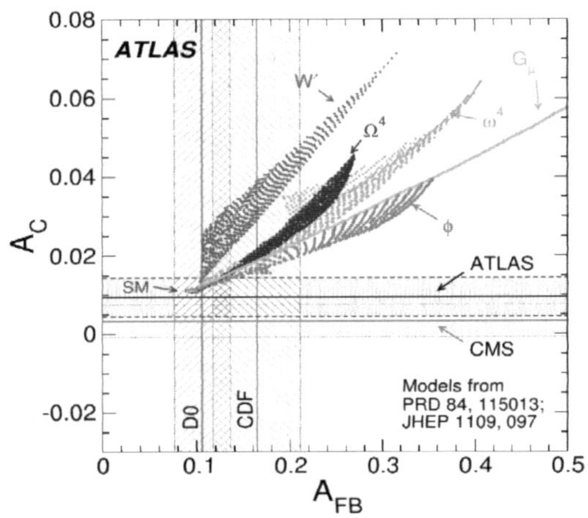

Figure 4. Allowed values from the SM for the two asymmetries A_C and A_{FB}. Other allowed regions for beyond the SM models are indicated. From [11].

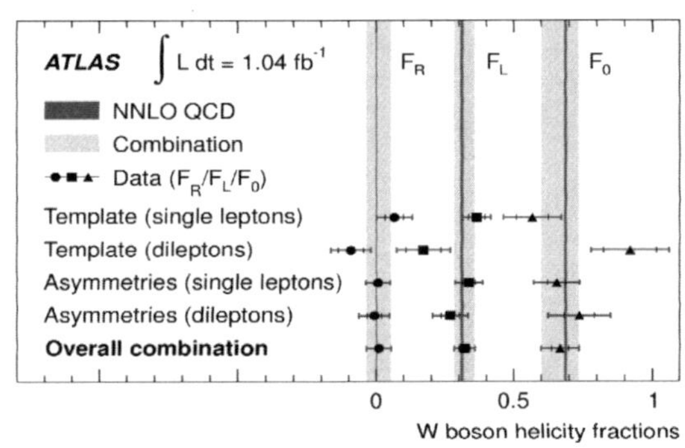

Figure 5. Measured values for helicity fractions, along with their statistical and systematic uncertainties. From [14].

4.3. Measurement of color flow in $t\bar{t}$ decays

The SM establishes that the b-quark decaying directly from the top quark via the emission of a W-boson (as shown in Figure 1) is color connected with its mother particle. This implies that when a couple of quarks are produced from the emerging W-boson, these are color connected as well. These two quarks originate a couple of jets, where a jet is a group of particles product of a quark hadronisation that are identified by the detector. In this context being color connected means that these two jets are not completely free from each other, as it is predicted by the SM. The MC simulation representing the SM keeps certain degree of connection between these two jets, which makes certain special observables get smaller values than what they would get if the jets are set to be free from each other. The degree of connection defines the different models for color connection or color flow. Similarly as with the previous analyses, two MC samples have been produced for this study: one keeping the predicted degree of connection between the produced jets by the SM and another one decreasing to zero the attraction or connection between the two jets from the W-boson, which is known as the *flipped model* [13]. To perform this study the quantity $\vec{r_i} = (\Delta y_i, \Delta\phi_i)$ is defined as the diference between the jet components coordinates (y_i, ϕ_i) and the jet coordinates (y_J, ϕ_J). This makes possible to define the pull vector of one of the jets emerging from the W-boson as:

$$\vec{v}_p^J = \sum_{i\in J} \frac{p_T^i |r_i|}{p_T^J} \vec{r_i} \tag{4}$$

This vector allows the definition of the pull angle $\theta_p(J_1, J_2)$ as the angle between the pull vector associated to one of the produced jets and the remainning jet number 2. The pull angle is sensitive to changes in the color connection between the two jets. If the color flow is flipped as mentioned before, its value gets smaller. Distributions of the pull angle θ_p from the SM and the flipped model are compared with the Data collected at $\sqrt{s} = 8$ TeV. After a statistical test is performed [13] it is concluded that the flipped model differs from the Data by 2.3 standard deviations, and no significant difference is found between the SM and the Data [13].

4.4. Measurement of W-boson polarization in $t\bar{t}$ decays

For this analysis the dilepton and semi-leptonic channels are used. The observable that is recorded at each event is $cos\theta^*$, where the angle θ^* is defined as the angle between the direction of the momentum of the charged lepton (in the W-boson rest frame) and the W boson momentum in the top quark rest frame. The shape of the distribution of $cos\theta^*$ is given by:

$$\frac{1}{N}\frac{dN}{dcos\,\theta^*} = \frac{3}{4}sin^2\theta^* F_0 + \frac{3}{8}(1 - cos\,\theta^*)^2 F_L + \frac{3}{8}(1 + cos\,\theta^*)^2 F_R \tag{5}$$

where F_0, F_L and F_R are known as *helicity fractions* and have a predicted value within the SM. As with the previous measurements here different MC samples are produced with all the posible values for the helicity fractions. The analysed Data corresponds to an integrated luminosity of $\int L\mathrm{d}t = 1.04$ fb^{-1} at 7 TeV. Figure 5 shows all the individual measurements of the helicity fractions performed by ATLAS using two different approaches and channels. After combining all the measurements in the figure, the final result is: $F_0 = 0.67 \pm 0.07$, $F_L = 0.32 \pm 0.04$ and $F_R = 0.01 \pm 0.05$, which are consistent with the SM and are in agreement with previous measurements as well [14].

4.5. Measurement of the top quark mass and width in $t\bar{t}$ decays

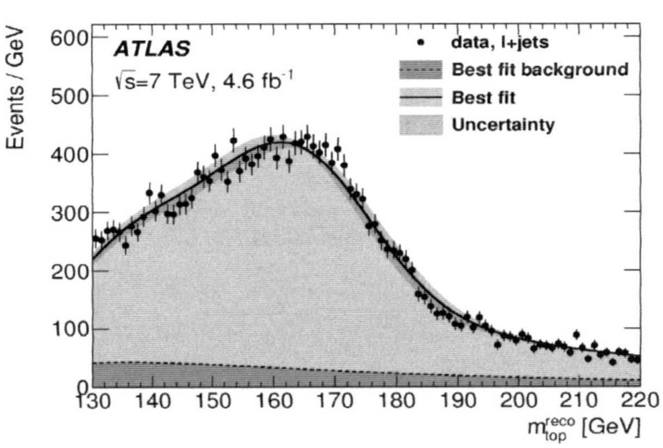

Figure 6. Reconstructed top mass distribution. From [15]

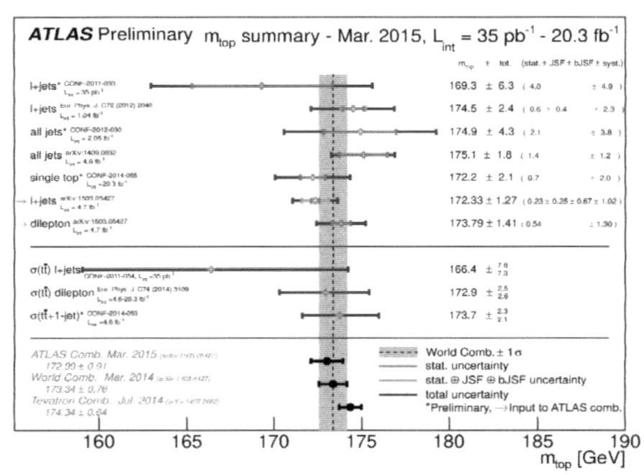

Figure 7. Recent measurements over m_{top}. From [15]

Finally, ATLAS has also measured the mass (m_{top}) and width (Γ_{top}) of the top-quark using a template fit in the same way as all the analyses previously mentioned in this contribution, producing MC samples with various values of the m_{top} and Γ_{top} parameters respectively. The most precise measurement of m_{top} was performed with a pp collision energy of 7 TeV using 4.6 fb^{-1} of Data. Figure 6 shows the distribution of the reconstructed top mass, here are included the histogram from the Data and the MC template that best fitted with it. The m_{top} measurement was performed using the semi-leptonic and dilepton channels separately, after the combination of both results the most precise mesurement obtained by ATLAS is $m_{top} = 172.99 \pm 0.91$ GeV [15]. Figure 7 shows this result along with other less precise results from ATLAS and the current world combination for m_{top}. A similar analysis is in progress to measure the Γ_{top} parameter, in this analysis a direct approach is followed as in the recent measurement from the CDF collaboration that sets a upper limit of $\Gamma_{top} < 6.38$ GeV at 95 % confidence level [16].

5. Conclusions and top physics measurements during LHC Run-II

All the measurements of properties of the top quark discussed in this contribution are in agreement with the SM predictions. The m_{top} and Γ_{top} measurements are very difficult, because even when having more than two orders of magnitude more Data than the Tevatron experiments, the LHC experiments deal with larger systematic uncertainties, situation that is expected to change during Run-II at $\sqrt{s}$ = 13 TeV, as all the detectors will be understood better. ATLAS went through important upgrades before the start of Run-II last December. Also in this second campaign the production of $t\bar{t}$ pairs is expected to increase by a factor of ~ 3.3 with respect to the production at 8 TeV. So higher precision for all the measurements discussed here is expected during Run-II.

References

[1] L. Evans and P. Bryant, LHC Technical Report, LHC Machine. 2008, JINST **3**, s08001
[2] ATLAS Experiment Luminosity Public Results,
URL: http//twiki.cern.ch/twiki/bin/view/AtlasPublic/LuminosityPublicResults,
Accessed: 24/07/2016
[3] Summary plots from the ATLAS Standard Model physics group,
URL: https://atlas.web.cern.ch/Atlas/GROUPS/PHYSICS/CombinedSummaryPlots/SM/,
Accessed: 24/07/2016
[4] ATLAS, CMS Collaborations, Combined Measurement of the Higgs Boson Mass in pp Collisions at s=7 and 8 TeV with the ATLAS and CMS Experiments, Phys. Rev. Lett. **114**, 191803 (2015), arXiv:1503.07589 [hep-ex]
[5] ATLAS summary plots for lower limits for supersymmetry,
URL: https://twiki.cern.ch/twiki/bin/view/AtlasPublic/SupersymmetryPublicResults,
Accessed: 24/07/2016
[6] Dr. Thorsten Wengler, Top quark angular distributions at LHC, TOP 2012 - 5th International Workshop on Top Quark Physics, 20 September, 2012, Winchester United Kingdom
[7] CKMfitter Group - J. Charles et al., Eur. Phys. J. C**41**, 1-131 (2005), hep-ph-0406184
[8] ATLAS collaboration, Search for charged Higgs bosons through the violation of lepton universality in $t\bar{t}$ events using pp collisions data at $\sqrt{s} = 7$ TeV with the ATLAS experiment, Eur. Phys. J. C, **73** 6 (2013) 2465, arXiv:1302.3694 [hep-ex].
[9] ATLAS Collaboration, Measurement of Spin Correlation in Top-Antitop Quark Events and Search for Top Squark Pair Production in pp Collisions at $\sqrt{s} = 8$ TeV Using the ATLAS Detector, Phys. Rev. Lett. **114**, 142001 (2015), arXiv:1412.4742 [hep-ex]
[10] CMS Collaboration, Inclusive and differential measurements of the $t\bar{t}$ charge asymmetry in pp collisions at $\sqrt{s} = 8$ TeV, Phys. Lett. B **757** (2016) 154, arXiv:1507.03119 [hep-ex]
[11] ATLAS Collaboration, Measurement of the charge asymmetry in top-quark pair production in the lepton-plus-jets final state in pp collision data at $\sqrt{s}$=8 TeV with the ATLAS detector, Eur. Phys. J. C**76** (2016) 87, arXiv:1509.02358 [hep-ex]
[12] CDF Collaboration, Measurement of the forward-backward asymmetry of top-quark and antiquark pairs using the full CDF Run II data set,Phys. Rev. D **93**, 112005 (2016), arXiv:1602.09015 [hep-ex]
[13] ATLAS collaboration, Measurement of colour flow with the jet pull angle in $t\bar{t}$ events using the ATLAS detector at $\sqrt{s} = 8$ TeV, Physics Letters B (2015) 475-493, arXiv:1506.05629 [hep-ex]
[14] ATLAS Collaboration, Measurement of the W boson polarization in top quark decays with the ATLAS detector, JHEP 1206 (2012) 088, arXiv:1205.2484 [hep-ex]
[15] ATLAS Collaboration, Measurement of the top quark mass in the $t\bar{t} \rightarrow$ lepton+jets and $t\bar{t} \rightarrow$ dilepton channels using $\sqrt{s} = 7$ TeV ATLAS data, Eur. Phys. J. C (2015) 75:330, arXiv:1503.05427 [hep-ex]
[16] CDF Collaboration, A Direct Measurement of the Total Decay Width of the Top Quark, Phys. Rev. Lett. **111**, 202001 (2013), arXiv:1308.4050[hep-ex]

Journal of Physics: Conference Series **761** (2016) 012070 doi:10.1088/1742-6596/761/1/012070

Some aspects on four quarks recombination

G. Toledo Sánchez and I. A. Toledano Juárez

Instituto de Física, Universidad Nacional Autónoma de México. México D. F. C.P. 04510

E-mail: toledo@fisica.unam.mx

Abstract.
We have performed a 3-D Monte Carlo simulation of a system composed of two identical light quarks (qq) and two identical antiquarks ($\bar{Q}\bar{Q}$) and determined whether it is energetically more favorable to form a tetraquark or two mesons, as a function of the interparticle separation distance which, for a fixed number of particles, can be identified as a particle density. In this proceedings, we highlight the main results and elaborate on the implications in properties like the correlation function for two-mesons and characterize the isolated diquark correlation function. We analize the four-body potential evolution and exhibit its linear behavior as a function of the invariant distance. We track the dynamical flipping among configurations to determine the recombination probability, exhibiting the importance of the tetraquark state.

1. Introduction

Recent experimental research has provided strong evidence on the formation of tetraquark states [1, 2, 3, 4]. A new era in hadron spectroscopy is just ahead [5], calling for a better understanding of the strong interaction in a scenario where conventional mesonic and baryonic systems try to cope with the experimental data. The early theoretical studies inquired on the existence and stability of the tetraquark[6, 7, 8, 9, 10, 11, 12, 13, 14], and how its mixing with a meson state can help us to understand the observed spectroscopy of states like the σ meson [14, 15].

A tetraquark state offers the opportunity to learn about the strong interaction features leading to its formation when two mesons are forced to approach each other as it could happen in a meson-meson collision or, on the opposite, when the four quarks are produced very close in space as in the WW decay, which eventually freeze out to two mesons [16, 17]. In these scenarios several questions can be asked: i) Under which conditions two mesons can turn into a tetraquark or mixed state? ii) Is there any property at the meson level that can reflect the presence of an intermediate tetraquark state? iii) What is the structure of a diquark? iv)How the many-body potential behaves? v) Can we turn it into a simpler form? vi) Can we distinguish between a molecular (meson-meson bound) and a tetraquark state? vii) Is the quark recombination relevant for hadronic states? viii) How is it modified by the presence of the tetraquark?, etc.
In a recent work [18], we addressed these questions using an effective model to mimic the strong interaction among quarks. In this proceedings we highlight the main results and elaborate on the implications. Let us recall the procedure followed:

We consider two identical light quarks qq (denoted by u) and two identical antiquarks $\bar{Q}\bar{Q}$ (denoted by d, s, c and b, depending on the constituent mass with respect to the light one). The seminal work by Lenz *et al.* [8] laid down a procedure to describe such a system in this context and found its general properties, as isolated objects, using a harmonic potential

Journal of Physics: Conference Series **761** (2016) 012070 doi:10.1088/1742-6596/761/1/012070

interaction. We perform a 3-D Monte Carlo simulation of the system including quantum correlations between particles to determine whether it is energetically more favorable to form a 4-body state (tetraquark) or two mesons, as a function of the interparticle separation which, for a fixed number of particles, can be identified as a particle density. We are interested in s-wave states where tensor and spin interaction effects are expected to be negligible in the gross features of the properties.

2. Meson variational wave function

Let us start with the description of the exact solution and the variational approach of a meson state, composed of a quark and an antiquark of mass m_1 and m_2 respectively. The strong interaction between this pair can be represented by a flux tube [7], which can be effectively described by a linear potential $V[\vec{r}_1, \vec{r}_2] = k|\vec{r}_1 - \vec{r}_2| = kr$ where k is an interaction constant and $r = |\vec{r}_1 - \vec{r}_2|$ is the relative distance between them. Lattice QCD studies have confirmed the linear behavior of the interaction among quarks at large distances [19]. The exact solution corresponds to the eigen-energies and eigen-functions:

$$E_n = \left[\frac{k^2}{2\mu}\right]^{1/3} |\xi_n|, \qquad \rho_n(r) = \frac{1}{r}\mathrm{Ai}\left(r\,[2\mu k]^{1/3} - \xi_n\right), \tag{1}$$

where ξ_n are the zeros of the Airy function (Ai) and μ is the reduced mass. The ground state corresponds to the first zero of the Airy function, with energy $E_0 = 2.3381$, in $m_1 = m_2 = m = k = 1$ units. A variational wave function that approaches the exact result for the ground state takes the following form:

$$F_\lambda(r) = \sqrt{\frac{3\lambda^2}{2\pi}} e^{-\lambda r^{3/2}}, \tag{2}$$

where λ is the variational parameter and r is the relative coordinate of the two-body system. We have found that the wave function reproduces the exact minimum energy value within a 1% error. This simple form allows to find analytical expressions for the expected variational parameter, energy and mean square radius. By matching them to their experimental values we can fix the physical units.

3. Four-body system

Using the previous results, we can now extend our study to a four-body system composed of two identical u-like quarks (qq) and two identical antiquarks ($\bar{Q}\bar{Q}$), which can be d, s, c and b-like. We will refer to this four-body system by qQ. At very low density, where the distance between quarks is large, the model must reproduce a system of two isolated mesons. In addition, as the density increases, the inter-particle separation becomes small allowing the interaction between the two mesons, which can be represented by quark exchange or a truly four-body interaction. All over, the Pauli blocking among identical particles must be enforced. We rely on the String-Flip Model [8, 20, 21] which has been also used to study dense matter systems [22, 23, 24], to describe the dynamical transition among these regimes.

The strong interaction of a many-body system can be well approached by considering that the quarks are connected by gluon flux tubes [7] according to a configuration producing the lowest energy state of the whole system. Let us denote the position of the particles by $\vec{r}_1$ and $\vec{r}_2$ and the antiparticles by $\vec{r}_3$ and $\vec{r}_4$. The way to link the four particles of the system consistent with the QCD restrictions of color neutrality are shown in Figure 1: i) The system is composed of two mesons (V_{m1} or V_{m2}). ii) The system is composed of a tetraquark (V_{4Q}). In this case, the shortest path linking the four particles is given by the Steiner-tree, which uses two auxiliary vectors (denoted by $\vec{k}$ and $\vec{l}$) to minimize the length of the configuration. The vector $\vec{k}$ links

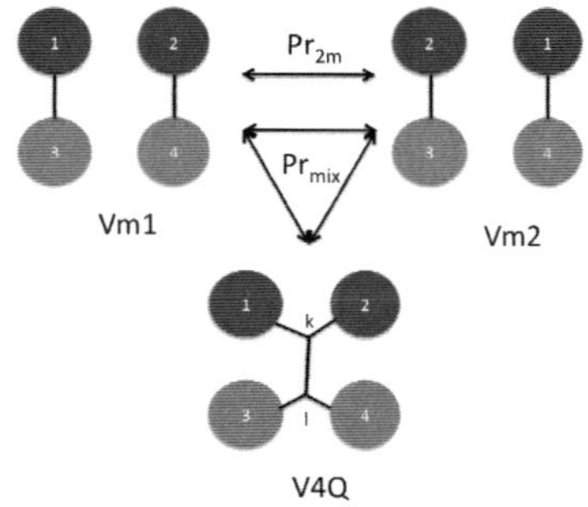

Figure 1. Potential configurations for the system (Vm1,Vm2,V4Q). 1 and 2 are the particles and 3 and 4 the antiparticles. Pr_{2m} and Pr_{mix} are the possible quark recombinations for the two mesons and mixed system, respectively.

the diquark sub-system and the vector $\vec{l}$ the anti-diquark one. When all these configurations are allowed, the potential of the system is chosen as the one producing the minimum potential energy in a given configuration:

$$V = min(V_{m1}, V_{m2}, V_{4Q}). \tag{3}$$

We propose the following variational wave function, which reproduces the two isolated mesons at very low densities, allows four-body correlations and incorporates the Fermi correlations as the density increases:

$$\Psi_\lambda = \Phi_{FG}\left(e^{-\lambda Q}\right), \tag{4}$$

where λ is the single variational parameter, Q is a function driven by the many-body potential, and Φ_{FG} is the free Fermi system wave function. For a four-body system, the Q function takes the following form, depending on the optimal potential at the given configuration, denoted by the sub-index:

$$Q_{m_1} = r_{13}^{3/2} + r_{24}^{3/2}, \qquad Q_{m_2} = r_{14}^{3/2} + r_{23}^{3/2} \tag{5}$$

or

$$Q_{4Q} = r_{1k}^{3/2} + r_{2k}^{3/2} + r_{kl}^{3/2} + r_{3l}^{3/2} + r_{4l}^{3/2}, \tag{6}$$

where the 3/2 power resembles the variational solution of single pairs linked by a linear potential, described in the previous section. Φ_{FG} is composed of single wave functions of a particle in a cubic box of side L. We can define a particle density parameter as a measure of the interparticle separation by: $\rho = N/V = 4/L^3$, where N is the number of particles of the system and V is the box volume. For a fixed number of particles, the change in the particle density corresponds to modify the box size and correspondingly to the inter-particle separation.

4. Results

The expectation value for the energy is computed using Monte Carlo techniques, in particular we use a Metropolis algorithm for the sampling, driven by the square of the wave function. We computed the properties of three systems:

- *Two mesons.* In this case, only meson configurations are allowed in the potential. We characterize the meson properties, comparing the results with the expected from the analytical solution in the isolated limit and the departure from them as density increases.

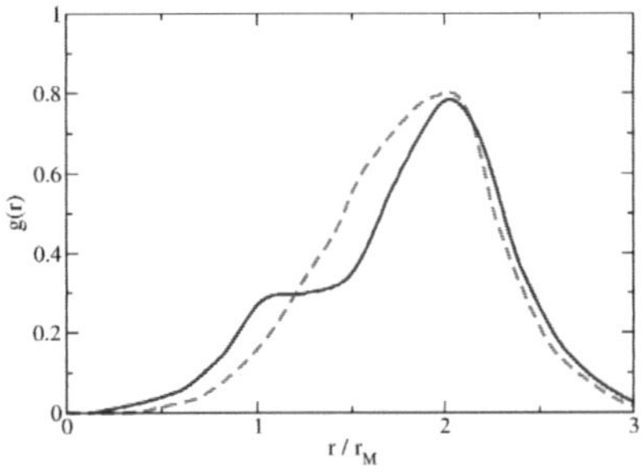

Figure 2. Two meson correlation function for ud at $\rho = 0.01$. Dashed (Solid) line corresponds to the two mesons (mixed) case.

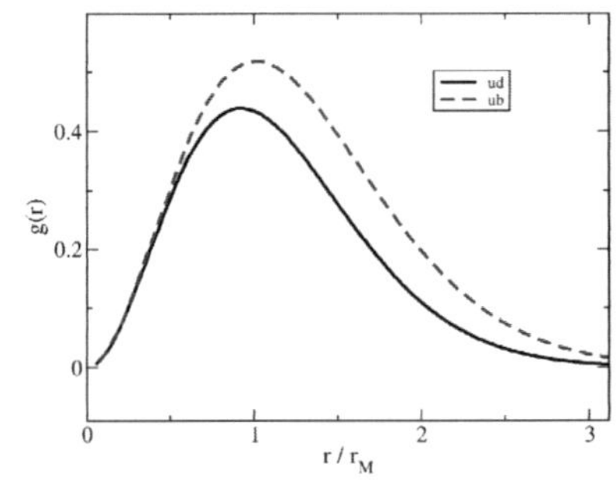

Figure 3. Diquark correlation function at zero density.

- *Tetraquark.* In this case, only tetraquark configurations are allowed in the potential. We characterize the tetraquark properties in the isolated limit and the departure from it as density increases.
- *Mixed.* In this case, all the possible configurations are allowed in the potential. We characterize the meson and tetraquark properties and their modifications due to the presence of the other configuration.

For all the above cases we explored the mass effect by considering the two quarks to be light (u)and the antiquarks having variable mass to resemble the $m_{d,s,c,b}/m_u$ ratios. Results are presented for several light-quarks to heavy-antiquarks mass ratios whenever they are found to be relevant.

4.1. Correlation functions

A useful observable to characterize the properties of the system is the two-particle correlation function. It measures the probability of finding two particles at a relative distance r from each other. In Figure 2, we show the meson-meson correlation function at a density $\rho = 0.01$ for the two-mesons (dashed line) and mixed case (solid line). We exhibit the results corresponding to a light-light (ud) system but a similar behavior is observed in the other mass ratio combinations. The comparison between both figures shows that there is a modification in the meson-meson correlation function by the presence of the tetraquark state at intermediate densities, a bump develops in the near tail of the correlation function, driven by the diquark formation and fades out as the density increases and all the quark correlations vanish. This may be an alternative observable to have indirect evidence of the tetraquark formation.

For the tetraquark case, we can explore the structure of the quark-quark (diquark) correlation function. In particular, in the zero density limit, the normalized tetraquark system correlation function for the diquark behaves as shown in Figure 3 for the light-light (solid line) and light-heavy (dashed line) cases, which can be parameterized by:

$$g(r)_{diquark} = A_0 r^2 e^{-r^{A_2}/A_1^2}. \tag{7}$$

In Table 1, we show the value of the parameters for each case. Note that the dependence on r of the diquark correlation function is close to others suggested in the literature [25, 26] but in

Table 1. Diquark correlation functions parameters for *ud* and *ub*-like mass ratios.

System	A_0	A_1	A_2
ud	0.64	1.24	1.51
ub	1.13	1.1	1.47

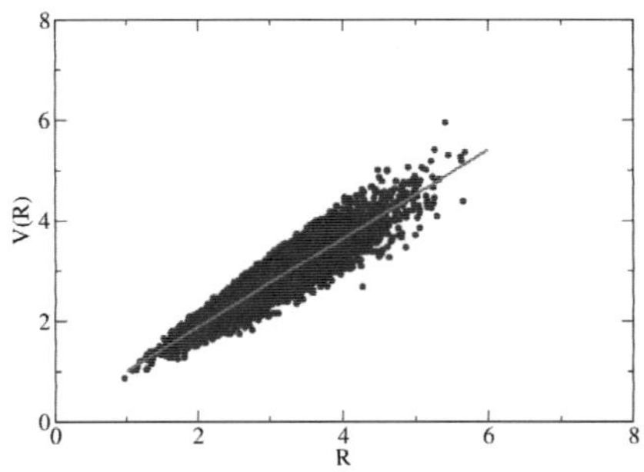

Figure 4. Tetraquark potential obtained in the simulation (symbols) as a function of R. The solid line is a linear fit to the data.

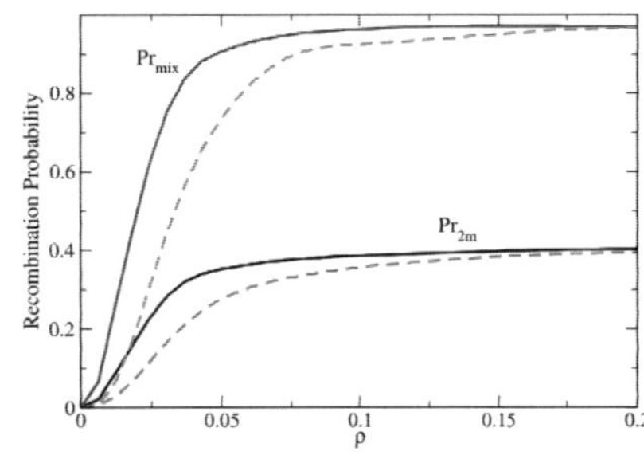

Figure 5. Recombination probabilities as a function of the density, for two mesons (Pr_{2m}) and mixed (Pr_{mix}) systems. The solid and dashed lines are for *ud* and *ub* systems respectively.

our case it is obtained in a dynamical way. The short distance region shows the competition between the Pauli blocking among the identical particles and the attractive potential.

4.2. Tetraquark potential

The tetraquark potential depends on the quarks positions and two auxiliary vectors, placed in such a way that the total length linking the quarks is the shortest one. These auxiliary vectors are modified in a non-trivial way whenever a single quark changes its position, making this a truly many-body interaction. An effective behavior of the potential can be set as linear respect to the invariant length, $R \equiv \sqrt{\sum r_{ij}^2}$:

$$V(R) = R_0 + BR, \tag{8}$$

where R_0 is the value of the potential at zero distance, which is expected to be modified by the short distance coulomb-like correction. In Figure 4, we show the tetraquark potential values as obtained from the simulation (symbols) as a function of R and a linear fit to them (solid line) when only tetraquarks are allowed. The simulation data clearly follows the linear fit, The slope shows small dependence on the density, and $B(\rho \approx 0) = 0.84 \pm 0.02$. In the mixed case there is a significant density dependence, starting below the corresponding value at zero density ($B(\rho \approx 0) = 0.67 \pm 0.02$) and as the density increases they approach to each other. This behavior is similar when considering different flavor systems.

An effective four-body contact potential, $V_4(contact)$, as a function of density, can be determined as the average potential irrespective of the length linking the particles.

4.3. Dynamical recombination

The four quark recombination can have important effects in systems where they are produced very close in space. An example can be found in the $WW \to qq\bar{Q}\bar{Q}$ decay, whose spatial separation at LEP2 energies is around 0.1 fm [16]. Although in the perturbative regime the recombination is small, in the non-perturbative regime the effect may be important. The typical scenarios to estimate the recombination are: Considering spherical or elongated bags color sources and the reconnection is proportional to the overlap of two color sources; strings considered as vortex lines where reconnection takes place when the cores of the two string pieces cross each other [17]. In our case, we would have, in addition, recombinations similar to tetraquark states which eventually freeze out to two mesons. The flipping from one configuration to another, driven by the minimal potential energy restriction, is a measure of the dynamical recombination property of the system. We define the recombination probability for the two mesons case by Pr_{2m}, denoting the number of flippings among meson configurations normalized to the total configurations. Considering the tetraquark configuration as contributing to the recombination, the probability Pr_{mix} counts the flipping among all of them, as depicted in Figure 1. The behavior of these probabilities as a function of density are shown in Figure 5. We observe a difference between them in size, which can be attributed to the normalization, and more subtle in shape, as the rising to the high density convergence value is faster approached in the mixed case.

We have presented some general features about a four quarks system recombination states but many questions are still open. The inclusion of more properties and refinement in the approaches are certainly important to be considered in future work.

Acknowledgments

We congratulate the Mexican Division of Particles and Fields for its 30th annual meeting, and thank the organizers for a very pleasant gathering.

References

[1] Z.Q. Liu et al. (Belle Collaboration), Phys. Rev. Lett. **110**, 252002 (2013).
[2] M. Ablikim et al. (BES III Collaboration), Phys. Rev. Lett. **110**, 252001 (2013).
[3] LHCb Collab. (R. Aaij et al.), Phys. Rev. Lett. **112**, 222002 (2014).
[4] V. M. Abazov, et al., (Dzero Collab.), Phys. Rev. Lett. 117, 022003 (2016)
[5] A. Hosaka, et al., PTEP 2016 (2016) no.6, 062C01
[6] R. L. Jaffe, Phys. Rev. D **15** 267(1977).
[7] N. Isgur and J. Paton Phys. Rev. D **31** 2910(1985).
[8] F. Lenz et al., Annals of Phys. **170** 65(1986).
[9] J. Carlson and V. R. Pandharipande Phys. Rev. D **43** 1652(1991).
[10] D. M. Brink and F. Stancu, Phys. Rev. D **49** 4665(1994).
[11] J. Vijande, A. Valcarce, J. M. Richard, Phys. Rev. D **87** 034040(2013).
[12] Pedro Bicudo, Nuno Cardoso, Marco Cardoso. Prog. Part. Nucl. Phys. **67** (2012). 440
[13] J. Ferretti, G. Galatà and E. Santopinto Phys. Rev. D **90** 054010(2014).
[14] A. Esposito et al, Int. Jour. Mod. Phys. A, **30** 1530002(2015).
[15] Ping Wang, Stephen R. Cotanch and Ignacio J. General, Eur. Phys. J. C. **55** 409(2008)
[16] G. Abbiendi et al, (The OPAL Collaboration), Eur. Phys. J. C. **45** 307(2006).
[17] T. Sjöstrand and V. A. Khoze, Z. Phys.C **62** 281(1994); Phys. Rev. Lett. **72** 28(1994).
[18] I. A. Toledano Juárez and G. Toledo Sánchez Phys. Rev. C **92**, 065204(2015)
[19] G. S. Bali and K. Schilling, Phys. Rev. D **47** 661(1993).
[20] C. J. Horowitz, E. J. Moniz and John W. Negele, Phys. Rev. D **31** 1689(1985).
[21] M. Oka and C. J. Horowitz, Phys. Rev. D **31** 2773(1985)
[22] C. J. Horowitz and J. Piekarewicz Nucl. Phys. A **536** 669(1992);
[23] G. Toledo Sánchez and J. Piekarewicz, Phys. Rev. C **65**, 045208 (2002).
[24] A. Ayala, J. Magnin, L. M. Montano and G.Toledo Sanchez Phys. Rev. C **80** 064905(2009).
[25] Samuel H. Blitz and Richard F. Lebed, Phys. Rev. D **91** 094025(2015).
[26] Pedro Bicudo and Marc Wagner, Phys. Rev. D **87**, 114511 (2013).

MWPF IOP Publishing
Journal of Physics: Conference Series **761** (2016) 012071 doi:10.1088/1742-6596/761/1/012071

Spectrum of Charmonia within a Contact Interaction

Marco Antonio Bedolla
Instituto de Fsica y Matemticas, Universidad Michoacana de San Nicolás de Hidalgo Edificio C-3, Ciudad Universitaria, C.P. 58040, Morelia, Michoacán, México.
E-mail: bedolla@ifm.umich.mx

Abstract. For the flavour-singlet heavy quark system of charmonia, we compute the masses of the ground state mesons in four different channels: pseudo-scalar ($\eta_c(1S)$), vector ($J/\Psi(1S)$), scalar ($\chi_{c_0}(1P)$) and axial vector ($\chi_{c_1}(1P)$), as well as the weak decay constants of the $\eta_c(1S)$ and $J/\Psi(1S)$. The framework for this analysis is provided by a symmetry-preserving Schwinger-Dyson equation (SDEs) treatment of a vector×vector contact interaction (CI). The results found for the meson masses and the weak decay constants, for the spin-spin combinations studied, are in fairly good agreement with experimental data and earlier model calculations based upon Schwinger-Dyson and Bethe-Salpeter equations (BSEs) involving sophisticated interaction kernels.

1. Introduction

First explorations for heavy mesons, both charmonia and bottomonia, with a consistent use of the rainbow-ladder truncation in the kernels of the gap and Bethe-Salpeter equations (BSEs), were undertaken by Jain and Munczek in Ref. [1]. They found the mass spectrum and the decay constants of pseudoscalar mesons in good agreement with experiments. This work was repeated with the Maris-Tandy model for $\bar{c}c$ bound states in Refs. [2–4]. We use a symmetry preserving vector-vector contact interaction (CI) [5–9]. This model provides a simple scheme to exploratory studies of the spontaneous chiral symmetry breaking and its consequenses like: dynamical mass generation, a quark condensate, the rise of goldstone bosons in the chiral limit and quark confinement. The results obtained from the CI model are quantitatively comparable to those obtained using sophisticated QCD model interactions, [10–12].

We employ this interaction for the analysis of the quark model heavy mesons for spins $J = 0, 1$ and study the mass spectrum and weak decay constants for charmonia. Without parameter readjustment, we find good agreement with charmonia masses. However, we need to modify the set of parameters to simultaneously account for the weak decay constants of the $\eta_c(1S)$ and $J/\Psi(1S)$, and the charge radius of $\eta_c(1S)$.

2. The Tools

2.1. The Bethe-Salpeter and the Gap Equations

Meson bound states appear as poles in a four-point function. The condition for the appearance of these poles in a particular J^{PC} channel is given by the homogeneous BSE [13–15]

$$[\Gamma_H(k;P)]_{tu} = \int \frac{d^4q}{(2\pi)^4}\, \chi(q;P)_{sr} K^{rs}_{tu}(q,k;P)\,, \tag{1}$$

Journal of Physics: Conference Series **761** (2016) 012071 doi:10.1088/1742-6596/761/1/012071

where $\chi(q;P) = S_f(q_+)\Gamma_H(q;P)S_g(q_-)$; $q_+ = q + \eta P$, $q_- = q - (1-\eta)P$; k (P) is the relative (total) momentum of the quark-antiquark system; S_f is the f-flavour quark propagator; $\Gamma_H(q;P)$ is the meson BSA, where $H = f\bar{g}$ specifies the flavour content of the meson; r, s, t, u represent colour, flavour, and spinor indices; and K is the quark-antiquark scattering kernel.

The f-flavour dressed-quark propagator, S_f, that enters Eq. (1) is obtained as the solution of the quark SDE, the so called gap equation [16–19]:[1]

$$S_f^{-1}(p) = i\gamma \cdot p + m_f + \Sigma_f(p)\,, \tag{2}$$

$$\Sigma_f(p) = \int \frac{\mathrm{d}^4 q}{(2\pi)^4}\, g^2 D_{\mu\nu}(p-q) \frac{\lambda^a}{2}\gamma_\mu S_f(q) \Gamma^a_\nu(p,q)\,, \tag{3}$$

where g is the strong coupling constant, $D_{\mu\nu}$ is the dressed-gluon propagator, Γ^a_ν is the dressed-quark-gluon vertex, and m_f is the bare f-flavour current-quark mass.

Both $D_{\mu\nu}$ and Γ^a_ν satisfy their own SDE, which in turn are coupled to higher n-point functions and so on *ad infinitum*. Therefore, the quark SDE, Eq. (2), is only one of the infinite set of coupled nonlinear integral equations. A tractable problem is defined once a truncation scheme has been specified, i.e., once the gluon propagator and the quark-gluon vertex are defined.

2.2. Axial-Vector Ward-Takahashi Identtiy

The phenomenological features of chiral symmetry and its dynamical breaking in QCD can be understood by means of the axWTI. In the chiral limit, it reads

$$-\,\mathrm{i}P_\mu \Gamma_{5\mu}(k;P) = S^{-1}(k_+)\gamma_5 + \gamma_5 S^{-1}(k_-)\,. \tag{4}$$

The axWTI relates the axial-vector vertex, $\Gamma_{5\mu}$, the pseudoscalar vertex, γ_5 and the quark propagator. This in turn implies a relationship between the kernel in the BSE and that in the quark SDE. It must be preserved by any viable truncation scheme of the SDE-BSE coupled system

$$\int \frac{\mathrm{d}^4 q}{(2\pi)^4} K_{tu;rs}(k,q;P)\, [\gamma_5 S(q_-) + S(q_+)\gamma_5]_{sr} = [\Sigma(k_+)\gamma_5 + \gamma_5 \Sigma(k_-)]_{tu}\,, \tag{5}$$

thus constraining the content of the quark-antiquark scattering kernel $K(p,q;P)$ if an essential symmetry of the strong interactions, and its breaking pattern, is to be faithfully reproduced.

2.3. Rainbow-Ladder truncation and the Contact Interaction

We employ a momentum-independent vector$\times$vector CI this interaction for the analysis of the quark model charmonia spectrum. Therefore, we use [7]

$$g^2 D_{\mu\nu}(k) = \frac{4\pi\alpha_{\mathrm{IR}}}{m_g^2}\delta_{\mu\nu} \equiv \frac{1}{m_G^2}\delta_{\mu\nu}\,, \tag{6}$$

in Eq. (3), where $m_g = 800$ MeV is a gluon mass scale which is in fact generated dynamically in QCD [20], and $\alpha_{\mathrm{IR}} = 0.93\pi$ is a parameter that determines the interaction strength. For the quark-gluon vertex, the rainbow truncation is used:

$$\Gamma^a_\mu(p,q) = \frac{\lambda^a}{2}\gamma_\mu\,. \tag{7}$$

[1] We work in a Euclidean metric where: $\{\gamma_\mu, \gamma_\nu\} = 2\delta_{mu\nu}$; $\gamma_\mu^\dagger = \gamma_\mu$; $\gamma_5 = \gamma_4\gamma_1\gamma_2\gamma_3$; $a \cdot b = \sum_{i=1}^4 a_i b_i$; and P_μ timelike $\Rightarrow P^2 < 0$.

Once the elements of the kernel in the quark SDE have been specified, we proceed to obtain and analyse its solution. Using Eq. (6) and Eq. (7), the quark equation, Eq. (2), takes the form

$$S_f^{-1}(p) = i\gamma \cdot p + m_f + \frac{4}{3}\frac{1}{m_G^2}\int \frac{\mathrm{d}^4 q}{(2\pi)^4}\gamma_\mu S_f(q)\gamma_\mu\,, \tag{8}$$

whose solution is of the form

$$S_f^{-1}(p) = \mathrm{i}\gamma \cdot p + M_f\,, \tag{9}$$

since the last term on the right-hand side of Eq. (8) is independent of the external momentum. The momentum-independent mass, M_f, is determined as the solution of

$$M_f = m_f + \frac{M_f}{3\pi^2 m_G^2}\int_0^\infty \mathrm{d}s\, s\frac{1}{s+M_f^2}\,. \tag{10}$$

Since Eq. (10) is divergent, we have to specify a regularization procedure. We employ the proper time regularization scheme, [21]

$$\begin{aligned}\frac{1}{s+M^2} &= \int_0^\infty \mathrm{d}\tau\, \mathrm{e}^{-\tau(s+M^2)} \to \int_{\tau_{\rm UV}^2}^{\tau_{\rm IR}{}^2} \mathrm{d}\tau\, \mathrm{e}^{-\tau(s+M^2)}\,,\\ &= \frac{\mathrm{e}^{-\tau_{\rm UV}^2(s+M^2)} - \mathrm{e}^{-\tau_{\rm IR}^2(s+M^2)}}{s+M^2}\,,\end{aligned} \tag{11}$$

where $\tau_{\rm IR}^2$ and $\tau_{\rm UV}^2$ are, respectively, infrared and ultraviolet regulators. A nonzero value for $\tau_{\rm IR} \equiv 1/\Lambda_{\rm IR}$ implements confinement by ensuring the absence of quark production thresholds [22]. Furthermore, since Eq. (6) does not define a renormalizable theory, $\tau_{\rm UV} \equiv 1/\Lambda_{\rm UV}$ cannot be removed, but instead plays a dynamical role and sets the scale for all dimensioned quantities. Thus

$$M_f = m_f + \frac{M_f}{3\pi^2 m_G^2}\mathcal{C}_{01}(M_f^2;\tau_{\rm IR},\tau_{\rm UV})\,, \tag{12}$$

where

$$\mathcal{C}_{\alpha\beta}(M^2;\tau_{\rm IR},\tau_{\rm UV}) = \frac{(M^2)^\nu}{\Gamma(\beta)}\Gamma(\beta-2,\tau_{\rm UV}^2 M^2,\tau_{\rm IR}^2 M^2)\,, \tag{13}$$

with $\nu = \alpha - (\beta - 2)$ and $\Gamma(a, z_1, z_2)$ is the generalized incomplete gamma function.

2.4. Classification of BSA in a Contact Interaction

We are interested in the static properties of several mesons. We begin with their classification and the general form of their BSA in the CI we are working with. Table 1 lists the spin quantum numbers of the quark model mesons under study.

L	J^{PC}	Type	L	J^{PC}	Type
0	0^{-+}	Pseudoscalars	1	0^{++}	Scalars
0	1^{--}	Vectors	1	1^{++}, 1^{+-}	Axial Vectors

Table 1. Quark model mesons

With the dependence on the relative momentum forbidden by the CI, the general form of the BSAs for the mesons listed in Table 1 are [23]:

$$\Gamma_{0^{-+}}(P) = \gamma_5\left[iE_{0^{-+}} + \frac{1}{2M}\gamma\cdot PF_{0^{-+}}\right], \tag{14}$$

$$\Gamma_{0^{++}}(P) = \mathbb{1}E_{0^{++}}, \tag{15}$$

$$\Gamma_{1^{--}\mu}(P) = \gamma_\mu^T E_{1^{--}} + \frac{1}{2M}\sigma_{\mu\nu}P_\nu F_{1^{--}}, \tag{16}$$

$$\Gamma_{1^{++}\mu}(P) = \gamma_5\left[\gamma_\mu^T E_{1^{++}} + \frac{1}{2M}\sigma_{\mu\nu}P_\nu F_{1^{++}}\right], \tag{17}$$

where M is a mass scale.

2.5. Normalization of the BSA

Since the BSE is a homogeneous equation, the BSA has to be normalized by a separate condition. In the Rainbow-Ladder truncation of the BSE, that condition takes a simple form ($\eta = 1$):

$$P_\mu = N_c\frac{\partial}{\partial P_\mu}\int\frac{d^4q}{(2\pi)^4}\,\mathrm{Tr}\left[\overline{\Gamma}_H(-Q)S(q_+)\Gamma_H(Q)S(q)\right], \tag{18}$$

at $Q = P$, with $P^2 = -m_H^2$, which ensures that the residue at the mass pole is unity. Here, Γ_H is the normalized BSA and $\overline{\Gamma}_H$ its charge conjugated version.

2.6. Pseudoscalar and Vector Decay Constants

Once the BSA has been normalized canonically, we can calculate observables from it. The pseudoscalar leptonic decay constant, $f_{0^{-+}}$, is defined by

$$P_\mu f_{0^{-+}} = N_c\int\frac{d^4q}{(2\pi)^4}\,\mathrm{Tr}\left[\gamma_5\gamma_\mu S(q_+)\Gamma_{0^{-+}}(P)S(q_-)\right]. \tag{19}$$

Similarly, the vector decay constant, $f_{1^{--}}$, is defined by

$$m_{1^{--}}f_{1^{--}} = \frac{N_c}{3}\int\frac{d^4q}{(2\pi)^4}\,\mathrm{Tr}\left[\gamma_\mu S(q_+)\Gamma_{\mu 1^{--}}S(q_-)\right], \tag{20}$$

where $m_{1^{--}}$ is the mass of the 1^{--} bound state, and the factor of 3 in the denominator comes from summing over the three polarizations of the spin-1 meson.

3. Results

The mass and BSA of a meson depend on its quantum numbers and can be found by solving Eq. (1). In order to do this, a fictitious eigenvalue, λ_H, is introduced to the bound state equation. Thus, the mass of the bound state in a particular channel, m_H, will be such that $\lambda_H(P^2 = -m_H^2) = 1$, where P is the meson's momentum. In any channel, the form of the homogeneous BSE for the CI is

$$K_H(m_H)\cdot\Gamma_H(m_H) = \lambda_H(m_H)\Gamma_H(m_H), \tag{21}$$

where K_H is a 2×2 matrix, and the subscript H indicates the dependence of the explicit expressions on the quantum numbers of the meson under consideration, see Eqs. (14-17). Equation (21) is an eigenvalue equation for the vector $\Gamma_H(m_H) = (E_H(m_H), F_H(m_H))^T$ with solutions for discrete values of $P^2 = -m_H^2$. Explicit expressions for every channel given in Table 1 in the reference [24].

	masses			
	$m_{\eta_c(1S)}$	$m_{J/\Psi(1S)}$	$m_{\chi_{c_0}(1P)}$	$m_{\chi_{c_1}(1P)}$
Experiment [29]	2.983	3.096	3.414	3.510
Contact Interaction	2.983*	2.979	3.412	3.442
	-	-	3.293	3.344
JM [1]	2.821	3.1	3.605	-
BK [25]	2.928	3.111	3.321	3.437
S1rp [4]	3.035	3.192	-	-
RB1 [26]	3.065	-	-	-
RB2 [26]	3.210	-	-	-
	decay constants			
Experiment [29]	0.361	0.416	-	-
Contact Interaction	0.084	0.080	-	-

Table 2. Ground state charmonia masses, BS amplitudes and decay constants obtained with the light sector parameter set: $m_g = 0.8\,\text{GeV}$, $\alpha_{IR} = 0.93\pi$, $\Lambda_{\text{IR}} = 0.24\,\text{GeV}$, $\Lambda_{\text{UV}} = 0.905\,\text{GeV}$. The current-quark mass is $m_c = 1.578^*\,\text{GeV}$, and the dynamically generated constituent-like mass is $M_c = 1.601\,\text{GeV}$. The value immediately below the CI results is obtained without a spin-orbit coupling $g_{so} = 0.24$. For a direct comparison, we quote values from other SDE approaches to calculate the masses of low lying charmonia. Dimensioned quantities are in GeV. (* = The current quark mass was fitted to obtain the mass of the pseudoscalar meson).

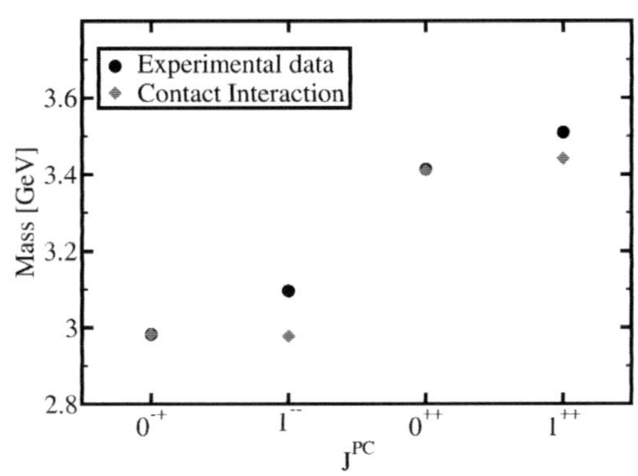

Figure 1. Contact interaction results for the $\bar{c}c$ mass spectrum using model parameters fitted in the light sector, see Table 2 PDG-labeled data is from [29].

3.1. Charmonia mass spectrum

The parameter set used in this calculation is the same as that obtained used in the light sector [5]. Only the current-quark mass for the charm quark is an input parameter, and it is fixed such that the experimental mass of the pseudoscalar is reproduced. The rest of the meson masses are predictions of the model. As can be seen from Table 2, the predictions for the masses of the remaining mesons are in good agreement with the results obtained from more sophisticated SDE-BSE model calculations [25,26], lattice QCD for the charm sector [27,28] as well as experimental values [29].

On the other side, Table 2 also shows that the pseudoscalar and vector decay constants, for the model parameters used, are strongly underestimated, in disagreement both with model calculations and experimental data. As noticed in Refs. [30–32], the decay constant is influenced by the high momentum tails of the dressed-quark propagator and the BSAs. This high momentum region probes the wave-function of quarkonia at origin. The CI, on the other hand, yields constant mass with no perturbative tail for large momenta. Therefore, this artefact of quarkonia has to be built into the model in an alternative manner. We know that with increasing mass of the heavy quarks, they become increasingly point-like in the configuration space. The closer the quarks get, the further the coupling strength between them decreases. Therefore, we cannot expect the decay constants to be correctly reproduced with the parameters of the light quark sector. The next step is to consider the possibility of extending the simple CI model to the heavy sector by reducing the effective coupling. However, the reduction in the strength of the kernel has to be compensated by increasing the ultraviolet cut-off. This makes sense by observing that the $\Lambda_{\rm UV}$ (highest energy scale associated with the system) used in the light quark sector is, in fact, less than the current charm quark mass. Therefore, it needs to be modified.

3.2. Contact Interaction Model for Charmonia

We look a balance between the effective coupling and the ultraviolet cut-off to describe the static properties of charmonia. For this purpose, we set out to redefine the parameters of the CI to study the masses and weak decay constants.

	masses			
	$m_{\eta_c(1S)}$	$m_{J/\Psi(1S)}$	$m_{\chi_{c_0}(1P)}$	$m_{\chi_{c_1}(1P)}$
Experiment [29]	2.983	3.096	3.414	3.510
Contact Interaction	2.950*	3.129	3.407	3.433
			3.194	3.254
JM [1]	2.821	3.1	3.605	-
BK [25]	2.928	3.111	3.321	3.437
RB1 [26]	3.065	-	-	-
RB2 [26]	3.210	-	-	-
	decay constants			
	f_{η_c}	$f_{J/\Psi}$		
Experiment [29]	0.361	0.416		
S1rp [4]	0.239	0.198		
S3ccp [4]	0.326	0.330		
BK [25]	0.399	0.448		
Contact Interaction	0.305	0.220		

Table 3. Ground state charmonia masses and decay constants obtained with the best-fit parameter set: $m_g = 0.8\,\text{GeV}$, $\alpha_{IR} = 0.93\pi/20$, $\Lambda_{\rm IR} = 0.24\,\text{GeV}$, $\Lambda_{\rm UV} = 2.788\,\text{GeV}$. The current-quark mass is $m_c = 0.956^*$ GeV, and the dynamically generated constituent-like mass is $M_c = 1.497\,\text{GeV}$. The value immediately below the CI results is obtained without a spin-orbit coupling $g_{so} = 0.24/3$. Dimensioned quantities are in GeV. (* = This parameter set was obtained from a best-fit to the mass and decay constant of the pseudoscalar and vector channels).

We retain the parameters m_g and $\Lambda_{\rm IR}$ of the light sector. Modern studies of the gluon propagator indicate that in the infrared, the dynamically generated gluon mass scale virtually remains unaffected by the introduction of heavy dynamical quark masses [33, 34]. The rest of

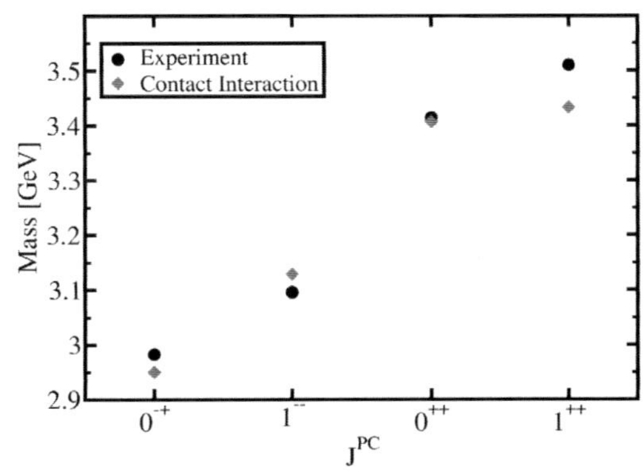

Figure 2. Contact interaction results for the $\bar{c}c$ mass spectrum, see Table 3. PDG-labelled data is taken from Ref. [29].

the parameters are obtained from a best-fit to the mass and decay constant of the pseudoscalar (η_c) and vector (J/Ψ) channels.

One can now readily calculate the masses of the ground state pseudo-scalar, vector, scalar, and axial vector mesons. The results are shown in Table 3 and Fig. 2. They are in very good agreement with experimental values and comparable to the best SDE results with refined truncations.

The decay constants for the $\eta_c(1S)$ and $J/\Psi(1S)$ channels are reported in Table 3. For the pseudoscalar meson, the result aligns nicely with the experimental value. However, this is not exactly the case for the vector channel. Furthermore, we note that the decay constant for $J/\Psi(1S)$ is smaller than that for $\eta_c(1S)$. The correct ordering can be recovered by reducing the interaction strength by a large factor. However, this is something we consider contrived and, therefore, not pursued further. Notice that one of the SDE results yields the $J/\Psi(1S)$ decay constant even smaller than our value [4].

4. Conclusions

We compute the quark model ground state spin-0 and spin-1 charmonia masses and decay constants using a rainbow-ladder truncation of the simultaneous set of SDE and BSE with a CI model of QCD, developed and tested for the light quark sector [5–9]. As the model is non-renormalizable, we employ proper time regularization scheme which ensures confinement is implemented through the absence of quark production threshold. Moreover, the relevant Ward identities and the low energy theorems such as Goldberger-Triemann relations are satisfied. Without parameter readjustment, we find that the masses of the studied mesons are in reasonably good agreement with experimental data and other model calculations. Moreover, the Gell-Mann–Oakes–Renner relation, valid for every current-quark mass in the pseudo-scalar channel, is always satisfied. However, the decay constants of pseudo-scalar as well as vector mesons are significantly underestimated.

We realize that the extension of the CI model to the heavy sector requires a reduction of the effective coupling, which mimics the high momentum tail of the quark mass function obtained in the SDE studies of QCD [30–32]. We only have to ensure that the reduction in the strength of the kernel is appropriately compensated by increasing the ultraviolet cut-off, a natural requirement for studying heavy quarks. We find that with a modified choice of two parameters, not only the masses of the ground state mesons, i.e., pseudo-scalar ($\eta_c(1S)$), vector ($J/\Psi(1S)$), scalar ($\chi_{c_0}(1P)$), and axial vector ($\chi_{c_1}(1P)$), but also their weak decay constants, are in much better agreement with the experiments [29] as well as earlier SDE calculations with QCD based refined truncations [25]. This is an encouraging first step towards a comprehensive study of heavy mesons in this approach. Further steps will involve flavored mesons and baryons. Our goal is to provide a unified description of light and heavy hadrons within the CI model.

Full discussion and details are found in our work [24].

5. Acknowledgments

I want to acknowledge the organizing committee for the financial support and my colaborators in [24] for fruitful discussions.

References

[1] P. Jain and H. J. Munczek, Phys.Rev. **D48**, 5403 (1993), hep-ph/9307221.
[2] A. Krassnigg and P. Maris, J.Phys.Conf.Ser. **9**, 153 (2005), nucl-th/0412058.
[3] M. Bhagwat, A. Krassnigg, P. Maris, and C. Roberts, Eur.Phys.J. **A31**, 630 (2007), nucl-th/0612027.
[4] N. Souchlas, Phys.Rev. **D81**, 114019 (2010).
[5] L. Gutierrez-Guerrero, A. Bashir, I. Cloet, and C. Roberts, Phys.Rev. **C81**, 065202 (2010), 1002.1968.
[6] H. Roberts, C. Roberts, A. Bashir, L. Gutierrez-Guerrero, and P. Tandy, Phys.Rev. **C82**, 065202 (2010), 1009.0067.
[7] C. Chen, L. Chang, C. D. Roberts, S. Wan, and D. J. Wilson, Few Body Syst. **53**, 293 (2012), 1204.2553.
[8] H. L. Roberts, L. Chang, I. C. Cloet, and C. D. Roberts, Few Body Syst. **51**, 1 (2011a), 1101.4244.
[9] H. Roberts, A. Bashir, L. Gutierrez-Guerrero, C. Roberts, and D. Wilson, Phys.Rev. **C83**, 065206 (2011b), 1102.4376.
[10] A. Bashir, L. Chang, I. C. Cloet, B. El-Bennich, Y.-X. Liu, *et al.*, Commun.Theor.Phys. **58**, 79 (2012b), 1201.3366.
[11] G. Eichmann, R. Alkofer, I. Cloet, A. Krassnigg, and C. Roberts, Phys.Rev. **C77**, 042202 (2008), 0802.1948.
[12] I. Cloet, A. Krassnigg, and C. Roberts, eConf **C070910**, 125 (2007), 0710.5746.
[13] F. Gross, *Relativistic quantum mechanics and field theory* (Wiley, New York, 1993), first ed.
[14] E. E. Salpeter and H. A. Bethe, Phys. Rev. **84**, 1232 (1951).
[15] M. Gell-Mann and F. Low, Phys. Rev. **84**, 350 (1951).
[16] C. D. Roberts, M. S. Bhagwat, A. Holl, and S. V. Wright, Eur. Phys. J. ST **140**, 53 (2007), 0802.0217.
[17] A. Holl, C. D. Roberts, and S. V. Wright (2006), nucl-th/0601071.
[18] P. Maris and C. D. Roberts, Int. J. Mod. Phys. **E12**, 297 (2003), nucl-th/0301049.
[19] R. Alkofer and L. von Smekal, Phys. Rept. **353**, 281 (2001), hep-ph/0007355.
[20] P. Boucaud, J. Leroy, A. L. Yaouanc, J. Micheli, O. Pene, *et al.*, Few Body Syst. **53**, 387 (2012), 1109.1936.
[21] D. Ebert, T. Feldmann, and H. Reinhardt, Phys.Lett. **B388**, 154 (1996), hep-ph/9608223.
[22] C. Roberts, Prog.Part.Nucl.Phys. **61**, 50 (2008), 0712.0633.
[23] C. H. Llewellyn-Smith, Ann. Phys. **53**, 521 (1969).
[24] M.A. Bedolla, J.J. Cobos-Martínezand A. Bashir, Phys.Rev. **D92**, 054031 (2015), 1101.4244.
[25] M. Blank and A. Krassnigg, Phys.Rev. **D84**, 096014 (2011), 1109.6509.
[26] E. Rojas, B. El-Bennich, and J. de Melo, Phys.Rev. **D90**(7), 074025 (2014), 1407.3598.
[27] E. Follana *et al.* (HPQCD Collaboration, UKQCD Collaboration), Phys.Rev. **D75**, 054502 (2007), hep-lat/0610092.
[28] M. Kalinowski and M. Wagner, PoS **LATTICE2013**, 241 (2014), 1310.5513.
[29] K. Nakamura and P. D. Group, Journal of Physics G: Nuclear and Particle Physics **37**(7A), 075021 (2010), URL `http://stacks.iop.org/0954-3899/37/i=7A/a=075021`.
[30] M. Bhagwat, A. Holl, A. Krassnigg, C. Roberts, and P. Tandy, Phys.Rev. **C70**, 035205 (2004), nucl-th/0403012.
[31] P. Maris and C. D. Roberts, Phys. Rev. **C56**, 3369 (1997), nucl-th/9708029.
[32] P. Maris and P. C. Tandy, Phys. Rev. **C60**, 055214 (1999), nucl-th/9905056.
[33] A. Ayala, A. Bashir, D. Binosi, M. Cristoforetti, and J. Rodriguez-Quintero, Phys.Rev. **D86**, 074512 (2012), 1208.0795.
[34] A. Bashir, A. Raya, and J. Rodriguez-Quintero, Phys.Rev. **D88**, 054003 (2013), 1302.5829.

Journal of Physics: Conference Series **761** (2016) 012072 doi:10.1088/1742-6596/761/1/012072

Long distance electromagnetic contribution to $B^{\pm} \to (K^{\pm}, \pi^{\pm})\ell^{+}\ell^{-}$ rare processes

S. L. Tostado[a,1]

[a] Departamento de Física, Centro de Investigación y de Estudios Avanzados del Instituto Politécnico Nacional, Apdo. Postal 14-740, 07000 México D.F., México

E-mail: [1]stostado@fis.cinvestav.mx

Abstract. We evaluate a long distance weak annihilation contribution to the decay rates of the $B^{\pm} \to (K^{\pm}, \pi^{\pm})\ell^{+}\ell^{-}$ rare processes. Despite a visible kinematic leptonic universality violation is obtained in the low q^2 region, its effect almost disappears in the region of interest for the R_K anomaly. We show that sizeable CP asymmetries are obtained when short and long distance prescriptions are taken into account in different energy intervals.

1. INTRODUCTION

Indirect evidence of new physics (NP) effects can be extracted by studying the most suppressed (rare) processes within the standard theory. The ones involving flavor changing neutral currents (FCNC) have triggered a lot of interest since some deviations from the standard model (SM) predictions have been observed experimentally. Short time ago, the LHCb collaboration reported the measurement of the ratio of muon to electron pairs produced in $B^{\pm} \to K^{\pm}$ decays, $R_K \equiv B(B^- \to K^-\mu^+\mu^-)/B(B^- \to K^-e^+e^-) = 0.745^{+0.090}_{-0.074} \pm 0.036$ [1], observing a deficit with respect to the SM value $R_K = 1.0003 \pm 0.0001$ [2, 3] in the $(1, 6)$ GeV^2 region for the squared invariant-mass of the lepton pair. Many models have been proposed in the literature to solve the anomaly [4], some of them involving non-universal leptonic interactions. In this work, we calculate a long-distance (LD) weak annihilation (WA) contribution to the $B^{\pm} \to (K^{\pm}, \pi^{\pm})\ell^{+}\ell^{-}$ decays, and study its effect on the branching ratios and the CP asymmetry for each of these rare processes.

2. SHORT DISTANCE APPROACH

In the SM, the $B^-(p_B) \to P^-(p_P)\ell^+(p_+)\ell^-(p_-)$ decay (with $P = \pi$ or K) involves contributions of penguin (Figure 1a), as well as W boson box (Figure 1b) diagrams, which are dominated by loops involving the top quark. This short distance picture (SD) is accurately described by the effective amplitude

$$\mathcal{M}_{QCDf} = \frac{G_F \alpha}{\sqrt{2}\pi} V_{tb} V_{tD}^{*} \xi_P(q^2) p_B^{\mu} \left[F_V(q^2) \bar{\ell}\gamma_{\mu}\ell + F_A(q^2) \bar{\ell}\gamma_{\mu}\gamma_5 \ell \right] , \tag{1}$$

where $q = p_+ + p_-$ is the momentum of the lepton pair, D stands for $d(s)$ in the case of $|\Delta S| = 0(1)$ transitions. The factors $F_V \approx C_9 = 4.214$ and $F_A = C_{10} = -4.312$, correspond to the vector and axial Wilson coefficients at NNLO [5], obtained from the weak

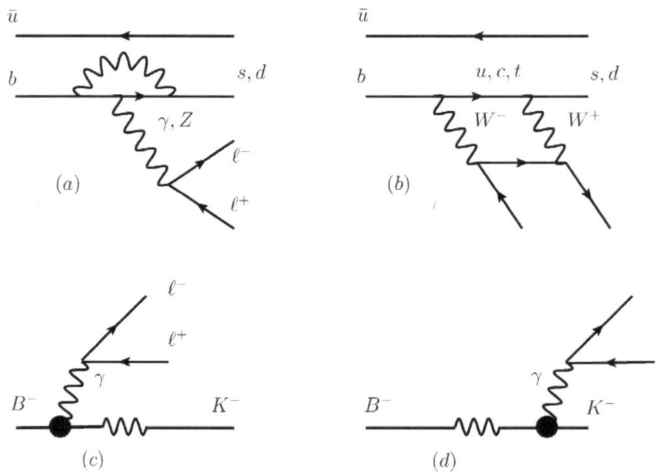

Figure 1. SD contributions to $B^- \to P^-\ell^+\ell^-$ decays are shown in (a) (penguin) and (b) (W box) diagrams. The LD WA contributions are shown in (c) and (d). The full circle denotes the electromagnetic form factor of the charged pseudoscalar mesons.

effective Hamiltonian for the $b \to q\ell^+\ell^-$ transition. The short-distance WA contributions induced by four quark operators can be absorbed into the effective F_V coefficient by replacing $\xi_P(q^2)F_V \to \xi_P(q^2)F_V + F^{WA}$. $\xi_P(q^2)$ stands for the weak form factor of the specific process, where we will use the expressions obtained from light-cone sum rules (LCSR) [2, 6] regarding the $\mathcal{P}_P(q^2)$ Gegenbauer moments as

$$\begin{aligned}\xi_\pi(q^2) &= \frac{0.918}{1-q^2/(5.32\text{ GeV})^2} - \frac{0.675}{1-q^2/(6.18\text{ GeV})^2} + \mathcal{P}_\pi(q^2),\\ \xi_K(q^2) &= \frac{0.0541}{1-q^2/(5.41\text{ GeV})^2} + \frac{0.2166}{[1-q^2/(5.41\text{ GeV})^2]^2} + \mathcal{P}_K(q^2).\end{aligned}$$

The SD WA diagrams have been studied in the QCD factorization (QCDf) framework [7], where the virtual photon exchange between hadrons and the dilepton pair and hard gluon scattering have been analysed [5, 8]. Along this paper we will refer to Eq. (1) indistinctly as the SD or QCDf contribution.

3. LONG DISTANCE WA CONTRIBUTION

The long distance WA diagrams that contribute to $B^-(p_B) \to P^-(p_P)$ transitions are depicted in (Figure 1c) and (Figure 1d). The decay amplitude for these diagrams can be written as

$$\begin{aligned}\mathcal{M}_{LD,WA} &= \sqrt{2}G_F(4\pi\alpha)V_{ub}V_{uD}^* f_B f_P \frac{1}{q^2(m_B^2-m_P^2)}\\ &\quad\times\left[m_B^2\left(F_P(q^2)-1\right) - m_P^2\left(F_B(q^2)-1\right)\right] p_B^\mu \bar{\ell}\gamma_\mu\ell\,, \qquad (2)\end{aligned}$$

where $F(q^2)$ denotes the electromagnetic form factor (FF) of the corresponding meson and f_M the meson decay constant. At first sight, it can be seen that the $D = s$ mode is CKM-suppressed compared to $D = d$, and that the second term between squared brackets, corresponding to photon emission from the B meson, is suppressed by the ratio of masses of the light meson to the heavy. In addition, we have found that the FF contribution of the B meson is suppressed

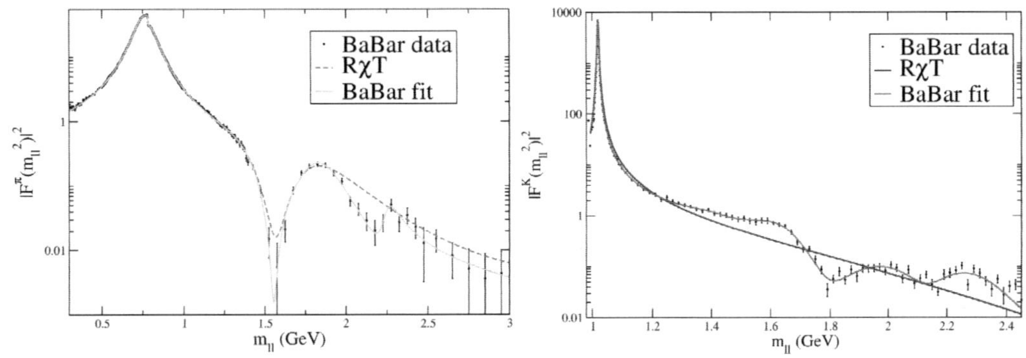

Figure 2. Comparisson of the RχT and BABAR fits of the electromagnetic pion (left) and kaon (right) FF as function of $m_{ll} = \sqrt{q^2}$ to experimental data [13].

itself due to the reduced charged radius of the heavy meson, such that its final contribution is negligible. We have analysed other kinds of structure dependent contributions, but these vanish because of the gauge invariance and/or are suppressed by $1/N_c$ [9]. The vector nature of the LD amplitude in Eq. (2) makes plausible its incorporation in a re-definition of the coefficients coming from operator O_9, namely

$$\xi_P(q^2)F_V \longrightarrow \xi_P(q^2)F_V + \kappa_P m_B^2 \left[\frac{F_P(q^2)-1}{q^2}\right], \tag{3}$$

where

$$\kappa_P = -8\pi^2 \frac{V_{ub}V_{uD}^*}{V_{tb}V_{tD}^*}\frac{f_B f_P}{m_B^2 - m_P^2}. \tag{4}$$

To complete our analysis [10], we will use the electromagnetic form factors (FF) of the light pseudoscalar mesons [11] obtained from Resonance Chiral Theory (RχT) [12], which are saturated by the lowest-lying light flavor resonances, ρ and ω for $P = \pi$, and the extremely narrow $\phi(1020)$ meson for $P = K$ [13]. For the sake of completeness, we will also use the fit to the experimental data provided by the BaBar collaboration to estimate the uncertainties owing to changes in the FF [13]. In order to avoid double counting of WA contributions, we will consider in the interval from threshold to 2 GeV^2 the RχT approach since we expect that chiral Lagrangians give an adequate description for low q^2 values, while QCDf will be more appropriate for describing the higher q^2 region. The matching of C_9^{eff} obtained with both descriptions at this intermediate energy scale is considered in detail in Ref. [10].

4. Results

The comparison between both approaches used for the FF can be found in Fig. 2, where good agreement was found in the low lepton invariant mass region, with minor differences above 1.5 GeV; these differences are taken into account in the final quoted errors. Figure 3 shows the spectra of the lepton pair, normalized to the B^- meson decay width, for low values of the lepton-pair energies. A marked violation of lepton universality is observed in this limited region of the q^2 range owing to kinematical effects.

In Table 1 we show separately the contribution of QCDf and LD prescriptions for the WA diagrams for different rages of q^2. We have restricted our analysis to regions below 8 GeV^2 (above this value the charm effects are expected to dominate). We can observe that the LD term contributes at the same level of the SD part when integrated branching ratios are computed from

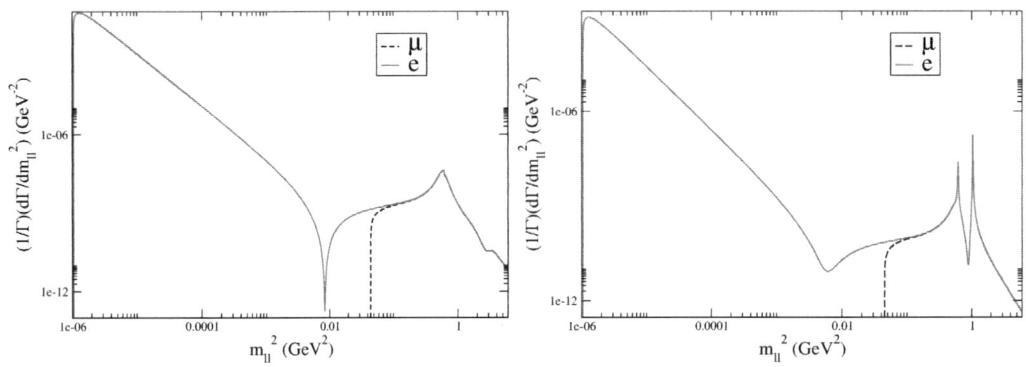

Figure 3. Normalized spectrum of the lepton-pair for the LD WA contribution, $P = \pi$ (left) and $P = K$ (right), in the region going from threshold up to 1.5 GeV^2.

the threshold, $q^2 \sim 0.05$ GeV^2. In the region above 1 GeV^2 the QCDf contribution dominates, these confirms the feasibility of NP physics searches in the $1 \leq q^2 \leq 6$ GeV^2 interval in the kaon mode, and $1 \leq q^2 \leq 8$ GeV^2 for the pion mode, where the hadronic pollution can be well under control.

It is possible to generate a direct CP asymmetry because the LD and SD WA contributions to the amplitude of the $B^\pm \to P^\pm \ell^+ \ell^-$ decays have different weak and strong phases. In this way, a (partially integrated over a finite q^2 range) CP asymmetry,

$$A_{CP}(P) = \frac{\Gamma(B^+ \to P^+\ell^+\ell^-) - \Gamma(B^- \to P^-\ell^+\ell^-)}{\Gamma(B^+ \to P^+\ell^+\ell^-) + \Gamma(B^- \to P^-\ell^+\ell^-)}\,, \tag{5}$$

is obtained from the interference of the diagrams shown in Figure 1. By inserting the amplitudes of LD WA and QCDf contributions in the above expression, and considering the matching of the LD and QCDf descriptions of the WA contributions described in the previous section, Eq. (5) leads to:

$$A_{CP}(P) = \begin{cases} (16.1 \pm 1.9)\%, & \text{for } P = \pi,\ 0.05 \leq q^2 \leq 8 \text{ GeV}^2, \\ (7.8 \pm 2.9)\%, & \text{for } P = \pi,\ 1 \leq q^2 \leq 8 \text{ GeV}^2, \\ (-1.0 \pm 0.3)\%, & \text{for } P = K,\ 1 \leq q^2 \leq 6 \text{ GeV}^2. \end{cases} \tag{6}$$

In the π meson case, we have found that the CP asymmetry for the first range is widely dominated by the LD WA contribution, while within the second range is dominated by QCDf effects. For kaons, the CP asymmetry is basically obtained from LD WA contributions (which are almost 70% within this range). The uncertainties in our results stem from the systematic error owing to parametrizations of the light meson electromagnetic form factors. For $P = \pi$, the CP asymmetries have been computed in [14, 15] considering the QCDf approach. These results and ours agree within the experimental error with the recent determination made by the LHCb collaboration for the integrated CP asymmetry, namely $A_{CP} = 0.11 \pm 0.12 \pm 0.01$ [16].

5. Concluding Remarks

We have calculated the effects of the WA diagrams in the LD regime to the $B^\pm \to (K^\pm, \pi^\pm)\ell^+\ell^-$ rare processes and shown their effect over the CP asymmetries. Despite a visible kinematic leptonic universality violation is obtained in the low q^2 region, the effect of LD WA contributions almost disappears in the region of interest for the R_K anomaly. This confirms the feasibility of the $1 \leq q^2 \leq 6$ GeV^2 range for NP searches in the $P = K$ case, and suggests to take

Table 1. Branching ratios of $B^- \to P^- \ell^+ \ell^-$ decays for $P = \pi$ (left hand side) and $P = K$ (right hand side) integrated in different q^2 ranges.

	$B^- \to \pi^- \ell^+ \ell^-$		$B^- \to K^- \ell^+ \ell^-$
	$0.05 \leq q^2 \leq 8$ GeV^2	$1 \leq q^2 \leq 8$ GeV^2	$1 \leq q^2 \leq 6$ GeV^2
LD	$(9.06 \pm 0.15) \cdot 10^{-9}$	$(4.74 \pm 0.05) \cdot 10^{-10}$	$(1.70 \pm 0.21) \cdot 10^{-9}$
interf.	$(-2.57 \pm 0.13) \cdot 10^{-9}$	$(-2^{+2}_{-1}) \cdot 10^{-10}$	$(-6 \pm 2) \cdot 10^{-11}$
QCDf	$(9.57^{+1.45}_{-1.01}) \cdot 10^{-9}$	$(8.43^{+1.31}_{-0.87}) \cdot 10^{-9}$	$(1.90^{+0.69}_{-0.41}) \times 10^{-7}$
Total	$(1.61^{+0.15}_{-0.11}) \cdot 10^{-8}$	$(8.69^{+1.31}_{-0.87}) \cdot 10^{-9}$	$(1.92^{+0.69}_{-0.41}) \times 10^{-7}$

$1 \leq q^2 \leq 8$ GeV^2 for NP searches when $P = \pi$. Similarly, we have obtained that measurable CP asymmetries are generated. Our results for the CP asymmetry in the $P = \pi$ case agree with the recent experimental measurements.

Acknowledgements

The author would like to thank Conacyt (México) for partial financial support under project 236394.

References

[1] R. Aaij *et al.* [LHCb Collaboration] 2014 Phys. Rev. Lett. **113** 151601.
[2] C. Bobeth, G. Hiller and G. Piranishvili 2007 J. High Energy Phys. **12** 040.
[3] C. Bobeth, G. Hiller, D. van Dyk and Ch. Wacker 2012 J. High Energy Phys. **01** 107; G. Hiller and F. Krüger 2004 Phys. Rev. D **69** 074020; G. Hiller and M. Schmaltz 2014 Phys. Rev. D **90** 054014.
[4] R. Alonso, B. Grinstein and J. Martín Camalich 2014 Phys. Rev. Lett. **113** 241802; D. Ghosh, M. Nardecchia and S. A. Renner 2014 JHEP **1412** 131; S. Biswas, D. Chowdhury, S. Han and S. J. Lee 2015 JHEP **1502** 142; T. Hurth, F. Mahmoudi and S. Neshatpour 2014 JHEP **1412** 053; S. L. Glashow, D. Guadagnoli and K. Lane 2015 Phys. Rev. Lett. **114** 091801; W. Altmannshofer and D. M. Straub Eur. Phys. J. C **75** (2015) no.8, 382; G. Hiller and M. Schmaltz 2015 JHEP **1502** 055; B. Gripaios, M. Nardecchia and S. A. Renner JHEP **1505** (2015) 006; B. Bhattacharya, A. Datta, D. London and S. Shivashankara 2015 Phys. Lett. B **742** 370; A. Crivellin, G. D'Ambrosio and J. Heeck Phys. Rev. Lett. **114** (2015) 151801; S. Sahoo and R. Mohanta Phys. Rev. D **91** (2015) no.9, 094019; I. d. M. Varzielas and G. Hiller JHEP **1506** (2015) 072 A. Crivellin, G. D'Ambrosio and J. Heeck Phys. Rev. D **91** (2015) no.7, 075006; C. Niehoff, P. Stangl and D. M. Straub, Phys. Lett. B **747** (2015) 182; D. A. Sierra, F. Staub and A. Vicente, Phys. Rev. D **92** (2015) no.1, 015001; S. M. Boucenna, J. W. F. Valle and A. Vicente Phys. Lett. B **750** (2015) 367.
[5] M. Beneke, Th. Feldmann and D. Seidel 2001 Nucl. Phys. B **612** 25; 2005 Eur. Phys. J. C**41** 173.
[6] P. Ball and R. Zwicky 2005 Phys. Rev. D **71** 014015.
[7] M. Beneke, G. Buchalla, M. Neubert and C. T. Sachrajda 1999 Phys. Rev. Lett. **83** 1914; 2000 Nucl. Phys. B **591** 313.
[8] G. Buchalla, A. J. Buras and M. E. Lautenbacher 1996 Rev. Mod. Phys. **68** 1125.
[9] G. Ecker, A. Pich and E. de Rafael 1987 Nucl. Phys. B **291** 692.
[10] A.Guevara, G. López Castro, P. Roig and S. L. Tostado 2015 Phys. Rev. D **92** 054035.
[11] E. Arganda, M. J. Herrero and J. Portolés, JHEP **0806** (2008) 079. D. Gómez Dumm and P. Roig, Eur. Phys. J. C **73** (2013) no.8, 2528. O. Shekhovtsova, T. Przedzinski, P. Roig and Z. Was, Phys. Rev. D **86** (2012) 113008.
[12] G. Ecker, J. Gasser, H. Leutwyler, A. Pich and E. de Rafael 1989 Phys. Lett. B **223** 425; G. Ecker, J. Gasser, A. Pich and E. de Rafael 1989 Nucl. Phys. B **321** 311.
[13] J. P. Lees *et al.* [BaBar Collaboration], Phys. Rev. D **86** (2012) 032013; **88** (2013) 3, 032013.
[14] W. S. Hou, M. Kohda and F. Xu 2014 Phys. Rev. D **90** 013002.
[15] C. Hambrock, G. Hiller, S. Schacht and R. Zwicky Phys. Rev. D **89** 074014.
[16] R. Aaij *et al.* [LHCb Collaboration] 2015 JHEP **10** 034.

Journal of Physics: Conference Series **761** (2016) 012073 doi:10.1088/1742-6596/761/1/012073

Lepton Flavour Violating τ decays in a 2HDM with SU(3) Yukawa matrices

L. López-Lozano

Área Académica de Matemáticas y Física, Universidad Autónoma del Estado de Hidalgo, Carr. Pachuca-Tulancingo Km. 4.5, C.P. 42184, Pachuca, Hidalgo, Mexico

E-mail: lao_lopez@uaeh.edu.mx

Abstract. Using the Lepton Flavor Violating (LFV) decay of τ with production of a pseudo-scalar neutral meson P^0, we calculate the parameters space of a version of 2HDM where the Yukawa matrix generated through a unitary transformation of the diagonal mass matrix. This unitary transformation corresponds to a particular invariant subgroup of SU(3). We study the channel $\tau \to eK_s^0$ and demonstrate that it is possible to have non small couplings an at the same time a LFV decay rate very small. This analysis shed light on the flavor structure in the frame of such group. This model is a particular case of the Partially Aligned 2HDM and motivates the analysis of more channels using this model for Yukawas.

1. Introduction

After the discovery of a scalar particle with all features of the Higgs boson, the Standard Model (SM) is the best description of particles at electroweak scale, nevertheless this discovery also leaves many open questions. Most of the lack of answers to these questions is related with the so called Flavor Problem where one important task is to understand the relations between the masses of fermions and the mixing angles. Likewise, many extensions of the SM have been proposed in order to parametrize possible new physics effects, although this has the disadvantage of introducing additional free parameters that have to be determined by experiments.

The 2HDM is an economical extension of SM that has been amply studied because it has many new phenomenology and could represent low energy limit of more general theories, for instance the low energy versions of SUSY models. Different versions of the 2HDM with NFC [1, 2], that introduce a Z_2 symmetry in the scalar sector to prevent Flavour Changing Neutral Currents (FCNC), have captured the attention due to its small number of free parameters. Also there are popular versions that are similar to low energy limits of models with extras symmetries, for instance the MSSM model [3]. The 2HDM with CP conserving VEV's contains two additional charged scalars, two neutral scalars and one neutral pseudo-scalar resulting in a rich phenomenology extending the SM. A review of this model is [4], and references therein.

The main drawback in versions that do not contain NFC, for instance the version III [5, 6], is the big number of free parameters. Usually, to reduce these parameters and to control the FCNC, some textures have been used for the Yukawa matrix shedding interesting results [7, 8, 9]. Also some efforts to reduce the free parameters have been done using experimental measurements of mixing angles. An example in the quark sector is given in [10].

From the phenomenological point of view, experiments with fermions determine directly or indirectly mixing matrices, namely for quarks the CKM matrix and for the leptons the PMNS matrix. Likewise in the mass basis, the diagonal mass matrix is determined through resonances. Although it is known that there exists a relation between masses and mixings, at this level these parameters can be considered as inputs of an extended model. There exist many approaches that build mass matrices. One of the most important ones is the 4-zero texture that has been proved in several contexts. One remarkable feature of this texture is that only 4 parameters are enough to describe both, quarks sector and the leptonic sector [11].

Although a general version of 2HDM has a rich phenomenology, experimental determination of free parameters is very difficult and highly dependent of the model. On the other hand a physical interpretation of this parameters is needed in terms of an underlying yet unknown complete flavor theory. In this sense, here we propose a model to rewrite the Yukawa matrices as the result of a unitary-like transformation generated by Gell-Mann matrices in a particular representation of the flavor basis as a form to inquire a flavor structure. In order to reduce the free parameters of the model we assume that only an invariant subgroup of SU(3) relates up-type quarks with down-type quarks. As a result of this parametrization there is one less parameter than in other successful approaches as the 4-zero textures.

As an application and to demonstrate that this model is phenomenologically viable, we calculate the lepton flavor violating couplings of semi-leptonic τ decays into one pseudo-scalar meson and a lepton. This decays with LFV are considered in a version of the 2HDM where the main contribution to the new physics comes from a pseudo-scalar interaction, that is, the interchange of the pseudo-scalar A^0. For the channel $\tau \to eK_s^0$ it is shown that the unitary property of CKM matrix suppresses naturally this processes, as seen in the experimental measurements even for non small couplings with scalar A^0.

2. Yukawa matrices generated by SU(3)

To generate a texture matrix using the fermion masses as inputs we make a flavor transformation of the form

$$T^{f'} = \xi A_L^{f\dagger} D^f A_R^f \qquad (1)$$

where $D^f \equiv \text{diag}(m_1^f, m_2^f, m_3^f)$ is the diagonal matrix, containing the fermion masses in normal order, for the $f = u, d, \ell, \nu$ flavors. On the other hand ξ is in general a complex number and A_L, A_R are 3×3 complex matrices. Also we have that in general $f' \neq f$. This transformation is general enough to construct any phenomenological texture because it has a big number of free parameters. This type of transformations are the starting point to introduce several hypothesis about the flavor structure for texture matrices. Transformations of the form (1) can be written such as the FCNC as in the Partially Aligned 2HDM (PA-2HDM) [12]. In this work we make the following assumptions:

- The matrices A_L^f and A_R^f are $SU(3)$ matrices that belong to a particular invariant subgroup, i.e. U-Spin, V-Spin and S-Spin.
- In order to reduce the number of free parameters we impose $A_L^f = A_R^f$, which means that the fermions transformation is unitary.
- For simplicity the scale parameter is set by $\xi = 1$ and f' is the up-type flavor (u) and f is the down-type flavor (d, ℓ).

We rewrite the decay rates in terms of the parameters of a unitary transformation in order to explore whether the phenomenology of τ decays can say something about the flavor structure in the frame of the SU(3) group. The second assumption says that the matrix that diagonalizes the mass matrix has any of the forms showed in the Table 2. The observables with less experimental error are the masses and mixings of fermions, thus we take these values as inputs

Subgroup	Generators
S-Spin (SU(2))	$\lambda_1, \lambda_2, \lambda_3$
U-Spin	$\lambda_4, \lambda_5, \frac{1}{4}(\sqrt{3}\lambda_3 + \lambda_8)$
V-Spin	$\lambda_6, \lambda_7, \frac{1}{4}(\sqrt{3}\lambda_3 - \lambda_8)$

Table 1. Generators of the invariant subgroups of SU(3).

of the model. Once we fix a unitary matrix that diagonalizes the mass matrix, we can find the rest of parameters for different types of fermions using the relation

$$U_L^u = U_L^d V_{\mathrm{CKM}}^\dagger, \tag{2}$$

that comes from the definition of V_{CKM}. Here U_L^u and U_L^d diagonalize the mass matrices when are written in the flavor basis. Thus we can see that in general the matrices that corresponds to every group have the forms in the Table 2.

S-Spin(SU(2))	$\begin{pmatrix} a_0 + B_S & A_S^* & 0 \\ A_S & a_0 - B_S & 0 \\ 0 & 0 & a_0 \end{pmatrix}$
U-Spin	$\begin{pmatrix} a_0 + B_U & 0 & A_U^* \\ 0 & a_0 & 0 \\ A_U & 0 & a_0 - B_U \end{pmatrix}$
V-Spin	$\begin{pmatrix} a_0 & 0 & 0 \\ 0 & a_0 + B_V & A_V^* \\ 0 & A_V & a_0 - B_V \end{pmatrix}$

Table 2. General form of matrix representation of invariant subgroups of SU(3), a_0, B's are real and A's are complex parameters.

As we can see, the matrices have 4 real free parameters. In terms of textures definitions, these matrices corresponds to 2-zero texture matrices and can be used to diagonalize mass matrices.

3. SU(3) Yukawa interactions in the 2HDM

In the context of a 2HDM, the Yukawa matrices in the mass basis takes the form

$$D^f = \frac{1}{\sqrt{2}}\left(v_1 \tilde{Y}_1^f + v_2 \tilde{Y}_2^f\right) \tag{3}$$

where $v_1 = v\cos\beta$ and $v_2 = v\sin\beta$. Once a Yukawa matrix is determined in the mass basis, the second one can be calculated. As an hypothesis we assume that Yukawa matrices have the form of (1), so we can write

$$\tilde{Y}_{X2}^d = (\sqrt{2}G_F)^{\frac{1}{2}} U_{XL}^d D^d U_{XL}^{d\dagger} \tag{4}$$

$$\tilde{Y}_{X2}^u = (\sqrt{2}G_F)^{\frac{1}{2}} U_{XL}^d V_{\mathrm{CKM}}^\dagger D^u V_{\mathrm{CKM}} U_{XL}^{d\dagger} \tag{5}$$

and for the leptonic sector

$$\tilde{Y}_{X'2}^\ell = (\sqrt{2}G_F)^{\frac{1}{2}} U_{X'L}^{\ell\dagger} D_\ell U_{X'L}^\ell \tag{6}$$

where U^f_{XL} has one of the forms of the Table 2. As we can see, all scalar interactions are given by 4 parameters for every type of fermion, i.e. a^f_o, B^f_X, $|A^f_X|$ and $\phi^f_{A_X}$ with $X, X' = S, U$ or V and $f = u, d, \ell$. Here $\phi^f_{A_X}$ is the phase of A^f_X. In principle only 8 parameters are needed to describe all semileptonic decays of mesons. We do not deal with Yukawa matrix for the neutrino sector because the phenomenology of meson decays has a scale of energy where the mass of neutrinos can be approximated to zero.

The Yukawa Lagrangian for charged leptons and quarks in the flavor basis for the 2HDM is given by

$$-\mathcal{L}^{\rm NC}_{\rm 2HDM} = \sum_{ij}\sum_{a=1}^{2}\left[\bar{Q}_{Li}\left(Y^u_{aij}\tilde{\Phi}_a u_{Rj} + Y^d_{aij}\Phi_a d_{Rj}\right) + Y^\ell_{aij}\bar{L}_{Li}\Phi_a \ell_{Rj} + \text{H.c.}\right] \tag{7}$$

In the mass basis the neutral interactions for the charged leptons can be cast as

$$\begin{aligned}\mathcal{L}^\ell_{Y,X} &= (\sqrt{2}G_F)^{\frac{1}{2}}\bar{\ell}_i\left[\left(m_i\frac{\cos\alpha}{\cos\beta}\delta_{ij} + \frac{\sin(\alpha-\beta)}{\sqrt{2}\cos\beta}(\tilde{Y'}^\ell_{X2})_{ij}\right)H^0\right.\\ &\quad + \left(-m_i\frac{\sin\alpha}{\cos\beta}\delta_{ij} + \frac{\cos(\alpha-\beta)}{\sqrt{2}\cos\beta}(\tilde{Y'}^\ell_{X2})_{ij}\right)h^0\\ &\quad \left.+ \left(-m_i\tan\beta\delta_{ij} + \frac{1}{\sqrt{2}\cos\beta}(\tilde{Y'}^\ell_{X2})_{ij}\right)\gamma^5 A^0\right]\ell_j\\ &\quad +\text{H.c.}\end{aligned} \tag{8}$$

The interactions of the scalar with quarks are

$$\begin{aligned}\mathcal{L}^d_{Y,X} &= (\sqrt{2}G_F)^{\frac{1}{2}}\bar{d}_i\left[\left(m_{d_i}\frac{\cos\alpha}{\cos\beta}\delta_{ij} + \frac{\sin(\alpha-\beta)}{\sqrt{2}\cos\beta}(\tilde{Y'}^d_{X2})_{ij}\right)H^0\right.\\ &\quad + \left(-m_{d_i}\frac{\sin\alpha}{\cos\beta}\delta_{ij} + \frac{\cos(\alpha-\beta)}{\sqrt{2}\cos\beta}(\tilde{Y'}^d_{X2})_{ij}\right)h^0\\ &\quad \left.+ \left(-m_{d_i}\tan\beta\delta_{ij} + \frac{1}{\sqrt{2}\cos\beta}(\tilde{Y'}^d_{X2})_{ij}\right)\gamma^5 A^0\right]d_j\\ &\quad +\text{H.c}\end{aligned} \tag{9}$$

$$\begin{aligned}\mathcal{L}^u_{Y,X} &= (\sqrt{2}G_F)^{\frac{1}{2}}\bar{u}_i\left[\left(m_{u_i}\frac{\sin\alpha}{\sin\beta}\delta_{ij} + \frac{\sin(\alpha-\beta)}{\sqrt{2}\sin\beta}(\tilde{Y'}^u_{X2})_{ij}\right)H^0\right.\\ &\quad + \left(m_{u_i}\frac{\cos\alpha}{\sin\beta}\delta_{ij} - \frac{\cos(\alpha-\beta)}{\sqrt{2}\sin\beta}(\tilde{Y'}^u_{X2})_{ij}\right)h^0\\ &\quad \left.+ \left(-m_{u_i}\cot\beta\delta_{ij} + \frac{1}{\sqrt{2}\sin\beta}(\tilde{Y'}^u_{X2})_{ij}\right)\gamma^5 A^0\right]u_j\\ &\quad +\text{H.c}\end{aligned} \tag{10}$$

where we have defined $\tilde{Y'}^f_{X2} = \frac{\tilde{Y}^f_{X2}}{(\sqrt{2}G_F)^{1/2}}$ in order to scale the parameters of the group to the electroweak scale. In this sense $\tilde{Y'}^f_{X2} = U^{f\dagger}_{XL}D^f U^f_{XL}$ and has units of mass.

4. Low energy flavor violation in τ decay with production of a pseudoscalar mesons

In order to make our phenomenological analysis we will work in the formalism of effective Lagrangian. In this case the LFV τ decay with production of neutral pseudo-scalar meson $\tau \to P^0\ell$ can be calculated from the expressions (8),(9) and (10). The advantage of channels with only one pseudo-scalar meson in the final state and with CP conservation at couplings, is that only the scalar A^0 contributes to decay reducing drastically the number of free parameters involved. In the low energy regime the effective Lagrangian is given by

$$-\mathcal{L}_{\text{eff}}^{XX'} = \sqrt{2}G_F \frac{M_W^2}{M_{A^0}^2} g^X_{A^0\ell_i\ell_j}\left(\bar{\ell}_i\gamma^5\ell_j\right)\left[\sum_{q_n,q_m} g^{X'}_{A^0 q_m q_n}\bar{q}_n\gamma^5 q_m\right] \tag{11}$$

where the effective dimensionless couplings are given by

$$g^X_{A^0\ell_i\ell_j} = \frac{1}{M_W}\left[-m_i\tan\beta\delta_{ij} + \frac{1}{\sqrt{2}\cos\beta}\left(U^{\ell\dagger}_{XL}D^dU^{\ell}_{XL}\right)_{ij}\right] \tag{12}$$

$$g^X_{A^0d_id_j} = \frac{1}{M_W}\left[-m_{d_i}\tan\beta\delta_{ij} + \frac{1}{\sqrt{2}\cos\beta}\left(U^{d\dagger}_{XL}D^dU^{\ell}_{XL}\right)_{ij}\right] \tag{13}$$

$$g^X_{A^0u_iu_j} = \frac{1}{M_W}\left[-m_i\cot\beta\delta_{ij} + \frac{1}{\sqrt{2}\sin\beta}\left(U^{d}_{XL}V^{\dagger}_{\text{CKM}}D^uV_{\text{CKM}}U^{d\dagger}_{XL}\right)_{ij}\right]. \tag{14}$$

Table 3. Updated experimental upper bounds of τ decays with LFV. In the third column we show the upper bounds on the couplings scaled with the mass of the pseudo-scalar A^0.

Process	Upper limit on BR [13]	$\frac{\lvert g^X_{A^0\tau\ell}\rvert}{M_A^2}\left\lvert\langle P\rvert\sum_{q_i,q_j}(g^{X'}_{A^0q_iq_j})\bar{q}_i\gamma^5q_j\lvert 0\rangle\right\rvert$
$\tau\to e^-\pi^0$	$< 8.0\times10^{-8}$	$< 2.13\times10^{-8}$
$\tau\to \mu^-\pi^0$	$< 1.1\times10^{-7}$	$< 2.67\times10^{-8}$
$\tau\to e^-K^0_s$	$< 2.6\times10^{-8}$	$< 1.31\times10^{-8}$
$\tau\to \mu^-K^0_s$	$< 2.3\times10^{-8}$	$< 1.32\times10^{-8}$
$\tau\to e^-\eta$	$< 9.2\times10^{-8}$	$< 2.52\times10^{-8}$
$\tau\to \mu^-\eta$	$< 6.5\times10^{-8}$	$< 2.27\times10^{-8}$
$\tau\to e^-\eta'$	$< 1.6\times10^{-7}$	$< 4.24\times10^{-8}$
$\tau\to \mu^-\eta'$	$< 1.3\times10^{-8}$	$< 1.32\times10^{-8}$

The decay width of τ with a neutral pseudo-scalar meson using the effective Hamiltonian (11) is [14],

$$\Gamma_{XX'}(\tau\to\ell P^0) = \frac{G_F^2}{8\pi}\left(\frac{M_W}{M_{A^0}}\right)^4\left[(m_\tau-m_\ell)^2-m_P^2\right]\frac{\lambda^{1/2}(m_\tau^2,m_\ell^2,m_P^2)}{m_\tau^3}$$
$$\times\lvert g^X_{A^0\tau\ell}\rvert^2\left\lvert\langle P\rvert\sum_{q_i,q_j}(g^{X'}_{A^0q_iq_j})\bar{q}_i\gamma^5q_j\lvert 0\rangle\right\rvert^2. \tag{15}$$

Using the experimental upper limit on the branching ratios we can calculate the parameters space for particular cases as the channel $\tau\to eK^0_s$.

Writing the couplings in terms of the parameters of the invariant transformation of $SU(3)$ we get several cases.

4.1. Couplings with the S-Spin subgroup

Using S-Spin transformation to generate the Yukawa matrix elements, we can calculate the couplings of the pseudo-scalar A^0 with fermions. In this work we consider only the processes at tree level with LFV in tau decay that produce a pseudo-scalar meson (see Table 3), thus in the lepton sector only the couplings $g^S_{A^0\tau e}$, $g^S_{A^0\tau\mu}$ are needed. From the expression (6), it is easy to see that LFV in these processes are identically zero leading to a plausible scenario of null LFV.

For completeness we write the couplings of A^0 to quarks that are implied in the LFV tau decays.

$$g^S_{A^0uu} = \frac{m_t}{\sqrt{2}M_W\sin\beta}\left\{(a^d_{S0})^2V^2_{ub} + \left[\left(V_{ud}(a^d_{S0}+B^u_S)+V_{us}\right)\left(V_{ud}(a^d_{s0}+B^d_S)+V_{us}A^{d*}_S\right)\right]\left(\frac{m_u}{m_t}\right) + \left(V_{us}(a^d_{s0}-B^d_S)+V_{ud}A^d_S\right)^2\left(\frac{m_c}{m_t}\right)\right\} \quad (16)$$

$$g^S_{A^0dd} = \frac{m_s}{\sqrt{2}M_W\cos\beta}\left[|A^d_S|^2 + \left((a^d_{S0}+B^d_S)^2-\sqrt{2}\sin\beta\right)\left(\frac{m_d}{m_s}\right)\right] \quad (17)$$

$$g^S_{A^0ss} = \frac{m_s}{\sqrt{2}M_W\cos\beta}\left[(a^d_{S0}-B^d_S)^2-\sqrt{2}\sin\beta+|A^d_S|^2\left(\frac{m_d}{m_s}\right)\right]. \quad (18)$$

For the couplings with flavor violation we have

$$g^S_{A^0uc} = \frac{m_t}{\sqrt{2}M_W\sin\beta}\left\{(a^d_{S0})^2V_{ub}V_{sb} + \left(V_{us}(a^d_{s0}-B^d_S)+V_{ud}A^d_S\right)\left(\frac{m_c}{m_t}\right) + \left(V_{ud}(a^d_{S0}+B^d_S)+V_{us}A^d_S\right)\left(V_{cd}(a^d_0+B^d_S)+V_{cs}A^{u*}_S\right)\left(\frac{m_u}{m_t}\right)\right\} \quad (19)$$

$$g^S_{A^0sd} = \frac{m_s}{\sqrt{2}M_W\cos\beta}\left[A^d_S(a^d_{S0}-B^d_S)+(a^d_{S0}+B^d_S)A^{d*}_S\left(\frac{m_d}{m_s}\right)\right]. \quad (20)$$

4.2. U-Spin couplings

In this subgroup we have that $g_{A^0\tau\mu}=0$ thus for the decay that we are considering, the LFV appears only in the decays with electron with the coupling

$$g_{A^0\tau e} = \frac{m_\tau}{\sqrt{2}M_W\cos\beta}\left[A^\ell_U(a^\ell_{U0}-B^\ell_U)+(a^\ell_{S0}+B^\ell_U)A^{\ell*}_U\left(\frac{m_e}{m_\tau}\right)\right]. \quad (21)$$

On the other hand, for the quark sector the couplings are

$$g^U_{A^0uu} = \frac{m_t}{\sqrt{2}M_W\sin\beta}\left\{\left[V_{ub}(a^d_{U0}-B^d_U)+A^d_UV_{ud}\right]^2+(a^d_0)^2V^2_{us}\left(\frac{m_c}{m_t}\right) + \left(\left[V_{ud}(a^d_{U0}+B^d_U)+A^d_UV_{ub}\right]\left[V_{ud}(a^d_{U0}+B^d_U)+V_{ub}A^d_U\right]-\sqrt{2}\cos\beta\right)\left(\frac{m_u}{m_\tau}\right)\right\} \quad (22)$$

$$g^U_{A^0dd} = \frac{m_b}{\sqrt{2}M_W\cos\beta}\left[A^{d2}_U+\left((a^d_{U0}-B^d_U)^2-\sqrt{2}\sin\beta\right)\left(\frac{m_d}{m_b}\right)\right] \quad (23)$$

$$g^U_{A^0ss} = \frac{m_s}{\sqrt{2}M_W\cos\beta}\left[(a^d_{U0})^2-\sqrt{2}\sin\beta\right] \quad (24)$$

$$g^U_{A^0uc} = \frac{m_t}{\sqrt{2}M_W\sin\beta}\left\{\left[V_{ub}(a^d_{U0}-B^d_U)+A^d_UV_{ud}\right]\left[V_{cb}(a_{U0}-B)+A^d_UV_{cd}\right] + \left[V_{ud}(a^d_{0U}+B^d_U)+A^d_UV_{ub}\right]\left[V_{cd}(a^d_{U0}+B^d_U)+V_{cb}A^{d*}_U\right]\left(\frac{m_u}{m_t}\right)+(a^d_{U0})^2V_{ud}V_{cs}\left(\frac{m_c}{m_t}\right)\right\} \quad (25)$$

$$g^U_{A^0sd} = \frac{m_s}{\sqrt{2}M_W\cos\beta}(a^d_{U0})^2. \quad (26)$$

(27)

4.3. V-Spin couplings

In this case $g^V_{A^0\tau e} = 0$ the LFV for decay with muons is given by

$$g_{A^0\tau\mu}\frac{m_\tau}{\sqrt{2}M_W\cos\beta}\left[A^d_V(a^d_{V0} - B^d_V) + (a^d_{V0} + B^d_V)A^{d*}_V\left(\frac{m_\mu}{m_\tau}\right)\right] \tag{28}$$

The couplings with quark sector is

$$g^V_{A^0uu} = \frac{m_u}{\sqrt{2}M_W\sin\beta}\left((a^d_{V0}) - \sqrt{2}\cos\beta\right) \tag{29}$$

$$\begin{aligned} g^V_{A^0dd} &= \frac{m_b}{\sqrt{2}M_W\cos\beta}\Big\{[V_{ub}(a_0 - B) + V_{us}A]^2 \\ &+ [V_{us}(a_0 + B) + AV_{ub}]\,[V_{us}(a_0 + B) + V_{ub}A^*]\left(\frac{m_s}{m_b}\right) \\ &+ \left((a_0)^2V^2_{ud} - \sqrt{2}\sin\beta\right)\left(\frac{m_d}{m_b}\right)\Big\} \end{aligned} \tag{30}$$

$$g^V_{A^0ss} = \frac{m_b}{\sqrt{2}M_W\cos\beta}\left[A^2 + \left((a_0 + B)^2 - \sqrt{2}\sin\beta\right)\left(\frac{m_s}{m_b}\right)\right] \tag{31}$$

$$\begin{aligned} g^V_{A^0uc} &= \frac{m_t}{\sqrt{2}M_W\sin\beta}\{[V_{ub}(a_0 - B) + V_{us}A]\,[V_{cb}(a_0 - B) + AV_{cs}] \\ &+ [V_{us}(a_0 + B) + AV_{ub}]\,[V_{cs}(a_0 + B) + V_{cb}A^*]\left(\frac{m_c}{m_t}\right) + (a_0)^2V_{ud}V_{cd}\left(\frac{m_u}{m_t}\right)\Big\} \end{aligned} \tag{32}$$

$$g^V_{A^0sd} = 0\,. \tag{33}$$

As we can seen, using the S-Spin invariant subgroup there are no flavor violating τ decays. On the other hand U-Spin and V-Spin exclude flavor violating decays with μ and e respectively. Although the couplings $g^X_{Af_1f_2}$ are quite complicated even with this simple approximation there is a decay particularly interesting because depend on only 2 parameters, namely $\tau \to eK^0_s$ for the U-Spin case. Here the condition $m_e << m_\tau$ leads to $g^U_{A\tau e} \simeq \frac{m_\tau\zeta^\ell_U}{\sqrt{2}M_W\cos\beta}$, with $\zeta^\ell_U = A^\ell_U(a^\ell_{0U} - B^\ell_U)$. On the other hand $g^U_{Asd} = \frac{m_s(a^d_{U0})^2}{\sqrt{2}M_W\cos\beta}$ depends only on one parameter.

$$\Gamma(\tau \to eK^0_s) \simeq (8.606 \times 10^{-15}\text{GeV}^5)\ \times\ \frac{|\zeta^\ell_U|^2(a^d_{0U})^4}{M^4_A}(1 + \tan^2\beta)^2\,. \tag{34}$$

Now using the upper bound of Table 3 we get the parameter space shown in the Figure 1.

5. Conclusions

In this work we have calculated the LFV couplings in the particular case where the flavor transformations correspond to a simple invariant subgroup $SU(3)$ transformation in a 2HDM with CP conservation at the tree level scalar interactions. We demonstrate that with this form the couplings are naturally suppressed controlling the FCNC and LFV processes. In particular we observed that the U-Spin flavor transformation leads to a small decay rate for $\tau \to eK^0_s$ although the couplings $g^U_{Af_1f_2}$ are not small. An important case is the decays with $\tau \to e\eta$ and $\tau \to e\eta'$ that allows to find information of the parameter a^d_{0U} in terms of $\tan\beta$, B^d_U, and A^d_U, but with similar behavior for $|\zeta^\ell_U|$.

Nevertheless in order to reduce the space of parameters a global analysis is needed that includes the use of decays $\tau \to \ell P_1P_2$ with P_1 and P_2 pseudo-scalars mesons. In these processes we get at tree level the contributions of H^0 and h^0 that introduce the mixing angle for scalars

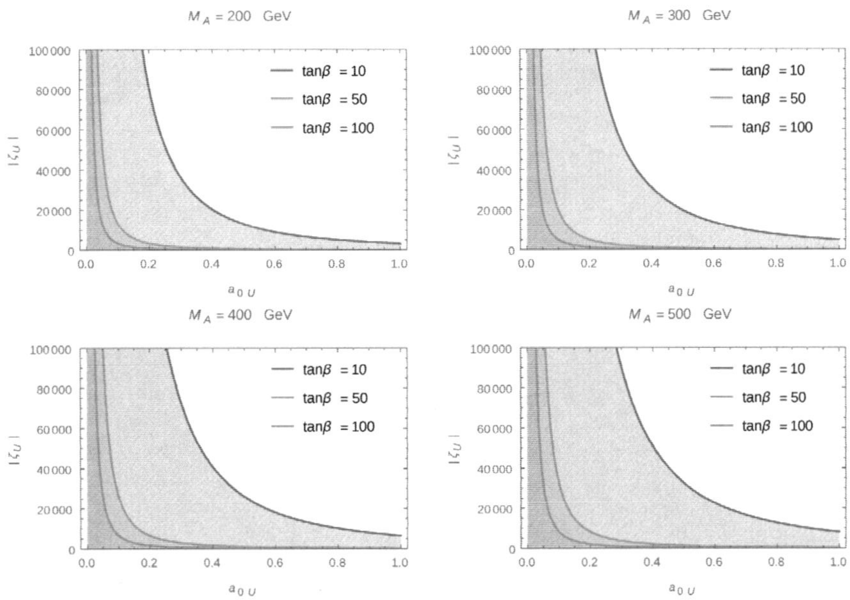

Figure 1. Here it is shown the parameters space calculated with the upper bound of the decay $\tau \to e K_s^0$. We can observe that it is not too sensitive to the values of M_A and $\tan\beta$ and for small values of the parameter a_{0U} the parameter $|\zeta_U^\ell|$ is essentially arbitrary. This plot shows that the parameters of the U-Spin flavor transformation can take a big interval of values and at the same time reproduce small LFV tau decay rates.

α as the new parameter. This forces us to use more than τ decays to bound properly the space of parameters and might reduce drastically the range of values that could take $|\zeta_U^\ell|$.

In this work we have shown how the introduction of suitable matrices coming from invariant subspaces of $SU(3)$ can suppress the LFV processes, in particular S-Spin eliminate at all this kind of decays obtaining the same result at tree level as the SM. In the case of U-Spin and V-Spin there are LFV τ decays but only with electron and μ respectively, suppressed by the hierarchy of masses and CKM matrix elements. Thus Yukawa matrices coming from this kind of flavour transformations are viable in a phenomenological sense.

Acknowledgment
The author thanks to P. Roig for his useful comments that helped to broaden the focus of this work.

References

[1] Glashow S L and Weinberg S 1977 *Phys.Rev.* **D15** 1958
[2] Paschos E 1977 *Phys.Rev.* **D15** 1966
[3] Crivellin A, Kokulu A and Greub C 2013 *Phys. Rev.* **D87** 094031 (*Preprint* 1303.5877)
[4] Branco G, Ferreira P, Lavoura L, Rebelo M, Sher M *et al.* 2012 *Phys.Rept.* **516** 1–102 (*Preprint* 1106.0034)
[5] Davidson S and Haber H E 2005 *Phys.Rev.* **D72** 035004 (*Preprint* hep-ph/0504050)
[6] Barger V D, Hewett J and Phillips R 1990 *Phys.Rev.* **D41** 3421–3441
[7] Barradas Guevara J, Cazarez Bush F, Cordero Cid A, Félix Beltran O, Hernández Sánchez J *et al.* 2010 *J.Phys.* **G37** 115008 (*Preprint* 1002.2626)
[8] Díaz-Cruz J, Noriega-Papaqui R and Rosado A 2005 *Phys.Rev.* **D71** 015014 (*Preprint* hep-ph/0410391)
[9] Díaz-Cruz J, Noriega-Papaqui R and Rosado A 2004 *Phys.Rev.* **D69** 095002 (*Preprint* hep-ph/0401194)

Journal of Physics: Conference Series **761** (2016) 012073 doi:10.1088/1742-6596/761/1/012073

[10] Félix-Beltrn O, González-Canales F, Hernández-Sánchez J, Moretti S, Noriega-Papaqui R *et al.* 2015 *Phys.Lett.* **B742** 347–352 (*Preprint* 1311.5210)
[11] Barranco J, Delepine D and López-Lozano L 2012 *Phys. Rev.* **D86** 053012 (*Preprint* 1205.0859)
[12] Hernández-Sanchez J, López-Lozano L, Noriega-Papaqui R and Rosado A 2012 *Phys. Rev.* **D85** 071301 (*Preprint* 1106.5035)
[13] Olive K *et al.* (Particle Data Group) 2014 *Chin.Phys.* **C38** 090001
[14] Gómez-Bock M, López Castro G, López-Lozano L and Rosado A 2009 *Phys.Rev.* **D80** 055017 (*Preprint* 0905.3351)

Journal of Physics: Conference Series **761** (2016) 012074 doi:10.1088/1742-6596/761/1/012074

Hairy Lovelock black holes and Stueckelberg mechanism for Weyl symmetry

Mariano Chernicoff [1], Gaston Giribet [2,3,4] and Julio Oliva[5]

[1]Departamento de Física, Facultad de Ciencias, Universidad Nacional Autónoma de México; A.P. 70-542, México D.F. 04510, México.

[2]Departamento de Física, Universidad de Buenos Aires and IFIBA - CONICET; Ciudad Universitaria, pabellón 1 (1428) Buenos Aires.

[3]Université Libre de Bruxelles and International Solvay Institutes; Campus Plaine C.P. 231 B-1050, Bruxelles, Belgium.

[4]Instituto de Física, Pontificia Universidad Católica de Valparaíso; Casilla 4950, Valparaíso, Chile.

[5]Departamento de Física, Universidad de Concepción; Casilla 160-C, Concepción, Chile.

E-mail: mchernicoff@ciencias.unam.mx, gaston@df.uba.ar, julioolivazapata@gmail.com

Abstract. Lovelock theory of gravity -and, in particular, Einstein theory- admits black hole solutions that can be equipped with a hair by conformally coupling the theory to a real scalar field. This is a secondary hair, meaning that it does not endow the black hole with new quantum numbers. It rather consists of a non-trivial scalar field profile of fixed intensity which turns out to be regular everywhere outside and on the horizon and, provided the cosmological constant is negative, behaves at large distance in a way compatible with the Anti-de Sitter (AdS) asymptotic. In this paper, we review the main features of these hairy black hole solutions, such as their geometrical and thermodynamical properties. The conformal coupling to matter in dimension $D > 4$ in principle includes higher-curvature terms. These couplings are obtained from the Lovelock action through the Stueckelberg strategy. As a consequence, the resulting scalar-tensor theory exhibits a self-duality under field redefinition that resembles T-duality. Through this field redefinition, the matter content of the theory transforms into a Lovelock action for a dual geometry. Since the hairy black holes only exist for special relations between the dual Lovelock coupling constants, it is natural to compare those relations with the causality bounds coming from AdS/CFT. We observe that, while the lower causality bound is always obeyed, the upper causality bound is violated. The latter, however, is saturated in the large D limit.

1. Introduction

Lovelock theory [1, 2, 3] is the most general torsionless metric theory of gravity that yields field equations involving a rank-2 symmetric tensor of second order and covariantly conserved. Being the natural generalization of general relativity, Lovelock theory exhibits nice properties such as the absence of ghosts around physically sensible backgrounds [4, 5]; it propagates a single massless graviton despite of being a higher-curvature theory. The action of Lovelock theory consists of the Einstein-Hilbert action augmented with higher-curvature terms $\mathcal{O}(R^k)$ that involve specific contractions of the Riemann tensors with no derivatives of it. While for spacetime dimension $D \leq 4$ the theory reduces to Einstein theory, for $D > 4$ it presents deviations from general relativity that typically emerge at distances comparable with the length scales defined by the dimensionful coupling constants of the higher-curvature terms.

Lovelock theory is relevant in high-energy physics for different reasons: the higher-curvature terms in its action resemble those appearing in the next-to-leading order of string theory low energy effective action and in M-theory compactifications. More generally, it provides a tractable example to study the consistency of more general higher-curvature models [6, 7]. Being of second order, this theory is the perfect setup to investigate the effects of introducing higher-curvature corrections in different scenarios, and this is the reason why in the last years it has become the favorite model to study higher-curvature corrections within the context of AdS/CFT correspondence [8, 9, 10]; see [11] and references therein.

In reference [12], it has been shown that Lovelock theory admits a natural generalization that leads to write down the most general conformally invariant single scalar field theory with second order field equations in D dimensions. This is in the same spirit as Horndeski and Galileon theories [13, 14]. In reference [15], hairy black hole solutions to Lovelock gravity coupled to such conformally invariant matter theory were found for special curves of the space of coupling constants. The scalar hair of these black holes is a *secondary hair*; that is, the scalar field does not endow the black hole with new quantum numbers, but its intensity turns out to be fixed in terms of some of the coupling constants in the action. The scalar field configuration is regular everywhere outside and on the horizon, and behaves at large distance in a way that respects the AdS asymptotic -here we will consider the case of negative cosmological constant-. In this paper, we will study different aspects of these hairy black hole solutions. We will review their main geometrical and thermodynamical properties. We will also study a duality that Lovelock theory conformally coupled to the scalar field exhibits under field redefinitions. Through such a field redefinition, the matter content of the theory transforms into a Lovelock action for a *dual* geometry. Since the hairy black holes exist only for special relations among the dual Lovelock coupling constants, we will compare those relations with the causality bounds coming from AdS/CFT. We will show that, while the lower causality bound is always obeyed, the upper causality bound is violated. Nevertheless, the latter is saturated in the large D limit.

2. Higher-curvature gravity and conformal couplings

2.1. Lovelock theory

Lovelock theory is defined by the action

$$I_{\{a\}}[g] = \int_{\mathcal{M}} d^D x\sqrt{-g} \sum_{k=0}^{[\frac{D-1}{2}]} a_k \frac{(2k)!}{2^k} \delta^{\mu_1}_{[\alpha_1} \delta^{\nu_1}_{\beta_1} \cdots \delta^{\mu_k}_{\alpha_k} \delta^{\nu_k}_{\beta_k]} R_{\mu_1\nu_1}{}^{\alpha_1\beta_1} \cdots R_{\mu_k\nu_k}{}^{\alpha_k\beta_k} \,, \tag{1}$$

where $R_{\nu\ \lambda\delta}^{\ \mu}$ is the Riemann curvature tensor, the symbol $[n]$ in the sum stands for the integer part of n, other brackets refer to antisymmetrization. The collection of real numbers $a_0, a_1, \ldots a_k$ in (1), to which the subindex $\{a\}$ on the left hand side of (1) refers, are coupling constants of the theory that are in principle independent. However, being dimensionful coupling constants, it is natural to think of them as being given by a unique fundamental length scale ℓ_P with $a_k \sim \ell_P^{2k-D}$. In (1), we are omitting boundary terms, cf. [16] (see also [17]). This action yields the most general theory of gravity whose fields equations are of second order in the metric. These field equations read

$$\mathcal{G}_\mu^\nu \equiv \sum_{k=0}^{[\frac{D-1}{2}]} a_k \frac{(2k)!}{2^{k+1}} \delta^{\nu}_{[\mu} \delta^{\alpha_1}_{\beta_1} \cdots \delta^{\alpha_{2k}}_{\beta_{2k}]} R^{\beta_1\beta_2}{}_{\alpha_1\alpha_2} \cdots R^{\beta_{2k-1}\beta_{2k}}{}_{\alpha_{2k-1}\alpha_{2k}} = 0 \,, \tag{2}$$

which, indeed, are of second order in the metric, although not of degree one in its second derivatives. Therefore, (2) can be regarded as the natural generalization of Einstein equations to higher dimensions.

2.2. Conformal couplings

Among the different ways to couple matter to Lovelock theory, there is one that results particularly interesting because it provides a tractable model that exhibits conformal symmetry. In dimension $D \geq 5$ a conformal invariant action for a scalar field may in principle include higher-curvature couplings. To write down the most general action of this sort, one may resort to the Stueckelberg strategy: Lovelock theory is not Weyl invariant. However, one can define the rescaled metric $\tilde{g}_{\mu\nu} = \varphi^{-2/s} g_{\mu\nu}$, with s being a real constant different from zero and φ being a real scalar field. Then, one considers the Riemann curvature tensor associated to $\tilde{g}_{\mu\nu}$,

$$S_{\mu\nu}{}^{\lambda\delta} \equiv \varphi^2 R_{\mu\nu}{}^{\lambda\delta} + \frac{4}{s}\varphi\delta^{[\lambda}_{[\mu}\nabla_{\nu]}\nabla^{\delta]}\varphi + \frac{4(1-s)}{s^2}\delta^{[\lambda}_{[\mu}\nabla_{\nu]}\varphi\nabla^{\delta]}\varphi - \frac{2}{s^2}\delta^{[\lambda}_{[\mu}\delta^{\delta]}_{\nu]}\nabla_\rho\varphi\nabla^\rho\varphi, \tag{3}$$

which happens to transform homogeneously under the Weyl transformation

$$g_{\mu\nu} \to e^{2\sigma} g_{\mu\nu}, \qquad \varphi \to e^{s\sigma}\varphi \tag{4}$$

with σ being a function of the spacetime coordinates. More precisely, under the rescaling (4) tensor (3) transforms as

$$S_{\mu\nu}{}^{\lambda\delta} \to e^{2(s-1)\sigma} S_{\mu\nu}{}^{\lambda\delta}. \tag{5}$$

This permits to construct a conformal invariant action for the compensator field φ by simply replacing $R_\nu{}^\mu{}_{\lambda\delta}$ by $S_\nu{}^\mu{}_{\lambda\delta}$ in (1). Then, the general action of Lovelock theory coupled to conformal matter takes the form [12]

$$I[g,\varphi] = \int_{\mathcal{M}} d^Dx\sqrt{-g} \sum_{k=0}^{[\frac{D-1}{2}]} \frac{(2k)!}{2^k}\delta^{\mu_1}_{[\alpha_1}\delta^{\nu_1}_{\beta_1}\dots\delta^{\mu_k}_{\alpha_k}\delta^{\nu_k}_{\beta_k]}\Big(a_k\, R_{\mu_1\nu_1}{}^{\alpha_1\beta_1}\dots R_{\mu_k\nu_k}{}^{\alpha_k\beta_k} + b_k\varphi^{m_k} S_{\mu_1\nu_1}{}^{\alpha_1\beta_1}\dots S_{\mu_k\nu_k}{}^{\alpha_k\beta_k}\Big), \tag{6}$$

where a_k and b_k represent coupling constants that are in principle arbitrary -only constrained by consistency conditions such as unitarity and causality-, and where $m_k = (2k-D)/s - 2k$. The parameter s corresponds to the conformal weight of the scalar field. The choice $s = 1 - D/2$ leads to the standard action of a conformally coupled scalar field theory

$$I[g,\varphi] = \int_{\mathcal{M}} d^Dx\sqrt{-g}\Big(\frac{1}{16\pi G}R - \frac{\Lambda}{8\pi G} - (\nabla\varphi)^2 - \frac{(D-2)}{4(D-2)}R\varphi^2 - \lambda\varphi^{2D/(D-2)} + \dots\Big)$$

where we have chosen $b_1 = -1$ and denoted $a_0 = \Lambda/(8\pi G)$, $a_1 = 1/(16\pi G)$, $b_0 = -\lambda$ as usual. The D-dimensional Planck length is given by $\ell_P = G^{1/(D-2)}$. The ellipses stand for terms with more than two derivatives on the metric and on φ. The latter, however, do not yield higher-order field equations. In fact, the new conformal couplings involve higher-curvature terms together with higher-derivatives of the scalar field, all arranged in a way that the field equations turn out to be conformally invariant as well as of second order [12]. The field equations coming from varying (6) with respect to the metric can be written in the form

$$\mathcal{G}_{\mu\nu} = 8\pi\ell_P^{D-2} T_{\mu\nu}\,, \tag{7}$$

with $\mathcal{G}_{\mu\nu}$ given by (2) and $T_{\mu\nu}$ given by

$$T^{\nu}_{\mu} = \sum_{k=0}^{[\frac{D-1}{2}]} \frac{(2k)!a_1 b_k}{2^k} \varphi^{m_k} \delta^{\nu}_{[\mu}\delta^{\alpha_1}_{\beta_1} \dots \delta^{\alpha_{2k}}_{\beta_{2k}]} S^{\beta_1\beta_2}{}_{\alpha_1\alpha_2} \dots S^{\beta_{2k-1}\beta_{2k}}{}_{\alpha_{2k-1}\alpha_{2k}} \,. \tag{8}$$

From this, we easily verify that, despite the presence of higher-curvature terms in the action, equations (7) are of second order in the fields. The same happens with the equation for the scalar field φ, which reads

$$\sum_{k=0}^{[\frac{D-1}{2}]} \frac{(2k-D)\, b_k}{2^k s} \varphi^{m_k - 1} \delta^{\mu_1}_{[\alpha_1} \delta^{\nu_1}_{\beta_1} \dots \delta^{\mu_k}_{\alpha_k} \delta^{\nu_k}_{\beta_k]} S_{\mu_1\nu_1}{}^{\alpha_1\beta_1} \dots S_{\mu_k\nu_k}{}^{\alpha_k\beta_k} = 0\,. \tag{9}$$

It is possible to show that the trace of $T_{\mu\nu}$ vanishes on-shell. This is consistent with the conformal invariance of the matter action.

3. Hairy black hole solution

In reference [15], explicit black hole solutions to field equations (7) have been found for arbitrary dimension $D > 4$, for horizons of either positive or negative curvature, and for arbitrary sign of the cosmological constant. Here, we will focus our attention to asymptotically AdS solutions with horizons of spherical topology.

There are special relations between the coupling constants $b_0, b_1, \dots b_{[(D-1)/2]}$ for which the static spherically symmetric solution to the field equations (7)-(9) is compatible with the ansatz

$$ds^2 = -F(r)dt^2 + \frac{1}{F(r)}dr^2 + r^2 d\Omega^2_{D-2}, \tag{10}$$

where $r \in \mathbb{R}_{>0}$, $t \in \mathbb{R}$, and where $d\Omega^2_{D-2}$ is the metric of a $(D-2)$-dimensional Euclidean space of constant curvature. Here, we will consider this space to be the unit $(D-2)$-sphere. In that case, the metric function $F(r)$ is root of the polynomial equation [18]

$$\sum_{k=0}^{[\frac{D-1}{2}]} a_k \frac{(D-1)!}{(D-2k-1)!} \Big(\frac{1-F(r)}{r^2}\Big)^k = \frac{M(D-1)}{\text{Vol}_\Omega r^{D-1}} + \frac{Q_0}{r^D}\,, \tag{11}$$

where $\text{Vol}_\Omega = 2\pi^{(D-1)/2}/\Gamma((D-1)/2)$ is the volume of the unit $(D-2)$-sphere.

Choosing $s = -1$, the solution for the scalar field takes the simple form

$$\varphi(r) = \frac{N}{r}, \tag{12}$$

with N satisfying the following constraints

$$\sum_{k=1}^{[\frac{D-1}{2}]} k\tilde{b}_k N^{2-2k} = 0\,, \quad \sum_{k=0}^{[\frac{D-1}{2}]} (D(D-1)+4k^2)\tilde{b}_k N^{-2k} = 0, \tag{13}$$

with $\tilde{b}_k = b_k(D-1)!/(D-2k-1)!$. These constraints also give a specific relation between the different coupling constants $b_0, b_1, \dots b_{[(D-1)/2]}$ for metric (10)-(11) to be a solution.

M in (11) is an arbitrary integration constant that corresponds to the black hole mass. In contrast, the constant Q_0 that weights the backreaction of the scalar field is fixed in terms of the coupling constants through the equation

$$Q_0 = \sum_{k=0}^{[\frac{D-1}{2}]} (D-2k-1)\tilde{b}_k N^{D-2k} . \tag{14}$$

This means that the conformal field φ corresponds to a secondary hair. The scalar field configuration remains finite everywhere outside and on the horizon, and it decays at large distance with a power law that depends on s (we considered in (12) the value $s = -1$). The particular dependence on Q_0 in the metric can be understood as the backreaction on the geometry produced by the scalar field energy density

$$T_0{}^0(r) = -\frac{Q_0}{(D-1)r^D}, \tag{15}$$

which is positive provided $Q_0 < 0$. Equation (11) receives additional contributions on the right hand side when the black hole is charged under $U(1)$ gauge fields; see (17) below.

For illustrative purposes it is enough to consider only up to $\mathcal{O}(R^2)$ terms in the purely gravitational part of the action; therefore, hereafter we set $a_{k>2} = 0$. In such case, equation (11) notably simplifies and yields

$$F(r) = 1 + \frac{r^2}{4\tilde{\alpha}}\left(1 - H(r)\right) , \tag{16}$$

with

$$\begin{aligned} H^2(r) &= 1 - \frac{8\tilde{\alpha}a_0}{(D-1)(D-2)a_1} + \frac{8\tilde{\alpha}M}{\mathrm{Vol}_\Omega(D-2)a_1r^{D-1}} + \\ & \frac{8\tilde{\alpha}Q_0}{(D-1)(D-2)a_1r^D} - \frac{8\tilde{\alpha}Q_1^2}{(D-2)(D-3)a_1\mathrm{Vol}_\Omega r^{2D-4}}, \end{aligned}$$

where we have defined $\tilde{\alpha} = a_2(D-4)(D-3)/(2a_1)$. This is a generalization of the Boulware-Deser solution [5] and of its electrically charged generalization [19]; see also [20]. Q_1 is the black hole charge under a minimally coupled Abelian gauge field with electric solution

$$A(r) = \frac{Q_1}{(D-3)r^{D-3}}dt. \tag{17}$$

The equation for $H^2(r)$ given above is quadratic, allowing in principle for the two signs of $H(r)$. This accounts for the two roots of $F(r)$ in the quadratic version of equation (11). Nevertheless, only the positive root corresponds to a physically sensible solution [5], yielding the following behavior for small $\tilde{\alpha}$

$$\begin{aligned} F(r) &\simeq 1 - \frac{16\pi GM}{\mathrm{Vol}_\Omega(D-2)r^{D-3}} + \frac{16\pi GQ_1^2}{\mathrm{Vol}_\Omega(D-2)(D-3)r^{2D-6}} - \\ & \frac{16\pi GQ_0}{(D-1)(D-2)r^{D-2}} - \frac{2\Lambda r^2}{(D-1)(D-2)} + \mathcal{O}(\tilde{\alpha}) , \end{aligned}$$

where again we have denoted $a_0 = -\Lambda/(8\pi G)$ and $a_1 = 1/(16\pi G)$. At first order in $\tilde{\alpha}$ this corresponds to the general relativity (GR) solution, with the second term on the right hand side being the Newtonian potential of the Schwarzschild-Tangherlini solution, the third and four being the Reissner-Nordström and the CFT matter contributions, respectively. The fifth term corresponds to the cosmological constant.

From (16), one can write the mass as a function of the horizon radius r_+, namely

$$M = \frac{\mathrm{Vol}_\Omega}{16\pi G}\Big[-\frac{2\Lambda r_+^{D-1}}{(D-1)} + (D-2) r_+^{D-3} + 2(D-2)\tilde{\alpha} r_+^{D-5} + \frac{16\pi G Q_1^2}{(D-3)\mathrm{Vol}_\Omega r_+^{D-3}} - \frac{16\pi G Q_0}{(D-1) r_+}\Big] \quad (18)$$

which, if $Q_1 = 0$ and $Q_0 > 0$, is not bounded from below due to the presence of the last term. This pathology compels to consider only black holes with $Q_0 \leq 0$. It is worth mentioning that, even if the $U(1)$ charge is present and screens the ultraviolet divergence due to Q_0, the solution with $Q_0 > 0$ still presents other pathologies, such as negative entropy configurations [16]. Therefore, hereafter we will restrict ourselves to the case $Q_0 < 0$.

4. Black hole thermodynamics

Now, let us discuss the thermodynamical properties of these hairy black holes. We will focus our attention on the effects produced by the presence of the contribution $Q_0 \neq 0$. Hereafter we consider $Q_1 = 0$. In this case, the Hawking temperature of the black hole solution (16) is given by

$$T = \frac{1}{4\pi(4\tilde{\alpha} + r_+^2)}\Big[-\frac{2\Lambda r_+^3}{(D-2)} + (D-3) r_+ + \frac{2(D-5)\tilde{\alpha}}{r_+} + \frac{16\pi G Q_0}{(D-1)(D-2) r_+^{D-3}}\Big] \quad (19)$$

The temperature presents different features depending on the parameters and it is necessary to consider different cases. At large r_+, the temperature correctly reproduces the behaviour of black holes in GR. The difference appears for small values of r_+. At short distances, provided $Q_0 \neq 0$, the leading term of (19) goes like like $T \sim Q_0/(4\tilde{\alpha} r_+^{D-3} + r_+^{D-1})$. However, it is important to take into account that physically sensible solutions correspond to $Q_0 < 0$. Therefore, there is a minimum value of r_+ for which $T \geq 0$. This is related to the fact that hairy black holes with sufficiently small radius exhibit positive specific heat. This leads to a slowed evaporation that yields remnants. This endpoint corresponds to extremal solutions with zero temperature.

From the Euclidean action one can compute the entropy

$$S = 4\pi\,\mathrm{Vol}_\Omega (D-2)\, a_1 \left(\frac{r_+^{D-2}}{D-2} + \frac{4\tilde{\alpha} r_+^{D-4}}{D-4} \right) + S_0 = \frac{A}{4G} + \mathcal{O}(\tilde{\alpha} r^{D-4}), \quad (20)$$

which obeys the first law of black hole thermodynamics with the mass and the temperature computed above. While the first term in (20) corresponds to the Bekenstein-Hawking area law, the entropy formula also have other contributions. On the one hand, there are corrections due to the Lovelock terms in the pure gravity part of the action. Notice also that there is a constant contribution S_0 which depends on the couplings $b_0, b_1, \ldots b_{[(D-1)/2]}$. Such a constant contribution, puzzling from the physical point of view, is a recurrent feature of theories that contain higher-curvature terms. In the model discussed here, the value of the constant S_0 depends on the value of Q_0, and also its sign does. As a consequence, in order to avoid negative values of entropy for positive r_+, it is necessary to restrict the analysis to the cases $Q_0 \leq 0$, $\tilde{\alpha} \geq 0$. The presence of S_0 in the entropy formula has implications for thermodynamical (in)stability of the black holes as well as for other features such as their inner black hole properties.

5. Dual geometry and causality bounds

Because the matter part of the action has been constructed by means of a Stueckelberg mechanism from the Lovelock action, the former can also be thought of as a theory of pure gravity for the dual metric $\tilde{g}_{\mu\nu}$. In other words, the action of the full theory (6) can be written as [21]

$$I[g,\varphi] = I_{\{a\}}[g] + I_{\{b\}}[\varphi^{-2/s}g], \tag{21}$$

or equivalently as

$$I[g,\varphi] = I_{\{b\}}[\tilde{g}] + I_{\{a\}}[\varphi^{+2/s}\tilde{g}]. \tag{22}$$

This means that for special choices of the couplings constants a_k and b_k the theory turns out to be self-dual under the transformation

$$g_{\mu\nu} \leftrightarrow \tilde{g}_{\mu\nu} = \varphi^{-2/s} g_{\mu\nu}\,, \qquad \varphi \leftrightarrow \tilde{\varphi} = \varphi^{-1}\,. \tag{23}$$

Constraints (13) admit particular solutions with $b_k = 0$ for all $k > 2$. This solution is

$$b_0 = -b_1 \frac{N^2}{2(D-3)(D-4)}, \qquad b_2 = -b_1 \frac{(D-2)(D^2-D-8)}{2DN^2}\,, \tag{24}$$

which yields

$$\frac{b_0 b_2}{b_1^2} = \frac{(D-2)(D^2-D-8)}{4D(D-3)(D-4)}\,. \tag{25}$$

These couplings b_k can be thought of as the Lovelock couplings $\tilde{a}_k$ of the dual quadratic gravity theory (22) for the metric $\tilde{g}_{\mu\nu}$. The dual theory admits asymptotically (A)dS solutions with curvature radius of order $|b_1/b_0|^{1/2}$. Therefore, it is natural to compare (25) with the causality constraint coming from AdS/CFT in D dimensions. For the dual CFT not to propagate perturbations with superluminal velocities, the couplings of the quadratic Lovelock action must lie within the window [9, 10, 22]

$$-\frac{(D-3)(3D-1)}{4(D+1)^2} \le \frac{\tilde{a}_0 \tilde{a}_2}{\tilde{a}_1^2} \le \frac{(D-3)(D-4)(D(D-3)+8)}{4(D^2-5D+10)^2}. \tag{26}$$

Comparing (26) with (25) in arbitrary dimension $D \ge 5$ one finds that the black hole solution exists for choices of the coupling constants b_k such that the dual quadratic gravity theory only obeys the lower bound in (26) but does not obey the upper bound. The upper bound, however, is met in the large D limit, when $b_0 b_2/b_1^2 \sim 1/4 + \mathcal{O}(1/D^2)$. In order to not to misinterpret this observation, it is worth pointing out that the violation of the upper causality constraint (26) does not imply a priori the instability of the hairy black hole solution. This is because the dual geometry $\tilde{g}_{\mu\nu}$ that corresponds to the original solution (10) is not the AdS$_D$ vacuum that the dual theory has, but a different solution that asymptotes to $\mathbb{R}^{1,1} \times S^{D-2}$ with the radius of the $(D-2)$-sphere given by N. It is also interesting that the hairy black hole solution with $\Lambda = 0$ corresponds to a dual geometry that asymptotes to AdS$_2 \times S^{D-2}$.

It may be also interesting to compare the expressions (25) and (26) with other special value for the combination $b_0 b_2/b_1^2$; namely,

$$\frac{b_0 b_2}{b_1^2} = \frac{(D-2)(D-1)}{4(D-4)(D-3)} \tag{27}$$

which corresponds to a curve on the space of the coupling constants for which the dual quadratic Lovelock theory has a single (A)dS vacuum and not two different effective cosmological constants. In $D = 5$ dimensions this coincides with the Chern-Simons theory (the special point $\lambda = 1/4$ in the notation employed in the AdS/CFT analysis, e.g. [8, 9, 10]), which exhibits peculiar properties. In the large D limit of (27) one finds again the behaviour $b_0 b_2/b_1^2 \sim 1/4 + \mathcal{O}(1/D)$. It would be interesting to further explore the properties that the solutions of this higher-curvature scalar-tensor theory exhibit in the large D limit [23].

Acknowledgments

The authors thank José Edelstein, Mario Galante, Andrés Goya, Matias Leoni, Guillem Pérez-Nadal, and Sourya Ray for discussions. This work has been supported by FONDECYT 1141073. and it has also been partially funded by FNRS-Belgium (convention FRFC PDR T.1025.14 and convention IISN 4.4503.15), by the Communauté Francaise de Belgique through the ARC program and by a donation from the Solvay family. MC is partially supported by Mexico's National Council of Science and Technology (CONACyT) grant 238734 and DGAPA-UNAM grant IN113115. GG is partially supported by CONICET of Argentina through grant PIP0595-13.

References

[1] C. Lanczos, *Annals Math.* **39**, 842 (1938).
[2] D. Lovelock, *J. Math. Phys.* **12**, 498 (1971).
[3] D. Lovelock, *J. Math. Phys.* **13**, 874 (1972).
[4] B. Zwiebach, *Phys. Lett.* B **156**, 315 (1985).
[5] D. G. Boulware and S. Deser, *Phys. Rev. Lett.* **55** (1985), 2656.
[6] X. O. Camanho, J. D. Edelstein, J. Maldacena and A. Zhiboedov, *JHEP* **1602**, 020 (2016).
[7] H. Reall, N. Tanahashi and B. Way, *Class. Quant. Grav.* **31**, 205005 (2014).
[8] M. Brigante, H. Liu, R. C. Myers, S. Shenker and S. Yaida, *Phys. Rev. Lett.* **100**, 191601 (2008).
[9] X. O. Camanho and J. D. Edelstein, *JHEP* **1004**, 007 (2010).
[10] A. Buchel, J. Escobedo, R. C. Myers, M. F. Paulos, A. Sinha and M. Smolkin, *JHEP* **1003**, 111 (2010).
[11] X. O. Camanho, [arXiv:1509.08129 [hep-th]].
[12] J. Oliva and S. Ray, *Class. Quant. Grav.* **29**, 205008 (2012).
[13] G. W. Horndeski, *Int. J. Theor. Phys.* **10**, 363 (1974).
[14] S. Dubovsky, T. Gregoire, A. Nicolis and R. Rattazzi, *JHEP* **0603**, 025 (2006).
[15] G. Giribet, M. Leoni, J. Oliva and S. Ray, *Phys. Rev. D* **89**, no. 8, 085040 (2014).
[16] M. Galante, G. Giribet, A. Goya and J. Oliva, *Phys. Rev. D* **92**, no. 10, 104039 (2015).
[17] R. A. Hennigar and R. B. Mann, Entropy **17**, no. 12, 8056 (2015)
[18] J. T. Wheeler, *Nucl. Phys. B* **268**, 737 (1986).
[19] D. L. Wiltshire, *Phys. Rev. D* **38**, 2445 (1988).
[20] C. Garraffo and G. Giribet, *Mod. Phys. Lett.* A **23**, 1801 (2008)
[21] M. Chernicoff, M. Galante, G. Giribet, A. Goya, M. Leoni, J. Oliva and G. Perez-Nadal, *JHEP* **1606**, 159 (2016).
[22] X. H. Ge and S. J. Sin, *JHEP* **0905**, 051 (2009)
[23] R. Emparan, R. Suzuki and K. Tanabe, *JHEP* **1306**, 009 (2013)

Confusing aspects regarding the symmetry of the vacuum state in theories with "spontaneous symmetry breaking"

Daniel Sudarsky
Instituto de Ciencias Nucleares, Universidad Nacional Autónoma de México, México D.F. 04510, México.
E-mail: sudarsky@nucleares.unam.mx

Abstract. The topic known as spontaneous symmetry breaking is afflicted with a high level of confusion. Here the topic is discussed in some detail focusing on the various sources of the confusion. A picture is presented, in which, with a small exception, and in contrast with standard presentations, the vacuum state is symmetric, and yet, the successful phenomenology usually associated with the subject is fully recovered.

1. Introduction

Spontaneous symmetry breaking (SSB) is regarded as a fundamental aspect in the understanding of multiple situations in various areas of theoretical physics. The general idea is often described as characterizing situations where the vacuum, or lowest energy state in the theory, does not share all the symmetries of the theory. When the symmetry is a continuous one that condition is then associated with the emergence of Nambu-Goldstone Bosons, or the Higgs mechanism, when Gauge Symmetries are broken.

Despite its importance, the topic is plagued by a serious level of confusion and by multiple misunderstandings, that can be found not only in research articles, but also in many leading textbooks. Moreover, the discussions found in the literature are often rather unclear about the general context in which they are carried on. In fact the SSB appellation is one that invokes things occurring in time: "everything was symmetric when suddenly and spontaneously a fundamental asymmetry simply appeared". If that was what one was referring to, one might then want to inquire about the timing of the symmetry breaking? and what exactly was then the cause of that occurrence? Here, again confusion can arise if we mix up the simple kind of classical statistical fluctuations, such as those associated with Brownian motion, and involve processes taking place in time, with the more fundamental quantum fluctuations, that are nothing more that the uncertainties that quantum mechanics normally associates with any given observable in an arbitrary quantum state, and as such refer to no explicit process in time. This line of inquire would rapidly take us into the realm of foundations of quantum mechanics, a topic that would require an extensive discussion on its own. Fortunately however, and despite the wording, the subject matter that concerns us here is something much simpler, namely, the question of under what conditions does the lowest energy state of a system exhibit the symmetries of the system's dynamics. It is also clear that the question itself is only well posed when one is dealing with situations in which the notion of energy is well defined, something that, as we know, requires stationarity of the space-time, and thus, from the onset, more general situations including the cosmological setting itself, would lie beyond the regime under consideration. Nonetheless confusion is abundant. Take the very important case of SSB in Gauge Theories which is, in fact, one of those associated with the most serious misunderstandings. Let us consider the following point: Gauge theories are theories with constrains (first order), and as it is well known, the Dirac rules for quantization of such theories require the physical states to be annihilated by the constraints. However the constrains are also the generators of infinitesimal gauge transformations, and thus, all physical states are necessary invariant. If the vacuum were not invariant, then it would follow that the vacuum could not be a physical state! Something seems clearly wrong in the standard accounts.

There are some people that understand these issues very well, but given the vast state of confusion that prevails in the literature on the subject it is also clear that there are many colleagues for whom this article could be helpful. In fact, before our original work on the subject [6] my colleague and I where quite confused ourselves. In order to help clarify the issues, in the present review I discuss the relevant ideas and sources of misunderstandings in the following contexts: Classical Mechanics, Quantum Mechanics, Global Symmetries in QFT for finite spatial extension, Global Symmetries in QFT in the limit of infinite spatial extension, Gauge Theories, and Statistical Mechanics. The goal would then be to identify and understand the fundamental differences regarding SSB in the various contexts, when they do exist, addressing the sources of confusion in each situation.

The manuscript is organized as follows: Section 2 deals with the simple case of discrete symmetries both at classical and quantum mechanical level, while section 3 extends the discussion for the continuous ones. Then we start addressing the case of infinite number of degrees of freedom in section 4, which is devoted to quantum field theory and global (or rigid) exact symmetries. The case of approximate symmetries in field theory is treated in section 5, while the situation in gauge theories is briefly discussed in section 6. Section 7 is devoted to the important differences between the previous treatments and those that one usually finds discussed in the statistical mechanical context. We end with a very brief discussion.

2. Classical and Quantum Mechanics, discrete symmetries.

Consider a system with a simple discrete symmetry P. In the classical theory there are two possibilities: either the lowest energy state is invariant under P, or there is a degeneracy in the state with lowest energy.

In the quantum theory the characterization of the situation is given by the existence of a unitary (or anti-unitary, a possibility we ignore here for simplicity) operator $\hat{P}$ acting on the Hilbert space $\mathcal{H}$ of the theory, $\hat{P} : \mathcal{H} \to \mathcal{H}$, which commutes with the Hamiltonian $PHP^{\dagger} = H$.

Consider for instance a particle in one dimension with a symmetric double well potential. Let the states $|R\rangle$ & $|L\rangle$, represent the localized wave-packets at the bottom of each well. The symmetry is reflected by

$$\hat{P}|R\rangle = |L\rangle, \qquad \hat{P}|L\rangle = |R\rangle. \tag{1}$$

The vacuum state must be of the form $|0\rangle = \frac{1}{\sqrt{2}}(|R\rangle + e^{i\alpha}|L\rangle)$. Now let us consider how do we determine α?

If $\langle R|H|L\rangle = c = |c|e^{i\beta} \neq 0$ then the minima of energy corresponds to $\alpha = \pi - \beta$. The point is that quantum interference can reduce the energy! By judiciously selecting the overall phase one can see that, of course, the vacuum is symmetric under P. However what happens if $c = 0$? Then α is undetermined and the vacuum is degenerate. Even some asymmetric states such as $|R\rangle$ or $|L\rangle$, are "vacuum states". In this case a system prepared in the state $|R\rangle$, has zero probability of tunneling to $|L\rangle$.

All this is very simple and clear, however we must note that, if c is fixed by the physical situation and the case $c = 0$ represents just an idealization, say, in the double well potential, the limit in which the barrier becomes infinite, the degeneration related with $c = 0$ is just a singular aspect of the limit: Regardless of how small $c \neq 0$ is, the vacuum is unique and symmetric, and it also determines, in general, a unique and symmetric vacuum in the limit. We should not confuse the **limit** $c \to 0$, with the **case** $c = 0$. In general there is no "continuity" in this limiting procedure. It is only when $c = 0$ represents the actual physical situation rather than an unphysical limit (for instance if the barrier is actually infinite) that we do have a true degeneration in the lowest energy states and the possibility of a non-symmetric vacuum. However, even then, among the multiplicity of vacua, there will be some that are symmetric states.

Of course the actual state of the system, will depend on how it was prepared. In this case energetic arguments do not allow one to select a unique quantum state. One cannot talk about the symmetry of " the vacuum" because there are symmetric and non-symmetric ones.

3. Mechanics Classical and Quantum: continuous symmetries

Consider now a system with two degrees of freedom and with a continuous $O(2)$ or $U(1)$ symmetry. Let the lagrangian of the theory be:

$$L = \frac{1}{2}m(\dot{X}_1^2 + \dot{X}_2^2) - \lambda(X_1^2 + X_2^2 - v^2)^2 . \tag{2}$$

The classical vacua form a degeneracy circle corresponding to $X_1+iX_2 = ve^{i\theta}$. Each one of these classical states breaks the symmetry. What happens at the quantum level? The Hamiltonian, in polar coordinates is:

$$H = \frac{1}{2m} P_r^2 + \frac{1}{2mr^2}P_\theta^2 + \lambda(r^2 - v^2)^2. \tag{3}$$

Note however that θ is not a good coordinate on all the configuration space manifold. The vacuum wave function is of the form $\Psi_0(r,\theta) = \frac{1}{\sqrt{2\pi}}\Phi_0(r)$, where $\Phi_0(r)$ is the vacuum of the "radial hamiltonian",

$$H^{(radial)} = \frac{1}{2m} P_r^2 + \lambda(r^2 - v^2)^2, \tag{4}$$

as all dependence on θ will increase the energy simply because the operator $\frac{1}{2mr^2}P_\theta^2$ is a positive definite operator. We thus have $H^{(radial)}|\Phi_0\rangle = E_0|\Phi_0\rangle$. Now, introducing the effective Hamiltonian for variable θ as :

$$H_{eff}(\theta, P_\theta) = \langle\Phi_0|H|\Phi_0\rangle = \langle\Phi_0|H^{(radial)}|\Phi_0\rangle + \langle\Phi_0|\frac{1}{2mr^2}P_\theta^2|\Phi_0\rangle \approx E_0 + \frac{1}{2mv^2}P_\theta^2, \tag{5}$$

we see that it corresponds to a free particle (or a harmonic oscillator with zero spring constant). This is the analog of a Nambu-Goldstone boson, a particle with zero mass. Note however that the vacuum state is symmetric, and in particular $\langle\Psi_0|X_1|\Psi_0\rangle = \langle\Psi_0|X_2|\Psi_0\rangle = 0$. There are Goldstone bosons even though there is no breakdown of the symmetry! However it is clear that $\langle\Psi_0|X_1^2|\Psi_0\rangle \neq \langle\Psi_0|X_1|\Psi_0\rangle^2$. That is, the state is characterized by strong quantum mechanical "correlations".

Consider next a system with three degrees of freedom to illustrate the analog of the (bosonic) mass generation mechanism usually associated with SSB. We take the system's Lagrangian to be:

$$L' = \frac{1}{2}m(\dot{X}_1^2 + \dot{X}_2^2) + (\mu/2)\dot{X}_3^2 - \lambda(X_1^2 + X_2^2 - v^2)^2 - \alpha(X_1^2 + X_2^2)X_3^2 . \tag{6}$$

We note that there is no quadratic term in X_3, so it is not a true harmonic oscillator.

The Hamiltonian (with the same change of variables as before) is:

$$H' = \frac{1}{2m}P_r^2 + \frac{1}{2mr^2}P_\theta^2 + \frac{1}{2\mu}P_3^2 + \lambda(r^2 - v^2)^2 + \alpha r^2 X_3^2. \tag{7}$$

The vacuum wave function is now: $\Psi_0'(r, \theta, X_3) = \frac{1}{2\pi}\Phi_0'(r, X_3)$, where $\Phi_0'(r, X_3)$ is the vacuum wave function for the two degrees of freedom system with Hamiltonian:

$$H^{(2)} = \frac{1}{2m}{P_r}^2 + \lambda(r^2 - v^2)^2 + \frac{1}{2\mu}P_3^2 + \alpha r^2 X_3^2. \tag{8}$$

Again, the vacuum is thus symmetric, as any dependence of the wavefunction on θ can only increase the energy. Again $\langle\Psi_0|X_1|\Psi_0\rangle = \langle\Psi_0|X_2|\Psi_0\rangle = 0$ and therefore there is no SSB!.

Comparing (7) and (3) we see that $H' = H + \frac{1}{2\mu}P_3^2 + \alpha r^2 X_3^2$ and from (8) and (4) we see that $H^{(2)} = H^{(radial)} + \frac{1}{2\mu}P_3^2 + \alpha r^2 X_3^2$.

Consider now the effective Hamiltonian for the X_3 degree of freedom: $H'_{eff}(X_3, P_3) = \langle\Psi_0|H'|\Psi_0\rangle$ where Ψ_0 is the vacuum state of the previous example, (i.e. the state described bellow eq. (3)). In this case we thus have,

$$H'_{eff}(X_3, P_3) = \langle\Psi_0|H|\Psi_0\rangle + \langle\Psi_0|(\frac{1}{2\mu}P_3^2 + \alpha r^2 X_3^2)|\Psi_0\rangle \approx E_0 + \frac{1}{2\mu}P_3^2 + \alpha v^2 X_3^2, \tag{9}$$

that is the X_3 degree of freedom (D.O.F) has now become a true harmonic oscillator with $K = 2\alpha v^2$. We note the analogy with the mass generation mechanism, again occurring without the breakdown of the symmetry by the vacuum state.

4. Quantum Field Theory and global symmetries

In this section we consider what part of the previous results extends to this situation? what does not ?, and then, why? Consider the paradigmatic case corresponding to a theory with n scalar fields $\Phi_i, i = 1, ...n$, and a global $O(n)$, symmetry described by the Lagrangian density:

$$\mathcal{L} = (1/2)\Sigma_{i=1}^n \partial_\mu\Phi_i\partial^\mu\Phi_i + V(\Sigma_{j=1}^n\Phi_j\Phi_j). \tag{10}$$

Let us focus on the case $n = 2$ where the potential has the standard Mexican Hat shape $V = \frac{\lambda}{4}(\Phi_1^2 + \Phi_2^2 - v^2)^2$. It is convenient to make a change of field variables so that $\Phi_1 = (\rho + v)\cos(\theta)$, and $\Phi_2 = (\rho + v)\sin(\theta)$. The conjugate momenta to these fields are: $\pi_\rho = \dot{\rho}$ and $\pi_\theta = (\rho + v)^2\dot{\theta}$, and the Hamiltonian density takes the form:

$$\mathcal{H} = \frac{1}{2}\pi_\rho^2 + \frac{1}{2(\rho + v)^2}\pi_\theta^2 + \frac{1}{2}(\partial_i\rho)^2 + \frac{1}{2}(\rho + v)^2(\partial_i\theta)^2 + \frac{\lambda}{4}(\rho^2 + 2v\rho)^2. \tag{11}$$

All this is standard [1], [2]. More nuanced treatments also involve discussions of stability under external perturbations and other considerations [3]. However, we want to focus on the simple and very specific issue of the symmetry of the vacuum of the theory in the absence of external factors, as a precise mathematical physics question. To this end let us now consider the situation in more detail.

In order to quantize the theory we separate into free and interaction Hamiltonian densities. The free part is:

$$\mathcal{H}_f = \frac{1}{2}\pi_\rho^2 + \frac{1}{2v^2}\pi_\theta^2 + \frac{1}{2}(\partial_i\rho)^2 + \frac{1}{2}v^2(\partial_i\theta)^2 + \frac{1}{2}m^2\rho^2 \tag{12}$$

with $m^2 = 2\lambda v^2$. We will now review the quantization procedure focusing on the treatment of the "zero mode"- the part which is position independent- of the field θ, which is the DOF which is supposed to acquire a non-vanishing expectation value in association with the SSB.

In order to proceed we consider the theory "in a box" with sides of length L with periodic boundary conditions. At the end we will explore the limit $L \to \infty$. We note that even with L finite, this is a theory with an infinite number of DOF, an issue that is sometimes mentioned as leading to a fundamental difference with Quantum Mechanics regarding the symmetry of the vacuum. It is well known that the question of number of DOF implies some differences between the QFT and Quantum Mechanics. The question we must consider here is whether there is a difference regarding the symmetry of the vacuum. Let us examine this in detail.

It is convenient at this point to make a canonical transformation to the new variables: $\phi \equiv v\theta$ and $\pi_\phi \equiv (1/v)\pi_\theta$, so that the free Hamiltonian is now:

$$H_f = \int d^3x[\frac{1}{2}\pi_\rho^2 + \frac{1}{2}\pi_\phi^2 + \frac{1}{2}(\partial_i\rho)^2 + \frac{1}{2}(\partial_i\phi)^2 + \frac{1}{2}m^2\rho^2]. \tag{13}$$

Making a Fourier expansion we have:

$$\rho(\vec{x},t) = \sum_{\vec{k}} \frac{1}{\sqrt{2\omega_k L^3}}(a_k(t)e^{-i\vec{k}\vec{x}} + a_k^\dagger(t)e^{i\vec{k}\vec{x}}), \pi_\rho(\vec{x},t) = -i\sum_{\vec{k}} \sqrt{\frac{\omega_k}{2L^3}}(a_k(t)e^{-i\vec{k}\vec{x}} - a_k^\dagger(t)e^{i\vec{k}\vec{x}}) \tag{14}$$

where $\omega_k = \sqrt{\vec{k}\cdot\vec{k} + m^2}$, the components of $\vec{k}$ are those compatible with boundary conditions and $[a_k, a_{k'}^\dagger] = \delta_{k,k'}$ (we use k instead of $\vec{k}$ in the indices to simplify the notation). In the same way we expand the field ϕ, while treating the zero mode separately. Thus we write,

$$\phi = \frac{1}{L^{3/2}}\phi_0(t) + \sum_{\vec{k}\neq 0} \frac{1}{\sqrt{2\omega_k' L^3}}(b_k(t)e^{-i\vec{k}\vec{x}} + b_k^\dagger(t)e^{i\vec{k}\vec{x}}) \tag{15}$$

$$\pi_\phi = \frac{\pi_0(t)}{L^{3/2}} - i\sum_{\vec{k}\neq 0} \sqrt{\frac{\omega_k'}{2L^3}}(b_k(t)e^{-i\vec{k}\vec{x}} - b_k^\dagger(t)e^{i\vec{k}\vec{x}}) \tag{16}$$

where $\omega_k' = |\vec{k}|$. The Hamiltonian is then,

$$H_f = \int \mathcal{H}_f\ d^3x = \frac{1}{2}\pi_0^2 + \sum_{\vec{k}}[\frac{1}{2}((\pi_\phi)_k)^2 + \frac{k^2}{2}(\phi_k)^2] + \sum_{\vec{k}}[\frac{1}{2}((\pi_\rho)_k)^2 + \frac{k^2+m^2}{2}(\rho_k)^2]. \tag{17}$$

This is the Hamiltonian for a free particle plus that of an infinite collection of harmonic oscillators. Note however that this is not exactly correct, because of the fact that that the field coordinate ϕ takes values in the interval $[0, 2\pi v]$ and it is periodical. We ignore the range restrictions on $\phi(x)$ (as those are irrelevant regarding the determination of the vacuum) and proceed with the analysis based on the Hamiltonian of the system which becomes:

$$H_f = \int \mathcal{H}_f d^3x = \frac{1}{2}\pi_0^2 + \sum_{\vec{k}}[\omega_k(a_k^\dagger a_k + \frac{1}{2}) + \omega_k'(b_k^\dagger b_k + \frac{1}{2})]. \tag{18}$$

The lowest energy state or vacuum $|0\rangle$ is then clearly characterized by $a_k|0\rangle = 0, b_k|0\rangle = 0$, & $\pi_0|0\rangle = 0$. The wave functional for the vacuum state is then the product of the wave functions for all the modes:

$$\Psi_0 = \otimes_k \psi_k^\phi[\phi_k] \otimes \psi_k^\rho[\rho_k], \tag{19}$$

where

$$\psi_k^\rho[\rho_k] = N_k \exp[-\frac{\sqrt{\vec{k}^2+m^2}}{2}\rho_k^2], \forall\vec{k}; \qquad \psi_k^\phi[\phi_k] = N_k' \exp[-\frac{\sqrt{\vec{k}^2}}{2}\phi_k^2], \forall\vec{k}\neq 0 \tag{20}$$

with N_k and N'_k normalization factors, while for the zero mode of the fieldϕ we have simply $\psi_0^\phi[\phi_0] = \frac{1}{(2\pi v L^{3/2})^{1/2}}$. This shows that the zero mode of the field has a constant wave function. Therefore there is no breakdown of the symmetry and in particular $\langle 0|\Phi_1|0\rangle = \langle 0|\Phi_2|0\rangle = 0$. Once more we have a situation where $\langle 0|\Phi_1^2|0\rangle \neq \langle 0|\Phi_1|0\rangle^2$, so the vacuum state is characterized by strong quantum correlations.

In fact, and independently of the composite nature of the operators: $\phi_1 = (\rho + v)\cos(\phi/v)$, y $\phi_2 = (\rho + v)\sin(\phi/v)$, one can see that the contribution of the zero mode to the vacuum expectation value of the field operators, vanish for any value of the "cut-off " (L). That is $\langle 0|\phi_i|0\rangle = 0; i = 1, 2$.

Thus, it is clear that this will remain valid also in the limit $L \to \infty$. Therefore this shows that the symmetry is not broken by the "vacuum". In fact one can show that: $\exp(i\epsilon \int \pi_\phi(x)dx)|0\rangle = |0\rangle$, where the operator in the l.h.s is rotation operator $U(1)$ with angle ϵ around the axis of the Mexican Hat potential.

4.1. Zero Mode, causality and "clustering"

In this subsection we briefly discuss a couple of objections to the overall picture we are presenting. The first one concerns the issue of "clustering". That is the often invoked requirement that, in the vacuum state, there should not be correlations of arbitrarily long range. It is clear, that in the examples we have considered, our vacuum fails to satisfy this condition. In our view this requirement is based on misconceptions about causality and is often presented with erroneous justifications. The usual argument is that unless the clustering requirement is enforced there would be violations of causality. One can think of simple examples of physical situations where there are long range correlations and where there are no violations of causality, such as those arising in standard EPR-B experiments. Of course there is then the nontrivial issue of how the state with such long correlations is prepared, or how did the system come into being set in such a state. We certainly acknowledge that this is an interesting question but it is also clear that this is a completely separate question from the identification of the vacuum or lowest energy state, and whether that state shares the symmetries of the theory.

We note that when the vacuum wave function has a "width" $\neq 0$ in the variable ϕ_0, the "clustering" property is violated. In fact the correlation Δ is

$$\Delta(\vec{x}, \vec{y}) \equiv \langle\phi(\vec{x}, t)\phi(\vec{y}, t)\rangle - \langle\phi(\vec{x}, t)\rangle\langle\phi(\vec{x}, t)\rangle = \langle\phi_0^2\rangle/L^3 + \delta(\vec{x}, \vec{y}) - \langle\phi_0\rangle^2/L^3 = \sigma_{\phi_0}^2/L^3 + \delta(\vec{x}, \vec{y}) \quad (21)$$

where σ_{ϕ_0} is the uncertainty on the value of the zero mode.

Thus, in the vacuum we do have long distance correlations: characterized by $\Delta(\vec{x}, \vec{y}) = \frac{\pi^2 v^2}{3} + \delta(\vec{x}, \vec{y})$ which are independent of our "universe' s size". Thus the clustering property is violated, independently of the size of the universe. If this was a real problem, it would not be resolved by taking the "vacuum" as a state sharply "picked" on ϕ_0 (as normally considered in the context of SSB), because such a state (which is not the true vacuum) would have violation of "clustering" property for the n-point functions for $\pi(x)$ instead of those of for $\phi(x)$. It is clear that, due Heisenberg's uncertainty relation between ψ_0 and π_0, one can not get rid of both. The violation of "clustering" is a manifestation of the quantum nature of the zero mode, and it survives in the $L \to \infty$ limit.

Another issue that is sometimes raised in the present context is weather one can simply ignore the zero mode, or just treat it classically. In fact in the compact case the zero mode must be treated quantum mechanically, otherwise one violates causality. Let us evaluate the commutator:

$$[\phi(x), \phi(y)] = \sum_{\vec{k}\neq 0} \frac{1}{2\omega'_k L^3} e^{(-ik\cdot(x-y))} - \sum_{\vec{k}\neq 0} \frac{1}{2\omega'_k L^3} e^{(ik\cdot(x-y))} + 2i\frac{(y^0 - x^0)}{L^3}, \quad (22)$$

where the last term comes from the zero mode. The point is that an explicit calculation shows that for $x-y$ space-like, the vanishing of the commutator arises due to cancellations between the contribution coming from the sum and that of the zero mode: The conclusion is that the zero mode **must** be treated quantum mechanically.

4.2. The $L=\infty$ case vs. the $L\to\infty$ limit.

For the case where L is finite, the energetic cost involved in localizing the zero mode of the field θ with an uncertainty $\delta\theta$ is $\delta E=\frac{1}{2v^2L^3(\delta\theta)^2}$ which tends to 0 as $L\to\infty$. Here we should recall the lessons from the Quantum Mechanical examples discussed at the beginning involving a potential barrier: As we saw there, it was essential to consider whether the barrier is very big and finite, or if it is truly infinite.

In the case $L=\infty$ the QFT construction faces a technical problem: We do not have a way to normalize all functions on R (or R^n). It is then argued in many treatments of the issue that we should only be interested on the functions of compact support as they characterize the excitations that "can be experimentally produced in a (finite) laboratory". Following this logic, one would, in effect, leave without a quantum treatment the zero mode of the field!. We should be careful here and avoid a potential confusion: It is **not** that the zero mode becomes classical, but that we do not have the mathematical tools to treat it quantum mechanically. However we should note that it would be incongruous to construct a theory for the excitations of compact support and use it to say something (like argue that there is SSB) about excitations which are not in this class.

If we justify this way of proceeding on our supposed limitation of interest to discuss only those excitations that we can produce in the laboratory, we could not then, come around and pretend to discuss the global behavior of the fields (including effects that might have cosmological causes). And particularly we would not be able to argue that those global modes that are left untreated determine something we do see in the laboratory, like the masses of elementary particles.

In fact, we must consider what is the true relevance of the $L=\infty$ case? Our universe is clearly not the infinite Minkowski space-time. That characterization is used only as an approximation for a region which is very large compared to the size of our measuring apparatus. What is relevant for such situations is then the limit, rather than the case of infinite Minkowski space-time.

When the limit is singular, as we saw, the appropriate thing to do is not "to consider the Hamiltonian in the limit", (among other things because we can't treat the zero mode) and then construct the theory, find the vacuum and do calculations, but rather, the appropriate procedure is: first construct the theory for finite L, then find the vacuum, do all calculations, and then take the limit $L\to\infty$. As we saw, in that case the treatment is transparent, the vacuum is always invariant, and there is nothing like SSB.

5. QFT , approximate symmetries. SU(2) & SU(3) flavor and the mesons.

The paradigmatic case is that of the effective strong interactions of light hadrons (see for instance [3]). The fundamental theory is QCD with a gauge group $SU(3)$ and global approximate symmetry $SU(2)_L\times SU(2)_R$: or Chiral symmetry (often one considers instead $SU(3)_L\times SU(3)_R$). The lagrangian density for this theory is :

$$\mathcal{L}=-\frac{1}{4}\mathrm{Tr}[F_{\mu\nu}F^{\mu\nu}]+\sum_{i=u,d}[\bar\psi^{(i)}_R D_\mu\gamma^\mu\psi^{(i)}_R+\bar\psi^{(i)}_L D_\mu\gamma^\mu\psi^{(i)}_L]+\sum_{i=u,d}m_i[\bar\psi^{(i)}_R\psi^{(i)}_L+\bar\psi^{(i)}_L\psi^{(i)}_R] \quad (23)$$

where $F_{\mu\nu}$ are the field strength tensors for the gauge fields (and are Lie-Algebra valued) and $\psi^{(i)}_{R,L}$ are the left and right components of the quark fields of various flavors i. Here D_μ are color-covariant derivatives γ^μ are the Dirac matrices, and m_i are the light masses for the quarks.

Journal of Physics: Conference Series **761** (2016) 012075 doi:10.1088/1742-6596/761/1/012075

The Lagrangian has various internal symmetries: $SU(3)$ color (indices not shown) an 2(3) $U(1)$ flavor symmetries.

If we set the 2 quark masses equal, we have an extra $SU(2)$ (flavor symmetry), and if we set them $m_i = 0$, we have an even larger symmetry: $SU(2)_L \times SU(2)_R$. In this case there should be two sets of $SU(2)$ conserved currents, denoted $J_V^{a\mu}(x)$ and $J_A^{a\mu}(x)$. The $J_V^{a\mu}(x)$ are associated with $SU(2)$ identical & simultaneous transformations of the left and right fermion field components, while, $J_A^{a\mu}(x)$, are related to opposite & simultaneous transformations of the left and right fermion field components.

The $J_A^{a\mu}(x)$ has the same quantum numbers of the pion, allowing the successful phenomenological identification known as (PCAC) [4]:

$$< 0|J_A^{a\mu}(x)|\pi(k), c >= ik^\mu f_\pi e^{-ikx}\delta^{ac} \tag{24}$$

where $|\pi(k), c>$ is the state with one pion of momentum k and flavor quantum number $SU(2) : c$. The proportionality constant f_π, turns out to be connected to the weak decay of the pion, and is thus measurable.

As far as we understand, the issue of the symmetry of the vacuum arises when considering objects which are obtained through formal manipulations such as:

$$f_\pi^2 \delta^{ab} = -\frac{i}{3}\int d^4x \langle 0|T(J_L^a(x)\cdot J_R^b(0))|0\rangle. \tag{25}$$

The point is that if the symmetry was 100% respected by the vacuum then the r.h.s should be 0, as that expression is not invariant under the $SU(2)$, axial transformations (note however that it is invariant under the vector ones). The question is then: what happens in the un-physical limit, in which the quark masses tend to zero ? The phenomenological argument is that the l.h.s. (i.e. f_π) can not be expected to tend to zero in this limit, because its value $130MeV$ is much larger than the value of $m_i \approx 5MeV$. This argument supports a point of view according to which, the symmetry would need to be broken even if $m_i = 0$, a situation where the Lagrangian is fully invariant.

The conclusion would be that the symmetry is spontaneously broken. This argument seems to ignore that in general, such limits are singular (as we saw in the QM example). Let us see a very clear example: Consider a free particle with mass M in 1-d: the state of lowest energy corresponds to a plane wave with zero momentum, i.e. a constant function. In such situation the expectation value $\langle 0|X|0\rangle$ is ill defined, but the state reflects the translational symmetry of the lagrangian, (a different issue is that of its normalization and the fact that $|\phi(X)|^2 X$ is even more divergent than $|\phi(X)|^2$).

Next let us consider a quadratic potential: $V(X) = \frac{K}{2}(X-5)^2$. The vacuum wave function is $\Phi(X) = Ne^{-\frac{(x-5)^2}{\sigma^2}}$, with $\sigma = \frac{1}{\sqrt{KM}}$. The expectation value is now well defined: $\langle 0|X|0\rangle = 5$ and the translational symmetry is clearly broken. However we note that the expectation value is independent of K, and thus we have: $\lim_{K\to 0}\langle 0|X|0\rangle = 5$ which **cannot be used** to conclude that, even in the **case** $K = 0$, the ground state should break the translational symmetry.

6. QFT Gauge Symmetries

This case has in fact been treated in detail by various authors [?, 8] and thus we will only devote a little space to them here. The central point is that these theories have, even at the classical level, a very serious problem: At the pure mathematical level they do not have a well-posed initial value formulation. This is easy to see by considering a given solution of the evolution equation with certain initial conditions at say the hypersurface $t = 0$ in Minkowski space-time. One can consider a gauge transformation which is the identity in a neighborhood

Journal of Physics: Conference Series **761** (2016) 012075 doi:10.1088/1742-6596/761/1/012075

of the initial hypersurface and becomes nontrivial elsewhere. The result will be again a solution of the equations of motion corresponding to the same initial conditions as the first one, and differing from it elsewhere in the space-time. The problem disappears once one identifies the gauge invariance as a redundancy in the description: Two solutions of the field equations that differ by a gauge transformation represent the same physical solution!

At the quantum level, we can think of the states of the system as wave functionals $\Psi[\phi, A_\mu]$. It is clear that for the theory to make sense, when evaluating the functional on two configurations that differ by a gauge transformation, the functional Ψ should give the same value. That is, all physical states must be gauge invariant.

The vacuum must be a physical state and therefore it needs to be gauge invariant. There is, in this sense, simply no room for SSB. However one finds statements affirming that in connection SSB non invariant fields acquire vacuum expectation values and thus the vacuum brakes the symmetry [1]. In fact in the standard model of particle physics the symmetry breaking by the vacuum characterized by a non-vanishing expectation value of the scalar fields in the theory is supposed to be responsible for generating masses for some of the gauge bosons W^+W^- and Z^0 and for the fermions in the theory.

The key to clarify the situation lies in the recognition that much of the treatment that is presented in many places is performed in schemes where one "fixes the gauge". That is, one is no longer working with the full theory but has instead chosen to work with a limited number of representatives of each equivalence class of field configurations (or to one such representative if in fact the gauge was fixed completely, something that is known to be generically impossible [5]).

As we have seen in various examples, the analogous of the mass generating mechanism works perfectly well without the need for the vacuum to break the symmetry, and that is exactly what happens in the present case. However in order to see this explicitly it is clear that one has to work in settings where the gauge has not been fixed. The reader might want to consult [6] to see this at work. The point is that it is incongruent to "fix the gauge" and then argue that somehow the gauge symmetry has been spontaneously broken.

7. Statistical Mechanics, Ferromagnetism.

One should not confuse the situations treated in Statistical Mechanics with those of Quantum Mechanics. They have many similarities but they are not the same. The starting point of the latter is the consideration of a large ensemble of identical systems. One is interested on ensemble averages of some physical quantities. Under certain conditions (ergodicity) one expects those to coincide with time averages for a particular system.

The standard applications involve some additional assumptions, particularly in the treatment of quantum systems. These include the interaction of the system with external D.O.F. often characterized as a thermal bath. Even at $T = 0$, the environment has a fundamental influence: It acts as the physical source of the ergodicity and of the **de-coherence**, which justifies the "random a priori random phases" hypothesis (see for instance [9]) for the elements of the ensemble, which is at the basis of most Quantum Statistical Mechanical treatments. This is required to justify various aspects of the treatments of the quantum system with the use of ensembles (micro-canonical, canonical, and gran canonical).

In contrast with the ordinary Quantum Mechanics, the treatment is fundamentally designed to deal with open systems which justifies in Stat. Phys. the considerations of, for instance, a system in equilibrium at a temperature T, which is represented by the Gibbs density matrix:

$$\rho(T) = e^{-\frac{\hat{H}}{kT}} = \sum_i e^{-\frac{E_i}{kT}} \, |\psi_i\rangle\langle\psi_i|, \tag{26}$$

instead of treatments designed to consider essentially pure states as is often the case in quantum theory. (In the above expression $\hat{H}$ is the system's hamiltonian, $\{|\psi_i\rangle\}$ a basis of its eigenstates with energies E_i and k is Boltzman's constant.) We should be careful not to forget about these when comparing or relating results of one with those of the other.

7.1. *Ferromagnetism.*

As a very illustrative example we consider a system of N atoms with spin $1/2$ in a fixed arrangement. We further assume that each atom interacts only with an external magnetic field $\vec{B}$ and with the nearest neighbors. The Hamiltonian is thus,

$$H = -J \sum_{a,\ b=1}^{N} \vec{s}_a \cdot \vec{s}_b - \mu \vec{B} \cdot \sum_{a=1}^{N} \vec{s_a}. \tag{27}$$

where J and μ are constants. The analysis is usually based on the canonical ensemble at temperature T: One computes the Helmholtz's free energy $A(T, B)$, and from it one obtains the magnetization trough the standard relation,

$$M(B,T) = -(\frac{\partial A}{\partial B})_{|T} \tag{28}$$

The so called spontaneous magnetization, often invoked as an analogy of SSB in QFT, corresponds to the case in which $M(0,T) \neq 0$ (i.e, magnetization without external B field) which happens when $T < T_c$ where T_c is called the critical temperature. The point is, however, that in this case the derivative does not exist !! This is hidden by the use of spherical coordinates which are bad coordinates at the origin. In fact in order to be precise, the equation above should be written as:

$$\vec{M}(\vec{B},T) = -(\vec{\nabla}_B A)_{|T} \tag{29}$$

where the B appearing in eq. (28) above, would correspond to the radial coordinate representing the magnitude of the vector $\vec{B}$. The feature that we are dealing with here is in fact an extreme

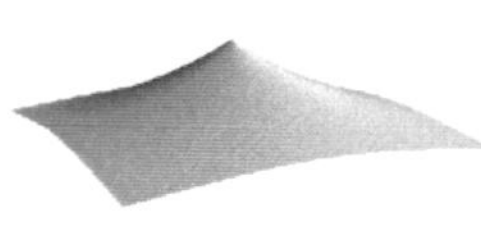

Figure 1: shows the actual shape of the function $A(T, \vec{B})$ in the thermodynamic limit as a function of $\vec{B}$ in a two dimensional case. It is clear that at the origin the function is not differentiable.

sensitivity to external fields that is reached in the thermodynamic limit $N \to \infty$. At any point different from the origin ($\vec{B} = 0$), the function A is differentiable and $\vec{M}$ is nonvanishing, nor does it vanish as $\vec{B} \to 0$. However the limiting behavior is dependent on the direction of approach to the origin, so strictly speaking the limit does not exist. It is nonetheless worth noting that in the realistic cases with finite N, the function A is smooth at the origin ($\vec{B} = 0$) and the symmetry of the situation then implies $\vec{M} = 0$. That is, we do not seem to have here something we might call spontaneous symmetry breaking.

However we must recall that the treatment above refers only to a statistical ensemble, while any consideration of an individual and finite system characterized by (27) leads, even when $\vec{B} = 0$ to a non-vanishing magnetization. Let us consider this last aspect of symmetry breaking, in more detail, in order to identify the differences with what we had faced before.

Journal of Physics: Conference Series **761** (2016) 012075 doi:10.1088/1742-6596/761/1/012075

The first observation is that the Hilbert space of each individual D.O.F. does not have symmetric states: If $|\vec{n},+\rangle$ is a spin pointing in the $+\vec{n}$ direction, we might try to built and rotationally invariant state as the superposition:

$$|\textbf{invariant}\rangle \equiv \int_{SU(2)} dg\ \mathbf{D}^{\frac{1}{2}}(g)|\vec{n},+\rangle, \tag{30}$$

where the matrix $\mathbf{D}^{1/2}(g)$ represents a Wigner rotation with g in $SU(2)$. The point however is that this gives as a result simply 0, that is the null vector in the Hilbert space. There is also no invariant state for the case of N spins when N is odd. That makes it hard to consider the limit $N \to \infty$. There are of course symmetric states for N even. In the case $N = 2$ these are the well known singlet states. However their misalignment implies an energetic cost according to 27. This seems to be a fundamental difference with the previously considered cases.

A second aspect worth noting refers to the statistical nature of the problem we faced and the fact that we deal in practice with an open quantum system. The ferromagnet is described in Statistical Mechanics by a density matrix (not a pure state), and this CAN be symmetric. For instance, for a single spin :

$$\rho_{inv} \equiv \int_{SU(2)} dg\ \mathbf{D}^{\frac{1}{2}}(g)|\vec{n},+\rangle(\mathbf{D}^{\frac{1}{2}}(g)|\vec{n},+\rangle)^{\dagger} = \frac{1}{2}I \tag{31}$$

where I is the unit matrix, which is in fact rotationally invariant. This illustrates that the conclusions regarding the issue of symmetry of the ground state of an individual quantum system can be very misleading if one relies on a standard statistical mechanical treatment. In particular it is worth noting that the treatment of individual systems composed of large number of spins can in fact be rather complex and cumbersome, as illustrated by works based on Majorana stars[10] or quantum constellations[11].

Analogously, for a system with N spins, the density matrix describing the ferromagnet (with $B = 0$) at low temperature, is spherically symmetric. However each of the possible states corresponding to an element in the ensemble, the one identified by observation, breaks the symmetry, and this corresponds to the selection of one of the degenerate vacua. This associated with the possibility, exemplified with the case $N = 1$, that the appropriate fully symmetric superposition of states, may not exist.

Moreover, when considering, in practice, the occurrence of "SSB" in this case, one in fact is looking for much more than in previous cases: Not only the state should not be symmetric, but it should be macroscopically non-symmetric. That is, somehow small domains of non-vanishing magnetization that could in principle lead to zero spatially averaged magnetization, should instead become correlated leading to a macroscopic magnetization of the individual sample. The fact that this is what one finds in the actual dealings with ferromagnets seems to connected with the EXTREME susceptibility to external fields which we saw in the evaluation of the magnetization.

Finally we should be mindful of the potential for confusions that arise due to the fact that a density matrix can be used to represent various things: i) The state of an ensemble, in which each of the elements is in a pure state, ii) A sub system which has no individual state (because it is part of a larger system) in which there are correlations between our subsystem and the rest. iii) We could have an ensemble of subsystems, each of which is as in ii). The reader is invited to see the discussion concerning *proper* and *improper* mixtures in [12], for further aspects of the importance of these distinction. The point is that our interest is establishing when an individual system's state is symmetric and the treatments involving ensembles tend to be obscure as this issue is concerned simply because, from the onset, one is dealing with a very large collection (generally infinite) of systems.

8. Discussion

As we have seen the treatments of the general subject of SSB and related topics generate a great deal of confusion, even at the text-book level. We touched on various representative aspects of those. In particular we showed that at the level of quantum filed theory, the most popular views are mistaken: In the case of rigid (non gauge) symmetries we saw that for the realistic situations involving finite spatial regions there is no SSB, while the considerations of infinite spatial extent required distinguishing the limit $L \to \infty$ (where again there is no SSB) from the physically irrelevant and mathematically problematic case $L = \infty$.

We showed that a treatment devoted to the case where L is allowed to be extended to the $L \to \infty$ limit, successfully accounts for the standard phenomenological aspects such as the emergence of Nambu-Goldstone bosons and the Higgs mechanism, while presenting a picture where the vacuum state is fully symmetric.

We briefly dealt with the case of gauge symmetries where again we hope to have clarified the situation and concluded there is simply no room for a vacuum state that breaks the symmetry. We ended considering the quantum statistical mechanical treatment case of ferromagnets, as one of the few examples where we truly face something we might call SSB.

Hopefully this manuscript will contribute to some extent in clarifying the various issues that tend to obscure many treatments of this very important topic in theoretical physics. For a more detailed discussion of all these issues we suggest consulting the work [6] and the references therein.

Acknowledgments

This work was supported in part by DGAPA-UNAM project IG100316 and by CONACyT project 101712.

References

[1] Zee A 2003 *Quantum Field Theory in a Nutshell* (Princeton University Press)
[2] Srednicki M 2007 *Quantum Field Theory* (Cambridge University Press)
[3] Weinberg S 1996 *The Quantum Theory of Fields Vol II ; Modern applications* (Cambridge University Press)
[4] Adler A 1964 *Phys. Rev.* B **135** 963
[5] Gribov V N 1978 *Nucl. Phys.* B **139** 1
[6] Perez A & Sudarsky D 2011 *IJMP-A* **26** 1493
[7] Eliutzur S 1975 *Phys. Rev.* D **12** 3978
[8] Strocchi F 2005 *Symmetry Breaking, Lecture Notes in Physics* (Berlin Heidelberg: Springer)
[9] Pathria R K & Beale P D 2011 *Statsitical Mechanics* (Elsevier) p 120
[10] Majorana E 1932 *Nuovo Cimento* **9** 43
[11] Qian N 2012 *Physics* **5** 65
[12] d'Espagnat B 1976 *Conceptual Foundations of Quantum Mechanics* (Addison Wesley)

Scalar fields in Cosmology: dark matter and inflation

L. Arturo Ureña-López
Departamento de Física, DCI, Campus León, Universidad de Guanajuato, 37150, León, Guanajuato, México
E-mail: lurena@ugto.mx

Abstract. Scalar fields have been widely used in cosmology during the last three decades, but it is until now that we have been able to fully understand their role as possible major matter components for the evolution of the Universe. Here we briefly present recent studies on scalar fields that show how their intrinsic properties are translated into observables related to the process of structure formation. In the first case, we consider a massive scalar field as a dark matter component, and present the corresponding mass power spectrum. It is confirmed that one of the distinctive features of the model is the suppression of structure at small cosmological scales. In the second case, we describe a generic method to find inflationary solutions without the need to be specific about the scalar potential. The method shows that single-field models of inflation can be classified in two groups according to their predictions of inflationary observables.

1. Introduction

What we actually call the standard model of Cosmology is a phenomenological description of the physics that seems to be at play to reproduce the observed Universe. The predictions of such model are in strikingly good agreement with a variety of observational constraints, and it is because of this reason that it is the fiducial model used in present and future observational forecasts. The model is also known as the ΛCDM model, for the initials of its main matter components: the cosmological constant Λ and Cold Dark Matter (CDM). The core of the model is based on a spatially-flat expanding Universe whose gravitational dynamics is governed by the equations of motion of General Relativity; in addition, the seeds of structure formation are Gaussian distributed adiabatic fluctuations with an almost scale-invariant spectrum[1].

One major concern in the present literature is the understanding of the formation of cosmological structure, and one principal matter component in this process is the CDM. The latter is as simple as it gets: a pressureless fluid of particles that have no direct, weak at most, interaction with other matter fluids apart from the gravitational one. These two basic properties are enough to provide a complex picture of structure formation through hierarchical assembling that fits data at different scales[2, 3].

Apart from the CDM component, the ΛCDM model also needs to set up the appropriate initial conditions so that the formation of structure can start as required by present observations of the distribution of galaxies. Such initial conditions, which we refer above as the seeds of structure formation, must have been produced by some separate process in the early Universe. It is because of this that the idea of inflation, which refers to an accelerated stage in the very early Universe, remains as one of the cornerstone ideas of modern Cosmology[4, 5, 6].

In this paper we will present the case of a scalar field as a model for dark matter and inflation. This is possible because of the flexible behavior that scalar fields exhibit within different cosmological settings (see for instance[7]). The results that will be presented in Secs. 2 and 3 below are part of recent studies on scalar fields[8, 9], with the aim to extract from them useful features that may help us to understand the subtleties behind the formation of cosmological structure.

2. Scalar field as dark matter

Here we present the equations of motion that must be solved in order to find the general behavior of cosmological scalar fields. For that, we shall consider the same features as that of the ΛCDM model: a spatially-flat Universe populated with standard matter fluids such as: radiation, neutrinos, baryons, and a cosmological constant Λ as the dark energy component. In addition, we consider a scalar field ϕ endowed with a generic potential $V(\phi)$, its Lagrangian density given by $\mathcal{L}_\phi = -(1/2)(\partial\phi)^2 - V(\phi)$.

If we use ρ_I and p_I to denote the energy density and pressure, respectively, for each matter fluid in the model apart from the scalar field, the full equations of motion for the background dynamics are:

$$H^2 = \left(\frac{\dot{a}}{a}\right)^2 = \frac{\kappa^2}{3}\left[\sum_I \rho_I + \frac{1}{2}\dot{\phi}^2 + V(\phi)\right], \quad \dot{H} = -\frac{\kappa^2}{2}\left[\sum_I (\rho_I + p_I) + 2\dot{\phi}^2\right], \tag{1}$$

$$\dot{\rho}_I = -3H(\rho_I + p_I), \quad \ddot{\phi} = -3H\dot{\phi} - \partial_\phi V, \tag{2}$$

where $\kappa^2 = 8\pi G$, a dot denotes derivative with respect to cosmic time t, and $H = \dot{a}/a$ is the Hubble parameter, with a the scale factor of the Universe.

In order to go beyond the background dynamics, we must take another step and calculate the linear perturbations to the background quantities; in this way, we will be able to study the influence of the scalar field on, for instance, the evolution of the CMB anisotropies. For that, we assume the synchronous gauge of metric perturbations with the line element $ds^2 = -dt^2 + a^2(t)(\delta_{ij} + h_{ij})dx^i dx^j$. The scalar field is given by $\phi(x,t) = \phi(t) + \varphi(x,t)$, where $\phi(t)$ is the background (homogeneous) field in Eqs. (1) to (2), and φ is its linear perturbation. Thus, the linearized KG equation for a Fourier mode $\varphi(k,t)$ reads[10, 11, 12, 13]:

$$\ddot{\varphi} = -3H\dot{\varphi} - (k^2/a^2 + \partial_\phi^2 V)\varphi - \frac{1}{2}\dot{\phi}\dot{h}, \tag{3}$$

where h is the trace of scalar metric perturbations (with $\dot{h}$ known as the metric continuity), and k is a comoving wavenumber.

We now apply the formalism above to the particular case of a scalar field as a dark matter component. For that, we need to consider a massive scalar field, and then the potential is:

$$V(\phi) = (1/2)m^2\phi^2, \tag{4}$$

where m is the mass scale of the scalar field. Other potentials have been considered in the literature, but their common feature is that they all can be approximated by the functional form (4), see for instance[14, 15, 16, 17].

The expected behavior at late times of the scalar field, under the influence of the quadratic potential (4), is that of rapid oscillations around the minimum of the potential located at $\phi = 0$. This is well known as a problematic regime which prevents an accurate solution of the scalar field equations of motion. However, as shown with more details in[8], such troublesome regime can be avoided by an appropriate change to polar variables for both the background and linear perturbations. Actually, it was in[8] that an accurate calculation of the full cosmological solutions was presented for the first time.

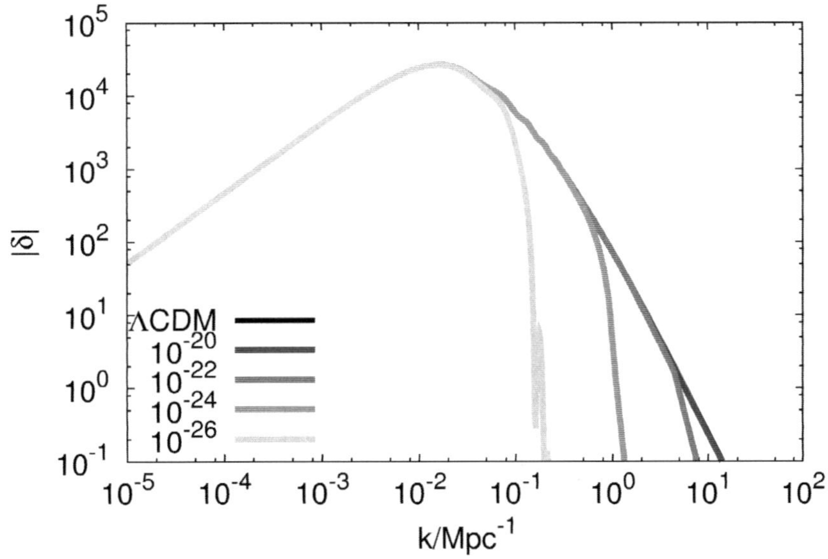

Figure 1. Mass power spectrum of scalar field density contrast calculated from the solution of Eq. (3) for the linear perturbation φ. The different colored lines denote the resultant power spectrum for different values, in units of eV, of the scalar field mass m (see Eq. (4)). We also show the case of the standard cosmological model ΛCDM (black solid line). The most prominent feature of the scalar field MPS is the sudden cut-off of the spectrum for large values of the wavenumber k, which means the suppression of cosmological structure at small scales.

As an example of the distinctive features of a scalar field as dark matter we present in figure 1 the mass power spectrum (MPS) of scalar field linear perturbations. The MPS was calculated with an amended version of the Boltzmann code CLASS[18] (the details of the amendments can be found in[8]). The most salient feature of the scalar field MPS is the sudden cut-off that appears for large values of the wavenumber k (here in units of Mpc^{-1}), which is interpreted as the suppression of cosmological structure at small scales $L < 1\,\mathrm{Mpc}$. The wavenumber at which the suppression starts is mainly determined by the scalar field mass m: the cosmological structure is suppressed at larger scales for smaller values of the scalar field mass, although the effect is only noticeable for masses $m < 10^{-20}\mathrm{eV}$.

3. Scalar fields as inflation

In the case of inflationary models with a single scalar field, the equations of motion can be simplified because of the assumption that the scalar field is the only matter field present in the early Universe. Another simplification is that we do not need to solve the equations of motion for linear perturbations, as there already exist general results for the features of the primordial perturbations that only require input from the dynamics in the background[4, 5, 6, 19].

Thus, the only equations we have to solve are:

$$H^2 = \frac{\kappa^2}{3}\left[\frac{1}{2}\dot{\phi}^2 + V(\phi)\right], \quad \dot{H} = -\kappa^2\dot{\phi}^2, \quad \ddot{\phi} = -3H\dot{\phi} - \partial_\phi V. \tag{5}$$

In order to have a more manageable system of differential equations, we now apply a *polar* change of variables in the form:

$$\frac{\kappa\dot{\phi}}{\sqrt{6}H} = \sin(\theta/2) \,, \quad \frac{\kappa V^{1/2}}{\sqrt{3}H} = \cos(\theta/2) \,, \quad y_1 = -2\sqrt{2}\frac{\partial_\phi V^{1/2}}{H} \,, \quad y_2 = -\frac{4\sqrt{3}}{\kappa}\frac{\partial^2_\phi V^{1/2}}{H} \,. \tag{6}$$

Thus, after straightforward manipulations, Eqs. (5) finally become:

$$\theta' = -3\sin\theta + y_1 \,, \tag{7}$$

$$y_1' = \frac{3}{2}\left(1-\cos\theta\right) y_1 + \sin(\theta/2) y_2 \,, \tag{8}$$

where a prime denotes a derivative with respect to the number of e-folds of expansion $N \equiv \ln(a)$. In principle, we would also require an equation of motion for the second potential variable y_2, but we will show that this is not strictly necessary for the description of the inflationary solutions.

In any given inflationary setting, we expect the scalar field to change slowly from a potential ($V(\phi) \gg \dot{\phi}^2$) to a kinetic dominated phase ($V(\phi) \ll \dot{\phi}^2$). This is also called de Sitter inflation, as the first part of the inflationary state is similar to that produced by a cosmological constant. Such change in the behavior of the scalar field can be parametrised more precisely in terms of the scalar field equation of state (EoS) w_ϕ, which is defined as:

$$w_\phi = \frac{\frac{1}{2}\dot{\phi}^2 - V(\phi)}{\frac{1}{2}\dot{\phi}^2 + V(\phi)} = -\cos\theta \,, \tag{9}$$

where the last expression on the rhs of Eq. (9) appears from the polar change of variables (6).

Hence, in terms of the EoS, the inflationary phase is represented by $w_\phi : -1 \to 1$; actually, it is only required that $w_\phi : -1 \to -1/3$ to span an accelerating phase. The latter can be written in terms of the angular variable as $\theta : 0 \to \theta_{end}$, where $\theta_{end} = \arccos(1/3) = 2\arcsin(1/\sqrt{3}) \simeq 1.2309\ldots$. This suggests that we can try a series solution of Eq. (8) with the following ansatz for the potential variable y_1:

$$y_1 = \sum_{j=1} k_{1j}\theta^j \,, \tag{10}$$

where k_{1j} are all constant coefficients. Once with the values of k_{1j} at hand we can make an expansion of Eq. (7) in the form:

$$\theta' = (k_{11} - 3)\theta + k_{12}\theta^2 + (k_{13} + 1/2)\theta^3 + \ldots \,. \tag{11}$$

Eq. (11) can then be solved at any order to provide a solution: $\theta_N = \theta_N(k_{1j}, \theta_{end}, N)$, where the subscript N denotes the number of e-foldings before the end of inflation.

For the calculation of inflationary quantities, we choose the so-called Hubble slow-roll (HSR) variables[20], under which various inflationary observables can be written in terms of the so-called HSR parameters ϵ_H and η_H. For instance, the spectral index n_S and the tensor-to-scalar ratio r, are given at first order in the HSR parameters as:

$$1 - n_S = 4\epsilon_H - 2\eta_H = 12\sin^2(\theta/2) - 6 + y_1\cot(\theta/2) \,, \tag{12}$$

$$r = 16\epsilon_H = 48\sin^2(\theta/2) \,, \tag{13}$$

where the last expressions on the rhs appear from the change of polar variables (6). Taking into account the expansions (10), Eqs. (13) can be written alternatively as:

$$1 - n_S \simeq 2(k_{11} - 3) + 2k_{12}\theta_N + (2k_{13} - k_{11}/6 + 3)\theta_N^2 \,, \quad r \simeq 12\theta_N^2 \,. \tag{14}$$

Journal of Physics: Conference Series **761** (2016) 012076 doi:10.1088/1742-6596/761/1/012076

In principle, Eqs. (11) and (14) is all what we need to find the values of the inflationary observables n_S and r, once the values of the expansion coefficients k_{1j} have been determined by other means. There is indeed a systematic, though a bit cumbersome, procedure to find the values of k_{1j} for any scalar field potential $V(\phi)$, the details of which can be found in[9].

One remarkable result of the above method is that de Sitter inflation in single-field models can be comprised in two general classes, which we called Class I and Class II, irrespective of the particular form of the scalar field potential $V(\phi)$. The key parameter for the new classification of inflationary models is the coefficient k_{12}. In the large N limit of the two classes we end up with only two typical behaviors in the solutions, namely[9]: $\theta_N \sim N^{-1/2}$ ($k_{12} = 0$: Class I), or $\theta_N \sim N^{-1}$ ($k_{12} \neq 0$: Class II), which in turn suggest that either $r \sim N^{-1}$, or $r \sim N^{-2}$, respectively, even though $1 - n_S \sim N^{-1}$ at the leading order in N for the two classes. Such classification of inflationary solutions was suggested before in[21, 22, 23], but the method in[9] is the first to give a general explanation of its existence.

To finish this section, we show in figures 2 and 3 the generic predictions for the observables n_S and r of the two classes of inflationary solutions, for $N = 60$, which is a typical number or e-folds before the end of inflation. It can be seen that it is Class I, rather than Class II, the one set of models that can provide a good fit to the observational data[9].

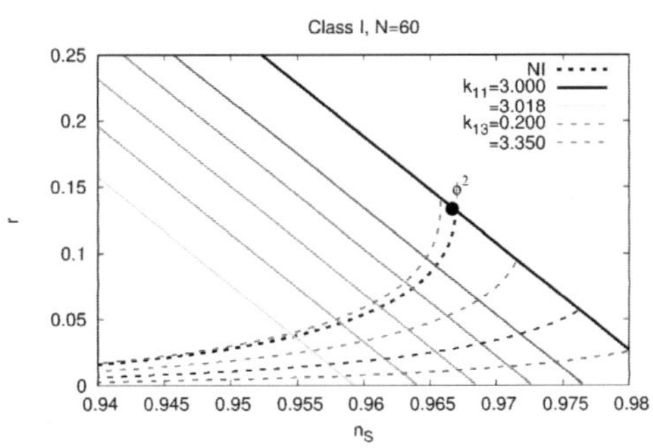

Figure 2. General results on the plane n_S vs r for the Class I of single field models of inflation, for different values of the coefficients k_{11} and k_{13}. The plots show the results for $N = 60$. Notice that Class I is not completely ruled out by observations, mostly because it can provide low enough values of the tensor-to-scalar ratio r. The predictions from the typical Large Field Inflation for the potential ϕ^2[24], and Natural Inflation (NI)[25], are indicated in the figure for comparison.

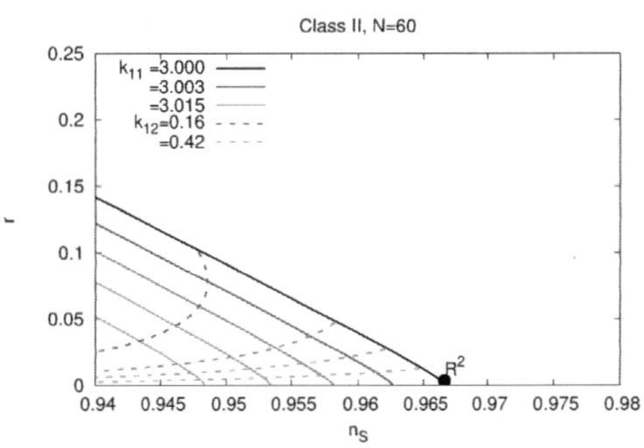

Figure 3. General results on the plane n_S vs r for Class II of single field models of inflation, for different values of the coefficients k_{11} and k_{12}. The plots show the results for $N = 60$. Notice that Class II covers a smaller area than Class I (see Fig. 2), and then it has less freedom to fit the observational data. The predictions from the typical Starobinsky's model (R^2)[26] model is indicated in the figure for comparison.

4. Conclusions and perspectives

We have presented two common instances in which scalar fields are used in Cosmology in order to provide models for unanswered riddles of the Universe.

In the first one, related to dark matter, we showed how the MPS, one of the most important quantities for the process of structure formation, is affected by the presence of a scalar field: the appearance of sudden cut-off at large wavenumbers, which is translated in the suppression of structure at small scales. This cut-off in the MPS has been the subject of study in many recent works about models of dark matter with scalar fields: from constraints coming from CMB anisotropies and other astrophysical probes[27, 28, 29, 14], to the most recent topic of the non-linear process of structure formation[30, 31], and the analysis of objects of galactic size[32, 33, 34].

The second instance was about single-field models of inflation and their generic predictions. Although there is already a large tradition on the study of inflationary models (see[19] and references therein), the method briefly presented in Sec. 3 provides a brand new way to show the intrinsic properties of scalar field models of inflation. Interestingly enough, the method gives a more solid confirmation of the existence of two generic classes of inflationary models. However, the question for the correct model, i.e. the particular $V(\phi)$ that could be the ultimate responsible for inflation, still remains open.

Acknowledgments

This work was partially supported by PRODEP, DAIP-UGTO research grant 732/2016 and 878/2016, PIFI, CONACyT México under grants 167335, 179881 and Fronteras 281, Fundación Marcos Moshinsky, and the Instituto Avanzado de Cosmología (IAC) collaboration.

References

[1] Ade P *et al.* (Planck Collaboration) 2015 (*Preprint* 1502.02114)
[2] Vogelsberger M, Genel S, Springel V, Torrey P, Sijacki D, Xu D, Snyder G F, Nelson D and Hernquist L 2014 *Mon. Not. Roy. Astron. Soc.* **444** 1518–1547 (*Preprint* 1405.2921)
[3] Genel S, Vogelsberger M, Springel V, Sijacki D, Nelson D, Snyder G, Rodriguez-Gomez V, Torrey P and Hernquist L 2014 *Mon. Not. Roy. Astron. Soc.* **445** 175–200 (*Preprint* 1405.3749)
[4] Liddle A R and Lyth D 2000 *Cosmological inflation and large scale structure*
[5] Mukhanov V 2005 *Physical foundations of cosmology*
[6] Bassett B A, Tsujikawa S and Wands D 2006 *Rev. Mod. Phys.* **78** 537–589 (*Preprint* astro-ph/0507632)
[7] Copeland E J, Sami M and Tsujikawa S 2006 *Int. J. Mod. Phys.* **D15** 1753–1936 (*Preprint* hep-th/0603057)
[8] Urea-Lpez L A and Gonzalez-Morales A X 2016 *JCAP* **1607** 048 (*Preprint* 1511.08195)
[9] Urena-Lopez L A 2015 (*Preprint* 1512.07142)
[10] Ratra B 1991 *Phys.Rev.* **D44** 352–364
[11] Ferreira P G and Joyce M 1997 *Phys.Rev.Lett.* **79** 4740–4743 (*Preprint* astro-ph/9707286)
[12] Ferreira P G and Joyce M 1998 *Phys.Rev.* **D58** 023503 (*Preprint* astro-ph/9711102)
[13] Perrotta F and Baccigalupi C 1999 *Phys.Rev.* **D59** 123508 (*Preprint* astro-ph/9811156)
[14] Matos T and Urena-Lopez L A 2001 *Phys.Rev.* **D63** 063506 (*Preprint* astro-ph/0006024)
[15] Sahni V and Wang L M 2000 *Phys.Rev.* **D62** 103517 (*Preprint* astro-ph/9910097)
[16] Beyer J and Wetterich C 2014 (*Preprint* 1407.0141)
[17] Marsh D J E and Silk J 2013 *Mon.Not.Roy.Astron.Soc.* **437** 2652 (*Preprint* 1307.1705)
[18] Lesgourgues J 2011 (*Preprint* 1104.2932)
[19] Vennin V, Koyama K and Wands D 2015 *JCAP* **1511** 008 (*Preprint* 1507.07575)
[20] Liddle A R, Parsons P and Barrow J D 1994 *Phys.Rev.* **D50** 7222–7232 (*Preprint* astro-ph/9408015)
[21] Roest D 2014 *JCAP* **1401** 007 (*Preprint* 1309.1285)
[22] Galante M, Kallosh R, Linde A and Roest D 2014 (*Preprint* 1412.3797)
[23] Mukhanov V 2013 *Eur.Phys.J.* **C73** 2486 (*Preprint* 1303.3925)
[24] Linde A D 1983 *Phys.Lett.* **B129** 177–181
[25] Freese K and Kinney W H 2014 (*Preprint* 1403.5277)
[26] Starobinsky A A 1980 *Phys.Lett.* **B91** 99–102
[27] Hlozek R, Grin D, Marsh D J E and Ferreira P G 2014 (*Preprint* 1410.2896)
[28] Marsh D J and Ferreira P G 2010 *Phys.Rev.* **D82** 103528 (*Preprint* 1009.3501)
[29] Magana J, Matos T, Suarez A and Sanchez-Salcedo F 2012 *JCAP* **1210** 003 (*Preprint* 1204.5255)
[30] Schive H Y, Chiueh T and Broadhurst T 2014 (*Preprint* 1406.6586)
[31] Woo T P and Chiueh T 2009 *Astrophys.J.* **697** 850–861 (*Preprint* 0806.0232)
[32] Du X, Behrens C and Niemeyer J C 2016 (*Preprint* 1608.02575)
[33] Schwabe B, Niemeyer J C and Engels J F 2016 *Phys. Rev.* **D94** 043513 (*Preprint* 1606.05151)
[34] Martinez-Medina L A, Bray H L and Matos T 2015 *JCAP* **1512** 025 (*Preprint* 1505.07154)

Identifying the nature of high energy Astroparticles

Karen Salomé Caballero Mora

Full time Professor, Facultad de Ciencias en Física y Matemáticas, Universidad Autónoma de Chiapas, Carretera Emiliano Zapata Km. 8 Rancho San Francisco, Ciudad Universitaria, Terán, Tuxtla Gutiérrez, Chiapas, Mexico

E-mail: karen.mora@unach.mx

Abstract. High energy Astroparticles include Cosmic Ray (CR), gamma ray and neutrinos, all of them coming from the universe. The origin and production, acceleration and propagation mechanisms of ultrahigh-energy CR (UHECR $\sim 10^{20}$ eV) are still unknown. Knowledge on particle interactions taking place at those energies, useful for studying current theories on particle physics, can be obtained only from measurements of high energy astroparticles. In the present document some techniques on data analysis of mass composition of UHECR with the Pierre Auger Observatory are described. The relevance of the muon component of air showers produced by the primary CR, as well as some low energy simulations of that component, are explained.

1. Introduction

The energy spectrum of CR shows the flux of particles as a function of energy. It extends from a few GeV to above 10^{20} eV. Acceleration mechanism together with propagation processes through the interstellar medium can explain qualitatively the whole energy range but the acceleration sites are still to be explained. The energy spectrum has some features which are easily distinguishable because of its similarity with a human leg, the knee and the ankle. Those features might be interpreted either as a change of the acceleration mechanism at sources, as a propagation effect or as a change of the hadronic interaction cross sections with increasing energy [1]. Figure 1 shows the spectrum measured by different experiments [2]. CR with energies up to some GeV arise predominantly from the Sun. From these energies up to a few 10^{15} eV CR could be originated by shock acceleration in galactic Supernova Remnants (SNRs) [3]. Although there is no experimental evidence, some theories point out that SNRs could accelerate cosmic rays up to energies of 10^{18} eV as well [5, 6]. The maximum energy that CR may obtain from a given astronomical object is limited by the combination of the magnetic field strength and shock size [7, 8]. On this basis, M. Hillas proposed a simple plot (Figure 2) to show that there are only few astrophysical objects (Active Galactic Neclei-AGNs and Gamma Ray Burst-GRBs, Intergalactic Medium-IGM is also shown) capable to accelerate nuclei and protons to energies of 10^{20} eV. There are other non-traditional theories trying to explain the origin of the extreme energetic particles, the so called *top-down* models. Some of those models include production of UHECRs from the decay of exotic particles such as Topological Defects (TD) [10], super heavy dark matter (SHDM) particles [11, 12, 13], or neutrino interactions with the relic neutrino background (Z-burst model) [14, 15]. All the top-down models suggest a large fraction of photons in the flux of UHECRs [16], which can be studied measuring the flux limits for photons and neutrinos. The Pierre Auger Collaboration has estimated upper limits on both of them [17, 18]. The upper limit set to the flux of photons allowed to constrain certain top-down models, which can not be considered as plausible scenarios for the production of UHECRs any more. The GZK

Journal of Physics: Conference Series **761** (2016) 012077 doi:10.1088/1742-6596/761/1/012077

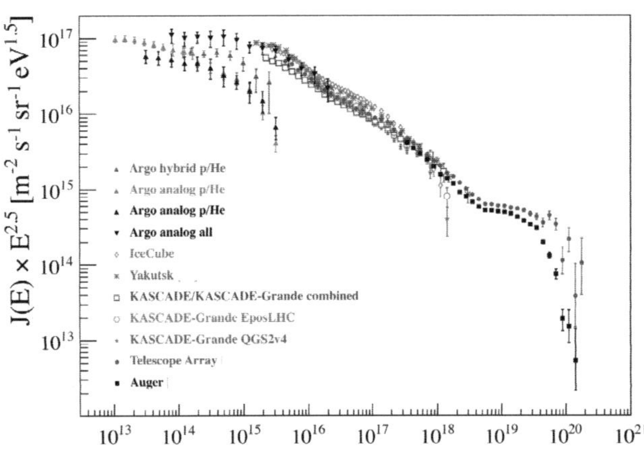

Figure 1. Cosmic ray energy spectrum as measured by many experiments over a wide range in energy, as presented at ICRC 2015 (modified from [2]).

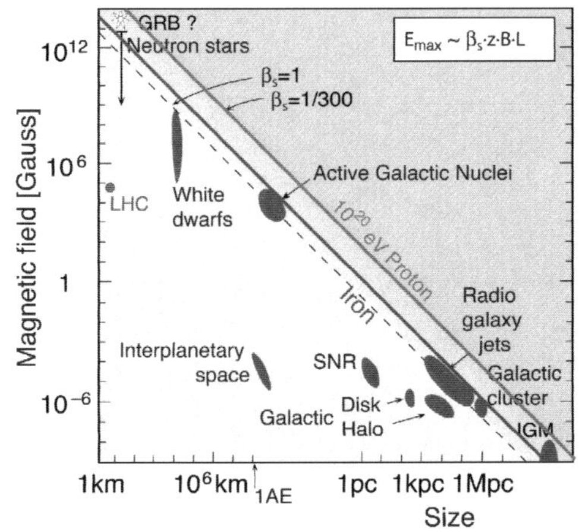

Figure 2. Hillas Plot showing the required size and magnetic field strength of possible UHECR sources. Objects below the corresponding lines are not able to accelerate proton or iron nuclei to an energy of 10^{20}eV [4].

effect [19, 20] also imposes a limit for the distance at which the sources of cosmic rays above energies of $5 \cdot 10^{19}$ eV can be located. Identification of sources of UHECRs by looking directly at their arrival directions is not easy because at relative low energies, the distribution of the arrival directions is isotropic. Nevertheless, a lot of effort has been concentrated to find significant anisotropies. Most of high energy particles in CR are protons; about 10% are helium nuclei, and 1% are neutrons or nuclei of heavier elements. Together, these account for 99% of the CR, and electrons and photons conform the remaining 1%. The number of neutrinos is estimated to be comparable to that of high energy photons, but it is very high at low energy because of the nuclear processes that occur in the Sun [21]. For UHECRs, studies on the depth of shower maximum performed by the Pierre Auger Observatory, have recently shown an unexpected non-trivial behavior in the mass composition above 10^{18} eV, more information on these results is described in this document. Interpreted with the leading LHC-tuned shower models, this behavior implies a gradual shift to a heavier composition. A number of fundamentally different astrophysical model scenarios have been developed to describe such evolution [22].
When CR arrive to the atmosphere, they produce an extensive air shower (EAS) of secondary particles. They reach a maximum and continues after the total energy of the primary cosmic ray is distributed to them [23]. The EAS has the components: *hadronic*, *muonic* and *electromagnetic*. In addition, there are particles with small contribution to the total energy balance, i.e. *UV-photons* and *radio emission*, or which are not detectable called *invisible component* (neutrinos and very low energy particles) [24]. Precision Auger measurements of shower properties, have revealed inconsistencies within present shower models, opening the possibility that the unexpected behavior is due to new hadronic interaction physics at energy scales beyond the reach of the LHC [22]. Gamma photons are photons of very high energy which can travel long distances without being deflected by galactic and extragalactic magnetic fields, hence they allow us to directly study their emission sources [21], neutrinos have the same situation. Gamma rays are involved in phenomena taking place in astrophysical objects such as AGNs and GRBs, as well as Dark Matter annihilation and decay, Lorentz Invariance Violation (LIV) and inter-galactic magnetic field measurement [26]. Such emissions can be meassured

with HAWC experiment at TeV-scale [27]. There is a feedback between CR physics and particle physics trying to understand the mechanisms of their interactions with matter and giving the possibility to develop fundamental theories of physics [28, 29, 30, 22]. UHERC are an alternative field of research to look for new physics beyond the Standard Model of elementary particles, in a kinematic region not accesible by terrestrial accelerators on the basic idea of the modification of effective cross sections at high energies. Examples of such theories are Supersymmetry (SUSY), Grand Unified Theories (GUT's) [31] and Extra Dimensions Models [32].

2. Mass Composition with the Pierre Auger Observatory

The Pierre Auger Observatory, is an international collaboration conformed by ~500 people from 82 institutions in 16 countries [33]. The scientific goal of Auger is to study UHECR (10^{18} to 10^{20} eV). The most important achievements of the experiment are described in [22]

2.1. Description and performance of the Observatory

Auger is a hybrid detector, it covers an area of 3000 km^2. The original components are a surface detector (SD) with a duty cycle of 100% and a fluorescence detector (FD) with a duty cycle of 15%. The **SD** has 1660 3.6 m diameter water Cherenkov tanks. Each station is full with 12,000 l of pure water. The light produced inside the tank is measured by three 9"Photonis photomultipliers tubes (PMTs). The PMTs are set on top of the tank symmetrically distributed, they look downwards into the water. The SD observes the particles of the EAS arriving to the earth, recording the lateral distribution. There are four **FD** stations located on the edges of the SD grid. Each of the stations contain six individual air-fluorescence light telescopes consisting of a spherical mirror with a spherical pixel camera. The camera is equiped with 440 hexagonal Photonis PMTs. The FD observes the longitudinal development of the EAS at the atmosphere. **Extensions** With the goal of studying the transition of galactic to extragalactic CR and to better understand the composition, Auger has included some enhancements including high elevetion fluorescence telescopes (HEAT), the AMIGA underground muon detectors and detectors of radio (AERA) and microwave emissions from air showers [22]. **Auger Prime (Primary cosmic Ray Identification with Muons and Electrons)** It consists of an upgrade of the experiment. Motivations: to get information on mass composition and the origin of the flux suppression at the highest energies, the search of a flux contribution of protons up to the highest energies, to study EAS and hadronic multiparticle production. The upgrades consist of a complementary Surface Scintilator Detector (SSD) above the existing SD stations, substitution of current electronics, the underground muon detector AMIGA and the extension of duty cycle of FD [22].

2.2. Mass Composition sensitive parameters

To know the mass composition of UHECR could give valuable information on the origin and acceleration mechanisms of such particles, as well as set constrains to the models. Interpreting the observed longitudinal shower profiles [34] with LHC-tuned interaction models, the Pierre Auger Observatory can conclude that there is a large fraction of protons present at 10^{18} eV, changing to a heavier composition, possibly dominated by elements of the CNO group, at $10^{19.5}$ eV [34]. In this section one of the first results (with data up to 2009) supporting this finding, based on mean values of X_{max} and Risetime mass composition sensitive parameters is described. **Risetime** ($\boldsymbol{t_{1/2}}$) is the time it takes for the integrated signal recorded by the SD stations, to rise from 10% to 50% of the final value. It is sensitive to the mass composition since its value changes for light initiated EAS against that of heavy initiated ones [8, 9]. Parameters derived from $t_{1/2}$ as its asymmetry [35], Δ_i [36] and Δ_{1000} are also mass sensitive. $\boldsymbol{X_{max}}$ is the most mass-sensitive observable recorded by the FD, it is the depth of the shower maximum, meassured by looking at the longitudinal shower profile [34].

2.3. Method for estimation of mass composition at the highest available energies

Observable Δ_{1000} is calculated for high quality hybrid events and calibrated with X_{max}. With this method one takes advantage of the high statistics provided by the SD to obtain a new estimation for X_{max}, X_{max}^{SD}. The energy range of the Observatory for measuring X_{max} was

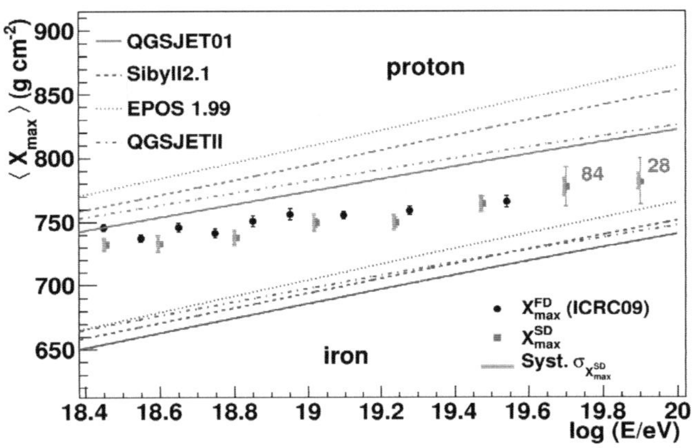

Figure 3. Comparison of the rate of change of mean X_{max}^{SD} with X_{max}^{FD} and with different hadronic interaction models, with energy. Numbers attached to the last two bins show the number of events at the given energy. Data remain in an intermediate composition between light and heavy [1].

Zenith Angle (o)	Energy x 10^{19} (eV)	InD (m)
36.56	2.29	765
44.77	1.37	910
57.57	1.62	1095

Figure 4. Examples of characteristic distances for $t_{1/2}$ for different zenith angles

extended to energies of 80 EeV. The mass composition corresponding to energies above $10^{19.6}$ eV is found to get heavier but still remains in a mixed mass composition, if the evolution of mean X_{max} with the energy is compared with models, as shown in Figure 3 [1] .

Improvement of $t_{1/2}$ parameter Typically $t_{1/2}$ is considered at 1000 m from the shower core per event. Nevertheless this parameter presents a large spread [1]. Since there is no physical reason to take that distance, it is worth to explore if there is other characteristic distance for each different event. Currently a study to find such a distance is being performed. It is found as the intersection distance (InD) of several parameterisations of $t_{1/2}$ as a function of the distance to the shower core for a given event. This is similar to the method used to find the 1000 m distance for the Lateral Distribution Function [37] . It is found, after analysing 15 high quality events, that the interesection changes for different zenith angles, preliminary representative results are shown in Figure 4. More studies are in progress [38].

2.4. Muon Component

The muonic component of air showers is sensitive to hadronic particle interactions at all stages in the EAS, and to many properties of hadronic interactions such as the multiplicity, elasticity, fraction of neutral secondary pions, and the baryon-to-pion ratio. Currently the number of muons can only be measured indirectly except at very large lateral distances and in very inclined showers, where muons dominate the shower signal at ground level [22]. The muon content in EAS also gives information on the composition of the primary UHECR, an iron primary may induce up to about 40% more muons than a proton primary of the same energy [39].

Inclined showers and Muon Component For estimating the energy of inclined showers, parameter N_{19}, representing the shower size is defined. It is based on the muon density model prediction to fit the signals recorded by the SD with respect to a reference profile for the muon density at ground of proton showers of 10^{19} eV simulated with several hadronic models. In order to test the effectiveness of N_{19} as estimator of the total number of muons relative to that in the reference distribution, a simulated N_{19} is compared to the true ratio $R_{\mu}^{MC} = N_{\mu}^{true}/N\mu, 19$, where N_{μ}^{true} is the true number of muons at ground. The mean value of the difference between N_{19}^{MC} and R_{μ}^{MC} is shown in Figure 5. There is a bias of less than 5% for showers with $R_{\mu}^{MC} > 0.6$, value above which the SD array is over 95% efficient. From this result, we conclude that N_{19} provides a direct measurement of the relative number of muons with respect to the reference distribution [40].

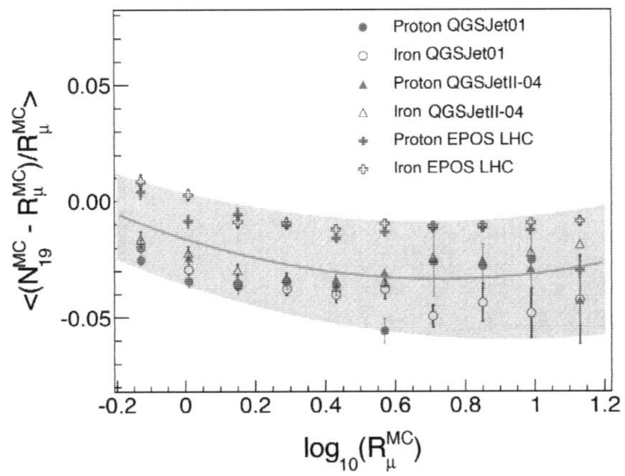

Figure 5. Average of the relative difference between N_{19}^{MC} and R_{μ}^{MC} for proton and iron showers simulated with QGSJet01, QGSJetII-04 and EPOS LHC. The band indicates the bias region and the solid line indicates the parameterised average bias [40].

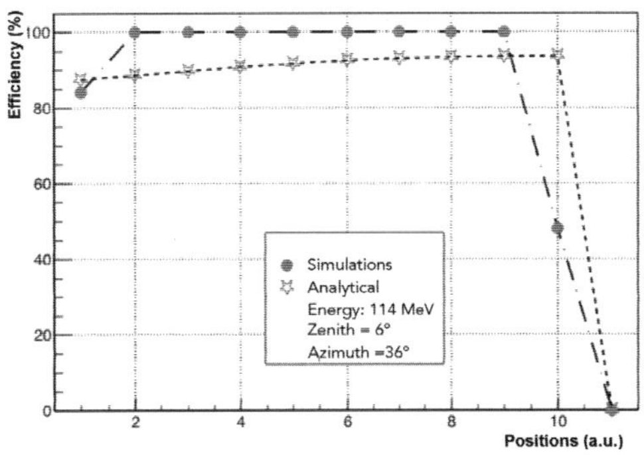

Figure 6. Comparison of simulations with expected results from analytical calculations for certain azimuth and zenith angles as example [41].

2.5. Low Energy muon studies

Burried Array Telescope at Auger (**BATATA**) is a muon telescope built to analyze the muon component of EAS and to quantify the electromagnetic contamination as a function of the depth, for energies above $6 \cdot 10^{15}$ eV. BATATA is conformed by plastic scintillators bars meant to validate the depth of the design of AMIGA counters. BATATA is also an independent muon detector. A simulation code has been developed to reproduce the data of the detector for comparisons with measurements. For testing the code, results obtained with it are compared with the expected results from analytical calculations for 9 positions on the detector (positions 1 to 9 in Figure 6). This study allowed to characterize the use of the simulation code (for particle interactions in the EAS at lower energies than the primary one) for getting the most reliable results. The maximum difference is $\sim 10\%$ [41].

2.6. Exotic Particles UHECRONS

Proposed exotic particle called UHECRON [42] is studied through simulations and compared with measurements of the Pierre Auger Observatory. UHECRON is a strongly interacting, long lived, massive and neutral particle, which do not has to interact with the Cosmic Background Radiation being able to reach the earth. Hybrid events are selected and compared with simulated events, those are mixtures of proton-UHECRON with 0.5, 2.5, 5 and 10 % of UHECRON content. Energies considered are from 10^{18} up to 10^{20} eV. The mixture with around 0.5 % of UHECRONs is selected as the one that best represents the experimental data. A Bayesian approach in order to determine the upper limit for the UHECRON composition fraction that may be present in the experimental data collected by the Observatory for energies above 10^{19} eV is performed. The upper limit for the UHECRON composition fraction determined by this method is 4.4% at 95% C.L. [44]. For more details on this kind of studies see [43, 44].

Acknowledgments

Many thanks to CONACyT Mexico for financial support through projects CB 243290 and CB 132197, and L'OREAL Mexico, CONACyT, UNESCO, CONALMEX and AMC for the *Fellowship for Women in Science 2014*. I want to thank also to all collaborators, who contributed to the work described in this document.

Journal of Physics: Conference Series **761** (2016) 012077 doi:10.1088/1742-6596/761/1/012077

References

[1] Caballero-Mora K.S. 2010, PhD thesis, pp 145 `http://digbib.ubka.uni-karlsruhe.de/volltexte/1000017918`, Karlsruhe, Germany
[2] Verzi V. 2015 *Proc of 34th Int. Cosmic Ray Conf., The Hague, The Netherlands* `PoS ICRC 2015 015`
[3] Aharonian F. A. *et al* 2004, *Nature* **432** pp 75-77
[4] Engel R. 2014 *Highest Energy Cosmic Rays: Auger and TA* talk at 19th ISCRA
[5] Ptuskin V. S. and Zirakashvili V. N. 2005, *A&A* pp 755-765
[6] Hillas A. M. 2005 *J. Phys.* **G31** pp R95-R131
[7] Drury L. O'C 1994 *Contemp. Phys* **35** pp 231-242 "doi10.1080/00107519408222090"
[8] Nagano M. and Watson A. 2000 *Rev. Mod. Phys.* **72** pp 689-732
[9] Barnhill D 2005 *Composition Analysis of Ultrahigh Energy Cosmic Rays Using the Pierre Auger Observatory Surface Detector* PhD thesis, UCLA, USA
[10] Hill C. T. 1983 *Nucl. Phys.* **B224** p 469
[11] Berezinsky V. Kachelriess M and Vilenkin A. 1997 *Phys. Rev. Lett* **79** pp 4302-4305
[12] Aloisio R. Berezinsky V. and Kachelriess M. 2004 *Phys. Rev.* **D69** p 094023
[13] Ellis J. R. Mayes V. E. and Nanopoulos D. V. 2006 *Phys. Rev.* **D74** p 115003
[14] Weiler J. T. 1999 *Astropart. Phys.* **11** pp 303
[15] Fargion D. Mele B. and Salis A. 1999 *Astrophys. J.* **517** pp 725-733
[16] Risse M. and Homola P. 2007 *Mod. Phys. Lett.* **A22** pp. 749-766
[17] Pierre Auger Collaboration, Abraham J. *et al* 2009 *Astropart. Phys.* **31** pp. 399-406
[18] Pierre Auger Collaboration, Aab A. *et al* 2015 *Phys. Rev. D* **91** pp. 092008
[19] Greisen K. 1966 *Phys. Rev. Lett* **16** p 748.
[20] Zatsepin G. T. and Kuz'min V. A. 1966 *Sov. Phys. JETP Lett.* **4** pp 78
[21] De Angelis A. and Pimenta M. J. Martins 2015, *Introduction to Particle and Astroparticle Physics Questions to the Universe* (Mailand:Springer), p XXIII 661.
[22] Pierre Auger CollaborationAab A. et al 2016 *Preliminary Design Report* `arXiv:1604.03637v1`
[23] Gaisser T. K. 1990 *Cosmic Rays and Particle Physics* (Cambridge:Cambridge University Press) p 279
[24] Ulrich R. M. 2008, PhD thesis, (Karslruhe:FZK GmbH) p 166 ISSN 0947-8620
[25] Grupen C. 2005 *Astroparticle Physics* (Berlin Heidelberg:Springer-Verlag) p XIV 442
[26] Lauer R. for the HAWC Collaboration 2016 talk at MACROS 2016
[27] Smith A. J. 2015 *Proc. of the 34th Int. Cosmic Ray Conf., The Hague, The Netherlands*, "arXiv:1508.05826v2http://pos.sissa.it/"
[28] Pierre Auger Collaboration, Abreu P. *et al* 2012 *Phys. Rev. Lett.* **109** p 062002
[29] Pierre Auger Collaboration, Farrar G. R. 2013 1108 *Proc. of 33rd Int. Cosmic Ray Conf., Rio de Janeiro, Brazil* `arXiv:1307.5059`
[30] Pierre Auger Collaboration, Aab A *et al* 2015 *Phys. Rev. D* **91** p 032003
[31] Mohapatra R. N. 2003 *Unification and Supersymmetry*, 3rd Edition (NewYork:Springer-Verlag)
[32] Pérez-Lorenzana 2005 *J. Phys. Conf. Ser.* **18**, 224-269
[33] `https://www.auger.org`
[34] Pierre Auger Collaboration Aab A *et al* 2014 *Phys. Rev. D90* **12** pp 122005 and 122006 `arXiv:1409.4809 and arXiv:1409.5083`
[35] Pierre Auger Collaboration, Abreu P. 2011 *Proc. of 32nd Int. Cosmic Ray Conf., Beijing, China* `arXiv:1107.4804`
[36] Smith B. 2008 *The Mass Composition of Cosmic Rays Above 1 EeV Inferred Using the Spread in Arrival Times of Air Shower Particles* PhD thesis, Leeds, UK
[37] Mariş I. 2005 Poster at Zeuthen Conference.
[38] Castellanos Valdés H. 2015 *Study of risetime as a function of the distance to the Shower Core in the Surface detector (SD) of the Pierre Auger Observatory*, VI School on Cosmic Rays and Astrophysics, Chiapas, Mex.
[39] Caballero-Mora K.S. *et al* 2008 *J. Phys.: Conf. Series* **116** 012007 pp 5
[40] Pierre Auger Collaboration, Valiño I. 2013 *Proc. of 33rd Int. Cosmic Ray Conf., Rio de Janeiro, Brazil* `CR-EX 0635 arXiv:1307.5059`
[41] Martínez Salazar I.A. 2016 *Caracterización de simulaciones de muones para el detector de Centelleo BATATA, ubicado en el Observatorio Pierre Auger* Bachelor thesis UNACH Chiapas, Mexico
[42] Albuquerque I. F. M. and Carvalho W. R. 2009 *Phys. Rev. D.* **80** 023006
[43] Schuster D 2011 *Searching for Exotic Particles at the Pierre Auger Observatory using bayesian inference* PhD thesis, Colorado School of Mines, USA
[44] Sánchez F. E. 2014 *Searching for UHECRONs above* $10^{18.4}eV$ *using Bayesian Inference* Master's thesis, CINVESTAV, Mexico

The Joint Physics Analysis Center: Recent results

César Fernández-Ramírez

Instituto de Ciencias Nucleares, Universidad Nacional Autónoma de México, Ciudad de México 04510, Mexico

E-mail: cesar.fernandez@nucleares.unam.mx

Abstract. We review some of the recent achievements of the Joint Physics Analysis Center, a theoretical collaboration with ties to experimental collaborations, that aims to provide amplitudes suitable for the analysis of the current and forthcoming experimental data on hadron physics. Since its foundation in 2013, the group is focused on hadron spectroscopy in preparation for the forthcoming high statistics and high precision experimental data from BELLEII, BESIII, CLAS12, COMPASS, GlueX, LHCb and (hopefully) PANDA collaborations. So far, we have developed amplitudes for πN scattering, $\bar{K}N$ scattering, pion and J/ψ photoproduction, two kaon photoproduction and three-body decays of light mesons (η, ω, ϕ). The codes for the amplitudes are available to download from the group web page and can be straightforwardly incorporated to the analysis of the experimental data.

1. Introduction

One of the many challenges of physics is to achieve a comprehensive understanding of the phenomena produced by strong coupling Quantum Chromodynamics (QCD), responsible for the binding of quarks inside hadrons and for more than 95% of the mass of the visible Universe. Hadrons are color singlets due to color confinement. Hence, the simplest structures are $q\bar{q}$ (quark-antiquark) for mesons and qqq (3-quarks) for baryons. However, there are many other ways we could build color neutral hadrons, for example: glueballs made out of a number of gluons, multi-quark states such as tetraquarks ($qq\bar{q}\bar{q}$) or pentaquarks ($qqqq\bar{q}$), hybrid mesons and baryons with gluonic components ($q\bar{q}g$ and $qqqg$) and so on. For a recent, brief and pedagogical summary of the current status of hadron physics we refer the reader to [1].

In the last years we have witnessed a dramatic advancement in detection techniques and accelerator technologies from the experimental side and algorithms for first principle QCD analyses on the theoretical side [2]. Consequently, many experimental facilities throughout the world have research programs on hadron spectroscopy, among them: LHCb [3], CMS [4] and COMPASS [5] at CERN (Geneve, Switzerland), CLAS12 [6] and GlueX [7] at Jefferson Lab (Newport News VA, USA), BESIII [8] at the Beijing Electron Positron Collider (Beijing, China), KLOE2 [9] at Laboratori Nazionali di Frascati (Frascati, Italy), BELLEII [10] at KEK (Tsukuba, Japan), the Hadron Experimental Facility [11] at J-PARC (Tokai, Japan), and (hopefully) the future PANDA [12] at FAIR (Darmstadt, Germany). As a result, several candidates for *exotic* hadrons beyond the $q\bar{q}$ and qqq picture have been discovered both in the meson [13] and baryon [14] sectors.

Given the experimental and theoretical interest on hadron spectroscopy, the Joint Physics Analysis Center (JPAC) [15] was set up to develop theoretical and phenomenological analysis

Journal of Physics: Conference Series **761** (2016) 012078 doi:10.1088/1742-6596/761/1/012078

methods to support hadron physics experiments. It started out in 2013 with 8 researchers, as a joint venture between Indiana University, George Washington University and Jefferson Lab. Currently it has expanded to 20 researchers distributed among the three founding institutions plus Rheinischen Friedrich-Wilhelms Universität Bonn and Johannes Gutenberg Universität Mainz in Germany, IFIC/CSIC-Universidad de Valencia in Spain, and Universidad Nacional Autónoma de México in Mexico.

The JPAC researchers are using S-matrix theory principles, –*i.e.* analyticity, crossing symmetry and unitarity [16]– to develop scattering amplitudes for several hadron reactions of theoretical and experimental interest In doing so, we work closely with experimentalists to implement the amplitudes in the existing data-analysis software.

The methods to write the amplitudes include the K-matrix [17] and N/D parametrizations [18]. These can be complemented with chiral symmetry constraints at low-energies [19] or Regge asymptotics [20] if needed. The amplitudes can be analytically continued from the real axis where the experimental data *live* to the complex-s plane and to the unphysical Riemann sheets, which allows us to unravel the existing resonances contributing to the physical reactions. It is important to note that the amplitudes developed by JPAC and the analytical continuation to the complex plane are also necessary to extract the resonances and their properties from first principle lattice gauge QCD [21].

In the next section we briefly summarize some of the results obtained by the group during the last two years.

2. Recent results

2.1. Three-body meson decays

Three-body decays play an essential role studying hadron dynamics and hadron spectroscopy. The extraction of CP violation phases often requires the amplitude analysis of three-body B and D decays [22] and several candidates for non-quark model states have been observed in heavy mesons decaying into three particles [14, 23, 24]. Heavy-meson decays have complicated phase spaces and their analysis can be cumbersome, hence, it is better to test our ideas on amplitude-building with simpler cases, like the decay of light mesons, *i.e.* η, ω, $\phi \to 3\pi$. Besides, these decays are interesting *per se* given the current and forthcoming precision data from various collaborations [25]. For these reasons, at JPAC we have developed an intensive work on dispersive approaches for three-particle final state interaction based on the Khuri-Treiman equations [26] and generalized Veneziano amplitudes [27], —this last work for $J/\psi \to 3\pi$ decay, although it can also be used for light-meson decays. Based on the dispersive approach we have developed amplitudes for η [28, 29], ω [30], and ϕ [30] decays to 3π.

The $\eta \to 3\pi$ decay is of particular interest because it is isospin-breaking and provides insight on the light-quark mass difference

$$\frac{1}{Q^2} = \frac{m_d^2 - m_u^2}{m_s^2 - \hat{m}^2}, \quad \hat{m} = \frac{(m_u + m_d)}{2}, \tag{1}$$

where m_u, m_d and m_s are the masses of u, d and s quarks respectively. In [29] we performed a simultaneous global fit to the KLOE-2 [31] and WASA-at-COSY data [32] with two real parameters and we determined $Q = 21.6 \pm 0.4$ [29] by matching our dispersive approach to nex-to-leading order Chiral Perturbation Theory (χPT) near the Adler zero [33].

2.2. Meson-baryon scattering

Meson-baryon scattering is one of the main resources to study the baryon spectrum. At JPAC, we have focused on πN [34] and $\bar{K}N$ [35] scattering because these are common final states detected in hadron experiments. In [34] we developed high-energy πN amplitudes using Regge physics [20] and connected them to the low-energy amplitudes (resonance region) from [36] using dispersion relations and Finite-Energy Sum Rules (FESRs). Consequently, we built a new set of πN amplitudes that cover the whole energy range for forward scattering.

In [35] we built a coupled-channel $\bar{K}N$ scattering model in the resonance region that incorporates up to 13 channels per partial wave and fulfills unitarity, has the correct analytical properties for the amplitudes in the resonance region and has the right threshold behavior for the partial waves. We also imposed analyticity of partial waves in the complex angular momentum plane [37]. The model was fitted to the Kent State University single-energy partial waves [38] using a genetic algorithm [39] and MINUIT [40]. We estimated the uncertainties in the parameters using bootstrap technique [41]. Once the model was fitted to the single-energy partial waves and the parameters established, we analytically continued the partial waves to the unphysical Riemann sheets and found the Λ^* and Σ^* poles (resonances) that constitute the low-lying hyperon spectrum [35, 42]. We expect to develop forward-scattering $\bar{K}N$ amplitudes [43] that cover the whole energy range combining high-energy $\bar{K}N$ Regge amplitudes with the low-energy amplitudes in [35] using FESR as was done in [34] for πN scattering.

The identification of baryons beyond the constituent quark model (either hybrids with glue as an essential constituent, molecules or pentaquarks) requires the identification of a whole flavor family, and not only the detailed study of one particular state [1]. Resonances of different angular momenta share the same Regge trajectory, what gives the spectrum an organized structure [20, 44]. The structure of the spectrum and the shape of the trajectories provide information on the composition of the states [35, 42, 45, 46, 48]. In [46] we combined Regge phenomenology [20, 37], the information on the hyperon spectrum from [35], and the $\Lambda(1405)$ pole positions from [47] to gain insight on the nature of the two $\Lambda(1405)$ states. We found that the higher-mass pole appears to be (mostly) a three-quark hyperon and the lower-mass pole either a molecule or a pentaquark [46, 48].

2.3. Meson photoproduction

Another front where JPAC is producing results is meson photoproduction. In particular, we have worked on two kaon photoproduction [49], high-energy pion photoproduction [50] and J/ψ photoproduction [51].

In [49] the double-Regge limit for the $\gamma p \to K^+K^-p$ reaction was computed. This reaction has been measured by CLAS collaboration (currently under analysis) and is of interest to study mesons with hidden strangeness that decay into two kaons. The double-Regge production mechanism constitutes a background to the reaction that contributes to the whole Dalitz plot and needs to be removed in order to isolate the mesons of interest.

There are many pion photoproduction models in the literature that describe the resonance region [52] but not recent ones for the high-energy regime. Considering that CLAS and GlueX collaborations will collect pion photoproduction data in the $E_\gamma \simeq 3-10$ GeV region we developed a high-energy neutral pion photoproduction [50] that covers the $E_\gamma > 4$ GeV region where the Regge poles dominate the amplitudes.

LHCb collaboration reported two possible candidates for pentaquarks in the $J/\psi p$ channel [14], however, the existence of these states is still to be confirmed. One of the most interesting proposals to detect the narrow $P_c(4450)$ pentaquark was made by Karliner and Rosner in [53]. They proposed to measure J/ψ photoproduction from the proton. If there is a pentaquark resonance that strongly couples to the $J/\psi p$ channel, it should be possible to detect it in the $\gamma p \to J/\psi p$ reaction. In [51] we built up a full model for this reaction and analyzed the available experimental data on $J/\psi p$ photoproduction from the proton. The resonance was introduced as a Breit–Wigner amplitude and the background was modeled with a Pomeron exchange. We found that data allow for the existence of a pentaquark whose $J/\psi p$ branching ratio has to be below 30% for spin-parity assignment $J^P = 3/2^-$ and below 17% for $J^P = 5/2^+$ at a 95%

confidence level. Jefferson Lab has just approved an experiment to measure this reaction [54].

3. Web page

We created an interactive website to make the exchange of information between JPAC and other theorists and experimentalist easier [55]. The JPAC web page at Indiana University [15] hosts downloadable versions of the codes to compute the $\eta \to 3\pi$ amplitude from [28, 29], the $\omega/\phi \to 3\pi$ amplitudes from [30], the πN amplitudes employing FESRs from [34], the $\bar{K}N$ partial waves and observables in the resonance region from [35] and the high-energy pion photoproduction observables from [50]. The codes can also be run on the web page and the outputs downloaded. Codes to compute the generalized Veneziano amplitude from [27] and the double-Regge limit for the $\gamma p \to K^+K^-p$ reaction using the model in [49] are available upon request. The codes for pentaquark photoproduction [51] will be available in the near future. We are currently in the process of building up a mirror of the JPAC web page with additional functionalities at Universidad Nacional Autónoma de México, hosted at the web servers of Instituto de Ciencias Nucleares [56].

4. Final remarks

The JPAC is a theoretical collaboration oriented to work with experimental collaborations and willing to cooperate with other theory groups. At present we have standing collaborations with BESIII, COMPASS, CLAS, GlueX, KLOE and LHCb. We are focused on hadron spectroscopy delivering amplitudes for three-body meson decays [26, 27, 28, 29, 30], meson-baryon scattering [34, 35, 42, 43], Regge phenomenology to gain insight on the internal structure of hadrons [46, 48] and meson photoproduction [49, 50, 51]. Codes for the amplitudes are available either from the JPAC web page [15, 55, 56] or upon request. We invite the hadron physics community to visit the website and send their comments and suggestions to the JPAC members.

Acknowledgments

We thank the Division of Particles and Fields of the Mexican Physical Society for their kind invitation to write this brief report on JPAC purpose and recent activities.

References

[1] Pennington MR 2016 *J. Phys. G: Nucl. Part. Phys.* **43** 054001
[2] Battaglieri M *et al* 2015 *Acta Phys. Polon.* B **46** 257
Briceño RA *et al* 2016 *Chin. Phys.* C **40** 042001
Dudek JJ, Shepherd MR and Mitchell RE 2016 *Nature* **534** 487
[3] `http://lhcb-public.web.cern.ch/lhcb-public/`
[4] `http://cms.web.cern.ch`
[5] `https://wwwcompass.cern.ch`
[6] `https://www.jlab.org/Hall-B/`
[7] `http://www.gluex.org/GlueX/`
[8] `http://www.ihep.ac.cn/english/E-Bepc/`
[9] `http://www.lnf.infn.it/kloe2/`
[10] `https://www.belle2.org`
[11] `http://j-parc.jp/Hadron/en/`
[12] `http://www.fair-center.eu/public/experiment-program/antiproton-physics/`
[13] Choi S *et al* (Belle collaboration) 2008 *Phys. Rev. Lett.* **100**, 142001
Aaij R *et al* (LHCb collaboration) 2014 *Phys. Rev. Lett.* **112** 222002
[14] Aaij R *et al* (LHCb collaboration) 2015 *Phys. Rev. Lett.* **115** 072001
Aaij R *et al* (LHCb collaboration) 2016 *Preprint* 1604.05708 [hep-ex]
[15] `http://www.indiana.edu/~jpac/`
[16] Eden RJ, Landshoff PV, Olive DI and Polkinghorne JC 2003 *The Analytic S-Matrix* (Cambridge: Cambridge University Press)
[17] Wigner EP 1946 *Phys. Rev.* **70** 15

Badalyan AM, Kok LP, Polikarpov MI and Simonov YA 1982 *Phys. Rep.* **82** 31
[18] Chew GF and Mandelstam S 1960 *Phys. Rev.* **119** 467
[19] Donoghue JF, Golowich E and Holstein BR 1992 *Cambridge Monographs in Particle Physics, Nuclear Physics and Cosmology Vol. 2: Dynamics of the Standard Model* (Cambridge: Cambridge University Press)
Gasser J, Sainio ME and Svarc A 1988 *Nucl. Phys.* B **307** 779
Bernard V, Kaiser N and Meißner U-G 1995 *Int. J. Mod. Phys.* E **4** 193
Gegelia J, Japaridze G and Wang XQ 2003 *J. Phys. G: Nucl. Phys.* **29** 2303
[20] Collins PDB 1977 *An Introduction to Regge Theory and High-Energy Physics* (Cambridge: Cambridge University Press)
[21] Dudek JJ, Edwards RG, Thomas CE and Wilson DJ 2014 *Phys. Rev. Lett.* **113** 182001
Wilson DJ, Briceño RA, Dudek JJ, Edwards RG and Thomas CE 2015 *Phys. Rev.* D **92** 094502
Dudek JJ, Edwards RG and Wilson DJ 2016 *Phys. Rev.* D **93** 094506
[22] Nakahama Y *et al* (Belle collaboration) 2010 *Phys. Rev.* D **82** 073011
Aaij R *et al* (LHCb collaboration) 2013 *Phys. Rev. Lett.* **111** 101801
Aaij R *et al* (LHCb collaboration) 2014 *Phys. Rev. Lett.* **112** 011801
Lees JP *et al* (BaBar collaboration) 2015 *Preprint* 1501.00705
Kränkl S, Mannel T and Virto J 2015 *Nucl. Phys.* B **899** 247
Niecknig F and Kubis B 2015 *JHEP* **1510** 142
Daub JT, Hanhart C and Kubis B 2016 *JHEP* **1602** 009
Nakamura SX 2016 *Phys. Rev.* D **93** 014005
[23] Swanson ES 2006 *Phys. Rep.* **429** 243
Esposito A, Guerrieri AL, Piccinini F, Pilloni A and Polosa AD 2015 *Int. J. Mod. Phys.* A **30** 1530002
Chen H-X, Chen W, Liu X and Zhu S-L 2016 *Phys. Rep.* **639** 1
[24] Ablikim M *et al* (BESIII collaboration) 2013 *Phys. Rev. Lett.* **110** 252001
Aaij *et al* (LHCb collaboration) 2014 *Phys. Rev. Lett.* **22** 222002
Olive KA *et al* (Particle Data Group) 2014 *Chin. Phys.* C **38** 090001
[25] Eugenio P *et al* (CLAS collaboration) 2003 *Search for new forms of hadronic matter in photoproduction*, Jefferson Lab experiment E04-005
Battaglieri M 2010 *Int. J. Mod. Phys.* E **19** 837
Adolph C *et al* (COMPASS collaboration) 2015 *Phys. Lett.* B **740** 303
Ablikim M *et al* (BESIII collaboration) 2015 *Phys. Rev.* D **92** 012014
[26] Guo P, Danilkin IV and Szczepaniak AP 2015 *Eur. Phys. J.* A **51** 135
[27] Szczepaniak AP and Pennington MR 2014 *Phys. Lett.* B **737** 283
[28] Guo P, Danilkin IV, Schott D, Fernández-Ramírez C, Mathieu V and Szczepaniak AP 2015 *Phys. Rev.* D **92** 054016
[29] Guo P, Danilkin IV, Fernández-Ramírez C, Mathieu V and Szczepaniak AP 2016 *Preprint* 1608.01447 [hep-ph]
[30] Danilkin IV, Fernández-Ramírez C, Guo P, Mathieu V, Shi M, Schott D and Szczepaniak AP 2015 *Phys. Rev.* D **91** 094029
[31] Adlarson P *et al* (WASA-at-COSY collaboration) 2014 *Phys. Rev.* C **90** 045207
[32] Anastasi A *et al* (KLOE2 collaboration) 2016 *JHEP* **1605** 019
[33] Colangelo G, Lanz S and Passemar E 2009 *PoS* **CD09** 047
Colangelo G, Lanz S, Leutwyler H and Passemar E 2011 *PoS* **EPS-HEP2011** 304
[34] Mathieu V, Danilkin IV, Fernández-Ramírez C, Pennington MR, Schott D, Szczepaniak AP and Fox G 2015 *Phys. Rev.* D **92** 074004
[35] Fernández-Ramírez C, Danilkin IV, Manley DM, Mathieu V and Szczepaniak AP 2016 *Phys. Rev.* D **93** 034029
[36] Workman RL, Arndt RA, Briscoe WJ, Paris MW and Strakovsky II 2012 *Phys. Rev.* C **86** 035202
[37] Gribov VN 2003 *The Theory of Complex Angular Momenta* (Cambridge: Cambridge University Press)
Gribov VN, Dokshitzer YL and Nyiri J 2009 *Strong Interactions of Hadrons at High Energies* (Cambridge: Cambridge University Press)
[38] Zhang H, Tulpan J, Shrestha M and Manley DM 2013 *Phys. Rev.* C **88**, 035204
[39] Fernández-Ramírez C, Moya de Guerra E, Udías A and Udías JM 2008 *Phys. Rev.* C **77** 065212
[40] James F and Roos M 1975 *Comput. Phys. Commun.* **10** 343
[41] Press WH, Teukolsky SA, Vetterling WT and Flannery BP 1992 *Numerical Recipes: The Art of Scientific Computing* (Cambridge: Cambridge University Press)
[42] Fernández-Ramírez C and Szczepaniak AP 2016 in *Hyperon Studies at JPAC —Mini-Proceedings of the Workshop on Physics with Neutral Kaon Beam at JLab (KL2016)*, ed. M. Amaryan *et al* *Preprint* 1604.02141 [hep-ph], p. 95

[43] Mathieu V 2016 in *Opportunities in the Hyperon Spectrum with Neutral Kaon Beam —Mini-Proceedings of the Workshop on Physics with Neutral Kaon Beam at JLab (KL2016)*, ed. M. Amaryan *et al Preprint* 1604.02141 [hep-ph], p. 152
[44] Chew GF and Frautschi SC 1962 *Phys. Rev. Lett.* **8** 41
[45] Londergan JT, Nebreda J, Peláez JR and Szczepaniak AP 2014 *Phys. Lett.* B **729** 9
Carrasco JA, Nebreda J, Peláez JR and Szczepaniak AP 2015 *Phys. Lett.* B **749** 399
Peláez JR and Rodas A 2016 *Preprint* 1607.03750 [hep-ph]
[46] Fernández-Ramírez C, Danilkin IV, Mathieu V and Szczepaniak AP 2016 *Phys. Rev.* D **93** 074015
[47] Roca L and Oset E 2013 *Phys. Rev.* C **87** 055201
Mai M and Meißner U-G 2015 *Eur. Phys. J.* A **51** 30
[48] Fernández-Ramírez C 2016 *J. Phys.: Conf. Ser.* **730** 012010
[49] Shi M, Danilkin IV, Fernández-Ramírez C, Mathieu V, Pennington MR, Schott D and Szczepaniak AP 2015 *Phys. Rev.* D **91** 034007
[50] Mathieu V, Fox G and Szczepaniak AP 2015 *Phys. Rev.* D **92** 074013
[51] Hiller Blin AN, Fernández-Ramírez C, Jackura A, Mathieu V, Mokeev VI, Pilloni A and Szczepaniak AP 2016 *Phys. Rev.* D **94** 034002
[52] Drechsel D, Hanstein O, Kamalov SS and Tiator L 1999 *Nucl. Phys.* A **645** 145
Fernández-Ramírez C, Moya de Guerra E and Udías JM 2006 *Ann. Phys., NY* **321** 1408
Drechsel D, Kamalov SS and Tiator L 2007 *Eur. Phys. J.* A **34** 69
Mariano A 2007 *Phys. Lett.* B **647** 253
Mariano A 2007 *J. Phys. G: Nucl. Phys.* **34** 1627
Juliá-Díaz B, Lee T-SH, Matsuyama A, Sato T and Smith LC 2008 *Phys. Rev.* C **77** 04520
Fernández-Ramírez C, Moya de Guerra E and Udías JM 2008 *Phys. Lett.* B **660** 188
Gasparyan A and Lutz MFM 2010 *Nucl. Phys.* A **848** 126
Workman RL, Briscoe WJ, Paris MW and Strakovsky II 2012 *Phys. Rev.* C **85** 025201
Anisovich AV, Beck R, Klempt E, Nikonov VA, Sarantsev AV and Thoma U 2012 *Eur. Phys. J.* A **48** 88
Kamano H, Nakamura SX, Lee T-SH and Sato T 2013 *Phys. Rev.* C **88** 035209
Rönchen D, Döring M, Huang F, Haberzettl H, Haidenbauer J, Hanhart C, Krewald S, Meißner U-G and Nakayama K 2014 *Eur. Phys. J.* A **50** 101
[53] Karliner M and Rosner JL 2016 *Phys. Lett.* B **752** 329
[54] Mezziani Z-E *et al* (Hall C collaboration) 2016 *A Search for the LHCb Charmed "Pentaquark" using Photoproduction of J/ψ at Threshold in Hall C at Jefferson Lab*, PAC44 approved experimental proposal PR12-16-007 at Jefferson Lab
[55] Mathieu V 2016 *AIP Conf. Proc.* **1735** 070004
[56] `http://nucleares.unam.mx/`

MWPF IOP Publishing
Journal of Physics: Conference Series **761** (2016) 012079 doi:10.1088/1742-6596/761/1/012079

Problems and Progress in Covariant High Spin Description

Mariana Kirchbach and Víctor Miguel Banda Guzmán

Instituto de Fisíca, Universidad Autónoma de San Luis Potosí, Av. Manuel Nava 6, Zona Universitaria, San Luis Potosí, SLP 78290, México

E-mail: `mariana@ifisica.uaslp.mx`, `vmbg@ifisica.uaslp.mx`

Abstract. A universal description of particles with spins $j \geq 1$, transforming in $(j, 0) \oplus (0, j)$, is developed by means of representation specific second order differential wave equations without auxiliary conditions and in covariant bases such as Lorentz tensors for bosons, Lorentz-tensors with Dirac spinor components for fermions, or, within the basis of the more fundamental Weyl-Van-der-Waerden $sl(2, C)$ spinor-tensors. At the root of the method, which is free from the pathologies suffered by the traditional approaches, are projectors constructed from the Casimir invariants of the spin-Lorentz group, and the group of translations in the Minkowski space time.

1. Introduction

High-spin $j \geq 1$ fields, both massive and massless, always have counted to the main topics in field theories. In particle physics such fields describe hadron resonances whose spins vary from 1/2 to 17/2 for baryons, and from zero to six for mesons [1]. At hadron colliders, intermediate high spin resonances can be produced which subsequently decay in a pair of photons, a process of high interest in the literature [2]. In gravity, high-spin bosons can couple to the metric tensor and cause its deformation [3], and are besides this fundamental to the physics of rotating black holes [4]. Also gravitational interactions between high-spin fermions are of interest [5]. The traditional approaches to high-spin fields have been developed between 1939 and 1964 (see [6] for a recent review, and [7] for a standard textbook) by Fierz and Pauli (FP) [8], Rarita and Schwinger [9], Bargmann and Wigner (BW) [10], Joos [11] and Weinberg [12]. In the next section we briefly highlight these four methods, comment on their problems, and report in section 3 on a new approach, suggested in [13]-[15], before closing with a brief summary.

2. The traditional methods for high-spin description and their problems

The methods for high-spin description are distinct through the type of the representation spaces (reps) used to embed the spin of interest. Below we list various possibilities to describe spin-j.

2.1. Multiple-spin valued representation spaces: the totally symmetric Lorentz tensors for bosons, and Lorentz–tensor–Dirac–spinors for fermions. The Fierz-Pauli and Rarita-Schwinger frameworks

Particles of spins, $j > \frac{1}{2}$, have been so far frequently described in terms of $so(1, 3)$ representation spaces of multiple spins and alternating parities, corresponding to the totally symmetric (**Sym**)

Lorentz tensors of rank-j,

$$\mathbf{Sym}\,\Phi_{\mu_1\ldots\mu_j} \simeq \left(\frac{j}{2},\frac{j}{2}\right), \tag{1}$$

for bosons (Fierz-Pauli framework [8]), or, the totally symmetric rank-$(j-1/2)$ Lorentz tensors with Dirac spinor components, ψ,

$$\mathbf{Sym}\,\Psi_{\mu_1\ldots\mu_{j-\frac{1}{2}}} \simeq \left(\frac{j-\frac{1}{2}}{2},\frac{j-\frac{1}{2}}{2}\right)\otimes\left[\left(\frac{1}{2},0\right)\oplus\left(0,\frac{1}{2}\right)\right], \tag{2}$$

for fermions (Rarita-Schwinger framework [9]), respectively. The particles have been associated with the degrees of freedom (d.o.f.) of the highest spins in the spaces under discussion, while the remaining degrees of freedom, corresponding to lower spins, had to be projected out by properly chosen auxiliary conditions for the sake of ensuring the correct number of physical d.o.f. required by spin-j. The wave equations in the Fierz-Pauli approach to high-spin- j bosons read,

$$(\partial^2+m^2)\,\mathbf{Sym}\,\Phi_{\mu_1\ldots\mu_i\ldots\mu_j}=0,\qquad \mathrm{tr}\,\mathbf{Sym}\,\Phi_{\mu_1\ldots\mu_i\ldots\mu_j}=0,\quad \partial^{\mu_i}\mathbf{Sym}\,\Phi_{\mu_1\ldots\mu_i\ldots\mu_j}=0, \tag{3}$$

while the Rarita-Schwinger approach to spin-j fermions is cast as,

$$(i\partial\!\!\!/-m)\mathbf{Sym}\,\psi_{\mu_1\ldots\mu_i\ldots\mu_{j-\frac{1}{2}}}=0,\ \gamma^{\mu_i}\mathbf{Sym}\,\psi_{\mu_1\ldots\mu_i\ldots\mu_{j-\frac{1}{2}}}=0,\qquad \partial^{\mu_i}\mathbf{Sym}\,\psi_{\mu_1\ldots\mu_i\ldots\mu_{j-\frac{1}{2}}}=0. \tag{4}$$

2.2. Single-spin valued $(j,0)\oplus(0,j)$ "bi-vector" representation spaces. The Joos-Weinberg framework

Alternatively, it has been considered that spin-j particles can also be described in terms of single-spin valued representation spaces of the even number of $2(2j+1)$ components,

$$\psi^{(j)}_B \;\simeq\; (j,0)\oplus(0,j),\quad B\in[1,2(2j+1)], \tag{5}$$

a scheme known as the Joos-Weinberg framework [11], [12]. Differently from the RS approach, the $2(2j+1)$-component wave function, $\mathbf{column}\left(\psi^{(j)}_1\ldots\psi^{(j)}_{2j+1},\psi^{(j)}_{(2j+1)+1}\ldots\psi^{(j)}_{2(2j+1)}\right)$, occasionally termed to as "bi-vector", satisfies one sole differential equation, which is however of a high-order,

$$\left(i^{2j}\left[\gamma_{\mu_1\mu_2\ldots\mu_{2j}}\right]_{AB}\partial^{\mu_1}\partial^{\mu_2}\ldots\partial^{\mu_{2j}}-m^{2j}\delta_{AB}\right)\psi^{(j)}_B(x) \;=\; 0. \tag{6}$$

Here, $\left[\gamma_{\mu_1\mu_2\ldots-\mu_{2j}}\right]_{AB}$ are the elements of the generalized Dirac Hermitean matrices of dimensionality $[2(2j+1)]\times[2(2j+1)]$, which transform as Lorentz tensors of rank-$2j$. The complete sets of such matrices have been extensively studied in the literature for the purpose of constructing all the possible field bi-linears needed in the definitions of the generalized currents, both transitional and diagonal [16]. Though well elaborated, the Weinberg-Joos formalism has attracted comparatively less attention than the linear Rarita-Schwinger framework not only because of the difficult to tackle high order of the differential equations and the high dimensionality of the generalized Dirac matrices. In addition, bi-vectors as a rule can not be equipped neither with Lorentz–, nor with Dirac indexes, a reason for which their couplings to spinorial, vectorial etc targets have to be described through uncomfortable rectangular matrices.

2.3. Single-spin valued totally symmetric Weyl–Van–der–Waerden tensor-spinor representation spaces. The Bargmann-Wigner framework

Another option for spin-j description is provided by the totally symmetric product, of $n = 2j$ copies of a Dirac spinor, $\psi \simeq (1/2,0)\oplus(0,1/2)$, where $(1/2,0)$, and $(0,1/2)$ are the right– and left-handed Weyl–Van–der-Waerden two-component spinors corresponding to the two nonequivalent fundamental $sl(2,C)$ representation spaces, also known as "spinor" (ξ), and "co-spinor" ($\dot{\eta}$),

$$\mathbf{Sym}\,\Psi^{(n)}_{b_1\ldots b_j\ldots b_n} = \mathbf{Sym}\,\psi_{b_1}\ldots\psi_{b_n} \simeq \mathbf{Sym}\left[\left(\frac{1}{2},0\right)\oplus\left(0,\frac{1}{2}\right)\right]_1 \otimes\ldots\otimes\left[\left(\frac{1}{2},0\right)\oplus\left(0,\frac{1}{2}\right)\right]_n, \quad (7)$$

where $b_i = 1,2,3,4$, a scheme known as the Bargmann-Wigner method [10]. The Bargmann-Wigner rank-n spinor satisfies also a high-order differential equation which reads,

$$(\gamma_\mu p^\mu - m)^{a_1b_1}\ldots(\gamma^\mu p^\mu - m)^{a_ib_i}\ldots.(\gamma_\mu p^\mu - m)^{a_nb_n}\,\mathbf{Sym}\,\Psi_{b_1\ldots b_i\ldots b_n} = 0. \quad (8)$$

All four schemes are known to be plagued by serious inconsistencies, which are however of distinct kinds. The Fierz-Pauli and the Rarita-Schwinger fields have multiple redundant components which are eliminated by the auxiliary conditions solely at the free-particle level. Upon couplings to external fields, the auxiliary conditions no longer serve their purpose, in consequence of which the wave fronts of the (classical) coupled solutions can propagate acausally (Velo-Zwanziger problem [17]). A further inconsistency has been revealed by Johnson and Sudarshan [18] who showed that the equal time anti-commutators between spin-3/2 fields, minimally coupled to the electromagnetic field, are not positive definite. In addition, the attempts to construct Lagrangians within these methods are becoming increasingly involved due to difficulties of incorporating the auxiliary conditions at the interaction level, a reason for which it is common to pursue alternative, though not fully covariant methods in data evaluations. Specifically in hadron collisions, where the calculations of the cross sections are carried out within the preferred Center of Mass frame, it is common to account for the contributions of intermediate states of high-spin-J to processes as, say, diphoton production, by parametrizing the corresponding polarized decay amplitude by a single constant and factorizing a rotational $SO(3)$ Wigner function, $d^J_{m,\lambda-\lambda'}(\theta)$, that encodes the angular dependence of the amplitude. Specifically in [2], the absence of high-spin Lagrangians has been compensated by suggesting effective Lagrangians that may be used to simulate some particular intermediate resonances of both low and high spins, through Monte Carlo generators.

On the other side, the Joos-Weinberg and the Bargmann-Wigner frameworks refer to a single wave equation without any need for auxiliary conditions, but they are of the order twice the particle's spin. Differential equations of the order higher than 2 are as a rule difficult to tackle because they lead to higher-order field theories which suffer severe inconsistencies such as ghost solutions of the bad type, kinetic terms of wrong signs, states of negative norms, violation of unitarity etc. Furthermore, particles in such theories can propagate non-locally. All the experimentally verified fundamental theories in modern physics are based on Lagrangians of second order in the momenta. Only in effective theories with out integrated fields, high order Lagrangians may appear. One of the reasons behind the problems caused by high order differential equations is the so Ostrogradskian instability [19] which, at the level of, say, classical mechanics, for concreteness, predicts phase spaces of unstable orbits, and needs a special effort to be handled [20]. In the light of the above discussion, it is desirable to search for a method which would allow for a consistent description of any spin,

- by one sole wave equation of maximally second order in the derivatives, and derivable from a Lagrangian,

- within Lorentz-, or, Weyl-Van-der-Waerden tensor-spinor bases.

Such a formalism for spin-j in single-spin– $(j,0)\oplus(0,j)$ reps has been designed in refs. [13],[14], [15], and is briefly highlighted below.

3. Progress in describing any spin-j in $(j,0)\oplus(0,j)$ in Lorentz-tensor–, Lorentz-tensor-Dirac-spinor–, or, Weyl-Van-der-Waerden tensor-spinor bases and wave equations of second order. The spin-Lorentz group projector method

The first goal of the references [13], [14], [15] has been to describe the pure spin $(j,0)\oplus(0,j)$ states, be it through Lorentz–, or through Weyl-Van-der-Waerden spinor-tensors, this for the sake of constructing by simple index contractions vertexes which involve interactions of high-spins with gauge fields, such as the photon, and/or spinorial targets, such as the proton, and thus avoid the introduction of the cumbersome index-matching rectangular matrices, typical for the Joos-Weinberg formalism. In order to illustrate the essentials of the method, we here bring as representative examples the two simplest cases, beginning with the description of the $(1,0)\oplus(0,1)$ field as a totally anti-symmetric Lorentz tensor of second rank, $B_{[\mu,\nu]}$, with the brackets denoting index anti-symmetrization. On these grounds, in a next step, pure spin-3/2 can be described by means of that very totally-antisymmetric second-rank Lorentz tensor but now with Dirac spinor, $\psi\simeq(1/2,0)\oplus(0,1/2)$, components, i.e. by the direct product, $B\otimes\psi$. Indeed, within the latter representation space, a reducible one, one finds the irreducible $(3/2,0)\oplus(0,3/2)$ sector contained as,

$$B\otimes\psi\longrightarrow\left[\left(\frac{3}{2},0\right)\oplus\left(0,\frac{3}{2}\right)\right]\oplus\left[\left(\frac{1}{2},1\right)\oplus\left(1,\frac{1}{2}\right)\right]\oplus\left[\left(\frac{1}{2},0\right)\oplus\left(0,\frac{1}{2}\right)\right]. \tag{9}$$

The latter equation shows that pure spin-3/2 transforming in $(3/2,0)\oplus(0,3/2)$ can be described in terms of a 24 component wave function, denoted by $\left[\Psi^{(3/2,0)\oplus(0,3/2)}_{[\mu,\nu]}\right]_a$, of two anti-symmetric Lorentz indexes, $[\mu,\nu]$, and a Dirac index, "a", taking the values, $a=1,2,3,4$, provided, one would be able to eliminate from them the redundant 16 components belonging to the $so(1,3)$ irreducible sectors, $\left(\frac{1}{2},1\right)\oplus\left(1,\frac{1}{2}\right)$, and $\left(\frac{1}{2},0\right)\oplus\left(0,\frac{1}{2}\right)$. Such an elimination can indeed be realized by the aid of momentum independent projectors constructed from the invariants of the spin-Lorentz group, a result due to [13],[14].

3.1. *The spin-Lorentz group and static covariant projectors on irreducible representation spaces*

The Lorentz group transforming the internal spin degrees of freedom, henceforth termed to as spin-Lorentz group, and denoted by, $\mathcal{L}$, is a subgroup of the complete Lorentz group, which acts besides on the spin- also on the external space-time degrees of freedom. The $\mathcal{L}$ generators, denoted by $S_{\mu\nu}$, are quadratic $d\times d$ constant matrices, where d fixes the finite dimensionality of the internal representation space, and encodes the spin value. For the special case of a pure spin, dimensionality and spin are related as $d=2(2j+1)$, while for representations of multiple spins, relations of the type $d=\sum_i(2j_i+1)$, or, $d=2\sum_i(2j_i+1)$, can hold valid. The algebra of the spin-Lorentz group, termed to as `homogeneous spin-Lorenz group` (HSLG) reads [21],

$$\mathcal{L}:\quad [S_{\mu\nu},S_{\rho\sigma}] = i(g_{\mu\rho}S_{\nu\sigma}-g_{\nu\rho}S_{\mu\sigma}+g_{\mu\sigma}S_{\rho\nu}-g_{\nu\sigma}S_{\rho\mu}). \tag{10}$$

It has two Casimir invariants, here denoted in their turn by, F and G, and defined as,

$$F_{AB}=\frac{1}{4}\left[S^{\mu\nu}\right]_{AD}\left[S_{\mu\nu}\right]_{DB},\qquad G_{AB}=\frac{1}{8}\epsilon_{\mu\nu}{}^{\alpha\beta}\left[S^{\mu\nu}\right]_{AC}\left[S_{\alpha\beta}\right]_{CB},\ A,B,C,D,...=1,...,d\tag{11}$$

The two operators in (11) identify unambiguously any <u>irreducible</u> finite dimensional $\mathcal{L}$ group representation space, here generically denoted by, $\psi_{(j_1,j_2)\oplus(j_2,j_1)} = \phi^R_{(j_1,j_2)} \oplus \phi^L_{(j_2,j_1)}$, where $\phi^R_{(j_1,j_2)}$ and $\phi^L_{(j_2,j_1)}$ are in their turn its left- and right-handed chiral components, according to,

$$\begin{aligned} F\,\psi_{(j_1,j_2)\oplus(j_2,j_1)} &= c_{(j_1,j_2)}\psi_{(j_1,j_2)\oplus(j_2,j_1)},\ c_{(j_1,j_2)} = \frac{1}{2}\left(K(K+2)+M^2\right), &(12)\\ G\,\phi^R_{(j_1,j_2)} &= r_{(j_1,j_2)}\phi^R_{(j_1,j_2)},\quad G\,\phi^L_{(j_2,j_1)} = r_{(j_2j_1)}\phi^L_{(j_2,j_1)}, &(13)\\ r_{(j_1,j_2)} &= -r_{(j_2j_1)} = i(K+1)M,\quad K=j_1+j_2,\quad M=|j_1-j_2|. &(14)\end{aligned}$$

The idea of [13] has been to employ the Casimir invariant F in the construction of a momentum independent (static) projectors on the irreducible sectors of the Lorentz-tensor-spinor in (9) and to explore the consequences. The Lorentz projector, denoted by $\mathcal{P}_F^{(3/2,0)\oplus(0,3/2)}$, that identifies the irreducible $(3/2,0)\oplus(0,3/2)$ representation space in (9), is constructed from F in (12) as,

$$\begin{aligned} \mathcal{P}_F^{(3/2,0)\oplus(0,3/2)} &= \left(\frac{F-c_{(1,1/2)}}{c_{(3/2,0)}-c_{(1,1/2)}}\right)\left(\frac{F-c_{(1/2,0)}}{c_{(3/2,0)}-c_{(1/2,0)}}\right),\\ c_{(1/2,0)} &= \frac{3}{4},\quad c_{(3/2,0)}=\frac{15}{4},\quad c_{(1,1/2)}=\frac{11}{4}. \end{aligned} \tag{15}$$

The equation (15) shows that the operator $\mathcal{P}_F^{(3/2,0)\oplus(0,3/2)}$ has the property to nullify any irreducible representation space which is different from $(3/2,0)\oplus(0,3/2)$. Instead, for $(3/2,0)\oplus(0,3/2)$, it acts as the identity operator, meaning that $\mathcal{P}_F^{(3/2,0)\oplus(0,3/2)}$ is a projector on this very space. In recalling the notation of the spin-3/2 wave function, $\left[\Psi^{(3/2,0)\oplus(0,3/2)}_{[\mu,\nu]}\right]_a$, we find

$$\left[\mathcal{P}_F^{(3/2,0)\oplus(0,3/2)}\right]_{[\mu,\nu]}{}^{[\eta,\rho]}{}_a{}^b\left[\Psi^{(3/2,0)\oplus(0,3/2)}_{[\eta,\rho]}\right]_b = \left[\Psi^{(3/2,0)\oplus(0,3/2)}_{[\mu,\nu]}\right]_a. \tag{16}$$

The simplest way to bring in kinematics is to set $\left[\Psi^{(3/2,0)\oplus(0,3/2)}_{[\mu,\nu]}\right]_a$ on its mass shell,

$$p^2\left[\mathcal{P}_F^{(3/2,0)\oplus(0,3/2)}\right]_{[\mu,\nu]}{}^{[\eta,\rho]}{}_a{}^b\left[\Psi^{(3/2,0)\oplus(0,3/2)}_{[\eta,\rho]}\right]_b = m^2\left[\Psi^{(3/2,0)\oplus(0,3/2)}_{[\mu,\nu]}\right]_a. \tag{17}$$

The symmetry of (17) is the direct product of the group of translations, $\mathcal{T}_{(1+3)}$, in $(1+3)$ space time, with the spin-Lorentz group, i.e. $\mathcal{T}_{(1+3)}\otimes\mathcal{L}$, some times termed to as the `inhomogeneous spin-Lorentz group`. In this way, it becomes possible to describe a particle residing in the pure spin-3/2 representation space, $(3/2,0)\oplus(0,3/2)$, of the $so(1,3)$ algebra of the spin-Lorentz group, by a Lorentz-tensor-Dirac-spinor wave function, and by means of the one sole second-order differential equation in (17). It has been shown in [13] that within this scheme, one reproduces the precise electromagnetic multipole moments obtained within the canonical Joos-Weinberg method, where the field is described by an eight component "bi-vector" [22]. It has been furthermore checked that also the properties of the remaining two sectors, known from elsewhere, are correctly reproduced, as it should be. Moreover, in [13] the proof that the wave equation in (17) is free from the Velo-Zwanziger problem has been delivered by demonstrating that the wave fronts of its (classical) solutions propagate always causally within an electromagnetic environment. The above presentation should leave it clear that the scheme does not restrict to product spaces of the type given in (9) but extends to any product spaces containing the pure spin $(j,0)\oplus(0,j)$ of interest. In particular, it applies to bases constructed as direct products of four-vectors, or, to general Weyl-Van-der-Waerden tensor-spinors, as used in the Bargmann-Wigner framework [15]. Moreover, the scheme also extends to high- spins carried by two-spin valued representation spaces of the type, $(1/2,j-1/2)\oplus(j-1/2,1/2)$, [14], in which case the second order wave equation is not obtained from the mass shell-condition alone but from combining it by another covariant projector, $\mathcal{P}^{(j,m)}_{\mathcal{W}^2} = (-\mathcal{W}^2/m^2 - j(j-1)p^2/m^2)/(2j)$, which projects besides on the mass shell, also on the highest of the two spin degrees of freedom. Here, $\mathcal{W}^2$ is the squared Pauli-Lubanski operator, the second Casimir invariant of the algebra of `inhomogeneous spin-Lorentz group`. Such a projector technique has first been employed by Aurilia and Umezawa in [23] at the free particle level, and independently, almost 40 years later by [24] at the interacting level. In the latter work the second order differential equation following from $\mathcal{P}^{(j,m)}_{\mathcal{W}^2}$ has been extended by the most general terms allowed by relativity and containing ∂_μ, ∂_ν commutators. These terms, identically vanishing at the free particle level, provide upon gauging essential contributions proportional to the electromagnetic field strength tensor, $F_{\mu\nu}$ and guarantee that the resulting wave equation is free from the Velo-Zwanziger problem for a g factor taking the value of $g = 2$. However, the solutions to second order equations, be them to the representation unspecific Klein-Gordon–, and Proca types, or to the representation specific equation in (17), describe arbitrary mixing of spin-j particles of opposite parities. This arbitrariness can be removed by the aid of the second Casimir invariant G in (11), the subject of the next subsection.

3.2. Separating left from right chiral degrees of freedom and identifying the parity states

In parallel to (15), also the G invariant of the spin-Lorentz group algebra can be employed in the construction of projector operators within any $(j_1, j_2) \oplus (j_2, j_1)$ representations space. Such operators, here denoted by $\mathcal{P}^{(j_1,j_2)}_G$ and $\mathcal{P}^{(j_2,j_1)}_G$, respectively, have the property to separate the reps under consideration into left (L)- and right (R)-handed degrees of freedom according to,

$$\mathcal{P}^{(j_1,j_2)}_G = \frac{1}{2}\frac{G + r_{(j_1,j_2)}}{r_{(j_1,j_2)}}, \qquad \mathcal{P}^{(j_1,j_2)}_G = -\frac{1}{2}\frac{G - r_{(j_1,j_2)}}{r_{(j_1,j_2)}}. \tag{18}$$

Therefore, from (18) one observes that the projectors $\mathcal{P}^{(j_1,j_2)}_G$ and $\mathcal{P}^{(j_2,j_2)}_G$ decompose the $2(2j_1 + 1)(2j_2 + 1)$ degrees of freedom residing in $(j_1, j_2) \oplus (j_2, j_1)$ into the two independent right-handed, (j_1, j_2), and left-handed, (j_2, j_1), irreducible sectors, each having half, i.e. $(2j_1 + 1)(2j_2 + 1)$, of the independent degrees of freedom of the initial representation space. Then the two equations,

$$\frac{p^2}{m^2}\mathcal{P}^{(j_1,j_2)}_G \mathcal{P}^{(j_1,j_2)\oplus(j_2,j_1)}_F \phi^R_{(j_1,j_2)} = \phi^R_{(j_1,j_2)}, \quad \frac{p^2}{m^2}\mathcal{P}^{(j_2,j_1)}_G \mathcal{P}^{(j_1,j_2)\oplus(j_2,j_1)}_F \phi^L_{(j_2,j_1)} = \phi^L_{(j_2,j_1)}, \tag{19}$$

uniquely fix the chiral components of the representation space under consideration. The chiral components are defined by the symmetric and anti-symmetric combinations of wave functions of positive and negative parity states, a reason for which the parity states spanning the representation space of interest can be recovered without ambiguities from the chiral states. Specifically for the Dirac spinor, $j_1 = 1/2, j_2 = 0$, where the spin-Lorentz group generators are $1/2\sigma_{\mu\nu}$, the G invariant calculates as, $-3/4i\gamma^5$, thus defining the corresponding projector operators as $\mathcal{P}^{(1/2,0)}_G = (1 - \gamma_5)/2$, and $\mathcal{P}^{(0,1/2)}_G = (1 + \gamma_5)/2$, leading to the respective Dirac's chiral spinors, $(u - v)/2$, and $(u + v)/2$.

4. Summary and conclusions

We suggested a universal recipe for calculating the states of particles with any high spin-j, covariantly transforming according to $(j, 0) \oplus (0, j)$, and in terms of Lorentz-tensor-Dirac-spinors (16), or, Weyl–Van–der-Waerden tensor-spinors [15], thus facilitating vertex constructions by simple contractions of indexes, and avoiding rectangular-matrix insertions. To avoid the problems of the high-order differential equations, we, similarly to the Fierz-Pauli method, approximated the kinematics through the mass-shell condition alone, and kept the reference frame specification of the reps (encoded by the boost) solely at the wave-function level (generated in any frame by construction), while dropping it from the wave equation. Boost incorporation into the wave equations necessarily rises their order. The wave equations in our method are derivable from a Lagrangian. To be specific, within the spin-Lorentz group projector method pure spin-3/2 transforming in the totally symmetric Lorentz tensor of second rank with Dirac spinor components, $\Psi_{\mu,\nu}^{(3/2,0)\oplus(0,3/2)}$, has been described by means of the following Lagrangian [13],

$$\begin{aligned}\mathcal{L}_{\text{free}}^{(3/2,0)\oplus(0,3/2)} &= \left(\partial^\mu [\overline{\Psi}^{(3/2,0)\oplus(0,3/2)}]^A\right)[\Gamma_{\mu\nu}^{(3/2,0)\oplus(0,3/2)}]_{AB}\partial^\nu[\psi^{(3/2,0)\oplus(0,3/2)}]^B \\ &- m^2[\overline{\Psi}^{(3/2,0)\oplus(0,3/2)}]^A[\Psi^{(3/2,0)\oplus(0,3/2)}]_A, \quad A=[\mu\nu], \quad B=[\gamma\delta], \quad (20)\end{aligned}$$

with the Γ-tensor being defined as,

$$\left[\Gamma^{(3/2,0)\oplus(0,3/2)\mu}{}_{\nu}\right]^{[\alpha\beta]}{}_{[\gamma\delta]} = 4\left[\mathcal{P}_F^{(3/2,0)\oplus(0,3/2)}\right]^{[\alpha\beta][\sigma\mu]}\left[\mathcal{P}_F^{(3/2,0)\oplus(0,3/2)}\right]_{[\sigma\nu][\gamma\delta]}$$

$$= \frac{1}{2}\left(\sigma^{\alpha\beta}\sigma^{\sigma\mu}\sigma_{\sigma\nu}\sigma_{\gamma\delta} + \sigma^{\sigma\mu}\sigma^{\alpha\beta}\sigma_{\gamma\delta}\sigma_{\sigma\nu} - 3\sigma^{\alpha\beta}\sigma^{\sigma\mu}\sigma_{\gamma\delta}\sigma_{\sigma\nu} - 3\sigma^{\sigma\mu}\sigma^{\alpha\beta}\sigma_{\sigma\nu}\sigma_{\gamma\delta}\right) + 4\sigma^{\sigma\mu}\sigma^{\alpha\beta}\sigma_{\sigma\nu}\sigma_{\gamma\delta}, \quad (21)$$

where $\sigma_{\mu\nu}$ stands for the standard totally anti-symmetric Dirac tensor of second rank. Along this line, Lagrangians for any spin can be constructed on the cost of increasing the number of the Lorentz indexes. We tested this approach in the evaluation of Compton scattering processes in [13], [14] and found that, compared to the Joos-Weinberg bi-vector basis, it notably speeds up the calculations by allowing employment of the FeynCalc software.

Acknowledgments: We thank the Editors for their kind invitation to contribute to this volume of historical relevance, in honor of the important impact of the DPyC on physics in Mexico.

References

[1] Olive K A *et al. Chinese Physics C* 2015 **38** 090001
[2] Panico G, Vecchi L, and Wulzer A 2016 *JHEP* **1606** 184
[3] Adamek J, Daverio D, Durrer R and Kunz M 2016 *Nature physics* **12** 346
[4] Campoleoni A, Gonzalez H A, Oblak B and Riegler M 2015 E-print Arxiv: 1512.03353 [hep-th]
[5] Henneaux M, Gómez G L and Rahman R 2014 *JHEP* **1401** 087
[6] Rahman R 2012 *PoS* **Modave VIII** 004; https://arxiv.org/abs/1512.03353
[7] Barut A and Raczka R 1980 *Group representation theory and its applications*, Vol. 2 (Moscow, Mir)
[8] Fierz M and Pauli W 1939 *Proc. Roy. Soc. Lond.* A **173** 211
[9] Rarita W and Schwinger J 1941 *Phys. Rev.* **60** 61
[10] Bargmann V and Wigner E D 1948 *Proc. Nat. Acad. Sci.* **34** 211
[11] Joos H 1962 *Fortschritte der Physik* **10** 65
[12] Weinberg S 1964 *Phys. Rev.* B **133** 1318
[13] Delgado Acosta E G, Banda Guzmán V M and Kirchbach M 2015 *Eur. Phys. J.* A **51** 35
[14] Delgado Acosta E G, Banda Guzmán V M and Kirchbach M 2015 *Int. J. Mod. Phys.* E **24**(7) 1550060
[15] Banda Guzmań V M and Kirchbach M 2016 E-Print arXiv:1604.06772[hep-ph]
[16] Sankaranarayanan A 1968 *Il Nuovo Cimento* LVI A **2** 459
[17] Velo G and Zwanziger D 1969 *Phys. Rev.* **188** 2218
[18] Johnson K and Sudarshan E C G 1961 *Ann. Phys. (N.Y.)* **13** 126
[19] Woodard R P 2003 *Phys. Rev.* A **67** 016102
[20] Tai-jun Chen *et al* 2013 *JCAP* **02** 042
[21] Kim Y S and Noz M 1986 *Theory and applications of the Poincaré group* (D. Reidel Publ. Comp., Dordrecht)
[22] Delgado Acosta E G, Kirchbach M, Napsuciale M and Rodriguez S 2012 *Phys. Rev.* D **85**) 116006
[23] Aurilia A and Umezawa H 1967 *Nuovo Cim.* A **51** 14
[24] Napsuciale M, Kirchbach M and Rodriguez S 2006 *Eur. Phys. J.* A **29** 289

From LHC physics to Dirac-Weyl materials

Alfredo Raya

Instituto de Física y Matemáticas, Universidad Michoacana de San Nicolás de Hidalgo. Edificio C-3, Ciudad Universitaria. Francisco J. Mújica s/n, Col. Felícitas del Río. C.P. 58040, Morelia, Michoacán, Mexico

E-mail: raya@ifm.umich.mx

Abstract. The quantum field theoretical description of particle physics under extreme conditions, namely, at finite temperature, density and in the presence of external magnetic fields, can naturally be extended to describe phenomenology in other branches of physics. In this contribution, I review some aspects of particle physics in the realm of condensed matter physics, particularly graphene and other Dirac-Weyl materials carried out in Mexico. I explore several features of the dynamics of fermions in (2+1)-dimensions which are relevant to heavy ion experiments, but that can be tested in table top experiments.

1. Introduction

Particle physics under extreme conditions –finite temperature and density, as well as in the presence of background magnetic fields– offers a natural opportunity to link ideas that have been developed in other fields, particularly condensed matter physics, and vice versa (see, for instance, Ref.[1]). These scenarios are realized either in heavy ion collision experiments in RHIC and ALICE and other natural sources, *e.g.*, compact stars, where several aspects of different phase transitions that hadron matter has experienced since the early instants after the Big-Bang can be tested from a quantum field theoretical point of view. Lattice QCD, Schwinger-Dyson equations and effective field theories are examples of frameworks that have been developed to describe the QCD phase diagram. Inclusion of temperature, density and/or magnetic fields results in a dimensional reduction of the corresponding quantum field theory under study.

On the other side of the bridge, a new era of materials science has emerged since the isolation of graphene membranes back in 2004 [2]. Although, theoretically speaking, graphene has been known since 1947 [3], the one-atom thickness of the membranes brings two-dimensional (2D) crystals to a natural platform for the development of nano-devices. Apart from the potential technological applications of graphene due to its high thermo-electric conductivity, stiffness yet flexibility and almost transparent for visible light, it is the nature of charge carriers in this material that makes it outstanding; at low energies, these quasi-particles behave as ultra-relativistic Dirac fermions. The difference is that these fermions move at a maximum, constant, energy independent Fermi velocity v_F which is some 300 times smaller than c. Therefore, graphene offers an opportunity to test aspects of quantum electrodynamics where relativistic effects are enhanced 300 times. After the isolation of graphene, a number of new materials were found which also exhibit relativistic charge carriers (Dirac-Weyl) [4].

In view of all the above, the study of fermions in (2+1)-dimensions naturally links the studies of particle physics under extreme conditions and condensed matter systems. A group of researchers, postdocs fellows, graduate and undergraduate students from UMSNH, UNAM, UNISON and UCOL have joined efforts to understand several aspects of these systems. In this contribution, I present the main results of our group in the past years to celebrate the XXX anniversary of the Division of Particles and Fields of the Mexican Physical Society.

2. Fermions in (2+1)-dimensions

Planar dynamics offers a number of interesting features, including generalized parity and time-reversal, fractional (anyonic) statistics and more (see, for instance, Ref. [5]). Particularly, for Dirac fermions, the Clifford Algebra

$$\{\gamma^\mu, \gamma^\nu\} = 2g^{\mu\nu} \tag{1}$$

can be realized in terms of 2×2 matrices, the Pauli σ matrices. There are, in fact, two irreducible representations, as can be seen from the product of all matrices

$$\Gamma = \gamma^0\gamma^1\gamma^2 = \pm i I_{2\times2} , \tag{2}$$

where $I_{2\times2}$ is the 2×2 identity matrix. A possible realization of these representations is

$$\gamma^0 = \sigma_z , \qquad \gamma^1 = i\sigma_x , \qquad \gamma^2 = \pm i\sigma_y . \tag{3}$$

Thus, at first glance, a chiral transformation cannot be defined. On top of that, the mass term $m\bar{\psi}\psi$ is odd under parity and time reversal transformations, and thus the spectrum for the massive Dirac equation,

$$(i\gamma^\mu\partial_\mu - m)\psi = 0 , \tag{4}$$

is not parity invariant. Notice, nevertheless, that the fermion propagator $S(p)$ is still of the form

$$iS(p) = \frac{i}{\gamma^\mu p_\mu - m + i\epsilon} , \tag{5}$$

regardless the representation we select to work with.

A fermion Lagrangian which is invariant under discrete transformations can be constructed by considering both irreducible representations simultaneously in the form

$$\mathcal{L} = \bar{\psi}(i\gamma^\mu\partial_\mu - m)\psi + \bar{\phi}(i\gamma^\mu\partial_\mu + m)\phi , \tag{6}$$

where ψ and ϕ are the fermion fields associated to each representation, but the γ-matrices are taken in a single irreducible one. We observe that the difference between the two fermion families can be realized in the sign of its mass term. This Lagrangian allows two chiral-like transformations, as has been discussed in [6]. Moreover the two irreducible representations can be merged into a single 4×4 reducible representation of the Dirac matrices (a detailed discussion can be found in Ref. [7], for instance) and hence we are led to the usual Dirac Lagrangian with the third spatial component set to zero, namely

$$\mathcal{L} = \bar{\psi}(i\gamma^\mu\partial_\mu - m)\psi , \qquad \mu = 0, 1, 2. \tag{7}$$

$\mathcal{L}$ above (7) is invariant under discrete symmetry transformations, describes fermions with two spin orientations and in its massless form, is invariant under the chiral transformation $\psi \to e^{i\alpha\gamma^5}\psi$ where $\gamma^5 = i\gamma^0\gamma^1\gamma^2\gamma^3$ as usual. The fermion propagator is still of the form (5), although the gamma matrices are now the usual 4×4. Furthermore, since γ^3 does not enter in (7), the massless reducible Lagrangian is invariant under two types of chiral transformations, namely

$$\psi \to e^{i\alpha\gamma^5}\psi\,, \qquad \text{and} \qquad \psi \to e^{i\beta\gamma^3}\psi\,, \tag{8}$$

and thus two mass terms can be considered in the Lagrangian,

$$m\bar{\psi}\psi \qquad \text{and} \qquad m_o\bar{\psi}\tau\psi\,, \qquad \tau = \frac{1}{2}[\gamma^3, \gamma^5]\,. \tag{9}$$

The former, which is the usual Dirac mass term, is invariant under discrete transformations, but breaks the chiral symmetries derived from the transformations in Eq. (8), while the latter, referred to in condensed matter physics literature as Haldane mass term [8], preserves both chiral symmetries, but is odd under parity and time-reversal transformations. Thus, the general form of the massive Dirac Lagrangian becomes

$$\mathcal{L} = \bar{\psi}(i\gamma^\mu\partial_\mu - m - m_o\tau)\psi\,. \tag{10}$$

Neither m nor m_o can be associated with poles of the corresponding propagator. By introducing the chiral projectors $\chi_\pm = (1 \pm \tau)/2$, along with the left- ad right-fields $\psi_\pm = \chi_\pm\psi$, the above Lagrangian (10) can be cast in a more familiar form

$$\mathcal{L} = \bar{\psi}_+(i\gamma^\mu\partial_\mu - m_+)\psi_+ \; + \; \bar{\psi}_-(i\gamma^\mu\partial_\mu - m_-)\psi_-\,, \tag{11}$$

where $m\pm = m \pm m_o$. This Lagrangian (11) describes two fermion species with well defined poles at $p^2 = m_\pm^2$, which are non-degenerate in mass. The corresponding (Euclidean) propagator is

$$\begin{aligned} S(p) &= -S_+(p)\chi_+ - S_-(p)\chi_- \\ &= -\frac{\gamma^\mu p_\mu + m_+}{p^2 + m_+^2}\chi_+ - \frac{\gamma^\mu p_\mu + m_-}{p^2 + m_-^2}\chi_-\,. \end{aligned} \tag{12}$$

We describe the electromagnetic interactions from this Lagrangian in different contexts below.

3. Quantum Electrodynamics

Quantum electrodynamics in (2+1)-dimensions (QED$_3$) is an interesting theory in several aspects. On one hand, it is a toy model for QCD under extreme conditions in the following sense: Any given quantum field theory at finite temperature is dimensionally reduced as the temperature $T \to \infty$. In the particular case of QCD, it is known that the non-Abelian corrections are suppressed as the inverse of the number of fermion families involved. Therefore, QED$_3$ can be considered as an effective description of QCD in these circumstances. The parallelism goes further by recalling that this theory exhibits dynamical chiral symmetry breaking and confinement. Thus, QED$_3$ resembles more to QCD than to ordinary QED.

Within the context of Schwinger-Dyson equations (SDE), QED$_3$ has been widely explored in Mexico [9, 10, 11, 12, 13, 14, 15, 16]. Early efforts were made to explore the possible gauge dependence of physical observables associated to the non-perturbative truncation of SDEs in quenched [9, 10, 11, 12] and unquenched [13, 14, 15] QED$_3$. The mathematical structure of the gap equation was deeply studied as well [16], establishing the criteria for which this equation could support multiple solutions. In these studies, the role of gauge invariance constraints like the Ward-Takahashi identities (WTI) and the Landau-Khalatnikov-Fradkin transformation (LKFT) of the fermion propagator played a crucial role. Proposals for the fermion-boson vertex satisfying these constraints were presented in [15, 17].

Special mention deserves the Mexican contribution to studies of LKFT for the fermion propagator and its implications to the perturbative and non-perturbative form of this two-point function. LKFT relates the fermion propagator in any covariant gauge, labeled by ξ, to the corresponding propagator in Landau gauge, $\xi = 0$, through the relation

$$S(x;\xi) = S(x;0)e^{-\alpha\xi x/2} , \tag{13}$$

where $\alpha = e^2/(4\pi)$ as usual. Perturbative constraints arising from LKFT (13) were explored in parity preserving [18] and parity non-preserving [19] QED_3. Because the transformation (13) is written in coordinate space, its usefulness in non-perturbative studies was not previously considered. After exploring the asymptotic form of the solution to the SDE in the light of LKFT [10], a significant contribution from Mexican participants was to establish the form of the said transformations in momentum space. Writing the propagator in the form

$$S(p;\xi) = \sigma_v(p;\xi)\gamma^\mu p_\mu + \sigma_s(p;\xi) , \tag{14}$$

the transformation (13) in momentum space is equivalent to the relations [11]

$$\begin{aligned} \sigma_v(p;\xi) &= \frac{\alpha\xi}{2\pi p^2}\int_0^\infty dk\; k^2\sigma_v(k;0)\left[\frac{1}{\lambda^-}+\frac{1}{\lambda^+}+\frac{1}{2kp}\ln\left|\frac{\lambda^-}{\lambda^+}\right|\right] , \\ \sigma_s(p;\xi) &= \frac{\alpha\xi}{2\pi p}\int_0^\infty dk\; k\sigma_s(k;0)\left[\frac{1}{\lambda^-}-\frac{1}{\lambda^+}\right] , \end{aligned} \tag{15}$$

with $\lambda^\pm = (\alpha\xi/2)^2 + (k \pm p)^2$. Out of these relations, it is possible to establish the gauge invariance of physical observables associated to a non-perturbative fermion propagator, like the chiral condensate [11] $-\langle\bar\psi\psi\rangle = Tr[S(x=0;\xi)]$, which is the order parameter for chiral symmetry breaking, the critical number of fermion families above which chiral symmetry is restored [13], the parameters associated to confinement [11, 15], and the position of the nodes of multiple solutions to the SDE [16]. To conclude this section, the one-loop corrections to the vacuum polarization tensor, fermion propagator and fermion-boson vertex were considered in [20] including the influence of a Chern-Simons term.

4. Fermions in External Magnetic Fields

The dynamics of fermions in a background magnetic field is also characterized by a dimensional reduction of the dynamics. Therefore, planar fermions are also participants in such scenarios found, for instance, during peripheral heavy-ion collisions. Solutions to the planar Dirac equation in a magnetic field perpendicularly aligned to the plane of motion of fermions were discussed in Ref. [21] in connection with the magnetic catalysis for the formation of the chiral condensate. Parity preserving and parity breaking condensates were studied. The corresponding fermion propagator was diagonalized in momentum space through the Ritus method [22] which was extended also to the case where the background magnetic field was not uniform, but varies with an exponential decay along one spatial direction [22, 23]. In this context, the condensate and electric current were found to follow the profile of the external field [23].

The combined effect of a magnetic field and a thermal bath on the dynamics of fermions induces interesting phase transitions. Whilst the magnetic field promotes the formation of the chiral condensate, a thermal bath at temperature T tries to melt the condensate. The impact of these two ingredients on parity preserving and parity breaking condensates for weak, intermediate and strong magnetic fields was explored in [24]. Magnetization and pair production rate through Schwinger mechanism were also explored in this scenario.

More recently, in an attempt to mimic the dynamics of quarks under extreme conditions, particularly under the influence of a magnetic field, a connection between the chiral magnetic effect (CME) in a condensed matter environment was established in Ref. [25]. The pseudo chiral magnetic effect is realized as the generation of a non-dissipative current by effects of a magnetic field *parallel* to the plane of motion of planar fermions in a medium characterized by a finite density in a thermal bath at temperature T. The effect of the external fields is to provide an imbalance between left and right reducible fermions with Dirac and Haldane mass gaps, resembling the imbalance of chiralities in the CME. Under these circumstances, it was established that an electric current is generated along the magnetic field which is proportional to the chiral number, in complete analogy to the CME.

5. Graphene

Low-energy dynamics of graphene can be effectively described by a version of massless QED$_3$ with photons living in 4 space-time dimensions. Inspired by brane-world scenarios, such a dynamics can be modeled by the *Reduced QED* action [26, 27]

$$I_{d_\gamma,d_e}[A_{\mu_\gamma},\psi_{(d_e)}] = \int d^{d_\gamma}x\ \mathcal{L}_{d_\gamma,d_e}, \tag{16}$$

where the Lagrangian

$$\mathcal{L}_{d_\gamma,d_e} = \bar{\psi}(x)i\gamma^{\mu_e}D_{\mu_e}\psi(x)\delta^{(d_\gamma-d_e)} - \frac{1}{4}F^{\mu_\gamma\nu_\gamma}_{\mu_\gamma\nu_\gamma} - \frac{1}{2\xi}(\partial_{\mu_\gamma}A^{\mu_\gamma})^2 \tag{17}$$

includes matter fields $\psi(x)$ restricted to a d_e-dimensional brane and gauge fields $A_{\mu_\gamma}(x)$ propagating in d_γ-bulk dimensions, with $d_\gamma > d_e$. Here D_μ represents the covariant derivative and $F_{\mu\nu}$ the field strength tensor. Gauge covariance of the associated fermion propagator was studied in Ref. [28] through LKFT.

The problem of light absorption in this material has been addressed by our group. From the action in Eq. (16), we can describe the propagation of electromagnetic waves through a graphene sample according to the modified Maxwell's equations

$$\partial_{\mu_\gamma}F^{\mu_\gamma\nu_\gamma} + \delta(z)\Pi^{\nu_\gamma\rho_\gamma}A_{\rho_\gamma} = 0\ , \tag{18}$$

subject to the conditions

$$\begin{aligned} A_{\mu_\gamma}\Big|_{z=0^+} - A_{\mu_\gamma}\Big|_{z=0^-} &= 0\ , \\ (\partial_z A_{\mu_\gamma})\Big|_{z=0^+} - (\partial_z A_{\mu_\gamma})\Big|_{z=0^-} &= \Pi^{\nu_\gamma}_{\mu_\gamma}A_{\nu_\gamma}\Big|_{z=0}\ , \end{aligned} \tag{19}$$

where A_{μ_γ} represents the incoming monochromatic plane wave and $\Pi^{\nu_\gamma\rho_\gamma}$ represents the vacuum polarization tensor. Considering that the incident wave is linearly polarized, with frequency ω, the reflected and transmitted waves can be described as

$$A = e^{-i\omega t}\begin{cases} \hat{e}_x e^{ik_z z} + (r_{xx}\hat{e}_x + r_{xy}\hat{e}_y)e^{-ik_z z}, & z<0, \\ (t_{xx}\hat{e}_x + t_{xy}\hat{e}_y)e^{ik_z z}, & z>0, \end{cases} \tag{20}$$

where $\hat{e}_{x,y}$ are the unit vectors along the directions x and y on the membrane. Then, the intensity of transmitted light $\mathcal{I}$ and the angle of polarization rotation θ_F (Faraday effect) can, respectively, be expressed as

$$\mathcal{I} = |t_{xx}|^2\ , \qquad \theta_F = \frac{1}{2}\arg\left(\frac{t_{xx}-it_{xy}}{t_{xx}+it_{xy}}\right)\ . \tag{21}$$

A calculation of $\Pi^{\nu_\gamma\rho_\gamma}$ for massless fermions in a weak magnetic field B perpendicular to the membrane [29] reveals that the 2.3% opacity of the membrane measured [30] gets corrected by a factor $-\alpha\pi(eB)^2/\omega^4$ and $\theta_F = 0$. Furthermore, it was observed that the Haldane mass term in absence of any magnetic field is capable of generating a finite value for $\theta_F = \alpha$ [31].

6. Final remarks

In this contribution we have reviewed several aspects of planar Dirac fermions as a bridge between the physics of LHC and the Dirac-Weyl materials carried out by a group of Mexican researchers, postdocs fellows and students. We would like to mention the independent work of another group regarding the influence of Chern-Simons terms in the dynamics of topological insulators [32] which are relevant for the studies carried out by ours. We hope this effort continues to shed light for fundamental physics in condensed matter environments.

Acknowledgments

The author thanks Saúl Hernández Ortiz and J. Javier Cobos Martínez for valuable discussions and careful reading of the manuscript. He also acknowledges support from CIC-UMSNH and CONACYT under Grants 4.22 and 256494 respectively.

References

[1] Rajagopal K and Wilczek F 2000, in *At the Frontier of Particle Physics / Handbook of QCD*, Shifman M, ed., (World Scientific).
[2] Novoselov K S *et al* 2005, *Nature* **438**, 197;
Zhang Y *et al.* 2005, *Nature* **438**, 201;
Gusynin V P and S.G. Sharapov S G 2005, *Phys. Rev. Lett.* **95**.
[3] Wallace P R 1947, *Phys. Rev.***71**, 622.
[4] Vafek O and Vishwanath A 2014, *Ann. Rev. Cond. Matt. Phys***5** 83.
[5] Khare A 2005, *Fractional Statistics and Quantum Theory*, 2nd. Edition (World Scientific, 2005), ISBN 981-256-160-9.
[6] Anguiano Ma de J and Bashir A 2005, *Few Body Syst.* **37**, 71.
[7] Raya A and Reyes E D 2008, *J. Phys. A***41** 355401;
[8] Haldane F D M 1988, *Phys. Rev. Lett.* **61**, 2015.
[9] Bashir A, Huet A, and Raya A 2002, *Phys. Rev. D***66**, 025029.
[10] Bashir A and Raya A 2005, *Nucl. Phys. B***709**, 307.
[11] Bashir A and Raya A 2007. *Few Body Syst.***41**, 185.
[12] Bashir A, Raya A, Cloet I C, and Roberts C D 2008, *Phys. Rev. C***78**, 055201.
[13] Bashir A, Raya A, Sánchez-Madrigal S, and Roberts C D 2009, *Few Body Syst.* **46** 229.
[14] Hofmann C P, Raya A, and Sánchez-Madrigal S 2010, *Phys. Rev. D***82** 096011;
Sánchez-Madrigal S, Hofmann C P, and Raya A 2011, *J. Phys. Conf. Ser.* **287** 012028.
[15] Bashir A, Raya A, Sánchez-Madrigal S 2011, *Phys. Rev. D***84** 036013.
[16] Raya K, Bashir A, Hernández-Ortiz S, Raya A, and Roberts C D 2013,*Phys. Rev. D***88**, 096003.
[17] Bashir A and Raya A 2001, *Phys. Rev. D***64** 105001.
[18] Bashir A and Raya A 2002, *Phys. Rev. D***66**, 105005.
[19] Bashir A, Raya A, and Sánchez-Madrigal S 2008, *J. Phys. A***41**, 505401.
[20] Concha-Sánchez Y, Raya A, and Tejeda-Yeomans M E 2013, *Phys. Rev. D***87** 035001.
[21] Anguiano Ma de J, Bashir A and Raya A 2007, *Phys. Rev. D***76** 127702.
[22] Murguía G, Raya A, Sanchez A, and Reyes E 2010, *Am. J. Phys.* **78**, 700.
[23] Raya A and Reyes E 2010, *Phys. Rev. D***82** 016004.
[24] Ayala A, Bashir A, Gutiérrez E, Raya A, and Sánchez A 2010, *Phys. Rev. D***82** 056011.
[25] Mizher A J, Raya A, and Villavicencio C 2013, *Int. J. Mod. Phys. B***30**, 1550257;
Mizher A J, Raya A, and Villavicencio C 2016, *Nucl. Part. Phys. Proc.* **270-272** 181.
[26] Gorbar E V, Gusynin V P, and Miransky V A 2001, *Phys.Rev D***64**, 105028
[27] Teber S 2012, *Phys. Rev. D***86**, 025005.
[28] Ahmad A, Cobos-Martńez J J, Concha-Sánchez Y, and Raya A 2016, *Phys. Rev. D***93**, 094035.
[29] Valenzuela D, Hernández-Ortiz S, Loewe M, and Raya A 2015, *J. Phys. A***48**, 065402.
[30] Nair R R *et al.* 2008, *Science* **320**, 1308.
[31] Hernández-Ortiz S, Valenzuela D, Raya A, and Sánchez-Madrigal S 2015, *Int. J. Mod. Phys.* **30**, 1650084.
[32] Martín-Ruiz A, Cambiaso M, and Urrutia L F 2015, *Phys. Rev. D***92**, 125015;
Martín-Ruiz A, Cambiaso M, and Urrutia L F 2016,*Europhys. Lett.* **113**, 60005;
Martín-Ruiz A, Cambiaso M, and Urrutia L F 2016, *Phys.Rev. D***93**, 045022.

Journal of Physics: Conference Series **761** (2016) 012081 doi:10.1088/1742-6596/761/1/012081

Z-dark search with the ATLAS detector

Elizabeth Castaneda-Miranda
Department of Physics, University of Johannesburg, South Africa
E-mail: elizabeth.castaneda.miranda@cern.ch

Abstract. The search of the"hidden sector" via new neutral light bosons Z-dark (Z_d) could be revealed by the study of the decay of the discovered Higgs-like boson or any other undiscovered Higgs boson. After the LHC concluded a successful first period of running, The ATLAS Collaboration published its latest results on the $H \to Z_dZ_d \to 4l$ analysis using up to 20 fb^{-1} of integrated luminosity at $\sqrt{s} = 8$ TeV. In this proceeding I present a summary of the recent results on the search of the Z_d in the signature $H \to Z_dZ_d \to 4l$ with the ATLAS detector at the LHC.

1. Introduction

The $H \to Z_dZ_d \to 4l$ is an interesting search compatible with one of the main decay channels used in the discovery of the SM Higgs boson, $H \to ZZ^* \to 4l$ [1]. This search is motivated by many extensions to the Standard Model (SM) which provide a candidate for the dark force mediator to explain the astrophysical observations of the positron excess [2], or a candidate mediator of the "hidden or dark sector" [3, 4, 5].

This document presents the results on the Z_d search, for both model independent and dependent cases. The study is based on the SM $H \to ZZ^* \to 4l$ [1, 6], with objective to find a significant excess in the dilepton mass region $15 \leq m_{Z_d \to ll} \leq 60$ GeV, with a Higgs mass fixed $m_H = 125$ GeV. Otherwise, 95% upper bound are set on the main parameters of interest as a function of the m_{Z_d}. The information presented here is reported and published by the ATLAS Collaboration [7].

2. The ATLAS detector

It is a multipurpose detector located at point 1 at the LHC. Generally speaking, the ATLAS detector consists of four major subsystems: Inner Detector (ID), Calorimeter, Muon Spectrometer (MS) and Magnet System. These subsystems are integrated with the following components: Trigger and Data Acquisition System and Computing System [8, 9]. The coordinate system used in ATLAS is the right-handed with origin at interaction point in the center of the detector, the z-axis along the beam line and the x-y plane perpendicular to the beam line. Cylindrical coordinates (r, ϕ) are used in the transverse plane, ϕ is the azimuthal angle around the beam line. Observables labelled "transverse" are projected into the x-y plane. The pseudorapidity is defined in terms of the polar angle θ as $\eta = -\ln \tan \theta/2$.

3. Signal and background simulation samples

The new neutral light boson is produced via the Higgs boson production ($H \to Z_dZ_d \to 4l$) in the gluon fusion (ggF). The benchmark model used is the hidden Abelian Higgs model

Journal of Physics: Conference Series **761** (2016) 012081 doi:10.1088/1742-6596/761/1/012081

Table 1. List of the background processes and their corresponding event generators, PDF's calculators, and their QCD and EW applied corrections. The references of the generators, used for this study, are found in [7].

Process	Generator	PDF	QCD corr.	EW corr.
ggFF, VBF	POWHEG+PYTHIA8	CT10	NLO and NNLO	NLO
WH, ZH	PYTHIA8	CT10	NNLO	NLO
$t\bar{t}$H	PYTHIA8	CTEQ6L1	NLO	—
$q\bar{q} \to$ ZZ*	POWHEG+PYTHIA8	CT10	NLO	—
gg $\to$ ZZ*	gg2ZZ+JIMMY	CT10	NLO	—
Z+jets	ALPGEN	CTEQ6L1	NLO and NNLO	—
$t\bar{t}$	MC@NLO+HERWIG, JIMMY	CT10	NLO	—
WZ and WW	SHERPA	CT10	NLO	—
ZJ/ψ, $Z\Upsilon$	PYTHIA8	CTEQ6L1	—	—

(HAHM) [3, 4, 5]. The samples are generated for $m_H = 125$ GeV and $15 \leq m_{Z_d} \leq 60$ GeV, in steps of 5 GeV, using MadGraph with CTEQ6L1 for the parton distribution functions (PDF). For taking into account the parton showering, hadronization, and initial- and final-state radiation PYTHIA8 and PHOTOS are used. In the $H \to Z_d Z_d \to 4l$ search, the $h \to 4l$ is normalized from data.

The background processes considered for this study are listed in Table 1. For the SM Higgs boson in ggF production, the QCD soft-gluon resummations, calculated in the next-to-next-to-leading-logarithmic (NNLL) approximation, are applied. The Z+jets production, modelled with up to five partons, is divided in two sources: Z+light-jets and $Zb\bar{b}$. For data and simulation comparison, the QCD cross-section calculations (see Table 1) are used to normalize the simulation for inclusive Z boson and $Zb\bar{b}$ production. The ZJ/ψ and $Z\Upsilon$ Monte Carlo (MC) samples are normalized using the ATLAS measurement described in [10]. The SM Higgs productions, SM ZZ^* and diboson processes are normalized with the theoretical cross-sections and decay branching ratios as well as their uncertainties. For taking into account the different conditions in the pileup, as a function of the instantaneous luminosity, the simulated minimun-bias overlayed events, generated with PYTHIA8 onto the hard-scattering process, were used. Such events were reweighted according to the distribution of the mean number of interactions observed in data. All the MC background samples generated are reconstructed using the "full detector simulation" ATLAS software [11] which is based on GEANT4 [12]. The signal samples are reconstructed with the "fast detector simulation" ATLAS software [13] which only parametrized the response of the electromagnetic and hadronic showers in the ATLAS calorimeter, and the rest of the systems are reconstructed as "full detector simulation".

4. Event selection

For this search, 20.3 fb^{-1} of integrated luminosity is used which was recorded in the optimal function of the detector. All the MC processes are normalized with integrated luminosity mentioned. The event selection applied is the same for both data and MC. There are three categories that are studied: 4−electrons ($4e$), 4−muons (4μ) and 2−electrons 2−muons ($2e2\mu$).

It is requested that the (data/MC) events pass the combination of single-lepton and dilepton triggers. Table 2 shows the list of single-lepton and dilepton triggers used, with their respective thresholds in transverse energy/momentum. All these events must contain a reconstructed primary vertex, defined as the vertex with highest $\sum p_T^2$ of the associated tracks, with at least three tracks of $p_T > 0.4$ GeV each one. The muon candidates selected are formed by matching reconstructed ID tracks with either a complete track or track-segment reconstructed in MS [14]. The acceptance is extended using tracks reconstructed in the forward region of the MS

Journal of Physics: Conference Series **761** (2016) 012081 doi:10.1088/1742-6596/761/1/012081

Table 2. Triggers used with their respective E_T or p_T thresholds.

Trigger	E_T [GeV]	p_T [GeV]
single-electron	25	—
single-muon	—	24
dielectron	12	—
dimuon (symm.)	—	13
dimuon (asymm.)	—	18 and 8
electron-muon	12 or 24	8

$(2.5 < |\eta| < 2.7)$, which is outside the ID coverage. If both an ID or a complete MS track are present, the two independent momentum measurements are combined; otherwise the information of the ID or the MS is used alone. In the case of the electrons candidates, they must have well-reconstructed ID track pointing to an electromagnetic calorimeter cluster and the cluster should satisfy a set of identification criteria [15]. Tracks associated with electromagnetic clusters are fitted using a Gaussian-Sum Filter [15], which allows for bremsstrahlung energy losses to be taken into account. Each electron (muon) must satisfy $p_T > 7$ GeV ($p_T > 6$ GeV) and be measured in the range $|\eta| < 2.47$ ($|\eta| < 2.7$). The longitudinal impact parameters of the leptons along the beam axis z_0 are required to be within 10 mm of the reconstructed primary vertex (muons without ID track are exempted of this requirement). To reject the cosmic rays, muon tracks are required to have a transverse impact parameter d_0 (closest approach to the primary vertex in the x-y plane) of less than 1mm. For avoiding double-counting of leptons, an overlap removal is applied. If two electrons share the same ID track or within in a cone $(\Delta R \equiv \sqrt{(\Delta\phi)^2 + (\Delta\eta)^2} < 0.1)$, the one with the highest transverse energy deposit in the calorimeter (E_T) is kept. An electron within in a muon cone ($\Delta R = 0.2$) is removed, and a calorimeter-based reconstructed muon within an electron cone ($\Delta R = 0.2$) is removed.

After selecting the leptons with proper quality, events with at least four leptons, with all possible combinations of four leptons containing two same-flavor and opposite-sign (SFOS) defined the "quadruplets", are kept. Each lepton is ordered by p_T, the three highest-p_T leptons should have: $p_{Tl1} > 20$ GeV, $p_{Tl2} > 15$ GeV and $p_{Tl3} > 10$ GeV. It is also required that at least one (two) of the lepton(s), from the quadruplet, must satisfy the single-lepton (dilepton) trigger requirements. Each dilepton invariant mass or each SFOS pair, from the quadruplet, are denoted by m_{12} (formed by leptons 1 and 2) and m_{34} (formed by leptons 3 and 4). For the $H \to Z_d Z_d \to 4l$ study, both Z_d boson are considered to be on-shell, because of that, there is not any distinction between the SFOS pairs. If in the event there are several "quadruplets", only the quadruplet that satisfies the minimum value of $\Delta m \equiv |m_{12} - m_{34}|$ is selected. Subsequently, the transverse impact parameter significance (IP) and the isolation requirements are imposed to the leptons of the quadruplet selected, and four final requirements that defined the two signal regions (SR) considered in this study are shown in Table 3. In Figs. 1 and 2 show the dilepton and four-lepton invariant masses after the IP requirements, respectively, in the dilepton mass the m_{12} and m_{34} are combined.

5. Background estimation

For the $H \to Z_d Z_d \to 4l$ analysis, the main background contributions are $ZZ^* \to 4l$ and $H \to ZZ^* \to 4l$ which are reduced considerably after applying the tight SR (see Table 3). Also small contribution from Z+jets, $t\bar{t}$, WW and WZ can be seen from Fig. 3. The $H \to ZZ^* \to 4l$, $ZZ^* \to 4l$, WW and WZ background processes are estimated from MC simulation, their normalization is based on theoretical cross-sections calculations, integrated

Journal of Physics: Conference Series **761** (2016) 012081 doi:10.1088/1742-6596/761/1/012081

Table 3. Requirements after selecting the one quadruplet, and the loose and tight SR definitions.

Requirements	
Track isolation	$\frac{\sum p_T^{tracks}}{p_T^{lepton}}(\Delta R = 0.2) < 0.15$
Calorimeter isolation	$\frac{\sum E_T^{clusters}}{E_T^{e}(p_T^{\mu})}(\Delta R = 0.2) <0.2$ (0.3 and μ's no ID 0.15)
IP	$\frac{d_0}{\sigma_0^{IP\ uncer.}} < 6.5$ for e's (3.5 for μ's)
Loose SR	$115 < m_4 < 130$ GeV, $\lvert m_{Z_d} - m_{ll}\rvert < \frac{m_H}{2}$, $m_H = 125$ GeV, and Z, $J/\psi - \Upsilon$ vetoes: $\lvert m_{Z_{PDG}} - m_{ll}\rvert < 10$ GeV, $m_{ll} < 12$ GeV (ll=12,34, and 23, 14)
Tight SR	Loose SR and $\lvert m_{Z_d} - m_{ll}\rvert < \delta m$ (5/3/4.5 GeV for $4e/4\mu/2e2\mu$)

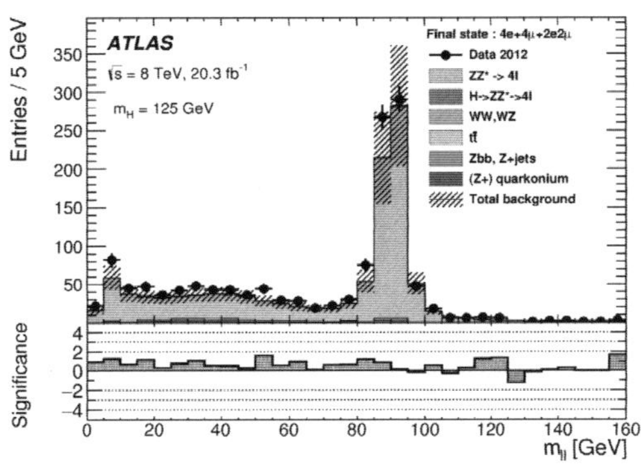

Figure 1. Dilepton invariant mass $m_{ll=12,34}$ for $m_H = 125$ GeV at the IP requirements.

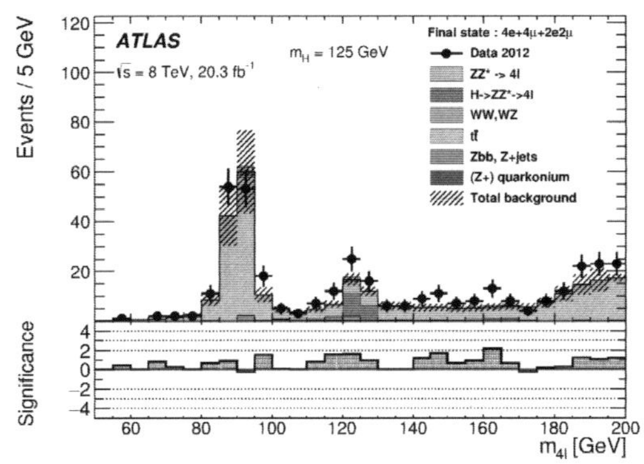

Figure 2. Four-lepton invariant mass for $m_H = 125$ GeV at the IP requirements.

luminosity recorded, and acceptance efficiency. In the case of zero MC background expected events, in the tight SR, 68% C.L. upper bound, which corresponds to 1.14 events [16], is estimated as $N_{background} < L \times \sigma \times \frac{1.14}{N_{total}}$, where L is the integrated luminosity recorded, σ the cross-section of the background processes and N_{tot} is the total number of weighted events simulated for the background processes. This background estimation is validated using a control region defined by reversing, the four-lepton invariant mass requirement ($m_{4l} < 115$ GeV or $m_{4l} > 115$ GeV), good agreement is found between the expectation and observation results [7].

6. Systematic uncertainties

The systematic uncertainties on the theoretical cross-section calculations as well as the detector uncertainties are taken into account. The PDF effects, α_s, renormalization and factorization scale uncertainties are applied to the ZZ^* backgrounds, and for all the SM Higgs productions applied on the total inclusive cross-sections, using values obtained from [17]. Uncertainties due to the limited number of MC events in the $t\bar{t}$, Z+jets, ZJ/ψ, $Z\Upsilon$ and WW/WZ background processes are estimated (see Sec. 5). The luminosity uncertainty [18] is applied to all signal and background yields. The detector systematic uncertainties on the electron and muons, within the acceptance on SR, are applied (see Table 4): uncertainties on electron/muon identification, reconstruction and trigger efficiency (summarized in Electron/Muon identification), on

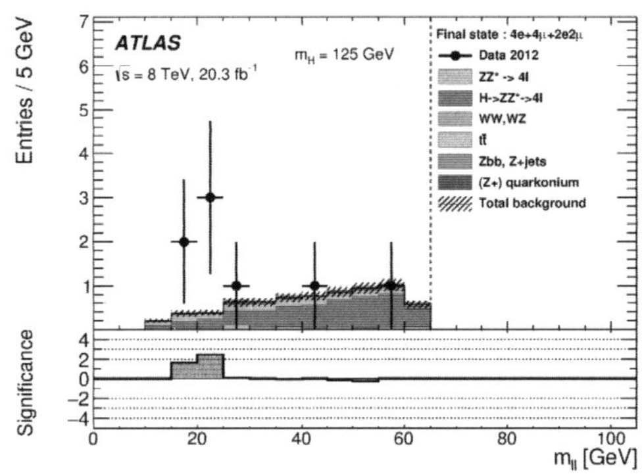

Figure 3. Dilepton invariant mass $m_{ll=12,34}$ for $m_H = 125$ GeV at the Loose SR.

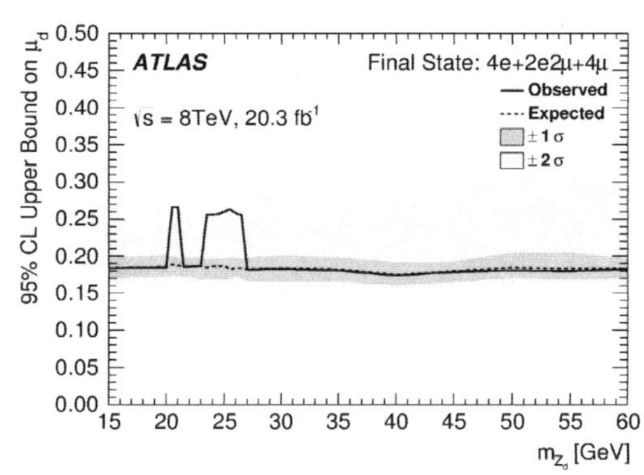

Figure 4. The 95% C.L. upper bound on the signal strength μ_d, see Eq 1.

Table 4. Relative systematic uncertainties on the event yields.

Source	4μ	$4e$	$2e2\mu$
Electron identification	—	6.7	3.2
Electron energy scale	—	0.8	0.3
Muon identification	2.6	—	1.3
Muon momentum scale	0.1	—	0.1
Luminosity	2.8	2.8	2.8
ggF QCD	7.8	7.8	7.8
ggF PDF's and α_s	7.5	7.5	7.5
ZZ* Normalization	5.0	5.0	5.0

the electron/muon energy/momentum scale, and on the electron/muon energy/momentum resolution (summarized in Electron/Muon energy/momentum scale). Also the uncertainties from the isolation and IP requirements are taken into account.

7. Results

Four data events pass after applying the loose SR requirements: one in the $4e$, two in the 4μ and one in the $2e2\mu$ channel. Only two of these events pass the tight SR: the event in the $4e$ and one of the events in the 4μ channel. The events in $4e$ and 4μ channels are consistent with the Z_d mass range $23.5 \leq m_{Z_d} \leq 26.5$ GeV and $20.5 \leq m_{Z_d} \leq 21.0$ GeV, respectively, as shown in Fig. 3. In the mass range of 15 to 60 GeV the interpolated histograms are used [19], generated, based on the MC signal simulations, in steps of 0.5 GeV to obtain the signal acceptance at the hypothesized m_{Z_d}. Table 5 shows the expected and observed number of events after applying the tight SR requirements. In the absence of any significant excess, the 95% C.L. upper bound is computed on the parameter of interest, the signal strength, which is defined as:

$$\mu_d = \frac{\sigma \times BR(H \to Z_d Z_d \to 4l)}{[\sigma \times BR(H \to ZZ^* \to 4l)]_{SM}}. \tag{1}$$

The limits are computed for each final state and their combination for the model independent case, following the CL_s modified frequentist formalism with the profile-likelihood test statistic [20]. The nuisance parameters associated with the systematic uncertainties described

Table 5. The expected and observed number of events in the tight signal region for each final state are shown. The results are for the mass regions $m_{Z_d} = 25$ and 20.5 GeV. The statistical and systematic uncertainties are given for the signal and background expected events. The $H \to ZZ^* \to 4l$ expected events are the combination of the ggF, VBF, ZH, WH and $t\bar{t}H$ processes.

Process	$4e$	4μ	$2e2\mu$
$H \to ZZ^* \to 4l$	$(1.5 \pm 0.3 \pm 0.2) \times 10^{-2}$	$(1.0 \pm 0.3 \pm 0.3) \times 10^{-2}$	$(2.9 \pm 1.0 \pm 2.0) \times 10^{-3}$
$ZZ^* \to 4l$	$(7.1 \pm 3.6 \pm 0.5) \times 10^{-4}$	$(8.4 \pm 3.8 \pm 0.5) \times 10^{-3}$	$(9.1 \pm 3.6 \pm 0.6) \times 10^{-3}$
WW, WZ	$< 0.7 \times 10^{-2}$	$< 0.7 \times 10^{-2}$	$< 0.7 \times 10^{-2}$
$t\bar{t}$	$< 3.0 \times 10^{-2}$	$< 3.0 \times 10^{-2}$	$< 3.0 \times 10^{-2}$
Zbb, Z+jets	$< 0.2 \times 10^{-2}$	$< 0.2 \times 10^{-2}$	$< 0.2 \times 10^{-2}$
ZJ/ψ, $Z\Upsilon$	$< 2.3 \times 10^{-3}$	$< 2.3 \times 10^{-3}$	$< 2.3 \times 10^{-3}$
Total background	$< 5.6 \times 10^{-2}$	$< 5.9 \times 10^{-2}$	$< 5.3 \times 10^{-2}$
Data	1	0	0
$H \to ZZ^* \to 4l$	$(1.2 \pm 0.3 \pm 0.2) \times 10^{-2}$	$(5.8 \pm 2.0 \pm 2.0) \times 10^{-3}$	$(2.6 \pm 1.0 \pm 0.2) \times 10^{-3}$
$ZZ^* \to 4l$	$(3.5 \pm 2.0 \pm 0.2) \times 10^{-3}$	$(4.1 \pm 2.7 \pm 0.2) \times 10^{-3}$	$(2.0 \pm 0.6 \pm 0.1) \times 10^{-2}$
WW, WZ	$< 0.7 \times 10^{-2}$	$< 0.7 \times 10^{-2}$	$< 0.7 \times 10^{-2}$
$t\bar{t}$	$< 3.0 \times 10^{-2}$	$< 3.0 \times 10^{-2}$	$< 3.0 \times 10^{-2}$
Zbb, Z+jets	$< 0.2 \times 10^{-2}$	$< 0.2 \times 10^{-2}$	$< 0.2 \times 10^{-2}$
ZJ/ψ, $Z\Upsilon$	$< 2.3 \times 10^{-3}$	$< 2.3 \times 10^{-3}$	$< 2.3 \times 10^{-3}$
Total background	$< 5.3 \times 10^{-2}$	$< 5.1 \times 10^{-2}$	$< 6.4 \times 10^{-2}$
Data	0	1	0

in Sec 6 are profiled. The systematic uncertainties on Table 4 are 100% correlated between the signal and background. In Figs. 4 and 5 show the 95% C.L. upper bound on the μ_d and on the branching ratio of $H \to Z_d Z_d \to 4l$ as a function of the mass m_{Z_d} for the combination of the three final states: $4e$, 4μ and $2e2\mu$; assuming the SM Higgs boson cross-section and $BR(H \to ZZ^* \to 4l)_{SM} = 1.25^{-4}$ [17].

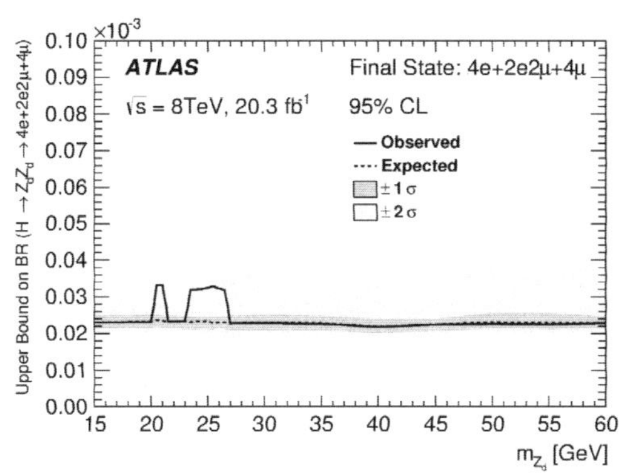

Figure 5. The 95% C.L. upper bound on the $BR(H \to Z_d Z_d \to 4l)$.

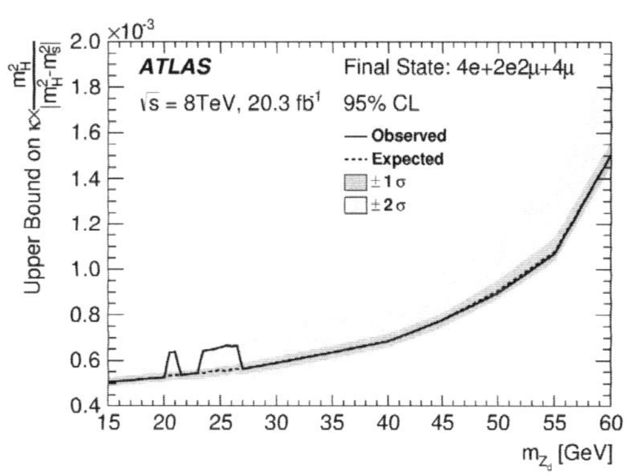

Figure 6. The 95% C.L. upper bound on the Higgs mixing parameter, see Eq. 2.

For the model dependent case, the simplest benchmark model is the SM plus a dark vector boson and a dark Higgs boson [3, 4, 5], where the branching ratio of the $Z_d \to ll$ is given as a function of the m_{Z_d}, which can be used to convert the upper bound measurement of the μ_d into an upper bound on the branching ratio BR$(H \to Z_d Z_d)$, assuming the SM Higgs production cross-section [7]. The $H \to Z_d Z_d$ decay can be used to obtain an m_{Z_d}–dependent limit on an

Higgs mixing parameter κ' [4]:

$$\kappa' = \kappa \times \frac{m_H^2}{|m_H^2 - m_s^2|}, \tag{2}$$

where κ is the size of the Higgs portal coupling and m_s is the mass of the dark Higgs boson. The partial width of the $H \to Z_d Z_d$ is given in terms of κ [4]. In the regime where the Higgs mixing parameter dominates ($\kappa \gg \epsilon$), $m_s > m_H/2$, $m_{Z_d} < m_H/2$ and $H \to Z_d Z^* \to 4l$ are negligible, the only relevant decay is $H \to Z_d Z_d$. The Higgs portal coupling parameter κ is [4]:

$$\kappa^2 = \frac{\Gamma_{SM}}{f(m_{Z_d})} \frac{BR(H \to Z_d Z_d)}{1 - BR(H \to Z_d Z_d)}. \tag{3}$$

The upper bound on the effective Higgs mixing parameter as a function of m_{Z_d} is shown in Fig. 6 for $m_H/2 < m_s < 2m_H$, this corresponds to an upper bound on the Higgs portal coupling in the range $\kappa \sim (1 - 10) \times 10^{-4}$.

8. Conclusions

The search of the Z_d light boson in a mass range of $15 \leq m_{Z_d} \leq 60$ GeV, via $H \to Z_d Z_d \to 4l$ with $m_H = 125$ GeV, is presented at $\sqrt{s} = 8$ TeV. Two data events in the tight SR are observed in $4e$ and 4μ final states: $4e$, $m_{Z_{d1,2}} = 21.8,\ 28.1$ GeV and 4μ, $m_{Z_{d1,2}} = 23.2,\ 18.0$ GeV. Since there is not significant excess observed, the 95% C.L. upper bound on the μ_d, and on the $BR(H \to Z_d Z_d \to 4l)$ are set, combining the three final states. For the mass range $20 \leq m_{Z_d} \leq 28$ GeV, in the model independent case, the upper bounds for μ_d are $(18{-}27)\times 10^{-2}$; in case of the model depending study, the upper bound for the $BR(H \to Z_d Z_d \to 4l)$ is $(2.1 - 3.2) \times 10^{-5}$, and the upper bound on the Higgs mixing parameter is $(5.0 - 6.5) \times 10^{-4}$ (see Fig. 4, 5 and 6).

8.1. Acknowledgments

It is a pleasure to thank the ATLAS Collaboration, to CONACyT for the support received as a postdoctoral fellow with the University of Johannesburg, to Professor Simon Connell, Ketevi Assamagan and the University of Cape Town for giving me the opportunity to be part of the SouthAfrican group in the ATLAS Collaboration and for working in this project.

References

[1] The ATLAS Collaborations 2012 *Phys. Lett.* B **716** 1
[2] The AMS Collaboration 2013 *Phys. Rev. Lett.* **110** 141102
[3] S. Gopalakrishna, S. Jung, and J. D. Wells 2008 *Phys. Rev.* D **78** 055002; J. D. Wells 2008 *arXiv:0803.1243*
[4] D. Curtin, R Essig, S. Gori, and J. Shelton 2015 *J. High Energy Phys.* **02** 157
[5] D. Curtin *et al.* 2014 *Phys. Rev.* D **90** 075004
[6] The ATLAS Collaborations 2015 *Phys. Rev.* D **91** 012006
[7] The ATLAS Collaboration 2015 *Phys. Rev.* D **92** 092001
[8] The ATLAS Collaboration 2008 *arXiv:0901.0512*;
[9] The ATLAS Collaboration 2012 *Eur. Phys. J.* C **72** 1849
[10] The ATLAS Collaboration 2015 *Eur. Phys. J.* C **75** 229
[11] The ATLAS Collaboration 2010 *Eur. Phys. J.* C **70** 823
[12] S. Agostinelli *et al.* 2003 *Nucl. Instrum. Methods Phys. Res.* Sect A **506** 250
[13] The ATLAS Collaboration 2010 *ATL-PHYS-PUB-2010-013*
[14] The ATLAS Collaboration 2011 *ATLAS-CONF-2011-063*
[15] The ATLAS Collaboration 2012 *Eur. Phys. J.* C **72** 1909; *ATLAS-CONF-2012-047*
[16] K. Olive *et al.* 2014 *Chin. Phys.* C **38** 090001
[17] LHC Higgs cross section working group *et al. arXiv: 1101.0593 arXiv:1201.3084*
[18] The ATLAS Collaboration 2013 *Eur. Phys. J.* C **73** 2518
[19] A. L. Read 1999 *Nucl. Instrum. Methods Phys. Res.* Sec. A **425** 357
[20] A. L. Read 2002 *J. Phys.* G **28** 2693; G. Cowan *et al.* 2011 *Eur. Phys. J.* C **71** 1554

Dark matter and the neutrino portal paradigm

V. Gonzalez Macías[1], J. Illana[2], J. Wudka[1]

[1]Department of Physics & Astronomy, University of California Riverside, Riverside, CA 92521-0413, USA

[2]CAFPE and Depto. de Física Teórica y del Cosmos, Universidad de Granada 18071 Campus de Fuentenueva, Granada, Spain

E-mail: vanniagm@fisica.ugto.mx, jillana@ugr.es, jose.wudka@ucr.edu

Abstract. A simple extension of the Standard Model (SM) that provides an explicit realization of the dark-matter (DM) neutrino-portal paradigm is presented. The leading interactions between the dark sector, containing scalars and relic fermions, and the SM involve neutrinos. This model meets all observational constraints.

1. Introduction

Dark Matter (DM) presents an unambiguous evidence of physics beyond the Standard Model (SM). The most compelling DM paradigm assumes that it consists of one or more particles with very weak couplings to the SM [1, 2], and having the correct abundance to explain the CMB observations [3]. This hypothesis has been probed extensively using direct [4–6] and indirect detection [7–12] experiments, and in collider processes [13–16]. To date, no evidence of DM effects in any of these experiments has been confirmed.

Many viable DM candidates have been proposed, spanning a large range in masses and interaction strengths (see, for example, [2, 17]). In particular, reference [18] describes a possible scenario that ensures naturally small direct and indirect detection signals, without compromising the relic abundance inferred from CMB experiments. This scenario is based on the assumption that interactions between the dark and SM sectors are mediated by one or more Dirac fermions $\mathcal{F}$, assumed neutral under all dark and SM symmetries, except fermion number. In addition, the dark sector is assumed to contain (at least) one fermion Ψ and one scalar Φ that have the same (non trivial) transformations under a symmetry group $\mathcal{G}_{\rm DM}$, whose nature is not necessarily specified; the only assumption is that all SM particles are singlets under $\mathcal{G}_{\rm DM}$, which ensures that the lightest dark-sector particle will be stable and so serve as a DM candidate.

Reference [19] shows the simplest model that realizes such a scenario and the implications of existing and projected experimental restrictions on the model parameters. Despite the high precision constraints available, there are significant regions of parameter space allowed. Given that the leading couplings between the DM and the SM sectors involve neutrinos, the model has a distinctive identifying feature: the presence of a monochromatic neutrino signal generated by DM annihilation in astrophysical objects.

2. Neutrino portal in effective theories

A dark sector that contains scalars Φ and fermions Ψ allows for the presence of an effective interaction with the SM of the form $\mathcal{O}^{(5)} = (\bar\Psi\Phi)(\tilde\phi^\dagger\ell)$ where ℓ and ϕ denote, respectively, the isodoublets for a left-handed SM lepton and SM scalar ($\tilde\phi = i\sigma^2\phi^*$); this dimension-5 operator can be generated at tree-level by the exchange of a neutral fermion $\mathcal{F}$. Within the $\mathcal{F}$-mediated paradigm this operator describes the strongest interactions between the SM and the dark sector, which always involve a neutrino: this is a neutrino portal scenario (neutrino portals have been studied in related contexts for example in [23]). The presence of a Ψ-Φ-ν coupling also implies that the heaviest of the dark particles will promptly decay into the lightest, so there will be a single DM relic despite having a dark sector with two (or more) particles.

In addition to the above $\mathcal{F}$-induced coupling, the presence of dark scalars allows for the usual Higgs portal coupling $|\Phi|^2|\phi|^2$. If the dark fermion is heavier than the dark scalar, $m_\Psi > m_\Phi$, then Φ constitutes the DM relic and the physics of the model is dominated by the effects of the Higgs portal coupling, which has been extensively studied in the literature. In contrast, if $m_\Psi < m_\Phi$, the Higgs portal coupling is secondary to (2) and the phenomenology is completely different; for example, the leading interactions relevant for direct detection are produced by the dimension 5: $|\phi|^2\bar\Psi\Psi$ and 6: $(\phi^\dagger \overleftrightarrow{D}_\mu \phi)\left(\bar\Psi\gamma^\mu P_{L,R}\Psi\right)$, $(\bar\ell\gamma_\mu\ell)(\bar\Psi\gamma^\mu P_{L,R}\Psi)$ effective operators that are generated at one loop[1] by the $\mathcal{F}$. The fact that the $\mathcal{F}$ create interactions at tree-level and at one loop is what allows for the required relic abundance to be obtained within the constraints of direct and indirect detection experiments, without fine-tuning.

3. UV completion: Neutrino portal DM model

The simplest model realization (fully described in [19]) of the above paradigm consists of the following: i) A dark sector which contains one scalar Φ and one fermion field Ψ, transforming under a global symmetry under which all SM fields are singlets such that $m_\Phi > m_\Psi$, so that the fermion is stable. ii) The model contains in addition three Dirac fermions $\mathcal{F}$, neutral under the dark and the SM symmetries. iii) Lepton number is conserved (except for possible instanton effects). The Lagrangian of the model is:

$$\begin{aligned}\mathcal{L} &= \bar\ell i\not{D}\ell + \overline{e_R} i\not{D} e_R + \bar\Psi(i\not\partial - m_\Psi)\Psi + \bar{\mathcal{F}}(i\not\partial - M)\mathcal{F} + |\partial\Phi|^2 - m_\Phi^2|\Phi|^2 \\ &\quad - \left(\bar\ell Y^{(e)} e_R\phi + \bar\ell Y^{(\nu)}\mathcal{F}\tilde\phi + \bar\Psi z^\dagger \mathcal{F}\Phi + \text{H.c.}\right) - \lambda_x|\Phi|^2|\phi|^2 \qquad (1)\end{aligned}$$

where ℓ_i and e_{Ri} denote, respectively, the left-handed SM lepton isodoublets and right-handed isosinglets ($i = 1, 2, 3$ is a family index); ϕ is the SM scalar isodoublet; Ψ and Φ are the dark fermion and scalar fields; and $\mathcal{F}_i$ are the (Dirac) neutral fermion mediators. M is the 3×3 Hermitian mass matrix for the $\mathcal{F}$, while the Yukawa couplings $Y^{(e)}, Y^{(\nu)}$ are general 3×3 complex matrices.

The fields $\mathcal{F}$ and ν are replaced by the fields N and n_L to diagonalize the mass matrix:

$$\mathcal{F} = U^\dagger(\mathcal{C}U_L N_L - \mathcal{S}n_L + U_R N_R), \qquad \nu = V(\mathcal{S}U_L N_L + \mathcal{C}n_L) \qquad (2)$$

where $Y^{(\nu)} = \sqrt{2}/v V\eta U M$ is redefined with a standard polar decomposition, and $\mathcal{C} = 1/(\mathbb{1}+\eta^2)^{1/2}$, $\mathcal{S} = \eta/(\mathbb{1}+\eta^2)^{1/2}$. The unitary matrices $U_{L,R}$ are chosen such that $U_R^\dagger U M U^\dagger \mathcal{C}^{-1} U_L = M_N =$ diagonal. In this basis the n_L are massless left-handed fermions that correspond to the SM neutrinos, and (N_L, N_R) form a set of Dirac fermions with mass matrix M_N. The interaction terms then become:

$$\begin{aligned}- \quad &\mathcal{L}_{\text{int}} = (H/v)\,\bar e m_e e + \lambda_x|\phi|^2|\Phi|^2 + \frac{g}{2c_W}\bar e \not{Z}(2s_W^2 - P_L)e \\ + \quad &\left[\bar\Psi z^T U^\dagger(\mathcal{C}U_L N_L - \mathcal{S}n_L + U_R N_R)\Phi + \text{H.c.}\right] + \frac{g}{\sqrt2}\left[\bar e \not{W} V(\mathcal{C}n_L + \mathcal{S}U_L N_L) + \text{H.c.}\right] \\ + \quad &(H/v)\left[\bar N_R M_N U_L^\dagger \mathcal{S}(\mathcal{C}n_L + \mathcal{S}U_L N_L) + \text{H.c.}\right] + \frac{g}{2c_W}(\bar n_L\mathcal{C} + \bar N_L U_L^\dagger\mathcal{S})\not{Z}(\mathcal{C}n_L + \mathcal{S}U_L N_L) \quad (3)\end{aligned}$$

The number of physical parameters is 3 for m_e, M, z and η each, plus 1 for m_Ψ, m_Φ, λ_x each, plus 4 for V and 8 for U, for a total of 27.

Note that the presence of the $\mathcal{F}$ in (3) modifies the couplings of the charged and light neutral leptons to the SM bosons, as well as couplings involving the heavy neutral leptons N. The DM-SM interactions arise from the fourth term in (3).

As emphasized above, the term $\bar\Psi z^T U^\dagger \mathcal{S} n_L \Phi$, represents the leading couplings between the dark and SM sectors, which justifies our denoting this a "neutrino portal" scenario. The presence of

[1] Current-current operators involving quarks or right-handed leptons are only generated at ≥ 2 loops.

the Ψ-Φ-n_L coupling implies that whenever $m_\Phi > m_\Psi$ the scalar field will decay promptly into the fermion and a neutrino:[2] this model, while having a multi-component dark sector, has a single component DM relic. However, the presence of the Φ is essential for generating the leading DM-SM interactions. Although Z-DM and H-DM couplings are not generated at tree level in (3), they are induced at one loop (fig. 3) and represent the leading coupling in direct detection processes and produce important resonant effects in the annihilation cross section when $m_\Psi \simeq m_Z/2$.

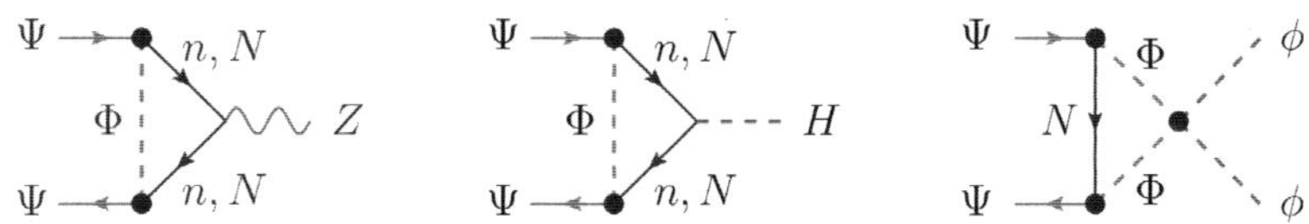

Figure 1. Interaction of the Z (left) and H bosons (center & right) with the DM fields induced at one loop.

In the special case, thoroughly revised in [19] , where $|\eta| \ll 1$ and the mass states N are almost degenerate: $M \simeq \Lambda[\mathbb{1} + \mathcal{O}(\eta^2)]$. This implies that, up to $\mathcal{O}(\eta^2)$ corrections, $U_L = U_R = U$, $\mathcal{C} \simeq \mathbb{1} - \eta^2/2$, $\mathcal{S} \simeq \eta$ and $M \simeq \Lambda(\mathbb{1} - \frac{1}{2}U^\dagger\eta^2 U)$. As a consequence, the observables of interest (the cross sections relevant for relic abundance and indirect detection calculations) will depend on λ_x, η, U and z only through the two real combinations $|\eta U z|$ and $\lambda_x|z|^2$.

4. Observational constraints

All the experimental constraints that are presented below are computed assuming the special case of quasi-degenerate heavy fermions. In this case the relevant parameters are the masses m_Ψ, m_Φ, a heavy mass scale Λ (the cut-off scale) and the coupling combinations $|\eta U z|$ and $\lambda_x|z|^2$. Two possibilities are analyzed for the mass spectrum in the dark sector, a quasi-degenerate spectrum $m_\Phi < m_\Psi + 10\,\text{GeV}$ and a non-degenerate spectrum $m_\Phi \geq m_\Psi + 10\,\text{GeV}$, with $m_\Phi > m_\Psi$ in either case, as required in the scenario considered here. Additionally, it is required that $|z| \leq 2$, which is slightly more conservative than the limits $|z_i| < \sqrt{4\pi}$, derived from tree-level unitarity[3] for each component of z, using the process $\Psi\mathcal{F} \to \Psi\mathcal{F}$. We used the aid of the public codes `MicrOmegas` [20] and `CALCHEP` [21, 22].

4.1. Electroweak constraints

The tightest restrictions on the model parameters are derived from the decays of the Z and H gauge bosons [4].

(i) The experimental result $\Gamma(Z \to \text{inv}) = 499.0 \pm 1.5\,\text{MeV}$ [24] for the invisible width of the Z generates the most stringent bound on the parameters of the model: $\sum_i \eta_i^2 < 0.014 \quad (3\sigma)$ where the η_i are the diagonal elements of the diagonal matrix η.

(ii) Latest results from the ATLAS experiment at the LHC [25–27] report an upper bound $\Gamma(H \to \text{inv}) < 2.2\,\text{MeV}$ at a 90% C.L. so that, for $m_H \gg m_\Psi$,

$$\frac{v}{\Lambda}\left||\eta U z|^2\frac{\Lambda^2}{v^2} + \lambda_x|z|^2 \ln\frac{\Lambda}{m_\Phi}\right| < 1.7\,. \tag{4}$$

4.2. Relic abundance

The leading DM-SM interaction is generated by the (tree-level) exchange of the dark scalars Φ (figure 2) and represent the most important reaction responsible for the equilibration between the dark and standard sectors in the early Universe. If the dark scalar is slightly heavier than the dark fermion (the so-called the quasi-degenerate case), even though it decays promptly to

[2] If $m_\Psi > m_\Phi$ it is the fermions that decay.

[3] We impose tree-level unitarity given our requirement that the model remains perturbative.

[4] W mediated decays of fermions or neutrino mixings are also modified but are weaker than those from the Z boson decay.

the dark fermion and a neutrino, coannihilation processes become important when computing the present density of the relic fermions since the temperature in this case at the equilibration process is higher than the difference in their masses. All of these effects are taken into account in the numerical calculations.[5] The remaining relevant interaction is generated by the one-loop induced Ψ-Ψ-H coupling, and consists of the s-channel exchange of the H boson (figure 2), dominant in the resonance region $m_\Psi \simeq m_H/2$. A similar interaction generated by the s-channel exchange of the Z boson is small but observable in the resonance region $m_\Psi \simeq m_Z/2$. The constraints imposed by Planck [3] are better illustrated in the $(m_\Psi, \Lambda_{\rm eff})$ plane (fig. 3), where $\Lambda_{\rm eff}$ defined as $\Lambda_{\rm eff} = \frac{v/\sqrt{2}}{|\eta U z|}(\frac{m_\Phi^2+m_\Psi^2}{m_\Psi^2})^{1/2}$.

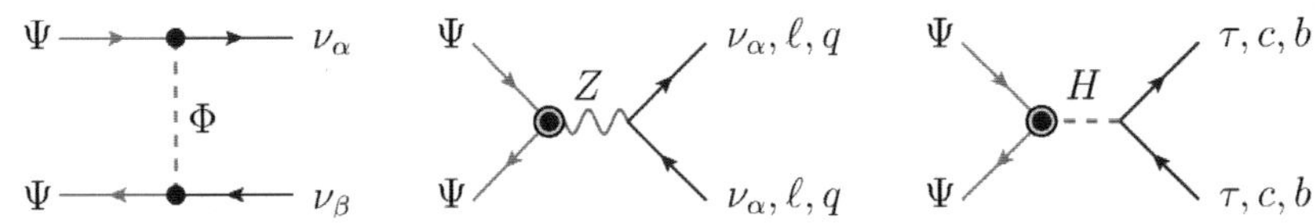

Figure 2. Leading DM-SM interactions in the annihilation channels.

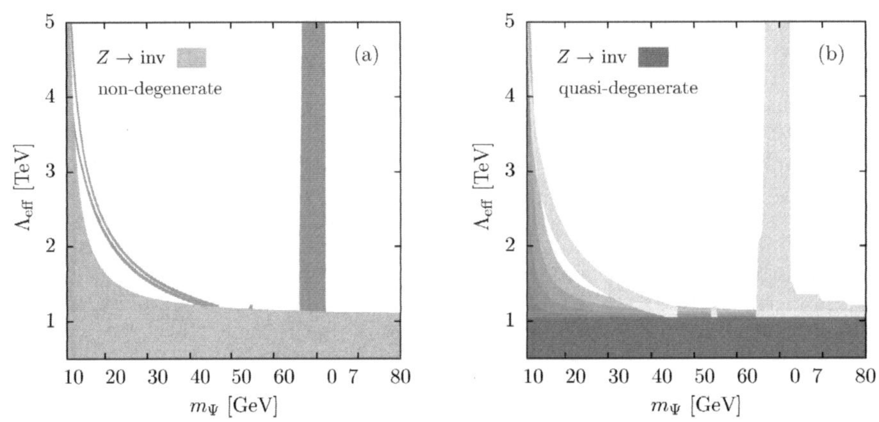

Figure 3. Constraints on the DM-ν portal model from the relic abundance 3σ bounds obtained by the Planck experiment: (a) Allowed regions for the non-degenerate case (blue); for comparison, the green band results from the analytic approximation valid outside the resonance regions. (b) The cyan areas denote the allowed regions for the quasi-degenerate case; in this case the invisible Z constraint is sensitive to m_Φ with the grey bands corresponding to $m_\Phi = m_\Psi + \{0, 1, 3, 6, 10\}$ GeV (dark to light grey, respectively). We set $|z| = 2$ for illustration.

4.3. Direct detection

The leading interactions between the dark matter and the neutral bosons Z and H are induced at one loop, generated by the diagrams in figure 3 hence, naturally suppressing the interactions of the dark matter with quarks. Both axial-vector and vector and scalar couplings are proportional to $|\eta U z|^2$ or $\lambda_x |z|^2 \ln(\Lambda/m_\Phi)$, where only the first parameter combination is affected by the relic abundance constraints. Hence both spin-dependent and spin-independent cross sections are roughly of the same order. Figure 4 shows the projection of the numerical results for xenon nuclei to the $(m_\Psi, \sigma_{\rm SI})$ plane, together with the present bounds from LUX [4][6] and the expected sensitivity from XENON1T [28].

[5] Coannihilation channels for the equilibration process, such as $\Psi\Phi \to We$, $Z\nu$, $H\nu$, and $\Phi\Phi \to HH, WW, ZZ$ become important when kinematically allowed; all are included in the numerical calculations below.

[6] We present here updated results using the latest LUX bounds (2016) from new analyzed data.

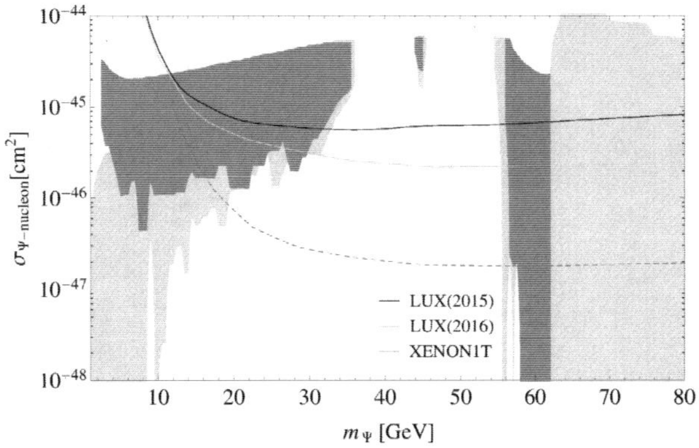

Figure 4. DM-nucleon spin-independent cross sections compatible with the relic abundance and electroweak constraints for the non-degenerate (blue) and quasi-degenerate (cyan) cases. The region above the solid (dashed) lines is (will be) excluded by the LUX (2016) (XENON1T) experiments.

4.4. Indirect detection

Accumulated DM particles in the core of astronomical objects like the Sun can then annihilate into neutrinos, or other SM particles, that can be detected, after traveling to Earth and interacting with the detectors, in astrophysical high energy neutrino experiments (see for example [29]). Given the small DM velocities, the neutrino spectrum of the $\Psi\bar{\Psi} \to \nu\bar{\nu}$ channel in our model is essentially a delta function centered around $E_\nu \simeq m_\Psi$ that would show up as a monochromatic line in a detector if the corresponding experimental sensitivity is reached. In figure 5 we show the predicted flux of muons produced from the interaction of the neutrinos inside a water Cherenkov detector (contained flux), and the flux of muons produced from the interaction of up-going neutrinos with the rocks surrounding the detector (upward flux), as predicted by our model and taking into account Planck and LUX constraints. There are no significant experimental constraints for neutrino final states in DM annihilation for DM masses below 100 GeV[7].

Other extragalactic unresolved point-sources, such as the galactic halo, galactic center, galaxy clusters, dwarf galaxy satellites are also sources of DM annihilation products that may be accessible to indirect detection experiments [17, 32]. In particular, gamma rays and neutrinos produced as primary or secondary products of DM annihilation will travel essentially undisturbed through space, so the flux of these particles is proportional to the (present time) thermally-averaged, annihilation cross section of non-relativistic DM relics. In figure 6 (right) we show the annihilation cross section of $\Psi\bar{\Psi}$ into neutrinos versus the DM mass m_Ψ for regions in parameter space that meet the relic abundance and direct detection constraints. There are no significant experimental constraints for neutrino final states in DM annihilation for DM masses below 100 GeV. Figure 6 (left) shows the annihilation cross sections of the process $\Psi\bar{\Psi} \to b\bar{b}$ in the non-relativistic limit generated by the Z boson exchange, versus the DM mass m_Ψ, with all points fulfilling the relic abundance and direct detection constraints. The recent Fermi-LAT limit [9], obtained by searching for $b\bar{b}$ annihilation products in several dwarf galaxies having a high ratio of DM to ordinary matter, is also displayed.

5. Conclusions

A simple model realization for DM interactions in the neutrino portal paradigm is revised taking into account updated LUX results [4]. The relatively large DM-neutrino couplings allow an annihilation cross section large enough to generate the expected relic density, while simultaneously

[7] Available limits by SuperKamiokande or Icecube [7, 30] depend on the DM annihilation final states, which are chosen usually to be the so-called soft ($b\bar{b}$) or hard channels ($\tau^+\tau^-$); also available are limits for the W^+W^- and direct neutrino production channels [31]

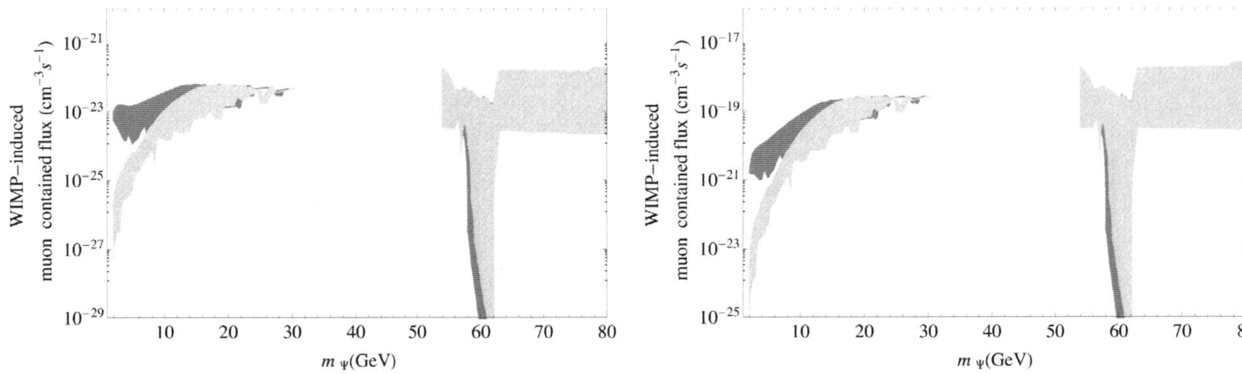

Figure 5. Induced muon rate by neutrinos produced from DM annihilation in the core of the Sun for the non-degenerate (blue) and quasi-degenerate (cyan) cases. The left figure shows the muons produced by neutrinos interacting within the detector (contained), and the right figure shows the induced muon flux by neutrinos interacting with the surrounding material (upward). These figures show the allowed region of parameter space from Planck, Lux (2016), and electroweak constraints.

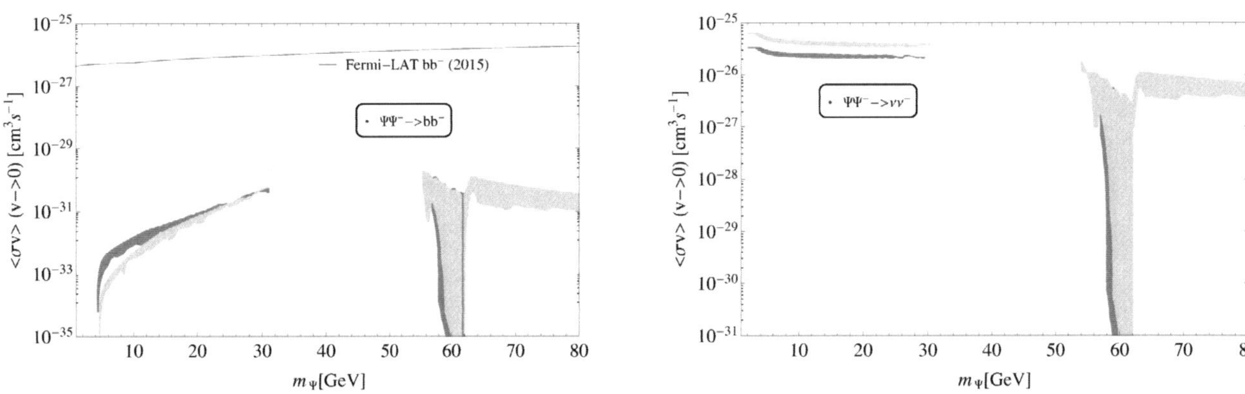

Figure 6. Annihilation cross section into b quarks (left) and ν (right) final states for the non-degenerate (blue) and quasi-degenerate (cyan) cases. These figures show the allowed region of parameter space from Planck, Lux (2016), and electroweak constraints at 3σ, together with the latest Fermi-LAT results.

obeying the direct-detection constraints, because of the suppressed couplings to the Z and H without fine tunning. The indirect detection constraints are also easily accommodated because in this scenario the main annihilation products are neutrinos, for which the available limits are weak. It is of interest that there are two distinctive scenarios depending on the mass spectrum in the dark sector. If the dark scalars are only sightly heavier than the fermions, coannihilation processes become important in generating the freeze-out of the DM fermions, the lightest and only stable particles, and wide regions of parameter space are allowed. In the case of non-degenerate dark particle states, electroweak constraints restrict the DM mass to lie in the range $2.3\,\text{GeV} \leq m_\Psi \leq 35\,\text{GeV}$ or near the H resonance region $m_\Psi \simeq m_H/2$. In contrast, for quasi-degenerate dark scalars and fermions, the electroweak constraints together with LUX constraints exclude only the relatively narrow range $35\,\text{GeV} \leq m_\Psi \leq 52\,\text{GeV}$.

References

[1] M. W. Goodman and E. Witten, Phys. Rev. D **31**, 3059 (1985).

[2] J. L. Feng, Ann. Rev. Astron. Astrophys. **48**, 495 (2010) [arXiv:1003.0904 [astro-ph.CO]].
[3] P. A. R. Ade *et al.* [Planck Collaboration], Astron. Astrophys. **571**, A16 (2014) [arXiv:1303.5076 [astro-ph.CO]].
[4] D. S. Akerib *et al.* [LUX Collaboration], Phys. Rev. Lett. **112**, 091303 (2014). Updated results: http://luxdarkmatter.org/ arXiv:1512.03506 [astro-ph.CO].
[5] E. Aprile *et al.* [XENON100 Collaboration], Phys. Rev. Lett. **109**, 181301 (2012) [arXiv:1207.5988 [astro-ph.CO]].
[6] Z. Ahmed *et al.* [CDMS-II Collaboration], Science **327**, 1619 (2010) [arXiv:0912.3592 [astro-ph.CO]].
[7] K. Choi *et al.* [Super-Kamiokande Collaboration], Phys. Rev. Lett. **114**, no. 14, 141301 (2015) [arXiv:1503.04858 [hep-ex]].
[8] M. G. Aartsen *et al.* [IceCube Collaboration], Eur. Phys. J. C **75**, no. 99, 20 (2015) [arXiv:1406.6868 [astro-ph.HE]].
[9] M. Ackermann *et al.* [Fermi-LAT Collaboration], arXiv:1503.02641 [astro-ph.HE].
[10] F. Aharonian *et al.* [HESS Collaboration], Phys. Rev. Lett. **97**, 221102 (2006) [astro-ph/0610509].
[11] M. Aguilar *et al.* [AMS Collaboration], Phys. Rev. Lett. **110**, 141102 (2013).
[12] O. Adriani *et al.* [PAMELA Collaboration], Nature **458**, 607 (2009) [arXiv:0810.4995 [astro-ph]].
[13] V. Khachatryan *et al.* [CMS Collaboration], Eur. Phys. J. C **75**, no. 5, 235 (2015) [arXiv:1408.3583 [hep-ex]].
[14] G. Aad *et al.* [ATLAS Collaboration], JHEP **1304**, 075 (2013) [arXiv:1210.4491 [hep-ex]].
[15] J. Goodman, M. Ibe, A. Rajaraman, W. Shepherd, T. M. P. Tait and H. B. Yu, Phys. Rev. D **82** (2010) 116010 [arXiv:1008.1783 [hep-ph]].
[16] Y. Bai, P. J. Fox and R. Harnik, JHEP **1012**, 048 (2010) [arXiv:1005.3797 [hep-ph]].
[17] G. Bertone, D. Hooper and J. Silk, Phys. Rept. **405**, 279 (2005) [hep-ph/0404175].
[18] V. G. Macías and J. Wudka, JHEP **1507**, 161 (2015) [arXiv:1506.03825 [hep-ph]].
[19] V. Gonzlez-Macas, J. I. Illana and J. Wudka, JHEP **1605**, 171 (2016) doi:10.1007/JHEP05(2016)171 [arXiv:1601.05051 [hep-ph]].
[20] G. Belanger, F. Boudjema, A. Pukhov and A. Semenov, Comput. Phys. Commun. **185**, 960 (2014) [arXiv:1305.0237 [hep-ph]].
[21] A. Belyaev, N. D. Christensen and A. Pukhov, Comput. Phys. Commun. **184**, 1729 (2013) [arXiv:1207.6082 [hep-ph]].
[22] A. Alloul, N. D. Christensen, C. Degrande, C. Duhr and B. Fuks, Comput. Phys. Commun. **185**, 2250 (2014) [arXiv:1310.1921 [hep-ph]].
[23] N. Cosme, L. Lopez Honorez and M. H. G. Tytgat, Phys. Rev. D **72**, 043505 (2005) [hep-ph/0506320]; H. An, S. L. Chen, R. N. Mohapatra and Y. Zhang, JHEP **1003**, 124 (2010) [arXiv:0911.4463 [hep-ph]];
[24] K. A. Olive *et al.* [Particle Data Group Collaboration], Chin. Phys. C **38**, 090001 (2014).
[25] G. Aad *et al.* [ATLAS Collaboration], arXiv:1509.00672 [hep-ex].
[26] G. Aad *et al.* [ATLAS Collaboration], arXiv:1508.07869 [hep-ex].
[27] CMS Collaboration [CMS Collaboration], CMS-PAS-HIG-15-012.
[28] E. Aprile *et al.* [XENON Collaboration], [arXiv:1512.07501 [physics.ins-det]].
[29] M. Cirelli, N. Fornengo, T. Montaruli, I. A. Sokalski, A. Strumia and F. Vissani, Nucl. Phys. B **727**, 99 (2005) [Nucl. Phys. B **790**, 338 (2008)] [hep-ph/0506298].
[30] M. G. Aartsen *et al.* [IceCube Collaboration], Phys. Rev. Lett. **110**, no. 13, 131302 (2013) [arXiv:1212.4097 [astro-ph.HE]].
[31] G. Belanger, J. Da Silva, T. Perrillat-Bottonet and A. Pukhov, JCAP **1512** (2015) no.12, 036 [arXiv:1507.07987 [hep-ph]].
[32] H. Yuksel, S. Horiuchi, J. F. Beacom and S. Ando, Phys. Rev. D **76**, 123506 (2007) [arXiv:0707.0196 [astro-ph]].

Fundamental physics with low-energy neutrons

Libertad Barrón-Palos

Instituto de Física, Universidad Nacional Autónoma de México, Apartado Postal 20-364, México

E-mail: `libertad@fisica.unam.mx`

Abstract. Low-energy neutrons are playing a prominent role in a growing number of fundamental physics studies. This paper provides a brief description of the physics that some of the experiments in the area are addressing.

1. Introduction

In the present days, many of the new advances in fundamental physics come from experiments in high energy physics, however experiments at low energies, particularly those with high intensities of low-energy neutrons, also can shed light on some of the most fundamental questions of the physics laws that govern the Universe. The unique properties of low-energy neutrons (with energies ranging from a few meV to hundreds of neV), that can be effectively formed into intense polarized beams, or stored in traps with boundaries that are material, magnetic, gravitational, or combination of these, make it possible to use them in a variety of studies that include the properties of the neutron itself (electric dipole moment), its decay (neutron lifetime and correlation parameters), the study of fundamental interactions (hadronic weak interaction), the violation of symmetries (parity and time reversal), or the search for possible new exotic interactions, just to mention some examples.

2. Parity violation and the hadronic weak interaction

As a consequence of the non-perturbative nature of QCD at low energies and the dominance in intensity of the effects of the strong force in the nuclear interactions, we do not have a first-principles description of the weak interaction between hadrons. The weak interaction is the only fundamental interaction that can change the flavor of quarks and can violate the parity symmetry, thus the weak interaction can be studied in flavor-changing mechanisms, like the strangeness-changing non-leptonic decays of mesons and baryons, or in hadronic and nuclear processes where parity is violated. The use of polarized cold neutron beams in the study of nuclear processes that exhibit parity violation (PV) is of interest for several reasons [1, 2]. At low energies the strong interaction between nucleons is dominated by the exchange of pions, which corresponds to isospin exchange of $\Delta I = 1$; charged currents are suppressed for this isospin channel and therefore low-energy nuclear weak interactions offer the possibility to study neutral currents. Since the range of the weak bosons is small compared to the nucleon size, these studies constitute a probe for quark-quark correlations. Also a better understanding of the nucleon-nucleon weak interaction would be of help in testing nuclear structure models using PV effects in nuclear and atomic systems. In addition to this, there are observed phenomena in the strangeness-changing sector of the hadronic weak interaction (HWI) which cannot be explained in the framework of the existing theories, like the dominance of the $\Delta I = 1/2$ channel in the decay of kaons or the relatively weak amplitudes in the non-leptonic decay of hyperons [2]; the understanding of these phenomena requires input from the flavor-conserving PV sector. The theoretical attempts to describe the HWI include the one-meson exchange model, proposed more than 30 years ago by Desplanques, Donoghue and Holstein [3], effective field theories (EFT) [4] and more recently lattice QCD calculations [5]. For the first two, the experimental determination of weak coupling constants that cannot be calculated is necessary; for the last, comparison of theoretical and experimental results can guide new theoretical predictions for other systems. For several years now, a program to determine the weak coupling constants in experiments using polarized cold neutrons and light targets has been taking place [6, 7]. One of the purposes of this program is to measure PV observables in few-nucleon systems, where nuclear structure uncertainties are smaller and a more reliable correlation with the weak coupling constants can be attained. In this context, so far three low-energy neutron experimental collaborations have taken data: NPDGamma, Neutron Spin Rotation (NSR) and n-^{3}He. In the capture of polarized cold neutrons by protons, the NPDGamma experiment measures the PV asymmetry, with respect to the neutron spin direction, in the emission of gamma rays that follows the capture ($\vec{n}+p \rightarrow d+\gamma$). This asymmetry is dominated by the $\Delta I = 1$ $^3S_1 - {}^3P_1$ parity-odd transition in the np system, therefore it is related to the weak coupling that characterizes the exchange of one pion in the HWI, h^1_π. The details of the experiment can be found at [1, 8, 9]. The experiment had a first stage at the Los Alamos National Laboratory Neutron Science Center (LANSCE) whose result was statistically limited [1]. A second stage of the experiment at the Spallation Neutron Source of the Oak Ridge National Laboratory (SNS-ORNL) concluded this year and results with statistical uncertainty in the 10^{-8} range will be published soon. The n-^{3}He collaboration finished data taking last year at the SNS-ORNL and analysis is in progress. This experiment measures the asymmetry, with respect to the neutron spin direction, in the direction of emission of the protons produced in the $\vec{n}+{}^3He \rightarrow t+p$ nuclear reaction. NSR is an experiment that measures the neutron spin rotation angle for polarized cold neutrons passing through unpolarized ^4He. The experiment has a statistically limited result [10] from its first stage at the Center for Neutron Research of the National Institute of Standards and Technology (NCNR), and the collaboration is preparing an improved apparatus for the second stage also to take place at the NCNR [11].

3. Probing the electroweak interaction in the neutron beta decay

The free neutron beta decay presents the possibility to measure a number of observables: the neutron lifetime τ_n, the electron-neutrino correlation coefficient a , the Fierz interference term b, and in the case of polarized neutrons, the correlation coefficients of the neutron spin and the momentum of the electron A and the neutrino B. An interesting aspect of the neutron beta decay is that in the electroweak theory three of these observables (a, A and B) depend on only one parameter, λ, the ratio of axial-vector to vector neutron coupling constants (g_A and g_V), while b should be zero. The experimental determination of these coefficients would over-constrain λ, allowing for the performance of different consistency checks of the electroweak theory and for a sensitive search for new physics. In addition, the precise determination of λ, together with τ_n, can be used to determine the V_{ud} term in the Cabibbo-Kobayashi-Maskawa (CKM) matrix. The accurate determination of the matrix elements is important for its unitarity consistency checks (weak universality), as well as to establish constraints to physics beyond the SM. At present the most precise determination of V_{ud} comes from super-allowed nuclear beta decays. Its extraction from free neutron beta decay, which would have the advantage of avoiding many-nucleon effects present in other nuclear decays, has however not been as precise due to experimental inconsistencies in λ and τ_n. While several experimental collaborations are trying to resolve the neutron lifetime puzzle [12, 13], neutron studies like Nab [14] and $abBA$ [15], to take place in the near future at the SNS-ORNL, are attempting to resolve these inconsistencies for λ, potentially making the free neutron beta decay the most accurate way to determine V_{ud}.

4. Search for possible exotic long-range spin-dependent interactions of the neutron

In the spontaneous symmetry breaking in theories with two or more Higgs doublets, a new vector boson associated with an extra axial generator acting on quark and lepton fields may appear, giving rise to an exotic interaction mediated by a light vector boson coupling to a fermion. In the non relativistic limit this interaction gives rise to potentials proportional to $g_A^2\vec{\sigma} \cdot (\vec{v} \times \vec{r})$ and $g_A g_V \vec{\sigma} \cdot \vec{v}$ [16]. These potentials are among the 16 different operator structures that were studied by Dobrescu and Mocioiu [17] and that are part of the possible non-relativistic fermion-fermion interactions that satisfy rotational invariance. Interactions of this type have been poorly studied due to the experimental challenges that they encompass, however setting limits to these possible exotic interactions is becoming more relevant, as many theories beyond the SM propose, in some cases as a way to explain dark matter and energy, the existence of new light weakly-coupled particles that would induce interactions of relatively long range. Neutrons are a very suitable tool to probe for such interactions since they can be formed into beams with high polarization and their energy and momentum transfers can access the mesoscopic scale. In the recent years polarized low-energy neutrons have been used to set the most stringent limits in some possible exotic interactions, like the limit at distances below 1 cm for any parity-odd long-range interaction of the neutron with matter mediated by spin 1 boson exchange, the upper bound on parity-odd components of possible in-matter gravitational torsion coupled to neutrons [18], and limits for possible parity-even exotic interactions of polarized neutrons with matter from spin 1 boson exchange with axial couplings [19]. Currently the NSR collaboration is conducting an experiment at LANSCE to improve the existing limits in the possible exotic axial-vector couplings of neutrons to matter for ranges below 1 cm [16].

5. Parity and time reversal violation in compound nuclei

One of the challenging questions for the SM is the explanation of the asymmetry between matter and anti-mater in the early stages of the Universe, which led to the predominance of matter that we observe in the present. Almost 50 years ago, Andrei Shakarov [20] proposed three necessary conditions that should have been satisfied in the early Universe so that an imbalance between matter and anti-matter could have been produced: violation of baryon number, violation of C and CP symmetries, and deviation from thermal equilibrium. Regarding the second of these conditions, CP violation has been observed in the decay of Kaons and B mesons [21, 22, 23], however additional sources of CP violation, as well as CP violation in strongly interacting systems, are needed in order to account for the matter/anti-matter imbalance. Since the CPT symmetry remains as one that has to be conserved in any physics interaction, the violation of time reversal symmetry is a via to search for sources of CP violation. Electric dipole moments (EDM) and T violation in resonances of compound nuclei are possible sources of T violation and therefore CP violation in strongly interacting systems. Regarding fundamental neutron physics, significant efforts to measure the neutron EDM are taking place [24] and future experiments to find evidence of T violation in resonances of compound nuclei that can be accessed with low-energy neutrons are in preparation [25]. In the past, large amplification of PV amplitudes in compound nuclei due to nuclear structure have been observed [26], and theory indicates that similar enhancement in T violation amplitudes are to be expected.

Acknowledgments

We gratefully acknowledge the support of PAPIIT-UNAM (Grant No. IG101016).

References

[1] M. T. Gericke *et al* 2011 *Phys. Rev.* C **83** 015505
[2] M. J. Ramsey-Musolf and S. A. Page 2006 *Annu. Rev. Nucl. Part. Sci.* **56** 2
[3] B. Desplanques, J. F. Donoghue and B. R. Holstein 1980 *Ann. Phys.* **124** 449
[4] Shi-Lin Zhu, C. M.Maekawa, B. R.Holstein, M. J. Ramsey-Musolf and U. van Kolck 2005 *Nucl. Phy.* A **748** 435–498
[5] J. Wasem 2012 *Phys. Rev.* C **85** 022501
[6] J. S. Nico and W. M. Snow 2005 *Annu. Rev. Nucl. Part. Sci.* **55** 27
[7] L. Barrón-Palos *et al* 2009 *Rev. Mex. Fís.* **55** (2) 18
[8] N. Fomin *et al* 2013 *AIP Conf. Proc.* **1560** 145
[9] K. B. Grammer *et al* 2015 *Phys. Rev.* B **91** 180301(R)
[10] W.M. Snow *et al* 2011 *Phys. Rev.* C **83** 022501 (R)
[11] W. M. Snow *et al* 2015 *Rev. Sci. Instrum.* **86** 055101
[12] G. L. Greene and P. Geltenbort 2016 *Sci. Am.* **314** 36-41
[13] F. E. Wietfeldt and G. L. Greene 2011 *Rev. Mod. Phys.* **83** 1173
[14] S. Baessler *et al* 2013 *AIP Conf. Proc.* **1560** 114
[15] L. Barrón-Palos *et al* 2010 *J. Phys. Conf. Ser.* **239** 012013
[16] W. M. Snow *et al* 2015 *JPS Conf. Proc.* **8** 026003
[17] B. Dobrescu and I. Mocioiu 2006 *J. High Energy Phys.* **11** 005
[18] H. Yan and W.M. Snow 2013 *Phys. Rev. Lett.* **110** 082003
[19] F.M. Piegsa and G. Pignol 2012 *Phys. Rev. Lett.* **180** 181801
[20] A. D. Sakharov 1967 *J. Exp. Theor. Phys.* **5** 24-27
[21] J. H. Christenson, J. W. Cronin, V. L. Fitch and R. Turlay 1964 *Phys. Rev. Lett.* **13** 138
[22] A. Alavi-Harati *et al* (KTeV Collaboration) 1999 *Phys. Rev. Lett.* **83** 22
[23] R. Aaij *et al* (LHCb Collaboration) 2013 *Phys. Rev. Lett.* **110** 221601
[24] J. M. Pendlebury *et al* 2015 *Phys. Rev.* D **92** 092003
[25] J. D. Bowman and V. Gudkov 2014 *Phys. Rev.* C **90** 065503
[26] G. E. Mitchell, J. D. Bowman, S. I. Penttilä , E. I. Sharapov 2001 *Phys. Rep.* **354** 157

Journal of Physics: Conference Series **761** (2016) 012084 doi:10.1088/1742-6596/761/1/012084

Towards an effective description for the J/ψ decays

FV Flores-Baez

Facultad de Ciencias Físico-Matemáticas, Universidad Autónoma de Nuevo León
Ciudad Universitaria, San Nicolás de los Garza, Nuevo León, 64450, México

E-mail: fflores@fcfm.uanl.mx

Abstract. In this work we propose a phenomenological Resonance Chiral Theory inspired vertex $J/\psi\rho\pi$, which is a basic ingredient for a description of the J/ψ phenomenology. This vertex is characterized by only one coupling constant which is computed through the analysis of the $J/\psi \to \gamma\pi^0$ decay. Assuming ρ meson dominance we compute the $\mathcal{BR}(J/\psi \to \pi^0\pi^+\pi^-)$, finding a good agreement with the experimental measurement.

1. Introduction

It is well known that Chiral Perturbation Theory (χPT) [1–3] is a rigorous effective field theory framework for a description of Quantum Chromodynamics (QCD) at low energies, well below the $\rho(770)$ meson mass. At energies $m_\rho \lesssim E \lesssim 2\,\text{GeV}$ the Resonance Chiral Theory (RχT) [4,5] emerges as a framework to describe the interactions between the light quark resonances and the chiral pseudo-goldstone bosons. The RχT lagrangian fulfills the discrete symmetries of QCD and chiral symmetry and contains an even sector [4,6] and an odd intrinsic-parity sector [7,8]. It also models a large-N_C expansion [9,10]. This framework has been applied for instance to the study hadronic tau decays [11–16] at leading order in the $N_C \to \infty$ limit and to a description [17] of the pion vector form factor up to the next to leading order in the large number of colours expansion.

However the low-energy couplings of both χPT and RχT cannot be determined from chiral symmetry alone, they are free parameters of the lagrangian. Nonetheless demanding a short distance behaviour in accordance with QCD and its Operator Product Expansion (OPE) [18,19], one is able to extract constraints between several couplings. This procedure has been applied in Refs. [7, 20] in order to impose restrictions on the couplings of the odd-intrinsic parity sector, obtaining a set of relations between them. Still this method does not provide full information on the couplings and some of them remain unknown. Alternatively one can rely on the phenomenology for their determination or at least to establish some relations between them.

One of the couplings of the odd-intrinsic-parity RχT Lagrangian (OIPL) that remains unknown after the first procedure above is the corresponding to an operator of the four that contribute to the vertex VVP (being V short for vector resonance and P for pseudoscalar meson, all of them of the lightest $SU(3)$ multiplets), namely d_4 in of Ref. [7] (we take this notation in the following) or κ_4^{VV} in Ref. [8]. It appears, for instance, in the leading order determination of the odd-intrinsic-parity $\mathcal{O}(p^6)$ chiral perturbation theory couplings C_{17}^W, C_{19}^W, C_{20}^W and C_{21}^W [8,21] (though with a factor that suppresses its contribution) and, more importantly, in processes like the radiative decay of the tau lepton into two pions [22], relevant for the determination of the muon anomalous magnetic moment. One way to determine the d_4 coupling is to study the $J/\psi(1S)$ (J/ψ for short) resonance phenomenology, in particular we can focus on the $J/\psi \to \omega(782)\pi^+\pi^-$ decay.

The phenomenology of this process extracted from the experimental analyses is still a matter of discussion. The BES Collaboration [23] has performed a fit to the decay data by considering a wide set of intermediate resonance states and as a result they reported a strong signal of resonances $f_0(500)$, $f_2(1270)$, $b_1(1235)$ in this particular J/ψ decay. However they did not include the virtual $\rho(770)$ transition, namely, $J/\psi \to \rho(770)\pi \to \omega(782)\pi^+\pi^-$ channel in their analysis. The inclusion of this contribution within the chiral unitary approach was performed in Ref. [24], where some additional final state interactions where also taken into account. The $\rho(770)\pi$ intermediate state was later considered in Ref. [25] where it was found that, depending on the choice of the parameters of the form factor, the contribution of the $\rho(770)$ to the total decay width of $J/\psi \to \omega(782)\pi^+\pi^-$ extends from 3.1% to 21%. The inclusion of the $\rho(770)$ was also found to give a solution to the problem put forward in Ref. [23] of the disagreement between the value obtained for the width of the $b_1(1235)$ and the PDG value [26].

In order to disentangle the $\rho(770)$ contribution in the $J/\psi \to \omega(782)\pi^+\pi^-$ decay one has to consider two vertices: $\rho - \omega - \pi$ and $J/\psi - \rho - \pi$. Here we focus in the latter, by investigating its participation in the $J/\psi \to \gamma\pi^0$ decay, assuming that RχT is reliable until $E \sim 2$ GeV and considering a phenomenological Lagrangian that describes the $J/\psi - \rho - \pi$ vertex. Subsequently with our result we give a prediction for the $\mathcal{B}(J/\psi \to \pi^+\pi^-\pi^0)$. Our result will be useful for the determination of the d_4 coupling, where the knowledge of both vertices is mandatory [27].

Another point of discussion is the coupling of the J/ψ vector resonance to the lightest hadrons, both vectors and pseudoscalar mesons. We will take the realization of the three-gluon-annihilation OZI rule put forward in Ref. [28] in the frame of the OIPL and where the J/ψ is considered as a $SU(3)$ singlet in the Proca representation and the lightest mesons obey the constraints of chiral $SU(3)$ symmetry. This complements nicely our framework.

In Section 2 we present the basis of RχT and our proposal for the $J/\psi VP$ vertex; in Section 3 we analyse the $J/\psi \to \gamma\pi^0$ decay and we estimate the $\mathcal{B}(J/\psi \to \pi^0\pi^+\pi^-)$ decay with our assumptions; finally we give our conclusions in Section 4.

2. Resonance Chiral Theory

Resonance Chiral Theory [4, 5] provides an appropriate setting for the study of resonance contributions to low-energy ($E \lesssim 2\,\text{GeV}$) hadrodynamics. Essentially it involves the construction of a phenomenological Lagrangian that describes the interaction of the lightest resonances ($N_f = 2, 3$, being N_f the number of flavours) with the pseudoscalar stable mesons, namely pions and kaons. A detailed account of the method is given in Ref. [29].

On the other hand, it is customary to use the antisymmetric formulation to describe the light quark spin-1 resonances. This is because in the even-intrinsic-parity sector the $\mathcal{O}(p^4)$ χPT low-energy constants, consistent with higher energy constraints, are obtained directly upon integration of the resonances if one uses that formalism. If instead we resort to the Proca formalism, for instance, the same goal would need to add local contributions to the pure resonance Lagrangian [5]. However we will use the latter representation to describe the $J/\psi(1S)$ field.

In the antisymmetric tensor field formalism, spin-1 resonances are described by a field that satisfies:

$$\langle 0|V_{\mu\nu}|V, p\rangle = \frac{i}{m_V}\left[p_\mu \eta_\nu(p) - p_\nu \eta_\mu(p)\right] , \tag{1}$$

where m_V is the mass of the resonance, p its momentum and η_ν its polarization vector. Its propagator in this formalism reads:

$$\Omega^{\alpha\gamma\tau\delta}(p) = 2\,i\,\left[\frac{\Omega_L^{\alpha\gamma\tau\delta}(p)}{m_V^2 - p^2} + \frac{\Omega_T^{\alpha\gamma\tau\delta}(p)}{m_V^2}\right]. \tag{2}$$

The transverse (T) and longitudinal (L) part of the propagator are given by:

$$\Omega_T^{\alpha\gamma\tau\delta}(p) = \frac{1}{2p^2}\left[p^2\, g^{\alpha\tau}\, g^{\gamma\delta} - g^{\alpha\tau}\, p^\gamma\, p^\delta - g^{\gamma\tau}\, p^\alpha\, p^\delta - (\tau \leftrightarrow \delta)\right],$$
$$\Omega_L^{\alpha\gamma\tau\delta}(p) = \frac{1}{2p^2}\left[g^{\alpha\tau}\, p^\gamma\, p^\delta - g^{\gamma\tau}\, p^\alpha\, p^\delta - (\tau \leftrightarrow \delta)\right]. \tag{3}$$

The leading and next-to-leading RχT phenomenological Lagrangian in the even-intrinsin-parity sector has already been worked out [4,6]. Here we collect the pieces relevant for our purposes:

$$\mathcal{L}_{\mathrm{R}\chi\mathrm{T}} = \mathcal{L}_2 + \mathcal{L}_{\mathrm{kin}}^{\mathrm{V}} + \mathcal{L}_{\mathrm{V}} + \mathcal{L}_{J/\psi\mathrm{VP}}\,. \tag{4}$$

Here $\mathcal{L}_2$ is the $\mathcal{O}(p^2)$ χPT Lagrangian:

$$\mathcal{L}_2 = \frac{F^2}{4}\langle u_\mu u^\mu + \chi_+ \rangle, \tag{5}$$

where $F \sim 92.2\,\mathrm{MeV}$ [26] is the pseudoscalar meson decay constant in the chiral limit. The brackets indicate a trace in the flavour space. The notation here and in the following is the one in Ref. [6]. The next piece involves kinetic term of the vector resonances $\mathcal{L}_{\mathrm{kin}}^{\mathrm{V}}$, including also the one of the J/ψ state. The leading interacting operators of the even-intrinsic-parity sector, linear in the light quark vector resonance fields, are given by:

$$\mathcal{L}_{\mathrm{V}} = \frac{F_V}{2\sqrt{2}}\langle V_{\mu\nu} f_+^{\mu\nu}\rangle + i\,\frac{G_V}{\sqrt{2}}\langle V_{\mu\nu} u^\mu u^\nu\rangle, \tag{6}$$

where $V_{\mu\nu}$ indicates the $U(3)$ nonet of the lightest vector meson resonances, namely $\rho(770)$, $K^*(892)$, $\omega(782)$, $\phi(1020)$, and F_V and G_V coupling constants not determined by the symmetries. As commented above the enforced symmetries do not give information on the couplings of the operators: F, F_V, G_V. Within the RχT scheme some information on them can be extracted by requiring that relevant Green functions, determined with our phenomenological Lagrangian in Eq. (4) satisfy, asymptotically (i.e. at large energies or transfers of momenta), the leading behaviour described by QCD in the large-N_C limit. To proceed one would need to determine those Green functions using our RχT Lagrangian in a model of the large-N_C limit in which we consider only one multiplet of light resonances. Usually the constraints translate into relations between the coupling constants. Though reducing the infinite tower of resonances (of the large-N_C limit) to only one could look, admittedly, a strong assumption, several works along this line indicate that this is a good approximation when spin-1 resonances are involved [7, 20, 30, 31]. See Ref. [32] for limitations on this approach. For instance, it was established [20] that $F_V^2 = 3F^2$, which is in good agreement with phenomenological determinations of the F_V coupling constant [12]. In the following we will assume this condition.

Hadron decays of the J/ψ involve a rich dynamics mainly driven by light-flavoured meson resonances. Accordingly those decays are a very important source of information on the effective couplings of the RχT framework. The study of the processes we are interested in requires to add the $\mathcal{L}_{J/\psi\mathrm{VP}}$ piece in Eq. (4) that corresponds to the vertex of a J/ψ vector resonance with one light ($N_f = 3$) vector nonet (V) and pseudoscalar (P) mesons. As the J/ψ is a pure $\bar{c}c$ state, the decay $J/\psi \to VP$ is OZI suppressed. The operators ruling the vertex will obey the chiral and unitary symmetry in the part that corresponds to the light quark hadrons. For the description of the J/ψ state we will take a Proca realization of the field and it will be a singlet under the light quark symmetries. With this proviso we have just one leading operator:

$$\mathcal{L}_{J/\psi \mathrm{VP}} = m_\psi \, H_1 \, \varepsilon_{\mu\nu\rho\sigma} \, \Psi^\mu \, \langle \, u^\nu \, V^{\rho\sigma} \, \rangle \, , \tag{7}$$

neglecting $SU(3)$ breaking and double OZI suppressed contributions [28]. Here H_1 is an unknown coupling constant and Ψ_μ is the field of the J/ψ vector meson; the fact that we have taken a Proca realization for the latter warrants that the number of operators, describing the vertex, is smaller than in the case of the antisymmetric realization, in fact only one in this case.

3. The $J/\psi \to \gamma\pi^0$ decay and the H_1 coupling .

As our main goal is to obtain information on the $J/\psi VP$ vertex given in Eq. (7), here we proceed to do a phenomenological determination of H_1 from the study of $J/\psi \to \pi^0\gamma$ and confirm its consistency in the $J/\psi \to \pi^+\pi^-\pi^0$ decay.

Within our framework the $J/\psi \to \pi^0\gamma$ decay is driven, at leading order, by the vertex in Eq. (7). Hence we can use the experimentally determined $\mathcal{B}(J/\psi \to \gamma\pi^0)$ [33] to determine the H_1 coupling. The only diagram that contributes is the one in Figure 1.

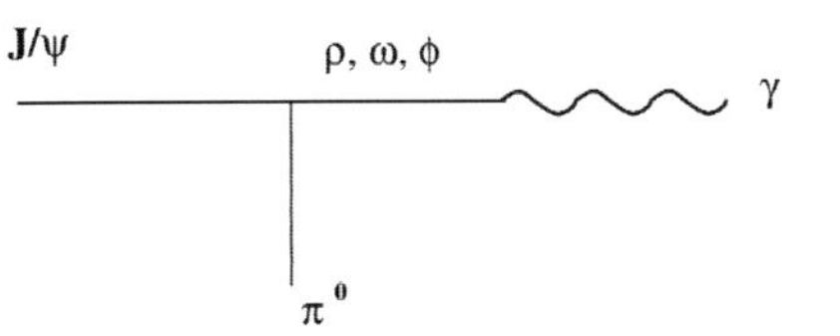

Figure 1. The $J/\psi \to \pi^0\gamma$ decay encloses the H_1 coupling in Eq. (7). Only the $\rho(770)$ contributes to this process and its coupling to the photon is given by Eq. (6).

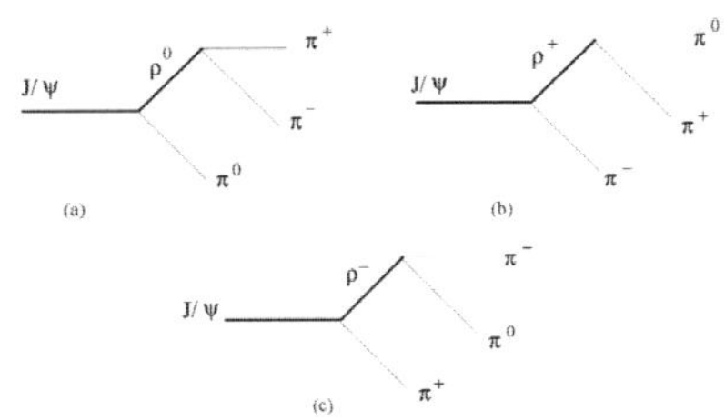

Figure 2. Feynman diagrams for the $J/\psi \to \pi\pi\pi$ decay. Notice that in diagram (a) there are no contributions of ω or ϕ.

The amplitude for this diagram, contributing to $J/\psi(\eta, P) \to \pi^0(q)\gamma(\varepsilon, k)$ reads:

$$\mathcal{M}_{\pi^0\gamma} = -i2\sqrt{2}eH_1 \frac{F_V m_\psi}{F m_\rho^2} \, \varepsilon_{\mu\sigma\rho\nu} \eta^\mu(P) \epsilon^\sigma(k) k^\rho q^\nu . \tag{8}$$

Taking $F_V = \sqrt{3}F$ and the central value of the experimental result $\mathcal{B}(J/\psi \to \pi^0\gamma) = (3.49^{+0.33}_{-0.30}) \times 10^{-5}$ [26], the absolute value of the coupling is found to be,

$$|H_1| \simeq 2.36 \times 10^{-5}. \tag{9}$$

On the other hand, the $J/\psi \to \pi^+\pi^-\pi^0$ decay has recently been measured at BESIII [34], showing a $\rho(770)$-dominance in the $\pi\pi$ invariant mass distribution. Indeed the known $\rho\pi$ *puzzle* that has challenged for long many theoretical determinations is precisely to understand why other higher-mass intermediate states are suppressed and the dynamics is $\rho\pi$ dominated [35–39]. The role of the $J/\psi\rho\pi$ vertex has widely been studied in different frameworks [24, 35, 40–43]. Here, in order to obtain information on the Lagrangian given in Eq. (7) we use this system.

The $J/\psi\rho\pi$ vertex is given by Eq. (7) that, for the fields of our interest, reads:

$$\mathcal{L}_{\psi\rho\pi} = -\sqrt{2}\frac{m_\psi}{F} \, H_1 \, \varepsilon_{\mu\nu\rho\sigma} \Psi^\mu \left[\partial^\nu \pi^+ \rho_-^{\rho\sigma} + \partial^\nu \pi^- \rho_+^{\rho\sigma} + \partial^\nu \pi^0 \rho_0^{\rho\sigma} \right] , \tag{10}$$

and preserves, by construction, the chiral symmetry.
Now we can compute the amplitude for the process $J/\psi \to \pi^0\pi^-\pi^+$ that proceeds through the diagrams in Figure 2. The description of the $\rho\pi\pi$ vertex is given by Eq. (6). The amplitude for the $J/\psi(\varepsilon) \to \pi^0(p_0)\pi^-(p_-)\pi^+(p_+)$ decay is:

$$\mathcal{M}_{3\pi} = -i4\sqrt{2}\frac{G_V H_1}{F^3 m_\psi}\varepsilon_{\nu\sigma\rho\mu}\, p_+^\nu\, p_-^\sigma\, p_0^\rho\, \varepsilon^\mu \left[\frac{m_\psi^2}{\hat{m}_\rho^2(s)-s}+\frac{m_\psi^2}{\hat{m}_\rho^2(u)-u}+\frac{m_\psi^2}{\hat{m}_\rho^2(t)-t}\right], \tag{11}$$

where the Mandelstam variables are defined as $s=(p_++p_-)^2$, $t=(p_0+p_+)^2$ and $u=(p_0+p_-)^2$. Here the finite width of the $\rho(770)$ resonance is taken into account by the mass term, $\hat{m}_\rho^2(s) = m_\rho^2 - im_\rho\Gamma_\rho(s)$, and where [44]:

$$\Gamma_\rho(s) = \frac{m_\rho s}{96\pi F^2}\left[\left(1-\frac{4m_\pi^2}{s}\right)^{3/2}\theta(s-4m_\pi^2)+\frac{1}{2}\left(1-\frac{4m_K^2}{s}\right)^{3/2}\theta(s-4m_K^2)\right]. \tag{12}$$

The decay width depends on two Mandelstam variables, we choose the pair (s,u), whose minimum and maximum values are given as follows,

$$\begin{aligned} 4m_\pi^2 \le s \le (m_\psi - m_\pi)^2 , \\ u_-(s) \le u \le u_+(s) , \end{aligned} \tag{13}$$

where

$$u_\pm(s) = \frac{1}{4s}\left[(m_\psi^2-m_\pi^2)^2 - \left(\lambda^{1/2}(s,m_\pi^2,m_\pi^2) \mp \lambda^{1/2}(m_\psi^2,s,m_\pi^2)\right)^2\right]. \tag{14}$$

Using our finding given in Eq. (9), our prediction is found to be,

$$\mathcal{B}(J/\psi \to \pi^+\pi^-\pi^0) = 1.7097\times 10^{-2} , \tag{15}$$

which is not far away from the value $(2.11\pm.07)\times 10^{-2}$ reported in Ref. [26].

4. Conclusions

Summarizing the work done, we have determined the value of H_1 except by a sign, and we have proved the reliability of our finding by estimating the $\mathcal{BR}(J/\psi \to \pi^+\pi^-\pi^0)$ which competes very well with the measurements. With this result we have a trustworthy model for an effective description of the $J/\psi VP$ vertex. More analysis in others J/ψ decays could shed light on the determination of the correct sign, the central value of this coupling and the truthful impact of the possible direct term $J/\psi\pi\pi\pi$ which is not considered here.

Acknowledgments

This research has been supported in part by Conacyt-SNI and by UANL through the grant PAICYT-2015. Flores-Baez thanks the hospitality and support from IFIC-Valencia where part of this work was done and also thanks to J Portoles for his valuable reading of this manuscript.

References

[1] Weinberg S 1979 *Physica* **A96** 327–340
[2] Gasser J and Leutwyler H 1984 *Annals Phys.* **158** 142
[3] Gasser J and Leutwyler H 1985 *Nucl. Phys.* **B250** 465–516
[4] Ecker G, Gasser J, Pich A and de Rafael E 1989 *Nucl. Phys.* **B321** 311–342
[5] Ecker G, Gasser J, Leutwyler H, Pich A and de Rafael E 1989 *Phys. Lett.* **B223** 425–432

[6] Cirigliano V, Ecker G, Eidemuller M, Kaiser R, Pich A and Portoles J 2006 *Nucl. Phys.* **B753** 139–177 (*Preprint* hep-ph/0603205)
[7] Ruiz-Femenia P D, Pich A and Portoles J 2003 *JHEP* **07** 003 (*Preprint* hep-ph/0306157)
[8] Kampf K and Novotny J 2011 *Phys. Rev.* **D84** 014036 (*Preprint* 1104.3137)
[9] 't Hooft G 1974 *Nucl. Phys.* **B72** 461
[10] Witten E 1979 *Nucl. Phys.* **B160** 57–115
[11] Gomez Dumm D, Pich A and Portoles J 2004 *Phys. Rev.* **D69** 073002 (*Preprint* hep-ph/0312183)
[12] Dumm D G, Roig P, Pich A and Portoles J 2010 *Phys. Lett.* **B685** 158–164 (*Preprint* 0911.4436)
[13] Dumm D G, Roig P, Pich A and Portoles J 2010 *Phys. Rev.* **D81** 034031 (*Preprint* 0911.2640)
[14] Escribano R, Gonzalez-Solis S and Roig P 2013 *JHEP* **10** 039 (*Preprint* 1307.7908)
[15] Guo Z H 2008 *Phys. Rev.* **D78** 033004 (*Preprint* 0806.4322)
[16] Roig P, Guevara A and López Castro G 2014 *Phys. Rev.* **D89** 073016 (*Preprint* 1401.4099)
[17] Rosell I, Sanz-Cillero J J and Pich A 2004 *JHEP* **08** 042 (*Preprint* hep-ph/0407240)
[18] Shifman M A, Vainshtein A I and Zakharov V I 1979 *Nucl. Phys.* **B147** 385–447
[19] Shifman M A, Vainshtein A I and Zakharov V I 1979 *Nucl. Phys.* **B147** 448–518
[20] Roig P and Sanz Cillero J J 2014 *Phys. Lett.* **B733** 158–163 (*Preprint* 1312.6206)
[21] Bijnens J, Girlanda L and Talavera P 2002 *Eur. Phys. J.* **C23** 539–544 (*Preprint* hep-ph/0110400)
[22] Flores-Tlalpa A, Lopez Castro G and Sanchez Toledo G 2005 *Phys. Rev.* **D72** 113003 (*Preprint* hep-ph/0511315)
[23] Ablikim M *et al.* (BES) 2004 *Phys. Lett.* **B598** 149–158 (*Preprint* hep-ex/0406038)
[24] Roca L, Palomar J E, Oset E and Chiang H C 2004 *Nucl. Phys.* **A744** 127–155 (*Preprint* hep-ph/0405228)
[25] Wu F Q and Zou B S 2006 *Phys. Rev.* **D73** 114008 (*Preprint* hep-ph/0603224)
[26] Olive K A *et al.* (Particle Data Group) 2014 *Chin. Phys.* **C38** 090001
[27] Flores-Baez F V, Morones-Ibarra J and Portoles J , work in progress
[28] Chen Y H, Guo Z H and Zou B S 2015 *Phys. Rev.* **D91** 014010 (*Preprint* 1411.1159)
[29] Portoles J 2010 *AIP Conf. Proc.* **1322** 178–187 (*Preprint* 1010.3360)
[30] Cirigliano V, Ecker G, Eidemuller M, Pich A and Portoles J 2004 *Phys. Lett.* **B596** 96–106 (*Preprint* hep-ph/0404004)
[31] Cirigliano V, Ecker G, Eidemuller M, Kaiser R, Pich A and Portoles J 2005 *JHEP* **04** 006 (*Preprint* hep-ph/0503108)
[32] Bijnens J, Gamiz E, Lipartia E and Prades J 2003 *JHEP* **04** 055 (*Preprint* hep-ph/0304222)
[33] Ablikim M *et al.* (BES) 2006 *Phys. Rev.* **D73** 052008 (*Preprint* hep-ex/0510066)
[34] Ablikim M *et al.* (BESIII) 2012 *Phys. Lett.* **B710** 594–599 (*Preprint* 1202.2048)
[35] Szczepaniak A P and Pennington M R 2014 *Phys. Lett.* **B737** 283–288
[36] Kisslinger L S, Parno D and Riordan S 2008 *Adv. High Energy Phys.* **2008** 982341 (*Preprint* 0805.1943)
[37] Suzuki M 2001 *Phys. Rev.* **D63** 054021 (*Preprint* hep-ph/0006296)
[38] Chen Y Q and Braaten E 1998 *Phys. Rev. Lett.* **80** 5060–5063 (*Preprint* hep-ph/9801226)
[39] Brodsky S J and Karliner M 1997 *Phys. Rev. Lett.* **78** 4682–4685 (*Preprint* hep-ph/9704379)
[40] Liu B, Buescher M, Guo F K, Hanhart C and Meissner U G 2009 *Eur. Phys. J.* **C63** 93–99 (*Preprint* 0901.1185)
[41] Liu X, Zeng X Q and Li X Q 2006 *Phys. Rev.* **D74** 074003 (*Preprint* hep-ph/0606191)
[42] Lahde T A and Meissner U G 2006 *Phys. Rev.* **D74** 034021 (*Preprint* hep-ph/0606133)
[43] Li X Q, Bugg D V and Zou B S 1997 *Phys. Rev.* **D55** 1421–1424
[44] Gomez Dumm D, Pich A and Portoles J 2000 *Phys. Rev.* **D62** 054014 (*Preprint* hep-ph/0003320)

Some aspects of large-N_c heavy baryon chiral perturbation theory

Rubén Flores-Mendieta
Instituto de Física, Universidad Autónoma de San Luis Potosí, San Luis Potosí, SLP, Mexico
E-mail: `ruben@ifisica.uaslp.mx`

Abstract. Large-N_c chiral perturbation theory is a calculational scheme for baryons that simultaneously exhibits both the m_q and $1/N_c$ expansions, where N_c is the number of colors of quarks. In this work an outline of the calculation of one-loop corrections to the leading baryon axial-vector and vector form factors in baryon semileptonic decays to order $\mathcal{O}(p^2)$ in the chiral expansion is provided in order to show how the expansion works.

1. Introduction
Chiral perturbation theory [1, 2] is a framework for studying strong-interaction processes at low energies. The global chiral symmetry $\mathrm{SU}(3)_L \times \mathrm{SU}(3)_R$ in QCD is spontaneously broken down to the vector subgroup $\mathrm{SU}(3)_V$ by the vacuum expectation value which yields eight massless and spinless Goldstone bosons. Those bosons correspond to the spontaneously broken internal symmetry generators, which have negative parity, baryon number zero, and transform as an octet under $\mathrm{SU}(3)_V$. The pseudoscalar octet mesons (π, K, η) qualify as candidates for Goldstone bosons. The finite masses of the physical mesons are interpreted as a consequence of the explicit symmetry breaking due to the finite masses of the light quarks m_q in the QCD Lagrangian [3].

The effective Lagrangian is structured in terms of a simultaneous expansion in powers of covariant derivatives and m_q terms so physical observables can be expanded order by order in powers of p^2/Λ_χ^2 and m_Π^2/Λ_χ^2, where p and m_Π are the meson momentum and mass, respectively, and $\Lambda_\chi \sim 1$ GeV is the scale of chiral symmetry breaking.

When chiral perturbation theory is extended to include baryons, it is convenient to introduce velocity-dependent baryon fields, so that the expansion of the baryon chiral Lagrangian in powers of m_q and $1/M$ (where M is the baryon mass) is manifest [4]. This approach is called *heavy baryon chiral perturbation theory*. It was used to compute the corrections to the axial-vector coupling defined in baryon semileptonic decays (BSD) due to meson loops [4]. While these corrections are large when only octet baryon intermediate states are retained, the inclusion of decuplet baryon intermediate states yields large cancellations [4].

Large-N_c QCD is the generalization of QCD from three colors $N_c = 3$ to $N_c \gg 3$ [5, 6]. The baryon sector in the large-N_c limit exhibits an exact $\mathrm{SU}(2N_f)$ spin-flavor symmetry, where N_f is the number of light quark flavors. The difference between the SU(3) invariant masses of the decuplet and octet baryons is $\Delta \propto 1/N_c$ so decuplet and octet baryon states become degenerate in the large-N_c limit. The spin-flavor symmetry allows one to classify large-N_c baryon states and matrix elements and to compute static properties of baryons in an expansion in $1/N_c$, for example masses [7, 8], axial-vector couplings, magnetic moments [9] and vector couplings [10].

Next, *large-N_c heavy baryon chiral perturbation theory* is a calculational scheme that simultaneously exhibits both the m_q and $1/N_c$ expansions. The $1/N_c$ chiral Lagrangian was introduced in Ref. [11] and used to compute flavor-**27** baryon mass splittings at leading order in chiral perturbation theory. Other baryon properties also computed are axial-vector couplings [12, 13, 14], magnetic moments [15, 16], vector couplings [17] and charge radii [18].

The aim is to provide an overview of the calculation of corrections to form factor of BSD.

2. The chiral Lagrangian for baryons in the $1/N_c$ expansion

The $1/N_c$ baryon chiral Lagrangian that implements nonet symmetry and contracted spin-flavor symmetry for baryons in the large-N_c limit is written as [11]

$$\mathcal{L}_{\text{baryon}} = i\mathcal{D}^0 - \mathcal{M}_{\text{hyperfine}} + \text{Tr}\left(\mathcal{A}^k\lambda^c\right) A^{kc} + \frac{1}{N_c}\text{Tr}\left(\mathcal{A}^k\frac{2I}{\sqrt{6}}\right) A^k + \ldots, \tag{1}$$

where $\mathcal{D}^0 = \partial^0\mathcal{I} + \text{Tr}\left(\mathcal{V}^0\lambda^c\right) T^c$.

The baryon operators contained in Eq. (1) have well-defined $1/N_c$ expansions; they can be expressed as polynomials in the SU(6) spin-flavor generators [8]

$$J^k = q^\dagger\frac{\sigma^k}{2}q, \qquad T^c = q^\dagger\frac{\lambda^c}{2}q, \qquad G^{kc} = q^\dagger\frac{\sigma^k}{2}\frac{\lambda^c}{2}q, \tag{2}$$

where $q^\dagger$ and q are operators that create and annihilate states in the fundamental representation of SU(6), and σ^k and λ^c are the Pauli and Gell-Mann matrices, respectively. The flavor indices in Eqs. (1)–(2) run from one to nine to account for the full meson nonet π, K, η and η'.

$\mathcal{M}_{\text{hyperfine}}$ denotes the spin splittings of the tower of baryon states with spins $1/2, \ldots, N_c/2$ in the flavor representations. The vector and axial vector combinations of the meson fields,

$$\mathcal{V}^0 = \frac{1}{2}\left(\xi\partial^0\xi^\dagger + \xi^\dagger\partial^0\xi\right), \qquad \mathcal{A}^k = \frac{i}{2}\left(\xi\nabla^k\xi^\dagger - \xi^\dagger\nabla^k\xi\right), \tag{3}$$

couple to baryon vector and axial-vector currents, respectively, $\xi = \exp[i\Pi(x)/f]$, where $\Pi(x)$ is the nonet of Goldstone boson fields and $f \approx 93$ MeV is the meson decay constant. The $\ell = 1$ flavor octet axial-vector pion combination couples to the baryon axial-vector current, A^{kc}. It is a spin-1 object, an octet under SU(3), and odd under time reversal. Its $1/N_c$ expansion is [8]

$$A^{kc} = a_1 G^{kc} + \sum_{n=2,3}^{N_c} b_n \frac{1}{N_c^{n-1}}\mathcal{D}_n^{kc} + \sum_{n=3,5}^{N_c} c_n \frac{1}{N_c^{n-1}}\mathcal{O}_n^{kc}. \tag{4}$$

The first terms in expansion (4) are

$$\mathcal{D}_2^{kc} = J^k T^c, \qquad \mathcal{D}_3^{kc} = \{J^k, \{J^r, G^{rc}\}\}, \qquad \mathcal{O}_3^{kc} = \{J^2, G^{kc}\} - \frac{1}{2}\{J^k, \{J^r, G^{rc}\}\}, \tag{5}$$

and higher order terms are obtained as $\mathcal{D}_n^{kc} = \{J^2, \mathcal{D}_{n-2}^{kc}\}$ and $\mathcal{O}_n^{kc} = \{J^2, \mathcal{O}_{n-2}^{kc}\}$ for $n \geq 4$. The unknown coefficients a_1, b_n, and c_n have expansions in powers of $1/N_c$ and are order unity at leading order. At the physical value $N_c = 3$ the series can be truncated as

$$A^{kc} = a_1 G^{kc} + b_2\frac{1}{N_c}\mathcal{D}_2^{kc} + b_3\frac{1}{N_c^2}\mathcal{D}_3^{kc} + c_3\frac{1}{N_c^2}\mathcal{O}_3^{kc}. \tag{6}$$

The matrix elements of the space components of A^{kc} in Eq. (6) between SU(6) states give the tree-level values of the axial-vector couplings.

Similarly, the $1/N_c$ expansion for the baryon mass operator $\mathcal{M}$ can be written as [7, 8]

$$\mathcal{M} = m_0 N_c \mathcal{I} + \sum_{n=2,4}^{N_c-1} m_n \frac{1}{N_c^{n-1}} J^n, \tag{7}$$

where m_n are unknown coefficients. The first term on the right-hand side of Eq. (7) is the overall spin-independent mass of the baryon multiplet and is removed from the chiral Lagrangian by the heavy baryon field redefinition [4]. The additional spin-dependent terms define $\mathcal{M}_{\text{hyperfine}}$.

3. One-loop corrections to the baryon axial-vector current

A^{kc} is renormalized by the one-loop diagrams displayed in Fig. 1, which are given by the product of a group theoretic structure times a loop integral [12, 13].

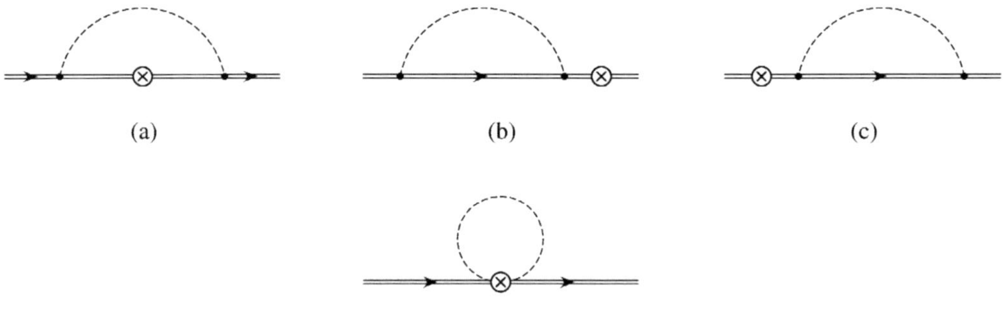

Figure 1. One-loop corrections to the baryon axial-vector current.

Specifically, the correction arising from the sum of the diagrams of Figs. 1(a,b,c) reads [12]

$$\begin{aligned}\delta A^{kc} &= \frac{1}{2}[A^{ja},[A^{jb},A^{kc}]]\Pi^{ab}_{(1)} - \frac{1}{2}\{A^{ja},[A^{kc},[\mathcal{M},A^{jb}]]\}\Pi^{ab}_{(2)} \\ &\quad + \frac{1}{6}\left([A^{ja},[[\mathcal{M},[\mathcal{M},A^{jb}]],A^{kc}]] - \frac{1}{2}[[\mathcal{M},A^{ja}],[[\mathcal{M},A^{jb}],A^{kc}]]\right)\Pi^{ab}_{(3)} + \ldots \end{aligned} \tag{8}$$

Here, A^{ia} and A^{jb} are used at the meson-baryon vertices and $\Pi^{ab}_{(n)}$ is a symmetric tensor which contains meson loop integrals $F^{(n)}(m_\Pi, \Delta, \mu)$ with the exchange of a single meson. It decomposes into flavor singlet, flavor **8**, and flavor **27** representations as [11]

$$\Pi^{ab}_{(n)} = F^{(n)}_{\mathbf{1}}\delta^{ab} + F^{(n)}_{\mathbf{8}} d^{ab8} + F^{(n)}_{\mathbf{27}}\left[\delta^{a8}\delta^{b8} - \frac{1}{8}\delta^{ab} - \frac{3}{5}d^{ab8}d^{888}\right], \tag{9}$$

where $F^{(n)}_{\mathbf{k}}$ are linear combinations of $F^{(n)}(m_\Pi, \Delta, \mu)$ [12].

δA^{kc} contains n-body operators,[1] with $n > N_c$, which are complicated commutators and/or anticommutators of the one-body operators J^k, T^c, and G^{kc}. All these structures should be reduced and rewritten as linear combinations of the operator basis, with $n \leq N_c$. Although the operator basis is complete and independent, the reduction could be rather difficult.

[1] An n-body operator is one with n q's and n $q^\dagger$'s and can be written as a polynomial of order n in J^i, T^a, and G^{ia} [8].

The N_c dependence of the matrix elements of J^i, T^a, and G^{ia} is nontrivial [8]. A simplified counting rule for baryons with spins of order unity is [12]

$$T^a \sim N_c, \qquad G^{ia} \sim N_c, \qquad J^i \sim 1. \tag{10}$$

Similarly, $f \propto \sqrt{N_c}$, so the functions $F^{(n)}(m_\Pi, \Delta, \mu)$ introduce a $1/N_c$ suppression.

Naively, the double commutator would be $\mathcal{O}(N_c^3)$: one factor of N_c from each A^{kc}. However, there are large-N_c cancellations between the diagrams of Figs. 1(a,b,c), *provided all baryon states in a complete multiplet of the large-N_c spin-flavor symmetry are included in the sum over intermediate states* [12]. The terms with one and two baryon mass insertions also contain large-N_c cancellations [12]. Thus, the correction δA^{kc} is $\mathcal{O}(1)$, or $1/N_c$ times the tree level value.

In a similar way, the one-loop correction to A^{kc} from the diagram of Fig. 1(d) is given by

$$\delta A^{kc} = -\frac{1}{2}[T^a, [T^b, A^{kc}]]\Pi^{ab}, \tag{11}$$

where Π^{ab} is a tensor similar to the one in Eq. (9) with a different loop integral [12].

The double commutator in Eq. (11) is proportional to A^{kc} [12] so the correction is at most $\mathcal{O}(1)$ and is of the same order as the one arising from the sum of Figs. 1(a,b,c).

Finally, in the limit $\Delta/m_\Pi = 0$ the one-loop correction to A^{kc} becomes

$$\delta A^{kc} = \frac{1}{2}[A^{ja}, [A^{jb}, A^{kc}]]\Pi_{(1)}^{ab} - \frac{1}{2}[T^a, [T^b, A^{kc}]]\Pi^{ab}. \tag{12}$$

The matrix elements between spin-$\frac{1}{2}$ baryon states of the space components of the renormalized baryon axial-vector current, $A^{kc} + \delta A^{kc}$, give the actual value of g_1, as defined in BSD, with a normalization such that $g_1 \approx 1.27$ for neutron decay.

A comparison can be done between the g_1 obtained in the combined formalism and the one obtained in heavy baryon chiral perturbation theory [4] through the relations between the SU(3) invariant couplings D, F, $\mathcal{C}$ and $\mathcal{H}$ and the operator coefficients of the expansion (6),

$$D = \frac{1}{2}a_1 + \frac{1}{6}b_3, \quad F = \frac{1}{3}a_1 + \frac{1}{6}b_2 + \frac{1}{9}b_3, \quad \mathcal{C} = -a_1 - \frac{1}{2}c_3, \quad \mathcal{H} = -\frac{3}{2}a_1 - \frac{3}{2}b_2 - \frac{5}{2}b_3. \tag{13}$$

Both approaches yield the same results.

4. One-loop corrections to the baryon vector current

The baryon vector current V^{0c} is a spin-0 object and an octet under SU(3). At $q^2 = 0$, V^{0c} is the generator of SU(3) symmetry transformations, so to all orders in the $1/N_c$ expansion [10]

$$V^{0c} = T^c. \tag{14}$$

The matrix elements of V^{0c} of Eq. (14) between SU(6) baryon states yield the form factors f_1 in the limit of exact SU(3) symmetry, $f_1^{\mathrm{SU(3)}}$. One-loop corrections to these form factors for $|\Delta S| = 1$ processes arise from the diagrams displayed in Fig. 2. Due to the Ademollo-Gatto theorem, these corrections appear to order $\epsilon^2 \sim (m_s - \hat{m})^2$. Details can be found in Ref. [17].

For instance, the contribution to V^{0c} arising from Fig. 2(a) can be written as

$$\delta V_{(\mathrm{a})}^{c} = \sum_{\mathrm{j}} A^{ia}\mathcal{P}_{\mathrm{j}} A^{ib} P^{abc}(\Delta_{\mathrm{j}}). \tag{15}$$

Here $\mathcal{P}_{\mathrm{j}}$ is the baryon propagator for spin $J = \mathrm{j}$ [11]

$$\frac{i\mathcal{P}_{\mathrm{j}}}{k^0 - \Delta_{\mathrm{j}}}, \tag{16}$$

Journal of Physics: Conference Series **761** (2016) 012085 doi:10.1088/1742-6596/761/1/012085

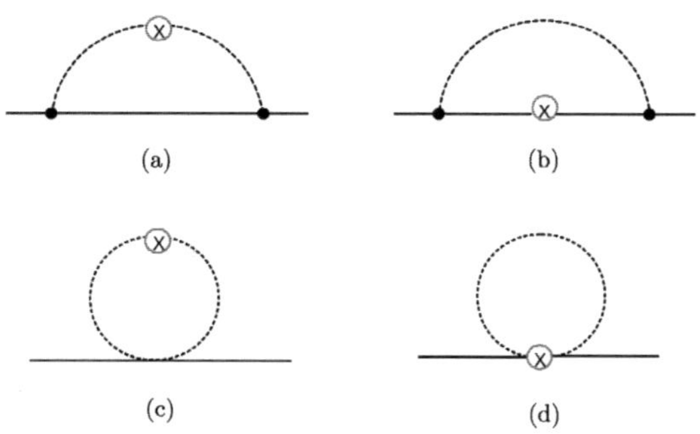

Figure 2. One-loop corrections to the vector current..

which satisfies by definition $\mathcal{P}_{\rm j}^2 = \mathcal{P}_{\rm j}$ and $\mathcal{P}_{\rm j}\mathcal{P}_{\rm j'} = 0$ for $\rm j \neq j'$. Similarly, $\Delta_{\rm j}$ stands for the difference of the hyperfine mass splitting between the intermediate baryon with spin $J = \rm j$ and the external baryon. The remaining corrections can be obtained without difficulty in the combined expansion in m_q and $1/N_c$. Details can be found in Ref. [17].

The one-loop corrections to order $\mathcal{O}(p^2)$ in the chiral expansion to f_1 for Λp process is [17]

$$\begin{aligned}\left[\frac{f_1}{f_1^{\rm SU(3)}}\right]_{\Lambda p} &= 1 + \left[\frac{3}{8} + \frac{17}{32}a_1^2 + \frac{3}{16}a_1b_2 + \frac{17}{48}a_1b_3 + \frac{1}{32}b_2^2 + \frac{1}{16}b_2b_3 + \frac{17}{288}b_3^2\right] H(m_\pi, m_K) \\ &+ \left[\frac{3}{8} + \frac{9}{32}a_1^2 + \frac{3}{16}a_1b_2 + \frac{3}{16}a_1b_3 + \frac{1}{32}b_2^2 + \frac{1}{16}b_2b_3 + \frac{1}{32}b_3^2\right] H(m_K, m_\eta) \\ &+ \left[-\frac{1}{4}a_1^2 - \frac{1}{4}a_1c_3 - \frac{1}{16}c_3^2\right] K(m_\pi, m_K, \Delta), \end{aligned} \tag{17}$$

where $H(m_1, m_2)$ and $K(m_1, m_2, \Delta)$ arise from loop integrals. When relations (13) are used in Eq. (17), the expression obtained in heavy baryon chiral perturbation theory [19] is recovered.

The chiral corrections to the vector form factor can be parametrized as

$$f_1 = f_1^{\rm SU(3)}(1 + \delta^{(2)} + \ldots), \tag{18}$$

where $\delta^{(2)}$ is the leading SU(3)-breaking loop correction $\mathcal{O}(p^2)$ and the dots stand for higher chiral corrections not computed yet within the present formalism.

5. Fit to data

A comparison of the theoretical expressions with the available experimental data [20] through a least-squares fit can be performed. For octet baryons the data are the decay rates R, the ratios g_1/f_1, the angular correlation coefficients $\alpha_{e\nu}$ and the spin-asymmetry coefficients α_e, α_ν, α_B, A and B. For decuplet baryons, the axial couplings g for the processes $\Delta \to N\pi$, $\Sigma^* \to \Lambda\pi$, $\Sigma^* \to \Sigma\pi$, $\Xi^* \to \Xi\pi$ are determined. The free parameters are the operator coefficients of A^{kc} plus a set of parameters d_i which are introduced by explicit SU(3) symmetry breaking Ref. [17].

The fit yields [17]

$$\begin{aligned} &a_1 = 0.95 \pm 0.14, \quad b_2 = -1.10 \pm 0.19, \quad b_3 = 1.10 \pm 0.09, \quad c_3 = 1.07 \pm 0.15, \\ &d_1 = 0.62 \pm 0.13, \quad d_2 = -0.57 \pm 0.24, \quad d_3 = 0.39 \pm 0.05, \quad d_4 = -0.06 \pm 0.08. \end{aligned} \tag{19}$$

with $\chi^2 = 5.5/4\,\text{dof}$. With the best-fit parameters, the predicted values of observables in BSD as well as the pattern of SU(3) symmetry breaking in f_1 are listed in Table 1 for completeness.

Table 1. Predicted observables in BSD. R is given in $10^{-3}\,\mathrm{s}^{-1}$ for np process and $10^{6}\,\mathrm{s}^{-1}$ for the rest.

	np	$\Sigma^+\Lambda$	$\Sigma^-\Lambda$	Λp	$\Sigma^- n$	$\Xi^-\Lambda$	$\Xi^-\Sigma^0$	$\Xi^0\Sigma^+$
R	1.128	0.232	0.387	2.949	6.284	2.844	0.454	0.820
$\alpha_{e\nu}$	−0.079	−0.406	−0.414	−0.026	0.340	0.559		
α_e	−0.087			0.014	−0.627			
α_ν	0.987			0.977	−0.357			
α_B				−0.586	0.668			
A			0.049			0.583		
B			0.888					
g_1/f_1	1.270			0.718	−0.340	0.254	1.217	1.217
$f_1/f_1^{\mathrm{SU(3)}}$				0.952	0.966	0.953	0.962	0.962

To order $\mathcal{O}(p^2)$ in the chiral expansion, corrections to the form factors f_1 produce a systematic decrease of around 5% with respect to their SU(3) symmetric values. Also, the values of the predicted observables are consistent with the measured ones.

6. Conclusions

The combined formalism in m_q and $1/N_c$ expansions is an outstanding approach to analyze the static properties of baryons in a systematic way. Here the analysis of leading axial-vector and vector form factors that appear in BSD has been presented to exemplify how the method works. A number of baryon properties can be studied with it, which constitutes an important advancement in our understanding of the low-energy consequences of QCD.

6.1. Acknowledgments

This work was supported in part by CONACYT and FAI (UASLP), Mexico.

References

[1] Weinberg S 1979 *Physica* A **96** 327
[2] Gasser J and Leutwyler H 1984 *Annals Phys.* **158** 142
[3] Gasser J and Leutwyler H 1982 *Phys. Rept.* **87** 77
[4] Jenkins E and Manohar A V 1991 *Phys. Lett.* B **255** 558; *Phys. Lett.* B **259** 353
[5] 't Hooft G 1974 *Nucl. Phys.* **B75** 461
[6] Witten E 1979 *Nucl. Phys.* **B160** 57
[7] Dashen R, Jenkins E, and Manohar A V 1994 *Phys. Rev.* D **49** 4713
[8] Dashen R, Jenkins E, and Manohar A V 1995 *Phys. Rev.* D **51** 3697
[9] Dai J, Dashen R, Jenkins E and Manohar A V 1996 *Phys. Rev.* D **53** 273
[10] Flores-Mendieta R, Jenkins E, and Manohar A V 1998 *Phys. Rev.* D **58** 094028
[11] Jenkins E 1996 *Phys. Rev.* D **53** 2625
[12] Flores-Mendieta R, Hofmann C P, Jenkins E and Manohar A V 2000 *Phys. Rev.* D **62** 034001
[13] Flores-Mendieta R and Hofmann C.P. 2006 *Phys. Rev.* D **74** 094001
[14] Flores-Mendieta R, Hernández-Ruiz M A, and Hofmann C P 2012 *Phys. Rev.* D **86** 094041
[15] Flores-Mendieta R 2009 *Phys. Rev.* D **80** 094014
[16] Ahuatzin G, Flores-Mendieta R, Hernández-Ruiz M A and Hofmann C P 2014 *Phys. Rev.* D **89** 034012
[17] Flores-Mendieta R and Goity J L 2014 *Phys. Rev.* D **90** 114008
[18] Flores-Mendieta R and Rivera-Ruiz A 2015 *Phys. Rev.* D **92** 094026
[19] Anderson J and Luty M A 1993 *Phys. Rev.* D **47** 4975
[20] Olive K A *et al.* [Particle Data Group Collaboration] 2014 *Chin. Phys. C* **38** 090001

Decay coupling constants sum rules for dibaryon octet into two baryon octets with λ_8 first order SU(3) symmetry breaking

E. N. Polanco-Euán[1], V. Gupta[2], G. Sánchez-Colón[3], and B. A. Bambah[4]

[1,2,3]Departamento de Física Aplicada.
Centro de Investigación y de Estudios Avanzados del IPN.
Unidad Mérida.
A.P. 73, Cordemex.
Mérida, Yucatán, 97310. MÉXICO.
[4]School of Physics.
University of Hyderabad.
Hyderabad, A.P. 500 046, India.

E-mail: [3]`gabriel.sanchez@cinvestav.mx`

Abstract. The SU(3) octet states with baryon number $B = 2$, hexaquark dibaryons, are considered. Strong decay coupling constants sum rules for dibaryon octet into two ordinary baryon octets with λ_8 first order SU(3) symmetry breaking are given. An SU(4) extension of the analysis is commented upon. Possibilities for the experimental observation of multibaryon and anti-multibaryon states are pointed out.

1. Introduction.

The first hexaquark or sextaquark state with baryon number $B = 2$, a dibaryon, was proposed in 1964 [1]. Hexaquarks received more attention after the proposal of the H-dibaryon ($uuddss$) by Jaffe [2]. The presence of hexaquark states have also been conjectured in models with colour flux tube (string) confinement, [3, 4, 5]. The theoretical case for the dibaryon (H-dibaryon) has recently been made stronger by lattice-QCD calculations in the SU(3) flavor limit by the NPLQCD and HALQCD collaborations [6, 7]. The presence of multiquark states including the hexaquark in hot-dense matter is predicted in certain statistical [8] and holographic models [9], and since strangeness enhancement is associated with the quark-gluon plasma, hexaquarks with strangeness may have an observable signature in the upcoming heavy ion experiments at LHC. More recently Bashkanov *et. al* [10] have suggested novel dibaryon states and their possible experimental signatures. These include exclusive measurements of the $np \to np\pi^0\pi^0$ cross section at the WASA detector in COSY, Julich, which have shown the presence of a $d^*(2348)$ resonance that can be interpreted as a dibaryon state [11]. Recently in Ref. [12], it was noted that the deuteron, a stable dibaryon, belongs to the $B = 2$, $\mathbf{10^*}$ representation in the SU(3) $\mathbf{8}\otimes\mathbf{8}$ decomposition of two $J^P = (1/2)^+$ baryon octets (denoted here as $B_\mathbf{8}$) and sum rules for strong decays of $\mathbf{10^*}$ and $\mathbf{10}$ dibaryons decuplets (denoted $D_{\mathbf{10^*},\mathbf{10}}$) into an antisymmetric final state of two octet baryons, $D_{\mathbf{10^*},\mathbf{10}} \to B_\mathbf{8} + B_\mathbf{8}$, were given. A recent high statistics search in inclusive

Upsilon decays was done for H-dibaryons in the symmetric SU(3) state at the Belle detector of KEKB e^+e collider [13], which put stringent constraints on the H-dibaryon mass and clearly showed that if a H-dibaryon exists with mass $\leq 2\,m_\Lambda$, then it must be different from a deuteron. M. Oka has also proposed the possibility of several dibaryon resonances examining flavor SU(3) octet dibaryon systems employing the quark cluster model [14]. Thus, it is important to study the various decay channels of a dibaryon belonging to the dibaryon octet.

In this paper we present sum rules for the strong decays of the dibaryon SU(3) octet (denoted $D_\mathbf{8}$) into two ordinary $J^P = (1/2)^+$ baryon octets with the usual $\lambda_8 = (1/\sqrt{3})\,\mathrm{diag}(1, 1, -2)$ first order SU(3) symmetry breaking (the SU(3) symmetry breaking interaction term transforms as the eigth component of the unitary spin, *i.e.* like the hypercharge [15]). This additional study of the symmetry breaking mechanism at SU(3) level sets the stage for how to treat the breaking of SU(4) to get reasonable estimates for couplings of charmed hexaquark states.

In Sec. 2 the baryonic content of the $D_\mathbf{8}$ states and their expected masses are presented. In Sec. 3 decay sum rules in broken SU(3) are determined. In Sec. 3.1.1 the extension to the SU(4) case is considered.

Finally we discuss some open problems for multiquark states.

2. Baryonic content and masses of dibaryon octet

The (Y, I) states (with hypercharge Y and isospin I) of the dibaryon octet, denoted by $D_\mathbf{8}(Y, I)$ are given in Table 1. The last column gives an estimate of the average mass of $D_\mathbf{8}(Y, I)$ depending on its baryonic content. From these we obtain:

$$3D_\mathbf{8}(0,0) + D_\mathbf{8}(0,1) = 9286.2\mathrm{MeV}, \tag{1}$$

and

$$2[D_\mathbf{8}(1,1/2) + D_\mathbf{8}(-1,1/2)] = 9132.0\mathrm{MeV}. \tag{2}$$

Table 1. The (Y, I) states $D_\mathbf{8}(Y, I)$ in the $D_\mathbf{8}$ dibaryon octet that one obtains in the reduction $\mathbf{8} \otimes \mathbf{8}$ of two ordinary baryon octets $B_\mathbf{8} + B_\mathbf{8}$. For the SU(3) Clebsch-Gordan coefficients see Refs. [15, 16]. The average mass (in MeV) is given in the last column.

$D_\mathbf{8}(Y,I)$	$B_\mathbf{8} + B_\mathbf{8}$	Average mass
$D_\mathbf{8}(1,\frac{1}{2})$	$\frac{1}{2}[(\Sigma N')_{\frac{1}{2}} + (N\Sigma')_{\frac{1}{2}} + (N\Lambda')_{\frac{1}{2}} - (\Lambda N')_{\frac{1}{2}}]$	2093.5
$D_\mathbf{8}(0,1)$	$\frac{1}{\sqrt{6}}[2(\Sigma\Sigma')_1 + (N\Xi')_1 - (\Xi N')_1]$	2514.9
$D_\mathbf{8}(0,0)$	$\frac{1}{\sqrt{2}}[(N\Xi')_0 + (\Xi N')_0]$	2251.5
$D_\mathbf{8}(-1,\frac{1}{2})$	$\frac{1}{\sqrt{2}}[(\Sigma\Xi')_{\frac{1}{2}} + (\Xi\Sigma')_{\frac{1}{2}} + (\Lambda\Xi')_{\frac{1}{2}} - (\Xi\Lambda')_{\frac{1}{2}}]$	2472.5

It is remarkable how well these estimates satisfy the first order Gell Mann-Okubo mass formula for the octet [17, 18].

3. Sum rules for $D_8 \to B_8 + B_8$ strong decays with first order symmetry breaking of SU(3).

We consider now dibaryon octet strong decays into a final state of two baryon octets: $D_\mathbf{8} \to B_\mathbf{8} + B_\mathbf{8}$. Following Ref. [15] we write $B_\mathbf{8}$ and $D_\mathbf{8}$ matrices as:

$$B_{\mathbf{8}} = \begin{pmatrix} \frac{1}{\sqrt{6}}\Lambda + \frac{1}{\sqrt{2}}\Sigma^0 & \Sigma^+ & p \\ \Sigma^- & \frac{1}{\sqrt{6}}\Lambda - \frac{1}{\sqrt{2}}\Sigma^0 & n \\ -\Xi^- & \Xi^0 & -\sqrt{\frac{2}{3}}\Lambda \end{pmatrix}, \tag{3}$$

$$D_{\mathbf{8}} = \begin{pmatrix} \frac{1}{\sqrt{6}}D_{\mathbf{8}}(0,0,0) + \frac{1}{\sqrt{2}}D_{\mathbf{8}}(0,1,0) & D_{\mathbf{8}}(0,1,+1) & D_{\mathbf{8}}(1,1/2,+1/2) \\ D_{\mathbf{8}}(0,1,-1) & \frac{1}{\sqrt{6}}D_{\mathbf{8}}(0,0,0) - \frac{1}{\sqrt{2}}D_{\mathbf{8}}(0,1,0) & D_{\mathbf{8}}(1,1/2,-1/2) \\ -D_{\mathbf{8}}(-1,1/2,-1/2) & D_{\mathbf{8}}(-1,1/2,+1/2) & -\sqrt{\frac{2}{3}}D_{\mathbf{8}}(0,0,0) \end{pmatrix}, \tag{4}$$

where the (Y, I, I_3) states of $D_{\mathbf{8}}$ are denoted by $D_{\mathbf{8}}(Y, I, I_3)$ with I_3 the isospin third component.

In general, for exact SU(3) the Yukawa interaction for $D_{\mathbf{8}} \to B_{\mathbf{8}} + B_{\mathbf{8}}$ ($\mathbf{8} \to \mathbf{8} \otimes \mathbf{8}$) is characterized by two parameters, $g^0_{\mathbf{8}}$ and $g^0_{\mathbf{8}'}$, since there are two octets in the decomposition:

$$\mathbf{8} \otimes \mathbf{8} = \mathbf{1} \oplus \mathbf{8} \oplus \mathbf{8}' \oplus \mathbf{10} \oplus \mathbf{10}^* \oplus \mathbf{27}. \tag{5}$$

The **1**, **8**, and **27** are symmetric under the interchange of the original two octets and $\mathbf{8}'$, **10**, and $\mathbf{10}^*$ are antisymmetric.

With λ_8 first order breaking of SU(3) we expect eight additional parameters in addition to $g^0_{\mathbf{8}}$ and $g^0_{\mathbf{8}'}$ since the one dimensional representation **1** appears eight times in the decomposition $\mathbf{8} \otimes \mathbf{8} \otimes \mathbf{8} \otimes \mathbf{8}$, four corresponding to couplings with flavor-antisymmetric final state representations and four corresponding to couplings with flavor-symmetric final state representations.

In this work we will analyze both cases, antisymmetric and symmetric final state, separately.

3.1. Decays of octet dibaryon into $B_{\mathbf{8}} + B_{\mathbf{8}}$ flavor-antisymmetric final state.
We consider strong decays of the dibaryon octet into a flavor-antisymmetric final state of two baryon octets. As described above, in this case we have five parameters: $g^0_{\mathbf{8}'}$ from exact SU(3) and g_k ($k = 1, 2, 3, 4$) from λ_8 first order SU(3) symmetry breaking. The $D_{\mathbf{8}} \to B_{\mathbf{8}} + B_{\mathbf{8}}$ Yukawa couplings in terms of $g^0_{\mathbf{8}'}$ and g_k ($k = 1, 2, 3, 4$) are (for sake of clarity, we put a prime on the baryon matrix in the second octet, $\overline{D}_{\mathbf{8}} = D^{\mathrm{T}}_{\mathbf{8}}$, and the traces are on the flavor space):

$$g^0_{\mathbf{8}'} \,\mathrm{Tr}\left\{\overline{D}_{\mathbf{8}}\left[B_{\mathbf{8}}B'_{\mathbf{8}} - B'_{\mathbf{8}}B_{\mathbf{8}}\right]\right\}, \tag{6}$$

$$g_1 \,\mathrm{Tr}\left\{\overline{D}_{\mathbf{8}}\lambda_8\left[B_{\mathbf{8}}B'_{\mathbf{8}} - B'_{\mathbf{8}}B_{\mathbf{8}}\right]\right\}, \tag{7}$$

$$g_2 \,\mathrm{Tr}\left\{\overline{D}_{\mathbf{8}}\left[B_{\mathbf{8}}\lambda_8 B'_{\mathbf{8}} - B'_{\mathbf{8}}\lambda_8 B_{\mathbf{8}}\right]\right\}, \tag{8}$$

$$g_3 \,\mathrm{Tr}\left\{\overline{D}_{\mathbf{8}}\left[B_{\mathbf{8}}B'_{\mathbf{8}} - B'_{\mathbf{8}}B_{\mathbf{8}}\right]\lambda_8\right\}, \tag{9}$$

$$g_4 \left\{\mathrm{Tr}\left[\overline{D}_{\mathbf{8}}B_{\mathbf{8}}\right]\mathrm{Tr}\left[B'_{\mathbf{8}}\lambda_8\right] - \mathrm{Tr}\left[\overline{D}_{\mathbf{8}}B'_{\mathbf{8}}\right]\mathrm{Tr}\left[B_{\mathbf{8}}\lambda_8\right]\right\}. \tag{10}$$

There are eight independent decay coupling constants to be determined. We take these eight constants to be:

$$G[D_{\mathbf{8}}(1,1/2,+1/2) \to p\,\Sigma^{0'}], \quad G[D_{\mathbf{8}}(1,1/2,+1/2) \to p\,\Lambda'], \tag{11}$$

$$G[D_{\mathbf{8}}(0,1,+1) \to p\,\Xi^{0'}],\; G[D_{\mathbf{8}}(0,1,+1) \to \Sigma^+\,\Lambda'],\; G[D_{\mathbf{8}}(0,1,+1) \to \Sigma^+\,\Sigma^{0'}], \tag{12}$$

$$G[D_{\mathbf{8}}(0,0,0) \to p\,\Xi^{-\prime}], \tag{13}$$

$$G[D_{\mathbf{8}}(-1,1/2,+1/2) \to \Sigma^+\,\Xi^{-\prime}], \quad G[D_{\mathbf{8}}(-1,1/2,+1/2) \to \Lambda\,\Xi^{0\prime}]. \tag{14}$$

The remaining decay coupling constants are given by:

$$\begin{aligned}\sqrt{2}\,G[D_{\mathbf{8}}(1,1/2,+1/2) \to p\,\Sigma^{0\prime}] &= -G[D_{\mathbf{8}}(1,1/2,+1/2) \to n\,\Sigma^{+\prime}] = \\ = \; G[D_{\mathbf{8}}(1,1/2,-1/2) \to p\,\Sigma^{-\prime}] &= -\sqrt{2}\,G[D_{\mathbf{8}}(1,1/2,-1/2) \to n\,\Sigma^{0\prime}],\end{aligned} \tag{15}$$

$$G[D_{\mathbf{8}}(1,1/2,+1/2) \to p\,\Lambda'] = G[D_{\mathbf{8}}(1,1/2,-1/2) \to n\,\Lambda'], \tag{16}$$

$$G[D_{\mathbf{8}}(0,1,+1) \to \Sigma^+\,\Lambda'] = G[D_{\mathbf{8}}(0,1,0) \to \Sigma^0\,\Lambda'] = G[D_{\mathbf{8}}(0,1,-1) \to \Sigma^-\,\Lambda'], \tag{17}$$

$$\begin{aligned}G[D_{\mathbf{8}}(0,1,+1) \to p\,\Xi^{0\prime}] &= \sqrt{2}\,G[D_{\mathbf{8}}(0,1,0) \to p\,\Xi^{-\prime}] = \sqrt{2}\,G[D_{\mathbf{8}}(0,1,0) \to n\,\Xi^{0\prime}] = \\ &= G[D_{\mathbf{8}}(0,1,-1) \to n\,\Xi^{-\prime}],\end{aligned} \tag{18}$$

$$G[D_{\mathbf{8}}(0,1,+1) \to \Sigma^+\,\Sigma^{0\prime}] = G[D_{\mathbf{8}}(0,1,0) \to \Sigma^+\,\Sigma^{-\prime}] = G[D_{\mathbf{8}}(0,1,-1) \to \Sigma^0\,\Sigma^{-\prime}], \tag{19}$$

$$G[D_{\mathbf{8}}(0,0,0) \to n\,\Xi^{0\prime}] = -G[D_{\mathbf{8}}(0,0,0) \to p\,\Xi^{-\prime}], \tag{20}$$

$$\begin{aligned}G[D_{\mathbf{8}}(-1,1/2,+1/2) \to \Sigma^+\,\Xi^{-\prime}] &= -\sqrt{2}\,G[D_{\mathbf{8}}(-1,1/2,+1/2) \to \Sigma^0\,\Xi^{0\prime}] = \\ = \; \sqrt{2}\,G[D_{\mathbf{8}}(-1,1/2,-1/2) \to \Sigma^0\,\Xi^{-\prime}] &= -G[D_{\mathbf{8}}(-1,1/2,-1/2) \to \Sigma^-\,\Xi^{0\prime}],\end{aligned} \tag{21}$$

$$G[D_{\mathbf{8}}(-1,1/2,+1/2) \to \Lambda\,\Xi^{0\prime}] = G[D_{\mathbf{8}}(-1,1/2,-1/2) \to \Lambda\,\Xi^{-\prime}]. \tag{22}$$

The eight independent decay coupling constants (11)-(14) are described by five parameters $g^0_{8'}$ and g_k ($k = 1,2,3,4$). Thus, three SU(3) broken sum rules between the Yukawa couplings may be deduced:

$$\begin{aligned}4\,G[D_{\mathbf{8}}(1,1/2,+1/2) \to p\,\Sigma^{0\prime}] &= 2\sqrt{3}\,G[D_{\mathbf{8}}(0,0,0) \to p\,\Xi^{-\prime}] \\ &- \sqrt{2}\,G[D_{\mathbf{8}}(0,1,+1) \to p\,\Xi^{0\prime}] + 2\,G[D_{\mathbf{8}}(0,1,+1) \to \Sigma^+\,\Sigma^{0\prime}] \\ &- 2\sqrt{2}\,G[D_{\mathbf{8}}(-1,1/2,+1/2) \to \Sigma^+\,\Xi^{-\prime}],\end{aligned} \tag{23}$$

$$\begin{aligned}2\sqrt{6}\,G[D_{\mathbf{8}}(1,1/2,+1/2) \to p\,\Lambda'] &= 3\sqrt{6}\,G[D_{\mathbf{8}}(0,0,0) \to p\,\Xi^{-\prime}] \\ &+ 5\,G[D_{\mathbf{8}}(0,1,+1) \to p\,\Xi^{0\prime}] - \sqrt{2}\,G[D_{\mathbf{8}}(0,1,+1) \to \Sigma^+\,\Sigma^{0\prime}] \\ &+ 2\sqrt{6}\,G[D_{\mathbf{8}}(-1,1/2,+1/2) \to \Xi^0\,\Lambda'],\end{aligned} \tag{24}$$

$$\begin{aligned}&\sqrt{6}\,G[\,D_{\mathbf{8}}(0,1,+1) \to \Sigma^+\,\Lambda'] = \\ &= 2\,G[D_{\mathbf{8}}(0,1,+1) \to p\,\Xi^{0\prime}] - \sqrt{2}\,G[D_{\mathbf{8}}(0,1,+1) \to \Sigma^+\,\Sigma^{0\prime}] \\ &+ 3\,G[D_{\mathbf{8}}(-1,1/2,+1/2) \to \Sigma^+\,\Xi^{-\prime}] + \sqrt{6}\,G[D_{\mathbf{8}}(-1,1/2,+1/2) \to \Xi^0\,\Lambda'],\end{aligned} \tag{25}$$

with identical relationships for $G[D_{\mathbf{8}}(Y,I,I_3) \to B'\,B] = -G[D_{\mathbf{8}}(Y,I,I_3) \to B\,B']$.

3.1.1. Comments on the extension to SU(4). Exact SU(4) symmetry is often used to make estimates of charmed baryon couplings. This gives rise to wrong results as SU(4) is badly broken. In this section we comment on the extension to SU(4) of the analysis of Sec. 3.1 [19].

The SU(4) group representation includes a new quantum number charm, which comes as a new quark. The baryons fit into the product:

$$\mathbf{4}\otimes\mathbf{4}\otimes\mathbf{4}=\mathbf{4}^{*}\oplus\mathbf{20'}\oplus\mathbf{20'}\oplus\mathbf{20}, \tag{26}$$

the $J^P=(1/2)^+$ baryons are assigned to the $\mathbf{20'}$ representations and the $J^P=(3/2)^+$ resonances are placed in the $\mathbf{20}$.

In Sec. 3.1 the decays of the dibaryon octet $D_{\mathbf{8}}$ into an antisymmetric final state $B_{\mathbf{8}}+B_{\mathbf{8}}$ were studied. Correspondingly, for charmed dibaryons the decays of the 64-plet, $D_{\mathbf{64}}$, into an antisymmetric final state $B_{\mathbf{20'}}+B_{\mathbf{20'}}$ have to be considered, as can be seen from the decomposition:

$$\mathbf{20'}\otimes\mathbf{20'}=\mathbf{6}\oplus\mathbf{10}\oplus\mathbf{10}^{*}\oplus\mathbf{50}\oplus\mathbf{64_A}\oplus\mathbf{64_S}\oplus\mathbf{70}\oplus\mathbf{126}. \tag{27}$$

For exact SU(4) the Yukawa interaction is characterized by only one coupling constant, since the antisymmetrical $\mathbf{64_A}$ representation appears only once in the decomposition of $\mathbf{20'}\otimes\mathbf{20'}$.

For SU(4) breaking it can be assumed, by using the spurion method, that the symmetry breaking operator transforms like the $Y=I=0$ component of the SU(4) $\mathbf{15}$ representation. Denote the symmetry breaking interaction as a spurion, $S_{\mathbf{15}}$, the $(Y,I)=(0,0)$ state of the SU(4) $\mathbf{15}$, and consider the reaction

$$S_{\mathbf{15}}+D_{\mathbf{64}}\to B_{\mathbf{20'}}+B_{\mathbf{20'}} \tag{28}$$

as SU(4) invariant. Consider the decomposition

$$\mathbf{15}\otimes\mathbf{64}=\mathbf{6}\oplus\mathbf{10}\oplus\mathbf{10}^{*}\oplus\mathbf{50}\oplus 3(\mathbf{64})\oplus\mathbf{70}\oplus\mathbf{70}^{*}\oplus\mathbf{126}\oplus\mathbf{126}^{*}\oplus\mathbf{300}, \tag{29}$$

and the decomposition of $\mathbf{20'}\otimes\mathbf{20'}$, Eq. (27), where the representations $\mathbf{6}$, $\mathbf{50}$, $\mathbf{64_A}$, and $\mathbf{70}$ are antisymmetric. For the two final baryon $\mathbf{20'}$ be in a flavor antisymmetric state, we find that there are six SU(4) breaking terms to first order, namely, $\langle\mathbf{6}|\mathbf{6}\rangle$, $\langle\mathbf{50}|\mathbf{50}\rangle$, $\langle\mathbf{64}|\mathbf{64_A}\rangle$, $\langle\mathbf{64'}|\mathbf{64_A}\rangle$, $\langle\mathbf{64''}|\mathbf{64_A}\rangle$, and $\langle\mathbf{70}|\mathbf{70}\rangle$.

Sum rules for SU(4) breaking along with exact SU(3) could be calculated at this first stage. As a second part, combined SU(4) and SU(3) first order symmetry breaking, it has to be considered that in the spurion method the SU(3) breaking interaction is assumed to transform as the $Y=I=0$, component of a SU(3) octet which is contained in the SU(4) $\mathbf{15}$ representation. Since the breaking of SU(4) symmetry should be on a similar scale as the masses of the considered baryons is hard to say that linear breaking is enough for phenomenological applications so that the second order breaking could be more important than the mixed SU(3) SU(4) breaking. Due to its extension, in this work we will not proceed further with the analysis to find the sum rules for the decays $D_{\mathbf{64}}\to B_{\mathbf{20'}}+B_{\mathbf{20'}}$ with SU(4) symmetry breaking for the cases described.

3.2. Decays of octet dibaryon into $B_{\mathbf{8}}+B_{\mathbf{8}}$ flavor-symmetric final state.
Now we consider decays of the dibaryon octet $D_{\mathbf{8}}$ into two baryon octets flavor-symmetric final state. In principle, with λ_8 first order breaking of SU(3) there are five additional parameters g'_i $(i=1,2,3,4,5)$ in addition to $g^0_{\mathbf{8}}$:

$$g^0_{\mathbf{8}}\,\mathrm{Tr}\left\{\overline{D}_{\mathbf{8}}\left[B_{\mathbf{8}}B'_{\mathbf{8}}+B'_{\mathbf{8}}B_{\mathbf{8}}\right]\right\}, \tag{30}$$

$$g'_1\,\mathrm{Tr}\left\{\overline{D}_{\mathbf{8}}\lambda_8\left[B_{\mathbf{8}}B'_{\mathbf{8}}+B'_{\mathbf{8}}B_{\mathbf{8}}\right]\right\}, \tag{31}$$

$$g_2' \,\mathrm{Tr}\left\{\overline{D}_{\mathbf{8}}\left[B_{\mathbf{8}}\lambda_8\, B_{\mathbf{8}}' + B_{\mathbf{8}}'\lambda_8\, B_{\mathbf{8}}\right]\right\}, \tag{32}$$

$$g_3' \,\mathrm{Tr}\left\{\overline{D}_{\mathbf{8}}\left[B_{\mathbf{8}}B_{\mathbf{8}}' + B_{\mathbf{8}}'B_{\mathbf{8}}\right]\lambda_8\right\}, \tag{33}$$

$$g_4' \left\{\mathrm{Tr}\left[\overline{D}_{\mathbf{8}}B_{\mathbf{8}}\right]\mathrm{Tr}\left[B_{\mathbf{8}}'\lambda_8\right] + \mathrm{Tr}\left[\overline{D}_{\mathbf{8}}B_{\mathbf{8}}'\right]\mathrm{Tr}\left[B_{\mathbf{8}}\lambda_8\right]\right\}. \tag{34}$$

$$g_5' \,\mathrm{Tr}\left[\overline{D}_{\mathbf{8}}\lambda_8\right]\mathrm{Tr}\left[B_{\mathbf{8}}B_{\mathbf{8}}'\right]. \tag{35}$$

However, the terms in Eqs. (30)-(35) are not linearly independent, there is the mathematical identity:

$$\begin{aligned}
&\mathrm{Tr}\left\{\overline{D}_{\mathbf{8}}\lambda_8\left[B_{\mathbf{8}}B_{\mathbf{8}}' + B_{\mathbf{8}}'B_{\mathbf{8}}\right]\right\} + \mathrm{Tr}\left\{\overline{D}_{\mathbf{8}}\left[B_{\mathbf{8}}\lambda_8\, B_{\mathbf{8}}' + B_{\mathbf{8}}'\lambda_8\, B_{\mathbf{8}}\right]\right\} \\
&+\mathrm{Tr}\left\{\overline{D}_{\mathbf{8}}\left[B_{\mathbf{8}}B_{\mathbf{8}}' + B_{\mathbf{8}}'B_{\mathbf{8}}\right]\lambda_8\right\} = \\
&\mathrm{Tr}\left[\overline{D}_{\mathbf{8}}B_{\mathbf{8}}\right]\mathrm{Tr}\left[B_{\mathbf{8}}'\lambda_8\right] + \mathrm{Tr}\left[\overline{D}_{\mathbf{8}}B_{\mathbf{8}}'\right]\mathrm{Tr}\left[B_{\mathbf{8}}\lambda_8\right] + \mathrm{Tr}\left[\overline{D}_{\mathbf{8}}\lambda_8\right]\mathrm{Tr}\left[B_{\mathbf{8}}B_{\mathbf{8}}'\right].
\end{aligned} \tag{36}$$

Using (36) to eliminate the g_2' term, the five independent terms are (with new parameters g_i $(i = 1, 2, 3, 4)$): (30), (31), and (33)-(35).

As described at the beginning of this section, we have five parameters in this case: $g_{\mathbf{8}}^0$ from exact SU(3) and g_i $(i = 1, 2, 3, 4)$ from λ_8 first order SU(3) symmetry breaking. There are nine independent coupling constants to be determined, we take these to be:

$$G[D_{\mathbf{8}}(1, 1/2, +1/2) \to p\,\Sigma^{0'}], \quad G[D_{\mathbf{8}}(1, 1/2, +1/2) \to p\,\Lambda'], \tag{37}$$

$$G[D_{\mathbf{8}}(0, 1, +1) \to p\,\Xi^{0'}], \quad G[D_{\mathbf{8}}(0, 1, +1) \to \Sigma^+\,\Lambda'], \tag{38}$$

$$G[D_{\mathbf{8}}(0, 0, 0) \to p\,\Xi^{-'}],\ G[D_{\mathbf{8}}(0, 0, 0) \to \Sigma^+\Sigma^{-'}],\ G[D_{\mathbf{8}}(0, 0, 0) \to \Lambda\Lambda'], \tag{39}$$

$$G[D_{\mathbf{8}}(-1, 1/2, +1/2) \to \Sigma^+\,\Xi^{-'}], \quad G[D_{\mathbf{8}}(-1, 1/2, +1/2) \to \Lambda\,\Xi^{0'}]. \tag{40}$$

The remaining decay coupling constants are given by the same relations, Eqs. (15)-(22), with $G[D_{\mathbf{8}}(0, 1, +1) \to \Sigma^+\,\Sigma^{0'}] = 0$ and $G[D_{\mathbf{8}}(0, 0, 0) \to \Sigma^+\Sigma^{-'}] = -G[D_{\mathbf{8}}(0, 0, 0) \to \Sigma^0\Sigma^{0'}]$.

The nine coupling constants (37)-(40) are described by the five parameters $g_{\mathbf{8}}^0$ and g_i. Thus, four SU(3) broken sum rules between the decay couplings constants may be deduced:

$$\begin{aligned}
&\sqrt{6}\,G[D_{\mathbf{8}}(1, 1/2, +1/2) \to p\,\Lambda'] = -\sqrt{2}\,G[D_{\mathbf{8}}(1, 1/2, +1/2) \to p\,\Sigma^{0'}] \\
&\quad - \sqrt{6}\,G[D_{\mathbf{8}}(-1, 1/2, +1/2) \to \Lambda\,\Xi^{0'}] + G[D_{\mathbf{8}}(-1, 1/2, +1/2) \to \Sigma^+\,\Xi^{-'}],
\end{aligned} \tag{41}$$

$$\begin{aligned}
&\sqrt{6}\,G[D_{\mathbf{8}}(0, 1, +1) \to \Sigma^+\,\Lambda'] = -2\,G[D_{\mathbf{8}}(0, 1, +1) \to p\,\Xi^{0'}] \\
&\quad - \sqrt{6}\,G[D_{\mathbf{8}}(-1, 1/2, +1/2) \to \Lambda\,\Xi^{0'}] + G[D_{\mathbf{8}}(-1, 1/2, +1/2) \to \Sigma^+\,\Xi^{-'}],
\end{aligned} \tag{42}$$

$$\sqrt{6}\,G[D_{\mathbf{8}}(0,0,0) \to p\,\Xi^{-\prime}] = 2\sqrt{2}\,G[D_{\mathbf{8}}(1,1/2,+1/2) \to p\,\Sigma^{0\prime}] + G[D_{\mathbf{8}}(0,1,+1) \to p\,\Xi^{0\prime}] + \sqrt{6}\,G[D_{\mathbf{8}}(0,0,0) \to \Sigma^+\,\Sigma^{-\prime}] + 2\,G[D_{\mathbf{8}}(-1,1/2,+1/2) \to \Sigma^+\,\Xi^{-\prime}], \quad (43)$$

$$3\sqrt{3}\,G\,[D_{\mathbf{8}}(0,0,0) \to \Lambda\,\Lambda'] = -8\,G[D_{\mathbf{8}}(1,1/2,+1/2) \to p\,\Sigma^{0\prime}] - 2\sqrt{2}\,G[D_{\mathbf{8}}(0,1,+1) \to p\,\Xi^{0\prime}] - 3\sqrt{3}\,G[D_{\mathbf{8}}(0,0,0) \to \Sigma^+\,\Sigma^{-\prime}] - 6\sqrt{3}\,G[D_{\mathbf{8}}(-1,1/2,+1/2) \to \Lambda\,\Xi^{0\prime}] - \sqrt{2}\,G[D_{\mathbf{8}}(-1,1/2,+1/2) \to \Sigma^+\,\Xi^{-\prime}]. \quad (44)$$

With identical relationships for $G[D_{\mathbf{8}}(Y,I,I_3) \to B'\,B] = G[D_{\mathbf{8}}(Y,I,I_3) \to B\,B']$.

4. Concluding remarks

Earlier, strong decays of sextaquark dibaryon decuplets (**10*** and **10** representations) into a flavor-antisymmetric final state of two ordinary baryon octets were considered [12]. In this paper we have given sum rules with λ_8 first order SU(3) symmetry breaking for strong decays of dibaryon octet $D_{\mathbf{8}}$ into two ordinary baryon octets $B_{\mathbf{8}} + B_{\mathbf{8}}$. The sum rules given here and in earlier papers [12, 20] can be applied into $B(\mathbf{8}) + \bar{B}(\mathbf{8})$ with appropiate changes.

Very recently LHCb has reported that a multiquark state $Z(4430)^-$ (tetraquark) made of charm, anti-charm, down and anti-up quarks [21] exists. This opens up a whole new area of multiquark states with charm and thus an SU(4) extension of our results is called for.

Apart from the heavy ion experiments, which we mentioned in the introduction, in our view e^+e^- annihilation reaction is best suited to find multibaryon plus antimultibaryon pairs, since the signal for resonance will be a clear signal. However LHCb will also be a good place for observing these states.

4.1. Acknowledgments

V. Gupta, E. N. Polanco-Euán, and G. Sánchez-Colón would like to thank CONACyT (México) for partial support. B. A. Bambah would like to acknowledge Ms. A. Aiswarya for checking some calculations.

References

[1] F. J. Dyson and N.-H. Xuong, *Phys. Rev. Lett.* **13**, 815 (1964); Erratum *Phys. Rev. Lett.* **14**, 339 (1965).
[2] R. L. Jaffe, *Phys. Rev. Lett.* **38**, 195 (1977); Erratum *Phys. Rev. Lett.* **38**, 617 (1977).
[3] Y. Nambu and B. Bambah, *Phys. Rev. D* **26**, 2871 (1982).
[4] M. Crawford and B. Bambah, *Phys. Rev. D* **29**, 1334 (1984).
[5] J. Vijande, A. Valcarce and J. M. Richard, *Phys. Rev. D* **85**, 014019 (2012).
[6] T. Inoue et al. (HAL QCD), *Phys. Rev. Lett.* **106**, 162002 (2011).
[7] S. R. Beane *et al.* [NPLQCD Collaboration], *Phys. Rev. D* **85**, 054511 (2012).
[8] S. Cho et al. [ExHIC Collaboration], **Phys. Rev. Lett.** 106, 212001 (2011).
[9] P. Burikham and E. Hirunsirisawat, *Adv. High Energy Phys.* **2011**, 123184 (2011).
[10] M. Bashkanov, Stanley J. Brodsky and H. Clement, *Phys. Lett. B* **727**, 438 (2013).
[11] P. Alderson *et.al.*, *Phys. Rev. Lett.* **112**, 202301 (2014).
[12] V. Gupta and G. Sánchez-Colón, *Mod. Phys. Lett. A* **30**, 1550010 (2015).
[13] B. H. Kim *et al.* [Belle Collaboration], *Phys. Rev. D* **86**, 010001 (2012).
[14] M. Oka, *Phys. Rev. D* **38**, 298 (1988).
[15] P. Carruthers. *Introduction to Unitary Symmetry.* New York: Interscience (1966).
[16] J. J. de Swart, *Rev. Mod. Phys.* **35**, 916 (1963).
[17] M. Gell-Mann, *"The Eightfold Way: A Theory of Strong Interaction Symmetry"*, California Institute of Technology Synchrotron Laboratory Report CTSL-20 (1961), unpublished.

[18] S. Okubo, *Prog. Theor. Phys.* **27**, 949 (1962).
[19] V. Gupta, *Pramana* **6**, 259 (1976).
[20] V. Gupta and V. Singh, *Phys. Rev.* **135**, B1442 (1964).
[21] R. Aaij *et al.* [LHCb Collaboration], *Phys. Rev. Lett.* **112**, no. 22, 222002 (2014).

Interpretation of topologically restricted measurements in lattice σ-models

Irais Bautista[a], **Wolfgang Bietenholz**[a], **Urs Gerber**[a,b], **Christoph P. Hofmann**[c], **Héctor Mejía-Díaz**[a] **and Lilian Prado**[a]

[a] Instituto de Ciencias Nucleares, Universidad Nacional Autónoma de México
A.P. 70-543, C.P. 04510 Ciudad de México, Mexico
[b] Instituto de Física y Matemáticas
Universidad Michoacana de San Nicolás de Hidalgo
Edificio C-3, Apdo. Postal 2-82, C.P. 58040, Morelia, Michoacán, Mexico
[c] Facultad de Ciencias, Universidad de Colima
Bernal Díaz del Castillo 340, Colima C.P. 28045, Mexico

E-mail: wolbi@nucleares.unam.mx

Abstract. We consider models with topological sectors, and difficulties with their Monte Carlo simulation. In particular we are concerned with the situation where a simulation has an extremely long auto-correlation time with respect to the topological charge. Then reliable numerical measurements are possible only within single topological sectors. The challenge is to assemble such restricted measurements to obtain an approximation for the full-fledged result, which corresponds to the correct sampling over the entire set of configurations. Under certain conditions this is possible, and it provides in addition an estimate for the topological susceptibility χ_t. Moreover, the evaluation of χ_t might be feasible even from data in just one topological sector, based on the correlation of the topological charge density. Here we present numerical test results for these techniques in the framework of non-linear σ-models.

1. Topological sectors in quantum physics

We are going to consider quantum physics in the functional integral formulation with Euclidean time. For a number of models, the paths — or configurations — to be summed over split into *topological sectors*. A simple example is the quantum rotor, *i.e.* a quantum mechanical scalar particle moving freely on a circle with periodic boundary conditions in time (say with period T). This corresponds to the 1d O(2) model with the action, partition function and expectation values

$$S[\varphi] = \frac{\beta}{2}\int_0^T \dot{\varphi}^2\, dt \ , \quad Z = \int D\varphi\ \exp(-S[\varphi]) \ , \quad \langle \mathcal{O} \rangle = \frac{1}{Z}\int D\varphi\ \mathcal{O}[\varphi]\ \exp(-S[\varphi]) \ . \qquad (1.1)$$

(β represents the inverse coupling in the O(2) model, or the moment of inertia of the rotor.) The functional integral $\int D\varphi$ sums over all closed paths $\varphi(t) \in S^1$, $\varphi(0) = \varphi(T)$. These paths occur in disjoint subsets, which can be labelled by their winding number, or *topological charge*

$$Q = \frac{1}{2\pi}\int_0^T \dot{\varphi}\ dt \in \mathbb{Z} \ . \qquad (1.2)$$

A continuous deformation of a path does not change Q, hence it remains in the same subset, *i.e.* in the same topological sector.

This can be generalised to quantum field theory on a torus, namely to O(N) models in $d = N - 1$ dimensions, where the field is a classical spin $\vec{S}(x) \in S^{N-1}$. For the models in this class, the configurations are structured in topological sectors. We are going to address the cases $N = 2$ and 3. Further models with topological sectors are *e.g.* the 2d CP($N - 1$) models, 2d U(1) gauge theory, 4d Yang-Mills theories and QCD. In all these cases, the configurations can only be deformed continuously within a fixed topological sector (at finite action). The functional integral splits into disjoint integrals over these sectors, which are characterised by a topological charge $Q \in \mathbb{Z}$.

2. Lattice regularisation and Monte Carlo simulation

We proceed to the lattice formulation, which reduces the (periodic) volume to a set of discrete sites x, say with a cubic structure and lattice spacing a. Matter fields live on these sites, Φ_x, while gauge fields can be given by compact link variables $U_{x,\mu} \in \{\text{gauge group}\}, \mu = 1 \dots d$. This represents a gauge invariant UV regularisation.

On the lattice regularised level, there are *a priori* no topological sectors; all configurations can be continuously deformed into one another. Still, something similar exists: the action may have local minima, and their vicinities are separated by regions of high action. Hence transitions between these sectors are statistically suppressed, where

$$p[\Phi] = \frac{1}{Z} \exp(-S[\Phi]) \tag{2.1}$$

is interpreted as the probability of some configuration $[\Phi]$. In such cases, it is often convenient to introduce a topological charge Q even on the lattice. In the presence of chiral fermions (with a lattice Dirac operator that obeys the Ginsparg-Wilson Relation), this can be done best by the fermion index [1]. For ($N - 1$)-dimensional O(N) models, the optimal formulation maps local patches of the lattice configuration onto S^{N-1}, and the (hyper-)spherical area represents the topological charge density ("geometric definition") [2]. These definitions both have the virtue that they provide integer Q values for each configuration (up to a subset of measure zero), even on the lattice.

Monte Carlo simulations generate a sequence of configurations (a "Monte Carlo history", usually a Markov chain) randomly with the probability given in eq. (2.1). Summing over this set provides an approximation for expectation values of observables, such as n-point functions. There are still statistical errors (the set of configurations is finite), and systematic errors (for instance due to the finite lattice spacing $a > 0$), which can be reduced with increased computational effort. On the other hand, we stress that this method is fully *non-perturbative*.

3. Topologically frozen numerical measurements

Most popular algorithms generate configurations by performing a long sequence of small updates, until a new, (quasi-)independent configuration emerges, to be used for the numerical measurement. This includes in particular the Hybrid Monte Carlo (HMC) algorithm [3], which is standard in QCD simulations with dynamical quarks.

Regarding theories with topological sectors, however, such small-update algorithms are plagued by the problem that they have a hard time to change the topological sector. This requires the sequence of update steps to pass through a region of low probability. However, in order to sample the entire space of configurations correctly, the algorithm is supposed to change Q frequently. As a striking counter-example, we mention the QCD studies with chiral quarks reported in Refs. [4], where the extensive HMC histories were entirely confined to the sector with $Q = 0$. For Wilson-type quarks the problem is less severe so far, *i.e.* in the range of lattice

Journal of Physics: Conference Series **761** (2016) 012087 doi:10.1088/1742-6596/761/1/012087

spacings $0.05\,\text{fm} \lesssim a \lesssim 0.15\,\text{fm}$ which have typically been used up to now. However, once we push for even finer lattices — in order to suppress further the systematic lattice artifacts — the problem will become more severe: it is welcome for many reasons that the system is more continuum-like, but this also makes topological transitions harder.

Therefore Monte Carlo histories which are trapped in one topological sector for a very long (computing) time, *i.e.* over a huge number of update steps, are a serious issue. The question arises whether we can still perform correct numerical measurements of n-point functions, or *e.g.* of the topological susceptibility

$$\chi_{\rm t} = \frac{1}{V}(\langle Q^2 \rangle - \langle Q \rangle^2) \ . \tag{3.1}$$

As a remedy, the use of open boundary conditions has been advocated [5]; then Q is not integer anymore, so it can vary continuously. Here we consider attempts to deal with this problem while keeping the boundary conditions periodic. We test two procedures for this purpose, with numerical data for the 1d O(2) and the 2d O(3) model, and we refer to the geometric definition of the topological charge of a lattice configuration. Its contributions from some local patches (nearest neighbour sites in $d = 1$, triangles in $d = 2$) form the lattice topological charge density q.

4. Correlation of the topological charge density

First we consider a method suggested in Ref. [6] to determine $\chi_{\rm t}$ even if only one single topological sector has been explored in the simulation. We denote the restricted expectation value within one sector as $\langle \ldots \rangle_Q$. Ref. [6] derived an approximate formula for the correlation of the topological lattice charge density at large Euclidean time separation t,[1]

$$\lim_{t\to\infty} \langle \, q_0 \, q_t \, \rangle_{|Q|} \approx -\frac{1}{V}\chi_{\rm t} + \frac{Q^2}{V^2} \ . \tag{4.1}$$

The derivation assumes the charges Q to be Gauss distributed around zero, $\langle Q^2 \rangle = V\chi_{\rm t}$ to be large, and $|Q|/\langle Q^2 \rangle$ to be small. The explicit meaning of these requirements has to be investigated numerically.[2] Here we give results that we obtained while preparing an extensive paper on this subject [9]. For the 1d O(2) model we consider the standard lattice action (we will call it simply "standard action") and the Manton action [10]. For the sum over all topological sectors, analytical results are given in Ref. [11], and additional numerical results in Ref. [12]. For the 2d O(3) model we simulated the standard action (in lattice units)

$$S[\vec{S}] = \beta \sum_{x,\mu} (1 - \vec{S}_x \cdot \vec{S}_{x+\hat{\mu}}) \ , \tag{4.2}$$

where $\hat{\mu}$ is a lattice unit vector in μ-direction.[3] We used an efficient cluster algorithm [13], which is not restricted to small update steps, and which does therefore change the topological sector frequently. Hence, in these tests, correctly distributed configurations are available, and they provide a check for the indirect methods. The latter are relevant for instance in gauge theories, where no efficient cluster algorithm is known.

Our results for $\langle \, q_0 \, q_t \, \rangle_{|Q|}$ are shown in Figures 1 and 2. In all cases we observe excellent agreement with the values corresponding to the approximation formula (4.1). However, in the

[1] On the right-hand-side there is another correction term, $(\langle Q^4 \rangle - 3\langle Q^2 \rangle^2)/(2V\chi_{\rm t}^2)$, for deviations from Gauss-distributed charges Q ("kurtosis"), but we checked that it is negligible in the examples that we considered.

[2] Actually there is already a successful determination of $\chi_{\rm t}$ in 2-flavour QCD along these lines [7], but it used the flavour singlet pseudo-scalar density, as suggested in Ref. [8].

[3] The Manton action replaces the function $1 - \cos \Delta\varphi_{x,\mu}$ of the relative angle (in the last term) by $\frac{1}{2}(\Delta\varphi_{x,\mu})^2$. For the quantum rotor this reduces the lattice artifacts drastically; this action turns out to be classically perfect [11].

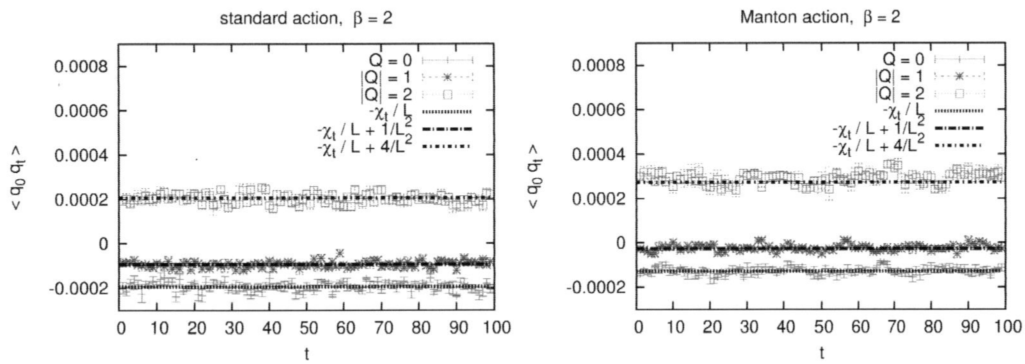

Figure 1. The correlation function of the topological charge density over Euclidean time separation t for the 1d O(2) model on a spin chain of length $L = 100$ and inverse coupling $\beta = 2$. We show results for the standard action (with $\langle Q^2 \rangle = 1.94$) on the left, and for the Manton action (with $\langle Q^2 \rangle = 1.27$) on the right. The black dotted lines correspond to the approximation (4.1) for $|Q| = 0, 1, 2$, if we insert the analytically known values for χ_t [11].

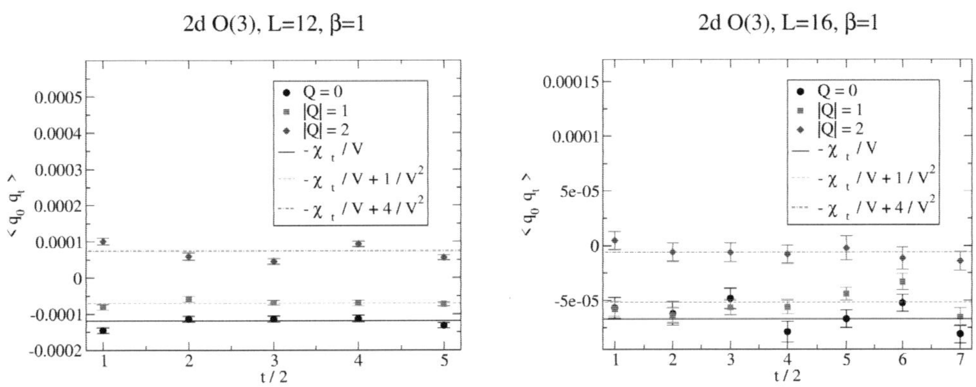

Figure 2. The correlation function of the topological charge density over Euclidean time separation t for the 2d O(3) model at $\beta = 1$. We show results for the standard action in volume $V = 12 \times 12$ (with $\langle Q^2 \rangle = 2.46$) on the left, and $V = 16 \times 16$ (with $\langle Q^2 \rangle = 4.39$) on the right. The horizontal lines are the values for $|Q| = 0$, 1 and 2 obtained from the approximation (4.1), if we insert the directly measured topological susceptibility χ_t.

field theoretic example of Figure 2 the correlation length is short ($\xi \simeq 1.3$). To move closer to the continuum limit, we have to increase β, and — in order to keep $\langle Q^2 \rangle$ relatively large — also the volume V. Consequently the predicted plateaux for $\langle q_0 \, q_t \rangle_{|Q|=0,1,2}$ move rapidly towards zero, and it becomes difficult to disentangle them in the numerical data from each other and from zero. However, this distinction is crucial for the evaluation of χ_t. The question how far this is still feasible is discussed in detail in Ref. [9], including applications to 4d SU(2) Yang-Mills theory. The crucial remedy is the computation of all-to-all correlators.

5. Approximate topological summation of observables

Now we proceed to a more ambitious goal: the computation of some expectation value $\langle \mathcal{O} \rangle$, if only a few topologically restricted measurements $\langle \mathcal{O} \rangle_{|Q|}$ are available. An approximate formula for this purpose was first derived in Ref. [14], referring to the pion mass in QCD simulations (for later considerations, see Refs. [15–18]). It can also be applied to other observables $\mathcal{O}$, and other models with topological sectors. The relevant approximation formula reads

$$\langle \mathcal{O} \rangle_{|Q|} \approx \langle \mathcal{O} \rangle + \frac{c}{2V\chi_t}\Big(1 - \frac{Q^2}{V\chi_t}\Big) , \tag{5.1}$$

where c is a constant, which (like χ_t and $\langle \mathcal{O} \rangle$) stabilises in large volumes. Together with $\langle \mathcal{O} \rangle$ and χ_t, there are three unknown terms on the right-hand-side. If we employ numerical results for the left-hand-side, in various $|Q|$ and V, they can in principle be determined by a fit (this requires results in at least two different volumes).

The assumptions in the derivation are (qualitatively) the same as for eq. (4.1), hence we need again a reasonably large $\langle Q^2 \rangle = V\chi_t$, and we should only use rather small charges $|Q|$. The tests of this approximation, which have been carried out with numerical data, were performed in the Schwinger model [15, 17, 18], for the quantum rotor [16, 18] and for 4d SU(2) gauge theory [18].

First we present results for the 1d O(2) model, where we measured the restricted correlation length $\xi_{|Q|}$ for $|Q| = 0, 1, 2$. Actually the (connected) correlation function exhibits a neat exponential decay — resp. a cosh shape — only if all sectors are included. Topological restriction deforms this decay somewhat, hence the evaluation of $\xi_{|Q|}$ is less obvious [14, 16–18]. To illustrate this property, we show in Figure 3 the correlation function in a small volume of $L = 14$ at $\beta = 2$, for $|Q| = 0$ and 1, and in total. We see that only the total (unrestricted) correlation matches a cosh function very well. (The formula for the correlation at fixed Q is given in Ref. [16].)

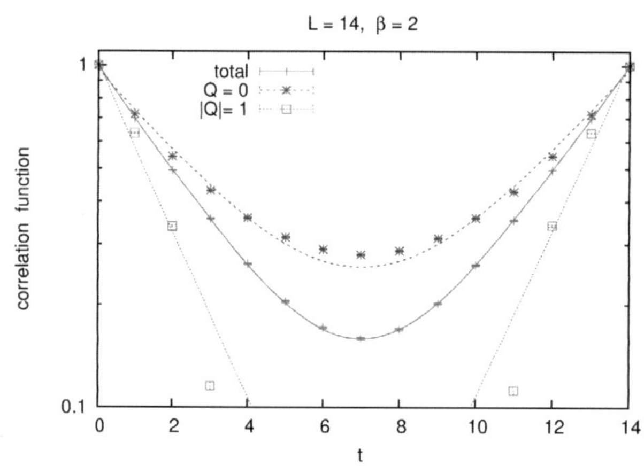

Figure 3. The correlation function for the 1d O(2) model with $L = 14$, $\beta = 2$ for the sectors $|Q| = 0$ and 1, and in total. Only the complete correlation follows well a fit to a cosh function.

Now we increase the inverse coupling to $\beta = 4$, and we consider larger sizes in the range $L = 150 \dots 400$. Once we have evaluated $\xi_{|Q|}$ as well as possible, we use ξ_0 as an input for the summation formula (5.1) to obtain approximations $\xi_{1,\text{approx}}$ and $\xi_{2,\text{approx}}$, which are then compared to the directly measured values. Figure 4 shows good agreement over a broad range of L, both for the standard action and for the Manton action. In particular for the smaller sizes L, this observation is highly non-trivial since there is a considerable splitting between the lengths $\xi_{|Q|}$. Of course, in very large volumes all values coincide, so this agreement becomes trivial.

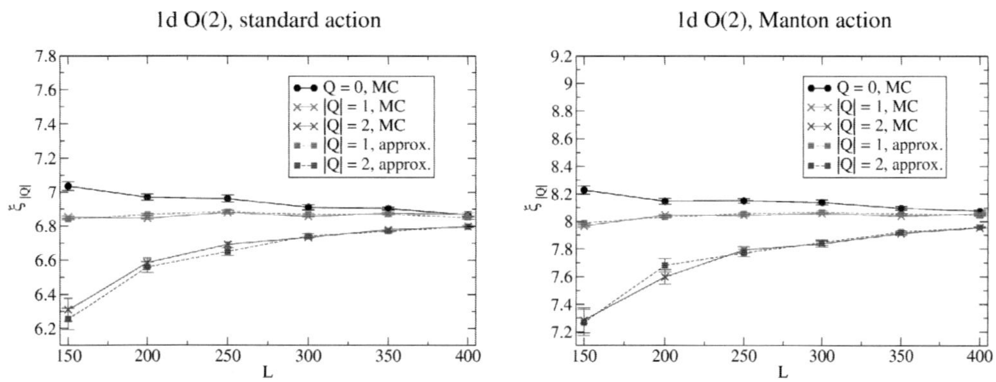

Figure 4. The topologically restricted correlation lengths ξ_0, ξ_1, ξ_2 measured directly in Monte Carlo (MC) simulations of the 1d O(2) model at $\beta = 4$. For ξ_1 and ξ_2 these results are compared to the values obtained from the approximation formula (5.1), if ξ_0 is used as an input. We observe good agreement in all cases.

Now we address the true correlation length ξ, which describes the decay of the correlation function measured over all topological sectors. Generally, in very large volumes all restricted values also coincide with the complete result, as eq. (5.1) shows, hence also this agreement becomes trivial. However, such large volumes are often inaccessible in Monte Carlo simulations. Therefore we explore whether in somewhat smaller volumes the correction terms on the right-hand-side of eq. (5.1) can help to still extract the correct result. The volume should not be too small, however, otherwise additional corrections, which are missing in approximation (5.1), become significant.[4]

Figure 5 shows the exact correlation length at $\beta = 4$, which is known analytically [11] (dashed lines), along with the numerically measured values for ξ_0, ξ_1, ξ_2 in sizes $L = 150 \dots 400$. Table

	Standard action			Manton action		
fitting range for L	250 – 400	300 – 400	theory	250 – 400	300 – 400	theory
ξ	6.77(5)	6.79(2)	6.815	7.95(5)	7.88(6)	8.000

Table 1. Fitting results for the correlation length ξ in the 1d O(2) model at $\beta = 4$, based on the formula (5.1). We insert the numerically measured values for ξ_0, ξ_1, ξ_2 in some range for the size L, and determine the optimal approximation for ξ by a least-square fit. We arrive at results which are close to the theoretical value.

1 gives the results based on eq. (5.1), if we insert the numerical and the topologically restricted values in some range of sizes L. This is the procedure to handle a situation where the complete result is not known. Now the three unknown parameters are strongly over-determined, and we perform a least-square fit to obtain the estimates in Table 1. The fact that there are only small fitting errors confirms the consistency of this approach. Moreover, the resulting values are close to the exact values, which is again non-trivial regarding the significant splitting of ξ_0, ξ_1 and ξ_2 (see Figure 5).

[4] An extension to higher-order corrections is discussed in Ref. [16], and Refs. [19] include also ordinary finite size effects (not related to fixed topology).

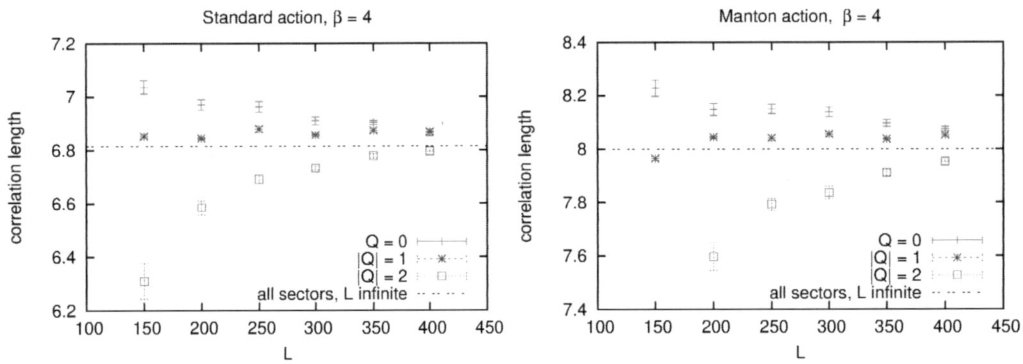

Figure 5. The analytic value of the correlation length ξ in the 1d O(2) model (dashed lines), and numerical measurements for the topologically restricted correlation lengths ξ_0, ξ_1, ξ_2 in sizes $L = 150 \dots 400$. We see a clear splitting both for the standard action (on the left) and for the Manton action (on the right). Hence it is non-trivial that these measured values provide good estimate for ξ, when we insert them into approximation (5.1) and perform a least-square fit for the three unknown terms. Corresponding results are given in Table 1.

Finally we add an application of formula (5.1) to the 2d O(3) model. Here our observable is the action density, *i.e.* the mean value of the action divided by the volume, $\langle S \rangle / V$. Figure 6 shows results for the sectors $|Q| = 0,\ 1,\ 2$ and for all configurations, in $L \times L$ volumes with $L = 8 \dots 32$. We see that they all converge to the same value for increasing size L.

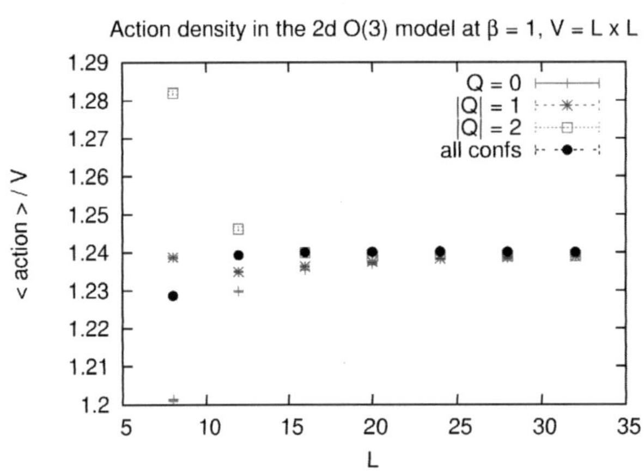

Figure 6. Numerical results for the action density of the 2d O(3) model with the standard action at $\beta = 1$ in various $L \times L$ volumes. We show the complete result, as well as the values restricted to $|Q| = 0,\ 1$ or 2.

If we were not able to measure the result over all configurations reliably — *i.e.* if we only had Monte Carlo histories that hardly ever change the topological sector — we could now use the restricted measurements in some range of L, insert them into formula (5.1) and estimate $\langle S \rangle / V$, as well as χ_t, with a least-square fit. Such results are displayed in Table 2. We see that even the use of modest volumes provides estimates, which agree well with the results of a direct complete measurement (this is feasible in this case thanks to the cluster algorithm).

fitting range for L	16 – 24	16 – 28	16 – 32	directly measured in all sectors at $L = 32$
$\langle S\rangle/V$	1.24038(12)	1.24027(8)	1.24015(5)	1.24008(5)
χ_t	0.0173(6)	0.0169(5)	0.0164(5)	0.01721(4)

Table 2. Estimates for the action density $\langle S\rangle/V$ and for the topological susceptibility χ_t in the 2d O(3) model at $\beta = 1$, based on topologically restricted results in some range of L. We find good agreement with the directly measured values, without topological restriction (last column).

6. Summary

Most Monte Carlo simulations of lattice field theory still have to be performed with small-update algorithms. For theories with topological sectors, this often implies that the Monte Carlo history is confined to one sector over an enormous number of update steps — a problem, which is getting worse for decreasing lattice spacing. This concerns in particular QCD simulations with (light) dynamical quarks.

In this case, it may happen that observables can be measured well only within a fixed sector, *i.e.* one is limited to topologically restricted expectation values $\langle \mathcal{O}\rangle_Q$. On the conceptual side, this rises algorithmic questions related to ergodicity. If we trust the restricted measurements $\langle \mathcal{O}\rangle_Q$, the next issue is to extract physical information from them.

An approximate formula derived in Ref. [14] allows us in principle to obtain an estimate for the physical value $\langle \mathcal{O}\rangle$, based on restricted results in various topological sectors and in various volumes. Moreover, if the Monte Carlo history rarely changes Q, a direct measurement of the topological susceptibility χ_t is hardly possible. A fit to the formula (5.1) of Ref. [14] provides in addition an estimate for χ_t. That specific quantity can also be estimated from the correlation of the topological charge density, even within just one sector, based on another approximation formula put forward in Ref. [6], eq. (4.1).

Here we have tested both formulae in non-linear σ-models with topological sectors. In this case, the full-fledged results (involving all topological sectors) are available for comparison, either analytically or numerically, hence we could check how far the approximations work. In Section 4 we tested eq. (4.1) to determine χ_t, and in Section 5 eq. (5.1) for the determination of expectation values $\langle \mathcal{O}\rangle$ and χ_t.[5]

Our results show that these formulae do work well in a suitable regime, where the assumptions for their derivations are decently satisfied. In particular, we need $\langle Q^2\rangle \gtrsim 1.5$, and we should only involve sectors with $|Q| \leq 2$. These properties have been further explored in σ-models and in Yang-Mills gauge theory [9, 18], in order to probe the prospects for applications in QCD.

Acknowledgments

We thank Christopher Czaban, Arthur Dromard, Lukas Gonglach, Ivan Hip and Marc Wagner for helpful communication and collaboration. This work was supported by the Mexican *Consejo Nacional de Ciencia y Tecnología* (CONACyT) through projects CB-2010/155905 and CB-2013/22281, as well as DGAPA-UNAM, grant IN107915. The simulations were performed on the cluster of the Instituto de Ciencias Nucleares, UNAM.

[1] Hasenfratz P, Laliena V and Niedermayer F 1998 *Phys. Lett.* B **427** 125
[2] Berg B and Lüscher M 1981 *Nucl. Phys.* B **190** 412
[3] Duane S, Kennedy A D, Pendleton B J and Roweth D 1987 *Phys. Lett.* **195** B 216
[4] Fukaya H *et al* 2007 *Phys. Rev. Lett.* **98** 172001, 2007 *Phys. Rev.* D **76** 054503. Aoki S *et al* 2008 *Phys. Rev.* D **78** 014508. Borsanyi Sz *et al* arXiv:1510.03376 [hep-lat]

[5] Very recently, yet another method to measure χ_t based on configurations in a single topological sector — denoted as the *slab method* — has been tested very recently and successfully [20].

Journal of Physics: Conference Series **761** (2016) 012087 doi:10.1088/1742-6596/761/1/012087

[5] Lüscher M 2010 *JHEP* **1008** 071
[6] Aoki S, Fukaya H, Hashimoto S and Onogi T 2007 *Phys. Rev.* D **76** 054508
[7] Aoki S *et al* 2008 *Phys. Lett.* B **665** 294
[8] Fukaya H and Onogi T 2004 *Phys. Rev.* D **70** 054508
[9] Bautista I, Bietenholz W, Dromard A, Gerber U, Gonglach L, Hofmann C P, Mejía-Díaz H and Wagner M 2015 *Phys. Rev.* D **92** 114510
[10] Manton N 1980 *Phys. Lett.* B **96** 328
[11] Bietenholz W, Brower R, Chandrasekharan S and Wiese U-J 1997 *Phys. Lett.* B **407** 283
[12] Boyer T, Bietenholz W and Wuilloud J 2007 *Int. J. Mod. Phys.* C **18** 1497
[13] Wolff U 1989 *Phys. Rev. Lett.* **62** 361
[14] Brower R, Chandrasekharan S, Negele J W and Wiese U-J 2003 *Phys. Lett.* B **560** 64
[15] Bietenholz W and Hip I 2008 *PoS LATTICE2008* 079; 2012 *J. Phys. Conf. Ser.* **378** 012041
Bietenholz W, Hip I, Shcheredin S and Volkholz J 2012 *Eur. Phys. J.* C **72** 1938
[16] Dromard A and Wagner M, 2013 *PoS LATTICE2013* 339, 2014 *Phys. Rev.* D **90** 074505
[17] Czaban C and Wagner M, 2013 *PoS LATTICE2013* 465
[18] Bietenholz W, Czaban C, Dromard A, Gerber U, Hofmann C P, Mejía-Díaz H and Wagner M 2016 *Phys. Rev.* D **93** 114516
[19] Dromard A, Bietenholz W, Gerber U, Mejía-Díaz H and Wagner M 2015 *Acta Phys. Polon. Supp.* **8** 391, 2015 arXiv:1510.08809 [hep-lat]
[20] Bietenholz W, de Forcrand P and Gerber U, 2015 *JHEP* **1512** 070. Dromard A, Bietenholz W, Cichy K and Wagner M 2016 arXiv:1605.08637 [hep-lat]

Journal of Physics: Conference Series **761** (2016) 012088 doi:10.1088/1742-6596/761/1/012088

$W^*\gamma^*\pi$ form factors in the $\tau^- \to \pi^-\ell^+\ell^-\nu_\tau$ decays

Adolfo Guevara

Departamento de Física, Centro de Investigación y de Estudios Avanzados, Avenida IPN 2508, Col. San Pedro Zacatenco, 07360 México DF, México

E-mail: aguevara@fis.cinvestav.mx

Abstract. We study the disintegration $\tau^- \to \pi^-\ell^+\ell^-\nu_\tau$, where l denotes either an electron or a muon. This process requires knowledge about the $W^*\gamma^*\pi$ interaction, which is modeled using Resonance Chiral Theory. Within this framework we study the sensitivity of the observables of the process, such as the branching ratios and the dilepton mass spectra, to the model dependent contributions. The information of the hadronization of the quark currents is encoded in form factors which are evaluated using the lightest vector and axial-vector multiplets as degrees of freedom. We determine the couplings by matching the asymptotic QCD behavior of the form factors in the limit of an infinite number of colours. The branching ratios we obtain show that the process with $l = e$ should be measured soon in B factories.

1. Introduction

It is well known that the interaction between quarks and gluons becomes stronger in low-energy processes [1]. Due to this non-perturbative characteristic of QCD it is advisable to appeal to effective theories. To study this process it is necessary to have precise information about the $W^*\gamma^*\pi$ interaction. The relevant form factors have been studied before with Chiral Perturbation Theory (χPT) [2] in the $\pi \to e^+e^-e\ \nu_e$ decays[3, 4, 5], where the photon is almost on-shell. The knowledge of this interaction can be used to compute radiative corrections and the couplings needed to calculate the lightest pseudoscalar contributions to the hadronic light-by-light piece of the anomalous magnetic moment of the muon[6].

The difference between the $\tau^- \to \pi^-\ell^+\ell^-\nu_\tau$ process and the pion decay is that the process we study may have an energy above the chiral symmetry breaking scale, where the χPT is no longer valid; this is why we employ Resonance Chiral Theory (RχT)[7, 8]. RχT introduces meson resonances as dynamical fields preserving the chiral symmetry properties and carrying out the required calculation by means of an expansion in powers of the inverse of the number of colours, $\frac{1}{N_C}$[9].

2. Decay amplitude

The $\tau^- \to \pi^-\ell^+\ell^-\nu_\tau$ [10] decays have not been measured yet, and in order to compute the branching ratios and invariant mass spectra we follow the work done in [11], where the process with a real photon is analyzed and the participating vector and axial-vector form factors are obtained. As a consequence of the virtuality of the photon, a diagram that vanished in [11] now contributes to the amplitude, giving rise to a third contributing form factor, B in eq. (4), proportional to the vector two-pion form factor, which was implemented following [12].

The contributions to the process decay amplitude are the bremsstrahlung off the τ, off the π and the vector and axial-vector contributions; respectively[10]

$$iM_{IB\tau} = iG_F V_{ud} e^2 f_\pi \frac{1}{k^2} P_\mu l_\nu L^{\mu\nu} \ , \tag{1}$$

$$iM_{IB\pi} = iG_F V_{ud} e^2 f_\pi \frac{1}{k^2} l^\nu \left(\frac{2P_\nu (P+k)_\mu}{(P+k)^2 - m_\pi^2} - g_{\mu\nu} \right) L^\mu \ , \tag{2}$$

$$M_V = -G_F V_{ud} e^2 \frac{1}{k^2} F_V \left(p\cdot k, k^2\right) \varepsilon_{\mu\nu\rho\sigma} l^\nu k^\rho p^\sigma L^\mu, \tag{3}$$

$$\begin{aligned} M_A \ = \ & iG_F V_{ud} \frac{2e^2}{k^2} l^\nu \left\{ F_A\left(p\cdot k, k^2\right) \left[g_{\mu\nu}\left(p\cdot k - k^2\right) - P_\nu k_\mu \right] + \right. \\ & \left. B(k^2) k^2 \left[g_{\mu\nu} - \frac{(P+k)_\mu P_\nu}{k^2 + 2P\cdot k} \right] \right\} L^\mu \ . \end{aligned} \tag{4}$$

Where we have defined

$$l^\nu = \bar{u}_{\ell^-}\left(P_-\right) \gamma_\nu v_{\ell^+}\left(P_+\right), \tag{5a}$$

$$L^{\mu\nu} = \bar{u}_{\nu_\tau}(q) \gamma^\mu \left(1-\gamma_5\right) \left(\frac{\gamma\cdot\left(P_\tau - k\right) + M_\tau}{\left(P_\tau - k\right)^2 - {M_\tau}^2} \right) \gamma^\nu u_\tau\left(P_\tau\right), \tag{5b}$$

$$L^\mu = \bar{u}_{\nu_\tau}(q) \gamma^\mu \left(1-\gamma^5\right) u_\tau\left(P_\tau\right). \tag{5c}$$

3. Form factors

Since the mass of the τ lepton is larger than the χPT expansion parameter, the perturbative approach does no longer converge for this process. At this energy scale the lightest resonances are used as dynamical fields, because the masses of these resonances are smaller or of the order of the τ lepton mass. This is done with RχT, using as expansion parameter $\frac{1}{N_C}$ to compute the form factors.

Thus, in the framework of RχT we find that

$$\begin{aligned} F_V(t,k^2) = & -\frac{N_C}{24\pi^2 F_\pi} + \frac{2\sqrt{2}F_V}{3F_\pi M_V}[(c_2 - c_1 - c_5)t + (c_5 - c_1 - c_2 - 8c_3)m_\pi^2 + 2(c_6 - c_5)k^2] \\ & + \frac{2\sqrt{2}F_V}{3F_\pi M_V} D_\rho(t)[(c_1 - c_2 - c_5 + 2c_6)t + (c_5 - c_1 - c_2 - 8c_3)m_\pi^2 + (c_2 - c_1 - c_5)k^2] \\ & + \frac{4F_V^2}{3F_\pi} D_\rho(t)[d_3(t + 4k^2) + (d_1 + 8d_2 - d_3)m_\pi^2], \end{aligned} \tag{6}$$

$$F_A(t,k^2) = \frac{F_V^2}{F_\pi}\left(1 - \frac{2G_V}{f_V}\right) D_\rho(k^2) - \frac{F_A^2}{F_\pi} D_{a_1}(t) + \frac{F_A F_V}{\sqrt{2}F_\pi} D_{a_1}(t) D_\rho(k^2)(-\lambda'' t + \lambda_0 m_\pi^2), \tag{7}$$

where

$$D_R(s) = \frac{1}{M_R^2 - s - iM_R\Gamma_R(s)}. \tag{8}$$

Table 1. The central values of the different contributions to the branching ratio of the $\tau^- \to \pi^- \ell^+ \ell^- \nu_\tau$ decays are given on the left hand side of the table. The error bands due to variations of the couplings are displayed in the right hand side of the table. The error band of the IB-IB contributions stems from the uncertainties on the F_π decay constant and the τ lepton lifetime [1].

	$l = e$	$l = \mu$	$l = e$	$l = \mu$
IB-IB	1.461×10^{-5}	1.600×10^{-7}	$\pm 0.006 \times 10^{-5}$	$\pm 0.007 \times 10^{-7}$
IB-V	-2×10^{-8}	1.4×10^{-8}	$[-1 \times 10^{-7}, 1 \times 10^{-7}]$	$[-4 \times 10^{-9}, 4 \times 10^{-8}]$
IB-A	-9×10^{-7}	1.01×10^{-7}	$[-3 \times 10^{-6}, 2 \times 10^{-6}]$	$[-2 \times 10^{-7}, 6 \times 10^{-7}]$
V-V	1.16×10^{-6}	6.30×10^{-7}	$[4 \times 10^{-7}, 4 \times 10^{-6}]$	$[1 \times 10^{-7}, 3 \times 10^{-6}]$
A-A	2.20×10^{-6}	1.033×10^{-6}	$[1 \times 10^{-6}, 9 \times 10^{-6}]$	$[2 \times 10^{-7}, 6 \times 10^{-6}]$
V-A	2×10^{-10}	-5×10^{-11}	$\sim 10^{-10}$	$\sim 10^{-10}$
Total	1.710×10^{-5}	1.938×10^{-6}	$(1.7^{+1.1}_{-0.3} \times 10^{-5})$	$[3 \times 10^{-7}, 1 \times 10^{-5}]$

and

$$B(k^2) = F_\pi \frac{F_V^{\pi^+\pi^-}|_\rho(k^2) - 1}{k^2}, \tag{9}$$

where $F_V^{\pi^+\pi^-}|_\rho(k^2)$ is the $I = 1$ part of the $\pi^+\pi^-$ vector form factor.

Given the rather wide widths of the $\rho(770)$ and $a_1(1260)$ resonances, their off-shell behavior becomes important and has to be considered in eq. (8). A consistent and formalism independent definition for the energy-dependent spin-one resonance widths was given in [13], which we follow for the $\rho(770)$ case. The simplified expression that we use for the $a_1(1260)$ meson was obtained in [14].

4. Branching ratios

The form factors are completely determined using short distance constraints [7, 15], which allows us to compute the branching ratio and the $\ell^+\ell^-$ invariant mass spectra. After numerical integration, we obtained the contributions to the total branching ratio of the different parts. These different contributions come from the squared moduli of the three amplitudes described in the previous section and their interferences. However, in our approach there is an uncertainty in this short distance constraints which is $\lesssim 20\%$. Taking this into account, the contributions vary as noted in Table 1.

The normalized invariant-mass distribution of the lepton pair,

$$\frac{d\Gamma\left(\tau^- \to \pi^- \ell^+ \ell^- \nu_\tau\right)}{ds_{34}} \frac{1}{\Gamma_\tau},$$

where Γ_τ is the inverse of the τ lifetime and s_{34} the invariant mass of the $\ell^+\ell^-$ pair, is shown in Figure 1 for $\ell = e$, and in Figure 2 for $l = \mu$. It can be observed in Figure 1 that the structure independent contribution dominates the spectrum for values of $s_{34} \lesssim 0.1$ GeV^2, but for larger values, the axial-vector contribution dominates the rest of the spectrum apart from the ρ peak. The interference terms are negative for most of the spectrum and do not appear in the figure because of their smallness. In Figure 2, at low invariant mass the A contribution is comparable to the IB contribution, but for relatively low values and in the higher energy

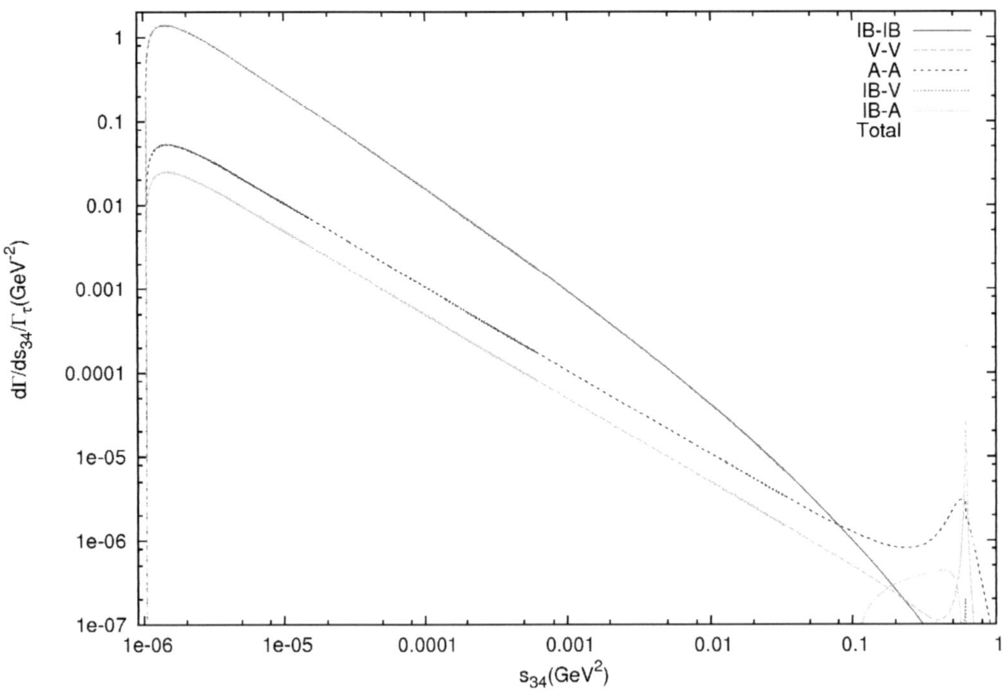

Figure 1. Contributions to the invariant mass distribution of the different parts for $l = e$.

region the A contribution dominates the spectrum, except around the ρ mass. The branching fractions shown in Table 1 were obtained by integrating numerically these invariant-mass distributions and checked from a direct integration over the five independent kinematic variables.

The fact that in both cases the structure dependent contribution dominates at $s_{34} \gtrsim 0.1\ \text{GeV}^2$ justifies our assumption of including only one multiplet of vector and axial-vector resonances. It can be seen in both figures that the A-A contribution is peaked in the ρ mass region, this is due to the inclusion of the B form factor, because it is proportional to the I $= 1$ component of the electromagnetic di-pion form factor, showing how essential this contribution is.

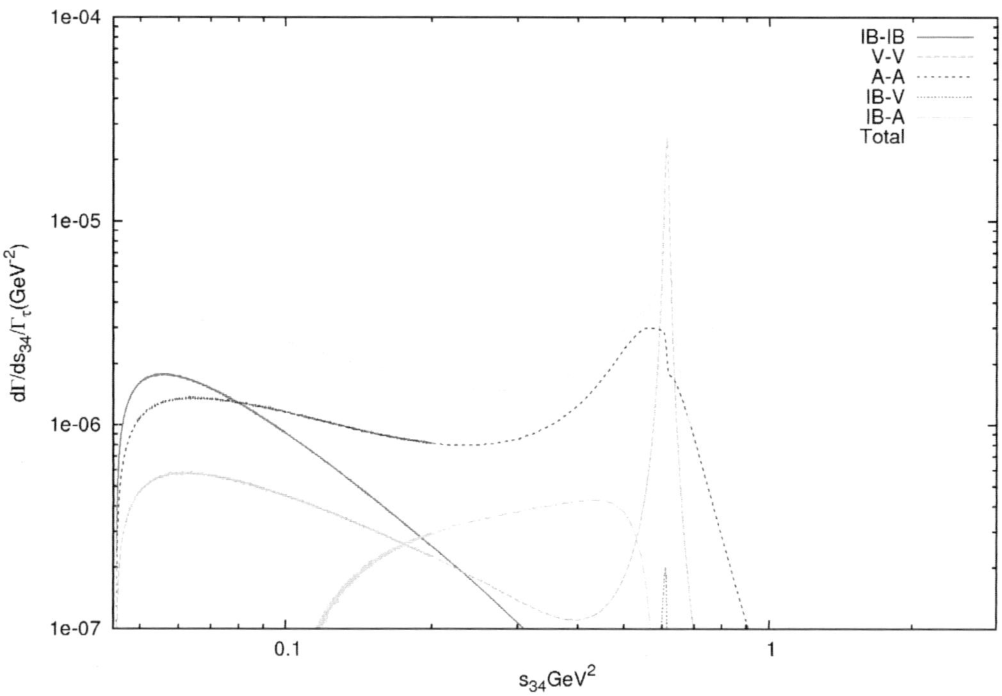

Figure 2. Contributions to the invariant mass distribution of the different parts for $l = \mu$.

5. Conclusions

We have studied the $\tau^- \to \pi^- \ell^+ \ell^- \nu_\tau$ process for the first time, describing in detail the $W^* \gamma^* \pi$ interaction using RχT. We have been able to predict all involved couplings using short-distance QCD constraints. The uncertainty of this procedure has been considered in our error estimates.

As a result, we predict the branching ratios $BR(\tau^- \to \pi^- e^+ e^- \nu_\tau) = \left(1.7^{+1.1}_{-0.3}\right) \times 10^{-5}$ and $BR(\tau^- \to \pi^- \mu^+ \mu^- \nu_\tau) = \left[3 \times 10^{-7}, 10^{-5}\right]$, where the hadronic error dominates our result for the $\ell = \mu$ case. From these figures, we conclude that the $\ell = e$ decay mode should be measured soon in the B-factories. This may also be true for the $\ell = \mu$ case only if the branching fraction happens to be close to the upper limit of the range we have given.

6. Acknowledgements

This work has been partially supported by the Spanish grant FPA2011-25948 and Conacyt (México).

References

[1] Beringer J *et al.* (Particle Data Group) 2012 *Phys. Rev.* D **86** 010001.
[2] Weinberg, S 1979 *Physica* A **96** 327.
Gasser J and Leutwyler H 1984 *Annals Phys.* **158** 142, 1985 *Nucl.Phys.* B**250** 465.
[3] Bryman D, Depommier P and Leroy C 1982 *Phys. Reports* **88** 151.
DeBaenst P and Pestieau J 1968 *Nuovo Cim.* A**53** 137.
[4] Kersch A and Scheck F 1986 *Nucl. Phys.* B **263** 475.
[5] Vaks V and Ioffe B 1958 *Nuovo Cim.* **10** 342.
[6] Roig P, Guevara A and López Castro G 2013 1401.4099 [hep-ph].
[7] Ecker G *et. al.* 1989 *Nucl.Phys.* B **321** 311, *Phys.Lett.* B **223** 425.
[8] Portolés J 2010*AIP Conf.Proc.* 1322.
[9] 't Hoooft G 1974 *Nucl.Phys.* B **72** 461, **75** 461.
Witten E 1979 *Nucl.Phys.* B**160** 57.
[10] Roig P, Guevara A and López Castro G 2012*Phys. Rev.* D**88**.
[11] Guo Z and Roig P 2010 *Phys. Rev.* D **82** 113016.
[12] Gómez Dumm D and Roig P 2013 *Eur. Phys. Jour.* C **73** 2528.
[13] Gómez Dumm D, Pich A and Portolés J 2000 *Phys. Rev.* D**62** 054014.
[14] Gómez Dumm D *et. al.* 2010 *Phys. Lett.* B **685** 158.
[15] Ruiz-Femenía P, Pich A and Portolés J 2003 *JHEP* **0307** 003.
Gómez Dumm D, Pich A and Portolés J 2004 *Phys. Rev.* D **69** 073002.
Cirigliano, V *et. al.* 2004 *Phys. Lett.* B **596** 96, 2006 *Nucl.Phys.* B **753** 139.
Guo Z H 2008 *Phys. Rev.* D **78** 033004.
Gómez Dumm D *et. al.* 2010 *Phys. Rev.* D **81** 034031.
Kampf, K and Novotny, J 2011 *Phys. Rev.* D **84** 014036.
Roig P and Sanz-Cillero J J *Preprint* hep-ph/1312.6206.

A calculation of the $Z_H \to \gamma H$ decay in the Littlest Higgs Model

J I Aranda[(a)], **I Cortés-Maldonado**[(b)], **F Ramírez-Zavaleta**[(a)], **E S Tututi**[(a)]

[(a)]Facultad de Ciencias Físico Matemáticas, Universidad Michoacana de San Nicolás de Hidalgo, Avenida Francisco J. Mújica S/N, 58060, Morelia, Michoacán, México.
[(b)]Departamento de Física, CINVESTAV, Apartado Postal 14-740, 07000, México, D. F., México.

E-mail: icortesm@cern.ch

Abstract. New heavy neutral gauge bosons are predicted in many extensions of the Standard Model, those new bosons are associated with additional gauge symmetries. We present a preliminary calculation of the branching ratio decay for heavy neutral gauge bosons (Z_H) into γH in the most popular version of the Little Higgs models. The calculation involves the main contributions at one-loop level induced by fermions, scalars and gauge bosons. Preliminary results show a very suppressed branching ratio of the order of 10^{-6}.

1. Introduction

The standard model of fundamental interactions (SM) is a non-abelian gauge theory, based on the gauge group $SU(3)_C \times SU(2)_L \times U(1)_Y$. The model is used to describe the interactions of fundamental particles and fundamental forces. The theory agrees with experimental data. Moreover, the discovery of the Higgs boson at the Large Hadron Collider (LHC) in 2012 was an exciting moment on particle physics. The mechanism proposed by Francois Englert and Peter Higgs ingeniously explains how the Higgs field gives mass to both particles that constitute matter as well as particles that mediate forces among particles [1]. These scientists answered the fundamental question: *Why do particles have mass?* in the middle sixties. After nearly 50 years Englert and Higgs won the nobel prize in 2013 for this contribution. To search for the Higgs boson, the physicists needed to build a bigger collider than Tevatron at Fermilab, with higher center energy of mass. The Tevatron was operating at a center of mass energy of 1.96 TeV, corresponding to an integrated luminosity of 9.45 fb^{-1} in proton anti-proton collisions. Even though the LHC was operating at half capacity (7 TeV), it was able to collect enough data to observe a resonance around 125 GeV corresponding to the evidence of the presence of the Higgs boson [2].

Recently, the LHC has been improved and the four detectors that are within it have been reconditioned. With these improvements it will be possible to explore the region of TeV's. At the TeV's scale, there exist theories beyond the SM which are based on non-linear realization of some global symmetry G broken down to H. In this sense, the so-called Littlest Higgs (LH) model [3, 4] represents a new attempt to solve the problem of quadratic divergences on the mass of the Higgs boson. The LH model predicts new heavy particles, such as the Z_H gauge boson.

Journal of Physics: Conference Series **761** (2016) 012089 doi:10.1088/1742-6596/761/1/012089

In this work we are specially interested on the $Z_H \to H\gamma$ decay mediated by the SM particles and the new heavy particles.

2. The Littlest Higgs Model

In the SM, the Higgs boson mass receives corrections at one-loop level, the most significant correction is due to the top quark loop. The contributions coming from loops of the electroweak gauge bosons and of the Higgs boson itself are smaller than the top quark contributions. It means that the Higgs boson mass is sensitive to high energy scales. If we assume the SM is valid up to Fermi scale, the parameters in the theory need to be carefully fine-tuned to keep the Higgs mass at an acceptable value of at most a one hundred GeV. The problem above is known as *hierarchy problem*. In order to solve this problem, there are a number of scenarios where the quadratic divergences of the Higgs boson are cancelled by means of new heavy particles. The most famous one is the supersymmetric theory.

The Little Higgs models (LHM) have been proposed as alternative to solve the hierarchy problem [3]. These models are characterized by a set of global and gauge symmetries which are spontaneously broken to the Standard Model gauge group. In these models there is a set of new particles that play the role of partners of the SM gauge bosons and the SM fermions. These new particles contributions cancel the quadratic divergences to the Higgs boson mass, m_H, arising at one-loop from the exchange of SM particles. The most economical version of these models is called *Littlest Higgs model* [4]. The LH model is a nonlinear sigma model with a global symmetry under the $SU(5)$ group, the global symmetry is broken down to $SO(5)$ by a vacuum expectation value (VEV) f of the order of TeV's. Since the model should be an extension of the SM, the unbroken symmetry group should contain the $SU(2) \times U(1)$ group.

The LH model contain a lot of free parameters, below we list the relevant parameters for our calculation [5]:

- $SU(2)_H$ couplings: $\tan\theta = s/c = g_1/g_2$.
- $U(1)_H$ couplings: $\tan\theta' = s'/c' = g'_1/g'_2$.
- Scalar triplet VEV Φ : v' $(v' < v^2/4f)$.
- Top partner mass m_T.

The particle content is presented in Table 1.

Particle	Spin	Squared mass
Φ^0, Φ^P, Φ^+, Φ^-, Φ^{++}, Φ^{--}	0	$\frac{2m_H^2 f^2}{v^2\left[1-\left(\frac{4v'f}{v^2}\right)^2\right]}$
T	$\frac{1}{2}$	$\frac{v^2}{m_t^2}\left(\lambda_1\lambda_2 f\right)^2$
A_H	1	$m_Z^2 s_W^2\left(\frac{f^2}{5s'^2c'^2v^2}-1\right)$
Z_H	1	$m_W^2\left(\frac{f^2}{s^2c^2v^2}-1\right)$
W_H^+, W_H^-	1	$m_W^2\left(\frac{f^2}{s^2c^2v^2}-1\right)$

Table 1. The new particles predicted by the LH model.

3. The $Z_H \to \gamma H$ decay

At the tree level, the extra neutral gauge boson Z_H decay into γ and Higgs boson is absent in the theory. The process is induced at one-loop level. In this work we present a preliminary result of the branching ratio for the $Z_H \to \gamma H$ decay in the context of the LH model. We have considered the main contributions at one-loop level coming from fermions, scalars and charged gauge bosons from the SM. The heavy masses of the particles previously mentioned are also considered. We can appreciate their contributions in Fig. 1.

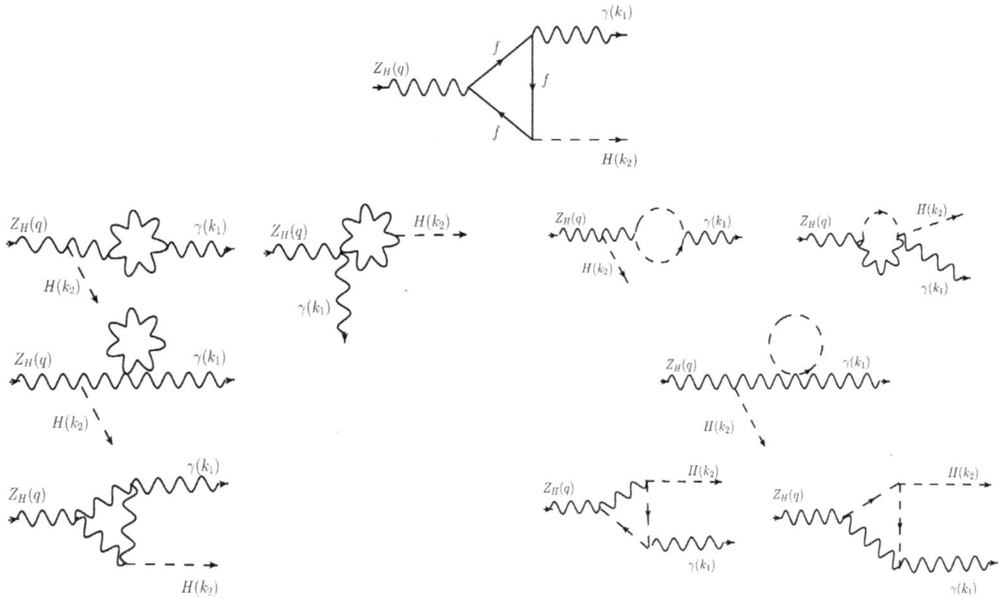

Figure 1. Feynman diagrams contributing to the $Z_H \to \gamma H$ decay at one-loop level in the LH model. (a) Fermionic contribution (solid lines). (b) Charged gauge bosons contribution (undulating lines). (c) Scalar contribution (dashed lines).

The associated decay amplitude is given by:

$$\mathcal{M}(Z_H \to \gamma H) = \mathcal{M}_T^{\mu\nu} \epsilon_\mu(q) \epsilon_\nu(k_1), \tag{1}$$

where

$$\mathcal{M}_T^{\mu\nu} = \mathcal{A}_T g^{\mu\nu} + \mathcal{B}_T \hat{k}_1^\mu \hat{q}^\nu, \tag{2}$$

being $\hat{k}_1 \equiv k_1/m_{Z_H}$ and $\hat{q} \equiv q/m_{Z_H}$. $\mathcal{M}_T^{\mu\nu}$ is the tensorial amplitude corresponding to three sets of Feynman diagrams shown in Fig. 1. The polarization vectors associated to the gauge bosons Z_H and γ are given by $\epsilon_\mu(q)$ and $\epsilon_\nu(k_1)$, respectively. The $\mathcal{A}_T$ and $\mathcal{B}_T$ form factors are given in terms of Passarino-Veltman scalar functions. Explicitly, $\mathcal{A}_T = \sum_f \mathcal{A}_f + \sum_{i=1}^{3} \mathcal{A}_{G_i} + \sum_{i=1}^{2} \mathcal{A}_{S_i}$ and $\mathcal{B}_T = \sum_f \mathcal{B}_f + \sum_{i=1}^{3} \mathcal{B}_{G_i} + \sum_{i=1}^{2} \mathcal{B}_{S_i}$. Here, f includes all charged fermions, G_i represents charged gauge bosons (W, W_H), and S_i symbolizes charged scalars (ϕ^+, ϕ^-, ϕ^{++} and ϕ^{--}) [6]. The associated decay width is given by:

$$\Gamma(Z_H \to \gamma H) = \frac{\mathcal{A}_T^2 \left(1 - \frac{m_H^2}{m_{Z_H}^2}\right)}{8\pi m_{Z_H}}. \tag{3}$$

It is very important to mention that all the $\mathcal{A}_i$ form factors in $\mathcal{A}_T$ are free of ultraviolet divergences and the Lorentz structure in Eq. (2) satisfies the Ward identity $k_{1\nu}\mathcal{M}_T^{\mu\nu} = 0$. For more details of the calculation and the corresponding numerical evaluation consult Ref. [6]. Preliminary results on the evaluation of the branching ratio for the $Z_H \to \gamma H$ decay tell us that around of $m_{Z_H} \sim 5$ TeV, the branching ratio is of the order of 10^{-5}.

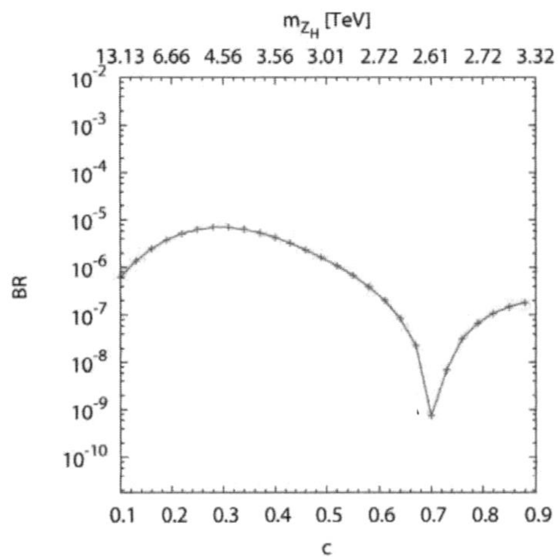

Figure 2. Branching ratio for the $Z_H \to \gamma H$ decay as a function of the c parameter for $f = 4$ TeV.

4. Conclusions

We have presented a brief description of the LH model, which predicts the existence of a new neutral massive gauge boson identified as the Z_H gauge boson. Since the existence of an extra neutral massive gauge boson is being tested at the LHC it provides one of many variants to study the presence of the Z_H gauge boson via the $Z_H \to \gamma H$ decay. The preliminary numerical analysis suggests that this process is suppressed as it has an associated branching ratio of the order of 10^{-5} at most.

Acknowledgments

This work has been partially supported by SNI and CIC-UMSNH (México).

References

[1] Higgs P W, Phys. Rev. Lett. **13**, 508 (1964).
[2] The Atlas Collaboration, Phys. Lett. **B725**, 120 (2013).
[3] Arkani-Hamed N, Cohen A G and Georgi H, Phys. Lett. **B513**, 232 (2001).
[4] Arkani-Hamed N, Cohen A G, Katz E and Nelson A E, JHEP**07**, 034 (2002).
[5] Han Tao *et al.*, Phys. Rev. D**67**, 095004 (2003).
[6] Aranda J I, Cortés-Maldonado I, Ramírez-Zavaleta F and Tututi E S, J. Phys. G: Nucl. Part. Phys. **43**, 025002 (2016).
[7] Hahn T and Perez-Victoria M, Comput. Phys. Commun. **118**, 153 (1999).

Journal of Physics: Conference Series **761** (2016) 012090 doi:10.1088/1742-6596/761/1/012090

Constraining the $Z'bs$ coupling from the $B \to X_s\gamma$ decay

J I Aranda[(a)], **D Espinosa-Gómez**[(b)], **F Ramírez-Zavaleta**[(a)], **E S Tututi**[(a)]

[(a)]Facultad de Ciencias Físico Matemáticas, Universidad Michoacana de San Nicolás de Hidalgo, Avenida Francisco J. Mújica S/N, 58060, Morelia, Michoacán, México.

E-mail: feramirez@umich.mx

Abstract. The Z' one-loop effects on the flavor-violating $b \to s\gamma$ process are analyzed. We employed the sequential Z model in order to establish a bound for the flavor-violating $Z'bs$ coupling, $|\Omega_{bs}|^2$. We found a bound of the order of 10^{-1} for the mentioned coupling in the mass range 2 TeV$< m_{Z'} <$3 TeV.

1. Introduction

The presence of a new neutral massive gauge boson (Z') is predicted in the context of numerous extensions of the Standard Model (SM). The $SU_C(3) \times SU_L(2) \times U_Y(1) \times U'(1)$ extended electroweak gauge group is the simplest model that predicts the existence of the Z' boson [1, 2]. Although absent in the SM, the phenomenon of flavor violation can arise in many of its well-motived extensions. One interesting feature of most models beyond the SM is the presence of generalized Yukawa sectors or generalized current sectors, which favor nondiagonal transitions mediated by neutral massive particles of spin-0 or spin-1 [1, 2]. Currently, experimental data have proven the existence of neutrino oscillations, which in simple words, tell us that lepton flavor conservation is violated in nature. However, the only signal of lepton flavor violation comes from transitions between neutral leptons. Thus, flavor violating transitions between charged fermions constitute an interesting subject of study, since if they occur in nature are additional evidence of flavor violation.

The purpose of this work is to study the effect of quark flavor violation mediated by the Z' boson on the $b \to s\gamma$ process, exploiting the rich phenomenology that offers the meson weak decays. Theoretically, it has been shown that this decay is highly sensitive to possible new physics effects in different contexts [3]. Thus, the main goal in this work is to estimate numerically the strength of the $Z'bs$ coupling from the $b \to s\gamma$ decay. However, it is well known that this process can not be tackled in isolation, but must be considered as part of the meson decay $B \to X_s\gamma$.

2. The flavor-violating Lagrangian

We consider the most general renormalizable Lagrangian, which includes the lepton flavor violation mediated by a new neutral massive gauge boson, coming from any extended or grand unification model [2]. This Lagrangian can be written as

$$\mathcal{L}_{NC} = \sum_{i,j} \Big[\bar{f}_i \gamma^\alpha (\Omega_{Lfifj} P_L + \Omega_{Rfifj} P_R) f_j + \bar{f}_j \gamma^\alpha (\Omega^*_{Lfjfi} P_L + \Omega^*_{Rfjfi} P_R) f_i \Big] Z'_\alpha, \tag{1}$$

where f_i is any fermion of the SM and Z'_α is the neutral gauge boson predicted by several extensions of the SM [1, 4, 5, 6]. The parameters $\Omega_{Lq_iq_j}$ and $\Omega_{Rq_iq_j}$ represent the strength of the coupling $Z'q_iq_j$, where q_i is any quark of the SM.

In the context of sequential Z model, the diagonal couplings of the Z' boson with pairs of fermions are parameterized by $Q^{f_i}_{L,R}$ [7, 8], and are related to the coupling parameters Ω as follows: $\Omega_{Lf_if_i} = -g_2 Q^{f_i}_L$ and $\Omega_{Rf_if_i} = -g_2 Q^{f_i}_R$. Here, g_2 is the gauge coupling of the Z' boson given as

$$g_2 = \frac{g}{\cos\theta_W}, \tag{2}$$

where θ_W is the weak mixing angle.

The vertices associated with flavor-violating ($Z'f_if_j$) and flavor-conserving ($Z'f_if_i$) couplings can be extracted from the Lagrangian given in Eq. (1):

$$\gamma^\alpha (\Omega_{Lfifj} P_L + \Omega_{Rfifj} P_R) \tag{3}$$

and

$$-g_2 \gamma^\alpha (Q^{f_i}_L P_L + Q^{f_i}_R P_R). \tag{4}$$

3. The $b \to s\gamma$ decay

The contribution of the flavor-violating coupling, $Z'bs$, to the $b \to s\gamma$ decay is calculated from Feynman diagrams shown in Figure 1. The tensorial amplitude of the $b \to s\gamma$ process was obtained through FeynCalc program by using the Passarino-Veltman reduction scheme. Once applied the Gordon identities and contracting with the photon polarization vector, the tensorial amplitude for the $b \to s\gamma$ decay is reduced to

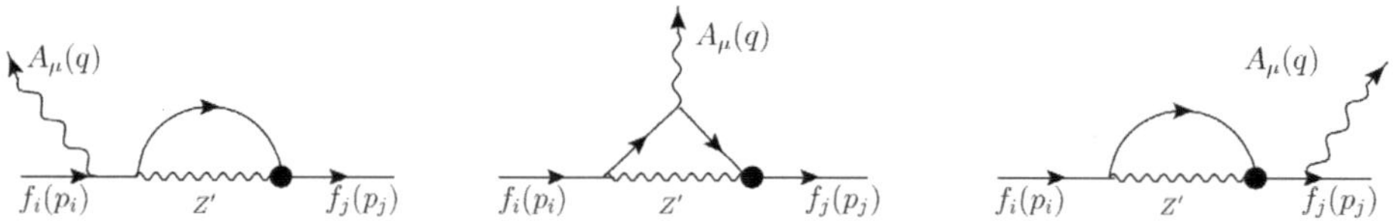

Figure 1. Feynman Diagrams contributing to the $b \to s\gamma$ decay.

$$\mathcal{M}^\mu_{\text{Tot}} = \frac{ieg_2Q_{f_i}}{64\pi^2 m_i} \bar{u}(p_j)\, \sigma^{\mu\alpha} q_\alpha \epsilon_\mu(q) \Big[F_1 (Q^{f_i}_L - Q^{f_i}_R)(\Omega_{Lf_if_j} - \Omega_{Rf_if_j}) + F_2 (Q^{f_i}_L \Omega_{Lf_if_j} + Q^{f_i}_R \Omega_{Rf_if_j}) + \Big(F_1 (Q^{f_i}_L - Q^{f_i}_R)(\Omega_{Lf_if_j} + \Omega_{Rf_if_j}) + F_2 (Q^{f_i}_L \Omega_{Lf_if_j} - Q^{f_i}_R \Omega_{Rf_if_j}) \Big) \gamma^5 \Big] u(p_i),$$

where

$$F_1 = \frac{m_i^2}{m_{Z'^2}} - 6\Big(B_0(0) - B_0(1)\Big),$$

$$F_2 = 2\left[1 + 2\left(C_0(1)m_i^2 + \frac{m_{Z'}^2}{m_i^2}\Big(B_0(0) - B_0(1)\Big)\right)\right],$$

with $Q_{f_i} = 1/3$. Also, $B_0(0) \equiv B_0(0, m_i^2, m_{Z'}^2)$, $B_0(1) \equiv B_0(m_i^2, m_i^2, m_{Z'}^2)$ y $C_0(1) \equiv C_0(m_i^2, 0, 0, -m_i^2, m_{Z'}^2, m_i^2)$ are the Passarino-Veltman scalar functions. The tensorial amplitude presented above is free of ultraviolet divergences because the functions B_0 are mutually subtracted. Moreover, it becomes evident the explicit gauge invariance, since this amplitude is of dipolar type and whenever it is contracted with the photon momentum tensor, q_μ, the simple Ward identity is satisfied.

4. Calculation of the $Z'bs$ coupling

In order to obtain the strength of the $Z'bs$ coupling, we use the discrepancy between the SM theoretical prediction and experimental measurement on the $B \to X_s\gamma$ process to quantify possible effects of new physics that might be in this gap [9]. Such a mismatch is identified by the following hypothesis

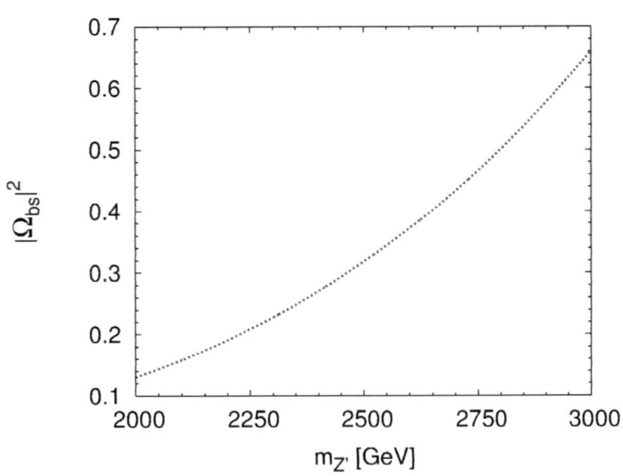

Figure 2. Strength of the $Z'bs$ coupling as a function of the Z' boson mass.

$$R_{TOT-ME} \approx R_{EXP-ME} = \frac{\Gamma_{ME+NF} - \Gamma_{ME}}{\Gamma_{ME}} = \frac{Br_{ME+NF}}{Br_{ME}} - 1, \quad (5)$$

where Γ_{EXP} is the experimental decay width of the process and Γ_{ME} is the theoretical prediction of the SM [1]. Also, Br_{EXP} and Br_{ME} are the associated decay fractions, respectively. By making use of the experimental results found by the collaborations BABAR, Belle and CLEO, along with the analysis of the Heavy Flavor Averaging Group [9], we have that $Br(B \to X_s\gamma) = (3.52 \pm 0.23 \pm 0.09) \times 10^{-4}$, which should be contrasted with the theoretical

[1] Since this contribution was presented, an updated evaluation of this observable at NNLO in QCD became available [14], making the bound more restrictive and thus favouring heavier LFV Z' bosons than here considered.

prediction in the SM: $Br(B \to X_s\gamma) = (3.15 \pm 0.23) \times 10^{-4}$ [10]. This implies that the discrepancy between theory and experiment is $R_{EXP-ME} = 0.117 \pm 0.113$ [11]. To restrict the strength of the $Z'bs$ coupling we assume that the total theoretical prediction agrees with the experimental value. Details of calculation of the contributions of the $b \to s\gamma$ process on the decay $B \to X_s\gamma$ can be found in Ref. [12].

The graph of the intensity of the $Z'bs$ coupling as a function of the mass of the Z' boson is shown in Fig. 2. It can be seen that the intensity of the parameter $|\Omega_{bs}|^2$ increases as Z' boson mass grows up; the mass range studied corresponds to 2 Tev $< m_{Z'} <$ 3 TeV, which is consistent with the experimental bound on the Z' mass reported by experimental collaborations ATLAS and CMS [13]. It is clear that this is not a restrictive bound, since $|\Omega_{bs}|^2$ is of order 10^{-1} and it is an unfavorable scenario to search for new physics.

5. Conclusions

We calculated the analytical expression of the amplitude associated with the $b \to s\gamma$ process that violates flavor, which is mediated by a neutral boson called Z'. We obtained a result which is free of ultraviolet divergences and consistent with the decoupling theorem. We estimated the intensity of the $Z'bs$ coupling as a function of the Z' boson mass. By making use of the experimental measurement for the branching ratio of the $B \to X_s\gamma$ decay it was calculated a bound for the $|\Omega_{bs}|^2$ parameter, which is about 10^{-1} for the mass range 2 Tev$< m_{Z'} <$3 Tev. The bound estimated results poorly restrictive.

Acknowledgments

This work has been partially supported by SNI and CIC-UMSNH (México).

References

[1] Durkin L S and Langacker P, Phys. Lett. **B166**, 436 (1986); Cvetic M and Langacker P, in *Proceedings of Ottawa 1992: Beyond the Standard Model III*, edited by Godfrey S and Kalyniak P (World Scientific, Singapore, 1992), p. 454; Chiang C-W, Lin Y-F and Tandean J, JHEP**11**, 083 (2011).
[2] Langacker P and Luo M, Phys. Rev. D**45**, 278 (1992).
[3] Hewett J L, hep-ph/9406302.
[4] Langacker P and Plümacher M, Phys. Rev. D**62**, 013006 (2000); He X-G and Valencia G, Phys. Rev. D**74**, 013011 (2006); Chiang C-W, Deshpande N G and Jiang J, JHEP**08**, 075 (2006).
[5] Pati J C and Salam A, Phys. Rev. D**10**, 275 (1974); Mohapatra R N and Pati J C, Phys. Rev. D**11**, 566 (1975).
[6] Pisano F and Pleitez V, Phys. Rev. D**46**, 410 (1992); Frampton P H, Phys. Rev. Lett. **69**, 2889 (1992).
[7] Robinett R W and Rosner J L, Phys. Rev. D**26**, 2396 (1982).
[8] Arhrib A, Cheung K, Chiang C-W and Yuan T-C, Phys. Rev. D**73**, 075015 (2006).
[9] Barberio E *et al.* (Heavy Flavor Averaging Group), arXiv:0808.1297.
[10] Misiak M *et al.*, Phys. Rev. Lett. **98**, 022002 (2007); Misiak M and Steinhauser M, Nucl. Phys. **B764**, 62 (2007).
[11] Aranda J I, Montaño J, Ramírez-Zavaleta F, Toscano J J and Tututi E S, Phys. Rev. D**86**, 035008 (2012).
[12] Espinosa Gómez David, Bachelor Thesis, Facultad de Ciencias Físico-Matemáticas, Universidad Michoacana de San Nicolás de Hidalgo, Noviembre 2013.
[13] Chatrchyan S *et al.* (CMS Collaboration), JHEP**05**, 093 (2011); The CMS Collaboration, Report No. CMS-PAS-EXO-11-019; Chatrchyan S *et al.* (CMS Collaboration), Phys. Lett. B**704**, 123 (2011); The CMS Collaboration, Report No. CMS-PAS-EXO-11-006; The ATLAS Collaboration, Report No. ATLAS-CONF-2012-007.
[14] M. Misiak *et al.*, Phys. Rev. Lett. **114** (2015) no.22, 221801.

AUTHOR INDEX

AUTHOR INDEX